W9-BWZ-506

ITALIAN FOLKTALES

ITALIAN FOLKTALES

Selected and Retold by

ITALO CALVINO

Translated by George Martin

A Harvest Book

A Helen and Kurt Wolff Book

Harcourt, Inc.

San Diego   New York   London

Copyright © 1956 Giulio Einaudi editore, s.p.a., Torino
English translation copyright © 1980 by Harcourt, Inc.

All rights reserved.
No part of this publication may be reproduced or transmitted
in any form or by any means, electronic or mechanical, including
photocopy, recording, or any information storage and retrieval system,
without permission in writing from the publisher.

Requests for permission to make copies of any
part of the work should be mailed to:
Permissions Department, Harcourt, Inc.,
6277 Sea Harbor Drive, Orlando, Florida 32887-6777.

The woodcut illustrations are reproduced from
Proverbi milanesi, Proverbi siciliani, and *Proverbi del Veneto*
by kind permission of Aldo Martello-Giunti Editore, S.p.A., Milan.

Library of Congress Cataloging-in-Publication Data
Calvino, Italo, 1923–1985
Italian folktales.
Translation of Fiabe italiane.
"A Helen and Kurt Wolff book."
I. Tales, Italian.   I. Title.
GR176.C3413   398.2'1'0945   80-11879
ISBN 0-15-145770-0
ISBN 0-15-645489-0 (pbk.)

Printed in the United States of America
H J L N P O M K I G

Contents

Contents

Translator's Acknowledgments

My thanks, first of all, to Willard R. Trask and Ines Delgado de Torres, for certain thoughtful and judicious remarks to me that are actually responsible for my getting launched in the translation of these folktales. Next, I am deeply grateful to Italo Calvino and to Helen Wolff for their encouragement at every turn. I feel especially fortunate to have had so painstaking—and patient—an editor as Sheila Cudahy, from whose expertise in literature, in translation, and in Italian I have profited immeasurably. My father, G. W. Martin, also deserves special thanks for his useful comments on portions of the manuscript. Finally, I would like to express my appreciation to the Translation Center at Columbia University for an award made possible by a grant from the National Endowment for the Arts.

GEORGE MARTIN

Introduction

A Journey Through Folklore

The writing of this book was originally undertaken because of a publishing need: a collection of Italian folktales to take its rightful place alongside the great anthologies of foreign folklore. The problem was which text to choose. Was there an Italian equivalent of the Brothers Grimm?

It is generally accepted that Italian tales from the oral tradition were recorded in literary works long before those from any other country. In Venice, as early as the middle of the sixteenth century, tales of wizardry and enchantment (some of them in dialect) as well as realistic novellas written in a Boccaccio-like style were collected by Straparola in his *Piacevoli Notti*. These tales imparted to his book a flavor of magic—part gothic, part oriental—suggestive of Carpaccio. In Naples, in the seventeenth century, Giambattista Basile wrote fairy tales in Neapolitan dialect and in baroque style and gave us the *Pentameron or Entertainment for the Little Ones* (which in our century was translated into Italian by no less a personage than the philosopher Benedetto Croce). Basile's work resembles the dream of an odd Mediterranean Shakespeare, obsessed with the horrible, for whom there never were enough ogres or witches, in whose far-fetched and grotesque metaphors the sublime was intermingled with the coarse and the sordid. And in the eighteenth century, again in Venice, to countervail Goldoni's middle-class comedies, Carlo Gozzi, a surly conservative, deeming that the public deserved no better, brought to the stage folktales in which he mingled fairies and wizards with the Harlequins and Pantaloons of the Commedia dell'Arte.

But it was no longer a novelty: ever since the seventeenth century in France, fairy tales had flourished in Versailles at the court of the Sun King, where Charles Perrault created a genre and set down in writing a refined version of simple popular tales which, up to then, had been transmitted by word of mouth. The genre became fashionable and lost its artlessness: noble ladies and *précieuses* took to transcribing and inventing fairy stories. Thus dressed up and embellished, in the forty-one volumes of the *Cabinet des fées*, the folktale waxed and waned in French literature along with a taste for elegant fantasy counterbalanced by formal Cartesian rationalism.

Thanks to the Brothers Grimm it flourished again, somber and earthy, at the beginning of the nineteenth century in German Romantic literature, this time as the anonymous creation of the Volksgeist, which had its roots in a timeless medieval period. A patriotic cult for the poetry of the common people spread among the *littérateurs* of Europe: Tommaseo and other scholars sought out Italian popular poetry but the tales waited in vain for an Italian Romantic to discover them.

Through the diligent efforts of the folklorists of the positivistic generation, people began to write down tales told by old women. These folklorists looked upon India, as did Max Muller, as the source of all stories and myths, if not of mankind itself. The solar religions impressed them as being so complex that they had to invent Cinderella to account for the dawn, and Snow White for the spring. But meantime, after the example first set by the Germans (Widter and Wolf in Venice, Hermann Knust in Leghorn, the Austrian Schneller in Trentino, and Laura Gonzenbach in Sicily), people began collecting "novelline"—Angelo De Gubernatis in Siena, Vittorio Imbriani in Florence, in Campania, and in Lombardy; Domenico Comparetti in Pisa; Giuseppe Pitrè in Sicily. Some made do with a rough summary, but others, more painstaking, succeeded in preserving and transmitting the pristine freshness of the original stories. This passion communicated itself to a host of local researchers, collectors of dialectal oddities and minutiae, who became the contributors to the journals of folkloristic archives.

In this manner huge numbers of popular tales were transmitted by word of mouth in various dialects, especially during the last third of the nineteenth century. The unremitting efforts of these "demo-psychologists," as Pitrè labeled them, were never properly acknowledged and the patrimony they had brought to light was destined to remain locked up in specialized libraries; the material never circulated among the public. An "Italian Grimm" did not emerge, although as early as 1875 Comparetti had attempted to put together a general anthology from a number of regions, publishing in the series "Poems and Tales of the Italian People," which he and D'Ancona edited, one volume of *Popular Italian Tales*, with the promise of two additional volumes which, however, never materialized.

The folktale as a genre, confined to scholarly interests in learned mono-graphs, never had the romantic vogue among Italian writers and poets that it had enjoyed in the rest of Europe, from Tieck to Pushkin; it was taken over, instead, by writers of children's books, the master of them all being Carlo Collodi, who, some years before writing *Pinocchio*, had translated from the French a number of seventeenth-century fairy tales. From time to time, some famous writer such as Luigi Capuana, the major novelist of the Sicilian naturalist school, would do as a book for children a collection of tales having its roots both in fantasy and popular sentiment.

But there was no readable master collection of Italian folktales which would be popular in every sense of the word. Could such a book be assembled now? It was decided that I should do it.

For me, as I knew only too well, it was a leap in the dark, a plunge into an unknown sea into which others before me, over the course of 150 years, had flung themselves, not out of any desire for the unusual, but because of a deep-rooted conviction that some essential, mysterious element lying in the ocean depths must be salvaged to ensure the survival of the race; there was, of course, the risk of disappearing into the deep, as did Cola Fish in the Sicilian and Neapolitan legend. For the Brothers Grimm, the salvaging meant bringing to light the fragments of an ancient religion that had been preserved by the common people and had lain dormant until the glorious day of Napoleon's defeat had finally awakened the German national consciousness. In the eyes of the "Indianists," the essential element consisted of the allegories of the first Aryans who, in trying to explain the mystery of the sun and the moon, laid the foundations for religious and civil evolution. To the anthropologists it signified the somber and bloody initiation rites of tribal youths, rites that have been identical from time immemorial, from paleolithic hunters to today's primitive peoples. The followers of the Finnish school, in setting up a method for tracing migrations among Buddhist countries, Ireland, and the Sahara, applied a system similar to that used for the classification of coleoptera, which, in their cataloging process, reduced findings to algebraic sigla of the Type-Index and Motif-Index. What the Freudians salvaged was a repertory of ambiguous dreams common to all men, plucked from the oblivion of awakenings and set down in canonical form to represent the most basic anxieties. And for students of local traditions everywhere, it was a humble faith in an unknown god, rustic and familiar, who found a mouthpiece in the peasantry.

I, however, plunged into that submarine world totally unequipped, without even a tankful of intellectual enthusiasm for anything spontaneous and primitive. I was subjected to all the discomforts of immersion in an almost formless element which, like the sluggish and passive oral tradition, could never be brought under conscious control. ("You're not even a Southerner!" an uncompromising ethnologist friend said to me.) I could not forget, for even an instant, with what mystifying material I was dealing. Fascinated and perplexed, I considered every hypothesis which opposing schools of thought proposed in this area, being careful not to allow mere theorizing to cloud the esthetic pleasure that I might derive from these texts, and at the same time taking care not to be prematurely charmed by such complex, stratified, and elusive material. One might well ask why I undertook the project, were it not for the one bond I had with folktales—which I shall clarify in due course.

Meanwhile, as I started to work, to take stock of the material available, to classify the stories into a catalog which kept expanding, I was gradually possessed by a kind of mania, an insatiable hunger for more and more versions and variants. Collating, categorizing, comparing became a fever. I could feel myself succumbing to a passion akin to that of entomologists, which I thought characteristic of the scholars of the Folklore Fellows Communications of Helsinki, a passion which rapidly degenerated into a mania, as a result of which I would have given all of Proust in exchange for a new variant of the "gold-dung donkey." I'd quiver with disappointment if I came upon the episode of the bridegroom who loses his memory as he kisses his mother, instead of finding the one with the ugly Saracen woman, and my eye became so discerning—as is the wont with maniacs—that I could distinguish at a glance in the most difficult Apulian or Friulian text a "Prezzemolina" type from a "Bellinda" type.

I was unexpectedly caught in the spiderlike web of my study, not so much by its formal, outward aspect as by its innermost particularities: infinite variety and infinite repetition. At the same time, the side of me that remained lucid, uncorrupted, and merely excited about the progression of the mania, was discovering that this fund of Italian folklore, in its richness, limpidity, variety, and blend of the real and the unreal, is unsurpassed by even the most famous folktales of Germanic, Nordic, and Slavic countries. This is true not only where the story is recorded from the words of an outstanding narrator—more often than not a woman—or when the story is laid in a region noted for brilliant storytelling, but the very essence of the Italian folktale is unparalleled grace, wit, and unity of design. Its composition and genius for synthesizing the essence of a type is unique. Thus, the longer I remained steeped in the material, the fewer became my reservations; I was truly exalted by the expedition, and meanwhile the cataloging passion— maniacal and solitary—was replaced by a desire to describe for others the unsuspected sights I had come upon.

Now my journey through folklore is over, the book is done. As I write this preface I feel aloof, detached. Will it be possible to come down to earth again? For two years I have lived in woodlands and enchanted castles, torn between contemplation and action: on the one hand hoping to catch a glimpse of the face of the beautiful creature of mystery who, each night, lies down beside her knight; on the other, having to choose between the cloak of invisibility or the magical foot, feather, or claw that could meta- morphose me into an animal. And during these two years the world about me gradually took on the attributes of fairyland, where everything that happened was a spell or a metamorphosis, where individuals, plucked from the chiaroscuro of a state of mind, were carried away by predestined loves, or were bewitched; where sudden disappearances, monstrous transformations occurred, where right had to be discerned from wrong, where paths bristling with obstacles led to a happiness held captive by dragons. Also in the lives of peoples and nations, which until now had seemed to be at a standstill, anything seemed possible: snake pits opened up and were transformed into rivers of milk; kings who had been thought kindly turned out to be brutal parents; silent, bewitched kingdoms suddenly came back to life. I had the impression that the lost rules which govern the world of folklore were tumbling out of the magic box I had opened.

Now that the book is finished, I know that this was not a hallucination, a sort of professional malady, but the confirmation of something I already suspected—folktales are real.

Taken all together, they offer, in their oft-repeated and constantly varying examinations of human vicissitudes, a general explanation of life preserved in the slow ripening of rustic consciences; these folk stories are the catalog of the potential destinies of men and women, especially for that stage in life when destiny is formed, i.e., youth, beginning with birth, which itself often foreshadows the future; then the departure from home, and, finally, through the trials of growing up, the attainment of maturity and the proof of one's humanity. This sketch, although summary, encompasses everything: the arbitrary division of humans, albeit in essence equal, into kings and poor people; the persecution of the innocent and their subsequent vindication, which are the terms inherent in every life; love unrecognized when first encountered and then no sooner experienced than lost; the common fate of subjection to spells, or having one's existence predetermined by complex and unknown forces. This complexity pervades one's entire existence and forces one to struggle to free oneself, to determine one's own fate; at the same time we can liberate ourselves only if we liberate other people, for this is a sine qua non of one's own liberation. There must be fidelity to a goal and purity of heart, values fundamental to salvation and triumph. There must also be beauty, a sign of grace that can be masked by the humble, ugly guise of a frog; and above all, there must be present the infinite possibilities of mutation, the unifying element in everything: men, beasts, plants, things.

Criteria for My Work

The method of transcribing folktales "from the mouths of the people" was started by the Brothers Grimm and was gradually developed during the second half of the century into "scientific" canons scrupulously faithful to the dialect of the narrator. The Grimms' approach was not "scientific" in the modern sense of the word, or only halfway so. A study of their manuscripts confirms what is abundantly plain to an experienced eye perusing *Kinder- und Hausmärchen*, namely that the Grimms (Wilhelm in particular) had added their own personal touch to the tales told by little old women, not only translating a major part from German dialects, but integrating the variants, recasting the story whenever the original was too crude, touching up expressions and images, giving stylistic unity to the discordant voices.

The foregoing serves as an introduction and justification (if I may take refuge behind names so famous and remote) for the hybrid nature of my work, which likewise is only halfway "scientific," or three-quarters so; as for the final quarter, it is the product of my own judgment. The scientific portion is actually the work of others, of those folklorists who, in the span of one century, patiently set down the texts that served as my raw material. What I did with it is comparable to the second part of the Grimms' project: I selected from mountains of narratives (always basically the same ones and amounting altogether to some fifty types) the most unusual, beautiful, and original texts. I translated them from the dialects in which they were recorded or when, unfortunately, the only version extant was an Italian translation lacking the freshness of authenticity, I assumed the thorny task of recasting it and restoring its lost originality. I enriched the text selected from other versions and whenever possible did so without altering its character or unity, and at the same time filled it out and made it more plastic. I touched up as delicately as possible those portions that were either missing or too sketchy. I preserved, linguistically, a language never too colloquial, yet colorful and as derivative as possible of a dialect, without having recourse to "cultivated" expressions—an Italian sufficiently elastic to incorporate from the dialect images and turns of speech that were the most expressive and unusual.

As the notes at the end of the book testify, I worked on material already collected and published in books and specialized journals, or else available in unpublished manuscripts in museums and libraries. I did not personally hear the stories told by little old women, not because they were not available, but because, with all the folklore collections of the nineteenth century I already had abundant material to work on. Nor am I sure that attempts on my part to gather any of it from scratch would have appreciably improved my book.

My work had two objectives: the presentation of every type of folktale, the existence of which is documented in Italian dialects; and the representation of all regions of Italy.

As for the real and genuine fairy tale (*fiaba*), that is, the wonderful, magical story that tells of kings of imprecise realms, all its "types" of any significance are represented by one or more versions that struck me as being the most characteristic, the least stereotyped, and the most steeped in local color (a concept I shall clarify presently). The book is also interspersed with religious and local legends, short stories, animal fables, jokes, and anecdotes— in short, popular narrative components of various kinds which I came across in my search and which held me by their beauty or else served to represent regions for which I lacked other material.

I drew very little on local legends concerning place origins or customs or historical records; this is a field entirely different from that of the folktale: the narratives are short, undeveloped, and their anthologies, with very few exceptions, do not reproduce the speech of the people; they only evoke it in a nostalgic, romantic style: in short, it was material I found unusable.

As for Italian dialects, I have taken all those that make up the Italian linguistic area, but not all those in Italy as a country. Thus I dealt with folktales from the French coast of Nice, whose dialect is closer to the Ligurian than to the Provençal, and passed over those from the Italian Aosta valley where a French dialect is spoken: I included some of the Venetian dialects of Jugoslav Dalmatia, and none from the German-speaking South Tyrol province of Italy. I made an exception for the small settlements of Greek-speaking sections of Calabria, two of whose folktales I included (since their narrative folklore is well integrated with that of the rest of Calabria; in any case, I am happy to have them in the book).

In parentheses at the end of each folktale in the book is the name of a locality or region. In no instance does it signify that the folktale in question originated in that particular area. Folktales are the same the world over. To say "from where" a folktale comes makes little sense; thus, scholars of the "Finnish" or historical-geographic school who seek to determine the zone of origin of each type of folktale come up with rather dubious results, placing them anywhere between Asia and Europe. But international circulation of common tales does not exclude their diversity, which is expressed, according to an Italian scholar, "through the choice or rejection of certain motifs, the preference for certain kinds, the creation of particular characters, the atmosphere suffusing the narration, the stylistic traits that reflect a formal, defined culture." Folktales are labeled "Italian" insofar as they are narrated by the people of Italy, tales that have come into our narrative folklore via the oral tradition; but we also classify them as Venetian, Tuscan, Sicilian. Since the folktale, regardless of its origin, tends to absorb something of the place where it is narrated—a landscape, a custom, a moral outlook, or else merely a very faint accent or flavor of that locality—the degree to which a tale is imbued with that Venetian, Tuscan, or Sicilian something is what led me to choose it.

The notes at the end of the volume account for the name of the locality I have assigned to each story, and they also list the versions I read in other Italian dialects. So it will be quite clear that the designation "Monferrato" or "Marche" or "Terra d'Otranto" does not mean that the folktale itself had its origin in Monferrato, Marche, or Terra d'Otranto; but that as I recorded that particular tale I kept foremost in mind the version of it from one of those regions. Because of the various texts at my disposal, this particular one struck me as not only the most beautiful or the richest or the most skillfully narrated, but also as the one which, rooted in its native heath, had drawn from it the most pith, thereby becoming typical of Monferrato, Marche, or Otranto.

It must be noted that with many of the first folklorists, the urge to collect and publish was stimulated by the "comparatist" passion peculiar to the literary culture of the period, in which similarity rather than diversity was stressed, and when evidence of the universal diffusion of a motif rather than the distinction of a particular place, time, and narrative personality was emphasized. The geographic designations of my book are, in certain cases indisputable (in many of the Sicilian tales, for instance), while in others they will appear arbitrary, justified solely by the bibliographical reference in the note.

In all this I was guided by the Tuscan proverb dear to Nerucci: "The tale is not beautiful if nothing is added to it"—in other words, its value consists in what is woven and rewoven into it. I too have thought of myself as a link in the anonymous chain without end by which folktales are handed down, links that are never merely instruments or passive transmitters, but—and here the proverb meets Benedetto Croce's theory about popular poetry—its real "authors."

The Folklore Anthologies

The work of documentation of Italian narrative done over nearly a century by folklorists has a very uneven geographical distribution. For some regions, I found a mine lode of material; for others, almost nothing. There are full, good collections for two regions in particular: Tuscany and Sicily.

For Sicily my most important source is Giuseppe Pitrè's *Fiabe, novelle e racconti popolari Siciliani* (*Sicilian Fables, Stories and Popular Tales*, 1875). It consists of four volumes containing 300 narratives classified according to type, written in all the dialects of Sicily: it is a scholarly work, painstakingly documented, replete with footnotes of "variants and collations" and lexical comparatist notes.

Giuseppe Pitrè (1841–1916) was a medical doctor dedicated to the study of folklore, who had a large team of collectors working for him.

The secret of Pitrè's work is that it gets us away from the abstract notion of "people" talking; instead, we come into contact with narrators having distinct personalities, who are identified by name, age, occupation. This makes it possible to uncover through the strata of timeless and faceless stories and through crude stereotyped expressions, traces of a personal world of more sensitive imagination, whose inner rhythm, passion, and hope are expressed through the tone of the narrator.

Pitrè's collection dates from 1875; in 1881, Verga wrote *I Malavoglia* (*The House by the Medlar Tree*). Contemporaneously, two Sicilians, the novelist and the scholar, listened, each with a different purpose, to the fishermen and gossips, so as to transcribe their speech. We may compare the ideal catalog of voices, proverbs, and customs which each of them sought to put together, the novelist coordinating it with his own inner lyrical and choral rhythm, the folklorist with a carefully labeled museum which can be inspected in the twenty-five volumes of the *Biblioteca delle tradizioni popolari siciliane* (1871–1913), the twenty-four years of his journal, *Archivio per lo studio delle tradizioni popolari,* 1882–1906 ("Archives for the Study of Popular Traditions"), the sixteen volumes of the series, *Curiosità popolari tradizionali* ("Popular Traditional Oddities"), and even in his collection of folk art and craftsmanship now housed in the Pitrè Museum in Palermo. Pitrè did in folklore what Verga had done in literature; he was the first folklorist to transcribe not only traditional motifs or linguistic usages, but the inner poetry of the stories.

With the advent of Pitrè, the folklore movement began taking into account, in the very existence of a storytelling tradition, the part played by poetic creativity. This is entirely different from the field of folksong, where the song is forever fixed by its lines and rhymes, anonymously repeated in its choruses and with a very limited possible range of individual variations. The folktale must be re-created each time. At the core of the narrative is the storyteller, a prominent figure in every village or hamlet, who has his or her own style and appeal. And it is through this individual that the timeless folktale is linked with the world of its listeners and with history.

The protagonist of Pitrè's collection is an illiterate old woman, Agatuzza Messia, a former domestic of Pitrè's, a quilt maker in Borgo (a section of Palermo) living at 8 Largo Celso Nero. She is the narrator for a number of Pitrè's best tales and I have freely chosen among them (see stories 148 through 158). This is how Pitrè, in the preface to his anthology, describes his model narrator:

She is far from beautiful, but is glib and eloquent; she has an appealing way of speaking, which makes one aware of her extraordinary memory and talent. Messia is in her seventies, is a mother, grandmother, and great-grandmother; as a little girl she heard stories from her grandmother, whose own mother had told them, having herself heard countless stories from one of her grandfathers. She had a good memory so never forgot them. There are women who hear hundreds of stories and never remember one; there are others who remember them but haven't the knack of storytelling. Her friends in Borgo thought her a born storyteller; the more she talked, the more they wanted to listen.

Messia can't read, but she knows lots of things others don't, and talks about them so picturesquely that one cannot help but appreciate her. I call my readers' attention to this picturesqueness of speech. If the setting of the story is aboard a ship due to sail, she speaks, apparently unconsciously, with nautical terms and turns of phrase characteristic of sailors or seafaring people. If the heroine of a story turns up penniless and woebegone at the house of a baker, Messia's language adapts itself so well to that situation that one can see her kneading the dough and baking the bread—which in Palermo is done only by professional bakers. No need to mention domestic situations, for this is where Messia is in her element; inevitably, for a woman who, like all her neighbors, has brought up her children and the children of her children "to serve the home and the Lord," as they say.

Messia saw me come into the world and held me in her arms; this is how I heard from her lips the many beautiful stories that bear her imprint. She repeated to the young man the tales she had told the child thirty years before; nor has her narration lost one whit of its original purity, ease and grace.

Messia, like a typical Sicilian storyteller, fills her narrative with color, nature, objects; she conjures up magic, but frequently bases it on realism, on a picture of the condition of the common people; hence her imaginative language, but a language firmly rooted in commonsensical speech and sayings. She is always ready to bring to life feminine characters who are active, enterprising, and courageous, in contrast to the traditional concept of the Sicilian woman as a passive and withdrawn creature. (This strikes me as a personal, conscious choice.) She passes completely over what I should say was the dominant element in the majority of Sicilian tales: amorous longing, a predilection for the theme of love as exemplified in the lost husband or wife motif, so widespread in Mediterranean folklore and dating back to the oldest written example, the Hellenistic tale of *Amor and Psyche* told in Apuleius' *Metamorphosis* (second century A.D.) and repeated through the ages in hundreds and hundreds of stories about encounters and separations, mysterious bridegrooms from the nether regions, invisible brides, and horse- or serpent-kings who turn into handsome young men at night. Or else the fragile, delicate genre that is neither myth, short story, nor ballad, epitomized by that sigh of melancholy, sensual joy, *La sorella del Conte* ("The Count's Sister").

In contrast to the Sicilian folktale which unfolds in a somewhat limited gamut of themes and often from a realistic starting point (how many hungry families go out into the country looking for plants to make soup!), the Tuscan folktale proves a fertile ground for the most varied cultural influences. My favorite Tuscan source is Gherardo Nerucci's *Sessanta novelle popolari Montanesi,* 1880 (*Sixty Popular Tales from Montale*—a village near Pistoia), written in an odd Tuscan vernacular. In one of these a certain Pietro di Canestrino, a laborer, told, in *La Regina Marmotta* ("The Sleeping Queen"), the most Ariosto-like tale to come from the mouth of a man of the people. The story is an uncertain byproduct of the sixteenth-century epic, not in its plot (which in its broad outline is reminiscent of a well-known folktale) nor in its fanciful geography (which also dates back to the ballads of chivalry) but in its manner of narrating, of creating magic through the wealth of descriptions of gardens and palaces (much more complete and literary in the original text than in my own highly abridged reconstruction, where I sought to avoid any appreciable divergence from the general tone of the present book). The original description of the queen's palace even includes a list of famous beauties of the past, introduced as statues: ". . . and these statues represented many famous women, alike in dress, but different in countenance, and they included Lucretia of Rome, Isabella of Ferrara, Elizabeth and Leonore of Mantua, Varisilla Veronese with her beautiful face and unusual features; the sixth, Diana of Regno Morese and Terra Luba, the most renowned for her beauty in Spain, France, Italy, England, and Austria, and whose royal blood was the purest . . ." and so forth.

The book of the sixty Montale stories appeared in 1880, when many of the most important Italian collections of folktales had already been published, but the lawyer Gherardo Nerucci (1828–1906, somewhat older than the other folklorists of the "scientific" generation) had begun collecting tales much earlier, in 1868. Many of the sixty stories were already in the anthologies of his colleagues; some of the most beautiful tales in the Imbriani and Comparetti collections are his. Nerucci was not concerned with comparatist storytelling (his interest in popular tales was linguistic), but it is plain that the Montale stories are based, often, on literary sources. To be sure, this village also produced its share of obscure and prehistoric tales, such as *Testa di Bufala* ("Buffalo Head") which simply clamors for ethnological interpretation. There were also tales that appear strangely modern and "invented," like *La novellina delle scimmie* ("The Little Tale of the Monkeys") but a great many of the stories have the same themes and plots as popular poems (going back to the fourteenth to sixteenth centuries) and as the *Arabian Nights*; these (with only the settings transposed) are so faithful to the French eighteenth-century translation by Galland (allowing for an adaptation to suit Western taste) as to exclude the possibility of influences stemming from long ago, via who knows what oral paths from the Orient. These were unquestionably directly plucked from literature and transposed into folklore quite recently. Thus, when the widowed Luisa Ginanni repeats from beginning to end the plot of Boccaccio's *Andreuccio da Perugia*, I believe that it derives not from the popular tradition that was the source of Boccaccio's story, but from a direct version in dialect of the most picaresque story in the *Decameron*.

Thus, with Boccaccio, we come close to defining the spirit in which Pistoia country people told stories. It would appear that in this region the link has been established (or that Nerucci perceived it) between a *fiaba* (fairy tale) and a *novella* (short story); the transition between the narrative of magic and the narrative of fortune or individual bravery has been pinpointed. The tale of magic flows smoothly into a mundane, "bourgeois" story, short story or novel of adventure, or tear-jerking account of a damsel in distress. Let us take, for example, *Il figliuolo del mercante di Milano* ("The Son of the Merchant of Milan") which belongs to a very old and obscure type of folktale: the youth who draws from his adventures—always the same, in which a dog, poisoned food, birds play a role—a riddle in nonsense rhyme. He puts it to a princess who is reputed to be a riddle-solver, and thus wins her hand. In Montale, the hero is not the usual predestined character, but a young man of initiative, ready to run risks; he knows how to capitalize on his winnings and profit from his losses. The proof is that—and it is very odd behavior in a fairy-tale hero—instead of marrying the princess, he releases her from all obligation to him in exchange for economic gain. This happens not just once, but twice in succession: the first time in exchange for a magic object (more exactly, the permission to win it himself), and the second, even more practical, in exchange for a steady income. The supernatural origin of Menichino's success is quite overshadowed by his true native ability to make the most of these magic powers and retain all profits for himself. But Menichino's outstanding trait is sincerity, the ability to win people's trust: the hallmark of a businessman.

Nerucci's favorite storyteller is the widow Luisa Ginanni. Of all the Montale storytellers, she knows the greatest number (three-quarters of the collection originate with her); often her imagery is quite striking, but there is no great difference between her voice and that of others. In a style full of verbal invention Nerucci intended to show us the richness of the unusual vernacular that results when the people of Montale speak in Italian—a harsh, mangled, violent Tuscan. Whereas with most of the other texts my task was somehow to enhance the stylistic color, with Nerucci's my rewriting had to tone it down; there were, consequently, unavoidable losses.

Rewriting Tuscan texts from a vernacular not so different from current Italian was for me a difficult task. The odds were stacked against me. And the hardest ones—for the simple reason that they are the most beautiful and already have a distinct style—were those fifteen or so tales I singled out from

Nerucci. (On the other hand, in the case of the Sicilian texts from Pitrè's collection, the more beautiful they were, the easier my task turned out to be; I translated them literally or freely, as the text required.)

As I have already indicated, Tuscany and Sicily have the choicest selection of folktales, both in quantity and quality. And immediately after, with its own special interpretation of a world of fantasy, is Venice, or rather, the entire range of Venetian dialects. The outstanding name here is Domenico Giuseppe Bernoni, who published (in 1873, 1875, 1893) several booklets of Venetian tales. These fables are remarkable for their limpidity and poetic power; although well-known types recur in them they always somehow evoke Venice, her spaces and light, and in one way or another they all impart an aquatic flavor; the sea, canals, voyages, ships, or the Levant figure in them. Bernoni does not give the names of the narrators, nor do we know how faithful he was to the original tales; but there is a distinct harmony overlying the gentle tones of the dialect and the atmosphere pervading various folktales, qualities which I hope are reflected in my transcription of the seven tales I singled out from his collection (stories 29 through 35).

In the same period, Bologna's contribution, through Carolina Coronedi-Berti, was a copious, first-rate anthology (1874), written in a dialect brimming over with zest and consisting of well-rounded, well-told stories shrouded in a somewhat hallucinatory ambience and set in familiar landscapes. Although the names of the storytellers are not given, one is often conscious of a feminine presence, who tends, at times, to be sentimental, at others to be dashing.

In Giggi Zanazzo's Roman collection (1907), the tale becomes a pretext for verbal entertainment. It is based upon knavish and suggestive modes of speech and makes for rewarding and pleasant reading.

The Abruzzi have to their credit two very fine collections: the two volumes by Gennaro Finamore (1836–1923), a teacher and medical doctor. He collected texts in dialect from various localities and transcribed them with great linguistic precision; a vein of melancholy poetry occasionally emerges from these texts. The other Abruzzi anthology is the work of the archeologist Antonio De Nino (1836–1907), a friend of D'Annunzio's, who recast the tales in Italian in very brief stories interspersed with short ballads and refrains in dialect in a playful and childlike style—a method of doubtful value from the scientific point of view as well as from mine. But the book contains many unusual stories (although a number of them are culled from the *Arabian Nights*) and curious ones (see my *Gobba, zoppa e collotorto*, "Hunchback Wryneck Hobbler"), that have an underlying irony and playfulness.

Eight of the best tales I ran into are in Apulian dialect, in Pietro Pellizzari's book, *Fiabe e canzoni popolari del contado di Maglie in terra d'Otranto* (*Popular Fables and Songs from the Country of Maglie in Terra d'Otranto*, 1881). They are familiar types, but written in a language so witty, in so racy a style, with such enjoyment of grotesque malformation, that they give the impression of having been conceived in that very language, like the excellent

I cinque capestrati ("Five scapegraces") the plot of which can be found in its every detail in Basile.

In Calabria (mainly in the village of Palmi) Letterio Di Francia, the scholarly author of a history of storytelling, transcribed a collection, *Fiabe e novelle calabresi* (*Calabrian Fables and Short Stories*) published in 1929 and 1931, complete with the fullest and most precise notes ever recorded in Italian. The author names the different outstanding narrators, among whom there is a certain Annunziata Palermo. Calabrian storytelling exhibits a rich, colorful, complex imagination, but within it the logic of the plot is often lost, leaving only the unraveling of the enchantment.

Characteristics of the Italian Folktale

Can one speak of the Italian folktale? Or must the question of the folktale be dealt with in terms of a remote age that is not only prehistoric but also pregeographic?

The disciplines concerned with studying the relationships between the folktale and the rites of primitive societies yield surprising, and for me, convincing results, as in Propp's *Historical Roots of Russian Fairy Tales* (1946), and it seems to me that the origins of the folktale are to be found in these rites. But having arrived at this conclusion there are still many unanswered questions. Was the birth and development of folklore a parallel and similar phenomenon throughout the world as the proponents of polygenesis claim? In view of the complexity of certain types that explanation may be too simple. Can ethnology explain every motif, every narrative complex throughout the world? Evidently not. Therefore, quite apart from the question of the ancient sources of folktales, the importance of the life which every folktale has had during a historical period must be recognized: storytelling as entertainment means the passage of the tale from narrator to narrator, from country to country, often by means of a written version, a book, until the story has spread over the entire area where it is to be found today.

Between the fourteenth and sixteenth centuries, Tuscany, through its ballads and popular verses, often imitative of folklore motifs, must have defined and diffused the most successful categories. The ballad has its own history, distinct from that of the folktale, but the two cross: the ballad draws its motifs from the tale and in turn adapts the tale to suit its motif.

We must be careful not to "medievalize" the folktale too much. The ethnological view plucks the fable from the décor given it by a romantic taste, and has accustomed us to see the castle as the hut where initiations to the hunt took place, to regard the princess as a sacrificial offering to the dragon for agricultural necessities, the wizard as the sorcerer of the clan. Moreover, one need only glance at any collection faithful to the oral tradition to understand that the people (namely those of the nineteenth century unfamiliar either with the illustrations of children's books or with Disney's *Snow White*) do not visualize the folktales in terms of images which seem natural to us. In these stories description is almost always minimal, the terminology is general. Italian folklore tells of palaces, not castles. It rarely speaks of a prince or princess but rather of the son or daughter of the king. The names of supernatural beings such as ogres or witches are drawn from the most ancient pagan background of the locality. The names of these beings are not classified with any precision, not only because of the diversity of the dialects—for example, in Piedmont the *masca* (witch) is the *mamma-draga* (mother dragon) in Sicily, and in Romagna the *om salbadgh* (wild man) is the *nanni-orcu* (orca) in Puglia—but also because of the confusion that arises within the confines of a dialect; for example, in Tuscany *mago* (sorcerer) and *drago* (dragon) are often confused and used interchangeably.

Nevertheless, the medieval stamp on the popular tale remains strong and enduring. These stories abound in tournaments to win the hands of princesses, with knightly feats, with devils, and with distortions of hallowed traditions. Therefore, one must examine as a prime occurrence in the historical life of the fable that moment of osmosis between folktale and the epic of chivalry, the probable source of which was Gothic France, whence its influence spread into Italy via the popular epic. That substratum of pagan and animistic culture which at the time of Apuleius had taken on the trappings and names of classical mythology, subsequently fell under the influence of the feudal and chivalrous imagination, of the institutions, ethics, and religious beliefs of the Middle Ages.

At a certain point this amalgam blends with yet another, a wave of images and transfigurations of oriental origin which had spread from the south when contacts with and threats from the Saracens and Turks were at their height. In the numerous sea stories I have included, the reader will see how the notion of the world divided between Christian and Muslim takes the place of the arbitrary geography of the folktale. The folktale clothes its motifs in the habits of diverse societies. In the West the imprint of feudalism prevailed (notwithstanding certain nineteenth-century touches in the south such as an English lord) while in the Orient the bourgeois folktale of the fortunes of Aladdin or of Ali Baba dominated.

One of the few folktales, perhaps—according to Stith Thompson the only one—that can be considered of "probable Italian origin," is that of the love of three oranges (as in Gozzi) or of the three lemons (as in Basile) or of the three pomegranates (as in the version I chose). The tale abounds in metamorphoses in baroque (or Persian?) taste altogether worthy of Basile's inventive powers or of those of a visionary weaver of rugs. It consists of a series of metaphors strung into a story—ricotta and blood, fruit and girl; a Saracen woman who looks at her image in a well, a girl in a tree who turns into a dove, the dove's drops of blood from which a tree suddenly grows and, completing the circle, the fruit out of which the girl emerges. I should like to have given this folktale greater prominence but having examined numerous popular versions I did not find one that could be considered the frame story. I have included two versions, one (no. 107) integrated with some others is from Abruzzi and represents the most classic form of the tale, the other (no. 8) is a curious variation from Liguria. However, I must say in this instance, Basile is unrivaled and I refer the reader to his tale, the last in the *Pentamerone*.

In this very mysterious story of transformations, with its precise rhythm, its joyous logic at work, I think I perceive a characteristic of the popular elaboration of the Italian folktale. There is a genuine feeling for beauty in the communions or metamorphoses of woman and fruit, of woman and plant in the two beautiful companion pieces of the *Ragazza mela* ("Apple Girl") from Florence (no. 85) and the *Rosmarina* ("Rosemary") from Palermo (no. 161). The secret lies in the metaphorical link: the image of the freshness of the apple and of the girl or of the pears beneath which the girl is hidden in order to increase the weight of the basket (no. 11) in the story "The Little Girl Sold with the Pears."

The natural cruelties of the folktale give way to the rules of harmony. The continuous flow of blood that characterizes the Grimms' brutal tales is absent. The Italian folktale seldom displays unbearable ferocity. Although the notion of cruelty persists along with an injustice bordering on inhumanity as part of the constant stuff of stories, although the woods forever echo with the weeping of maidens or of forsaken brides with severed hands, gory ferocity is never gratuitous; the narrative does not dwell on the torment of the victim, not even under pretense of pity, but moves swiftly to a healing solution, a part of which is a quick and pitiless punishment of the male-factor—or more often the malefactress—to be tarred and burned in the grim tradition of witches' pyres, or, as in Sicily, being thrown from a window and then burned.

A continuous quiver of love runs through Italian folklore. In speaking of the Sicilian tales, I mentioned the popularity of the Cupid and Psyche type found not only in Sicily but also in Tuscany and more or less everywhere. The supernatural bridegroom is joined in an underground dwelling. Neither his name nor his secret must be divulged lest he vanish. A lover is produced by sorcery from a basin of milk; a bird in flight is wounded by an envious rival who puts ground glass in a basin or tacks on a window ledge where the bird will alight. There is the serpent- or swine-king who at night turns into a handsome youth for the bride who respects him, while the wax of the candle lighted by curiosity thrusts him back under the evil spell. In the story of Bellinda and the monster a curious sentimental relationship develops between them. In those cases where the suffering partner is the man, it is the bewitched bride who comes silently at night to join him in the deserted palace; it is the fairy love of Liombruno who must remain a secret, it is the girl-dove who recovers her wings and flies. These stories differ but they tell of a precarious love that unites two incompatible worlds, and of a love tested by absence; stories of unknowable lovers who unite only in the moment in which they are lost to one another.

Fairy tales are rarely structured along the simple and basic lines we associate with a love story in which the characters fall in love and encounter obstacles to their marriage. This theme is developed only occasionally in some melancholy tales from Sardinia, a land where girls used to be courted from their windows. The countless tales of the conquest or liberation of a princess always deal with someone never seen, a victim to be released by a test of valor, or a stake to be won in a joust to fulfill a destiny, or else someone falls in love with a portrait or the sound of a name or envisages the beloved in a drop of blood on a white portion of ricotta. These are abstract or symbolic romantic attachments that savor of witchcraft or of malediction. However, the most positive and deeply felt loves in folklore are not these, but rather those in which the beloved is first possessed and then conquered.

Propp, in *Historical Roots of Russian Fairy Tales*, gives the Cupid and Psyche type of story a suggestive interpretation. Psyche is the girl who lives in a house where youths are segregated during the final phase of their initiation. She comes into contact with young men in the guise of animals, or in the dark, since they must be seen by no one. Hence it is as if only one invisible youth loved her. Once the period of initiation is over, the young men return home and forget the girl who lived segregated with them. They marry and begin new families. The story grows out of this crisis. It describes a love born during the initiation and doomed to destruction by religious laws, and shows how a woman rebels against this law and recovers her young lover. Although the customs of millennia are disregarded, the plot of the story still reflects the spirit of those laws and describes every love thwarted and forbidden by law, convention, or social disparity. That is why it has been possible, from prehistory to the present, to preserve, not as a fixed formula but as a flowing element, the sensuality so often underlying this love, evident in the ecstasy and frenzy of mysterious nocturnal embraces.

This eroticism of tales that we now consider as a part of children's literature proves that oral tradition was not intended for any particular age level; it was simply an account of marvels, a full expression of the poetic needs at that cultural stage.

Folktales especially intended for children exist, to be sure, but as a separate genre, neglected by more ambitious storytellers and carried on in a humbler and more familiar tradition, having the following characteristics: a theme of fear and cruelty, scatological or obscene details, lines of verse interpolated into the prose and slipping into nonsense rhymes (see story no. 37, "Petie Pete versus Witch Bea-Witch"), characteristics of coarseness and cruelty which would be considered wholly unsuitable in children's books today.

The tendency to dwell on the wondrous remains dominant, even when closely allied with morality. The moral is always implicit in the folktale in the victory of the simple virtues of the good characters and in the punishment of the equally simple and absolutely perverse wrongdoers: rarely is it sententiously or didactically presented. No doubt the moral function of the tale, in the popular conception, is to be sought not in the subject matter but in the very nature of the folktale, in the mere fact of telling and listening. That too can be interpreted as prudent and practical moralism, such as the tale of "The Parrot" seems to suggest (see no. 15). This is a tale within a tale; Comparetti and Pitrè both published it at the beginning of their anthologies as a kind of prologue. The parrot, by telling an interminable story, manages to save the virtue of a girl. It is a symbolic defense of the narrative art against those who accuse it of being profane and hedonistic. The suspense of the story keeps the fascinated listener from transgression. This is its minimal and conservative justification, but something more profound is revealed in the very narrative construction of "The Parrot": the art of storytelling which the narrator displays and which is humorously exemplified in the parody of tales that "never end." Therein lies, for us, its real moral: the storyteller, with a kind of instinctive skillfulness, shies away from the constraint of popular tradition, from the unwritten law that the common people are capable only of repeating trite themes without ever actually "creating"; perhaps the narrator thinks that he is producing only variations on a theme, whereas actually he ends up telling us what is in his heart.

A regard for conventions and a free inventiveness are equally necessary in constructing a folktale. Once the theme is laid out there are certain steps required to reach a solution; they are interchangeable ingredients—the horse hide carried up in flight by an eagle, the well that leads to the netherworld, dove-maidens whose clothes are stolen while they bathe, magic boots and cloak purloined by thieves, three nuts that must be cracked, the house of the winds where information is given about the path to be followed, and so on. It is up to the narrator to organize these, to pile them up like the bricks in a wall, hurrying over the dull places, all this depending upon the degree of the narrator's talent and what he puts into his story, mixing his own mortar of local color and personal tribulations and expectations.

Certainly the greater or lesser ease with which one picks one's way through a fantasy world has its grounds in one's actual experience and culture: we notice, for instance, the different ways Sicilian and Tuscan folktales refer to kings. As a rule, the court of kings in popular tales is a general and abstract concept, a vague symbol of power and wealth; in Sicily, however, king, court, and nobility are distinct and concrete institutions, with their hierarchy, protocol, and moral code—a whole world and terminology, mostly invented but with which the illiterate old women narrators are familiar down to the last detail. "There was a king of Spain who had a left-hand squire and a right-hand squire." It is characteristic of Sicilian folktales that kings never make an important decision without the advice of their counselors: "Gentlemen, what is your advice?" or, more briefly, the king shouts, "Counselors! Counselors!" and they advise him.

But Tuscany, although more cultivated in many areas, has never had a king; "king" is here a generic term with no institutional implication; it evokes no more than the condition of affluence; storytellers say "that king" just as they would say "that gentleman," without any royal association or any notion of a court, of an aristocratic hierarchy, or even of a real land. It is thus possible to find one king living next door to another, looking out the window at each other or paying visits to one another, just like two good country burghers.

In contrast to this world of kings is that of the peasants. The "realistic" foundation of many folktales, the point of departure spurred by dire need, hunger, or unemployment is typical of a large number of Italian popular narratives. I have already noted as a prime motif of numerous and particularly southern folktales that of the *cavoliccidaru* (cabbage picker): the cupboard is bare, so father or mother, along with daughters, scour the countryside for plants with which to make soup; pulling up a cabbage larger than the others, they come upon a passage into an underground world where a supernatural husband waits; or there may be a witch who will hold the girl prisoner, or a Bluebeard who feeds upon human flesh. Or else, especially in seaside localities, in place of the farmer who has neither land nor work, there is a hapless fisherman, who one day nets a big talking fish.

But the "realistic" state of destitution is not merely a starting point for the folktale, a sort of springboard into wonderland, a foil for the regal and the supernatural. There are folktales that deal with peasants from start to finish, with an agricultural laborer as hero, whose magic powers are merely complements to natural human strength and persistence. These folktales appear like fragments of an epic of laborers that never took shape and which on occasion borrows its themes from episodes of chivalry, replacing deeds and tournaments to win princesses with mounds of earth to be moved by plow or spade. Examples of these are the remarkable Sicilian tale, "Out in the World" and the Abruzzi "Joseph Ciufolo, Tiller-Flutist," or "The North Wind's Gift" from Tuscany, and "Fourteen" from Le Marche; and on the subject of women's work and tribulations, "Misfortune" and "The Two Cousins" (both Sicilian).

Those who know how rare it is in popular (and nonpopular) poetry to fashion a dream without resorting to escapism, will appreciate these instances of a self-awareness that does not deny the invention of a destiny, or the force of reality which bursts forth into fantasy. Folklore could teach us no better lesson, poetic or moral.

I. C.

Translated by Catherine Hill

ITALIAN FOLKTALES

1

Dauntless Little John

There was once a lad whom everyone called Dauntless Little John, since he was afraid of nothing. Traveling about the world, he came to an inn, where he asked for lodgings. "We have no room here," said the innkeeper, "but if you're not afraid, I will direct you to a certain palace where you can stay."

"Why should I be afraid?"

"People shudder at the thought of that palace, since nobody who's gone in has come out alive. In the morning the friars go up with the bier for anyone brave enough to spend the night inside."

So what did Little John do but pick up a lamp, a bottle, and a sausage, and march straight to the palace.

At midnight he was sitting at the table eating, when he heard a voice in the chimney. "Shall I throw it down?"

"Go ahead!" replied Little John.

Down the chimney into the fireplace fell a man's leg. Little John drank a glass of wine.

Then the voice spoke again. "Shall I throw it down?"

"Go ahead!" So another leg dropped into the fireplace. Little John bit into the sausage.

"Shall I throw it down?"

"Go ahead!" So down came an arm. Little John began whistling a tune.

"Shall I throw it down?"

"By all means!" And there was another arm.

"Shall I throw it down?"

"Yes!"

Then came the trunk of a body, and the arms and legs stuck onto it, and there stood a man without a head.

"Shall I throw it down?"

"Throw it down!"

Down came the head and sprang into place atop the trunk. He was truly a giant, and Little John raised his glass and said, "To your health!"

The giant said, "Take the lamp and come with me."

Little John picked up the lamp, but didn't budge.

"You go first!" said the giant.

"No, after you," insisted Little John.

"After you!" thundered the giant.

"You lead the way!" yelled Little John.

So the giant went first, with Little John behind him lighting the way, and they went through room after room until they had walked the whole length of the palace. Beneath one of the staircases was a small door.

"Open it!" ordered the giant.

"You open it!" replied Little John.

So the giant shoved it open with his shoulder. There was a spiral staircase.

"Go on down," directed the giant.

"After you," answered Little John.

They went down the steps into a cellar, and the giant pointed to a stone slab on the ground. "Raise that!"

"You raise it!" replied Little John, and the giant lifted it as though it were a mere pebble.

Beneath the slab were three pots of gold. "Carry those upstairs!" ordered the giant.

"You carry them up!" answered Little John. And the giant carried them up one by one.

When they were back in the hall where the great fireplace was, the giant said, "Little John, the spell has been broken!" At that, one of his legs came off and kicked its way up the chimney. "One of these pots of gold is for you." An arm came loose and climbed up the chimney. "The second pot of gold is for the friars who come to carry away your body, believing you perished." The other arm came off and followed the first. "The third pot of gold is for the first poor man who comes by." Then the other leg dropped off, leaving the giant seated on the floor. "Keep the palace for yourself." The trunk separated from the head and vanished. "The owners of the palace and their children are now gone forever." At that, the head disappeared up the chimney.

As soon as it was light, a dirge arose: *"Miserere mei, miserere mei."* The friars had come with the bier to carry off Little John's body. But there he stood, at the window, smoking his pipe!

Dauntless Little John was a wealthy youth indeed with all those gold pieces, and he lived happily in his palace. Then one day what should he do but look behind him and see his shadow: he was so frightened he died.

4

2

The Man Wreathed in Seaweed

A king had his crier announce in the town squares that whoever found his missing daughter would be rewarded with a fortune. But the announcement brought no results, since no one had any idea of the girl's whereabouts. She had been kidnapped one night, and they had already looked the world over for her.

A sea captain suddenly had the thought that since she wasn't on land she might well be on the sea, so he got a ship ready to go out in search of her. But when the time came to sign up the crew, not one sailor stepped forward, since no one wanted to go on a dangerous expedition that would last no telling how long.

The captain waited on the pier, but fearful of being the first to embark, no one approached his ship. Also on the pier was Samphire Starboard, a reputed tramp and tippler, whom no ship captain was ever willing to sign on.

"Listen," said our captain, "how would you like to sail with me?"

"I'd like to very much."

"Come aboard, then."

So Samphire Starboard was the first to embark. After that, other sailors took heart and boarded the ship.

Once he was on the ship, Samphire Starboard did nothing but stand around all day long with his hands in his pockets and dream about the taverns he had left behind. The other sailors cursed him because there was no knowing when the voyage would end, provisions were scarce, and he did nothing to earn his keep. The captain decided to get rid of him. "See that little island?" he asked, pointing to an isolated reef in the middle of the sea. "Get into a rowboat and go explore it. We'll be cruising right around here."

Samphire Starboard stepped into the rowboat, and the ship sailed away at full speed, leaving him stranded in the middle of the sea. He approached the reef, spied a cave, and went in. Tied up inside was a very beautiful maiden, who was none other than the king's daughter.

"How did you manage to find me?" she asked.

"I was fishing for octopi," explained Samphire.

"I was kidnapped by a huge octopus, whose prisoner I now am," said the king's daughter. "Flee before it returns. But note that for three hours a day it changes into a red mullet and can be caught. But you have to kill the mullet at once, or it will change into a sea gull and fly away."

Samphire Starboard hid his boat and waited out of sight on the reef. From the sea emerged the octopus, which was so large that it could reach clear around the island with its tentacles. All its suckers shook, having smelled a man on the reef. But the hour arrived when it had to change into a fish, and suddenly it became a red mullet and disappeared into the sea. Samphire Starboard lowered fishing nets and pulled them back up full of gurnard, sturgeon, and dentex. The last haul produced the red mullet, shaking like a leaf. Samphire raised his oar to kill it, but instead of the red mullet he struck the sea gull flying out of the net and broke its wing. The gull then changed back into an octopus, whose wounded tentacles spurted dark red blood. Samphire was upon it instantly and beat it to death with the oar. The king's daughter gave him a diamond ring as a token of the gratitude she would always feel toward him.

"Come and I'll take you to your father," he said, showing her into his boat. But the boat was tiny and they were out in the middle of the sea. After rowing and rowing they spied a ship in the distance. Samphire signaled to it with an oar draped with the king's daughter's gown. The ship spotted them and took them aboard. It was the same ship that had earlier discharged and abandoned Samphire. Seeing him back with the king's daughter, the captain said, "Poor Samphire Starboard! Here we thought you were lost and now, after looking all over for you, we see you return with the king's daughter! That calls for a real celebration!" To Samphire Starboard, who'd not touched a drop of wine for months on end, that seemed too good to be true.

They were almost in sight of their home port when the captain led Samphire to a table and placed several bottles of wine before him. Samphire drank and drank until he fell unconscious to the floor. Then the captain said to the king's daughter, "Don't dare tell your father that drunkard freed you. Tell him I freed you myself, since I'm the captain of the ship and ordered him to rescue you."

The king's daughter neither agreed nor disagreed. "I know what I'll tell him," she answered.

To be on the safe side, the captain decided to do away with Samphire Starboard once and for all. That night, they picked him up, still as drunk as could be, and threw him into the sea. At dawn the ship was in sight of port. With flags they signaled they were bringing home the king's daughter safe and sound. A band played on the pier, where the king waited with the entire court.

A date was chosen for the king's daughter to wed the captain. On the day of the wedding, the mariners in port saw a man emerge from the water. He was covered from head to foot with seaweed, and out of his pockets and the holes in his clothes swam fish and shrimps. It was none other than Samphire Starboard. He climbed out of the water and went ambling through the city streets, with seaweed draping his head and body and dragging along behind him. At that very moment the wedding procession was moving through the street and came face to face with the man wreathed in seaweed. Everyone stopped. "Who is this?" asked the king. "Seize him!" The guards came up, but Samphire Starboard raised a hand and the diamond on his finger sparkled in the sunlight.

"My daughter's ring!" exclaimed the king.

"Yes," said the daughter, "this man was my rescuer and will be my bridegroom."

Samphire Starboard told his story, and the captain was imprisoned. Green though he was with seaweed, Samphire took his place beside the bride clad in white and was joined to her in matrimony.

(Riviera ligure di ponente)

◆ 3 ◆

The Ship with Three Decks

Once there was a poor couple who lived way out in the country. A baby boy was born to them, but there was no one anywhere around to be his godfather. They went into town, but they didn't know a soul there and couldn't have the child baptized without a godfather. They saw a man wrapped in a black cloak on the church doorstep and asked, "Kind sir, would you please be this boy's godfather?" The man agreed, and the child was baptized.

When they came out of the church, the stranger said, "I now must give my godson his present. Take this purse, which is to be used to raise and educate him. And give him this letter when he has learned to read." The father and mother were thunderstruck, but before they could find words of thanks and ask the man his name, he had disappeared.

The purse was full of gold crowns, which paid for the boy's education. Once he could read, his parents gave him the letter, which said:

7

Dear Godson,

I am going back to repossess my throne after a long exile, and I need an heir. As soon as you read this letter, set out on a journey to your dear godfather, the king of England.

P.S. Along the way, beware of a cross-eyed man, a cripple, and a mangy character.

The youth said, "Father, Mother, farewell. I must go to my godfather." After a few days of walking, he met a traveler who asked, "Where are you going, my lad?"

"To England."

"So am I. We shall travel together."

The youth noticed the man's eyes: one of them looked east, and the other west, so the boy realized this was the cross-eyed man he must avoid. He found a pretext for stopping, then took another road.

He met another traveler sitting on a stone. "Are you going to England? We'll therefore travel together," said the stranger, who got up and limped along, leaning on a stick. He's the cripple, thought the youth, and changed roads again.

He met a third traveler, whose eyes, like his legs, bespoke perfect health. As for any scalp disease, this man had the thickest and cleanest head of black hair you ever saw. As the stranger was also on his way to England, they traveled together. They stopped for the night at an inn, where the youth, wary of his companion, handed over his purse and the letter for the king to the innkeeper for safekeeping. During the night while everybody was sleeping, the stranger rose and went to the innkeeper for purse, letter, and horse. In the morning the young man found himself alone, penniless, on foot, and with no letter for the king.

"Your servant came to me in the night," explained the innkeeper, "for all your belongings. Then he left. . . ."

The youth set out on foot. At a bend in the road he spied his horse tethered to a tree in a field. He was about to untie it, when from behind the tree rushed last night's companion armed with a pistol. "If you don't want to die on the spot," he said, "you must become my servant and pretend I'm the king of England's godson." As he spoke, he removed his black wig, revealing a scalp completely covered with mange.

They set out, the mangy one on horseback, the youth on foot, and at last reached England. With open arms the king welcomed the mangy one, taking him for his godson, while the real godson was assigned to the stables as stable boy. But the mangy one couldn't wait to get rid of his companion, and the opportunity soon presented itself. The king one day said to the false godson, "If you could free my daughter from the spell that holds her prisoner on a certain island, I'd give her to you in marriage. The only difficulty is that nobody who has attempted to free her has ever come back alive." The mangy one lost no time in replying. "Try sending my servant, who is surely capable of setting her free."

The king summoned the youth at once and asked, "Can you set my daughter free?"

"Your daughter? Tell me where she is, Majesty!"

The king would only say, "I warn you that you'll lose your head if you come back to me without her."

The youth went to the pier and watched the ships sail away. He had no idea how to reach the princess's island. An old sailor with a beard down to his knees approached him and said, "Ask for a ship with three decks."

The youth went to the king and had a ship with three decks rigged. When it was in port and ready to weigh anchor, the old sailor reappeared. "Now have one deck loaded with cheese rinds, another with bread crumbs, and the third with stinking carrion."

The youth had the three decks loaded.

"Now," said the old man, "when the king says, 'Choose all the sailors you want,' you will reply, 'I need only one,' and select me." That he did, and the whole town turned out to watch the ship sail off with that strange cargo and a crew of one, who also happened to be on his last legs.

They sailed for three months straight, at the end of which time they spied a lighthouse in the night and entered a port. All they could make out on shore were low, low houses and stealthy movement. At last a voice asked, "What cargo do you carry?"

"Cheese rinds," replied the old sailor.

"Fine," they said on shore. "That's what we need."

It was the Island of Rats, where all the inhabitants were rats, who said, "We'll buy the entire cargo, but we have no money with which to pay you. But any time you need us, you have only to say, 'Rats, fine rats, help us!' and we'll be right there to help you."

The youth and the sailor dropped the gangplank, and the rats came aboard and unloaded the cheese rinds in a flash.

From there the men sailed to another island. It was also night and they could make out nothing at all in port. It was worse than the other place, with not a house or a tree anywhere in sight. "What cargo do you bring?" asked voices in the dark.

"Bread crumbs," replied the sailor.

"Fine! That's just what we need!"

It was the Island of Ants, where all the inhabitants were ants. Nor did they have any money either, but they said, "Whenever you need us, you have only to say, 'Ants, fine ants, help us!' and we'll be right there, no matter where you are."

The ants carried all the bread crumbs down the fore and aft moorings, and the ship cast off again.

It came to an island of rocky cliffs that dropped straight down to port. "What cargo do you bring?" cried voices from above.

"Stinking carrion!"

"Excellent! That's just what we need," and huge shadows swooped down on the ship.

It was the Island of Vultures, inhabited entirely by those greedy birds. They flew off with every ounce of carrion, promising in return to help the men whenever they called, "Vultures, fine vultures, help us!"

After several more months of sailing, they landed on the island where the king of England's daughter was a prisoner. They disembarked, walked through a long cave, and emerged before a palace in a garden. A dwarf walked out to meet them. "Is the king of England's daughter here?" asked the youth.

"Come in and ask Fairy Sibiana," replied the dwarf, showing them into the palace, which had gold floors and crystal walls. Fairy Sibiana sat on a throne of crystal and gold.

"Kings and princes have brought entire armies to free the princess," said the fairy, "and every last one of them died."

"All I have are my will and my courage," said the youth.

"Well, then, you must undergo three trials. If you fail, you'll not get away from here alive. Do you see that mountain shutting out the sun from my view? You must level it by tomorrow morning. When I wake up I want the sunlight streaming into my room."

The dwarf came out with a pickax and led the youth to the foot of the mountain. The young man brought the pickax down once, and the blade snapped in two. "Now how am I going to dig?" he wondered, then remembered the rats on the other island. "Rats, fine rats, help me!"

He'd not got the words out of his mouth before the mountain was swarming with rats from top to bottom. They dug and gnawed and clawed, while the mountain dwindled and dwindled and dwindled....

Next morning Fairy Sibiana was awakened by the first rays of sun streaming into her room. "Congratulations!" she said to the youth, "but you're not done yet." She led him to the palace's underground vaults, in the center of which was a room with a ceiling as high as a church's and containing one big heap of peas and lentils that reached the ceiling. "You have this whole night to separate the peas from the lentils into two distinct piles. Heaven help you if you leave one single lentil in the pea pile, or one single pea in the lentil pile."

The dwarf left him a candle wick and went off with the fairy. As the wick burned down to nothing, the youth continued to stare at the huge pile, wondering how any human could ever accomplish so intricate a task. Then he remembered the ants on the other island. "Ants, fine ants," he called, "help me!"

No sooner had he said those words than the entire cellar teemed with those tiny insects. They converged on the heap and, with order and patience, made two separate piles, one team of ants carrying peas and the other lentils.

"I'm still not defeated," said the fairy when she saw the task completed. "A far more difficult trial now awaits you. You have from now till dawn to fetch me a barrel of the water of long life."

The spring of long life was at the top of a steep mountain infested with savage beasts. Scaling the mountain was out of the question, much less while carrying a barrel. But the youth called, "Vultures, fine vultures, help me!" and the sky darkened with vultures circling down to earth. The youth attached a phial to the neck of each, and the vultures soared in a grand formation straight to the spring on the mountaintop, filled their phials, and flew back with them to the youth, who poured the water into the barrel he had waiting.

When the barrel was full, hoofbeats were heard retreating. Fairy Sibiana was fleeing for dear life, followed by her dwarfs, while out of the palace ran the king of England's daughter, cheering: "I'm safe at last! You set me free!"

With the king's daughter and the water of long life, the youth returned to his ship, where the old sailor was all ready to weigh anchor.

The king of England scanned the sea every day through his telescope. Seeing a ship approach that was flying the English flag, he ran to port overjoyed. When the mangy one beheld the youth safe and sound and escorting the king's daughter, he was fit to be tied and resolved to have him killed.

While the king was celebrating his daughter's return with a grand banquet, two grim-looking fellows came to get the youth, saying it was a matter of life and death. Puzzled, he followed them. When they got to the woods, the two fellows, who were assassins hired by the mangy one, drew their knives and cut the youth's throat.

Meanwhile at the banquet, the king's daughter was more and more worried, since the youth had gone off with that sinister pair and not returned. She went out looking for him and, reaching the woods, found his body covered with wounds. But the old sailor had brought along the barrel containing the water of long life, in which he immersed the youth's body, only to see him jump right back out as sound as ever and so handsome that the king's daughter threw her arms around his neck.

The mangy one was livid with rage. "What's in that barrel?" he asked. "Boiling oil," replied the sailor.

So the mangy one had a barrel of oil heated to boiling and announced to the princess: "If you don't love me I'll kill myself." He stabbed himself with his dagger and leaped into the boiling oil. He was instantly scalded to death. Also his black wig had flown off when he leaped, revealing his mangy head.

"Ah, the mangy one!" exclaimed the king of England. "The cruelest of all my enemies. He finally got what was coming to him. So you, valiant youth, are my godson! You shall marry my daughter and inherit my kingdom!" And so it was.

(Riviera ligure di ponente)

◈ 4 ◈

The Man Who Came Out Only at Night

Long ago there lived a poor fisherman with three marriageable daughters. A certain young man asked for the hand of one of them, but people were wary of him since he came out only at night. The oldest daughter and then the middle daughter both said no to him, but the third girl said yes. The wedding was celebrated at night, and as soon as the couple was alone, the bridegroom announced to his bride: "I must tell you a secret: I am under an evil spell and doomed to be a tortoise by day and a man at night. There's only one way to break the spell: I must leave my wife right after the wedding and travel around the world, at night as a man and by day as a tortoise. If I come back and find that my wife has remained loyal to me all along and endured every hardship for my sake, I'll become a man again for good."

"I am willing," said the bride.

The bridegroom slipped a diamond ring on her finger. "If you use it to a good end, this ring will help you in whatever situation you find yourself."

Day had dawned, and the bridegroom turned into a tortoise and crawled off to begin his journey around the world.

The bride went about the city in search of work. Along the way, she came across a child crying and said to his mother, "Let me hold him in my arms and calm him."

"You'd be the first person to do that," answered the mother. "He's been crying all day long."

"By the power of the diamond," whispered the bride, "may the child laugh and dance and frolic!" At that, the child started laughing, dancing, and frolicking.

Next, the bride entered a bakery and said to the woman who owned it, "You'll have no regrets if you hire me to work for you." The owner hired her, and she began making bread, saying under her breath, "By the power of the diamond, let the whole town buy bread at this bakery as long as I work here!" From then on, people poured in and out with no sign of a letup. Among the customers were three young men who saw the bride and fell in love with her.

"If you let me spend a night with you," one of them said to her, "I'll give you a thousand francs."

"I'll give you two thousand," said another.

"And I'll make it three thousand," said the third.

She collected the three thousand francs from the third man and smuggled him into the bakery that very night.

"I'll be with you in a minute," she told him, "after I've put the yeast into the flour. While you're waiting, would you please knead the dough a little bit for me?"

The man began kneading, and kneaded and kneaded and kneaded. By the power of the diamond, he couldn't for the life of him take his hands out of the dough, and therefore went on kneading till daylight.

"So you finally finished!" she said to him. "You really took your time!"

And she sent him packing.

Then she said yes to the man with the two thousand francs, brought him in as soon as it grew dark, and told him to blow on the fire a moment so that it wouldn't go out. He blew and blew and blew. By the power of the diamond, he had to keep right on blowing up to the next morning, with his face bulging like a wineskin.

"What a way to behave!" she said to him in the morning. "You come to see me, but spend the night blowing on the fire!"

And she sent him packing.

The next night she brought in the man with the thousand francs. "I have to add the yeast," she told him. "While I'm doing that, go shut the door."

The man shut the door, which by the power of the diamond came open again right away. All night long he closed it only to see it immediately reopen, and in no time the sun was up.

"Did you finally close this door? Well, you may now open it again and get out."

Seething with rage, the three men denounced her to the authorities. In that day and time there were, in addition to policemen, women officers who were called whenever a woman was to be brought into custody. So four women officers went to apprehend the bride.

"By the power of the diamond," said the bride, "let these women box one another's ears until tomorrow morning."

The four women officers began boxing one another's ears so hard that their heads swelled up like pumpkins, and they still went on striking each other for all they were worth.

When the women officers failed to return with the culprit, four male officers were sent out to look for them. The bride saw them coming and said, "By the power of the diamond, let those men play leapfrog." One of the male officers dropped down at once on all fours; a second one moved up, put his hands on the officer's back, and leaped over him, with the third and fourth following in his tracks. Thus began a game of leapfrog.

Right at that point, a tortoise came crawling into view. It was the husband returning from his trip around the world. He saw his wife, and behold! He was again a handsome young man, and a handsome young man he remained, by his wife's side, up to a ripe old age.

(Riviera ligure di ponente)

◈ 5 ◈

And Seven!

A woman had a daughter who was big and fat and so gluttonous that when her mother brought the soup to the table she would eat one bowl, then a second, then a third, and keep on calling for more. Her mother filled her bowl, saying, "That makes three! And four! And five!" When the daughter asked for a seventh bowl of soup, her mother, instead of filling the bowl, whacked her over the head, shouting, "And seven!"

A well-dressed young man was passing by just then and saw the mother through the window hitting the girl and crying, "And seven!"

As the big fat young lady captured his fancy immediately, he went in and asked, "Seven of what?"

Ashamed of her daughter's gluttony, the mother replied, "Seven spindles of hemp! I have a daughter so crazy about work that she'd even spin the wool on the sheep's back! Can you imagine that she's already spun seven spindles of hemp this morning and still wants to spin? To make her stop, I have to beat her."

"If she's that hard-working, give her to me," said the young man. "I'll try her out to see if you're telling the truth and then I'll marry her."

He took her to his house and shut her up in a room full of hemp waiting to be spun. "I'm a sea captain, and I'm leaving on a voyage," he said. "If you've spun all this hemp by the time I return, I'll marry you."

The room also contained exquisite clothes and jewels, for the captain happened to be very rich. "When you become my wife," he explained, "these things will all be yours." Then he left her.

The girl spent her days trying on dresses and jewels and admiring herself in the mirror. She also devoted much time to planning meals, which the household servants prepared for her. None of the hemp was spun yet, and in one more day the captain would be back. The girl gave up all hope of ever marrying him and burst into tears. She was still crying when through the window flew a bundle of rags and came to rest on its feet: it was an old woman with long eyelashes. "Don't be afraid," she told the girl. "I've come to help you. I'll spin while you make the skein."

You never saw anyone spin with the speed of that old woman. In just a quarter of an hour she had spun every bit of hemp. And the more she spun, the longer her lashes became; longer than her nose, longer than her chin, they came down more than a foot; and her eyelids also grew much longer.

When the work was finished, the girl said, "How can I repay you, my good lady?"

"I don't want to be repaid. Just invite me to your wedding banquet when you marry the captain."

"How do I go about inviting you?"

"Just call 'Columbina' and I'll come. But heaven help you if you forget my name. It would be as though I'd never helped you, and you'd be undone."

The next day the captain arrived and found the hemp all spun. "Excellent!" he said. "I believe you're just the bride I was seeking. Here are the clothes and jewels I bought for you. But now I have to go on another voyage. Let's have a second test. Here's twice the amount of hemp I gave you before. If you spin it all by the time I return, I'll marry you."

As she had done before, the girl spent her time trying on gowns and jewels, eating soup and lasagna, and got to the last day with all the hemp still waiting to be spun. She was weeping over it when, lo and behold, something dropped down the chimney, and into the room rolled a bundle of rags. It came to rest on its feet, and there stood an old woman with sagging lips. This one too promised to help, began spinning, and worked even faster than the other old woman. The more she spun, the more her lips sagged. When the hemp was all spun in a half-hour, the old woman asked only to be invited to the wedding banquet. "Just call 'Columbara.' But don't forget my name, or my help will have been in vain and you will suffer."

The captain returned and asked before he even got into the house, "Did you spin it all?"

"I just now finished!"

"Take these clothes and jewels. Now, if I come back from my third voyage and find you've spun this third load of hemp, which is much bigger than the other two, I promise we'll get married at once."

As usual, the girl waited until the last day without touching the hemp. Down from the roof's gutter fell a bundle of rags, and out came an old woman with buckteeth. She began spinning, spinning ever faster, and the more she spun, the longer grew her teeth.

"To invite me to your wedding banquet," said the old woman, "you must call 'Columbun.' But if you forget my name, it would be better if you'd never seen me."

When the captain came home and found the hemp all spun, he was completely satisfied. "Fine," he said, "now you will be my wife." He ordered preparations made for the wedding, to which he invited all the nobility in town.

Caught up in the preparations, the bride thought no more of the old women. On the morning of the wedding she remembered that she was supposed to invite them, but when she went to pronounce their names, she found they had slipped her mind. She cudgeled her brains but, for the life of her, couldn't recall a single name.

From the cheerful girl she was, she sank into a state of bottomless gloom. The captain noticed it and asked her what the matter was, but she would say nothing. Unable to account for her sadness, the bridegroom thought, This is perhaps not the right day. He therefore postponed the wedding until the day after. But the next day was still worse, and the day following we won't even mention. With every day that passed, the bride became gloomier and quieter, with her brows knit in concentration. He told her jokes and stories in an effort to make her laugh, but nothing he said or did affected her.

Since he couldn't cheer her up, he decided to go hunting and cheer himself up. Right in the heart of the woods he was caught in a storm and took refuge in a hovel. He was in there in the dark, when he heard voices:

"O Columbina!"

"O Columbara!"

"O Columbun!"

"Put on the pot to make polenta! That confounded bride won't be inviting us to her banquet after all!"

The captain wheeled around and saw three crones. One had eyelashes that dragged on the ground, another lips that hung down to her feet, and the third teeth that grazed her knees.

Well, well, he thought to himself. Now I can tell her something that will make her laugh. If she doesn't laugh over what I've just seen, she'll never laugh at anything!

He went home and said to his bride, "Just listen to this. Today I was in the woods and went into a hovel to get out of the rain. I go in and what should I see but three crones: one with eyelashes that dragged on the ground, another with lips that hung down to her feet, and the third with teeth that grazed her knees. And they called each other: 'O Columbina,' 'O Columbara,' 'O Columbun!' "

The bride's face brightened instantly, and she burst out laughing, and laughed and laughed. "Order the wedding banquet right away. But I'm asking one favor of you: since those three crones made me laugh so hard, let me invite them to the banquet."

Invite them she did. For the three old women a separate round table was set up, but so small that what with the eyelashes of one, the lips of the other, and the teeth of the third, you no longer knew what was what.

When dinner was over, the bridegroom asked Columbina, "Tell me, good lady, why are your lashes so long?"

"That's from straining my eyes to spin fine thread!" said Columbina.

"And you, why are your lips so thick?"

"That comes from always rubbing my finger on them to wet the thread!" said Columbara.

"And you, how on earth did your teeth get so long?"

"That's from biting the knot of the thread!" said Columbun.

"I see," said the bridegroom, and he turned to his wife. "Go get the spindle." When she brought it to him, he threw it into the fire. "You'll spin no more for the rest of your life!"

So the big, fat bride lived happily ever after.

(Riviera ligure di ponente)

◈ 6 ◈

Body-without-Soul

There was a widow with a son named Jack, who at thirteen wanted to leave home to seek his fortune. His mother said to him, "What do you expect to do out in the world? Don't you know you're still a little boy? When you're able to fell that pine tree behind our house with one kick, then you can go."

Every day after that, as soon as he rose in the morning, Jack would get a running start and jump against the trunk of the tree with both feet, but the pine never budged an inch and he fell flat on his back. He would get up again, shake the dirt off, and go back inside.

At last one fine morning he jumped with all his might, and the tree gave way and toppled to the ground, his roots in the air. Jack ran and got his mother who, surveying the felled tree, said, "You may now go wherever you wish, my son." Jack bid her farewell and set out.

After walking for days and days he came to a city whose king had a horse named Rondello that no one had ever been able to ride. People constantly tried, but were thrown just when it appeared they would succeed. Looking on, Jack soon realized that the horse was afraid of its own shadow, so he volunteered to break Rondello himself. He began by going up to the horse in the stable, talking to it and patting it; then he suddenly jumped into the saddle and rode the animal outside straight into the sun. That way it couldn't see any shadow to frighten it. Jack took a steady hold of the reins, pressed his knees to the horse, and galloped off. A quarter of an hour later Rondello was as docile as a lamb, but let no one ride him after that but Jack.

From then on, Jack served the king, who was so fond of him that the other servants grew jealous and plotted to get rid of him.

Now the king had a daughter who had been kidnapped in her infancy by the sorcerer Body-without-Soul, and no one had heard of her since. The servants went to the king claiming Jack had boasted to everybody he would free her. The king sent for him. Jack was amazed and said this was the first he had even heard of the king's daughter. But the fact that anyone had dared make light of the episode concerning his daughter so infuriated the king that he said, "Either you free her, or I'll have you beheaded."

Since there was no calming the king now, Jack asked for a rusty sword they kept hanging on the wall, saddled Rondello, and rode off. Crossing a forest, he saw a lion motioning him to stop. Although a bit uneasy, Jack disliked the idea of running away, so he dismounted and asked what the lion wanted.

"Jack," said the lion, "as you can see, there are four of us here: myself, a dog, an eagle, and an ant. We have a dead donkey to parcel among us. Since you have a sword, carve the animal and give us each a portion." Jack cut off the donkey's head and gave it to the ant. "Here you are. This will make you a nice home and supply you with all the food you'll ever want." Next he cut off the hoofs and gave them to the dog. "Here's something to gnaw on as long as you like." He cut out the entrails and gave them to the eagle. "This is your food, which you can carry to the treetops where you perch." All the rest he gave to the lion, which as the biggest of the four deserved the largest portion. He got back on his horse and started off, only to hear his name called. "Dear me," he thought, "I must have made some mistake in dividing the parts." But the lion said to him, "You did us a big favor and you were very fair. As one good deed deserves another, I'm giving you one of my claws which will turn you into the fiercest lion in the world when you wear it." The dog said, "Here is one of my whiskers, which will turn you into the fastest dog on earth, whenever you place it under your nose." The eagle said, "Here is a feather from my wings which can change you into the biggest and strongest eagle in the sky." The ant said, "I'm giving you one of my tiny legs. Put it on and you will become an ant so small that no one can see you, even with a magnifying glass."

Jack took his presents, thanked the four animals, and departed. As he was uncertain whether the gifts were magic or not, thinking the animals might have played a joke on him, he stopped as soon as he was out of sight to test them. He became lion, dog, eagle, and ant; next ant, eagle, dog, and lion; then eagle, ant, lion, and dog; finally dog, lion, ant, and eagle. Yes, everything worked like a charm! All smiles, he moved onward.

Beyond the forest was a lake, on whose shore stood the castle of Body-without-Soul. Jack changed into an eagle and flew straight to the edge of a closed window. Then he changed into an ant and crawled into the room. It was a beautiful bedchamber where, beneath a canopy, lay the king's daughter asleep. Still an ant, Jack went crawling over her cheek until she awakened. Then he removed the tiny ant leg, and the king's daughter suddenly beheld a handsome youth at her side.

"Don't be afraid," he said, signaling silence. "I've come to free you. You must get the sorcerer to tell you what could kill him."

When the sorcerer returned, Jack changed back into an ant. The king's daughter made a big to-do over Body-without-Soul, seating him at her feet and drawing his head onto her lap. Then she began: "My darling sorcerer, I know you're a body without a soul and therefore incapable of dying. But I live in constant fear of someone finding your soul and putting you to death."

"I can tell *you* the secret," replied the sorcerer, "since you're imprisoned here and can't possibly betray me. To slay me would require a lion mighty enough to kill the black lion in the forest. Out of the belly of the dead lion would leap a black dog so swift that only the fastest dog on earth could catch it. Out of the belly of the dead black dog would fly a black eagle that could withstand every eagle under the sun. But if by chance that eagle were slain, a black egg would have to be taken out of its craw and cracked over my brow for my soul to fly away and leave me dead. Does all that seem easy? Do you have any real grounds for worry?"

With his tiny ant ears, Jack took in every word, then crawled back under the window to the ledge, where he again turned into an eagle and soared into the forest. There he changed into a lion and stalked the underbrush until he came face to face with the black lion. The black lion jumped him, but Jack, being the strongest lion in the world, tore it to bits. (Back at the castle, the sorcerer felt his head spin.) The lion's belly was slit open, and out bolted a swift-footed black dog, but Jack turned into the fastest dog on earth, caught him, and they rolled together in a ball, biting each other until the black dog lay dead. (Back at the castle, the sorcerer had to take to his bed.) The dog's belly was slit open and out flew a black eagle, but Jack became the most powerful eagle under the sun and they soared through the sky pecking and clawing each other until the black eagle folded its wings and fell to earth. (At the castle, the sorcerer ran a high fever and curled up under the bedclothes.)

Jack changed back into a man, opened the eagle's craw, and removed the black egg. He returned to the castle and gave it to the king's daughter, who was overjoyed.

"How on earth did you do it?" she asked.

"Nothing to it," replied Jack. "The rest is now up to you."

The king's daughter entered the sorcerer's bedchamber, asking, "How do you feel?"

"Woe's me! I've been betrayed . . ."

"I brought you a cup of broth. Drink some."

The sorcerer sat up in his bed and bent over to drink the broth.

"Here, let me break an egg into it and give it more body." At that, the king's daughter broke the black egg over his brow and Body-without-Soul died on the spot.

Jack took the king's daughter home to her father. Everyone was overjoyed, and the young couple was married forthwith.

(Riviera ligure di ponente)

❖ 7 ❖

Money Can Do Everything

There was once a prince as rich as cream, who took it into his head to put up a palace right across the street from the king's, but a palace far more splendid than the king's. Once it was finished, he put on its front in bold lettering: MONEY CAN DO EVERYTHING.

When the king came out and saw that, he sent immediately for the prince, who was new in town and hadn't yet visited the court.

"Congratulations," the king said. "Your palace is a true wonder. My house looks like a hut compared with it. Congratulations! But was it your idea to put up the words: Money can do everything?"

The prince realized that maybe he had gone too far.

"Yes it was," he answered, "but if Your Majesty doesn't like it, I can easily have the letters stripped off."

"Oh, no, I wouldn't think of having you do that. I merely wanted to hear from your own lips what you meant by such a statement. For instance, do you think that, with your money, you could have me assassinated?"

The prince realized he had got himself into a tight spot.

"Oh, Majesty, forgive me. I'll have the words removed at once. And if you don't like the palace, just say so, and I'll have it torn down too."

"No, no, leave it the way it is. But since you claim a person with money can do anything, prove it to me. I'll give you three days to try to talk to my daughter. If you manage to speak to her, well and good; you will marry her. If not, I'll have you beheaded. Is that clear?"

The prince was too distressed to eat, drink, or sleep. Day and night, all he thought of was how he might save his neck. By the second day he was certain of failure and decided to make his will. His plight was hopeless, for the king's daughter had been closed up in a castle surrounded by one hundred guards. Pale and limp as a rag, the prince lay on his bed waiting to die, when in walked his old nurse, a decrepit old soul now who had nursed him as a baby and who still worked for him. Finding him so haggard, the old woman asked what was wrong. Hemming and hawing, he told her the whole story.

"So?" said the nurse. "And you're giving up, like that? You make me laugh! I'll see what I can do about all this!"

Off she wobbled to the finest silversmith in town and ordered him to make a solid silver goose that would open and close its bill. The goose was to be as big as a man and hollow inside. "It must be ready tomorrow," she added.

"Tomorrow? You're crazy!" exclaimed the silversmith.

"Tomorrow I said!" The old woman pulled out a purse of gold coins and continued, "Think it over. This is the down payment. I'll give you the rest tomorrow when you deliver the goose."

The silversmith was dumbfounded. "That makes all the difference in the world," he said. "I'll do my best to have the goose tomorrow."

The next day the goose was ready, and it was a beauty.

The old woman said to the prince, "Take your violin and get inside the goose. Play as soon as we reach the road."

They wound their way through the city, with the old woman pulling the silver goose along by a ribbon and the prince inside playing his violin. The people lined the streets to watch: there wasn't a soul in town that didn't come running to see the beautiful goose. Word of it reached the castle where the king's daughter was shut up, and she asked her father to let her go and see the unusual sight.

The king said, "Time's up for that boastful prince tomorrow. You can go out then and see the goose."

But the girl had heard that the old woman with the goose would be gone by tomorrow. Therefore the king had the goose brought inside the castle so his daughter could see it. That's just what the old woman was counting on. As soon as the princess was alone with the silver goose and delighting in the music pouring from its bill, the goose suddenly opened and out stepped a man.

"Don't be afraid," said the man. "I am the prince who must either speak to you or be decapitated by your father tomorrow morning. You can say you spoke to me and save my life."

The next day the king sent for the prince. "Well, did your money make it possible for you to speak to my daughter?"

"Yes, Majesty," answered the prince.

"What! Do you mean you spoke to her?"

"Ask her."

The girl came in and told how the prince was hidden in the silver goose which the king himself had ordered brought inside the castle.

The king, at that, removed his crown and placed it on the prince's head. "That means you have not only money but also a fine head! Live happily, for I am giving you my daughter in marriage."

(Genoa)

◈ 8 ◈

The Little Shepherd

There was once a shepherd boy no bigger than a mite and as mean as could be. On his way out to pasture one day, he passed a poultry dealer carrying a basket of eggs on her head. So what did he do but throw a stone into the basket and break every single egg. Enraged, the poor woman screamed a curse: "You shall get no bigger until you've found lovely Bargaglina of the three singing apples!"

From that time on, the shepherd boy grew thin and puny, and the more his mother attended to him, the punier he became. Finally she asked, "What on earth has happened to you? Have you done a bad turn for which someone placed a curse on you?" He then told her about his meanness to the poultry dealer, repeating the woman's words to him, "You shall get no bigger until you've found lovely Bargaglina of the three singing apples!"

"In that case," said his mother, "you've no choice but to go in search of this lovely Bargaglina."

The shepherd set out. He came to a bridge, on which a little lady was rocking to and fro in a walnut shell.

"Who goes there?"

"A friend."

"Lift my eyelids a little, so I can see you."

"I'm seeking lovely Bargaglina of the three singing apples. Do you know anything about her?"

"No, but take this stone; it will come in handy."

The shepherd came to another bridge, where another little lady was bathing in an eggshell.

"Who goes there?"

"A friend."

"Lift my eyelids a little, so I can see you."

"I'm seeking lovely Bargaglina of the three singing apples. Have you any news of her?"

"No, but take this ivory comb, which will come in handy."

The shepherd put it in his pocket and walked on until he came to a stream where a man was filling a bag with fog. When asked about lovely Bargaglina, the man claimed to know nothing about her, but he gave the shepherd a pocketful of fog, which would come in handy.

Next he came to a mill whose miller, a talking fox, said, "Yes, I know who lovely Bargaglina is, but you'll have difficulty finding her. Walk straight ahead until you come to a house with the door open. Go inside and you'll see a crystal cage hung with many little bells. In the cage are the singing apples. You must take the cage, but watch out for a certain old woman. If her eyes are open, that means she's asleep. If they're closed, she's surely awake."

The shepherd moved on. He found the old woman with her eyes closed and realized she was awake. "My lad," said the old woman, "glance down in my hair and see if I've any lice."

He looked, and as he was delousing her, she opened her eyes and he knew she had fallen asleep. So he quickly picked up the crystal cage and fled. But the little bells on the cage tinkled, and the old woman awakened and sent a hundred horsemen after him. Hearing them almost upon him, the shepherd dropped the stone he had in his pocket. It changed instantly into a steep, rocky mountain, and the horses all fell and broke their legs.

Now horseless, the cavalrymen returned to the old woman, who then sent out two hundred mounted soldiers. Seeing himself in new peril, the shepherd threw down the ivory comb. It turned into a mountain as slick as glass, down which horses and riders all slid to their death.

The old woman then sent three hundred horsemen after him, but he pulled out the pocketful of fog, hurled it over his shoulder, and the army got lost in it. Meanwhile, the shepherd had grown thirsty and, having nothing with him to drink, removed one of the three apples from the cage and cut into it. A tiny voice said, "Gently, please, or you'll hurt me." Gently, he finished cutting the apple, ate one half, and put the other in his pocket. At length he came to a well near his house, where he reached into his pocket for the rest of the apple. In its place was a tiny, tiny lady.

"I'm lovely Bargaglina," she said, "and I like cake. Go get me a cake, I'm famished."

The well was one of those closed wells, with a hole in the center, so the shepherd seated the lady on the rim, telling her to wait there until he came back with the cake.

Meanwhile, a servant known as Ugly Slave came to the well for water. She spied the lovely little lady and said, "How come you're so little and beautiful while I'm so big and ugly?" And she grew so furious that she threw the tiny creature into the well.

The shepherd returned and was heartbroken to find lovely Bargaglina gone.

Now his mother also went to that well for water, and what should she find in her bucket one day but a fish. She took it home and fried it. They ate it and threw the bones out the window. There where they fell, a tree grew up and got so big that it shut out all the light from the house. The shepherd therefore cut it down and chopped it up for firewood, which he brought inside. By that time his mother had died, and he lived there all by himself, now punier than ever, since no matter what he tried, he couldn't grow any bigger. Every day he went out to the pasture and came back home at night. How great was his amazement upon finding the dishes and pans he'd used in the morning all washed for him when he came home! He couldn't imagine who was doing this. At last he decided to hide behind the door and find out. Whom should he then see but a very dainty maiden emerge from the woodpile, wash the dishes, sweep the house, and make his bed, after which she opened the cupboard and helped herself to a cake.

Out sprang the shepherd, asking, "Who are you? How did you get in?"

"I'm lovely Bargaglina," replied the maiden, "the girl you found in your pocket in place of the apple half. Ugly Slave threw me into the well, and I turned into a fish, then into fishbones thrown out the window. From fishbones I changed into a tree seed, next into a tree that grew and grew, and finally into firewood you cut. Now, every day while you're away, I become lovely Bargaglina."

Thanks to the rediscovery of lovely Bargaglina, the shepherd grew by leaps and bounds, and lovely Bargaglina along with him. Soon he was a handsome youth and married lovely Bargaglina. They had a big feast. I was there, under the table. They threw me a bone, which hit me on the nose and stuck for good.

(Inland vicinity of Genoa)

◈ 9 ◈

Silver Nose

There was once a widowed washerwoman with three daughters. All four of them worked their fingers to the bone washing, but they still went hungry. One day the oldest daughter said to her mother, "I intend to leave home, even if I have to go and work for the Devil."

"Don't talk like that, daughter," replied the mother. "Goodness knows what might happen to you."

Not many days afterward, they received a visit from a gentleman attired in black. He was the height of courtesy and had a silver nose.

"I am aware of the fact that you have three daughters," he said to the mother. "Would you let one come and work for me?"

The mother would have consented at once, had it not been for that silver nose which she didn't like the looks of. She called her oldest girl aside and said, "No man on earth has a silver nose. If you go off with him you might well live to regret it, so watch out."

The daughter, who was dying to leave home, paid no attention to her mother and left with the man. They walked for miles and miles, crossing woods and mountains, and finally came in sight of an intense glow in the distance like that of a fire. "What is that I see way down there in the valley?" asked the girl, growing uneasy.

"My house. That's just where we are going," replied Silver Nose.

The girl followed along, but couldn't keep from trembling. They came to a large palace, and Silver Nose took her through it and showed her every room, each one more beautiful than the other, and he gave her the key to each one. When they reached the door of the last room, Silver Nose gave her the key and said, "You must never open this door for any reason whatever, or you'll wish you hadn't! You're in charge of all the rooms but this one."

He's hiding something from me, thought the girl, and resolved to open that door the minute Silver Nose left the house. That night, while she was sleeping in her little room, in tiptoed Silver Nose and placed a rose in her hair. Then he left just as quietly as he had entered.

The next morning Silver Nose went out on business. Finding herself alone with all the keys, the girl ran and unlocked the forbidden door. No sooner had she cracked it than smoke and flames shot out, while she caught sight of a crowd of damned souls in agony inside the fiery room. She then realized that Silver Nose was the Devil and that the room was Hell. She screamed, slammed the door, and took to her heels. But a tongue of fire had scorched the rose she wore in her hair.

Silver Nose came home and saw the singed rose. "So that's how you obey me!" he said. He snatched her up, opened the door to Hell, and flung her into the flames.

The next day he went back to the widow. "Your daughter is getting along very well at my house, but the work is so heavy she needs help. Could you send us your second daughter too?" So Silver Nose returned home with one of the girl's sisters. He showed her around the house, gave her all the keys, and told her she could open all the rooms except the last. "Do you think," said the girl, "I would have any reason to open it? I am not interested in your personal business." That night after the girl went to sleep, Silver Nose tiptoed in and put a carnation in her hair.

When Silver Nose went out the next morning, the first thing the girl did was go and open the forbidden door. She was instantly assailed by smoke, flames, and howls of the damned souls, in whose midst she spotted her sister. "Sister, free me from this Hell!" screamed the first girl. But the middle girl grew weak in the knees, slammed the door, and ran. She was now sure that Silver Nose was the Devil, from whom she couldn't hide or escape. Silver Nose returned and noticed her hair right away. The carnation was withered, so without a word he snatched her up and threw her into Hell too.

The next day, in his customary aristocratic attire, he reappeared at the washerwoman's house. "There is so much work to be done at my house that not even two girls are enough. Could I have your third daughter as well?" He thus returned home with the third sister, Lucia, who was the most cunning of them all. She too was shown around the house and given the same instructions as her sisters. She too had a flower put in her hair while she was sleeping: a jasmine blossom. The first thing Lucia did when she got up next morning was arrange her hair. Looking in the mirror, she noticed the jasmine. "Well, well!" she said. "Silver Nose pinned a jasmine on me. How thoughtful of him! Who knows why he did it? In any case I'll keep it fresh." She put it into a glass of water, combed her hair, then said, "Now let's take a look at that mysterious door."

She just barely opened it, and out rushed a flame. She glimpsed countless people burning, and there in the middle of the crowd were her big sisters. "Lucia! Lucia!" they screamed. "Get us out of here! Save us!"

At once Lucia shut the door tightly and began thinking how she might rescue her sisters.

By the time the Devil got home, Lucia had put her jasmine back in her hair, and acted as though nothing had happened that day. Silver Nose looked at the jasmine. "Oh, it's still fresh," he said.

"Of course, why shouldn't it be? Why would anyone wear withered flowers in her hair?"

"Oh, I was just talking to be talking," answered Silver Nose. "You seem like a clever girl. Keep it up, and we'll never quarrel. Are you happy?"

"Yes, but I'd be happier if I didn't have something bothering me."

"What's bothering you?"

"When I left my mother, she wasn't feeling too well. Now I have no news at all of her."

"If that's all you're worried about," said the Devil, "I'll drop by her house and see how she's doing."

"Thank you, that is very kind of you. If you can go tomorrow, I'll get up a bag of laundry at once which my mother can wash if she is well enough. The bag won't be too heavy for you, will it?"

"Of course not. I can carry anything under the sun, no matter how heavy it is."

When the Devil went out again that day, Lucia opened the door to Hell, pulled out her oldest sister, and tied her up in a bag. "Keep still in there, Carlotta," she told her. "The Devil himself will carry you back home. But any time he so much as thinks of putting the bag down, you must say, 'I see you, I see you!'"

The Devil returned, and Lucia said, "Here is the bag of things to be washed. Do you promise you'll take it straight to my mother?"

"You don't trust me?" asked the Devil.

"Certainly I trust you, all the more so with my special ability to see from a great distance away. If you dare put the bag down somewhere, I'll see you."

"Yes, of course," said the Devil, but he had little faith in her claim of being able to see things a great distance away. He flung the bag over his shoulder. "My goodness, this dirty stuff is heavy!" he exclaimed.

"Naturally!" replied the girl. "How many years has it been since you had anything washed?"

Silver Nose set out for the washerwoman's, but when he was only halfway there, he said to himself, "Maybe . . . but I shall see if this girl isn't emptying my house of everything I own, under the pretext of sending out laundry." He went to put the bag down and open it.

"I see you, I see you!" suddenly screamed the sister inside the bag.

"By Jove, it's true! She can see from afar!" exclaimed Silver Nose. He threw the bag back over his shoulder and marched straight to Lucia's mother's house. "Your daughter sends you this stuff to wash and wants to know how you are. . . ."

As soon as he left, the washerwoman opened the sack, and you can imagine her joy upon finding her oldest daughter inside.

A week later, sly Lucia pretended to be sad once more and told Silver Nose she wanted news of her mother.

She sent him to her house with another bag of laundry. So Silver Nose carried off the second sister, without managing to peep inside because of the "I see you, I see you!" which came from the bag the instant he started to open it. The washerwoman, who now knew Silver Nose was the Devil, was quite frightened when he returned, for she was sure he would ask for the clean wash from last time. But Silver Nose put down the new bag and said, "I'll get the clean wash some other time. This heavy bag has broken my back, and I want to go home with nothing to carry."

When he had gone, the washerwoman anxiously opened the bag and embraced her second daughter. But she was more worried than ever about Lucia, who was now alone in the Devil's hands.

What did Lucia do? Not long afterward she started up again about news of her mother. By now the Devil was sick and tired of carrying laundry, but he had grown too fond of this obedient girl to say no to her. As soon as it grew dark, Lucia announced she had a bad headache and would go to bed early. "I'll prepare the laundry and leave the bag out for you, so if I don't feel like getting up in the morning, you can be on your way."

Now Lucia had made a rag doll the same size as herself. She put it in bed under the covers, cut off her own braids, and sewed them on the doll's head. The doll then looked like Lucia asleep, and Lucia closed herself up in the bag.

In the morning the Devil saw the girl snuggled down under the covers and set out with the bag over his shoulder. "She's sick this morning," he said to himself, "and won't be looking. It's the perfect time to see if this really is nothing but laundry." At that, he put the bag down and was about to open it. "I see you, I see you!" cried Lucia.

"By Jove, it's her voice to a tee, as though she were right here! Better not joke with such a girl." He took up the bag again and carried it to the washerwoman. "I'll come back later for everything," he said rapidly. "I have to get home right away because Lucia is sick."

So the family was finally reunited. Since Lucia had also carried off great sums of the Devil's money, they were now able to live in comfort and happiness. They planted a cross before the door, and from then on, the Devil kept his distance.

(Langhe)

❖ 10 ❖

The Count's Beard

The town of Pocapaglia was perched on the pinnacle of a hill so steep that its inhabitants tied little bags on the tail feathers of their hens to catch each freshly laid egg that otherwise would have gone rolling down the slopes into the woods below.

All of which goes to show that the people of Pocapaglia were not the dunces they were said to be, and that the proverb,

> In Pocapaglian ways
> The donkey whistles, the master brays,

merely reflected the malicious grudge the neighboring townspeople bore the Pocapaglians for their peaceful ways and their reluctance to quarrel with anyone.

"Yes, yes," was all the Pocapaglians would reply, "but just wait until Masino returns, and you will see who brays more, we or you."

Everybody in Pocapaglia loved Masino, the smartest boy in town. He was no stronger physically than anybody else; in fact, he even looked rather puny. But he had always been very clever. Concerned over how little he was at birth, his mother had bathed him in warm wine to keep him alive and make him a little stronger. His father had heated the wine with a red-hot horseshoe. That way Masino absorbed the subtlety of wine and the endurance of iron. To cool him off after his bath, his mother cradled him in the shell of an unripened chestnut; it was bitter and gave him understanding.

At the time the Pocapaglians were awaiting the return of Masino, whom no one had seen since the day he went off to be a soldier (and who was now most likely somewhere in Africa), strange things started happening in Pocapaglia. Every evening as the cattle came back from pasture in the plain below, an animal was whisked away by Micillina the Witch.

The witch would hide in the woods at the foot of the hill, and all she needed to do was give one hearty puff, and she had herself an ox. When the farmers heard her steal through the thicket after dark, their teeth would chatter, and everyone would fall down in a swoon. That became so common that people took to saying:

> Beware of Micillina, that old witch,
> For all your oxen she will filch,
> Then train on you her crossèd-eye,
> And wait for you to fall and die.

At night they began lighting huge bonfires to keep Micillina the Witch from venturing out of the woods. But she would sneak up on the solitary farmer watching over cattle beside the bonfire and knock him out in one breath. In the morning upon awaking, he'd find cows and oxen gone, and his friends would hear him weeping and moaning and hitting himself on the head. Then everybody combed the woods for traces of the stolen cattle, but found only tufts of hair, hairpins, and footprints left here and there by Micillina the Witch.

Things went from bad to worse. Shut up all the time in the barn, the cows grew as thin as rails. A rake instead of a brush was all that was needed to groom them, from rib to rib. Nobody dared lead the cattle to pasture any more. Everyone stayed clear of the woods now, and the mushrooms that grew there went unpicked and got as big as umbrellas.

Micillina the Witch was not tempted to plunder other towns, knowing full well that calm and peace-loving people were to be found only in Pocapaglia. There the poor farmers lit a big bonfire every night in the town square, while the women and children locked themselves indoors. The men sat around the fire scratching their heads and groaning. Day after day they scratched and groaned until a decision was finally reached to go to the count for help.

The count lived high above the town on a large circular estate surrounded by a massive wall. The top of the wall was encrusted with sharp bits of glass. One Sunday morning all the townsmen arrived, with hats in hand. They knocked, the door swung open, and they filed into the courtyard before the count's round dwelling, which had bars at all the windows. Around the courtyard sat the count's soldiers smoothing their mustaches with oil to make them shine and scowling at the farmers. At the end of the courtyard, in a velvet chair, sat the count himself with his long black beard, which four soldiers were combing from head to foot.

The oldest farmer took heart and said, "Your Honor, we have dared come to you about our misfortune. As our cattle go into the woods, Micillina the Witch appears and makes off with them." So, amid sighs and groans, with the other farmers nodding in assent, he told the count all about their nightmare.

The count remained silent.

"We have come here," said the old man, "to be so bold as to ask Your Honor's advice."

The count remained silent.

"We have come here," he added, "to be so bold as to ask Your Honor to help us. If you assigned us an escort of soldiers, we could again take our cattle down to pasture."

The count shook his head. "If I let you have the soldiers," he said, "I must also let you have the captain. . . ."

The farmers listened, hardly daring to hope.

"But if the captain is away in the evening," said the count, "who can I play lotto with?"

The farmers fell to their knees. "Help us, noble count, for pity's sake!" The soldiers around the courtyard yawned and stroked their mustaches.

Again the count shook his head and said:

> I am the count and I count for three;
> No witch have I seen,
> So, no witch has there been.

At those words and still yawning, the soldiers picked up their guns and, with bayonets extended, moved slowly toward the farmers, who turned and filed silently out of the courtyard.

Back in the town square and completely discouraged, the farmers had no idea what to do next. But the senior of them all, the one who had spoken to the count, said, "There's nothing left to do but send for Masino!"

So they wrote Masino a letter and sent it to Africa. Then one evening, while they were all gathered around the bonfire as usual, Masino returned. Imagine the welcome they gave him, the embraces, the pots of hot, spiced wine! "Where on earth have you been? What did you see? If you only knew what we have been going through!"

Masino let them have their say, then he had his. "In Africa I saw cannibals who ate not men but locusts; in the desert I saw a madman who had let his fingernails grow twelve meters long to dig for water; in the sea I saw a fish with a shoe and a slipper who wanted to be king of the other fish, since no other fish possessed shoe or slipper; in Sicily I saw a woman with seventy sons and only one kettle; in Naples I saw people who walked while standing still, since the chatter of other people kept them going; I saw sinners and I saw saints; I saw fat people and people no bigger than mites; many, many frightened souls did I see, but never so many as here in Pocapaglia."

The farmers hung their heads in shame, for Masino had hit a sensitive spot in suggesting they were cowards. But Masino was not cross with his fellow townsmen. He asked for a detailed account of the witch's doings, then said, "Let me ask you three questions, and at the stroke of midnight I'll go out and catch the witch and bring her back to you."

"Let's hear your questions! Out with them!" they all said.

"The first question is for the barber. How many people came to you this month?"

The barber replied:

> "Long beards, short beards,
> Fine beards, coarse beards,
> Locks straight, locks curly,
> All I trimmèd in a hurry."

"Your turn now, cobbler. How many people brought you their old shoes to mend this month?"

"Alas!" began the cobbler:

> "Shoes of wood, shoes of leather,
> Nail by nail I hammered back together,
> Mended shoes of satin and shoes of serpent.
> But there's nothing left to do,
> All their money is spent."

"The third question goes to you, rope maker. How much rope did you sell this month?"

The rope maker replied:

> "Rope galore of every sort I sold:
> Hemp rope, braided, wicker, cord,
> Needle-thin to arm-thick,
> Lard-soft to iron-strong . . .
> This month I couldn't go wrong."

"Very well," said Masino, stretching out by the fire. "I'm now going to sleep for a few hours, I'm very tired. Wake me up at midnight and I'll go after the witch." He put his hat over his face and fell asleep.

The farmers kept perfectly quiet until midnight, not even daring to breathe, for fear of awaking him. At midnight Masino shook himself, yawned, drank a cup of mulled wine, spat three times into the fire, got up without looking at a soul, and headed for the woods.

The farmers stayed behind watching the fire burn down and the last embers turn to ashes. Then, whom should Masino drag in by the beard but the count! A count that wept, kicked, and pleaded for mercy.

"Here's the witch!" cried Masino, and asked, "Where did you put the mulled wine?"

Beneath the farmers' amazed stares, the count tried to make himself as small as possible, sitting on the ground and shrinking up like a cold-bitten fly.

"The thief could have been none of you," explained Masino, "since you had all gone to the barber and had no hair to lose in the bushes. Then there were those tracks made by big heavy shoes, but all of you go barefoot. Nor could the thief have been a ghost, since he wouldn't have needed to buy all that cord to tie up the animals and carry them away. But where is my mulled wine?"

Shaking all over, the count tried to hide in that beard of his which Masino had tousled and torn in pulling him out of the bushes.

"How did he ever make us faint by just looking at us?" asked one farmer.

"He would smite you on the head with a padded club. That way you would hear only a whir. He'd leave no mark on you, you'd simply wake up with a headache."

"And those hairpins he lost?" asked another.

"They were used to hold his beard up on his head and make it look like a woman's hair."

Until then the farmers had listened in silence, but when Masino said, "And now, what shall we do with him?" a storm of shouts arose: "Burn him! Skin him alive! String him up for a scarecrow! Seal him in a cask and roll him down the cliff! Sew him up in a sack with six cats and six dogs!"

"Have mercy!" said the count in a voice just above a whisper.

"Spare him," said Masino, "and he will bring back your cattle and clean your barns. And since he enjoyed going into the woods at night, make him go there every night and gather bundles of firewood for each of you. Tell the children never to pick up the hairpins they find on the ground, for they belong to Micillina the Witch, whose hair and beard will be disheveled from now on."

The farmers followed the suggestion, and soon Masino left Pocapaglia to travel about the world. In the course of his travels, he found himself fighting in first one war and another, and they all lasted so long that this saying sprang up:

> Soldier fighter, what a hard lot!
> Wretched food, the ground for a cot.
> You feed the cannon powder:
> Boom-BOOM! Boom-BOOM! Boom louder!

(Bra)

◈ 11 ◈

The Little Girl Sold with the Pears

Once a man had a pear tree that used to bear four baskets of pears a year. One year, though, it only bore three baskets and a half, while he was supposed to carry four to the king. Seeing no other way out, he put his youngest daughter into the fourth basket and covered her up with pears and leaves.

The baskets were carried into the king's pantry, where the child stayed in hiding underneath the pears. But having nothing to eat, she began nibbling on the pears. After a while the servants noticed the supply of pears dwindling and also saw the cores. "There must be a rat or a mole gnawing on the pears," they said. "We shall look inside the baskets." They removed the top and found the little girl.

"What are you doing here?" they asked. "Come with us and work in the king's kitchen."

They called her Perina, and she was such a clever little girl that in no time she was doing the housework better than the king's own maidservants. She was so pretty no one could help loving her. The king's son, who was her age exactly, was always with Perina, and they became very fond of each other.

As the maiden grew up, the maidservants began to envy her. They held their tongues for a while, then accused Perina of boasting she would go and steal the witches' treasure. The king got wind of it and sent for the girl. "Is it true you boasted you would go and steal the witches' treasure?"

"No, Sacred Crown, I made no such boast."

"You did so," insisted the king, "and now you have to keep your word." At that, he banished her from the palace until she should return with the treasure.

On and on she walked until nightfall. Perina came to an apple tree, but kept on going. She next came to a peach tree, but still didn't stop. Then she came to a pear tree, climbed it, and fell asleep.

In the morning there stood a little old woman under the tree. "What are you doing up there, my daughter?" asked the old woman.

Perina told her about the difficulty she was in. The old woman said, "Take these three pounds of grease, three pounds of bread, and three pounds of millet and be on your way." Perina thanked her very much and moved on.

She came to a bakery where three women were pulling out their hair to sweep out the oven with. Perina gave them the three pounds of millet, which they then used to sweep out the oven and allowed the little girl to continue on her way.

On and on she walked and met three mastiffs that barked and rushed at anyone coming their way. Perina threw them the three pounds of bread, and they let her pass.

After walking for miles and miles she came to a blood-red river, which she had no idea how to cross. But the old woman had told her to say:

> "Fine water so red,
> I must make haste;
> Else, of you would I taste."

At those words, the waters parted and let her through.

On the other side of the river, Perina beheld one of the finest and largest palaces in the world. But the door was opening and slamming so rapidly that no one could possibly go in. Perina therefore applied the three pounds of grease to its hinges, and from then on it opened and closed quite gently.

Inside, Perina spied the treasure chest sitting on a small table. She picked it up and was about to go off with it, when the chest spoke: "Door, kill her, kill her!"

"I won't, either, since she greased my hinges that hadn't been looked after since goodness knows when."

Perina reached the river, and the chest said, "River, drown her, drown her!"

"I won't, either," replied the river, "since she called me 'Fine water so red.' "

She came to the dogs, and the chest said, "Dogs, devour her, devour her!"

"We won't, either," replied the dogs, "since she gave us three pounds of bread."

She came to the bakery oven. "Oven, burn her, burn her!"

But the three women replied, "We won't, either, since she gave us three pounds of millet, so that now we can spare our hair."

When she was almost home, Perina, who had as much curiosity as the next little girl, decided to peep into the treasure chest. She opened it, and out came a hen and her brood of gold chicks. They scuttled away too fast for a soul to catch them. Perina struck out after them. She passed the apple tree, but they were nowhere in sight. She passed the peach tree, where there was still no sign of them. She came to the pear tree, and there stood the little old woman with a wand in her hand and hen and chicks feeding around her. "Shoo, shoo!" went the old woman, and the hen and chicks reentered the treasure chest.

Upon her arrival, the king's son came out to meet her. "When my father asked what you want as a reward, tell him that box filled with coal in the cellar."

On the doorstep of the royal palace stood the maidservants, the king, and the entire court. Perina handed the king the hen with the brood of gold chicks. "Ask for whatever you want," said the king, "and I will give it to you."

"I would like the box of coal in the cellar," replied Perina.

They brought her the box of coal, which she opened, and out jumped the king's son, who was hiding inside. The king was then happy for Perina to marry his son.

(Monferrato)

◈ 12 ◈

The Snake

A farmer went out mowing every day, and at noon one or the other of his three daughters would bring him his lunch. On a certain day it fell to the oldest girl to go. By the time she reached the woods, though, she was tired and sat down on a stone to rest a minute before proceeding to the meadow. No sooner had she taken a seat than she felt a strong thud underneath, and out crawled a snake. The girl dropped the basket and ran home as fast as her legs would carry her. That day the father went hungry and when he came in from the field he scolded his daughters angrily.

The next day the middle girl started out. She too sat down on the stone, and the same thing occurred as the day before. Then the third girl said, "It's my turn now, but I'm not afraid." Instead of one lunch basket, she prepared two. When she felt the thud and saw the snake, she gave it one of the baskets of food, and the snake spoke. "Take me home with you, and I will bring you luck." The girl put the snake in her apron and then went on to her father with his lunch. When she got back home, she placed the snake under her bed. It grew so rapidly that soon it was too big to fit under the bed, so it went away. Before leaving, however, it bestowed three charms on the girl: weeping, she would shed tears of pearl and silver; laughing, she would see golden pomegranate seeds fall from her head; and washing her hands, she would produce fish of every kind.

That day there was nothing in the house to eat, and her father and sisters were weak from hunger, so what did she do but wash her hands and see the basin fill up with fish! Her sisters became envious and convinced their father that there was something strange behind all this and that he would be wise to lock the girl up in the attic.

From the attic window the girl looked into the king's garden, where the king's son was playing ball. Running after the ball, he slipped and fell, sending the girl into peals of laughter. As she laughed, gold pomegranate seeds rained from her head on the garden. The king's son had no idea where they came from, for the girl had slammed the window.

Returning to the garden next day to play ball, the king's son noticed that a pomegranate tree had sprung up. It was already quite tall and laden with fruit. He went to pick the pomegranates, but the tree grew taller right before his eyes, and all he had to do was reach for a pomegranate and the branches would rise a foot beyond his grasp. Since nobody managed to pluck so much as one leaf of the tree, the king assembled the wise men to explain the magic spell. The oldest of them all said that only one maiden would be able to pick the fruit and that she would become the bride of the king's son.

So the king issued a proclamation for all marriageable girls to come to the garden, under pain of death, to try to pick the pomegranates. Girls of every race and station showed up, but no ladders were ever long enough for them to reach the fruit. Among the contestants were the farmer's two older daughters, but they fell off the ladder and landed flat on their backs. The king had the houses searched and found other girls, including the one locked up in the attic. As soon as they took her to the tree, the branches bent down and placed the pomegranates right in her hands. Everyone cheered, "That's the bride, that's the bride!" with the king's son shouting loudest of all.

Preparations were made for the wedding, to which the sisters, as envious as ever, were invited. They all three rode in the same carriage, which drew to a halt in the middle of a forest. The older girls ordered the younger one out of the carriage, cut off her hands, gouged out her eyes, and left her lying unconscious in the bushes. Then the oldest girl dressed in the wedding gown and went to the king's son. He couldn't understand why she'd become so ugly, but since she faintly resembled the other girl, he decided he'd been mistaken all along about her original beauty.

Eyeless and handless, the maiden remained in the forest weeping. A carter came by and had pity on her. He seated her on his mule and took her to his house. She told him to look down: the ground was strewn with silver and pearls, which were none other than the girl's tears. The carter took them and sold them for more than a thousand crowns. How glad he was to have taken the poor girl in, even if she was unable to work and help the family.

One day the girl felt a snake wrap around her leg: it was the snake she had once befriended. "Did you know your sister married the king's son and became queen, since the old king died? Now she's expecting a baby and wants figs."

The girl said to the carter, "Load a mule with figs and take them to the queen."

"Where am I going to get figs this time of year?" asked the carter. It happened to be winter.

But the next morning he went into the garden and found the fig tree laden with fruit, even though there wasn't a leaf on the tree. He filled up two baskets and loaded them onto his donkey.

"How high a price can I ask for figs in winter?" said the carter.

"Ask for a pair of eyes," replied the maiden.

That he did, but neither the king nor the queen nor her other sister would have ever gouged out their eyes. So the sisters talked the matter over. "Let's give him our sister's eyes, which are of no use to us." With those eyes they purchased the figs.

The carter returned to the maiden with the eyes. She put them back in place and saw again as well as ever.

Then the queen had a desire for peaches, and the king sent to the carter asking if he couldn't find some peaches the way he'd found figs. The next morning the peach tree in the carter's garden was laden with peaches, and he took a load to court at once on his donkey. When they asked him what he wanted for them, he replied, "A pair of hands."

But nobody would cut off their hands, not even to please the king. Then the sisters talked the matter over. "Let's give him our sister's."

When the girl got her hands back, she reattached them to her arms and was as sound as ever.

Not long afterward, the queen went into labor and brought forth a scorpion. The king nonetheless gave a ball, to which everybody was invited. The girl went dressed as a queen and was the belle of the ball. The king fell in love with her and realized she was his true bride. She laughed golden seeds, wept pearls, and washed fish into the basin, as she told her story from start to finish.

The two wicked sisters and the scorpion were burned on a pyre sky-high. On the same day the grand wedding banquet took place.

> They put on the dog and high did they soar;
> I saw, I heard, I hid behind the door.
> Then to dine repaired I to the inn,
> And there my story draws to an end.

(Monferrato)

◈ 13 ◈

The Three Castles

A boy had taken it into his head to go out and steal. He also told his mother.

"Aren't you ashamed!" said his mother. "Go to confession at once, and you'll see what the priest has to say to you."

The boy went to confession. "Stealing is a sin," said the priest, "unless you steal from thieves."

The boy went to the woods and found thieves. He knocked at their door and got himself hired as a servant.

"We steal," explained the thieves, "but we're not committing a sin, because we rob the tax collectors."

One night when the thieves had gone out to rob a tax collector, the boy led the best mule out of the stable, loaded it with gold pieces, and fled.

He took the gold to his mother, then went to town to look for work. In that town was a king who had a hundred sheep, but no one wanted to be his shepherd. The boy volunteered, and the king said, "Look, there are the hundred sheep. Take them out tomorrow morning to the meadow, but don't cross the brook, because they would be eaten by a serpent on the other side. If you come back with none missing, I'll reward you. Fail to bring them all back, and I'll dismiss you on the spot, unless the serpent has already devoured you too."

To reach the meadow, he had to walk by the king's windows, where the king's daughter happened to be standing. She saw the boy, liked his looks, and threw him a cake. He caught it and carried it along to eat in the meadow. On reaching the meadow, he saw a white stone in the grass and said, "I'll sit down now and eat the cake from the king's daughter." But the stone happened to be on the other side of the brook. The shepherd paid no attention and jumped across the brook, with the sheep all following him.

The grass was high there, and the sheep grazed peacefully, while he sat on the stone eating his cake. All of a sudden he felt a blow under the rock which seemed to shake the world itself. The boy looked all around but, seeing nothing, went on eating his cake. Another blow more powerful than the first followed, but the shepherd ignored it. There was a third blow, and out from under the rock crawled a serpent with three heads. In each of its mouths it held a rose and crawled toward the boy, as though it wanted to offer him the roses. He was about to take them, when the serpent lunged at him with its three mouths all set to gobble him up in three bites. But the little shepherd proved the quicker, clubbing it with his staff over one head and the next and the next until the serpent lay dead.

Then he cut off the three heads with a sickle, putting two of them into his hunting jacket and crushing one to see what was inside. What should he find but a crystal key. The boy raised the stone and saw a door. Slipping the key into the lock and turning it, he found himself inside a splendid palace of solid crystal. Through all the doors came servants of crystal. "Good day, my lord, what are your wishes?"

"I wish to be shown all my treasures."

So they took him up crystal stairs into crystal towers; they showed him crystal stables with crystal horses and arms and armor of solid crystal. Then they led him into a crystal garden down avenues of crystal trees in which crystal birds sang, past flowerbeds where crystal flowers blossomed around crystal pools. The boy picked a small bunch of flowers and stuck the bouquet in his hat. When he brought the sheep home that night, the king's daughter was looking out the window and said, "May I have those flowers in your hat?"

"You certainly may," said the shepherd. "They are crystal flowers culled from the crystal garden of my solid crystal castle." He tossed her the bouquet, which she caught.

When he got back to the stone the next day, he crushed a second serpent head and found a silver key. He lifted the stone, slipped the silver key into the lock and entered a solid silver palace. Silver servants came running up saying, "Command, our lord!" They took him off to show him silver kitchens, where silver chickens roasted over silver fires, and silver gardens where silver peacocks spread their tails. The boy picked a little bunch of silver flowers and stuck them in his hat. That night he gave them to the king's daughter when she asked for them.

The third day, he crushed the third head and found a gold key. He slipped the key into the lock and entered a solid gold palace, where his servants were gold too, from wig to boots; the beds were gold, with gold sheets, pillows, and canopy; and in the aviaries fluttered hundreds of gold birds. In a garden of gold flowerbeds and fountains with gold sprays, he picked a small bunch of gold flowers to stick in his hat and gave them to the king's daughter that night.

Now the king announced a tournament, and the winner would have his daughter in marriage. The shepherd unlocked the door with the crystal key, entered the crystal palace and chose a crystal horse with crystal bridle and saddle, and thus rode to the tournament in crystal armor and carrying a crystal lance. He defeated all the other knights and fled without revealing who he was.

The next day he returned on a silver horse with trappings of silver, dressed in silver armor and carrying his silver lance and shield. He defeated everyone and fled, still unknown to all. The third day he returned on a gold horse, outfitted entirely in gold. He was victorious the third time as well, and the princess said, "I know who you are. You're the man who gave me flowers of crystal, silver, and gold, from the gardens of your castles of crystal, silver, and gold."

So they got married, and the little shepherd became king.

> And all were very happy and gay,
> But to me who watched they gave no thought nor pay.

(Monferrato)

❖ 14 ❖

The Prince Who Married a Frog

There was once a king who had three sons of marriageable age. In order to avoid any dispute over their choice of three brides, he said, "Aim as far as you can with the sling. There where the stone falls you will get your wife."

The three sons picked up their slings and shot. The oldest boy sent his stone flying all the way to the roof of a bakery, so he got the baker girl. The second boy released his stone, which came down on the house of a weaver. The youngest son's stone landed in a ditch.

Immediately after the shots, each boy rushed off to his betrothed with a ring. The oldest brother was met by a lovely maiden as fresh as a newly baked cake, the middle brother by a fair girl with silky hair and skin, while the youngest, after looking and looking, saw nothing but a frog in that ditch.

They returned to the king to tell him about their betrothed. "Now," said the king, "whoever has the best wife will inherit the kingdom. Here begin the tests." He gave them each some hemp to be spun and returned within three days, to see which betrothed was the best spinner.

The sons went to their betrothed and urged them to spin their best. Highly embarrassed, the youngest boy took the hemp to the rim of the ditch and called:

"Frog, frog!"

"Who calls?"

"Your love who loves you not."

"If you love me not, never mind. Later you shall, when a fine figure I cut."

The frog jumped out of the water onto a leaf. The king's son gave her the hemp, telling her he'd pick up the spun thread three days later.

Three days later the older brothers anxiously hastened to the baker girl and the weaver girl to pick up their spun hemp. The baker girl produced a beautiful piece of work; the weaver girl, who was an expert at this sort of thing, had spun hers to look like silk. But how did the youngest son fare? He went to the ditch and called:

"Frog, frog!"

"Who calls?"

"Your love who loves you not."

"If you love me not, never mind. Later you shall, when a fine figure I cut."

She jumped onto a leaf holding a walnut in her mouth. He was somewhat embarrassed to give his father a walnut while his brothers brought spun hemp. He nevertheless took heart and presented the king with the walnut. The king, who had already scrutinized the handiwork of the baker and the weaver girls, cracked open the walnut as the older brothers looked on, snickering. Out came cloth as fine as gossamer that continued to unroll until the throne room was covered with it. "But there's no end to this cloth!" exclaimed the king. No sooner were the words out of his mouth than the cloth came to an end.

But the father refused to accept the idea of a frog becoming queen. His favorite hunting bitch had just had three puppies, which he gave the three sons. "Take them to your betrothed and go back for them a month later. The one who's taken the best care of her dog will become the queen."

A month later, the baker girl's dog had turned into a big, fat mastiff, having got all the bread he could eat. The weaver's dog, not nearly so well supplied, was now a half-starved hound. The youngest son came in with a small box. The king opened it and out jumped a tiny, beribboned poodle, impeccably groomed and perfumed, that stood on its hind legs and marched and counted.

"No doubt about it," said the king, "my youngest son will be king, and the frog will be queen."

The wedding of all three brothers was set for the same day. The older brothers went for their brides in garlanded carriages drawn by four horses, and the brides climbed in, decked with feathers and jewels.

The youngest boy went to the ditch, where the frog awaited him in a carriage fashioned out of a fig leaf and drawn by four snails. They set out. He walked ahead while the snails followed, pulling the fig leaf with the frog upon it. Every now and then he stopped for them to catch up with him, and once he even fell asleep. When he awakened, a gold carriage had pulled up beside him. It was drawn by two white horses, and inside on velvet upholstery, sat a maiden as dazzling as the sun and dressed in an emerald-green gown.

"Who are you?" asked the youngest son.

"I am the frog."

He couldn't believe it, so the maiden opened a jewel case containing the fig leaf, the frog skin, and four snail shells. "I was a princess turned into a frog, and the only chance I had of getting my human form back was for a king's son to agree to marry me the way I was."

The king was overjoyed and told his two older sons, who were consumed with envy, that whoever picked the wrong wife was unworthy of the crown. So the youngest boy and his bride became king and queen.

(Monferrato)

The Parrot

Once upon a time there was a merchant who was supposed to go away on business, but he was afraid to leave his daughter at home by herself, as a certain king had designs on her.

"Dear daughter," he said, "I'm leaving, but you must promise not to stick your head out the door or let anyone in until I get back."

Now that very morning the daughter had seen a handsome parrot in the tree outside her window. He was a well-bred parrot, and the maiden had delighted in talking with him.

"Father," she replied, "it just breaks my heart to have to stay home all by myself. Couldn't I at least have a parrot to keep me company?"

The merchant, who lived only for his daughter, went out at once to get her a parrot. He found an old man who sold him one for a song. He took the bird to his daughter, and after much last-minute advice to her, he set out on his trip.

No sooner was the merchant out of sight than the king began devising a way to join the maiden. He enlisted an old woman in his scheme and sent her to the girl with a letter.

In the meantime the maiden got into conversation with the parrot. "Talk to me, parrot."

"I will tell you a good story. Once upon a time there was a king who had a daughter. She was an only child, with no brothers or sisters, nor did she have any playmates. So they made her a doll the same size as herself, with a face and clothes exactly like her own. Everywhere she went the doll went too, and no one could tell them apart. One day as king, daughter, and doll drove through the woods in their carriage, they were attacked by enemies who killed the king and carried off his daughter, leaving the doll behind in the abandoned carriage. The maiden screamed and cried so, the enemies let her go, and she wandered off into the woods by herself. She eventually reached the court of a certain queen and became a servant. She was such a clever girl that the queen liked her better all the time. The other servants grew jealous and plotted her downfall. 'You are aware, of course,' they said, 'that the queen likes you very much and tells you everything. But there's one thing which we know and you don't. She had a son who died.' At that, the maiden went to the queen and asked, 'Majesty, is it true you had a son who died?' Upon hearing those words, the queen almost fainted. Heaven help anyone who recalled that fact! The penalty for mentioning that dead son was no less than

45

death. The maiden too was condemned to die, but the queen took pity on her and had her shut up in a dungeon instead. There the girl gave way to despair, refusing all food and passing her nights weeping. At midnight, as she sat there weeping, she heard the door bolts slide back, and in walked five men: four of them were sorcerers and the fifth was the queen's son, their prisoner, whom they were taking out for exercise."

At that moment, the parrot was interrupted by a servant bearing a letter for the merchant's daughter. It was from the king, who had finally managed to get it to her. But the girl was eager to hear what happened next in the tale, which had reached the most exciting part, so she said, "I will receive no letters until my father returns. Parrot, go on with your story."

The servant took the letter away, and the parrot continued. "In the morning the jailers noticed the prisoner had not eaten a thing and they told the queen. The queen sent for her, and the maiden told her that her son was alive and in the dungeon a prisoner of four sorcerers, who took him out every night at midnight for exercise. The queen dispatched twelve soldiers armed with crowbars, who killed the sorcerers and freed her son. Then she gave him as a husband to the maiden who had saved him."

The servant knocked again, insisting that the young lady read the king's letter. "Very well. Now that the story is over, I can read the letter," said the merchant's daughter.

"But it's not finished yet, there's still some more to come," the parrot hastened to say. "Just listen to this: the maiden was not interested in marrying the queen's son. She settled for a purse of money and a man's outfit and moved on to another city. The son of this city's king was ill, and no doctor knew how to cure him. From midnight to dawn he raved like one possessed. The maiden showed up in man's attire, claiming to be a foreign doctor and asking to be left with the youth for one night. The first thing she did was look under the bed and find a trapdoor. She opened it and went down into a long corridor, at the end of which a lamp was burning."

At that moment the servant knocked and announced there was an old woman to see the young lady, whose aunt she claimed to be. (It was not an aunt, but the old woman sent by the king.) But the merchant's daughter was dying to know the outcome of the tale, so she said she was receiving no one. "Go on, parrot, go on with your story."

Thus the parrot continued. "The maiden walked down to that light and found an old woman boiling the heart of the king's son in a kettle, in revenge for the king's execution of her son. The maiden removed the heart from the kettle, carried it back to the king's son to eat, and he got well. The king said, 'I promised half of my kingdom to the doctor who cured my son. Since you are a woman, you will marry my son and become queen.' "

"It's a fine story," said the merchant's daughter. "Now that it's over, I can receive that woman who claims to be my aunt."

"But it's not quite over," said the parrot. "There's still some more to come. Just listen to this. The maiden in doctor's disguise also refused to marry that king's son and was off to another city whose king's son was under a spell and speechless. She hid under his bed; at midnight, she saw two witches coming through the window and remove a pebble from the young man's mouth, whereupon he could speak. Before leaving, they replaced the pebble, and he was again mute."

Someone knocked on the door, but the merchant's daughter was so absorbed in the story that she didn't even hear the knock. The parrot continued.

"The next night when the witches put the pebble on the bed, she gave the bedclothes a jerk and it dropped on the floor. Then she reached out for it and put it in her pocket. At dawn the witches couldn't find it and had to flee. The king's son was well, and they named the maiden physician to the court."

The knocking continued, and the merchant's daughter was all ready to say "Come in," but first she asked the parrot, "Does the story go on, or is it over?"

"It goes on," replied the parrot. "Just listen to this. The maiden wasn't interested in remaining as physician to the court, and moved on to another city. The talk there was that the king of this city had gone mad. He'd found a doll in the woods and fallen in love with it. He stayed shut up in his room admiring it and weeping because it was not a real live maiden. The girl went before the king. 'That is my doll!' she exclaimed. 'And this is my bride!' replied the king on seeing that she was the doll's living image."

There was another knock, and the parrot was at a total loss to continue the story. "Just a minute, just a minute, there's still a tiny bit more," he said, but he had no idea what to say next.

"Come on, open up, it's your father," said the merchant's voice.

"Ah, here we are at the end of the story," announced the parrot. "The king married the maiden, and they lived happily ever after."

The girl finally ran to open the door and embrace her father just back from his trip.

"Well done, my daughter!" said the merchant. "I see you've remained faithfully at home. And how is the parrot doing?"

They went to take a look at the bird, but in his place they found a handsome youth. "Forgive me, sir," said the youth. "I am a king who put on a parrot's disguise, because I am in love with your daughter. Aware of the intentions of a rival king to abduct her, I came here beneath a parrot's plumage to entertain her in an honorable manner and at the same time to prevent my rival from carrying out his schemes. I believe I have succeeded in both purposes, and that I can now ask for your daughter's hand in marriage."

The merchant gave his consent. His daughter married the king who had told her the tale, and the other king died of rage.

(Monferrato)

◈ 16 ◈

The Twelve Oxen

There were twelve brothers who fell out with their father, and all twelve of them left home. They built themselves a house in the woods and made their living as carpenters. Meanwhile their parents had a baby girl, who was a great comfort to them. The child grew up without ever meeting her twelve brothers. She had only heard them mentioned, and she longed to see them.

One day she went to bathe at a fountain, and the first thing she did was remove her coral necklace and hang it on a twig. A raven came by, grabbed the necklace, and flew off with it. The girl ran into the woods after the raven and found her brothers' house. No one was at home, so she cooked the noodles, spooned them onto the brothers' plates, and hid under a bed. The brothers returned and, finding the noodles ready and waiting, sat down and ate. But then they grew uneasy, suspecting the witches had played a joke on them, for the woods were full of witches.

One of the twelve kept watch the next day and saw the girl jump out from under the bed. When the brothers learned she was not a witch but their own little sister, they made a great to-do over her and insisted that she remain with them. But they cautioned her to speak to no one in the woods, because the place was full of witches.

One evening when the girl went to prepare supper, she found that the fire had gone out. To save time, she went to a nearby cottage to get a light. An old woman at the cottage graciously gave her the light, but said that, in exchange, she would come to the girl on the morrow and suck a bit of blood from her little finger.

"I can't let anyone in the house," said the girl. "My brothers forbid it."

"You don't even have to open the door," replied the old woman. "When I knock, all you have to do is stick your little finger through the keyhole, and I'll suck it."

So the old woman came by every evening to suck the blood from her, while the girl grew paler and paler. Her brothers noticed it and asked her so many questions that she admitted going to an old witch for a light and having to pay for it with her blood. "Just let us take care of her," said the brothers.

The witch arrived, knocked, and when the girl failed to stick her finger through the keyhole, she poked her head through the cat door. One of the brothers had his hatchet all ready and chopped off her head. Then they pitched the remains into a ravine.

One day on the way to the fountain, the girl met another old woman, who was selling white bowls.

"I have no money," said the girl.

"In that case I'll make you a present of them," said the old woman.

So when the brothers came home thirsty, they found twelve bowls filled with water. They pitched in and drank, and instantly changed into a herd of oxen. Only the twelfth, whose thirst was slight, barely touched the water and turned into a lamb. The sister therefore found herself alone with eleven oxen and one lamb to feed every day.

A prince out hunting went astray in the woods and, turning up at the girl's house, fell in love with her. He asked her to marry him, but she replied that she had to think of her oxen brothers and couldn't possibly leave them. The prince took her to his palace along with all the brothers. The girl became his princess bride, and the eleven oxen and the lamb were put into a marble barn with gold mangers.

But the witches in the woods did not give up. One day the princess was strolling under the grape arbor with her lambkin brother that she always carried with her, when an old woman walked up to her.

"Will you give me a bunch of grapes, my good princess?"

"Yes, dear old soul, help yourself."

"I can't reach up that high, please pick them for me."

"Right away," said the princess, reaching up for a bunch.

"Pick that bunch there, they're the ripest," said the old woman, pointing to a bunch above the cistern.

To reach it, the princess had to stand on the rim of the cistern. The old woman gave her a push, and the princess fell in. The lamb started bleating, and bleated all around the cistern, but nobody understood what it was bleating about, nor did they hear the princess moaning down in the well. Meanwhile the witch had taken the princess's shape and got into her bed. When the prince came home, he asked, "What are you doing in bed?"

"I'm sick," said the false princess. "I need to eat a morsel of lamb. Slaughter me that one out there that won't stop bleating."

"Didn't you tell me some time ago," asked the prince, "that the lamb was your brother? And you want to eat him now?"

The witch had blundered and was at a loss for words. The prince, sensing that something was amiss, went into the garden and followed the lamb that was bleating so pitifully. It approached the cistern, and the prince heard his wife calling.

"What are you doing at the bottom of the cistern?" he exclaimed. "Didn't I just leave you in bed?"

"No, I've been down here ever since this morning! A witch threw me in!"

The prince ordered his wife pulled up at once. The witch was caught and burned at the stake. While the fire burned, the oxen and also the lamb slowly turned back into fine, strapping young men, and you'd have thought the castle had been invaded by a band of giants. They were all made princes, while I've stayed as poor a soul as ever.

(Monferrato)

◈ 17 ◈

Crack and Crook

In a distant town there was a famous thief known as Crack, whom nobody had ever been able to catch. The main ambition of this Crack was to meet Crook, another notorious thief, and form a partnership with him. One day as Crack was eating lunch at the tavern across the table from a stranger, he went to look at his watch and found it missing. The only person in the world who could have taken it without my knowing, he thought, is Crook. So what did Crack do but turn right around and steal

Crook's purse. When the stranger got ready to pay for his lunch, he found his purse gone and said to his table companion, "Well, well, you must be Crack."

"And you must be Crook."

"Right."

"Fine, we'll work together."

They went to the city and made for the king's treasury, which was completely surrounded by guards. The thieves therefore dug an underground tunnel into the treasury and stole everything. Surveying his loss, the king had no idea how he might catch the robbers. He went to a man named Snare, who had been put in prison for stealing, and said, "If you can tell me who committed this robbery, I'll set you free and make you a marquis."

Snare replied, "It can be none other than Crack or Crook, or both of them together, since they are the most notorious thieves alive. But I'll tell you how you can catch them. Have the price of meat raised to one hundred dollars a pound. The person who pays that much for it will be your thief."

The king had the price of meat raised to one hundred dollars a pound, and everybody stopped buying meat. Finally it was reported that a friar had gone to a certain butcher and bought meat. Snare said, "That had to be Crack or Crook in disguise. I'll now disguise myself and go around to the houses begging. If anybody gives me meat, I'll make a red mark on the front door, and your guards can go and arrest the thieves."

But when he made a red mark on Crack's house, the thief saw it and went and marked all the other doors in the city with red, so there was no telling in the end where Crack and Crook lived.

Snare said to the king, "Didn't I tell you they were foxy? But there's someone else foxier than they are. Here's the next thing to do: put a tub of boiling pitch at the bottom of the treasury steps. Whoever goes down to steal will fall right into it, and his dead body will give him away."

Crack and Crook had run out of money in the meantime and decided to go back to the treasury for more. Crook went in first, but it was dark, and he fell into the tub. Crack came along and tried to pull his friend's body out of the pitch, but it stuck fast in the tub. He then cut off the head and carried it away.

The next day the king went to see if he had caught the thief. "This time we got him! We got him!" But the corpse had no head, so they were none the wiser about the thief or any accomplices he might have had.

Snare said, "There's one more thing we can do: have the dead man dragged through the city by two horses. The house where you hear somebody weeping has to be the thief's house."

In effect, when Crook's wife looked out the window and saw her husband's body being dragged through the street, she began screaming and crying. But Crack was there and knew right away that would be their undoing. He therefore started smashing dishes right and left and thrashing the poor woman at the same time. Attracted by all that screaming, the guards came in and found a man beating his wife for breaking up all the dishes in the house.

The king then had a decree posted on every street corner that he would pardon the thief who had robbed him, if the thief now managed to steal the sheets out from under him at night. Crack came forward and said he could do it.

That night the king undressed and went to bed with his gun to wait for the thief. Crack got a dead body from a gravedigger, dressed it in his own clothes, and carried it to the roof of the royal palace. At midnight the cadaver, held by a rope, was dangling before the king's windows. Thinking it was Crack, the king fired one shot and watched him fall, cord and all. He ran downstairs to see if he was dead. While the king was gone, Crack slipped into his room and stole the sheets. He was therefore pardoned, and so that he wouldn't have to steal any longer, the king married his daughter to him.

(Monferrato)

◈ 18 ◈

The Canary Prince

There was a king who had a daughter. Her mother was dead, and the stepmother was jealous of the girl and always spoke badly of her to the king. The maiden defended herself as best she could, but the stepmother was so contrary and insistent that the king, though he loved his daughter, finally gave in. He told the queen to send the girl away, but to some place where she would be comfortable, for he would never allow her to be mistreated. "Have no fear of that," said the stepmother, who then had the girl shut up in a castle in the heart of the forest. To keep her company, the queen selected a group of ladies-in-waiting, ordering them never to let the girl go out of the house or even to look out the windows.

Naturally they received a salary worthy of a royal household. The girl was given a beautiful room and all she wanted to eat and drink. The only thing she couldn't do was go outdoors. But the ladies, enjoying so much leisure time and money, thought only of themselves and paid no attention to her.

Every now and then the king would ask his wife, "And how is our daughter? What is she doing with herself these days?" To prove that she did take an interest in the girl, the queen called on her. The minute she stepped from her carriage, the ladies-in-waiting all rushed out and told her not to worry, the girl was well and happy. The queen went up to the girl's room for a moment. "So you're comfortable, are you? You need nothing, do you? You're looking well, I see; the country air is doing you good. Stay happy, now. Bye-bye, dear!" And off she went. She informed the king she had never seen his daughter so content.

On the contrary, always alone in that room, with ladies-in-waiting who didn't so much as look at her, the princess spent her days wistfully at the window. She sat there leaning on the windowsill, and had she not thought to put a pillow under them, she would have got calluses on her elbows. The window looked out on the forest, and all day long the princess saw nothing but treetops, clouds and, down below, the hunters' trail. Over that trail one day came the son of a king in pursuit of a wild boar. Nearing the castle known to have been unoccupied for no telling how many years, he was amazed to see washing spread out on the battlements, smoke rising from the chimneys, and open casements. As he looked about him, he noticed a beautiful maiden at one of the upper windows and smiled at her. The maiden saw the prince too, dressed in yellow, with hunter's leggings and gun, and smiling at her, so she smiled back at him. For a whole hour, they smiled, bowed, and curtsied, being too far apart to communicate in any other way.

The next day, under the pretext of going hunting, the king's son returned, dressed in yellow, and they stared at each other this time for two hours; in addition to smiles, bows, and curtsies, they put a hand over their hearts and waved handkerchiefs at great length. The third day the prince stopped for three hours, and they blew each other kisses. The fourth day he was there as usual, when from behind a tree a witch peeped and began to guffaw: "Ho, ho, ho, ho!"

"Who are you? What's so funny?" snapped the prince.

"What's so funny? Two lovers silly enough to stay so far apart!"

"Would you know how to get any closer to her, ninny?" asked the prince.

"I like you both," said the witch, "and I'll help you."

She knocked at the door and handed the ladies-in-waiting a big old book with yellow, smudgy pages, saying it was a gift to the princess so the young lady could pass the time reading. The ladies took it to the girl, who opened it at once and read: "This is a magic book. Turn the pages forward, and the man becomes a bird; turn them back, and the bird becomes a man once more."

The girl ran to the window, placed the book on the sill, and turned the pages in great haste while watching the youth in yellow standing in the path. Moving his arms, he was soon flapping wings and changed into a canary, dressed in yellow as he was. Up he soared above the treetops and headed straight for the window, coming to rest on the cushioned sill. The princess couldn't resist picking up the beautiful canary and kissing him; then remembering he was a young man, she blushed. But on second thought she wasn't ashamed at all and made haste to turn him back into a youth. She picked up the book and thumbed backward through it; the canary ruffled his yellow feathers, flapped his wings, then moved arms and was once more the youth dressed in yellow with the hunter's leggings, who knelt before her, declaring, "I love you!"

By the time they finished confessing all their love for one another, it was evening. Slowly, the princess leafed through the book. Looking into her eyes the youth turned back into a canary, perched on the windowsill, then on the eaves, then trusting to the wind, flew down in wide arcs, lighting on the lower limb of a tree. At that, she turned the pages back in the book and the canary was a prince once more who jumped down, whistled for his dogs, threw a kiss toward the window, and continued along the trail out of sight.

So every day the pages were turned forward to bring the prince flying up to the window at the top of the tower, then turned backward to restore his human form, then forward again to enable him to fly away, and finally backward for him to get home. Never in their whole life had the two young people known such happiness.

One day the queen called on her stepdaughter. She walked about the room, saying, "You're all right, aren't you? I see you're a trifle slimmer, but that's certainly no cause for concern, is it? It's true, isn't it, you've never felt better?" As she talked, she checked to see that everything was in place. She opened the window and peered out. Here came the prince in yellow along the trail with his dogs. "If this silly girl thinks she is going to flirt at the window," said the stepmother to herself, "she has another thought coming to her." She sent the girl for a glass of water and some sugar, then hurriedly removed five or six hairpins from her own hair and concealed them in the pillow with the sharp points sticking straight up. "That will teach her to lean on the windowsill!" The girl returned with the water and sugar, but the queen said, "Oh, I'm no longer thirsty; you drink it, my dear! I must be getting back to your father. You don't need anything, do you? Well, goodbye." And she was off.

As soon as the queen's carriage was out of sight, the girl hurriedly flipped over the pages of the book, the prince turned into a canary, flew to the window, and struck the pillow like an arrow. He instantly let out a shrill cry of pain. The yellow feathers were stained with blood; the canary had driven the pins into his breast. He rose with a convulsive flapping, trusted himself to the wind, descended in irregular arcs, and lit on the ground with outstretched wings. The frightened princess, not yet fully aware of what had happened, quickly turned the pages back in the hope there would be no wounds when he regained his human form. Alas, the prince reappeared dripping blood from the deep stabs that had rent the yellow garment on his chest, and lay back surrounded by his dogs.

At the howling of the dogs, the other hunters came to his aid and carried him off on a stretcher of branches, but he didn't so much as glance up at the window of his beloved, who was still overwhelmed with grief and fright.

Back at his palace, the prince showed no promise of recovery, nor did the doctors know what to do for him. The wounds refused to heal over, and constantly hurt. His father the king posted proclamations on every street corner promising a fortune to anyone who could cure him, but not a soul turned up to try.

The princess meanwhile was consumed with longing for her lover. She cut her sheets into thin strips which she tied one to the other in a long, long rope. Then one night she let herself down from the high tower and set out on the hunters' trail. But because of the thick darkness and the howls of the wolves, she decided to wait for daylight. Finding an old oak with a hollow trunk, she nestled inside and, in her exhaustion, fell asleep at once. She woke up while it was still pitch-dark, under the impression she had heard a whistle. Listening closely, she heard another whistle, then a third and a fourth, after which she saw four candle flames advancing. They were four witches coming from the four corners of the earth to their appointed meeting under that tree. Through a crack in the trunk the princess, unseen by them, spied on the four crones carrying candles and sneering a welcome to one another: "Ah, ah, ah!"

They lit a bonfire under the tree and sat down to warm themselves and roast a couple of bats for dinner. When they had eaten their fill, they began asking one another what they had seen of interest out in the world.

"I saw the sultan of Turkey, who bought himself twenty new wives."

"I saw the emperor of China, who has let his pigtail grow three yards long."

"I saw the king of the cannibals, who ate his chamberlain by mistake."

"I saw the king of this region, who has the sick son nobody can cure, since I alone know the remedy."

"And what is it?" asked the other witches.

"In the floor of his room is a loose tile. All one need do is lift the tile, and there underneath is a phial containing an ointment that would heal every one of his wounds."

It was all the princess inside the tree could do not to scream for joy. By this time the witches had told one another all they had to say, so each went her own way. The princess jumped from the tree and set out in the dawn for the city. At the first secondhand dealer's she came to, she bought an old doctor's gown and a pair of spectacles, and knocked at the royal palace. Seeing the little doctor with such scant paraphernalia, the servants weren't going to let him in, but the king said, "What harm could he do my son who can't be any worse off than he is now? Let him see what he can do." The sham doctor asked to be left alone with the sick man, and the request was granted.

Finding her lover groaning and unconscious in his sickbed, the princess felt like weeping and smothering him with kisses. But she restrained herself because of the urgency of carrying out the witch's directions. She paced up and down the room until she stepped on a loose tile, which she raised and discovered a phial of ointment. With it she rubbed the prince's wounds, and no sooner had she touched each one with ointment than the wound disappeared completely. Overjoyed she called the king, who came in and saw his son sleeping peacefully, with the color back in his cheeks, and no trace of any of the wounds.

"Ask for whatever you like, doctor," said the king. "All the wealth in the kingdom is yours."

"I wish no money," replied the doctor. "Just give me the prince's shield bearing the family coat-of-arms, his standard, and his yellow vest that was rent and bloodied." Upon receiving the three items, she took her leave.

Three days later, the king's son was again out hunting. He passed the castle in the heart of the forest, but didn't deign to look up at the princess's window. She immediately picked up the book, leafed through it, and the prince had no choice but change into a canary. He flew into the room, and the princess turned him back into a man. "Let me go," he said. "Isn't it enough to have pierced me with those pins of yours and caused me so much agony?" The prince, in truth, no longer loved the girl, blaming her for his misfortune.

On the verge of fainting, she exclaimed, "But I saved your life! I am the one who cured you!"

"That's not so," said the prince. "My life was saved by a foreign doctor who asked for no recompense except my coat-of-arms, my standard, and my bloodied vest!"

"Here are your coat-of-arms, your standard, and your vest! The doctor was none other than myself! The pins were the cruel doing of my stepmother!"

The prince gazed into her eyes, dumbfounded. Never had she looked so beautiful. He fell at her feet asking her forgiveness and declaring his deep gratitude and love.

That very evening he informed his father he was going to marry the maiden in the castle in the forest.

"You may marry only the daughter of a king or an emperor," replied his father.

"I shall marry the woman who saved my life."

So they made preparations for the wedding, inviting all the kings and queens in the vicinity. Also present was the princess's royal father, who had been informed of nothing. When the bride came out, he looked at her and exclaimed, "My daughter!"

"What!" said the royal host. "My son's bride is your daughter? Why did she not tell us?"

"Because," explained the bride, "I no longer consider myself the daughter of a man who let my stepmother imprison me." And she pointed at the queen.

Learning of all his daughter's misfortune, the father was filled with pity for the girl and with loathing for his wicked wife. Nor did he wait until he was back home to have the woman seized. Thus the marriage was celebrated to the satisfaction and joy of all, with the exception of that wretch.

(Turin)

◈ 19 ◈

King Crin

Once there was a king who, for a son, had a pig named King Crin. King Crin would saunter through the royal chambers and usually behave beautifully, as befits anybody of royal birth. Sometimes, though, he was cross. On one such occasion, his father asked, while stroking his back, "What is the matter? Why are you so cross?"

"Oink, oink," grunted King Crin. "I want a wife. Oink, oink, I want the baker's daughter!"

The king sent for the baker, who had three daughters, and asked if his oldest daughter was willing to marry his pig-son. Torn between the thrill of wedding the king's son and the horror of marrying a pig, the daughter made up her mind to accept the proposal.

Tickled pink, King Crin went wallowing in the town thoroughfares on his wedding night and got all muddy. He returned to the bridal chamber, where his bride was waiting for him. Intending to caress her, he rubbed against her skirt. The bride was disgusted and, instead of caressing him, gave him a kick. "Get away from here, you nasty pig!"

King Crin moved away, grunting. "Oink! You'll pay for that!"

That night the bride was discovered dead in her bed.

The old king was quite distressed, but a few months later when his son was again as cross as could be and clamoring for a wife, he sent for the baker's second daughter, who accepted.

The evening of the wedding King Crin went back out and wallowed in the muddy roads, only to return and rub against his bride, who drove him out of the room. "Scram, you nasty pig!" In the morning she was found dead. This incident gave the court a bad name, being the second of its kind.

More time went by, and King Crin began acting up again. "Would you have the nerve," said his father, "to ask for the baker's third daughter?"

"Oink, oink, I certainly would. Oink, oink, I must have her!"

So they sent for the third girl to see if she would marry King Crin. She was obviously quite happy to do so. On his wedding night, as usual, King Crin went out to wallow, then ran back inside all muddy to caress his wife. She responded with caresses of her own and dried him off with fine linen handkerchiefs, murmuring, "My handsome Crin, my darling Crin, I love you so." King Crin was overjoyed.

Next morning at the court everybody expected to hear that the third bride had been found dead, but out she came in higher spirits than ever. That was a grand occasion for celebration in the royal house, and the king gave a reception.

The next night the bride became curious to see King Crin as he slept, because she had her suspicions. She lit a taper and beheld a youth handsome beyond all stretches of the imagination. But as she stood there rapt with admiration, she accidentally dropped the taper on his arm. He woke up and jumped out of bed, furious. "You broke the spell and will never see me again, or only when you have wept seven bottles of tears and worn out seven pairs of iron shoes, seven iron mantles, and seven iron hats looking for me." At that, he vanished.

So deep was her distress that the bride had no choice but to go in search of her husband. She had a blacksmith forge seven pairs of iron shoes, seven iron mantles, and seven iron hats for her, then departed.

She walked all day long until night overtook her on a mountain, where she saw a cottage and knocked on the door. "My poor girl," said an old woman, "I can't give you shelter, since my son is the Wind who comes home and turns everything upside down, and woe to anyone in his way!"

But she begged and pleaded until the old woman brought her in and hid her. The Wind soon arrived and sniffed all around, saying:

"Human, human, I smell a human."

But his mother quieted him down with food. In the morning she rose at daybreak and softly awakened the young lady, advising, "Flee before my son gets up and take along this chestnut as a souvenir of me, but crack it open only in a serious emergency."

She walked all day long and was overtaken by night on top of another mountain. She spied a cottage, and an old lady on the doorstep said, "I would gladly lodge you, but I'm Lightning's mother, and poor you if my son came home and caught you here!" But then she took pity on her and hid her. Lightning arrived soon afterward:

"Human, human, I smell a human."

But he didn't find her and, after supper, went to bed.

"Flee before my son wakes up," said Lightning's mother in the morning, "and take along this walnut, which might come in very handy."

She walked all day long and was overtaken by night on top of another mountain. There stood the house of Thunder's mother, who ended up hiding her. Thunder too came in saying:

"Human, human, I smell a human."

But neither did he find her, and in the morning she went off with a hazelnut as a present from Thunder's mother.

After walking for miles and miles she reached a city whose princess, she learned, would soon marry a handsome young man staying at her castle. The young lady was sure that was her own husband. What could she do to prevent the marriage? How could she get into the castle.

She cracked open the chestnut and out poured diamonds and other jewels, which she went off to sell under the princess's windows. The princess looked out and invited her inside. The young lady said, "I'll let you have all these gems for nothing, if you allow me to spend one night in the bedchamber of the young man staying at your palace."

The princess was afraid the young lady would talk to him and maybe persuade him to flee with her, but her maid said, "Leave everything to me. We'll give him a sleeping potion and he won't wake up." They did just that, and as soon as the handsome youth went to sleep, the maid took the young lady into his bedchamber and left her. With her own eyes, the young lady saw that his was none other than her husband.

"Wake up, my love, wake up! I've walked all over for you, wearing out seven pairs of iron shoes, seven iron mantles, and seven iron hats: and I've wept seven bottles of tears. Now that I've finally found you, you sleep and don't hear me!"

And that went on till morning, when, at her wit's end, she cracked the walnut. Out rolled exquisite gowns and silks, each lovelier than the other. At the sight of all these wonderful things, the maid called the princess, who simply had to have them all and therefore granted the young lady another night with the youth. But the young lady was taken into the bedchamber later than the last time and brought out earlier in the morning.

Nor was this second night any more fruitful than the first. The poor girl cracked the hazelnut and out came horses and carriages. To acquire them, the princess again let her spend the night with the young man.

But by this time he had grown tired of drinking what they brought him every night, so he only pretended to swallow it while actually emptying the glass over his shoulder. When the young lady began talking to him, he made out as if he were sleeping, but the moment he was sure it was his wife, he jumped to his feet and embraced her. With all those horses and carriages they had no problem getting away and back home, where there was a grand celebration.

> They put on the dog and high did they soar,
> They saw me not, I stood behind the door.

(*Colline del Po*)

◈ 20 ◈

Those Stubborn Souls, the Biellese

A farmer was on his way down to Biella one day. The weather was so stormy that it was next to impossible to get over the roads. But the farmer had important business and pushed onward in the face of the driving rain.

He met an old man, who said to him, "A good day to you! Where are you going, my good man, in such haste?"

"To Biella," answered the farmer, without slowing down.

"You might at least say, 'God willing.' "

The farmer stopped, looked the old man in the eye, and snapped, "God willing, I'm on my way to Biella. But even if God isn't willing, I still have to go there all the same."

Now the old man happened to be the Lord. "In that case you'll go to Biella in seven years," he said. "In the meantime, jump into this swamp and stay there for seven years."

Suddenly the farmer changed into a frog and jumped into the swamp.

Seven years went by. The farmer came out of the swamp, turned back into a man, clapped his hat on his head, and continued on his way to market.

After a short distance he met the old man again. "And where are you going, my good man?"

"To Biella."

"You might say, 'God willing.' "

"If God wills it, fine. If not, I know the consequence and can now go into the swamp unassisted."

Nor for the life of him would he say one word more.

(Biellese)

◈ 21 ◈

The Pot of Marjoram

There was once an apothecary, who was a widower with a dear and beautiful daughter named Stella Diana. Every day Stella Diana went to a sewing mistress to learn how to sew. The mistress had a terrace full of potted flowers and plants, and Stella Diana would go out each afternoon to water a pot of marjoram she particularly liked. Facing the terrace was a balcony where a young nobleman stood and gazed at her. One day he spoke:

> "Stella Diana, Stella Diana,
> How many leaves show on your *maggiorana?*"

The girl replied:

> "Noble, handsome knight,
> How many stars twinkle in the night?"

He said:

"Stars cannot be counted, so many are there."

Then she said:

"See my marjoram you must not dare."

The nobleman disguised himself as a fishmonger and went to hawk his fish under the sewing mistress's windows. The mistress sent Stella Diana down to buy a fish for supper. The girl picked out a fish and asked the fishmonger how much it was. He told her the price, but it was so high she said she would not take the fish. Then he said, "For a kiss, I would let you have it for nothing."

Stella Diana gave him a quick kiss, and he gave her the fish for the sewing mistress's supper.

In the afternoon when Stella Diana appeared in the midst of the plants on the terrace, the nobleman said to her from his balcony:

"Stella Diana, Stella Diana,
How many leaves show on your *maggiorana*?"

She replied:

"Noble, handsome knight,
How many stars twinkle in the night?"

He said:

"Stars cannot be counted, so many are there."

Then she said:

"See my marjoram you must not dare."

Then the nobleman said:

"For only one tiny fish
You gave me a nice little kiss."

Realizing he had made a fool of her, Stella Diana angrily withdrew and immediately went to work to make a fool of him. She dressed as a man, buckling a jewel-studded belt of rare beauty around her waist. Then she mounted a mule and sauntered up and down the street past the nobleman. He spied the belt and exclaimed, "What an exquisite belt! Would you sell it to me?" Mimicking a man, she told him she would sell it at no price. He said he would do anything to have the belt. "Well," she said, "plant a kiss on my mule's tail, and I'll *give* you the belt." The gentleman really liked that belt, so he looked all around to make sure no one was watching, kissed the mule's tail, and went away with the belt.

The next time they saw each other, the usual words were exchanged between the terrace and the balcony.

"Stella Diana, Stella Diana,
How many leaves show on your *maggiorana*?"

"Noble, handsome knight,
How many stars twinkle in the night?"

"Stars cannot be counted, so many are there."

"See my marjoram you must not dare."

"For only one tiny fish,
You gave me a nice little kiss."

"To get a belt without fail,
You kissed my mule right on his tail."

That quip cut the nobleman to the quick, so he put a word in the ear of the sewing mistress and received permission to hide under her steps. When Stella Diana came down the steps, the youth reached up and grabbed her by the skirt. The maiden screamed:

"Mistress, mistress,
The steps do grab me by my dress!"

In the afternoon this dialogue took place, with Stella on the terrace and the nobleman on the balcony:

"Stella Diana, Stella Diana,
How many leaves show on your *maggiorana*?"

"Noble, handsome knight,
How many stars twinkle in the night?"

"Stars cannot be counted, so many are there."

"See my marjoram you must not dare."

"For only one tiny fish,
You gave me a nice little kiss."

"To get a belt without fail,
You kissed my mule right on his tail."

"Mistress, mistress,
The steps do grab me by my dress!"

This time Stella Diana was cut to the quick. She thought to herself, Now I'll fix you! By bribing his manservant, she got into the youth's house one night and appeared before him completely veiled in a sheet and carrying a torch and an open book. At the sight of that ghost, the youth shook with fear:

> "Death, my love, I am young and patient,
> Go instead to auntie who's cross and ancient."

Stella Diana put out her torch and left. The next day the duet resumed:

> "Stella Diana, Stella Diana,
> How many leaves show on your *maggiorana?*"

> "Noble, handsome knight,
> How many stars twinkle in the night?"

> "Stars cannot be counted, so many are there."

> "See my marjoram you must not dare."

> "For only one tiny fish,
> You gave me a nice little kiss."

> "To get a belt without fail,
> You kissed my mule right on his tail."

> "Mistress, mistress,
> The steps do grab me by my dress!"

> "Death, my love, I am young and patient,
> Go instead to auntie who's cross and ancient."

At this latest jest, the youth thought to himself, Enough is enough! I'll get even with her once and for all. No sooner said than done: he went to the apothecary and asked for Stella Diana's hand in marriage. The apothecary was delighted, and they drew up the marriage contract at once. As the wedding day drew nigh, Stella Diana began to fear that her bridegroom would take revenge on her for all her pranks. She thus decided to make a life-size pastry doll that resembled her to a tee. In place of the heart, she put a bladder filled with whipped cream. On retiring to her room after the wedding, she placed the doll in bed wearing her own nightcap and gown; then she hid.

The bridegroom came in. "Ah, here we are by ourselves at last! The time has come to avenge myself for all the humiliations I suffered at your hands." He unsheathed a dagger and thrust it into the doll's heart. The bladder burst, and whipped cream spurted everywhere, even into the bridegroom's mouth.

"Poor me! My Stella Diana's blood tastes so sweet! How could I have killed her? Ah, if only I could bring her back to life!"

Out jumped Stella Diana, as sound as a bell. "Here is your Stella Diana. I'm not dead, God forbid!"

Overjoyed, the bridegroom embraced her, and from that time on they lived in perfect harmony.

(Milan)

◈ 22 ◈

The Billiards Player

There was once a young man who spent his days in the cafés challenging people to a game of billiards. One day in a certain café he met a foreign gentleman.

"Shall we play billiards?" proposed the youth.

"Let's do."

"For what stakes shall we play?"

"If you win, I'll give you my daughter in marriage," replied the stranger.

They played their game, which the young man won.

The stranger said, "Fine. I am the king of the Sun. I'll write to you without delay." Then he left.

From one day to the next the youth expected the mailman to bring him a letter from the king of the Sun, but no letter ever came. He therefore set out in search of the king. Every Sunday he stopped in a different city, waited for the people to come out of Mass, and asked the oldtimers if they knew where the king of the Sun kept himself. No one had any idea, but once an old man said to him, "I'm sure he exists, but his whereabouts are a complete mystery to me."

The youth journeyed for another week, and at last met a second old man coming out of Mass in another city. This old man directed him to still another city, where he arrived on Sunday and asked an old man coming out of Mass if he knew where the king of the Sun lived. "Just a stone's throw from here," said the old man. "At the end of this street, on the right, you will see his palace. You can't miss it, for it has no doors."

"How do you get in?"

"How would I know? I suggest you wait in yonder grove. You'll see a pool there where the king of the Sun's three daughters swim every day at noon."

The youth hid in the grove. Precisely at noon, here came the king of the Sun's three daughters. They undressed, dove into the pool, and began their swim. Meanwhile the young man sneaked up to the clothes of the prettiest girl and made off with them.

The three girls came out of the water and went to put their clothes back on. But the prettiest girl's clothes were missing.

"Hurry up," said her sisters, "we're all dressed. What's keeping you?"

"I can't find my clothes. Wait for me!"

"Look more carefully. We have to go." And they went off without her. The maiden started to cry.

Out jumped the youth. "If you take me to your father, I'll give you your clothes back."

"Who are you?"

"I beat the king of the Sun at billiards, so now I'm supposed to marry his daughter."

The two young people looked at each other and fell in love. The girl said, "You must marry me. But my father will blindfold you and tell you to choose one of us. You can pick me out by touching the hands of all three of us: one of my fingers was cut off." Then she took him into the king of the Sun's palace.

"I am here to marry your daughter," said the young man to the king of the Sun.

"Very well, you will marry her tomorrow," replied the king. "Meanwhile select the one you want." And he had him blindfolded.

The first girl came in. He touched her hands and said, "This one doesn't suit me."

The king sent him the second girl. The youth felt her hands and said, "Neither will this one do."

In came the third girl. The young man touched her hands just to be sure she was the one missing a finger. He said, "This is the maiden I wish to marry."

The wedding was celebrated, and the bride and groom retired to their own room in the palace. At midnight the bride said, "I can't hide it from you any longer that my father is planning to have you killed."

"Let us flee, then," he replied.

They got up early, took a horse apiece, and galloped away. The king also rose, entered the bridal chamber, and found the couple gone. He rushed to the stable and discovered his two finest horses missing. Then he sent out a troop of mounted soldiers after the newlyweds.

In the midst of their flight the king of the Sun's daughter, hearing hoofbeats behind them, looked around and saw a troop of soldiers advancing. So she took the comb out of her hair and flung it to the ground. The comb changed into a forest, in which a man and a woman were busy digging up tree stumps.

The soldiers asked them, "Did you see the king of the Sun's daughter go by with her husband?"

The man and the woman replied, "We're getting up stumps. When nighttime comes, we'll go home."

The soldiers raised their voices. "We asked if you saw the king of the Sun's daughter and her husband go by!"

The couple said, "Yes, when we have a cartload we go home."

Exasperated, the soldiers returned to the king, who asked, "Did you find them?"

"We were just about to lay hold of them, when all of a sudden we found ourselves in a forest and face to face with a man and a woman who gave us only foolish answers."

"You should have seized them. They were the newlyweds!"

So the pursuit resumed. The soldiers had almost caught up with the couple, when the king of the Sun's daughter again threw down her comb, which changed into a garden where a man and a woman were busy gathering chicory and radishes. The soldiers asked, "Did you see the king's daughter go by with her husband?"

"The radishes are a dime a bunch, and the chicory a nickel."

The soldiers repeated their question, while those two chattered on about radishes and chicory. The troops gave up and went home.

"We were within two feet of them," they told the king, "when all of a sudden we found ourselves in a garden and face to face with a man and a woman who gave us the most foolish of answers."

"You should have seized them! They were the newlyweds!"

After a mad chase, the soldiers were again on the heels of the couple, when the girl once more threw down her comb. The men found themselves before a church, where two sacristans were ringing the bells. The soldiers asked if they had seen the king of the Sun's daughter.

The sacristans said, "We're now ringing the second bell, next we'll ring the third, then comes Mass."

The soldiers gave up.

"You should have seized them! They were the newlyweds!" screamed the King. Then he too gave up.

(Milan)

Animal Speech

A rich merchant had a son named Bobo, who was both quick-witted and eager to learn. The father therefore put the boy in the charge of a learned teacher, who was to teach him all the languages.

When his studies were completed, Bobo came home. One evening he was walking with his father in the garden, where the sparrows were twittering so loudly in one of the trees that you couldn't hear yourself think. "These sparrows shatter my eardrums every evening," said the merchant, sticking his fingers in his ears.

"Shall I tell you what they are saying?" asked Bobo.

His father looked at him in amazement. "How would you know what the sparrows are saying? You're not a soothsayer, are you?"

"No, but my teacher taught me the languages of the various animals."

"Don't tell me that's where my money went!" said the father. "What was that teacher thinking of? I meant for him to teach you the languages of men, not of dumb beasts!"

"The tongues of animals are harder, so the teacher decided to start with them."

At that moment the dog ran up barking, and Bobo asked, "Shall I tell you what he's saying?"

"No, indeed! Don't let me hear another word about your dumb beasts' talk! Oh, the money I've thrown away!"

They were walking alongside the moat by now, and heard the frogs croaking. "The frogs also get on my nerves," grumbled the father.

"Father let me tell you what they . . ."

"The devil take you and the man who taught you!"

Angry over having thrown away his money to educate his son and associating this knowledge of animal speech with witchcraft, the father called in two servants and gave them secret instructions for the next morning.

Bobo was awakened at dawn, when one of the servants put him into the carriage and climbed in beside him. The other servant took a seat on the box, cracked his whip, and off they galloped. Bobo had no idea where they were going, but he noticed the sorrowful and swollen eyes of the servant beside him. "Where are we going?" asked Bobo. "Why are you so sad?" But the servant made no reply.

Then the horses began neighing, and Bobo understood what they were saying. "Gloomy is our trip, we are carrying young master to his death."

And the other horse answered, "Cruel indeed was his father's order."

"So you were ordered by my father to take me out and kill me!" said Bobo to the servants.

They shuddered. "How did you know?"

"The horses told me so," said Bobo. "Go ahead and kill me right now. Why torture me with a long wait?"

"We don't have the heart to kill you," said the servants. "How can we get around it?"

While they were talking, the dog ran up barking. He had chased the carriage all the way from home. Bobo listened to what he was saying. "I would give my life to save young master!"

"Even if my father is cruel," said Bobo, "there are still loyal beings such as you, dear servants, and this dog who declares himself ready to die for me."

"In that case," replied the servants, "we will kill the dog and carry his heart back to master. Flee for your life, young master."

Bobo embraced the servants and the faithful dog and wandered off. At nightfall he came to a farmhouse and asked for shelter. As everybody sat around the supper table, a dog began barking outside. Bobo went to the window to listen, then said, "Hurry, send the women and children off to bed and arm yourselves to the teeth and stay on the alert. A band of robbers will strike at midnight."

The people thought he had lost his mind. "How can you say such a thing? Who told you?"

"I learned it from the dog who was just now barking a warning. Poor animal, if I wasn't here, he'd only be wasting his breath. Listen to me and you'll be safe."

The farmers took their guns and hid behind a hedge, while their wives and children locked themselves in the house. At midnight there was a whistle, than another, then a third, followed by the sound of rushing feet. From the hedge came a volley of gunfire, and the thieves took to their heels. Two were killed and lay in the mud clutching their knives.

A big to-do was made over Bobo, and the farmers wanted him to stay on with them, but he said goodbye and continued on his way.

After miles and miles he came to another farmhouse in the evening. As he debated whether to knock, he heard frogs croaking in the ditch. He listened closely and heard: "Come on, throw me the Host! Throw it to *me*! If you don't I won't play any more! You won't catch it, and it will break in two! We've kept it whole all these years!" He went up and peered into the ditch: the frogs were playing ball with a consecrated wafer. Bobo made the sign of the cross.

"For six years now it's been in the ditch!" said one frog.

"Ever since the farmer's daughter was tempted by the Devil. Instead of swallowing the Host at communion, she hid it in her pocket and then threw it into the ditch here on her way home from church."

Bobo knocked on the door and was invited in to supper. Speaking with the farmer, he learned that the man had a daughter who had been sick for the last six years, but no doctor knew what ailed her and now she was dying.

"I should think so!" exclaimed Bobo. "She's being punished by God. Six years ago she threw the sacred Host into the ditch. You must find that Host and have her make a devout communion, at which she will get well."

The farmer was amazed. "Who told you all that?"

"Frogs," replied Bobo.

Doubtful, the farmer nevertheless searched the ditch, found the Host, and had his daughter receive communion, at which she got well. They had no idea how to repay Bobo for what he had done for them. But wanting nothing, he said goodbye and left.

One very hot day he met two men resting under a chestnut tree. He asked if he might join them, and stretched out beside them.

"Where are you gentlemen going?"

"To Rome. Haven't you heard the Pope is dead and a new one is being elected?"

Overhead, a flight of sparrows lit in the chestnut tree. "These sparrows are also going to Rome," said Bobo.

"How do you know?" asked the two men.

"I understand their speech." He listened closely and added, "Guess what they are saying."

"What?"

"They say that one of us three will be elected Pope."

In those days they chose the Pope by letting a dove loose in St. Peter's Square, where crowds of people waited. The man on whose head the dove lit would be the new Pope. The three men reached the packed square and made their way into the crowd. Round and round flew the dove and finally lit on Bobo's head.

In the midst of cheers and hymns of joy, he was lifted onto a throne and vested in rich robes. He stood to bless the crowd, when the hush that fell over the square was suddenly pierced by a cry. An old man had fallen unconscious to the ground, where he lay like a corpse. The new Pope rushed up to him and recognized his father. The old man was dying of remorse and just had time to ask his son's forgiveness before expiring in his arms.

Bobo forgave him and turned out to be one of the best popes the church has ever had.

(Mantua)

◈ 24 ◈

The Three Cottages

A poor woman who was dying called her three daughters to her bedside and said, "Dear daughters, it won't be long now before I die and leave you all by yourselves. When I'm gone, call on your uncles to build you each a little house. Love one another. Farewell." She then drew her last breath, and the three girls burst into tears.

They went out on the street, where they happened to meet one of their uncles who wove mats. Catherine, the oldest daughter, said, "Uncle, our mother has just died. Since you are so kind-hearted, will you build me a cottage out of rushes?"

So the uncle who wove mats built her a cottage out of rushes.

The other two sisters walked on until they met another uncle, who was a carpenter. Julia, the middle girl, said, "Uncle, our mother has just died. Since you are so kind-hearted, will you build me a wooden cottage?"

So the carpenter uncle built her a wooden cottage.

Now there was only Marietta, the youngest girl, who continued down the street until she met her uncle who was a blacksmith. "Uncle," she said, "my mother has just died. Since you are so kind-hearted, will you build me an iron cottage?"

So the blacksmith uncle built her an iron cottage.

At dusk the wolf came out. He went to Catherine's cottage and knocked on the door.

"Who is it?" asked Catherine.

"A poor little thing drenched to the bone. Please let me in."

"Get away from here. You're the wolf and want to eat me."

The wolf gave the rushes a push, walked in, and gobbled up Catherine. The next day the two sisters called on Catherine. They found the rushes pushed in and the cottage empty. "Oh, how awful!" they exclaimed. "Our big sister has surely been eaten by the wolf."

Toward evening the wolf came back and went to Julia's cottage. He knocked and she asked, "Who is it?"

"A poor little thing that's lost its way. Please give me shelter."

"No, you're the wolf and you would eat me next."

The wolf gave the wooden cottage one punch, flung open the door, and gobbled up Julia.

In the morning Marietta called on Julia, found her gone, and said to herself, "The wolf has eaten her up too! Poor me, now I'm all by myself in the world."

At dusk the wolf went to Marietta's cottage.

"Who is it?"

"A poor little thing half frozen to death. Please let me in."

"Get away from here, wolf! You ate my sisters and now you want to eat me, I know!"

The wolf threw himself against the door, but the door was iron like the rest of the cottage, so he only broke his shoulder. Howling with pain, he ran to the blacksmith.

"Fix my shoulder," he ordered.

"I fix iron, not bones," said the blacksmith.

"But I broke my bones on iron," argued the wolf, "so you're the one that has to fix them now."

The blacksmith therefore took his hammer and nails and fixed the wolf's shoulder.

The wolf went back to Marietta and called through the closed door, "Oh, Marietta dear, you caused me to break my shoulder, but I love you all the same. If you'll come with me tomorrow morning, we'll go for peas in a patch near here."

"Fine. Come by for me when you're ready."

But, smart girl that she was, she realized the wolf was only trying to get her out of the house so he could eat her. The next day she got up before dawn, went to the pea patch, picked a mess of peas, and carried them home in her apron. She put the peas on to cook and threw the pods out the window. At nine o'clock the wolf arrived. "Marietta dear, let's go for the peas."

"No, indeed, you dummy, I'm not going. I've already picked peas. Can't you see the pods under the window? Take a deep breath and you'll smell them cooking and lick your lips."

The wolf was fit to be tied, but he replied, "Oh, that's all right. I'll come by for you tomorrow morning, and we'll go out for lupins."

"Fine," said Marietta, "I'll expect you at nine."

But this time too she rose early, went to the lupin patch, picked an apronful, and took them home to cook. When the wolf appeared, she showed him the pods under the window.

The wolf swore to himself he would take revenge, but he said to her, "Naughty girl, you fooled me! And to think I'm so fond of you! Why don't you come with me tomorrow to a certain patch I'm familiar with. There we'll find wonderful pumpkins and have a real feast."

"Of course I'll come," said Marietta.

The next morning she ran to the pumpkin patch before daybreak, but this time the wolf didn't wait for nine o'clock. He too ran to the pumpkin patch to gobble up Marietta.

As soon as Marietta saw the wolf in the distance, she rapidly hollowed out a large pumpkin and squeezed inside, for there was absolutely nowhere else to hide or to flee. Smelling human flesh, the wolf went up and down and back and forth sniffing the pumpkins, but no Marietta could he find. He then thought, She must be back home already. I'll feast on the pumpkins by myself. And he began eating pumpkins right and left.

Marietta shuddered as the wolf approached her pumpkin, almost certain he would eat it too with her inside. But when he reached Marietta's pumpkin, the wolf was no longer hungry. "I'll take this big one home to Marietta," he said, "so that she will be my friend." He sank his teeth in the pumpkin and ran all the way with it to the iron cottage, where he threw it through the window.

"Marietta, my dear!" he called. "Look what a fine present I've brought you."

Back in safety, Marietta slipped out of the pumpkin, slammed the window, and made faces through the panes at the wolf. "Thank you, my friend," she said. "I was hiding in the pumpkin, and you carried me all the way home."

When he heard that, the wolf beat his head on the rocks.

That evening it snowed. Marietta was keeping warm at the fireside when she heard a noise in the chimney. That's the wolf coming to eat me, she thought to herself. She filled a kettle with water and hung it over the fire to boil. Little by little the wolf lowered himself through the chimney, then made a bound for what he thought was Marietta, but landed in the boiling water instead and scalded to death. So sly Marietta was rid of her enemy at last and lived in peace for the rest of her life.

(Mantua)

The Peasant Astrologer

A king had lost a precious ring. He looked all over for it, but nowhere was it to be found. He issued a proclamation stating that the astrologer who could tell him where it was would be rich for the rest of his life. Now there was a peasant by the name of Gàmbara, who was penniless and could neither read nor write. "Would it be so hard to play the astrologer?" he wondered. "I think I'll try." So he went to the king.

The king took him at his word, and shut him up in a room to study. There was nothing in the room but a bed and a table with a great big astrology book on it, and paper, pen, and ink. Gàmbara sat down at the table and began leafing through the book without understanding a word. Every now and then he made marks on the paper with the pen. As he didn't know how to write, he produced some very strange marks indeed, and the servants bringing him his lunch and his dinner got the idea he was an extremely wise astrologer.

Those servants had been the very ones to steal the ring. With their guilty conscience, they imagined from the knowing looks Gàmbara gave them whenever they went in that he suspected them, although the astrologer was only trying to look like an authority in his field. Fearful of being found out, they couldn't bow and scrape enough. "Yes, honorable astrologer! Your least wishes, honorable astrologer, are orders!"

Gàmbara, who was no astrologer, but a peasant and therefore cunning, suspected right away the servants knew something about the ring. So he set a trap for them.

One day, at the hour they brought in his lunch, he hid under the bed. The head servant came in and found no one in the room. Under the bed Gàmbara said in a loud voice, "That's one of them!" The servant put the dish down and withdrew in fright.

The second servant came in and heard a voice that seemed to come from underground. "That's two of them!" He too ran off.

Then the third came in. "That's three of them!"

The servants talked things over. "We have been found out, and if the astrologer accuses us to the king, we are done for."

So they decided to go to the astrologer and confess their theft. "We are poor men," they began; "if you tell the king what you have learned, we are lost. Please take this purse of gold and don't betray us."

Gàmbara took the purse and replied, "I won't betray you, provided you do as I say. Take the ring and make that turkey out in the farmyard swallow it. Then leave everything to me."

The next day Gàmbara went to the king and said that after much study he had learned where the ring was.

"Where is it?"

"A turkey has swallowed it."

They cut the turkey open and discovered the ring. The king heaped riches on the astrologer and honored him with a banquet attended by all the counts, marquis, barons, and grandees in the kingdom.

Among the many dishes served was a platter of *gamberi*, which means crayfish. Now crayfish were unknown to that country. Those served at the banquet were a present from the king of another country, and it was the first time people here had seen them.

"Since you are an astrologer," said the king to the peasant, "you must know the name of these things on the platter here."

The poor soul, who'd never seen or heard of them, mumbled to himself, "Ah, Gàmbara, Gàmbara, you're done for at last."

"Bravo!" said the king, who didn't know the peasant's real name. "You guessed it, the name is *gamberi*! You're the greatest astrologer in the world."

(Mantua)

◈ 26 ◈

The Wolf and the Three Girls

Once there were three sisters who worked in a certain town. Word reached them one day that their mother, who lived in Borgoforte, was deathly ill. The oldest sister therefore filled two baskets with four bottles of wine and four cakes and set out for Borgoforte. Along the way she met the wolf, who said to her, "Where are you going in such haste?"

"To Borgoforte to see Mamma, who is gravely ill."

"What's in those baskets?"

"Four bottles of wine and four cakes."

"Give them to me, or else—to put it bluntly—I'll eat you."

The girl gave the wolf everything and went flying back home to her sisters. Then the middle girl filled her baskets and left for Borgoforte. She too met the wolf.

"Where are you going in such haste?"

"To Borgoforte to see Mamma, who is gravely ill."

"What's in those baskets?"

"Four bottles of wine and four cakes."

"Give them to me, or else—to put it bluntly—I'll eat you."

So the second sister emptied her baskets and ran home. Then the youngest girl said, "Now it's my turn." She prepared the baskets and set out. There was the wolf.

"Where are you going in such haste?"

"To Borgoforte to see Mamma, who is gravely ill."

"What's in those baskets?"

"Four bottles of wine and four cakes."

"Give them to me, or else—to put it bluntly—I'll eat you."

The little girl took a cake and threw it to the wolf, who had his mouth open. She had made the cake especially for him and filled it with nails. The wolf caught it and bit into it, pricking his palate all over. He spat out the cake, leaped back, and ran off, shouting, "You'll pay for that!"

Taking certain short cuts known only to him, the wolf ran ahead and reached Borgoforte before the little girl. He slipped into the sick mother's house, gobbled her up, and took her place in bed.

The little girl arrived, found her mother with the sheet drawn up to her eyes, and said, "How dark you've become, Mamma!"

"That's because I've been sick so much, my child," said the wolf.

"How big your head has become, Mamma!"

"That's because I've worried so much, my child."

"Let me hug you, Mamma," said the little girl, and the wolf gobbled her up whole.

With the little girl in his belly, the wolf ran out of the house. But the townspeople, seeing him come out, chased him with pitchforks and shovels, cornered him, and killed him. They slit him open at once and out came mother and daughter still alive. The mother got well, and the little girl went back and said to her sisters, "Here I am, safe and sound!"

(Lake of Garda)

The Land Where One Never Dies

One day a young man said, "This tale about everybody having to die doesn't set too well with me. I will go in search of the land where one never dies."

He bid father, mother, uncles, and cousins goodbye and departed. For days and months he walked, asking everybody he met if they could direct him to the place where one never dies. But no one knew of any such place. One day he met an old man with a white beard down to his chest, pushing a wheelbarrow full of rocks. The boy asked him, "Could you direct me to that place where one never dies?"

"You don't want to die? Stick with me. Until I've finished carting away that entire mountain rock by rock, you shall not die."

"How long will it take you to level it?"

"One hundred years at least."

"And I'll have to die afterward?"

"I'm afraid so."

"No, this is no place for me. I will go to the place where one *never* dies."

He said goodbye to the old man and pushed onward. He walked for miles and came to a forest so vast that it seemed endless. There he saw an old man with a beard down to his navel pruning branches with a pruning hook. The young man asked, "Could you kindly tell me of a place where one never dies?"

"Stick with me," replied the old man. "Until I've trimmed all the trees in this forest with my pruning hook, you shall not die."

"How long will that take?"

"Who knows? At least two hundred years."

"And afterward I'll still have to die?"

"Indeed you will. Isn't two hundred years enough for you?"

"No, this is no place for me. I'm seeking a place where one *never* dies."

They said goodbye, and the youth continued onward. A few months later he reached the seashore. There he saw an old man with a beard down to his knees watching a duck drink seawater.

"Could you kindly tell me of a place where one never dies?"

"If you're afraid to die, stick with me. See that duck? Until it has drunk the sea dry, there's no danger at all of your dying."

"How long will it take?"

"Roughly three hundred years."

"And afterward I'll have to die?"

"What else do you expect? How much longer would you even want to live?"

"No, no, no. Not even this place is for me. I must go where one *never* dies."

He resumed his journey. One evening he came to a magnificent palace. He knocked, and the door was opened by an old man with a beard all the way down to his feet. "What is it you look for, young man?"

"I'm looking for the place where one never dies."

"Good for you, you've found it! This is the place where one never dies. As long as you stay with me, you can bet your boots you won't die."

"At last, after all the miles I've trudged! This is just the place I was seeking! But are you sure I'm not imposing on you?"

"Absolutely. I'm delighted to have company."

So the youth moved into the palace with the old man and lived like a lord. The years went by so fast and so pleasantly that he lost all track of time. Then one day he said to the old man, "There's no place on earth like here, but I really would like to pay my family a little visit and see how they're getting along."

"What family are you talking about? The last of your relatives died quite some time ago."

"I'd still like to go on a little journey, if only to revisit my birthplace and possibly run into the sons of my relatives' sons."

"If you're bent on going, follow my instructions. Go to the stable and get my white horse, which gallops like the wind. But once you're on him, never, never dismount for any reason whatever, or you will die on the spot."

"Don't worry, I'll stay in the saddle. You know how I hate the very idea of dying!"

He went to the stable, led out the white horse, got into the saddle, and was off like the wind. He passed the place where he had met the old man with the duck. There where the sea used to be was now a vast prairie. On the edge of it was a little pile of bones, the bones of the old man. "Just look at that," said the youth. "I was wise not to tarry here, or I too would now be dead."

He moved on and came to what was once the vast forest where the old man had to prune every single tree with his pruning hook. Not one tree was left, and the ground was as bare as a desert. "How right I was not to stop here, or I too would now be long gone, like the old soul in the forest."

He passed the place where the huge mountain had stood, which an old man was to cart away rock by rock. Now the ground was as level as a billiard table.

"Nor would I have fared any better here!"

On and on he went, finally reaching his town, but it had changed so much he no longer recognized it. Not only was his house gone, but even the street it had stood on. He inquired about his relatives, but no one had ever heard his family name. That was the end of it. "I might as well go back at once," he decided.

He turned his horse around and started back, but was not halfway home before he met a carter with a cart full of old shoes and drawn by an ox. "Sir," said the carter, "please be so kind as to dismount for a moment and help me dislodge this wheel sticking in the mud."

"I'm in a hurry and can't get out of the saddle," replied the youth.

"Please help me. I'm all by myself, as you can see, and night is coming on."

Moved to pity, the youth dismounted. He had only one foot on the ground and the other still in the stirrup, when the carter grabbed him by the arm and said: "I have you at last! Know who I am? Yes, I am Death! See all those old shoes in the cart? They're all the pairs you caused me to wear out running after you. Now you've fallen into my hands, from which no one ever escapes!"

So the poor young man had to die the same as everybody else.

(Verona)

◈ 28 ◈

The Devotee of St. Joseph

Once there was a man devoted exclusively to St. Joseph. He addressed all his prayers to St. Joseph, lit candles to St. Joseph, gave alms in the name of St. Joseph; in short, he recognized no one but St. Joseph. His dying day came, and he went before St. Peter. St. Peter refused to let him in, since the only thing to his credit were all those prayers he had said during his lifetime to St. Joseph. He had performed no good works to speak of, and behaved as if the Lord, our Lady, and all the other saints simply did not exist.

"Since I've come all the way here," said the devotee of St. Joseph, "let me at least see him."

So St. Peter sent for St. Joseph. St. Joseph came and, finding his devotee there, said, "Bravo! I'm really pleased to have you with us. Come on in right now."

"I can't. He won't let me."

"Why not?"

"Because he says I prayed only to you and to none of the other saints."

"Well, I'll be! What difference does that make? Come on in all the same."

But St. Peter continued to bar the way. A mighty squabble ensued, and St. Joseph ended up saying to St. Peter, "Either you let him in, or I'm taking my wife and my boy and moving Paradise somewhere else."

His wife was our Lady, his boy our Lord. St. Peter thought it wiser to give in and admit the devotee of St. Joseph.

(Verona)

◈ 29 ◈

The Three Crones

There were once three sisters who were all young. One was sixty-seven, another seventy-five, and the third ninety-four. Now these girls had a house with a nice little balcony, in the very middle of which was a hole for looking down on people passing along the street. The ninety-four-year-old sister, seeing a handsome young man approach, grabbed her finest scented handkerchief and sent it floating to the street just as the youth passed under the balcony. He picked it up, noticed the delightful scent, and concluded, "It can only belong to a very beautiful maiden." He walked on a way, then came back and rang the doorbell of that house. One of the three sisters answered the door, and the young man asked, "Would you please tell me if a young lady lives in this mansion, by chance?"

"Yes, indeed, and not just one."

"Would you do me a favor and allow me to see the one who lost this handkerchief?"

"No, that is impossible. A girl can't be seen before she's married. That's the rule at our mansion."

The youth was already so thrilled just imagining the girl's beauty that he said, "That's not asking a bit too much. I'll marry her sight unseen. Now I'm going to tell my mother I've found a lovely maiden whom I intend to marry."

He went home and told his mother all about it. She said, "Dear son, take care and don't let those people trick you. You must think before you act."

"They're not asking a bit too much. I've given my word, and a king must keep his promise," said the young man, who happened to be a king.

He returned to the bride's house and rang the doorbell. The same crone answered the door, and he asked, "Are you her grandmother?"

"That's right, I'm her grandmother."

"Since you're her grandmother, do me a favor and show me at least a finger of the girl."

"No, not now. You'll have to come back tomorrow."

The youth said goodbye and left. As soon as he was gone, the crones made an artificial finger out of the finger of a glove and a false fingernail. In the meantime his eagerness to see the finger kept him awake all night long. The sun came up at last, and he dressed and ran to the house.

"Madam," he said to the crone, "I've come to see my bride's finger."

"Yes, yes," she replied, "right away. You'll see it through the keyhole of this door."

The bride pushed the false finger through the keyhole. Bewitched by its beauty, the young man kissed the finger and slipped a diamond ring onto it. Head over heels in love by then, he said to the crone, "I must marry her forthwith, Granny; I can't wait any longer."

"You can marry her tomorrow, if you like."

"Perfect! I'll marry her tomorrow, on my honor as a king!"

Being rich, the three old women were able to get everything ready overnight for the wedding, down to the tiniest detail. The next day the bride dressed with the help of her two little sisters. The king arrived and said, "I'm here, Granny."

"Wait a minute, and we'll bring her to you."

Here she came at last, arm in arm with her sisters and covered with seven veils. "Remember," said the sisters, "you may not look at her face until you are in the bridal chamber."

They went to church and got married. Afterward the king wanted them all to go to dinner, but the crones would not allow it. "The bride, mind you, isn't used to such foolishness." So the king had to keep quiet. He was dying for night to come when he could be alone with the bride. The crones finally took her to her room, but made him wait outside while they undressed her and put her to bed. At last he went in and found the bride under the covers and the two old sisters still busying about the room. He undressed, and the old women went off with the lamp. But he'd brought along a candle in his pocket. He got it, lit it, and what should he see but an old withered crone streaked with wrinkles!

For an instant he was speechless and paralyzed with fright. Then in a fit of rage he seized his wife and hurled her through the window.

Under the window was a vine-covered trellis. The old crone went crashing through the trellis, but the hem of her nightgown caught on a broken slat and held her dangling in the air.

That night three fairies happened to be strolling through the gardens. Passing under the trellis, they spied the dangling crone. At that unexpected sight, all three fairies burst out laughing and laughed until their sides hurt. But when they had laughed their fill, one of them said, "Now that we've had such a good laugh at her expense, we must reward her."

"Indeed we must," agreed another. "I will that you become the most beautiful maiden in the world."

"I will," said the second fairy, "that you have the most handsome of husbands and that he love you with his whole heart."

"I will," said the third fairy, "that you be a great noblelady your whole life long."

At that, the fairies moved on.

At dawn the king awakened and remembered everything. To make sure it wasn't just a bad dream, he opened the window in order to see the monster he'd thrown out the night before. But there on the trellis sat the loveliest of maidens! He put his hands to his head.

"Goodness me, what have I done!" He had no idea how to draw her up, but finally took a sheet off the bed, threw her an end to grab hold of, then pulled her up into the room. Overjoyed to have her beside him once more, he begged her to forgive him, which she did, and they became the best of friends.

In a little while a knock was heard on the door. "It must be Granny," said the king. "Come in, come in!"

The old woman entered and saw in bed, in place of her ninety-four-year-old sister, the loveliest of young ladies, who said, as though nothing were amiss, "Clementine, bring me my coffee."

The old crone put a hand over her mouth to stifle a cry of amazement. Pretending everything was just as it should be, she went off and got the coffee. But the minute the king left the house to attend to his business, she ran to his wife and asked, "How in the world did you become so young?"

"Shhhhh!" cautioned the wife. "Lower your voice, please! Just wait until you hear what I did! I had myself planed!"

"Planed! Planed? Who did it for you? I'm going to get planed too."

"The carpenter!"

The old woman went running to the carpenter's shop lickety-split. "Carpenter, will you give me a good planing?"

"Oh, my goodness!" exclaimed the carpenter. "You're already dead-wood, but if I plane you, you'll go to kingdom come."

"Don't give it a thought."

"What do you mean, not give it a thought? After I've killed you, what then?"

"Don't worry, I tell you. Here's a thaler."

When he heard "thaler," the carpenter changed his mind. He took the money and said, "Lie down here on my workbench, and I'll plane you all you like," and he proceeded to plane a jaw.

The crone let out a scream.

"Now, now! If you scream, we won't get a thing done."

She rolled over, and the carpenter planed the other jaw. The old crone screamed no more: she was dead as dead can be.

Nothing more was ever heard of the other crone. Whether she drowned, had her throat slit, died in bed or elsewhere, no one knows.

The bride was the only one left in the house with the young king, and they lived happily ever after.

(Venice)

◈ 30 ◈

The Crab Prince

There was once a fisherman who never could catch enough fish to buy food for his family. One day though, when he went to pull up his nets, he felt a weight almost too heavy to move, but he tugged and tugged and found a crab so enormous that one pair of eyes was not enough to take it all in. "Oh, what a haul at last! Now I can buy food for my children!"

He took the crab home on his back and told his wife to put the pot on the fire, for he would return shortly with food. Then he carried the crab to the king's palace.

"Your Highness," he said to the king, "I've come to see if you will kindly buy this crab from me. My wife has put the pot on the fire, but I have no money to buy anything to go in it."

The king replied, "But what would I do with a crab? Can't you sell it to someone else?"

Just then the king's daughter came in. "Oh, what a fine crab, what a fine crab! Please buy it for me, Papa, please! We'll put it in the fishpond with the mullets and the goldfish."

The king's daughter was fascinated by fish and would sit for hours on the rim of the fishpond in the garden watching the mullets and the goldfish swim about. Her father could refuse her nothing, so he bought the crab. The fisherman put it into the fishpond and received a purse of gold coins that would feed his children for a whole month.

The princess never tired of watching the crab and spent all her time by the fishpond. She had become thoroughly familiar with him and his ways, noticing that from noon until three o'clock he always disappeared and went off goodness knows where. One day the king's daughter was there studying her crab, when she heard the doorbell ring. She looked down from her balcony, and there was a poor tramp asking for alms. She threw down a purse of money, but it flew past him into a ditch. The tramp went into the ditch after it, plunged under water, and began to swim. The ditch connected with the king's fishpond by an underground canal which continued on to no telling where. The tramp followed it and came out in a beautiful basin in the middle of a large underground hall hung with tapestries and containing a table sumptuously laid. The tramp stepped from the basin and hid behind the tapestries. At the stroke of noon, up popped a fairy in the middle of the basin, seated on the back of a crab. She and the crab jumped out of the water into the hall, the fairy tapped the crab with her wand, and there emerged from the crab shell a handsome youth. The young man took a seat at the table and the fairy tapped her wand, producing food in the dishes and wine in the bottles. When the youth had finished eating and drinking, he reentered the crab shell, which the fairy touched with her wand, and the crab took her onto his back once more, jumped into the basin, and disappeared underwater with her.

Then the tramp came out from behind the tapestries, dove into the water, and swam back to the king's fishpond. The king's daughter was there looking at her fish and, seeing the vagabond's head bob up, she asked, "What are you doing here?"

"Princess," said the tramp, "I have a wonderful thing to relate to you." He came out of the pond and told her the whole story.

"Now I understand where the crab goes from noon to three o'clock!" exclaimed the king's daughter. "Fine, tomorrow at noon we shall go together and see."

So the next day they both swam the underground canal from the fishpond to the underground hall and hid behind the tapestries. Exactly at noon, up popped the fairy on the crab's back. She tapped her wand and out stepped the handsome young man from the crab shell and took his place at the table. The princess, who already liked the crab, was charmed with the young man and immediately fell in love with him.

Seeing the empty crab shell right there next to her, she hid inside it.

When the youth got back into the shell he found the beautiful maiden there. "What have you done?" he whispered. "If the fairy learns of this, she will put us both to death."

"But I want to free you from the spell!" whispered the king's daughter. "Tell me what I must do."

"Impossible," said the young man. "Only a maiden who loved me enough to die for me could break the spell."

"I am that maiden," said the princess.

While this conversation was taking place inside the crab shell, the fairy seated herself on the crab's back, and the youth, working the crab claws as usual, carried her through the underground waterways to the open sea, without her suspecting that hidden inside with him was the king's daughter. On the way back to the fishpond after leaving the fairy at her destination, the young man, who happened to be a prince, explained to his beloved close beside him in the crab shell what to do to free him. "You must climb up on a rock on shore and play and sing. The fairy is enthralled by music and will emerge from the sea to listen to you and say, 'Play on, lovely maiden, your music is so delightful.' And you will reply, 'I certainly shall, if you give me the flower in your hair.' When you have that flower in your hand, I will be free, since the flower is my life."

Meanwhile the crab had reached the fishpond, and he let the king's daughter out of the shell.

The tramp had swum back by himself and, finding no princess, saw himself in serious trouble. But the maiden emerged from the fishpond, thanked him, and gave him a handsome reward. Then she went to her father and told him she wanted to study music and singing. The king, who never refused her anything, sent for the finest musicians and singers to give her lessons.

As soon as she had learned music, the daughter said to the king, "Papa, I want to go and play my violin on a rock by the sea."

"On a rock by the sea? Have you lost your mind?" But, as usual, he gave in to her and let her go with eight maids of honor dressed in white. As a precaution, he had her followed at a distance by a few armed soldiers.

Seated on a rock, with her eight maids of honor in white dresses on eight rocks around her, the king's daughter played her violin. From the waves rose the fairy. "How beautifully you play!" she said. "Play on, play on, it delights me so to hear you!"

The king's daughter said, "Indeed I shall, if you make me a present of that flower you are wearing in your hair, for I love flowers to distraction."

"I will give it to you if you can fetch it from where I throw it."

"I will fetch it," she assured the fairy, and started to play and sing. When the song was over, she said, "Now give me the flower."

"Here you are," said the fairy, and threw it as far as she could out to sea.

The princess dove into the sea and swam toward the flower floating on the waves. "Princess, princess! Help! Help!" screamed the eight maids of honor standing up on the rocks, with their white veils billowing in the wind. But the princess swam on and on, disappearing in the waves and coming back up; she was beginning to doubt whether she would reach the flower, when a big wave swept it right into her hand.

In that instant she heard a voice beneath her, saying, "You have given me back my life, and will be my bride. Now don't be afraid. I am under you and will carry you to shore. But say nothing of this, not even to your father. I must go and tell my parents, and within twenty-four hours I'll come and ask for your hand."

"Yes, yes, I understand" was all she could answer, since she was out of breath, while the crab underwater carried her to shore.

So when she got back home, all the princess told the king was that she had enjoyed herself immensely.

The next day at three, there was a roll of drums, a flourish of trumpets, a prancing of horses, and in walked a majordomo saying the son of his king requested an audience.

The prince put the customary request to the king for the princess's hand and then told the whole story. The king was somewhat taken aback, for he had been in the dark about everything. He sent for his daughter, who came running in and threw herself into the prince's arms, exclaiming, "This is my bridegroom, this is my bridegroom!" The king realized there was nothing to do but conclude the marriage as soon as possible.

(Venice)

Silent for Seven Years

There was once a mother and father with two little boys and a girl. The father was often away from home traveling and one day when he was away the two little boys said to their mother, "We are going to meet Papa!" Their mother replied, "Yes, yes, go ahead."

When they reached the woods the children stopped to play. Shortly afterward, they saw their father approaching and ran up and grabbed him around the legs, saying, "Papa! Papa!"

The father was in a bad humor that day and replied, "Don't bother me! Go away!" But the boys paid no attention and went on pulling on his legs.

Thoroughly irritated, the father yelled, "The Devil take you both!" In that moment the Devil came out and took them away before the father knew what had happened to them.

When the mother saw the father return without the children, she became worried and started crying. Her husband first told her he didn't know where they were, then he admitted cursing them, after which they disappeared from sight.

At that, their little sister spoke up. "Even if it means losing my own life, I'm going out to look for them." Ignoring her parents' protests, she got together a little food and departed.

Coming to a palace with an iron door, she went in and found herself before a gentleman, whom she asked, "Have you by any chance seen my brothers who were kidnapped by the Devil?"

"I can't say that I have. But go through that door into a room with twenty-four beds and see if the boys are there."

In effect, the maiden found her brothers in bed and was overjoyed. "So you are here, little brothers! That means you're safe after all!"

"Take a closer look," replied the brothers, "and see whether we are safe."

She peered beneath the bedclothes and beheld countless flames. "Oh, my brothers! What can I do to save you?"

"If you do not speak for seven years you will save us. But in that time you must go through fire and water."

"Don't worry, you can count on me."

She left them and walked back through the other room past the gentleman sitting there. He motioned to her to approach, but she shook her head, made the sign of the cross, and left the palace.

After walking and walking she found herself in a forest. Exhausted, she lay down and went to sleep. A king out hunting passed by and saw her sleeping. "What a beautiful girl!" he exclaimed, then woke her up to ask whatever brought her to the forest. With her head she made a sign that she was not there by her will. The king then asked, "Would you like to come with me?" and she nodded yes. Taking her at first for a deaf-mute, the king spoke loudly, but shortly realized she could hear even a whisper.

He got home and took her out of the carriage, telling his mother he had found a speechless maiden asleep in the forest, whom he was going to marry.

"I'll never consent to it!" exclaimed his mother.

"But here, *I* make the decisions," he snapped, and the wedding took place.

The mother-in-law was wicked-hearted and treated her daughter-in-law shamefully, but the daughter-in-law endured all in silence. Meanwhile she found herself with child. The mother-in-law forged a letter to her son calling him to a certain city where he was supposedly being swindled. The king said goodbye to his expectant wife and went off to attend to the matter. The wife gave birth to a baby boy, but the mother-in-law, in league with the midwife, placed a dog in bed beside the new mother and took the baby stuffed in a box to the palace roof. The poor young woman looked on frantically, but then remembered her condemned brothers and bit her tongue.

The mother-in-law wrote her son immediately that his wife had given birth to a dog. The king replied that he wished to hear no more about his wife. He ordered that she be given a little money for food and turned out of the palace before he got home.

But the old woman told a servant to take the young wife off, kill her, throw her body into the sea, and bring back her clothes.

When they reached the seashore, the servant said, "Please bow your head now, madam, as I'm obliged to kill you." With tears in her eyes, the young woman sank to her knees and joined her hands. Moved to pity, the servant merely cut off her hair and took all her clothes, leaving her his own shirt and trousers to put on.

Alone on the deserted shore, the young woman at last spotted a ship at sea and signaled to it. The ship carried soldiers who asked her who she was, never once suspecting she was a girl. In sign language she explained she was a sailor from a shipwrecked vessel and its sole survivor. The soldiers said, "Even if you can't talk, you can still help us wage war."

There was a battle, and the young woman fired her share of cannon shots. Because of her bravery, her comrades in arms made her a corporal right away. Once the war was over she requested a discharge, which was granted.

Back on land, she didn't know which way to turn. At night she spied a tumbledown house and went inside. Hearing footsteps at midnight, she peeped out and saw thirteen murderers go out the back door. She let them get well out of sight, then went to the rear of the house and found a large table laid for a feast. Thirteen places were set, and she went around the table taking just a tiny bit of food at each place, so that the murderers would find nothing missing when they came back. Then she returned to her hiding place, but forgot to remove her spoon from one of the plates before she left. The murderers came home in the middle of the night, and one of them noticed the spoon at once. "Look! Some stranger has been in here meddling."

"Well," replied another murderer, "let's go back out while one of us stays behind to keep watch." And so they did.

Thinking they had all left, the girl jumped out and the murderer grabbed her. "I have you now, you rogue! You just wait!"

Thoroughly frightened, she explained by signs that she was a mute and had come in because she was lost. The murderer comforted her and gave her food and drink. The others came home, heard the tale, and said, "Now that you are here you shall remain with us. Otherwise we'd have to kill you."

Nodding her agreement, she stayed on with them.

The murderers never left her by herself. One day the ringleader said to her, "Tomorrow night we're all going to descend on the palace of a certain king and steal all his valuables. You shall come with us."

He told her the name of the king, who happened to be her own husband, whom she wrote and warned of the danger. As a result, when the murderers started through the front door of the palace at midnight, the servants barricaded there in the dark hall slew them one by one. Thus died the ringleader and five others, while all the rest fled in every direction, leaving the young woman, who was also dressed as a murderer, at the mercy of the servants. What did they do but seize her, bind her hand and foot, and carry her off to prison. From her cell she could see them constructing the gallows in the town square. Only one more day, and her seven years of silence would be up. In sign language she begged them to put off her execution until tomorrow, to which the king consented. The next day they led her to the scaffold. On the first step she asked them in signs if, instead of executing her at three o'clock, they would wait one more hour. The king agreed to this also. Four o'clock struck and she was moving a step higher, when two warriors came forward, bowed to the king, and begged permission to speak.

"Speak," said the king.

"Why is that young man being sent to his death?"

The king explained why.

"That is no man, mind you, but our sister!" And they told the king why she had not uttered a word for the last seven years. Then they said to her, "Speak up, the danger is over, and we are safe."

They removed her shackles and, in the presence of the whole city, she said, "I'm the king's wife, and my wicked mother-in-law killed my baby. Go to the roof, get that box, and see whether I gave birth to a dog or to a baby boy." The king sent his servants for the box, and there inside lay a baby's skeleton.

At that, the whole city shouted, "String up the queen and the midwife in place of this courageous soul!" And so died the two old women, while the young wife returned to the palace with her husband, and the two brothers became prime ministers of the king.

(Venice)

◈ 32 ◈

The Dead Man's Palace

There was once a king who had a daughter. One day the girl was on the balcony with her maids of honor, when an old woman came walking by.

"My little lady," said the old woman, "be kind and give me something, won't you, please?"

"Yes, of course, good soul," replied the maiden, and threw her down a bag of coins.

"My little lady, that's not much money. . . . Won't you give me something more?"

The king's daughter threw down another bag. The old woman again said, "Little lady, can't you give me still a little bit more?"

At that, the king's daughter lost her temper. "You know what you are? You're a nuisance, that's what! I've given you money twice, and you're not getting a cent more!"

The old woman then faced her and said, "So that's how it is! I therefore call on heaven to prevent you from marrying until you have found the Dead Man!"

The king's daughter went back inside and burst into tears.

When her father learned why she was crying, he told her, "Never give any of those old wives' tales a thought!"

"I don't know what will become of me," she sobbed, "But I intend to depart in search of the Dead Man!"

"As you like! I give you up for lost!" replied the king, who then burst into tears himself. The girl paid no attention and set out.

After days and days of walking, she came to a marble palace. The door stood open and all the rooms were brightly lit. The girl walked in and called out, "Anybody home?"

There was no answer.

She went into the kitchen and found a pot of stew cooking on the stove. She opened the cupboard: it was full of provisions. "Since I'm already here, I may as well stay," she said, and sat down and ate, being quite hungry after all those days of traveling. When she had finished, she opened a door and saw a comfortable bed. "I'm going to bed. We'll see what tomorrow brings."

The next morning she awakened and resumed her tour of the palace. She opened every door until she finally found herself in a room in which a dead man lay. At his feet a note said:

> Whoever watches over me
> A year, three months, and a week,
> My beloved bride shall be.

"So I've found him I was seeking," murmured the girl. "All I have to do now is stick by him day and night." Nor did she budge from there except to prepare her meals.

Thus a year went by while she watched the dead man in solitude. Then one day a cry came up from the canal: "Slave girls for sale! Slave girls for sale!"

"That's an idea!" said the maiden. "I'll buy a slave girl right now. That way I'll at least have company, and every now and then I can lie down and sleep for a few minutes. I'm so tired now I could drop."

She went to the window, called to the slave dealer, and purchased one of his girls. She then had the girl come upstairs and stay by her side at all times.

Three more months went by, and the maiden was so tired that she said to the slave girl, "I'm going to bed now. Let me sleep for three days, no more. Wake me up the fourth day. Be careful and don't mistake the day!"

"Don't worry, I won't," replied the slave girl.

The maiden went to sleep, while the slave girl remained day and night with the dead man. Three days passed, then one more, and the maiden slept on. The slave girl said to herself, "Do you think I'd awaken her? Let her sleep, let her sleep!"

Time was up at last, and the dead man opened his eyes, saw the slave girl, embraced her, and said, "You shall be my beloved bride!"

At those words the whole palace sprang back to life. Servants came running from one wing, maids of honor, cooks, and coachmen from the other. In short, there were people all over the place.

The noise they made awakened the maiden, who realized she had slept a whole week. "I've been cheated!" she exclaimed. "That evil soul failed to call me, and now I've lost my fortune! Cursed was the hour and the minute when I bought that slave!"

The Dead Man happened to be a king. He asked the slave girl, "Did you watch over me all the time by yourself?"

"I brought in another woman for a few minutes every day, but she slept all the time and was of little use to me."

"Where is she now?" asked the king.

"In her room, sleeping as usual."

So the king married the slave girl. But despite all the royal finery he had her dressed in, despite all the gold and diamonds, she was still just as homely as homely could be. The king held open house for a week. After dinner it was his wish for the servants to join everybody else at the table for dessert, and he told his wife to be sure and invite the maid who had shared in her vigil.

"I'm not about to do that," said the bride. "In the first place she wouldn't come, for all she does is sleep."

On the contrary, all the poor maiden did was sigh and weep, day and night, over having slept one day too many and lost her fortune.

At the end of the week of open house, the king announced he had to go away to look after his property and that it was his custom on such occasions to bring back a present to every servant in his household. He sent for them all and asked them each what they wanted. Some said a handkerchief, some an outfit, others a pair of breeches, others a cutaway coat, and he wrote down everything so as to forget nothing. He said to his wife, "Call that maid of yours and let me find out what she wants, because I will bring her something too." The maiden was called in. The king found her so beautiful and genteel of face and speech that he was altogether charmed with her. "Tell me, my dear, what you want me to bring you."

"Please be so good," sighed the maiden, "as to bring me a tinderbox, a black taper, and a knife."

The king was dumbfounded by such a request. "Fine, fine, have no fear, I'll remember all three things."

He left, attended to his business, and then went to buy his servants' presents. Loaded down with all those purchases, he boarded his ship to sail back home. The ship weighed anchor, but could move neither forward nor backward. The sailors asked, "Venerable Majesty, did you possibly forget something?"

"No, not a thing," he replied, but then he went over his notes and saw that he had forgotten the maiden's three presents. He disembarked at once, entered a shop, and requested the three items.

The merchant looked him in the eye. "Excuse me for asking, but who will receive these things?"

"They are for one of my servants," replied the king.

"Listen carefully," said the merchant. "When you get home, don't give her anything, but make her wait three days. Then enter her room and say, 'Go get me a drink of water, and I'll give you the three presents.' As soon as she leaves the room, put them on her dresser and hide under the bed or in some other place where you can watch what she does."

"I'll do just that," said the king.

Upon his arrival all the servants ran out to meet him, and he gave them each the present he'd promised them. The last one in line was the maiden, who asked if he'd bought her three presents.

"You worrisome girl!" he replied. "I bought them and will give them to you later. . . ."

She returned to her room and cried, sure that he'd brought her nothing.

Three days later she heard a knock on the door, and there stood the king. "I'm here with your presents, but go get me a drink of water first, as I am thirsty."

The girl ran off. The king put everything on the dresser and hid under the bed. When she returned and found the king gone, she said, "He's gone off once more without giving me anything." She put the water down on the dresser and spied the presents.

Then she bolted the door, undressed, struck a light, and lit the black candle, which she placed on a little table. She picked up the knife and thrust it into the tabletop, saying, "Do you remember when I was back home with His Majesty my father and an old woman told me I wouldn't marry until I found the Dead Man?"

The knife replied, "Indeed I do."

"Do you remember my traveling the world over and finding a palace with the Dead Man inside?"

"Indeed I do."

"And you remember my watching him for a year and three months and then buying, to keep me company, that ugly slave I instructed to let me sleep for three days, since I was tired, and who let me sleep for a whole week only to have the Dead Man awaken, embrace her, and take her for his wife?"

"I remember only too well, alas!"

"Who should have been rewarded, myself who toiled a whole year and three months, or that woman who was there only a few days?"

"You."

"Since you remember everything and say I deserved the prize, fly from the little table and lodge yourself in my heart."

When the king heard the knife tearing loose from the table, he jumped out from under the bed, embraced the maiden, and said, "I heard everything! You shall be my wife! In the meantime stay in your room and leave everything to me."

He went to the slave girl and said, "Now that I'm back from my trip, I shall hold a week of open house."

"Just so you don't squander too much money on it," replied the slave.

"But that has always been my custom every time I've returned from a journey."

There was a grand banquet, and the king said to the slave, "I want all my servants to come in for dessert, and I want you to invite your maid too."

"Can't you let that woman alone? She's so disagreeable."

"If you don't invite her, I will." Thus the maiden came to the table, as tearful as ever.

When dinner was over, the king gave an account of his journey. He had visited a city, he said, whose king had been under a spell like the one he had been under himself. A maiden had watched him for a year and three months, then bought a slave to keep her company. Exhausted, the maiden went to sleep, and the slave girl failed to awaken her. The Dead Man woke up, saw the slave girl, and married her.

"Now tell me, everybody, which one should have been the king's bride, the one who watched for a week, or the one who watched for a year and three months?"

"The one who watched for a year and three months," agreed everyone.

The king continued. "Here, ladies and gentlemen, is the lady who watched for a year and three months, and here is the slave girl she bought. Tell me how this ugly Moor should die for her shameful treachery."

They all jumped to their feet clamoring, "Burn her! In the middle of the town square, in a caldron of flaming pitch!"

It was done. The king married the maiden, and they lived happily ever afterward, nor was any more ever said about them.

(Venice)

◈ 33 ◈

Pome and Peel

There was once a noble couple that longed to have a son, but alas, they had none. One day the lord was abroad and encountered a wizard. "Sir Wizard," he said, "please tell me what I can do to have a son."

The wizard gave him an apple and said, "Have your wife eat it, and at the end of nine months she will give birth to a fine baby boy."

The husband took the apple home to his wife. "Eat this apple, and we will have a fine baby boy. A wizard told me so."

Overjoyed, the wife called her maidservant and told her to peel the apple. The maidservant did so, but kept the peeling and ate it herself.

A son was born to the lady, and on the same day a son was born to her maidservant. The maidservant's son was as ruddy as an apple skin; the lady's son was as white as apple pulp. The lord looked on them both as his sons and reared and schooled them together.

Growing up, Pome and Peel loved each other like brothers. Out walking one day, they heard about a wizard's daughter as dazzling as the sun; but no one had ever seen her, as she never went abroad or even looked out her window. Pome and Peel had a large bronze horse built with a hollow belly, and they hid in it with a trumpet and a violin. The horse moved on wheels the boys turned from inside, and in that manner they rolled up to the wizard's palace and began to play. The wizard looked out and, seeing that wonderful bronze horse making music all by itself, invited it inside to entertain his daughter.

The maiden was delighted. But the minute she was left alone with the horse, out stepped Pome and Peel, and she was quite frightened. "Don't

95

be afraid," they said to her. "We heard how beautiful you are, and we just had to see you. If you want us to leave, we will. But if you like our music and want us to keep playing, we'll do so, then depart without letting anyone know we were ever here."

So they stayed on, playing and having a good time, and after a while the wizard's daughter didn't want them to leave. "Come with us," Pome told her, "and I'll marry you."

She accepted. They all hid in the horse's belly and off they rolled. No sooner had they gone than the wizard returned home, called his daughter, looked for her, questioned the guard at the gate: there was no sign of her anywhere. Then he realized he had been tricked, and he was furious. He went to the balcony and screamed three curses on the girl: "Let her come upon three horses—one white, one red, one black—and loving horses the way she does, let her leap on the white one, and let this horse be her undoing.

"Or else: Let her come upon three pretty little dogs—one white, one red, one black—and loving little black dogs the way she does, let her pick up the black one, and let this dog be her undoing.

"Or else: On the night she goes to bed with her spouse, let a giant snake come through the window, and let this snake be her undoing."

While the wizard was screaming those three curses from the balcony, three old fairies happened by on the street below and heard everything.

In the evening, weary from their long trip, the fairies stopped at an inn. As soon as they were inside, one of them remarked, "Just look at the wizard's daughter! She wouldn't be sleeping so soundly if she knew about her father's three curses!"

For there asleep on a bench in the inn were Pome, Peel, and the wizard's daughter. Peel wasn't actually asleep; perhaps he wasn't sleepy, or maybe he just considered it always wiser to sleep with one eye open and thus know what was going on around him.

So he overheard one fairy say, "It's the wizard's will for her to come upon three horses—one white, one red, one black—and leap on the white one, which will be her undoing."

"But," put in the second fairy, "if some far-seeing soul were present, he would cut off the horse's head at once, and nothing would happen."

The third fairy added, "Whoever breathes a word of this will turn to stone."

"Then it's the wizard's will for her to come upon three pretty little dogs," said the first fairy, "and pick up the very one that will be her undoing."

"But," commented the second fairy, "if some far-seeing soul were present, he would cut off the puppy's head at once, and nothing would happen."

"Whoever breathes a word of this," said the third fairy, "will turn to stone."

"It's finally his will, the first night she sleeps with her husband, for a giant snake to come through the window and destroy her."

But if some far-seeing soul were present, he would cut off the snake's head, and nothing would happen," chimed in the second fairy.

"Whoever breathes a word of this will turn to stone."

So Peel found himself in possession of three terrible secrets which he could not reveal without turning to stone.

The next morning they set out for a post house, where Pome's father had three horses waiting for them—one white, one red, one black. The wizard's daughter immediately jumped into the saddle on the white one, but Peel promptly unsheathed his sword and cut off the horse's head.

"What are you doing? Have you lost your mind?"

"Forgive me, I am not at liberty to explain."

"Pome, this Peel has a wicked heart!" said the wizard's daughter. "I will travel no further in his company."

But Peel admitted having cut off the horse's head in a moment of madness. He begged her to forgive him, which she ended up doing.

They reached the home of Pome's parents, and three pretty little dogs ran out to meet them—one white, one red, and one black. She bent down to pick up the black one, but Peel drew his sword and cut off the dog's head.

"Away with him at once, this crazy, cruel man!" screamed the bride.

At that moment Pome's parents came out. They heartily welcomed their son and his bride and, learning of the dispute with Peel, they persuaded her to pardon him once more. But at dinner, amidst the general merriment, Peel was pensive and aloof, nor could anyone make him say what was troubling him. "Nothing's the matter, absolutely nothing," he insisted, although he left the banquet early, under the pretext of being sleepy. But instead of going to his room, he entered the bridal chamber and hid under the bed.

The bride and bridegroom went to bed and fell asleep. Keeping watch, Peel soon heard the windowpane break, and in crawled a giant snake. Peel leaped out, bared his sword, and cut off the snake's head. At the commotion the bride awoke, saw Peel by the bed with his sword unsheathed, saw no snake (it had vanished), and screamed, "Help! Murder! Peel wants to kill us! I've pardoned him two times already, let him be put to death this time!"

Peel was seized, imprisoned, and three days later dressed for the gallows. Imagining himself now doomed in any event, he asked permission to tell Pome's wife three things before dying. She came to him in prison.

"Do you remember," Peel asked, "when we stopped at an inn?"

"Of course I do."

"Well, while you and your husband were sleeping, three fairies came in and said the wizard had placed three curses on his daughter: to come upon three horses and leap on the white horse, which would be her undoing. But, they added, should somebody quickly cut off the horse's head, nothing would happen. And whoever breathed a word of this would turn to stone."

As he said those words, poor Peel's feet and legs turned to marble.

The young woman understood. "That's enough, please!" she screamed. "Don't tell me any more!"

But he went on: "Doomed whether I speak or keep silent, I choose to speak. The three fairies also said the wizard's daughter would come upon three pretty little dogs . . ."

He related the curse regarding the little dogs and turned to stone up to his neck.

"I understand! Poor Peel, forgive me! Don't go on!" pleaded the bride.

But in a strained voice, since his throat was already marble, and stuttering, since his jaws were becoming marble, he told her about the curse with the snake. "But . . . whoever breathes a word of this . . . will turn to stone . . ." At that, he was silent, marble from head to foot.

"What have I done!" moaned the young wife. "This faithful soul is damned . . . unless . . . why, of course, the only person that can save him is my father." And she took paper, pen, and ink, and wrote her father, asking his forgiveness and begging him to come to her.

The wizard, whose child was the apple of his eye, came to her at breakneck speed. "Papa dear," she said as she kissed him, "I am asking you a favor. Look at this poor youth. After saving my life and protecting me from your three curses, he has turned to stone from head to foot."

Sighing, the wizard replied, "For the love I bear you, I will do this also." He drew a phial of balsam from his pocket, brushed Peel with it, and Peel sprang back to life as sound as ever.

Thus, instead of leading him to the gallows, they bore him home in triumph, amid music and singing, while the throngs around him shouted, "Long live Peel! Long live Peel!"

(Venice)

34

The Cloven Youth

A woman was expecting a baby and craved parsley. Next door to her lived a famous witch who had a whole garden of parsley. The garden gate was always open, since the parsley was so abundant, and all who wished could go in and help themselves. The woman with a craving for parsley went in, fell to, and didn't stop until she'd eaten half the garden. When the witch returned and found half the garden stripped, she said, "Ah-ha! They intend to eat me clean out. . . . I'll just keep watch tomorrow and find out who it is."

The woman showed up the next day to eat the rest of the parsley. She'd scarcely finished nibbling up the last plant when out jumped the witch and said, "Ah-ha! So you're the one who ate all my parsley!"

The woman was terrified. "Please let me go, I'm expecting a baby."

"Of course I'll let you go," replied the witch. "Only, the baby boy or girl you bear will be half mine and half yours when it is seven years old."

In great fright, the woman agreed, just to get away.

A boy was born. He grew, and when he was six, he met the witch on the street one day. Seeing him, she said, "Listen, remind your mother we have only one more year to go."

The child went home and said, "Mamma, an old woman told me we have only one more year to go."

"If she repeats that, tell her she's crazy."

When the boy was only three months away from his seventh birthday, the witch said to him, "Tell your mother we have only three months to go."

"My dear lady," he replied, "you're crazy!"

"Indeed! We'll see whether I'm crazy!

Three months later the old woman caught the boy on the street and took him to her house. She placed him on a table flat on his back, picked up a knife, and cut him lengthwise into two halves, starting at his head.

To one of the halves, she said, "You go home." To the other, "You stay with me."

One half remained, while the other went home and said, "You see, Mamma, what that old woman did to me? And you said she was crazy!" His mother threw up her hands, but what could she say now?

This half-boy grew up, but had no idea what calling to follow. He finally decided to be a fisherman. One day while eel fishing, he caught an eel as long as himself. When he pulled it up, the eel said, "Let me go and you'll fish me up again." He threw it back into the water, lowered his net once more, and pulled it up full of eels. He returned to shore with his boat brimming with eels and earned a bag of money.

The next day he caught the large eel again, which said, "Let me go, and whatever you wish will be granted, for the sake of the little eel." He let it go at once.

One day on his way out to fish as usual, he passed the king's palace. The king's daughter was on the balcony with her maids of honor. She saw this man with half a head, half a body, and just one leg, and burst out laughing. He looked up at her and said, "You laugh, do you? For the sake of the little eel, the king's daughter shall therefore have a son by me."

Not long after that, the king's daughter expected a child, and her parents became aware of it. "Just what is the meaning of this?" they asked.

"I'm as much in the dark as you are," replied the girl.

"What do you mean you're in the dark? Who's the father?"

"I really don't know. I know nothing about any of this." And she continued to say that, even though her parents went on questioning her, urging her to talk, and assuring her of their forgiveness. At last they lost patience and began insulting her and mistreating her.

The baby was born, a handsome son, but the girl's parents wept over the disgrace of having a fatherless child under their roof and called in a sorcerer to solve the riddle. The sorcerer said, "Wait until he's a year old."

After a year, the sorcerer said, "You must have a grand reception for all the noblemen in town. When they're all in the reception hall, let the child be carried past them with a gold apple and a silver apple. He will give his father the gold apple and his grandfather the silver apple."

The king sent out the invitations and had chairs set up around the walls of a large reception hall. When all the noblemen in town were seated, he sent for the nurse with the baby in her arms and put the two apples into his hands. "This one is for your father, and this one for your grandfather."

The nurse went around the hall and came back to the king, and the baby handed him the silver apple.

"I know only too well I'm your grandfather," said the king. "But I want to know who your father is."

But the baby was taken round and round the room, without giving the gold apple to a soul.

The sorcerer was called back and said, "Now give a reception for the town's poor men." So the king announced this reception.

When the cloven youth heard there was to be a reception at the palace for all the poor men in town, he said to his mother, "Get out my best half-shirt, my half-vest, my trouser, my pump, and my half-beret, for I'm invited to the king's."

The large hall was packed with paupers, fishermen, and beggars, who sat on benches the king had ordered placed against the walls. The nurse started around with the little boy holding the gold apple. "Go on," she told him, "give it to Papa." She continued around the room. The minute the child saw the cloven youth, he broke into a smile and threw his arms around the young man's neck, saying, "Here, Papa, take this apple!"

The poor men seated around the room on benches burst out laughing. "Ha, ha, ha! Just look who the king's daughter fell in love with!"

The only one among them to remain perfectly calm was the king, who said, "In that case, he shall become my daughter's husband!"

The wedding was performed at once. The newlyweds emerged from church, expecting to find a carriage waiting for them. Instead, there stood a barrel, a great big empty barrel. The cloven youth, his bride, and their child were put inside and sealed up, after which the barrel was thrown into the sea.

A storm was raging over the sea, and the barrel bobbed up and down on the waves until it finally disappeared from sight, and everyone at the king's palace said it had gone under for good.

It didn't sink, though, but floated out to sea. Sensing how frightened the king's daughter was, the cloven youth said to her, "My bride, would you like me to bring the barrel in to shore?"

The bride replied, scarcely above a whisper, "Yes, if you can."

No sooner said than done! For the sake of the little eel, the barrel came to rest on dry land. The cloven youth broke open the bottom, and they all three stepped out. It was mealtime, and for the sake of the little eel, a table appeared, set for three and laden with tasty dishes and beverages. When they had eaten and drunk their fill, the cloven youth asked, "Are you satisfied with me, my bride?"

"I would like you still better," she said, "if you were whole."

At that, he said to himself, "For the sake of the little eel, may I become whole and handsomer than ever." At once he became the handsomest of youths, completely whole, and dressed as a grand nobleman. "Are you satisfied?"

"Oh, yes, but I would be still more so if we were in a fine palace, instead of out here on this deserted shore."

He therefore thought to himself, For the sake of the little eel, may we find ourselves in a fine palace with two apple trees, one bearing gold apples and the other silver apples, and may we have maids, butlers, ladies-in-waiting, and everything one needs in a palace.

No sooner had he thought of all that than everything was there before him—palace, apples, and butlers.

A few days later, the cloven youth, who was now no longer cloven but whole, gave a reception for all the kings and queens in the vicinity, including his wife's father. The cloven youth, greeting them at the door, said, "Let me warn you of one thing: do not touch those gold apples and those silver apples. Heaven help you if you put your hands on them."

"Don't worry, don't worry!" replied the guests. "We'll keep our hands where they belong."

They sat down to eat and drink, while the cloven youth said to himself, "For the sake of the little eel, let a gold apple and a silver apple find their way into my father-in-law's pockets."

After dinner he took his guests into the garden for a stroll and noticed that two apples were missing. "Who took them?" he asked.

"Not I," answered all the kings. "I've not put my hands on a thing."

The cloven youth said, "I gave you warning those apples were not to be touched. Now I've no choice but to search Your Majesties."

He went down the line frisking them, king by king and queen by queen. No one had the apples. At last he came to his father-in-law, and there was an apple in each pocket. "Of all things! No one else dared touch them, but you stole two! You're now going to account to me!"

"But I know nothing about the apples," the king tried to explain. "I don't know how on earth . . . I didn't take them, I swear!"

"Even with all the evidence against you," said the cloven youth, "you still claim you're innocent?"

"Yes."

"Well, then, just as you are innocent, so was your daughter, and it's only fair that I do to you what you did to her."

In walked the young man's wife at that moment. "Never let it be said that my father had to suffer because of me. Even if he was cruel to me, he's still my father, and I beg you to be merciful to him."

Moved to pity, the cloven youth pardoned him. Happy to be reunited with his daughter, whom he'd given up for dead, and pleased to learn she was innocent, the king took them back to his palace, where they all lived in harmony from then on. Unless they have died in the meantime, they may well be there to this day.

(Venice)

35

Invisible Grandfather

There was once a mother with three daughters, and the family was as poor as poor could be. One day, one of the three girls said, "Look, rather than stay here and suffer, I'm going out in the world and seek my fortune." With that, she picked up and left.

After walking for miles and miles, she came to a palace. Finding the door open, she said, "I'll go in and see if they need a servant." She entered and called out, "Hello! Is anyone at home?" No one answered her. She walked into the kitchen and saw the pot boiling over the fire. Opening a cupboard, she found bread, rice, wine, a little bit of everything, and said, "Here's everything one could possibly need, and I am hungry, so I'm making myself some good soup right away."

As she uttered those words, she saw two hands setting the table. The hands put out a bowl of rice, and the girl said, "Now I'll eat," and sat down at once to the table. When she'd finished her rice, the hands brought her a cockerel, and the girl ate every bit of it. "Yes, indeed," she mused, "I was truly weak from hunger, but I feel better now."

She went through the palace and saw a beautiful reception hall, a breakfast room, and a bedroom with a canopied bed. "What a fine bed! I'm going to retire right away." She lay down and slept the whole night long.

The minute she woke up next morning, the same two hands appeared with coffee on a tray. She drank it, and the hands carried off the tray with the cup on it. After dressing, she passed into a large room containing a vast wardrobe full of dresses, shawls, skirts, and other wearing apparel. The girl cast off her rags and dressed in queenly attire. If she was beautiful before, no words can describe how lovely she was afterward.

Outside was an arbor, and she strolled under it at the very moment a king happened by. Catching sight of the beautiful maiden, he asked her under what conditions he could talk to her, for he was overwhelmed with admiration. The girl replied that she had neither father nor mother, but that if he would stop by another day, she would have an answer for him. The king bowed profusely, then rode off in his carriage.

The girl went back inside, approached the fireplace, and said, "Dear Sir, I ended up at this palace, but I've never seen a soul anywhere around, and now there's a king who's taken a liking to me. What must I tell him when he returns for an answer?"

From the chimney a voice answered her: "Beautiful you are and ever more beautiful will you be. I give you my blessing! Tell the king your poor, sick, solitary grandfather is glad for you to marry, provided you don't put off the wedding. Now go, my lovely one whose loveliness will increase." And the girl grew ever more beautiful.

The next day she appeared on her balcony just as the king was arriving, and the minute he saw her, he asked for her answer. She explained that she couldn't invite him inside, as her poor sick grandfather was there. But her grandfather was glad for them to marry if they would do so without delay, and meanwhile they could carry on their courtship on the balcony. The king was overjoyed.

They courted for a whole week, at the end of which the girl went up to the fireplace and said, "Grandfather, we've now courted for a week. Do you think that's long enough?"

He answered: "Go ahead and marry him and start carrying off everything in the house. Be sure you leave nothing behind. It is very important you take every single thing! Now go, my lovely one whose loveliness will increase." And she grew still more beautiful.

She went to the balcony, and the minute the king appeared, she told him to make arrangements for the wedding and to send carriages and horses in the meantime to haul away everything in her palace. It took them a good week to carry everything off. And the king said to his father, "Just look, Papa, at what fine things my bride has. Nothing our royal family has comes up to them. And just wait until you see what a beauty she is!"

In the meantime his fiancée had swept out the palace and thrown away brooms and brushes. It was now completely empty. All that remained was a golden necklace she intended putting on as soon as she departed, and she'd left it hanging on a nail. As she waited on the balcony, she saw the king approaching in his two-horse carriage, so she went to the fireplace and said, "Grandfather, I'm leaving now, since my bridegroom has come for me. Put your mind at rest, I've taken everything away and swept the palace clean."

"Good girl," said Grandfather. "I thank you. Beautiful you are, and ever more beautiful will you become."

More beautiful than ever, the girl climbed into the carriage, and the king embraced her; then they drove off. Halfway to the king's palace, she touched her neck and exclaimed, "Woe is me, I forgot my golden necklace.... Quick, let us go back for it!"

The king replied, "Don't give it a thought. We'll have a much finer one made for you."

But she insisted on going back at all costs. She got out of the carriage and went into the palace, while the king waited for her below. Approaching the fireplace, she said, "Grandfather?"

"What do you want?"

"Please forgive me, I forgot my golden necklace," and saying that, she took it off the nail.

"Be gone!" screamed the voice from the chimney. "Be gone, you hideous bearded woman!"

At that moment, as the girl slipped on her necklace, she felt hair touch her fingers. She looked in the mirror: she had a long beard that came halfway down her bosom.

Seeing her come out like that, the bridegroom put his hands to his head. "I told you we shouldn't have come back here! Now what will I tell my father after having praised your beauty to the skies? I can no longer take you home with me. But I have a cottage in the woods nearby, and I'll keep you there."

That he did, and called on her every day, for he still loved her and saw that she lacked nothing. Word got out and soon reached the king that his son was courting a bearded woman. At that, the king sent for his son and said to him, "What do you mean by courting a bearded woman? The dignity of the crown is at stake! Either you give her up, or I will put her to death!"

The youth went to the girl and said, "I must tell you something. My father has found out I'm courting a bearded woman, and said if I didn't leave you he would put you to death. What hope is there for us?"

"Do one thing for me," answered the girl. "Get someone to make me a black veil and a black velvet dress. Then take me to Grandfather, and we'll ask him to help us."

The prince brought her the dress and veil, and as soon as she was all wrapped up in them, they got into the carriage and drove off to the palace.

She approached the fireplace and said, "Grandfather?"

"Who's there?"

"It's me, Grandfather!"

"What do you want, you hideous, bearded woman?"

"Please listen, dear Grandfather. Because of you, I've been condemned to die . . ."

"Because of me? Didn't I tell you to take away everything, every single thing? If you hadn't left the golden necklace, I would now be free from my evil spell; but instead, I have to start my sentence all over again from the beginning!"

"Grandfather," said the girl, "I'm not asking for the beauty back I had in this palace, but I'd at least like my face to look as it did the first time I ever came here. Please, Grandfather, make me the way I used to be."

"Very well," said Grandfather; "you've not forgotten anything?"

"No, no," she replied. "I'm holding the necklace I left hanging on a nail."

Then the grandfather said to her, "Put it around your neck. Beautiful you were, and more beautiful will you become."

The girl put on her necklace, and the beard suddenly vanished.

"Grandfather! Thank you! Farewell!"

"Go, my lovely one, your loveliness will increase." And the girl became as dazzling as the sun.

She flew down the steps and into her bridegroom's carriage. The king's son was overjoyed to see her once more the way she used to be and even a hundred times more beautiful. He embraced her and said, "My father wouldn't dream of sentencing you to death now, and he wouldn't say it is undignified for the crown prince to court you."

As soon as they reached the royal palace, the father came out. "Here," said the son, "is the hideous bearded woman I have been courting."

"Ah," said the old king. "My son, you are exactly right. There couldn't be a lovelier girl under the sun." He embraced her and gave orders for the wedding, and in the meantime had her appear on the balcony, so that the whole town could see her. All the citizens immediately gathered before the balcony and, at the sight of the maiden, cried, "Long live our new queen!"

A few days later the two young people were married. At the wedding banquet they served radish preserves, peeled mice, skinned cats, and fried monkeys. They ate that, and enough was left over for tomorrow. To top off everything was a sprig of rosemary, token of remembrance, but nobody thought to say to me so much as "Have a glass of wine!"

(Venice)

◈ 36 ◈

The King of Denmark's Son

There was once a king and queen who seemed unable to have any children. At last, thanks to constant prayer to their idols, the queen gave birth to a baby girl. To learn the daughter's destiny, she called in twelve astrologers. Eleven of them she presented with a gold telescope apiece, but on the twelfth, the oldest astrologer, she bestowed only a silver telescope. The astrologers gathered round the girl. Some of them said she would be beautiful, others that she would be clever, others virtuous—in short, the usual things. Only the oldest astrologer remained silent. "Let us hear your prophecy too," said the king. At that, the old man, whose feelings had been hurt because of the telescope, answered that the predictions would all come true, but that the girl would fall in love with the first man she heard named.

"How can we prevent that?" asked the king.

"You will have to build a palace adjoining yours," explained the astrologer, "furnished as befits a king's daughter, and house her there with nurses and maidservants. But there mustn't be a single window in this palace, except a tiny little one way up at the top."

So was it done. The king called on his daughter once a month and saw her growing up and getting lovelier and lovelier, just as the astrologers had predicted she would. But the older she got, the more certain she became that she couldn't stay closed up there forever and that there had to be another world different from her prison.

One day while her maids were out in the garden, the girl went and stood under the little window way up at the top, then built a tower out of big tables, little tables, and chairs, piling one on top of the other until she reached the windowsill. She looked out and saw the sky with the sun and clouds; although the earth was not visible, she heard words and sounds floating up from it.

Two young men happened to be passing by. One of them said, "Just what is this palace next to the king's?"

"You don't know? They keep the king's daughter closed up in there, since it's been predicted she would fall in love with the first man she heard about."

"Is she beautiful?"

"They say she is, but no one has ever seen her."

"She can be as beautiful as beautiful can be, but never so fine-looking as the king of Denmark's son. Did you know that the king of Denmark's son is so dazzling he has to wear seven veils over his face? And he'll never marry until he finds a wife whose looks equal his own."

Hearing that conversation, the king's daughter was seized with frenzy and fell to the floor. Her maids of honor came running and found her weeping and raving. "I have to get out of here, I have to get out!"

"Calm down," answered the maids of honor; "wait and tell your father when he comes back to see you."

At the end of the month, her father came to call on her as usual, and she burst into tears and told him her imprisonment was senseless and that she had to get out. So the king took her to his palace and ordered that no man ever be mentioned in her presence. But the girl had the king of Denmark's son on her mind now, and she was always wistful. Her father repeatedly asked her what the matter was, but she replied, "Nothing, nothing at all!" Finally one day she took heart, entered her father's study, threw herself on her knees, and told him about the king of Denmark's son. "Please, Father, send to him and ask if he will have me for his wife."

"Get up from there and calm down," ordered the king. "I'll send ambassadors to him at once. I am more powerful than the king of Denmark, so he won't say no to me."

The ambassadors arrived at the king of Denmark's. The king called in his son, who entered with the seven veils over his face, and his father told him he was being sought in marriage.

At that, the boy lifted the first veil and asked the ambassadors, "Is she as fine-looking as I am?"

The ambassadors replied, "Your Highness, she is."

He lifted the second veil. "Is she as fine-looking as I am?"

"Your Highness, she is."

He thus lifted all the veils, one after the other, and when he had removed the last one, he asked, "Is she as fine-looking as I am?"

The ambassadors hung their heads. "Your Highness, no."

"Tell her, in that case, I don't want her."

"But she swore," insisted the ambassadors, "that if Your Majesty turned her down she would hang herself."

At that, the king of Denmark's son picked up a cord and threw it to the ambassadors. "Take her this rope and tell her to hang herself."

The ambassadors returned bearing the rope, and the king flew into a rage. But the girl cried, sighed, and pleaded until her father sent the ambassadors back to the king of Denmark.

This time as well the king of Denmark's son lifted all his veils down to the seventh and asked, "Is she as fine-looking as I am?"

"Your Highness, no."

"Then tell her I don't want her."

"She swore she'd take a knife and stab herself to death."

"Take this knife and tell her to stab herself to death."

When they returned with the knife, the king was ready to declare war on Denmark, but his daughter begged him to calm down, and a few months later she persuaded him to dispatch the ambassadors one more time.

The king of Denmark's son asked the same questions.

"Your Highness, no," replied the ambassadors when he lifted the last veil, "but if she's rejected again, she said she would take a pistol and shoot herself."

"Take this pistol and let her shoot herself."

They returned with the pistol. The king had another tantrum, and the daughter another crying spell. "Please, Father, make me an iron cask; close me up in it and send me out to sea."

Her father wouldn't hear of it, but she kept on begging until at last she was put into a cask dressed as a princess, with her crown on her head, and carrying cord, knife, pistol, and a few provisions for the crossing. Off she went over the sea.

After floating for days and days and days, she was washed ashore on an island where the palace of a queen stood. When the ladies-in-waiting opened the windows in the morning, they spied the cask on the beach. "Your Highness!" they exclaimed, "if you could only see what a beautiful little cask the sea has washed ashore!"

The queen ordered the cask brought in. They opened it as she looked on, and out stepped the beautiful maiden. "Why are you sailing the sea like that?" asked the queen, whereupon the girl explained.

"You have nothing to worry about, absolutely nothing," said the queen. "The king of Denmark's son is my brother. He comes to see me every month to drink a glass of seawater. He's expected here in just a few days."

So here came the king of Denmark's son. Instead of having the usual maid of honor take him his glass of seawater, his sister the queen sent that maiden to him. The minute he laid eyes on her he was in love. "Who is this beautiful lady?" he asked his sister.

"A friend of mine."

"Listen, my sister. From now on, instead of visiting you once a month, I'll come here every fortnight."

He returned in a fortnight, and the same maiden served him his glass of seawater.

"Listen, my sister, instead of every fortnight, I'll come here every week."

He was back the next week, but this time the original maid of honor took him his glass of seawater. The king of Denmark's son refused to drink it. "Is that other beautiful maiden no longer here?"

"She's not feeling very well."

"I shall go to her room and see her."

The maiden was in bed and, before her, on the sheet lay cord, knife, and pistol. But he looked only at her and paid no attention to the weapons. She said to him though, "Now, which one of these three things must I use?"

When he didn't understand, she explained that she was the daughter of the king who had sent ambassadors to him.

"I was misled!" he exclaimed. "Had I known you were so beautiful I would have said yes at once!"

At that, the maiden got up and wrote her father: "I am at the house of a certain queen, and here also is the king of Denmark's son, who wishes to marry me."

Overjoyed, her father came for her, and they all went to the king of Denmark's and celebrated the wedding.

(Venice)

◈ 37 ◈

Petie Pete versus Witch Bea-Witch

Petie Pete was a little boy just so tall who went to school. On the school road was a garden with a pear tree, which Petie Pete used to climb and eat the pears. Beneath the tree passed Witch Bea-Witch one day and said:

> "Petie Pete, pass me a pear
> With your little paw!
> I mean it, don't guffaw,
> My mouth waters, I swear, I swear!"

Petie Pete thought, Her mouth waters not for the pears but for me, and refused to come down the tree. He plucked a pear and threw it to Witch Bea-Witch. But the pear fell on the ground right where a cow had been by and deposited one of its mementos.

Witch Bea-Witch repeated:

> "Petie Pete, pass me a pear
> With your little paw!
> I mean it, don't guffaw,
> My mouth waters, I swear, I swear!"

But Petie Pete stayed in the tree and tossed down another pear, which fell on the ground right where a horse had been by and left a big puddle.

Witch Bea-Witch repeated her request, and Petie Pete thought it wiser to comply. He scampered down and offered her a pear. Witch Bea-Witch opened up her bag, but instead of putting in the pear, she put in Petie Pete, tied up the bag, and slung it over her shoulder.

After going a little way, Witch Bea-Witch had to stop and relieve herself; she put the bag down and went behind a bush. Meanwhile, with his little teeth as sharp as a rat's, Petie Pete gnawed the cord in two that tied up the bag, jumped out, shoved a heavy rock into the bag, and fled. Witch Bea-Witch took up the bag once more and flung it over her shoulder.

> "O Petie Pete,
> To carry you is a feat!"

she said, and wound her way home. The door was closed, so Witch Bea-Witch called her daughter:

> "Maggy Mag! Marguerite!
> Come undo the door;
> Then I ask you more:
> Put on the pot to stew Petie Pete."

Maggy Mag opened up, then placed a caldron of water over the fire. When the water came to a boil Witch Bea-Witch emptied her bag into it. *Splash*! went the stone and crashed through the caldron. Water poured into the fire and spattered all over the floor, burning Witch Bea-Witch's legs.

> "Mamma, just what do you mean
> By boiling stones in our tureen?"

cried Maggy Mag, and Witch Bea-Witch, dancing up and down in pain, snapped:

> "Child, rekindle the flame;
> I'll be back in a flash with something tame."

She changed clothes, donned a blond wig, and went out with the bag.

Instead of going on to school, Petie Pete had climbed back up the pear tree. In disguise, Witch Bea-Witch came by again, hoping he wouldn't recognize her, and said:

> "Petie Pete, pass me a pear
> With your little paw!
> I mean it, don't guffaw,
> My mouth waters, I swear, I swear!"

But Petie Pete had recognized her and dared not come down:

> "Pears I refuse old Witch Bea-Witch,
> Who would bag me without a hitch."

Then Witch Bea-Witch reassured him:

> "I'm not the soul you think, I swear,
> This morning only did I leave my lair.
> Petie, Pete, pass me a pear
> With your little paw so fair."

She kept on until she finally talked Petie Pete into coming down and giving her a pear. At once she shoved him down into the bag.

Reaching the bushes, she again had to stop and relieve herself; but this time the bag was tied too tight for Petie Pete to get away. So what did he do but call "Bobwhite" several times in imitation of quail. A hunter with his dog out hunting quail found the bag and opened it. Petie Pete jumped out and begged the hunter to put the dog into the bag in his place. When Witch Bea-Witch returned and shouldered the bag, the dog inside did nothing but squirm and whine, and Witch Bea-Witch said:

> "Petie Pete, there's nothing to help you,
> Bark like a dog is all you can do."

She got home and called her daughter:

> "Maggy Mag! Marguerite!
> Come undo the door;
> Then I ask you more:
> Put on the pot to stew Petie Pete."

But when she went to empty the bag into the boiling water, the angry dog slipped out, bit her on the shin, dashed into the yard, and gobbled up hens left and right.

> "Mamma, have you lost your mind?
> Is it on dogs you now want to dine?"

exclaimed Maggy Mag. Witch Bea-Witch snapped:

> "Child, rekindle the flame:
> I'll be back in a flash."

She changed clothes, donned a red wig, and returned to the pear tree. She went on at such length that Petie Pete fell into the trap once more. This time there were no rest stops. She carried the bag straight home where her daughter was waiting on the doorstep for her.

"Shut him up in the chicken coop," ordered the Witch, "and early tomorrow morning while I'm out, make him into hash with potatoes."

The next morning Maggy Mag took a carving board and knife to the henhouse and opened a little hen door.

> "Petie Pete, just for fun,
> Please lay your head upon this board."

He replied:

> "First show me how!"

Maggy Mag laid her neck on the board, and Petie Pete picked up the carving knife and cut off her head, which he put on to fry in the frying pan.

Witch Bea-Witch came back and exclaimed:

> "Marguerite, dear daughter,
> What have you thrown in the fryer?"

"Me!" piped Petie Pete, sitting on the hood over the fireplace.

"How did you get way up there?" asked Witch Bea-Witch.

"I piled one pot on top of the other and came on up."

So Witch Bea-Witch tried to make a ladder of pots to go after him, but when she got halfway to the top the pots came crashing down, and into the fire she fell and burned to ashes.

(Friuli)

Quack, Quack! Stick to My Back!

A king had a daughter as pretty as a picture whom all the princes and noblemen would have liked to marry, had it not been for the bargain she'd made with her father.

This king, mind you, had once given a big banquet, and while the guests all laughed and enjoyed themselves, his daughter remained serious and solemn-faced. "Why so glum?" asked her table companions. She answered them with total silence. They all tried to make her laugh, but failed.

"My daughter, are you angry?" asked her father.

"No, Father, I am not."

"Then why don't you laugh?"

"I wouldn't laugh even if my life depended on it."

Then the king had an idea. "Fine! Since you're so determined not to laugh, let's try something, rather let's make a bargain. Whoever would marry you must manage to make you laugh."

"Very well," said the princess. "But under this condition: whoever tries to make me laugh and fails will have his head cut off."

Thus was it agreed. All the guests witnessed the pact, and the royal word once given had to be kept.

The news spread to the four corners of the world, and all the princes and noblemen began competing for the hand of the lovely princess. But every single one who tried lost his life. Early each morning the princess would go out on her balcony to wait for a suitor to come by. Time was passing, and the king was more and more afraid his daughter would end up an old maid.

Now word of all this also reached a certain country village. You know how people sit around at night talking about all sorts of things, and so they got on the subject of the princess's bargain with her father. A boy with scalp disease, the son of a poor cobbler, listened open-mouthed. At length, he said, "I shall go myself and try!"

"Don't be silly, my son," answered his father.

"I'm serious, Father, I'm going. I shall set out tomorrow."

"Those people are serious too, and they'll put you to death."

"Father, I intend to become king!"

"Ha, ha, ha!" they all laughed. "A king with scalp disease!"

The next morning the father had forgotten all about it, when his son came in and announced, "Well, Father, I'm leaving. Here everybody looks down on me because of my scalp disease. Give me three loaves of bread, three gold florins, and a bottle of wine."

"But just think what you're letting yourself in for."

"I have," and with that, he departed.

He walked and walked and met a poor woman trudging along with the aid of a stick. "Are you hungry, madam?" asked the boy with the scalp disease.

"I certainly am, son. Could you give me something to eat?"

He gave her one of his three loaves, which she ate. But since she was still hungry, he gave her the second loaf as well. Feeling truly sorry for her, he ended up giving her the third one too.

On and on he went, until he met another woman in tatters.

"Could you give me a little money, my lad, to buy myself some sort of dress?"

He gave her a florin. Then he got to thinking one florin was perhaps not enough, so he gave her another one. But he felt so sorry for her that he handed her the third one too.

On and on he went until he met another woman, who was old, wrinkled, and panting with thirst.

"Dear boy, give me something to quench my thirst and you'll save a soul from Purgatory."

The boy with scalp disease handed her his bottle of wine. The old woman drank a little, and he kept on telling her to have more until she had drained the bottle dry. She looked up at last and was no longer an old woman, but a lovely blond maiden with a star in her hair.

"I know where you are going," she said, "and I know how kind-hearted you are, because the three women you met were all none other than myself. I shall now come to your assistance. Take this fine goose and carry it with you everywhere you go. Whenever anyone touches it, it will cry, 'Quack, quack!' and you must straightway say, 'Stick to my back!'" At that, the beautiful maiden vanished.

The youth continued on his way with the goose. He came to an inn at night and, having no money, took a seat outside on a bench. The innkeeper emerged and was going to drive him away, when his two daughters appeared, saw the goose, and said to their father, "Please don't send this stranger away, Father. Invite him in and offer him bed and board."

The innkeeper looked at the goose and, realizing what his daughters had in mind, said, "Very well, the young man will sleep in a nice room, and we'll put the goose in the barn."

"No, you won't," said the youth with scalp disease, "the goose goes where I go. It's too fine a goose to stay in a barn."

After dinner the youth retired for the night and put his goose under the bed. As he slept he thought he heard something stirring, and all of a sudden the goose went "Quack, quack!"

"Stick to my back!" he shouted and got up to see what was going on.

The innkeeper's daughter had crawled into the room in her nightgown, grabbed hold of the goose to steal its feathers, and now she was stuck in that position.

"Help! Sister! Come get me loose!" she cried. In came the sister, also in her nightgown, grabbed her sister around the waist to pull her loose from the goose. But the goose cried "Quack, quack!" and the young man added "Stick to my back!" so sister stuck to sister.

The youth looked out the window and saw that it was almost day. He dressed and left the inn, followed by the goose and the two girls stuck to it. Along the way he met a priest who, noticing the innkeeper's daughters in their nightgowns, exclaimed, "Shame on you two! Is that any way to be going about at this hour of day? I'll show you a thing or two!" At that, he began spanking them.

"Quack, quack!" went the goose.

"Stick to my back!" said the youth, so the priest stuck too.

They moved on with three persons now sticking to the goose. Whom should they meet but a coppersmith loaded down with pots and pans. "Oh, my goodness, what's this I see? A priest in a position like that? Just let me at him!" At that, the coppersmith whacked him hard.

"Quack, quack!" went the goose.

"Stick to my back!" added the youth with scalp disease, and the coppersmith stuck too, pots and all.

That particular morning the king's daughter was on her balcony as usual when the strange group came into view: the boy with scalp disease, the goose, the innkeeper's first daughter stuck to the goose, the innkeeper's second daughter stuck to the first, the priest stuck to the second girl, the coppersmith with pots and pans stuck to the priest. At that sight, the princess went into peals of laughter. Then she called her father, who also burst out laughing. The whole court looked out the windows and laughed until their sides hurt.

Right in the middle of all the mirth, the goose and everyone stuck to it disappeared.

There remained only the young man with scalp disease. He went up the steps and introduced himself to the king. The king glanced at him, noticed his scalp disease and his coarse old clothes all patched up, and had no idea what to do. "My good lad," he said to him, "I'll engage you as a servant. How's that?" But the boy with scalp disease refused the offer: he wanted to marry the princess.

To gain time, the king ordered him bathed from head to toe and clad in noble garb. When the youth reappeared, no one recognized him: he was so handsome now that the princess fell violently in love with him and had eyes for no one else.

Putting first things first, the young man insisted on fetching his father immediately. He pulled up in a carriage and found the poor cobbler on the doorstep grieving over the departure of his only son.

The youth took him back to the royal palace, introduced him to his father-in-law, the king, and his bride, the princess, and the marriage was celebrated at once.

(Friuli)

◈ 39 ◈

The Happy Man's Shirt

A king had an only son that he thought the world of. But this prince was always unhappy. He would spend days on end at his window staring into space.

"What on earth do you lack?" asked the king. "What's wrong with you?"

"I don't even know myself, Father."

"Are you in love? If there's a particular girl you fancy, tell me, and I'll arrange for you to marry her, no matter whether she's the daughter of the most powerful king on earth or the poorest peasant girl alive!"

"No, Father, I'm not in love."

The king tried in every way imaginable to cheer him up, but theaters, balls, concerts, and singing were all useless, and day by day the rosy hue drained from the prince's face.

The king issued a decree, and from every corner of the earth came the most learned philosophers, doctors, and professors. The king showed them the prince and asked for their advice. The wise men withdrew to think, then returned to the king. "Majesty, we have given the matter close thought and we have studied the stars. Here's what you must do. Look for a happy man, a man who's happy through and through, and exchange your son's shirt for his."

That same day the king sent ambassadors to all parts of the world in search of the happy man.

A priest was taken to the king. "Are you happy?" asked the king.

"Yes, indeed, Majesty."

"Fine. How would you like to be my bishop?"

"Oh, Majesty, if only it were so!"

"Away with you! Get out of my sight! I'm seeking a man who's happy just as he is, not one who's trying to better his lot."

Thus the search resumed, and before long the king was told about a neighboring king, who everybody said was a truly happy man. He had a wife as good as she was beautiful and a whole slew of children. He had conquered all his enemies, and his country was at peace. Again hopeful, the king immediately sent ambassadors to him to ask for his shirt.

The neighboring king received the ambassadors and said, "Yes, indeed, I have everything anybody could possibly want. But at the same time I worry because I'll have to die one day and leave it all. I can't sleep at night for worrying about that!" The ambassadors thought it wiser to go home without this man's shirt.

At his wit's end, the king went hunting. He fired at a hare but only wounded it, and the hare scampered away on three legs. The king pursued it, leaving the hunting party far behind him. Out in the open field he heard a man singing a refrain. The king stopped in his tracks. "Whoever sings like that is bound to be happy!" The song led him into a vineyard, where he found a young man singing and pruning the vines.

"Good day, Majesty," said the youth. "So early and already out in the country?"

"Bless you! Would you like me to take you to the capital? You will be my friend."

"Much obliged, Majesty, but I wouldn't even consider it. I wouldn't even change places with the Pope."

"Why not? Such a fine young man like you . . ."

"No, no, I tell you. I'm content with just what I have and want nothing more."

"A happy man at last!" thought the king. "Listen, young man. Do me a favor."

"With all my heart, Majesty, if I can."

"Wait just a minute," said the king, who, unable to contain his joy any longer, ran to get his retinue. "Come with me! My son is saved! My son is saved!" And he took them to the young man. "My dear lad," he began, "I'll give you whatever you want! But give me . . . give me . . ."

"What, Majesty?"

"My son is dying! Only you can save him. Come here!"

The king grabbed him and started unbuttoning the youth's jacket. All of a sudden he stopped, and his arms fell to his sides.

The happy man wore no shirt.

(Friuli)

◈ 40 ◈

One Night in Paradise

Once upon a time there were two close friends who, out of affection for each other, made this pledge: the first to get married would call on the other to be his best man, even if he should be at the ends of the earth.

Shortly thereafter one of the friends died. The survivor, who was planning to get married, had no idea what he should now do, so he sought the advice of his confessor.

"This is a ticklish situation," said the priest, "but you must keep your promise. Call on him even if he is dead. Go to his grave and say what you're supposed to say. It will then be up to him whether to come to your wedding or not."

The youth went to the grave and said, "Friend, the time has come for you to be my best man!"

The earth yawned, and out jumped the friend. "By all means. I have to keep my word, or else I'd end up in Purgatory for no telling how long."

They went home, and from there to church for the wedding. Then came the wedding banquet, where the dead youth told all kinds of stories, but not a word did he say about what he'd witnessed in the next world. The bridegroom longed to ask him some questions, but he didn't have the nerve. At the end of the banquet the dead man rose and said, "Friend, since I've done you this favor, would you walk me back a part of the way?"

"Why, certainly! But I can't go far, naturally, since this is my wedding night."

"I understand. You can turn back any time you like."

The bridegroom kissed his bride. "I'm going to step outside for a moment, and I'll be right back." He walked out with the dead man. They chatted about first one thing and then another, and before you knew it, they were at the grave. There they embraced, and the living man thought, If I don't ask him now, I'll never ask him. He therefore took heart and said, "Let me ask you something, since you are dead. What's it like in the hereafter?"

"I really can't say," answered the dead man. "If you want to find out, come along with me to Paradise."

The grave opened, and the living man followed the dead one inside. Thus they found themselves in Paradise. The dead man took his friend to a handsome crystal palace with gold doors, where angels played their harps for blessed souls to dance, with St. Peter strumming the double bass. The living man gaped at all the splendor, and goodness knows how long he would have remained in the palace if there hadn't been all the rest of Paradise to see. "Come on to another spot now," said the dead man, who led him into a garden whose trees, instead of foliage, displayed song birds of every color. "Wake up, let's move on!" said the dead man, guiding his visitor onto a lawn where angels danced as joyously and gracefully as lovers. "Next we'll go to see a star!" He could have gazed at the stars forever. Instead of water, their rivers ran with wine, and their land was of cheese.

All of a sudden, he started. "Oh, my goodness, friend, it's later than I thought. I have to get back to my bride, who's surely worried about me."

"Have you had enough of Paradise so soon?"

"Enough? If I had my choice . . ."

"And there's still so much to see!"

"I believe you, but I'd better be getting back."

"Very well, suit yourself." The dead man walked him back to the grave and vanished.

The living man stepped from the grave, but no longer recognized the cemetery. It was packed with monuments, statues, and tall trees. He left the cemetery and saw huge buildings in place of the simple stone cottages that used to line the streets. The streets were full of automobiles and streetcars, while airplanes flew through the skies. "Where on earth am I? Did I take the wrong street? And look how these people are dressed!"

He stopped a little old man on the street. "Sir, what is this town?"

"This city, you mean."

"All right, this city. But I don't recognize it, for the life of me. Can you please direct me to the house of the man who got married yesterday?"

"Yesterday? I happen to be the sacristan, and I can assure you no one got married yesterday!"

"What do you mean? I got married myself!" Then he gave an account of accompanying his dead friend to Paradise.

"You're dreaming," said the old man. "That's an old story people tell about the bridegroom who followed his friend into the grave and never came back, while his bride died of sorrow."

"That's not so, I'm the bridegroom myself!"

"Listen, the only thing for you to do is to go and speak with our bishop."

"Bishop? But here in town there's only the parish priest."

"What parish priest? For years and years we've had a bishop." And the sacristan took him to the bishop.

The youth told his story to the bishop, who recalled an event he'd heard about as a boy. He took down the parish books and began flipping back the pages. Thirty years ago, no. Fifty years ago, no. One hundred, no. Two hundred, no. He went on thumbing the pages. Finally on a yellowed, crumbling page he put his finger on those very names. "It was three hundred years ago. The young man disappeared from the cemetery, and the bride died of a broken heart. Read right here if you don't believe it!"

"But I'm the bridegroom myself!"

"And you went to the next world? Tell me about it!"

But the young man turned deathly pale, sank to the ground, and died before he could tell one single thing he had seen.

(Friuli)

◈ 41 ◈

Jesus and St. Peter in Friuli

I. How St. Peter Happened to Join Up with the Lord

There was once a poor man named Peter, who earned his living by fishing. One evening he went home tired, without having caught a single fish. To make matters worse, his wife had fixed no supper for him. "I spent all day looking around," she said, "but I found nothing, and you know we don't have any money."

"But how can I go to bed without any supper? Get me something to eat, and hurry!"

"There's nothing in the house, Peter. If you like, we can go to the field where those fine cabbages are and help ourselves to them."

"But I don't want to steal!"

"In that case we'll fast."

"Cabbages, you say? Can we get them? The two of us together . . ."

"Here's what we'll do: so as not to attract attention, one of us will go one way, the other the other."

Peter agreed, and they headed for the field, he by one road, and his wife by another. Along the way, Peter met a man with blond hair and gray eyes sitting on a post by the road. "What on earth is that stranger doing here?" wondered Peter, who addressed the man. "What's up, my good man?"

"I'm here to teach men not to do evil . . ." began the stranger.

Oh, my goodness, thought Peter, his remark seems directed at me in particular!

". . . and if they've done evil, to make them repent," concluded the stranger.

This kind of talk did not sit well with Peter. He cut the man short and moved on. But the stranger's words continued to ring in his ears. Upon reaching the field, he saw the shadowy figure of a woman moving about. "The owner's wife! The owner's wife! I'm getting out of here fast!" Peter took to his heels, overleaping rows of plants, ditches, and hedges. He ran all the way home, still thinking of the words, "make them repent," and as soon as he walked in, he grabbed the broom handle and thrashed his wife. "So you wanted me to become a thief, did you? You hussy! You lowdown wench!"

"Peter, for heaven's sake, forgive me!" cried the woman. "I wasn't able to steal a thing, since the man who owned the field arrived, and I had to flee for my life."

"And I was scared away by his wife! You hussy! You lowdown wench! The idea of making a thief of me! Just for that, I'm leaving you and going off to repent."

Off he ran and kept going until he overtook the stranger on the road and told him the whole story.

"Yes, Peter, you did the right thing by coming to me," said the stranger. "Now let me tell you that it wasn't the owner's wife you saw, but your very own; nor did she see the owner, but yourself. You scared each other, and your guilty consciences kept you from recognizing one another. Come along with me. You will be my foremost friend and my right arm. I am the Lord."

II. The Hare Liver

Once the Lord and St. Peter were going through a field, when out darted a hare from among the vegetable rows and bumped right into the Lord.

"Quick, Peter! Open your bag and put it in."

Peter bagged the hare and said, "We've not eaten in so long, Lord. We'll have to roast this one."

"Fine, Peter! This evening we'll have a good dinner. Since you really know how to cook, you can fix the hare for us."

They came to a town, saw the taverner's sign hanging above a certain door, and went in.

"Good evening, Mr. Innkeeper."

"Good evening, gentlemen."

"Bring us a half-bottle of wine," said the Lord. "And in the meantime, Peter, you go and cook the hare."

An expert all the way, St. Peter took a big knife, sharpened it on the billhook he always carried with him, skinned the hare, quartered it, and threw it into a pan. As it cooked, his mouth began to water. "How nice and fat it is! What an aroma! Just wait! I'll take a little taste to see if it's all right! It's delicious! Here's the liver done already: I believe I'll eat it with this bread crust. The Lord will never know the difference!" No sooner said than done! He stuck his fork into the liver and ate it. Then he called, "Lord, the hare is done! Shall I bring it to the table?"

"Done already? Bring it in, by all means."

Carrying the pan with one hand and wiping the grease from his whiskers with the other, Peter went back to the Lord. He put half of the hare on the Lord's plate and the other half on his own, and they began eating. But the Lord started looking around on his plate.

"Say, Peter, where is the liver?"

"Goodness, Lord, I don't know, I really didn't notice it. It's not on my plate either.... Maybe this hare had no liver ..."

"Maybe not," said the Lord smiling and taking up his fork again.

But Peter couldn't swallow a thing.

"Come on, Peter, don't you have any appetite? Do you have the liver on your stomach, perhaps?"

"Me, Lord?"

"But I'm not blaming you in the least. Go on, eat."

"I can't, Lord. I have sort of a lump here. I'll drink a glass of wine now."

That night Peter couldn't sleep a wink. He'd no sooner dozed off at dawn than the Lord awakened him so they'd get an early start and reach town before midday. Peter got up still worrying about the liver and fearing the Lord knew all about it.

In town all they saw were grave faces and lowered eyes; there was no sign of any of the dancing or merrymaking typical of all towns. "What's going on here?" wondered St. Peter, and the Lord said to him, "Just ask somebody, Peter."

Peter asked a soldier, who explained in a hushed voice that the king's daughter was so ill that the doctors had given up all hope of curing her, but the king had promised a bag of gold crowns to the person who made her well again.

"Listen, Peter," said the Lord, "I would like to see you win that bag of gold crowns. Go to the palace and say you're a famous doctor. Once you're alone with the king's daughter, take your pruning knife and cut off her head. Soak the head in water for one hour, then take it out and put it back on the king's daughter, who will be cured."

Peter marched straight to the king and asked to be left alone in his daughter's room for one hour. Once alone with her, he pulled out his knife and cut off her head, drenching the bed with blood. He threw the head into a pail of water and sat down to wait until an hour was up.

At the end of the hour, there was a mighty pounding on the door.

"Just a minute!" called Peter. He pulled the head out of water, set it on the girl's shoulders, but nothing happened! It wouldn't stick, and Peter became more frightened by the minute.

The knocking on the door grew louder. "Bam! Bam! BAM!"

"Open up, doctor!" ordered the king.

"What will I do? What will I do?"

BANG! They broke the door down, and in walked the king, who, seeing all the blood, cried, "What have you done, wretch? You've murdered my daughter! You'll hang for it! Guards! Bind him hand and foot and drag him to the gallows!"

"Majesty, forgive me, have mercy!"

"Away with him this instant!"

Only the Lord can get me out of this, thought Peter as he was hauled through the streets by the soldiers. Who should emerge from the crowd just then but the Lord. "Help, Lord, save me! Save me!"

"Where are you taking this man?" the Lord asked the soldiers.

"To the gallows."

"What has he done?"

"What has he done, you ask? He killed the king's daughter, that's what!"

"That's not so, let him go. Rather, take him to the king to get his bag of gold crowns. The king's daughter is as sound as a bell."

The soldiers went back to see if it was true. Reaching the royal palace, they saw the princess on the balcony in the highest of spirits, and the king walked straight up to Peter and handed him the bag of gold crowns.

Old though he was, Peter felt so unusually strong just then that he picked up the heavy bag as though it were a feather, slung it over his shoulder, and returned to the crossroads where the Lord was waiting for him.

"You see, Peter?"

"Now you're going to tell me, Lord, I'm good for nothing!"

"Hand me the money, which we'll divide up as usual."

Peter put the bag down, and the Lord began making the piles.

"Five for me, five for you, five for the other one . . ." and on and on in that manner. "Five for me, five for you, five for the other one . . ."

Peter looked on for a while, then asked, "But Lord, there're only two of us, why are you making three piles?"

"What, Peter?" And the Lord continued. "Here's for me, here's for you, here's for the other one . . ."

"Just who is the other one?"

"The one who ate the liver . . ."

"Lord, Lord," Peter hastened to say, "I'm the one who ate it!"

"So there, I've caught you. You did wrong, Peter, and the fear I've caused you has been your punishment. I forgive you, but don't do anything like that again."

Peter promised he wouldn't.

III. Hospitality

One evening after a long journey over the mountain roads, Jesus and St. Peter stopped at a certain woman's house and asked for shelter for that night. The woman looked them over and replied, "I'll have nothing to do with vagabonds."

"For pity's sake, madam!"

But the woman slammed the door in their faces.

Just as touchy as ever, Peter cast the Lord a look that said he, Peter, knew what was good for the woman. But the Lord ignored him and moved on to a humbler house black with soot. Inside a little woman was spinning at the fireside.

"Madam, would you be so kind as to give us lodgings for tonight? We've come a long way today and have no more strength to go any further."

"Why, of course! God's will be done! Do stop here, my good men.

Besides, where would you go, now that it's dark as pitch outdoors? I'll do what little I can to make you comfortable. Meanwhile, come up and warm yourselves at the fire. I bet you're hungry too."

"You're not far from the truth," replied Peter.

The little woman, whose name was Mistress Catín, tossed a few sticks on the fire and began to get supper—broth and the tenderest of beans, to Peter's delight, and a bit of honey she kept hanging from the rafters. Then she led them off to sleep on the hay.

"A good woman!" said Peter, stretching out contentedly.

Bright and early the next morning, upon bidding Mistress Catín good-bye, the Lord said, "Madam, whatever you start out doing this morning, you shall continue throughout the day." Then they left.

The little woman sat down at once to weave, and wove and wove and wove, the whole day long. The shuttle flew back and forth in the warp, and the house filled up with cloth, cloth, cloth; it poured out the door and windows, piling all the way up to the eaves of the house. At evenfall neighbor Giacoma called on Mistress Catín. Neighbor Giacoma was the woman who'd slammed the door in the face of Jesus and St. Peter. She saw all that cloth and wouldn't let Mistress Catín get a minute's peace until the little woman told her the whole story. Hearing that the two strangers she'd turned away were responsible for her neighbor's prosperity, she could have kicked herself. "Do you know whether those strangers intend to come back?" she asked.

"I believe so. They said they were only going down to the valley."

"Well, if they return, send them to my house, please, so they can do me a good turn too."

"Gladly, neighbor."

So the next evening when the two wayfarers reappeared at her door, Mistress Catín said, "To tell the truth, my house is too cluttered for me to put you up here tonight. But go to my neighbor Giacoma, in that house down there, and she'll do her utmost to make you comfortable."

Peter, who never forgot a thing, made a wry face and was about to speak his mind on neighbor Giacoma. The Lord, however, motioned to him to keep quiet, and they moved on to the other house. This time the woman made a great big to-do over them. "Why, good evening! Good evening! Did the gentlemen have a pleasant journey? Do come in, please! We're poor folks, but big-hearted all the same. Won't you come up to the fire and warm yourselves? I'm going to get dinner for you right away . . ."

So, amid all that fuss, the Lord and St. Peter ate and slept at neighbor Giacoma's house, and made ready to leave the next morning, with the woman still bowing and scraping before them.

"Madam," said the Lord, "whatever you begin doing this morning, you shall continue throughout the day." Then they left.

"Now I'll just show you what *I* can do!" chuckled the neighbor as she rolled up her sleeves. "I'll weave twice as much cloth as Mistress Catín." But before settling down to the loom, so as not to be interrupted later on, she decided to run in all haste to the dunghill and relieve herself. She got there and started—and it seemed to her she was doing it in a great big hurry—but she was unable to stop. "O mercy! What's the matter with me? How come I can't stop? Could I have eaten something that made me sick? Heavens above! But . . . but it couldn't be that . . ."

A half-hour later she tried to get away from there and back to her loom. Of course, she had to dash back to the dunghill at once. And there she spent the whole day. The result was a far cry from cloth! It's a wonder the Tagliamento didn't overflow its banks.

IV. Buckwheat

At sunset, three hot, sweaty, dusty wayfarers came into a village. In the courtyards people were threshing the last of their grain for the day, and chaff was still floating in the air.

"Mistress of this house!" said the three to a woman who was winnowing. The woman, a widow, invited them in, fed them, and said they could sleep in the hayloft if they would help her with the next day's threshing. The wayfarers, who were the Lord, St. John, and St. Peter, went to bed in the hayloft. At daybreak Peter heard the cock crow, and said, "Let's make haste and get up. We've eaten, and it's only fair that we go to work."

"Be quiet and sleep," replied the Lord, and St. Peter turned over on the other side. They'd no sooner gone back to sleep than the widow showed up with a stick in her hand. "Well! Do you think you are going to fritter away your time here in bed until Judgment Day? And after eating and drinking at my expense!" She gave Peter a whack on the back and stormed off, furious.

"You see how right I was?" said Peter, rubbing his back. "Come on, let's get up and go to work, or that confounded woman will thrash us for all she's worth."

Once more the Lord said, "Be quiet and sleep."

"That's all well and good. But if she comes back, *I'll* be the one to pay for it!"

"If you're so afraid of a woman," replied the Lord, "come over here and let John take your place."

They changed places and then all three went back to sleep.

Exasperated, the widow returned with her stick. "What! Still sleeping?" For the sake of fairness, she whacked the man sleeping in the middle this time, who was again Peter!

"It's always me!" groaned Peter, and to calm him down, the Lord changed places with him, saying, "Now you're the best protected of all. Be quiet and sleep."

Back came the widow. "It's your turn now!" Another whack went to Peter, who jumped out of the hay. "Let the Lord say what he will, I'm not staying." He ran into the courtyard, grabbed the thresher, and got to work as far away as possible from that awful woman.

In a little while here came the Lord and St. John. They also took up threshers, but the Lord said, "Bring me a firebrand." Motioning to the others to stay calm, he set fire to the four corners of the threshing floor. In a moment a great blaze engulfed the sheaves. You would have thought there'd be nothing but ashes when the fire burned down. Instead, on the right was hay; on the left, straw; in the air, chaff; and in the center, a mound of grain, threshed and shiny, as though already winnowed and sifted. The threshing was all done, and without one blow of the thresher.

The three didn't even wait to be thanked. They walked out of the courtyard and continued on their way. But the widow, still as haughty and greedy as ever, had the threshing floor cleared, the grain measured and put away, and a whole new load of sheaves brought out. As soon as the men untied the sheaves, the widow herself picked up a firebrand and set fire to the threshing floor. But this time the flames burned in earnest, and the grain crackled like fritters in a frying pan.

With her hands in her hair, the widow went flying out of the village after the wayfarers. The minute she caught up with them she fell to her knees and related her misfortune. Since she was now truly sorry, the Lord said to Peter, "Go and save what you can, showing her how to render good for evil."

St. Peter reached the threshing floor and made the sign of the cross. The fire went out, and the half-roasted grain all stuck together in a clot. Blackened, deformed, and cracked, it no longer looked like grain. But because of St. Peter's blessing, it was still full of flour, and those dark, tiny, pointed granules were the first buckwheat ever seen on the face of the earth.

(Friuli)

◈ 42 ◈

The Magic Ring

A poor boy said to his mother, "Mamma, I'm leaving home and going out into the world. Nobody in town esteems me, and if I stay here I'll never amount to a thing. I'm going out and make a fortune, and better days will come for you too."

At that, he departed. He came to a city, and was strolling through the streets when he saw a little old woman trudging up a hilly lane with two heavy buckets of water and gasping for breath. He approached her and said, "Let me carry the water. You'll never make it up the hill with such a load." He took the buckets, followed her to her cottage, went up the steps, and set the water down in the kitchen. In the room were many cats and dogs that all crowded around the old woman, purring or wagging their tails.

"How can I show you my appreciation?" asked the old woman.

"Don't mention it," he replied. "It was a pleasure to help you."

"Wait a minute," she said, and left the room. She came back with a little ring of no value and slipped it on his finger, saying, "This is a precious ring, mind you. Every time you turn it and ask for something, you will get your wish. Just be careful not to lose the ring, for that would be your undoing. To lessen that risk, I'm giving you one of my dogs and one of my cats that will follow you wherever you go. They are smart animals and sooner or later will prove quite useful."

The youth thanked her over and over, then left. But he set little store by her words. "Old wives' tales," he said to himself, not even thinking to give the ring a twist just to see what would happen. He left the city with the dog and cat trotting along beside him. He loved animals and was happy to have these two with him. Running and jumping with them, he entered a forest. Night fell, and he had to make his bed under a tree, with the dog and cat lying beside him. But he couldn't go to sleep because of the hunger gnawing at his stomach. He then remembered the ring he was wearing. It won't hurt to try it out, he thought to himself. He turned the ring and said, "I wish food and drink!"

He hadn't got the words out of his mouth before there appeared three chairs and a table laden with all kinds of food and drink. He took a seat and tied a napkin around his neck. In the other two chairs he seated the dog and the cat, tied napkins also around their necks, and they all three started eating with great relish. He now had confidence in the ring.

When the meal was over, he stretched out on the ground and considered all the wonderful things he would now be able to do. The hardest part was making a choice. For a while he imagined he would like piles of gold and silver, then he dreamed of horses and carriages, then castles and land, and the more he mused, the less he knew what he wanted. "But here I go losing my head," he finally said, exhausted from building castles in the air. "I've heard so often that people become perfect fools when they grow rich, but I intend to keep my wits about me at all times. That's enough for today; I'll give the matter more thought tomorrow." He turned over and went sound asleep, while the dog curled up at his feet and the cat at his head and kept watch over him.

When he woke up, the sun was already high in the sky and shining through the green treetops; a gentle wind blew, the birds sang, and he found himself completely refreshed by his night's rest. He thought of asking the ring for a horse, but the forest was so lovely he decided to cross it on foot. He thought of ordering breakfast, but such delicious wild strawberries grew in the forest, he made a meal of them instead. He then thought of calling for drink, but in the forest was a spring of the clearest water in the world, so he cupped his hands and drank from it. Wandering through fields and meadows he came at last to a large palace. A beautiful maiden happened to be looking out the window and smiled warmly at the carefree youth advancing so sprightly, with a dog and cat right behind him. He looked up and, although the ring was safe, his heart was lost. "Now is indeed the time to use the ring," he decided. Turning it, he said, "Let a palace rise opposite this one, outdoing it in beauty and containing every comfort."

In the twinkling of an eye the palace was there, far bigger and more beautiful than the other, and he was inside as though he'd always lived there, with the dog sleeping in his basket and the cat lying by the fire licking his paws. Opening the window, the youth found himself directly across from the beautiful maiden's window. They smiled at each other, sighed, and the young man realized the moment had come to go and ask for her hand in marriage. She was as pleased as her parents, and in a few days' time they were married.

Their first night together, after kisses, embraces, and caresses, she jumped up, saying, "But do tell me just how your palace sprouted all of a sudden like a mushroom?"

He wasn't sure whether he should tell her, but then he thought, She's my wife, and you don't keep secrets from your wife. He therefore told her about the ring. Then they fell asleep in bliss.

But while he was sleeping, his wife slipped the ring off his finger, then got up, called all the servants, and said, "Leave this palace at once; we are returning to my parents' house!" When she was safely home she turned the ring, saying, "Let my husband's palace be removed to the highest and steepest peak of the mountain in the distance!" And the palace vanished as though it had never existed. She looked at the mountain and saw the castle now poised on the peak.

The youth awakened in the morning, found his wife gone, threw open the window, and beheld the void. Taking a closer look, he made out deep ravines down below, and mountains all around him blanketed in snow. He went to touch the ring, and it too was gone. He called the servants, but no one answered. Instead, the dog and cat ran in. They had stayed there, because he'd told his wife only about the ring and said nothing of the two animals. Completely baffled at first, he gradually realized his wife had cruelly betrayed him, but the truth was little consolation to him. He peered out to see if he was going to be able to get down the mountain, but the doors and windows, alas, opened onto empty space, with the ravines plunging straight downward. The supplies in the palace would last only a few days, and the terrible thought struck him that he would then starve to death.

Seeing their master so downcast, the animals approached, and the dog spoke. "Don't give up hope yet, master. The cat and I will find our way down these cliffs, and then we'll get the ring for you."

"My dear pets," said the youth, "you're my only hope. Were it not for you, I'd jump off these cliffs rather than starve to death."

The dog and cat climbed out of the house, leaped down the cliffs, and came to the foot of the mountain. They raced across the plain to a river, where the dog took the cat on his back and swam to the other bank. When they arrived at the faithless wife's palace, it was already night and the whole house was sound asleep. In through the cat door they tiptoed, and the cat told the dog, "You stay here now and keep watch while I go upstairs and see what can be done."

Without a sound the cat rushed up the stairs and down the hall to the room where the false-hearted woman was sleeping, but the door was shut and he couldn't get in. While he thought frantically what to do, a rat ran by. The cat grabbed him. It was a big fat rat, who began begging and pleading with the cat to spare him. "I will," said the cat, "but you must gnaw a hole in this door big enough for me to crawl through."

The rat began to gnaw at once. He gnawed until he was blue in the face, but the hole was still too little, not only for the cat, but for the rat himself to get through.

So the cat said, "Do you have any little ones?"

"I should say so! I have seven or eight, each one as rambunctious as can be."

"Run get one," said the cat, "and if you don't come back, I'll catch you and eat you alive wherever you are."

The rat ran off and was back in a trice with a little rat. "Listen, little one," said the cat, "if you are clever you'll save your father's life. Go into this woman's room, crawl up on her bed, and pull off the ring she has on her finger."

The little rat ran inside, but was back in no time, quite upset. "She has no ring on," he said.

The cat, however, didn't lose heart. "That means she has it in her mouth," he said. "Go back, slap her nose with your tail and she'll sneeze, opening her mouth. The ring will drop out, you'll pick it up quickly, and run back here with it."

Everything happened the way the cat said. In a little while the rat returned with the ring. The cat took it and bounded down the steps.

"Do you have the ring?" asked the dog.

"Of course."

They plunged through the front door and headed back the way they had come. But down deep, the dog was consumed with jealousy because the cat had been the one to get the ring.

Arriving at the river, the dog said, "Give me the ring, and I'll ferry you across." The cat refused, and they started arguing. During the quarrel the cat let go of the ring, and it fell into the water, where a fish swallowed it. In a flash the dog grabbed the fish in his mouth, and then *he* had the ring. He carried the cat to the other bank, but they didn't make up and continued to argue all the way to their master's palace.

"Do you two have the ring?" he asked eagerly. The dog spit out the fish, the fish spit out the ring, but the cat said, "Don't believe him; I got the ring myself, and the dog stole it from me."

"But if I'd not caught the fish, the ring would have been lost forever."

Then the youth petted them both, saying, "My dears, don't argue so much, you're both very dear and precious to me." For a half-hour he caressed the dog with one hand and the cat with the other, until the animals were again as good friends as ever.

He took them into the palace, turned the ring on his finger, and said, "Let my palace take the place of my false-hearted wife's while she and her entire palace come here where I now am." At that, the two palaces sailed through the air and changed places with one another. His landed right in the middle of the plain; and hers, with her inside screaming like an eagle, perched on the sharp peak.

The youth sent for his mother and made her last years as joyous as he had promised. The dog and cat remained with him, always quarreling about something or other, but all in all everyone lived in harmony. And the ring? He used it occasionally, but not too much, thinking (and how truly!), It's not good for man to get everything he wants without effort.

When they went up the mountain, they found his wife cold and dead. She had starved to death. It was a bitter end, but she deserved none better.

(Trentino)

◈ 43 ◈

The Dead Man's Arm

It was the custom in a certain village, whenever a man died, for his sister to keep watch over his grave three nights in a row. If a girl should die, the watch would be kept by her brother. A certain maiden died, and her brother, a strapping youth who was afraid of nothing under the sun, went to the cemetery for the usual vigil.

At the stroke of midnight three dead men arose from their graves and asked, "How about a game?"

"Why not?" he answered. "But where do you want to play?"

"We always play in church."

They entered the church and showed him to an underground crypt packed with rotting coffins and a jumble of human bones. They picked up some of the bones and a skull and went back upstairs into church, where they stood the bones in a straight line on the floor. "These are our ninepins." They picked up the skull. "This is our ball." And they began bowling.

"Do you want to play for money?"

"Certainly!"

The young man bowled with the skull and was so good at it that he won each time and took every cent the dead men had. As soon as they ran out of money they carried ball and ninepins back to the crypt and retreated to their graves.

The second night the dead men wanted to play the return game, staking rings and gold teeth, and again the youth won everything. The third night they played still another round, at the end of which the men said, to the youth, "You've won again, and we have nothing left to give you. But since gaming debts are settled on the spot, we shall give you this dead man's arm, which is well preserved although a bit dry and will come in handier than a sword. No matter what enemy you touch with it, the arm will grab him around the chest and throw him down dead, even if he is a giant."

The dead men departed and left the young man standing there with that arm in his hand.

The next morning he took to his father the money and the gold won at ninepins and said, "Dear Father, I'm going out into the world and seek my fortune." His father gave him his blessing and the young man departed, with the dead man's arm hidden under his cloak.

He came to a large city, where the walls of the houses were draped in black crepe; the people all wore mourning and had even draped their horses and carriages in black. "What has happened?" he asked a sobbing passer-by, who explained, "Near that mountain, mind you, is a black castle occupied by sorcerers who exact from us a human being a day, and that is the end of the poor soul who goes to them. First they called for the girls, and the king was obliged to send them every last one of the chambermaids, housewives, baker girls, and weavers; then all the maids of honor at the court and all the noble ladies, and most recently his only daughter as well. And not a one of them has come back. Now the king is sending soldiers there three by three, but they fare no better. If only somebody could deliver us from the sorcerers, we would reward him with anything he wanted."

"I shall see what I can do about all this," said the youth and asked to be taken to the king at once. "Majesty, I will go to the castle all by myself."

The king looked him in the eye. "If you succeed, and if you free my daughter, I'll give her to you in marriage and you shall inherit my kingdom. You need only spend three nights at the castle for the spell to be broken and the sorcerers to vanish. On the battlements of the castle stands a cannon. If you're still alive tomorrow morning, fire one shot, day after tomorrow two, and the third morning three."

When night fell, the young man went to the black castle, with the dead man's arm under his cloak. He ascended the stairs and entered a room where a table had been set and laden with food, but the chairs were turned with their backs to the table. He left everything just as it was, entered the kitchen, lit the fire, and sat down next to the hearth, holding the dead man's arm ready. At midnight a chorus of voices cried down the chimney:

> "Many, many have we slain,
> You will be the next to wane!

Many, many have we slain,
You will be the next to wane!"

Then bang! out of the chimney dropped one sorcerer. Bang! another, and bang! a third. They all had frightfully ugly faces and long, long noses that wavered like octopus tentacles and clutched at the youth's arms and legs. Realizing it was vital to stay clear of those noses, he began brandishing the dead man's arm as though he were fencing. He tapped one sorcerer's chest with it, but nothing happened. He tapped a second one on the head, but still nothing happened. Then he tapped the third one on the nose, and the dead man's hand grabbed that nose and gave it a yank that left the sorcerer dead on the spot. Fully aware now that the noses were both dangerous and sensitive, the young man took aim. The dead man's arm seized the second sorcerer by the nose and finished him off. Then it took care of the third one. The young man rubbed his hands together in contentment and went off to bed.

In the morning he climbed to the battlements and fired the cannon: Boom! Down in the town below, where everyone anxiously waited, a crowd of handkerchiefs bordered in black waved in response.

When he went into the dining room in the evening, some of the chairs had been turned around and properly faced the table. Through other doors filed noble ladies and maids of honor, downcast and clad in mourning. They addressed the young man: "Please persevere and free us!" Then they sat down to the table and dined. After dinner they all bowed low and departed. He went into the kitchen and took a seat by the hearth to wait for midnight. When the twelfth chime had struck, voices were again heard in the chimney:

"Three of our brothers you slew,
Now we're coming after you!
Three of our brothers you slew,
Now we're coming after you!"

And bang! bang! bang! three sorcerers with long noses plummeted down the chimney. Brandishing the dead man's arm, the young man had them each by the nose in a flash, and in no time they were corpses themselves.

The next morning he fired two cannon shots: Boom! Boom! Down in the town a crowd of white handkerchiefs waved back: the black mourning strip had been removed from them.

The third evening he found still more chairs turned to the table in the dining room, and the black-clad maidens entered in greater numbers than the evening before. "Just one more night," they entreated, "and we'll all be free!" Then they dined with him and again departed. He sat down in his customary place in the kitchen. At midnight the voices set up a howl in the chimney like a whole choir:

> "Six of our brothers you slew,
> Now we're coming after you!
> Six of our brothers you slew,
> Now we're coming after you!"

And bang! bang! bang! bang! down rained sorcerers by the dozens, all with their long noses sticking out, but the youth whirled the dead man's arm round and round and killed them off as fast as they came. It was no trouble at all, since the only thing that shriveled paw had to do was grab them by the nose, and they were corpses themselves. He went to bed thoroughly satisfied, and the minute the cock crowed the whole castle came back to life. A procession of maidens and noble ladies dressed in gowns with trains entered the kitchen to thank him and pay him honor. In the middle of the procession came the princess. When she got up to the youth, she threw her arms around his neck and said, "I want you to be my husband!"

Three by three entered the freed soldiers and saluted.

"Go up to the battlements of the castle," commanded the youth, "and fire three cannon shots." They heard the thunder of the cannon down in the town and vigorously waved yellow, green, red, and blue handkerchiefs in response, accompanied by trumpets and bass drums.

The youth went down the mountain in the procession of free people and entered the town. The black crepe had disappeared, and all you saw were flags and colored streamers billowing in the wind. The king was there waiting for them, his crown entwined with flowers. The wedding was celebrated the same day and was such a grand event that people are still talking about it.

(Trentino)

The Science of Laziness

There was once an old Turk who had just one son, and the boy was dearly loved by his father. As everybody knows, the greatest scourge on earth for a Turk is work. Therefore, when the son turned fourteen, his father decided to send him to school to learn the science of laziness.

On the same street as the old Turk there lived a famous and highly respected professor. who had never done a lick of work in his life that he could get out of doing. The old Turk called on him and found him stretched out in the garden beneath a fig tree, with a cushion under his head, a cushion under his back, and a cushion under his buttocks. "Before talking to him I must first see how he does," said the old Turk to himself and hid behind a hedge to observe the man.

The professor lay as still as a corpse, with his eyes closed. The only time he moved was whenever he heard the thud of a ripe fig on the ground near where he lay: he would reach slowly out, bring the fruit to his mouth, and swallow it. Then he wouldn't stir again until another fig fell.

"This is just the professor my boy needs," decided the Turk. He came out of his hiding place, introduced himself, and asked if the professor would teach his son the science of laziness.

"Old man," answered the professor just above a whisper, "don't talk so much. It tires me to listen to you. If you want to bring up your son to be a true Turk, just send him to me."

The old Turk went home, took his son by the hand, thrust a feather pillow under his arm, and led him to that garden.

"I urge you," he told him, "to do everything you see this professor of idleness do."

The boy, who already had an inclination for that particular science, also stretched out under the fig tree. Observing his teacher, he saw him reach for every fig that fell and bring the fruit to his mouth. Why should I work myself to death reaching for figs? he thought, and lay there with his mouth wide open. Soon a fig fell into his mouth; he let it go down slowly, then reopened his mouth. Another fig fell, just missing his mouth. He kept perfectly still and murmured, "Why so wide of the mark? Fig, fall into my mouth!"

Seeing how wise the pupil was already, the professor said, "Go home. You have nothing to learn. You can even teach me something."

So the boy went home to his father, who thanked heaven for having given him such a smart son.

(Trieste)

◈ 45 ◈

Fair Brow

There was once a boy whose father said to him at the end of his schooling, "My son, now that you've finished your studies, the time is right for you to begin to travel. I will give you a ship so that you can get a start in loading, unloading, buying and selling. Work seriously, because I want you to learn to earn your living as soon as possible!"

He gave him seven thousand crowns with which to buy goods, and the boy set out. He had already sailed some distance without buying anything, when he came into a port and saw sitting on the shore a coffin, into which passers-by would all drop a small donation of money.

"Why are you keeping that corpse there?" he asked. "The dead wish to be buried."

"That man died saddled with debts," the boy was told. "It's the custom here to bury no one who has not paid his debts. We will not bury the man until his debts are paid up in full by charity."

"In that case let it be known that all his creditors should come to me to be paid. And take him away and bury him at once."

They made the announcement, and he paid every debt, without one penny left over for himself when he had finished. He therefore went back home, and his father asked, "What's the meaning of your returning so soon?"

"I sailed the sea and ran into pirates, who took all my capital."

"Don't worry, my son but be thankful they didn't take your life as well! I'll fit you out again, but don't venture into the same waters the next time." And he gave the boy another seven thousand crowns.

"You can be sure, Father, I'll change my course!" At that, he set out again.

Halfway across the sea, he saw a Turkish vessel and said to himself, "In this spot it's better to make friends than enemies: let's call on them and invite them to do likewise." He boarded the Turks' vessel and asked, "Where do you come from?"

"We come from the East!"

"And what do you carry?"

"Nothing but a beautiful maiden."

"To whom are you taking this maiden?"

"We will sell her to whoever wants to buy her. She's the daughter of our sultan and we kidnapped her on account of her great beauty."

"Let me have a look at her." When he saw her, he asked, "How much lo you want for her?"

"We are asking seven thousand crowns!"

So the youth gave the pirates all the money his father had given him and took the maiden to his ship. He had her baptized and married her, then went home to his father.

> "Welcome back, my son so fair,
> I can guess what prize you bring. . . .
>
> "Father, I bring a most precious gem,
> You will sing with joy when you see her!
> A maiden lovelier than you've e'er beheld:
> The daughter of the sultan of Turkey
> I bring as my first commodity!"

"Idiot! Is that all you've brought?" And the father angrily shook them both and threw them out of the house.

Poor things! They didn't know which way to turn.

"What will we do now?" he wondered. "I've nothing to my name."

But she said, "Listen, I can paint fine pictures. That's what I'll do and you'll go out and sell them. But beware of ever telling a soul they were done by me."

Meanwhile, back in Turkey, the sultan had dispatched ship after ship in search of his daughter. By chance, one of them arrived at the town where the young people were living. Many men disembarked, and the youth, seeing all those visitors in town, said to his wife, "Paint a lot of pictures, which we'll certainly sell today."

She did the pictures and said, "Here you are, but don't sell a one for less than twenty crowns."

He took them to the town square. The Turks arrived, glanced at the paintings, and said to one another, "Nobody but the sultan's daughter could have done these! She alone paints like that!" They moved closer and asked the young man how much he wanted for them.

"They are expensive," he replied. "I'm letting none of them go for less than twenty crowns."

"Fine, we'll buy them. But we'd like others as well."

"Come home and talk to my wife about it. She's the one who paints the pictures."

The Turks followed him home, and there was the sultan's daughter. They seized her, bound her, and carried her back to Turkey.

The husband was heartbroken. There he was with no wife, no trade, and no money. Every day he went to the harbor to look for a ship that might take him aboard, but he never found a one. Finally one day he saw an old man fishing from a little boat and said, "How much better off you are, good old soul, than I am!"

"Why do you say that, my boy?" replied the old man.

"How I would like to fish with you, good old soul!"

"If you wish to fish with me, come ahead! What with your pole and my boat, we might catch something of note!"

So the youth got in, and they made a pact to share everything, good and bad alike, that came their way. To begin, the old man divided his supper with the boy.

After eating, they went to sleep. Meanwhile a storm suddenly came up. The wind seized the boat, swept it over the waves, and finally grounded it on the shore of Turkey.

Seeing this boat land, the Turks took possession of it, made slaves of the two fishermen, and carried them before the sultan, who put them to work in the garden. The old man was to look after the vegetables, and the youth after the flowers. The two slaves made friends with the other gardeners and were very well off in the sultan's garden. The old man fashioned guitars, violins, flutes, clarinets, and piccolos, and the youth played them all and sang songs.

Now the sultan's daughter, for her punishment, had been imprisoned in a tall tower with her maids of honor. Hearing that fine playing and singing, she thought of her husband far away. "Only Fair Brow [as she called him] could play all instruments and sing in a voice far sweeter than any of them. Who is that playing and singing in the garden?"

Peeping through the slats in the blinds, which she was unable to open, she saw that the young musician was none other than her husband.

Every day the maids of honor took the gardeners a big basket to fill with flowers. The sultan's daughter therefore said to them, "Put that young man in the basket, cover him with flowers, and bring him up here!"

For a joke the gardeners put him in the basket, and the maids of honor carried him up in the tower. When they set the basket down, he bobbed up from under the flowers and found himself face to face with his wife! They hugged and kissed, telling each other everything. Then they began planning their escape.

They had a large ship loaded with pearls, precious stones, bars of gold, and jewels. Into the hold they lowered Fair Brow, then the sultan's daughter, then, one by one, all her maids of honor, after which the ship weighed anchor.

They were already on the open sea, when Fair Brow remembered the old man and said to his wife, "My dearest, I may lose my life for doing so, but I have to go back to shore. I cannot be unfaithful to my sworn word! I promised that old man we would always share everything, good and bad alike, that came our way!"

They turned back and found the old man on shore waiting for them. They brought him aboard and regained the open sea.

"Good old soul," said Fair Brow, "let us now divide things up. One half of all this treasure is for you, and the other half is for me."

"The same goes for your wife," said the old man. "One half of her is for you and the other half is for me!"

"Good old soul," replied the youth, "I am indebted to you, so I'll let you have all the treasure on this ship. But let me keep my wife all for myself."

"You are a generous youth. Note that I am the soul of the dead man for whose burial you arranged. All your luck stems from that good deed of yours."

He gave him his blessing and vanished.

The boat glided into its home port firing mighty cannon salutes: Fair Brow, the world's richest nobleman, was arriving with his wife. And who should be waiting on shore with open arms but his father.

> Happily from then on did they live,
> But nothing to me did they ever give.

(Istria)

The Stolen Crown

A king had three sons whom he loved very much. One day this king went hunting with his prime minister and, feeling very weary, lay down under a tree and went to sleep. Upon awakening, the first thing he did was look for his crown. It was neither on his head nor on the ground beside him, much less in the game bag, so where could it be? Right away he called to his prime minister, "Who took my crown?"

"Sacred Majesty, do you think I would so much as touch your crown? Nor have I seen anyone else who might have taken it!"

The king went home in a rage and had his prime minister sentenced to death, even though the poor man was completely innocent. The real culprit was Fairy Alcina, the queen of the fairies.

Ashamed to appear in public without his crown, the king shut himself up in his room and gave orders for no one to disturb him. That puzzled his sons, who couldn't imagine what had happened, and one day the oldest boy said to his two brothers, "Why would our father seclude himself like that and refuse to see anyone? He must have had an accident of some kind. I'll go in and try to cheer him up." But his father shouted him out of the room and would have surely struck him if the boy had not bolted.

"Let me try," said the middle boy. But he received the same welcome and retreated completely mortified.

It was now up to the youngest son, Benjamin, who was his father's favorite. Like his brothers, he went in and begged the king to say what the trouble was.

"I'd tell *you* everything," replied the king, "but this is too humiliating for words."

"If you won't tell me what's the matter, I'll kill myself rather than stand by and see you suffer." At that, he put a gun to his heart.

"Stop, my son!" cried the king. "I'll tell you everything!" And he went into detail about losing the crown, but entreated his son to say nothing about it to his brothers.

Benjamin listened attentively, then spoke. "The only person under the sun who could have made off with your crown is Fairy Alcina. She loves to torment people. I'll search the world over for her. Either I'll bring the crown back, or you'll never see me again."

He saddled a horse, filled a purse with money, and set out. At a certain point the road branched off in three different directions. A stone marker stood at the beginning of each new road. The first stone read: WHOEVER TAKES THIS ROAD WILL RETURN. The second stone read: GOODNESS KNOWS WHAT YOUR FATE WILL BE IF YOU TAKE THIS ROAD. The third stone read the opposite of the first: WHOEVER TAKES THIS ROAD WILL NEVER RETURN. He was about to start down the first road, but changed his mind and set foot in the second, only to backtrack and enter the third.

For a while the road was good, but then came brambles, stones, snakes, insects, and all kinds of wild animals. The horse could go no farther, so Benjamin dismounted, tethered the horse to a tree, kissed him goodbye, and said in a tearful voice, "We might never meet again." Then he continued his journey on foot.

After walking and walking he came to a cottage and knocked on the door, for he was quite hungry by this time. "Who is it?" asked a voice inside.

"A poor horseless knight requesting a little refreshment."

An old woman opened the door and asked in amazement, "What on earth are you doing in these parts, good lad? Don't come in, please! If my daughter should return and find you here, she'd kill you and eat you, upon my word. I am the mother of Bora the Northeast Wind. Wait right here and I'll bring you something to eat."

While he ate, Benjamin told the old woman why he was roaming the globe in search of Fairy Alcina. The old woman knew nothing about her, but nevertheless promised to help him, good old soul that she was. She brought him in and hid him under the bed; and when Bora arrived angry and ravenous, she fed her enough to quiet her hunger for a good while, then told her about the young man's plight and made her promise not to harm a hair on his head.

Bora, who was well fed by now, let Benjamin come out from under the bed and spoke to him as a friend. She told him that in the course of her routine journeys around the world she had seen his father's crown on Fairy Alcina's bed, together with a shawl of stars and a musical golden apple, both of which Alcina had stolen from two queens now imprisoned in a well by a magic spell. Finally she revealed the locations of Fairy Alcina's palace and the two queens' well.

"But how will I get inside the palace?" asked Benjamin.

"Take this potion," said Bora, "with which you will put the watchman to sleep. Then you can go in and find the gardener."

"How will I deal with the gardener?"

"Have no fear," said Bora. "Fairy Alcina's gardener is my father. Mother and I will recommend you to him."

After thanking mother and daughter profusely, Benjamin set out and didn't stop until he reached the fairy's palace. He put the watchman to sleep and found the gardener, who promised to help him. "The steps are guarded by two Moors who have orders to kill anyone who tries to pass, with the exception of myself when I take flowers to the fairy."

So Benjamin dressed as a gardener, picked up a large vase of tuberoses that hid his face, and went up the steps past the two Moors. He entered the boudoir of the sleeping fairy and picked up the crown which she had placed on the canopy, along with the shawl of stars and the golden apple. Then he turned and looked at the fairy: she was so beautiful he had the urge to kiss her as she slept. He was about to do so, when the golden apple sounded a few notes of music. Afraid the fairy would awaken, Benjamin fled, hiding his face in a vase of jasmines so the Moors wouldn't see him. It was a narrow escape, since whoever kissed Fairy Alcina turned to stone from head to foot.

Benjamin thanked the gardener and took the way back. After walking for six or seven hours he came to a well gone dry and so deep you couldn't see the bottom. Circling this well was a goose with wings wide enough to shelter several persons at a time. Realizing at once that Benjamin wished to go down into the well, it approached for him to get under a wing, then flew to the bottom.

There stood the two queens held prisoners by the magic spell. "Here are your shawl of stars and musical golden apple!" said Benjamin, jumping out from under the goose's wing. "You are now free! If you wish to leave with me, take your places here under the goose."

Overjoyed, the two queens took their places under the wings, and the goose soared from the well, passed over woods and mountains, and came to the spot where Benjamin's horse was tethered.

Benjamin bid the goose farewell, mounted the two queens on his horse, and returned to his father.

At the sight of his crown, the king was beside himself with joy. He made his son kneel and placed the crown on the boy's head. "It is yours, and you deserve it," said the king.

Benjamin married the lovelier of the two queens. There were celebrations galore and a beautiful life afterward, and on that note I bring my tale to a close.

(Dalmatia)

◈ 47 ◈

The King's Daughter Who Could Never
Get Enough Figs

A king issued a proclamation that whoever succeeded in giving his daughter her fill of figs would have her as his wife. One suitor then showed up with a whole basketful and didn't even have time to offer her the figs before she had eaten every one of them. When they were all gone, she said, "More!"

Three boys were out digging in a field. The oldest one said, "I don't feel like digging any longer. I shall go and try to give the king's daughter her fill of figs."

He climbed the fig tree with a large basket. When it was quite full he set out for the king's palace. Along the way he met a neighbor, who said, "Give me a fig."

"I can't," he replied. "I mean to give the king's daughter her fill of figs, and I may not have enough as it is." Then he moved on.

He reached the palace and was taken to the king's daughter, before whom he set the figs. Had he not picked up the basket the instant it became empty, she would have eaten that as well.

He went home, and the middle brother said, "I too have had enough of digging in the field. I shall try my luck at giving the king's daughter her fill of figs."

He climbed the tree, filled his basket, and off he went. He met the neighbor, who said, "Give me a fig."

The brother shrugged his shoulders and moved on. But he too had to grab up the empty basket, or the king's daughter would have eaten it as well.

Then the youngest boy announced his intention to go to the palace.

He was walking along with his basket, when the neighbor asked him also for a fig. "You may even take three," said the youth, holding out the basket.

The neighbor ate a fig, then gave him a magic wand, explaining, "When you get there all you have to do is strike the ground with this wand, and the basket will fill up again as soon as it becomes empty."

The king's daughter ate every single fig, but the youngest brother gave a tap with the wand, and the basket was full again. After two or three such taps, the king's daughter said to her father, "Figs! Ugh! I never want to see another one!"

The king said to the young man, "You've won all right, but if you want to marry my daughter, you must go to her aunt across the sea and invite her to the wedding."

Hearing that, the youngest brother went home in dismay. Along the way he met the neighbor on his doorstep and told him of his plight. The neighbor gave him a bugle. "Go to the seashore and blow this. The princess's aunt who lives across the sea will hear you and come over here. Then you can take her to the king."

The youth blew the bugle, and the aunt crossed the sea. Seeing her walk into the palace, the king said to the young man, "Bravo! But to wed my daughter you must have the gold ring now lying somehwere at the bottom of the sea."

The youth returned to the neighbor, who said, "Go back to the seashore and blow the bugle."

He did, and out of the water jumped a fish with the ring in its mouth. Seeing the ring, the king said, "In this bag are three hares for the wedding banquet, but they are too lean. Take them out to feed in the woods for three days and three nights and then bring them back in the same bag."

But who ever heard of letting hares loose in the woods and then recapturing them? When asked how you did it, the neighbor said, "When it gets dark blow the bugle and the hares will run back into the bag."

So the boy let the hares feed in the woods for three days and three nights. On the third day here came the aunt in disguise.

"What are you doing here in the woods, my boy?"

"I'm minding three hares."

"Sell me one."

"I can't."

"How much will you take for one?"

"One hundred crowns."

The aunt gave him one hundred crowns and left with the hare.

The young man waited until she was almost home, then blew the bugle. The hare slipped out of the aunt's hands and ran back to the woods and into the bag.

The king's daughter next went to the woods in disguise.

"What are you doing?"

"Minding three hares."

"Sell me one."

"I can't."

"How much will you take for one?"

"Three hundred crowns."

She paid him and left with the hare. But as she came in sight of home, the young man blew the bugle, and the hare slipped out of her hands, ran back to the woods, and into the bag.

Finally the king himself went to the woods in disguise.

"What are you doing?"

"Minding three hares."

"Sell me one."

"For three thousand crowns I will."

But this time too the hare slipped away and came back into the bag. The three days and three nights were up, so the young man returned to the king, who said, "One last test before you marry my daughter. You are to fill up this bag with the truth."

The neighbor was still on his doorstep, and told the boy, "You are well aware of all you did in the woods. Tell that, and the bag will fill up."

The youth went back to the king. The king held the bag open while the young man spoke. "The aunt came and bought a hare for one hundred crowns, but it got away from her and came back into the bag. Your daughter came and bought a hare for three hundred crowns, but it got away from her and came back into the bag. Finally you came, Majesty, and bought a hare for three thousand crowns, but it got away from you and came back into the bag."

All that was the truth, and the bag was now bulging.

The king realized at last that he had no choice but to give the young man his daughter.

(Romagna)

◈ 48 ◈

The Three Dogs

There was once an old farmer who had a son and a daughter. When time came for him to die, he called them to his bedside and said, "My children, I am about to die, and I have nothing to leave you except three little ewes in the barn. Try to live together in harmony and you'll never go hungry."

After his death, brother and sister continued to live together. The boy tended the flock and the girl stayed at home spinning and cooking. One day while the boy was out in the woods with the sheep, he met a little man with three dogs.

"Good day to you, my boy."

"Good day to you."

"What fine sheep you have there!"

"Those are also three fine dogs you have."

"Want to buy one?"

"For how much?"

"Give me one of the little sheep, and I'll give you one of my dogs."

"And what will my sister say?"

"What should she say? You really need a dog to guard the sheep!"

The boy let himself be talked into it: he gave the man a sheep, and received a dog. He asked what the dog's name was, and the little man said, "Crushiron."

When it was time to go home, he was very uneasy, knowing his sister would scold him. Indeed, when the girl went to the barn to milk the ewes and found only two sheep and a dog, she bawled him out and beat him.

"Will you please tell me what use we have of a dog? If you don't bring back all three ewes tomorrow, you'll be sorry!"

But he still felt he needed a dog to guard the sheep.

The next morning the boy returned to the same spot and again met the little man with two dogs and the ewe.

"Good day to you, my boy."

"Good day to you."

"My ewe is very lonesome," said the man.

"So is my dog," answered the boy.

"In that case, give me another ewe, and I'll give you another dog."

"Oh, my goodness! My sister was ready to kill me for parting with one sheep! Just imagine what she'd do if I traded another one!"

"Look, one dog by himself is no good. What would you do if two wolves turned up?"

So the boy agreed to a second exchange.

"What's this one's name?"

"Chewchains."

When he returned in the evening with one sheep and two dogs, his sister asked, "Did you bring all three ewes back?" But he had no idea what to tell her.

He said, "Yes, but there's no need for you to come to the barn. I'll milk them myself."

But the girl insisted on seeing for herself, and the brother consequently landed in bed without any supper. "I'll kill you," she told him, "if all three sheep are not back here by tomorrow night."

He was tending the animals in the woods the next day, when here came the little man with the two ewes and the last dog.

"Good day to you, my boy."

"Good day to you."

"Now this dog is dying of loneliness."

"So is my last sheep."

"Give me the sheep and take the dog."

"No, indeed. Don't even suggest such a thing."

"You have two dogs now. Why don't you want the third one? That way you would own three of the finest dogs on earth."

"What's his name?"

"Crashwall."

"Crushiron, Chewchains, Crashwall, come along with me."

When night fell, the boy didn't have the nerve to go home to his sister. "I'd better strike out on my own," he decided.

He had gone some distance, with the dogs right behind him, when it started to pour down rain. Night had fallen, and he had no idea where to go. Deep in the woods he spied a beautiful palace all lit up and surrounded by a high wall. He knocked, but no one opened. He called, but no one answered. Then he said, "Crashwall, help me!"

The words were not out of his mouth before Crashwall had torn a hole in the wall with his strong paws.

The boy and the dogs stepped through, but found themselves up against a thick iron gate. "Crushiron, it's your turn now!" said the boy, and in two bites Crushiron shattered the gate.

The palace door was held fast by a heavy chain. "Chewchains!" called the boy, and with one bite the dog had the chains off and the door open.

The dogs slipped in and up the steps, with the boy on their heels. Not a living soul was anywhere in sight. A fire burned brightly in the fireplace, and nearby stood a table laden with all kinds of delicious food. He sat down to eat and, under the table, discovered three bowls of soup for the dogs. When he had finished eating he entered another room and saw a bed turned down for the night, with three dog beds alongside it for his dogs. In the morning when he got up, there was a gun and horse all ready for him to take out hunting. He went hunting, and by the time he came home, the table had been set for lunch, the bed made, and whole house cleaned and put in order. That went on for days, and he never saw a soul, although his every wish was immediately satisfied. In sum, he lived like a gentleman of leisure. Then he got to thinking about his sister; there was no telling what a hard time she might be having, and he said, "I'll bring her here to live with me. Now that I'm so well off, she'd never dream of scolding me for failing to bring the sheep home."

The next morning he called the dogs, got on his horse and, dressed like a nobleman, rode to his sister's house. Seeing him from a distance, his sister, who sat on the doorstep spinning, wondered, "Who could that handsome nobleman be, riding toward my house?" But when she recognized her brother, still with those dogs instead of the sheep, she made one of her usual scenes.

"Wait a minute!" said her brother. "Why do you scream at me now that I am a gentleman of leisure and have come to take you to live with me? What use could we possibly have of the ewes now?"

He took her onto his horse and back to the palace, where she became a lady of leisure. Her least whim was always instantly satisfied. She continued to loathe the dogs, however, and every time her brother came home, she would resume her grumbling.

One day when her brother had gone hunting with the three dogs, she strolled into the garden and saw a beautiful orange on a tree at the end of the garden. She went to pick it, but as she pulled it from the branch, out rushed a dragon all set to devour her. She burst into tears, arguing that her brother, not she, was the real trespasser and that he should be eaten if the dragon had to devour somebody. The dragon claimed he couldn't eat her brother, who was always with those three dogs. The girl then said that to save her own life she would arrange for him to eat her brother, if he would tell her what to do. The dragon instructed her to chain up the dogs outside the garden wall. She promised to do so, and the dragon let her go free.

When the boy came home, his sister began complaining again about the dogs, saying she refused to have the mongrels around her any longer at mealtime, because they stank. So her brother, who had always obeyed her without protest, led them out and tied them up the way she said. Then she told him to go and pick that orange at the end of the garden, and he went. He was plucking the orange, when out rushed the dragon. Realizing he had been betrayed by his sister, he called, "Crushiron! Chewchains! Crashwall!" And Chewchains broke the chains, Crushirons shattered the bars of the gate, Crashwall burst through the wall with his paws. They charged the dragon and tore him to bits.

The youth returned to his sister and said, "Is that all you think of me? You meant for the dragon to devour me! I don't want to be with you another minute."

He got on his horse and rode away, followed by the three dogs. He came to the realm of a king who had an only daughter, who was shortly to be eaten by a dragon. He went to the king, asking for the girl's hand in marriage. "I can't give you my daughter," replied the king, "since she is to be devoured by a terrible beast. But if you set her free, she will be yours without fail!"

"Very well, Majesty, leave everything to me." He sought out the dragon, and the dogs tore him to bits. When he returned victorious, the king betrothed his daughter to him.

The wedding day arrived and the bridegroom, forgetting her past behavior, sent for his sister. Her grudge against him was as strong as ever, and she announced after the wedding, "Tonight I shall prepare my brother's bed." Taking that for a mark of sisterly devotion, everyone readily agreed she should make the bed. What did she do then but conceal under the sheets on the bridegroom's side of the bed a sharp saw. That night the brother lay down and was sliced in two. He was carried off to church amid great weeping, with the three dogs following the coffin. Then they locked the door, leaving the three dogs inside to guard the corpse, one on the right, one on the left, and one at the head.

When the dogs were sure everybody had gone, one of them spoke. "I'll go get it now."

"And I'll carry it," said another.

"And I'll do the anointing," said the third.

So two of the dogs went off and came back with a jar of ointment, and the one that had stayed behind to keep watch, greased the wound with the ointment, thus bringing the youth back to life as sound as ever.

The king ordered his men to find whoever had concealed that saw in the bed. When it was learned the sister had committed the crime, she was condemned to death.

The young people were now happily married, all the more so because the old king, grown weary, abdicated in favor of his son-in-law. But one thing marred the young man's happiness: the three dogs had disappeared without leaving a trace anywhere in the kingdom. He shed many a tear but, in the end, had to resign himself to the loss.

One morning an ambassador brought him tidings of three ships anchored at sea with three dignitaries aboard who wished to renew old ties of friendship with him. The young king smiled, because he had always been a simple country boy who had never known anyone important. He nevertheless followed the ambassador to those who claimed to be his friends. He found two kings and one emperor who made a big to-do over him, saying, "Don't you recognize us?"

"You must be mistaken," he replied.

"We never would have thought you'd forget your ever faithful dogs!"

"What!" he exclaimed. "Crushiron, Chewchains, and Crashwall? Transformed in this manner?"

"We were changed into dogs by a sorcerer and couldn't regain our true state until a country boy ascended the throne. Just as we helped you, so did you help us. From now on we'll be fast friends. No matter what happens, remember you can always count on two kings and an emperor to help you."

They stayed in the city for a few days amid grand celebrations. When time came to leave, they separated, wishing one another every blessing, and so they lived happily ever after.

(Romagna)

◈ 49 ◈

Uncle Wolf

There was once a greedy little girl. One day during carnival time, the schoolmistress said to the children, "If you are good and finish your knitting, I will give you pancakes."

But the little girl didn't know how to knit and asked for permission to go to the privy. There she sat down and fell asleep. When she came back into school, the other children had eaten all the pancakes. She went home crying and told her mother what had happened.

"Be a good little girl, my poor dear," said her mother. "I'll make pancakes for you." But her mother was so poor she didn't even have a skillet. "Go to Uncle Wolf and ask him if he'll lend us his skillet."

The little girl went to Uncle Wolf's house and knocked. Knock, knock.

"Who is it?"

"It's me!"

"For years and months, no one has knocked at this door! What do you want?"

"Mamma sent me to ask if you'll lend us your skillet to make pancakes."

"Just a minute, let me put my shirt on."

Knock, knock.

"Just a minute, let me put on my drawers."

Knock, knock.

"Just a minute, let me put on my pants."

Knock, knock.

"Just a minute, let me put on my overcoat."

Finally Uncle Wolf opened the door and gave her the skillet. "I'll lend it to you, but tell Mamma to return it full of pancakes, together with a round loaf of bread and a bottle of wine."

"Yes, yes, I'll bring you everything."

When she got home, her mother made her a whole stack of pancakes, and also a stack for Uncle Wolf. Before nightfall she said to the child, "Take the pancakes to Uncle Wolf together with this loaf of bread and bottle of wine."

Along the way the child, glutton that she was, began sniffing the pancakes. "Oh, what a wonderful smell! I think I'll try just one." But then she had to eat another and another and another, and soon the pancakes were all gone and followed by the bread, down to the last crumb, and the wine, down to the last drop.

Now to fill up the skillet she raked up some donkey manure from off the road. She refilled the bottle with dirty water. To replace the bread, she made a round loaf out of the lime she got from a stonemason working along the way. When she reached Uncle Wolf's, she gave him this ugly mess.

Uncle Wolf bit into a pancake. "Uck! This is donkey dung!" He uncorked the wine at once to wash the bad taste out of his mouth. "Uck! This is dirty water!" He bit off a piece of bread. "Uck! This is lime!" He glared at the child and said, "Tonight I'm coming to eat you!"

The child ran home to her mother. "Tonight Uncle Wolf is coming to eat me!"

Her mother went around closing doors and windows and stopping up all the holes in the house, so Uncle Wolf couldn't get in; but she forgot to stop up the chimney.

When it was night and the child was already in bed, Uncle Wolf's voice was heard outside the house. "I'm going to eat you now. I'm right outside!" Then a footstep was heard on the roof. "I'm going to eat you now! I'm on the roof!"

Then a clatter was heard in the chimney. "I'm going to eat you now. I'm in the chimney!"

"Mamma, Mamma! The wolf is here!"

"Hide under the covers!"

"I'm going to eat you now. I'm on the hearth!"

Shaking like a leaf, the child curled up as small as possible in a corner of the bed.

"I'm going to eat you now! I'm in the room!"

The little girl held her breath.

"I'm going to eat you now! I'm at the foot of the bed! Ahem, here I go!" And he gobbled her up.

So Uncle Wolf always eats greedy little girls.

(Romagna)

◈ 50 ◈

Giricoccola

A merchant who had three daughters was due to leave town on business. "Before going," he said to his daughters, "I shall give you a present, as I wish to leave you happy. Tell me what you want."

The girls thought it over and said they wanted gold, silver, and silk for spinning. Their father bought gold, silver, and silk, then departed, advising them to behave during his absence.

The youngest of the three sisters, whose name was Giricoccola, was the most beautiful, and her sisters always envied her. When their father had gone, the oldest girl took the gold to be spun, the second girl took the silver, thus leaving the silk for Giricoccola. After dinner they all three sat down by the window to spin. People passing by and glancing at the girls always stared at the youngest. That night the moon rose and looked in the window, saying:

> "Lovely is the one with gold,
> Lovelier still is the one with silver,
> But the one with silk surpasses them both.
> Good night, lovely girls and ugly girls alike."

Hearing that, the sisters were consumed with rage and decided to exchange threads. The next day they gave Giricoccola the silver and, after dinner, sat down by the window to spin. When the moon rose that night, she said:

> "Lovely is the one with gold,
> Lovelier still is the one with silk,
> But the one with silver surpasses them both,
> Good night, lovely girls and ugly girls alike."

Infuriated, the sisters taunted Giricoccola so much that only someone with that poor girl's patience could have stood it. The next afternoon when they went to the window to spin, they gave her the gold to see what the moon would say. But the minute the moon rose she said:

"Lovely is the one with silver,
Lovelier still is the one with silk,
But the one with gold surpasses them both.
Good night, lovely girls and ugly girls alike."

By now the sisters couldn't stand the sight of Giricoccola, so they locked her in the hayloft. The poor girl was there weeping, when the moon opened the little window with a moonbeam and said, "Come with me." She took her by the hand and carried her home with her.

The following afternoon the two sisters spun by themselves at the window. The moon rose in the evening and said:

"Lovely is the one with gold,
Lovelier still is the one with silver,
But the one at my house surpasses them both.
Good night, lovely girls and ugly girls alike."

At that, the sisters went flying to the hayloft to see what was up. Giricoccola was gone. They sent for a woman astrologer to find out where their sister was. The astrologer said that Giricoccola was at the moon's house and more comfortable than she had ever been.

"How can we bring about her death?" asked the sisters.

"Leave it all to me," replied the astrologer, who dressed as a gypsy and went to peddle her wares under the moon's windows.

Giricoccola looked out, and the astrologer said, "Would you like these handsome pins? I'll let you have them for a song!"

Now those pins truly delighted Giricoccola, and she invited the astrologer inside. "Here, let me put one in your hair," said the astrologer, and thrust the pin into her head. Giricoccola at once turned into a statue, and the astrologer ran off to report to the sisters.

When the moon came home from her trip around the world, she found the girl changed into a statue and said, "Didn't I tell you to let no one in? I should leave you just like that for disobeying me." But she finally relented and drew the pin from the girl's head. Giricoccola came back to life and promised never to let anyone else in.

A short time later the sisters returned to ask the astrologer if Giricoccola was still dead. The astrologer consulted her magic books and said that, for some strange reason, the girl was alive again and well. So the sisters once more urged the woman to put her to death. This time the astrologer took a box of combs to peddle under Giricoccola's windows. They were too much for the girl to resist, and she called the woman inside. But the minute the comb touched her head she changed back into a statue, and the astrologer ran off to the sisters.

The moon came home and, seeing the girl a statue once more, flew into a rage and called her every name under the sun. But when she had calmed down, she again forgave her, removed the comb from her head, and Giricoccola revived. "But if it happens one more time," warned the moon, "you are going to remain a statue!" Giricoccola solemnly promised to admit no one from that moment on.

But the sisters and the astrologer weren't about to give up! Here came the woman with an embroidered gown for sale, the most beautiful gown you ever saw. Giricoccola was so charmed with it that she had to try it on, and the minute she did, she became a statue. This time the moon washed her hands of the matter, selling the statue for three cents to a chimney sweep.

The chimney sweep took the beautiful statue about the city tied to his donkey's packsaddle, until one day the king's son saw it and fell in love with it. He bought the statue for its weight in gold and took it to his room, where he would spend hours adoring the stone maiden. Whenever he left the room, he would lock the door, wishing to be her sole admirer. Now his sisters were each anxious to have a gown like the statue's to wear to a gala ball, so they entered the room with a skeleton key while their brother was out and removed the maiden's gown.

No sooner was it off than Giricoccola stirred and came back to life. The sisters almost died of fright, but Giricoccola reassured them with her story. Then they had her hide behind a door to await their brother's return. The king's son was frantic upon discovering the statue missing, but out jumped Giricoccola and told him everything from beginning to end. The youth took her to his parents at once and introduced her as his bride. The wedding was celebrated immediately. Giricoccola's sisters learned of this from the astrologer and died of rage right then and there.

(Bologna)

51

Tabagnino the Hunchback

The hunchback Tabagnino was a poor cobbler who could no longer make ends meet, because people never brought him so much as one shoe to mend. He therefore set out in search of better luck elsewhere. It grew dark and he was wondering where he was going to sleep, when he spied a light in the distance. Going toward it, he came to a house and knocked. A woman answered the door, and he asked for shelter.

"But this is the home of the Wild Man," said the woman. "He eats everyone he meets. If I let you in, my husband will eat you too."

Tabagnino begged and pleaded until the woman took pity on him and said, "Well, come in, and I'll hide you under the ashes, if you don't mind."

That she did, and when the Wild Man arrived and went sniffing through the house, saying,

"The awful stench, it makes me leer,
There was or is a man in here,
My nose informs me, let him fear,
I know, I know, he's somewhere near,"

his wife replied, "Come on to the table, you're imagining things," and served him a big bowl of macaroni.

The couple both made a meal off the macaroni, and when the Wild Man had eaten all he could hold he said, "I've eaten my fill. You can give what's left over to anybody hiding in the house."

"As a matter of fact, there is a poor little man who asked me for shelter for tonight," said the wife. "If you promise not to eat him, I'll bring him out."

"Bring him out."

The woman pulled Tabagnino from the ashes and seated him at the table. Sitting across from the Wild Man, the poor hunchback cloaked in ashes shook like a leaf, but he took heart and ate the macaroni.

"Tonight I'm no longer hungry," said the Wild Man. "But I warn you, if you don't flee for your life early tomorrow morning, I'll swallow you whole."

With that said, they struck up a friendly conversation, and the hunchback, who was as crafty as the Devil himself, remarked, "That's a fine coverlet you have on your bed!"

"It is embroidered entirely in gold and silver," explained the Wild Man, "and fringed with solid gold."

"And that chest?"

"It contains two bags of money."

"And the wand behind the bed?"

"It brings good weather."

"And the voice we hear?"

"It's a parrot I keep in the henhouse and which talks like us."

"You really have some fine things!"

"Oh, they're not all here by any means! In the stable I have the most beautiful mare you ever saw, and she runs like the wind."

After supper the wife took Tabagnino back to his spot under the ashes and then went off to bed with her husband. When day broke, she called Tabagnino, "Get up, flee for your life before my husband rises!"

The hunchback thanked the woman and left.

On and on he traveled until he reached the king of Portugal's palace and asked for hospitality. The king wanted to see him and hear his story. Upon learning what was in the Wild Man's house, the king felt a strong desire to possess those wonderful things and said, "Listen carefully. You can stay here at the palace and do whatever you please, but I'm asking one thing in return."

"What's that, Majesty?"

"You said the Wild Man has a beautiful coverlet embroidered in gold and silver and fringed with solid gold. You must get it for me, or you'll lose your head."

"But how can I?" replied the hunchback. "The Wild Man devours everyone. You're sending me, for sure, to my death."

"That is not my concern. Think it over and manage the best you can."

The poor hunchback thought hard, then went to the king, saying, "Sacred Crown, give me a paper bag full of live hornets that have fasted for seven or eight days, and I'll bring you back the coverlet."

The king sent the army out to catch the hornets and gave them to Tabagnino. He also gave him a wand, saying, "Take this wand. It is magic and will come in very handy. When you have to cross water, strike the ground with the wand and have no fear. While you are gone I will be waiting for you in my palace across the sea."

The hunchback went to the Wild Man's house, where he listened at the door and heard them eating supper. He climbed up to the bedroom window, slipped through it, and hid under the bed. When the Wild Man and his wife retired and fell asleep, the hunchback thrust the paper bag full of hornets under the covers and sheets and opened it. Feeling that wonderful heat, the hornets came buzzing out, stinging right and left.

The Wild Man began tossing about and threw off the coverlet, which the hunchback rolled up under the bed. The hornets grew angry and stung for all they were worth. The Wild Man and his wife fled, screaming. Once Tabagnino was alone he too fled with the coverlet under his arm.

A few minutes later the Wild Man went to the window and called to the parrot in the henhouse, "Parrot, what is the hour?"

"The hour when the hunchback Tabagnino makes off with the beautiful coverlet!" replied the parrot.

The Wild Man ran into the bedroom and saw that the coverlet was missing. Then he took the mare and galloped after the hunchback, finally spotting him in the distance. But Tabagnino was already on the seashore striking the ground with the wand the king had given him: the waters divided, and he ran safely to the other shore, after which the waters again united. The Wild Man, halted on his shore, yelled:

> "O Tabagnino, you hateful thing,
> When will you come back to this shore?
> I mean to eat you before long,
> Or else not live any more."

At the sight of the coverlet, the king jumped for joy. He thanked the hunchback, then said, "Tabagnino, if you were sly enough to steal the coverlet, you can also make off with the wand which brings fine weather."

"But how could I ever do that, Sacred Crown?"

"Think hard, or you'll lose your head."

After some concentration, the hunchback asked the king for a little bag of walnuts.

He reached the Wild Man's house, listened at the door, and heard them going to bed. After climbing to the roof, he began pelting the tiles with walnuts. The clatter awakened the Wild Man, who said to his wife, "Just listen to that hailstorm! Take the wand and lay it on the roof at once, or the hail will ruin my wheat."

The woman got up, opened the window, and laid the wand on the roof, in perfect reach of Tabagnino, who picked it up immediately and fled.

Shortly after, the Wild Man rose, pleased that it had stopped hailing, and went to the window.

"Parrot, what is the hour?"

"It is the hour when the hunchback Tabagnino makes off with your good-weather wand."

The Wild Man jumped on the mare and galloped after the hunchback.

He was about to overtake him on the beach, when Tabagnino struck the ground with the wand. The sea opened, let Tabagnino through, then closed again. The Wild Man yelled:

> "O Tabagnino, you hateful thing,
> When will you come back to this shore?
> I mean to eat you before long,
> Or else not live any more."

At the sight of the wand, the king was overjoyed. But he said, "Now you must go and get me the two bags of money."

The hunchback thought it over, then had some woodchopper's tools readied, changed clothes, put on a false beard, and went off to the Wild Man's with hatchet, wedges, and sledgehammer. Now the Wild Man had never seen Tabagnino in the daylight, and Tabagnino had also been eating such hearty meals since he had been at the king's palace, that the Wild Man did not recognize him.

They greeted one another. "Where are you going?"

"For wood!"

"Oh, here in the forest there's all the wood you could possibly want!"

So Tabagnino took his tools and started working around a massive oak. Into the tree he hammered a wedge, then another, then a third, and proceeded to drive them in with the sledgehammer. In no time he grew impatient, pretending one wedge had gone in too far.

"Don't get upset," said the Wild Man. "I'll lend you a hand." And he stuck his hands in the opening to see if he could widen it and free the wedge. At that, Tabagnino struck the tree with the sledgehammer, knocking out all the wedges at once, and the crack in the trunk closed on the Wild Man's hands. "Help me, for heaven's sake, help me!" he began shouting. "Run to my house and tell my wife to give you those two large wedges of ours and set me free!"

Tabagnino ran to the house and said to the woman, "Quick, your husband wants you to give me those two bags of money out of the chest."

"How can I do that?" said the woman. "We have supplies to buy. If he'd said one bag, I could understand. But both of them, no!"

So Tabagnino opened the window and cried, "Is she to give me just one, or both of them?"

"Both of them, and be quick about it!" shouted the Wild Man.

"Did you hear that? He is very angry," said Tabagnino. He took the bags and fled.

With a good deal of struggling, the Wild Man pulled his hands out of the tree trunk, skinning them quite badly, and went home groaning. His wife asked right away, "But why did you ask me to give away the two bags of money?"

Her husband was stunned. He went to the parrot and asked: "What is the hour?"

"The hour the hunchback is carrying off the two bags of money!"

But this time the Wild Man was in too much pain for a chase and merely cursed the hunchback.

The king also ordered Tabagnino to go and steal the mare that ran like the wind. "How can I do it? The stable is locked, and the animal has many jingle-bells on its harness!" But then he thought it over and requested an awl and a little bag of cotton. With the awl he made a hole in the wall of the stable and squeezed through it. Then he began pricking the mare in the belly with the awl. The horse kicked, and the Wild Man, hearing the noise as he lay in bed, said, "Poor animal, she must not feel well tonight to be so restless."

Tabagnino waited a couple of minutes, then pricked her again with the awl. Tired of hearing the mare kick, the Wild Man went to the stable, led her out, and tethered her outdoors. Then he returned to his bed and fell asleep once more. The hunchback, who was hiding in the dark stable, came out through the hole he had made, stuffed the mare's bells with cotton, and muffled her hoofs. He untied the animal, climbed into the saddle, and galloped off in silence. A little later, the Wild Man woke up as usual and went to the window. "Parrot, what is the hour?"

"It is the hour the hunchback is riding your mare away!"

The Wild Man would have gone after him, but Tabagnino had the mare, and who could ever overtake that animal?

Overjoyed, the king said, "Now I want the parrot."

"But the parrot talks and screams!"

"Think of a way."

The hunchback ordered a few of the best trifles you ever tasted, candies, cookies, and all kinds of pastries. He put everything in a basket and left. "See what I brought you, parrot? You'll get things like this every day if you'll come with me."

The parrot ate the trifles and said, "Good!"

So by means of trifles, cookies, and candies, Tabagnino lured the bird away, and the next time the Wild Man called from the window, "Parrot, what is the hour?" there was no answer. "Didn't you hear me? I said, 'What is the hour?'" He ran to the henhouse and found it empty.

There was a big celebration when Tabagnino walked into the king's palace with the parrot. "Now after all this," said the king, "you have only one more thing to do."

"But there's nothing else to take!" replied the hunchback.

"What do you mean?" snapped the king. "The most important thing of all remains for you to bring me the Wild Man himself."

"I will try, Sacred Crown. Just give me a garment that hides my hump, and have my features well disguised."

The king called in his finest tailors and wigmakers, and thanks to the new clothes, blond wig and fine mustache they put on Tabagnino, no one would have ever recognized him as the hunchback from the king's court.

Thus disguised, the hunchback went off to the Wild Man and found him working in a field. Tabagnino doffed his hat in greeting.

"What are you looking for?" asked the Wild Man.

"I'm the coffin-maker," explained Tabagnino, "and I'm in search of planks for the coffin of Tabagnino the hunchback, who has died."

"So he finally dropped dead!" exclaimed the Wild Man. "That makes me so happy I'll give you the boards myself, and you can build the coffin right here."

"With pleasure," said the hunchback. "The only drawback to that is I won't be able to measure the body."

"That's no problem," answered the Wild Man. "The rascal was about my size. You may measure me."

Tabagnino began sawing the boards and nailing them together. When the coffin was done, he said, "Now let's see if this is the right size." The Wild Man stretched out inside. "Let's check the lid." Tabagnino put the lid in place and nailed it down. Then he took the coffin to the king.

All the noblemen in the vicinity came and carried the box to the middle of a meadow and set it afire. Then there was a grand celebration to mark the kingdom's liberation from the monster.

The king named Tabagnino his secretary and always held him in great esteem.

> Long tale, narrow way,
> I've finished, so now say your say.

(Bologna)

52

The King of the Animals

A man was left a widower with one daughter, the most beautiful girl you ever saw. Alone and with no one to look after his daughter, it wasn't long before he decided to remarry. But he got an awful wife who was as mean as could be to poor Stellina.

"Listen, Papa," said Stellina, "rather than stay here and suffer all the time, I'm going to the country to live."

"Be patient," replied her father.

But one day the stepmother slapped her for breaking a bowl, and Stellina left home, unable to endure any more. She went to the mountains to her aunt, a fairy of sorts, but very poor, who said, "Dear Stellina, I must send you out to tend the sheep, but I'm giving you my most precious possession." And she gave her a ring. "If you're ever in trouble, take this ring in your hand, and it will help you."

One morning when Stellina was in a meadow with her sheep, she saw a handsome youth approach. "You are too beautiful to tend sheep," he said to her. "Come, I'll marry you, and you'll live like a princess."

Stellina turned as red as a beet and was at a loss for an answer. But the young man finally talked her into going off with him. They climbed into a carriage waiting on the road and took off like lightning. They raced along for the better part of the day and drew to a halt before a handsome palace.

"Here you are, Stellina," said the young man, showing her inside. "This palace will be at your disposal. Ask for whatever you want and you will get it. I must now leave you to look after my business. We'll meet tomorrow morning." With that, he left.

Stellina was speechless with amazement. She felt herself being taken by the hand, but saw no one; and she let herself be led into a magnificent room where clothing and jewels had been laid out for her. She was undressed, then clothed as a noble lady. At every moment she was conscious of somebody busying about her, but she saw no one. When they had finished dressing her, they led her to another room where a steaming hot dinner was on the table. She sat down and began eating. The plates were changed and new courses brought in, but still she saw no one. After dinner she went to look around the palace. There were rooms decorated in yellow, rooms decorated in red, with divans and chairs and the loveliest objects imaginable. Behind the palace was a beautiful park, stocked with all sorts of animals, including dogs, cats, donkeys, hens, and even giant toads, and they sounded like a group of people all talking at once. Stellina was charmed at the sight and watched them until it grew dark. Then she decided to go to bed.

She entered a room furnished with a bed fit for a princess, and was helped out of her clothes. A lamp was brought to her along with night-gown and bedroom slippers. She went to bed. All was quiet. She fell asleep and didn't awaken until broad daylight.

"I'm going to ring the bell," she said, "and see if they come to serve me." But no sooner had she touched the bell than a silver tray appeared before her with coffee and cake. She drank the coffee and got up. She was dressed and had her hair done for her. In short, she was served like a princess.

Later on, the young man returned. "Did you sleep well? Are you happy?"

Stellina said yes, he squeezed her hand, and a few minutes later said goodbye and left. That was the daily visit he paid her, and the only time she ever saw him.

At the end of two months, Stellina was sick of such a routine. One morning after the young man had left, she said, "I should like to go out for a walk in the beautiful countryside and get some fresh air." She'd not got the words out before somebody brought her a beautiful bonnet, a fan, and a parasol. "So there's still somebody around who listens to me!" said Stellina. "Do let me see you, just once!" Those words fell on deaf ears, though; no one appeared.

In that instant Stellina remembered the ring her aunt had given her. Up to then, she'd not thought of using it, being in no need of anything. She went and got the ring out of the dresser drawer where she had put it. Once she had it in her hand, she said, "I ask to see who is here serving me!"

All of a sudden before her stood a beautiful maid of honor.

Stellina jumped for joy. "At last I'll have somebody to chat with!"

"Thank you, thank you," said the beautiful maid of honor. "You have finally made me visible after all the time I was under a spell and could neither speak nor be seen."

They became the best of friends and resolved to try to solve the mystery of the place. They went outside and started walking along a trail. On and on they went, and at last the trail passed between two columns. On one column was written ASK, on the other AND YOU WILL FIND OUT.

Facing the column with ASK written on it, Stellina said, "I want to know where I am."

The column that said AND YOU WILL FIND OUT replied: "You're in a place where you will fare well, but . . ."

"But what?" asked Stellina. "But what?" She asked first one column, then the other, but no answer came back.

That "but" struck the two friends as a bad sign, and they moved on, very worried. They soon reached the end of the park. On the other side of the fence a handsome knight was sitting on the ground.

When he saw them, the knight stood up, asking, "How do you happen to be in there? Be careful, you're in great danger, you're in the hands of the king of the animals. He brings everyone he catches to his treasure-filled palace here and devours them one by one."

Hearing that, Stellina was petrified. "How can we escape?" she asked the knight.

"I'll take you away," he replied. "I am the son of the king of India just passing by on my way around the world. I fell in love with you as soon as I saw you, and I'll take you to my father, who will give you and your maid of honor the welcome you deserve."

"Yes, I accept, I will come with you," said Stellina.

"But if the palace is full of treasure," said the maid of honor, "it's a pity to go off and leave it. Let's go get it while the king of the animals is away. And tomorrow morning we'll meet here again and flee."

"And how am I to spend the night out in the open?" asked the prince. "There's not even a hut here."

"I'll see to that!" announced Stellina. She pulled the ring out of her small purse, squeezed it, and said, "I will that a villa rise here at once, with servants, carriages, and all the necessaries for eating and sleeping." Right in the middle of the meadow rose a villa that was a marvel to behold. The prince said goodbye to the girls and went inside.

Stellina and her maid of honor returned to the palace and made a thorough search, from top to bottom. The cellar was full of innumerable boxes and trunks. "What is all this?" asked Stellina. "It looks more like a warehouse than a cellar. Let's see what's in these boxes." Taking off one lid after another they found silver, jewels, and money in abundance. Stellina squeezed the ring, saying, "Let all this be removed immediately to the prince of India's villa!" No sooner said than done, leaving the cellar completely bare.

Moving on through the palace, Stellina and the maid of honor came to a secret staircase. They went up and found themselves in the dark. A voice said over and over, "Poor me! Poor me!"

Stellina grew weak in the knees at that, then remembering the ring, she continued in the same direction. They were in a great hall, and on a table lay heads of men and women, along with arms and legs, while other members hung on the walls and chairs. The heads were saying, "Poor me! Poor me!" Terrified, the two girls realized these were the secret quarters of the king of the animals.

They found a barn full of hay, maize, and oats, and realized this was fodder for all the animals in the park, who were most likely men and women changed into animals by the king of the animals. He would surely eat them too.

That night Stellina had a hard time sleeping. In the morning, as usual, the king of the animals called on her and asked if she had slept well. Stellina replied politely, as she did every morning, hiding her excitement. "Goodbye, Stellina, stay happy," he said. "I'll see you again tomorrow." And he left.

Stellina and the maid of honor climbed to the loft of the barn at once, opened the window, and threw out hay, maize, and oats into the park, so that the animals would be too busy eating to notice them leave and alert the king of the animals. While they ate, Stellina and the maid of honor packed up and left.

When they got to the two columns, Stellina held the ring and said, "I order you to explain what you meant by 'but.'"

"That 'but' means you won't escape unless you bring about the death of the king of the animals," replied the column.

"How can I do that?"

"Go into the king's room and remove the walnut he keeps under the cushion in the armchair." With that pronouncement, the column crumbled to the ground.

Courageous to the end, Stellina went back and took the walnut from under the cushion. That very instant the king of the animals ran in, screaming, "Stellina, you have betrayed me!" Then he sank to the ground, dead.

He'd no sooner fallen than all the animals changed back to their true form. There were kings, queens, princes, and they all thanked Stellina and wanted to repay her with a kingdom or a royal marriage. "I'm sorry," she said, "I already have a bridegroom waiting for me." Together they all left the palace, which immediately burst into flames.

The noblemen went their separate ways, while Stellina journeyed to India with the prince and all the treasure. They got married and were happy forevermore.

(Bologna)

◈ 53 ◈

The Devil's Breeches

A man had a son, who was the handsomest boy you ever saw. The father fell ill and, one day, sent for his son. "Sandrino, my final hour has come. Please behave, and hold on to what little bit I'm leaving you."

He died, but instead of working and holding on to his inheritance, the son had a grand fling and in less than a year found himself penniless. So he went to the king of the city asking to be taken into service. Seeing what a handsome young man he was, the king engaged him as a footman. The minute the queen saw the youth, she took such a fancy to him that she insisted he be her personal footman. But when Sandrino realized that the queen was in love with him, he thought, I'd better be off before the king sees what's going on, and resigned. The king demanded an explanation, but Sandrino would only say that it was for urgent personal business, and departed.

He reached another city and again asked the king who was there to take him into service. Seeing what a tall, handsome youth Sandrino was, the king engaged him on the spot. Now the king had a daughter who, the instant she saw the young man, fell in love with him and became blind to all else around her. Sandrino was forced to resign before things got out of hand. When the king, who hadn't noticed what went on, requested an explanation for the departure, Sandrino stated it was for personal business, and the king could say no more.

He went to work for a prince, but the prince's wife fell in love with him, so he left that place too. He tried working for five or six other masters; but, every time, some lady fell in love with him and he would have to leave. The poor youth began to curse his good looks and ended up saying he would give his soul to the Devil just to be rid of them. No sooner had he made the statement than a young nobleman stood before him. "Why are you complaining so?" he asked, and Sandrino told him the reason.

"Listen," said the nobleman, "I'm going to give you this pair of breeches. Be sure to wear them all the time and never take them off. I'll return for them in exactly seven years. Meanwhile you must never wash so much as your face; and never trim your beard, hair, or fingernails. But you can do anything else you want to and be fully satisfied."

At that, he vanished, and the clock struck midnight.

Sandrino slipped into the breeches and threw himself upon the ground to sleep. He woke up in broad daylight, rubbed his eyes, and immediately remembered the breeches and what the Devil had told him. Rising to his feet, he felt the breeches weighting him down. He then took a few steps and what should be hear but a jingle of coins: the pants were full of gold pieces, and the more he removed, the more there came spilling out.

He went to a city, stopped at an inn, and rented the finest room they had. All day long he did nothing but pull money out of the breeches and pile it up. For every service rendered him, he gave a gold coin; every poor man who stretched out a hand also received a gold coin, so there was always a long line of men at his door.

One day he asked the manservant at the inn, "Would you know if there's a palace for sale?" The manservant mentioned one right across from the king's, but no one could afford it.

"See to the purchase of it," said Sandrino, "and I will pay you for your pains."

So the manservant got busy and arranged for the sale of the palace to Sandrino.

Sandrino had it completely refurnished and all the rooms on the ground floor lined with iron and the entrances walled up. Closed up inside the palace, he spent his days piling up money. When one room was full, he would move on to the next, and that way he filled up all the rooms on the ground floor. Time passed, and his hair and beard got so long that you no longer recognized him. His fingernails too were as long as combs for carding wool, while his toenails had grown to such length that he had to wear sandals like the friars', since he couldn't get into his shoes any more. A crust formed on his skin an inch thick. In short, he no longer looked like a man, but an animal. To keep his breeches clean, he covered them with white lead or with flour.

Now the king of that city had been drawn into war with a neighboring king and urgently needed money to continue the fight. One day he sent for his steward.

"What is it, Sacred Crown?"

"We are between the Devil and the deep blue sea," said the king. "I haven't another penny to wage war."

"Sacred Crown, there's the gentleman across the street who has more money than he knows what to do with. I can go and ask him for a loan of fifty million. The worst he can do is say no."

The steward went to Sandrino on the king's behalf, bowed and scraped before him, then explained his mission.

"Tell His Majesty I'm at his service," replied Sandrino, "on condition that he give me one of his three daughters in marriage, and I don't care which one."

"I will relay the message," answered the steward.

"I'll expect an answer in three days. Otherwise I'll consider myself released from all obligation."

Upon learning what strings were attached, the king exclaimed, "Oh, me! There's no telling what my daughters will say when they see this man who looks more like an animal. You might have at least asked for a portrait to prepare the girls for the shock."

"I'll go and ask him for one," said the steward.

When Sandrino was informed of the king's request, he called in a painter, had his portrait done, and sent it to the king. Beholding the brute, the king took a step backward, crying, "Could one of my daughters love a face like this?"

Just to see her reaction, he called in his oldest daughter and put the matter before her. The girl flatly rebelled. "You're proposing such things to me? Does he strike you as a man a maiden could possibly marry?" With that, she turned her back on the king and walked off.

The king sank into a black armchair he reserved for his bad days, and sat there completely disheartened. The next day he took heart again and sent for his middle daughter, but he was prepared for the worst. The girl came in and he made the same speech to her as to the first daughter, giving her to understand that the welfare of the kingdom depended on her reply. "Well, Father," said the girl, whose curiosity had been aroused, "let me see his picture."

The king handed her the portrait, but the minute she glanced at it, she hurled it away as if she had accidentally picked up a snake. "Father! I wouldn't have thought you capable of giving your daughter in marriage to a brute. Now I know how much you really love me!" With that, she stalked off, indignant and grumbling.

The king sighed to himself. "That's that! We're headed for sure destruction. If those two made such a fuss, I can just imagine the objections of the youngest, who is the most beautiful of the three." He sank into the black armchair and gave orders not to be disturbed the rest of the day. He failed to show up for dinner, but his daughters didn't deign to ask what the matter was with him. Only the youngest girl, without a word, slipped out of the room and went to find her father. She began wheedling him, saying, "But why are you so sad, Papa? Come on, get out of that chair, cheer up, or I'll break down and cry too."

She begged and pleaded so hard to know what the matter was, that the king told her. "Really?" replied the girl. "Show me the portrait, then. Come on, let's see it."

The king pulled the portrait out of a drawer and handed it to her. Zosa, as they called her, studied it from every angle and said, "Look, Father! Do you see what a beautiful forehead is hidden under this long, tousled hair? True, his skin is black; but washed, it would be something else entirely. Do you see how beautiful his hands would be, were it not for those awful fingernails? His feet too! And all the rest. Cheer up, Father, I'll marry him myself."

The king took Zosa in his arms and kissed her again and again. Then he called the steward and sent him to inform the gentleman that his youngest daughter was willing to marry him.

As soon as Sandrino heard it, he said, "Fine, we agree. Please tell His Majesty he can have fifty million, or better still, come get the money at once and fill up a little bag for yourself too, since I want to show my gratitude. Tell His Majesty not to worry about providing for the bride. I want to do it all myself."

When the sisters heard of Zosa's betrothal, they began teasing her, but she paid no attention and let them have their fun.

The steward went for the money, and Sandrino filled up a big bag of gold pieces for him. "I must now count them," said the steward, "because I have the impression there's more money here than the sum agreed on."

"That makes no difference," replied Sandrino, "if we run a little bit over."

Next he gave all the jewelers in the city an order for their finest earrings, necklaces, bracelets, brooches, and rings with diamonds as big as hazelnuts. He arranged the jewelry on a silver tray, which four of his valets presented to the bride.

The king was overjoyed, the daughter spent hours on end trying on the jewels, while her sisters suffered tortures from envy and said, "Everything would be perfect if only he had better looks."

"Just so long as he is kind, I am quite satisfied," replied Zosa.

Meanwhile Sandrino sent for the finest dressmakers, hatters, shoemakers, and seamstresses. He ordered everything needed for the trousseau, stating that it all had to be ready in a fortnight.

Now as money works every miracle, everything was ready in a fortnight: gowns as sheer as air, embroidered to the knees; petticoats trimmed with yards of damask; handkerchiefs so full of embroidery there was no place to blow your nose; dresses of many colored silks, of bejeweled gold and silver brocade, of red and deep blue velvet.

The evening before the wedding, Sandrino had four tubs filled with hot and cold water. When they were full he jumped into the one with the hottest water and soaked until the layers of filth on him loosened. Then he jumped into the other hot tub and proceeded to scrub; the dirt peeled off like chips from a carpenter's plane. He had not bathed for seven years! When the heaviest layers of dirt were gone, he jumped into another tub, this one full of perfumed water just barely tepid. There he lathered until his beautiful skin of former times was again recognizable. Then he jumped into the last tub containing cologne, where he lingered for some time in a final rinsing. "Send for the barber at once!" The barber came, sheared him like a sheep, applied curling irons and pomades, and finally cut his fingernails and toenails.

The next morning when he stepped from his carriage to fetch the bride, the sisters who were looking out the window to see the monster arrive, found themselves face to face with the handsomest of young men. "Who can he be? Probably someone sent by the bridegroom so as to avoid appearing in person."

Even Zosa thought it was one of his friends, and got into the carriage. Upon arriving at the palace, she said, "And the bridegroom?"

Sandrino took the portrait she had already seen of him and said, "Look carefully at those eyes, look at the mouth. Don't you recognize me?"

Zosa went wild with joy. "But how on earth did you ever sink to such a state?"

"Ask me nothing more," replied the bridegroom.

Upon learning that the bridegroom was none other than he, the sisters were consumed with envy. Throughout the wedding banquet they glared at the happy couple and whispered to one another, "We'll give our soul to the Devil for the sake of seeing their happiness end."

Now that very day the seven years set by the Devil were up, and he was expected at midnight to reclaim the breeches. At eleven o'clock the bridegroom bid all the guests good night, saying he wished to be alone. "Dear wife," he advised Zosa when they were by themselves, "you go on to bed, and I'll join you later." Zosa wondered, "What in the world is on his mind?" Still, assisted by her maids of honor, she undressed and went to bed.

Sandrino had bundled up the Devil's breeches, and sat waiting for him. The servants had all been sent off to bed, and he was alone. All of a sudden he noticed he had goose pimples and his heart was in his mouth. Midnight struck.

The house shook. Sandrino saw the Devil advancing toward him, and held out the bundle. "Here, take your breeches! Go on, take them!" he said.

"I ought to take your soul now," said the Devil.

Sandrino shuddered.

"But since you are responsible for my finding two other souls," continued the Devil, "I'll take those two instead of yours and leave you in peace!"

The next morning Sandrino was sleeping peacefully beside his wife, when the king came in to greet them and inquire if Zosa knew anything about her sisters, who had disappeared. They went into the sisters' bedroom, but it was empty. On the table was this note: "May you be cursed! We are damned because of you, and the Devil is taking us away."

Then Sandrino realized who the two souls were that the Devil had taken in place of him.

(Bologna)

◈ 54 ◈

Dear as Salt

There was once a king who had three daughters: a brunette, a redhead, and a blonde. The first was homely, the second so-so, while the youngest was the best-hearted and most beautiful of the three. The two older girls were, naturally, quite envious of her. Now the king had three thrones: a white one, a red one, and a black one. When he was happy he sat on the white one, when he was so-so on the red one, when he was angry on the black one.

One day he seated himself on the black throne, for he was angry with the two older girls. So they went in and made up to him. The oldest daughter asked, "Father, did you sleep well? You're on the black throne! That doesn't mean you're angry with me, does it?"

"It most certainly does. I am angry with *you*."

"But why, sir?"

"Because you don't love me at all."

"I do, Father. You are very dear to me."

"How dear?"

"As dear as bread."

The king gave a little snort, but said nothing more, for he was greatly pleased with the girl's answer.

Next came the middle girl. "Father, did you sleep well? Why are you on the black throne? You couldn't be angry with me, could you?"

"I certainly am."

"But why are you angry with *me*, sir?"

"Because you don't love me at all."

"I do love you, sir, you are quite, quite dear to me."

"Just how dear?"

"As dear as wine."

The king mumbled something to himself, but he was obviously delighted.

Now came the youngest daughter, as cheerful as cheerful could be. "Father! Did you sleep well? Why in the world are you on the black throne? Are you in any way angry with me?"

"I am indeed, for you don't love me either."

"I do love you, sir, you are exceedingly dear to me."

"How dear?"

"As dear as salt!"

Hearing that, the king was furious. "As dear as salt? Salt? You wretch! Out of my sight! I never want to lay eyes on you again!" And he ordered her taken to a forest and slain.

When her mother the queen, who doted on her, heard about that order, she racked her brains for a way to save her. In the royal palace was a silver candlestick big enough for Zizola—as they called the youngest daughter—to get inside its base, so the queen concealed her there. "Take this candlestick out and sell it," she said to her most trusted servant. "When anyone inquires how much you're asking for it, if they are poor people, say a fortune; but if it's a well-to-do gentleman, say he can have it for a song, and make sure he gets it." The queen kissed her daughter goodbye, giving her much parting advice, along with a store of figs, chocolate, and cookies.

The servant carried the candlestick to the town square, and to all those who wanted to know its price but whose looks he didn't like, he quoted a preposterous sum. At last the son of Hightower's king happened by, scrutinized the candlestick, then asked its price. The servant told him a ridiculously low figure, so the prince had the candlestick carried to his palace. It was placed in the dining room, and everybody who came to dinner marveled at it.

In the evening the prince always attended some social or other. As he wanted no one waiting up for him at home, the servants set out his supper and went off to bed. When Zizola was sure everyone had left the room, she jumped out of the candlestick, ate up everything on the table, and returned to her post inside the candlestick. The prince came home, found nothing out for him to eat, rang every bell in the house, and gave the servants a severe tongue-lashing. They swore they had set his supper out for him and that the dog or cat must have eaten it.

"If it happens again, I'm dismissing every one of you," stated the prince. He then ordered another supper, ate, and went off to bed.

Next evening, although everything was locked up, the same thing occurred. For a while it looked as if the prince would bring down the house with all his shouting. Then he said, "We'll just see what happens tomorrow night."

When tomorrow night came, what do you think he did? He hid under the table, which was covered by a cloth that came all the way to the floor. The servants set the table, putting out all the different dishes, then shooed the dog and cat out and locked the door behind them. No sooner was everyone gone than the candlestick opened and out stepped lovely Zizola. She went to the table and ate her fill. Out jumped the prince and grabbed her by the arm. She tried to get away, but he held her tight. Then Zizola fell to her knees and told him her whole story from beginning to end. The prince was, from the start, head over heels in love with her. He calmed her down and said, "You might as well know right now you will be my bride. For the time being, go back into the candlestick."

In bed, the prince couldn't shut his eyes the whole night long for thinking about lovely Zizola. In the morning he ordered the candlestick brought to his room; it was so beautiful he wanted it near him at night. The next thing he did was have his meals sent to him—double portions, for he was quite hungry. They brought him coffee, then a hot breakfast, then dinner, every meal with double servings. The minute they put the dishes down and left the room he locked the door, invited Zizola out, and the two of them ate together in glee.

The queen, who now had to take her meals in the dining room by herself, began sighing. "What on earth could my son have against me not to dine with me any more? Have I done anything to him?"

Again and again he told her to be patient, that he needed a little time to himself. Then one fine day he announced, "I am going to get married."

"And who is the bride?" asked the queen, cheered by the news.

The prince replied, "I am going to marry the candlestick!"

"Oh, goodness, my son has lost his mind!" said the queen, putting her hands over her eyes. He was serious, though. His mother tried to get him to see reason, reminding him of what people would say, but he wouldn't budge an inch: he ordered all wedding arrangements completed in a week.

On the appointed day, a long line of carriages left the palace. In the first one rode the prince, with the candlestick at his side. They reached the church, and the prince had the candlestick carried up to the altar. At exactly the right moment the candlestick opened, and out stepped Zizola in her brocaded dress, with countless gems adorning her neck and ears and sparkling all over. After the wedding they returned to the palace, where the queen heard their whole story. Being a very cunning lady, she said, "Leave everything to me, and I'll teach that father of yours a lesson."

So they had the wedding banquet and invited all the kings in the vicinity, including Zizola's father. For him, the queen had a special dinner prepared separately, without a grain of salt in any of the dishes. All the guests were informed the bride wasn't feeling well and couldn't come to the banquet. They began eating; but the king who got the tasteless soup started grumbling to himself. "That cook, that dumb cook forgot to salt the soup." What could he do but leave it.

The main course came to him just as saltless as the soup. The king put down his fork.

"Why aren't you eating, Majesty? Isn't it good?"

"Of course, of course. It's delicious!"

"Well, why don't you eat?"

"Uh, uh, I don't feel too well all of a sudden."

He tried putting a forkful of meat in his mouth, but no matter how much he chewed, the meat would not go down. Then he recalled what his daughter had told him, that he was as dear as salt to her. Overcome with remorse, he burst into tears. "Woe is me, what wrong have I done!"

The queen wanted to know what was the matter, and he told her all about Zizola. At that, the queen rose and summoned the little bride. Again and again her father embraced her; he couldn't help weeping and asking what she was doing there, as though she had risen from the dead. They sent for her mother too, renewed the festivities with a party every day, and I do believe everyone is there to this day and still dancing.

(Bologna)

The Queen of the
Three Mountains of Gold

There was once a poor man who had three sons. This man was sick and in great pain, and one day he called his sons to his bedside and said, "As you can see with your own eyes, I am about to die. I've nothing to leave you, but I ask you to be good boys and work as I've always done, and heaven will surely help you." He drew his last breath, and the three sons were left all alone.

The eldest said, "Let us now go out and seek work as our father advised." So they all three departed and went out into the world.

At nightfall they found themselves before a fine palace and knocked on the door to request shelter. They called and looked all around, but not a soul was anywhere in sight. They therefore went inside and found a table laden with a variety of good food. They stood there gaping, until the oldest boy said, "Since no one is home, let's sit down and eat. If anybody comes in, then we'll ask for permission." With that, they proceeded to eat and drink their fill.

Next, they strolled through the palace and came to a room with a nice bed surmounted by a canopy. Then they entered a room containing a bed with a garland of flowers above it; and last, a room where the bed was surmounted by a crown of golden leaves.

"These beds seem to have been prepared just for us," the boys agreed. "So let's go to bed!"

They each chose a room, and the oldest boy said, "Make sure you get up bright and early to leave, because I don't intend to wait for you."

In the morning the oldest boy rose quite early and left without a word, which was quite typical of him. When the middle boy awakened, he went into the room of his big brother and found him already gone, so he too dressed and departed.

Their little brother slept late. When he finally got up, he sought his brothers in vain. Breakfast was on the table, so he sat down and ate, then went to look out a window. Before him was a beautiful garden, which he decided to visit.

The youngest brother, whose name was Sandrino, was quite a handsome young man. As he strolled through the garden he spied a large pool at the end of one of the paths. Rising above the surface was the head of a very beautiful maiden immersed in water up to her neck and perfectly motionless.

Sandrino said, "What are you doing there, madam?"

"You are a real godsend, good youth," she replied. "I am the queen of the Three Mountains of Gold. A spell was cast over me and I must stay here in the water until I meet a man courageous enough to sleep in the palace three nights in a row."

"If that's all one has to do," answered Sandrino, "I'll sleep there myself."

"I will marry whoever succeeds in sleeping there, when the three days are up. But you must not be afraid if you hear a commotion and see all kinds of wild animals rush into the room. Stand your ground, and they will retreat without touching you."

"You can be sure I won't be afraid. I'll do exactly as you said."

At nightfall the boy went to bed. Then at the stroke of midnight he heard a tumult and recognized the roar of wild animals. "Here we go," said Sandrino, waiting to see what would happen.

Into the room rushed wolves, bears, eagles, serpents, and countless other beasts ferocious enough to make the Devil himself cringe. They circled the room and completely surrounded the bed, but Sandrino didn't cower the least bit, so the animals filed slowly out, and that was that.

In the morning the boy returned to the pool. The queen was now in water only up to her waist. She was happy and full of praise. At night the same animal music filled Sandrino's room, but he stood his ground and found the queen the next morning in water only up to her calves. She praised him to the skies, and Sandrino went to breakfast, all smiles.

He'd come to the last night. The animals roared mightily and closed in on the bed, but Sandrino still refused to flinch, and they finally left. In the morning only the queen's feet remained under water. He gave her his hand and led her out of the pool; maids of honor appeared and made a big fuss over her. They went to breakfast at once and set the wedding for three days thence.

On the morning of the wedding day, the queen said to Sandrino, "I must tell you something extremely important: when you kneel on the prayer bench, don't fall asleep, or I'll disappear and you'll see me no more."

"That's all we would need!" replied Sandrino. "How could I possibly fall asleep?"

They went to church, where he knelt on the *prie-dieu* and became so drowsy that he fell sound asleep, while the queen fled. Upon awaking a few minutes later, Sandrino looked and saw that the queen was no longer there. "Woe is me!" he said over and over. He returned to the palace and searched everywhere for her, but she was nowhere to be found. He therefore picked up a bag of money and set out in pursuit of her.

After walking all day long he entered an inn at night and asked the innkeeper if he had seen the queen of the Three Mountains of Gold. "I've not seen her myself," said the innkeeper, "but as I happen to be in charge of all the animals of the earth, I'll ask them if they've seen her." He whistled once, and here came dogs, cats, tigers, lions, monkeys, and other animals, and the innkeeper asked, "Have any of you seen the queen of the Three Mountains of Gold?"

"No," replied the animals, "we can't say that we have."

The innkeeper dismissed all the animals and said to Sandrino, "Look, tomorrow morning I'll send you to my brother who's in charge of all fish, and you can see what they have to say."

The next morning Sandrino gave the innkeeper a purse of money and left for his brother's.

When the man in charge of all fish heard that Sandrino had been sent by his brother, he invited him into his inn and said, "Just a minute and I'll call all the fish and ask them."

He whistled, and here came pike, tench, eels, sturgeon, dolphins, whales, and all the rest. "No, we've seen nothing," they all answered and were dismissed by the innkeeper, who then said to Sandrino, "Tomorrow I'll give you a letter of introduction to my brother, the one in charge of birds; they might have seen her."

Sandrino impatiently awaited the next day, and when it dawned he set out and walked and walked until he reached the third inn. "I'll oblige you immediately," said the innkeeper. He whistled, and all around them flew hens, owls, pheasants, birds of paradise, and falcons; only the eagle was missing. The innkeeper gave a second whistle, and the eagle appeared.

"I'm sorry I was late," said the eagle. "I was attending a banquet at the court of the king of Marone who's marrying the queen of the Three Mountains of Gold."

Hearing that news, Sandrino lost hope. But the man in charge of all birds said, "Cheer up, we'll see if we can do something about this." He turned to the eagle. "Will you carry this youth to the court of the king of Marone?"

"Right away!" said the eagle, "but I demand that every time I call for water he give me water, every time I call for bread he give me bread, and every time I call for meat he give me meat. Otherwise I'll throw him into the sea."

So the youth loaded up with two baskets of bread, two containers of water, and two pounds of meat. The eagle soared into the air with Sandrino astride. Every request the eagle made for bread, water, and meat was satisfied at once. But they still had a stretch of sea to cross, and the youth had run out of meat. The eagle called for meat, and Sandrino could think of nothing else to do but cut off flesh from his own leg and feed it to him. The queen had given him a magic salve, which he applied to the wound and healed it at once.

The eagle carried him right into the queen's room.

The minute they saw each other, they fell into each other's arms. They told each other everything, then the queen took him to the king, introducing Sandrino as her rescuer and bridegroom. The king thought it quite fitting for her to marry the young man and took great pleasure in proclaiming the wedding festivities, which lasted a month and one week.

(Bologna)

◈ 56 ◈

Lose Your Temper, and You Lose Your Bet

A poor man had three sons: Giovanni, Fiore, and Pírolo. Taken sick, he called his sons to his bedside. "As you can see with your own eyes, my sons, I am at death's door. All I have to leave you are three equal sums of money which I accumulated by hard work. Each of you take one and manage the best you can." No sooner had he said that than he heaved a deep sigh and died. The boys were heartbroken and wept; their poor father had left them forever.

They each took a bag of money, but Giovanni, the oldest son, said, "Brothers, we'll never make out if we don't work. What we have here won't last forever and we'll find ourselves out in the cold. One of us must begin looking around for work of some sort." The middle boy, Fiore, agreed. "You are quite right. I'll go out myself and see what I can find." Next morning he got up, washed, shined his boots, slung his bag of money over his shoulder, embraced his brothers, and set out.

He spent the whole day looking around and, toward evening, passed by a church and saw the archpriest outside getting some fresh air.

"Good evening, Father," said Fiore, doffing his hat.

"Good evening, young man, where are you going?"

"I'm going out into the world to seek my fortune."

"What have you there in that bag?"

"The share of money my poor father left me."

"How would you like to enter my household?"

"I'd like to."

"I too have a share of money, mind you. If you enter my service, we'll make a bargain: the first one to lose his temper will forfeit his share of money."

Fiore accepted the terms, and the archpriest took him out and showed him the plot of land to be tilled the next day, saying, "Once you begin working, there's no need to waste time going back and forth for breakfast and dinner. I'll send your meals out to you."

"As you wish, Father," replied Fiore. Then they sat down to supper and chatted awhile, after which the older servant woman showed the boy to his room.

Fiore got up bright and early the next morning and went out to dig up the field the archpriest had shown him the evening before. He dug until breakfast time, when he stopped and waited, expecting someone to show up any minute with food. When no one came, Fiore got upset and cursed. Since time was passing, he took up the spade again and went back to digging on an empty stomach in anticipation of dinnertime. At last it was dinnertime, and Fiore peered down the road to see if anyone was coming. Every time somebody approached, he was sure it must be the archpriest's servant and perked up; but it was always someone else, and he cursed a blue streak.

At last, around nightfall, the old woman arrived full of excuses: she'd been too busy with the laundry to come any sooner, and blah-blah-blah. . . . Although burning to call her every name under the sun, he controlled himself so as not to forfeit his sum of money to the priest. He dived into the old woman's basket and pulled out a pot and a bottle. He went to open the pot, but the lid seemed to have been cemented on and stuck fast. Screaming insults, Fiore sent pot and all flying. "But don't you realize," began the old woman as innocently as you please, "that we closed it up tight so the flies wouldn't get into it."

Fiore then grabbed the bottle, but it too was sealed up the same way. Cursing loud enough to awake the dead, he said, "Away with you! Go back and tell the archpriest he'll hear the rest from my own lips. He'll see if this is any way to treat a man!"

The servant went back to the archpriest, who was waiting at the door. "How did it go? How did it go?"

"It was perfect, Father, simply perfect! He's beside himself with rage!"

In a little while here came Fiore so long-faced you could have put a halter on him, and he hadn't shut the door before he launched out against the archpriest, calling him every name under the sun.

"Have you forgotten our agreement," said the archpriest, "that whoever flew off the handle first would forfeit his sum of money?"

"The Devil take that money too!" shouted Fiore, who packed up and left without the bag of money. The archpriest and his two servants laughed until they cried.

Half starved, exhausted, and angry, Fiore made his way home. His brothers, who were looking out the window as he came into view, knew right away from the expression on his face that he had fared badly.

Once he had satisfied his hunger and thirst, he told them what had happened. Giovanni said, "I bet if I go out I'll return with not only my money but the priest's and yours as well. Tell me where he lives and sit tight."

So Giovanni went to the archpriest, but he too became so enraged, what with hunger and thirst and that confounded pot and bottle, that he would have forfeited ten additional bags of money if he had had them. He came home as hungry and cross as a bear.

Pírolo, the youngest and most cunning of the three, said, "Let me go, brothers, and I'll be sure to return with your money and every cent of the archpriest's." The brothers were reluctant for him to go, lest the rest of their father's money be lost, but he begged and pleaded until they finally consented.

He reached the archpriest's house and entered his service. The usual bargain was made, and the archpriest added, "I have three bags of money, which I'm staking against your bag." They sat down to supper, and Pírolo wisely pocketed all the bread, meat, ham, and cheese he dared.

In the morning he was at work before sunrise. Naturally nobody showed up at breakfast time, so he took out his bread and cheese and ate. Then he went to a farmhouse, introducing himself as the archpriest's field hand, and asked for something to drink. The farmer and his family made a big to-do over him; they asked after the archpriest and chatted for a while, then took Pírolo to the cellar and drew a bowl of their finest wine, which lasted him until dinnertime. He thanked the people, promising to call on them again, and returned to his work in the best of spirits. Neither did anybody show up at dinnertime, but Pírolo had bread, ham, and other meat. Then he went back for more wine and returned to the field singing. As night began to fall, here came a little old woman down the road, the priest's old servant, bringing his dinner. And there was Pírolo singing!

"I'm sorry to be so late, young man . . ."

"Oh, don't give it a thought!" he replied. "It's never too late to eat."

At those words the old woman stood stock-still, then took out the pot with the sealed lid. He burst out laughing. "You clever souls! You fixed it so the flies wouldn't get in!" He pried off the lid with his hoe and ate the soup. Next he picked up the bottle, broke the bottleneck, again with his hoe, and drank the wine. When his hunger and thirst were satisfied, he said to the old woman, "You go on back, and I'll be home just as soon as I've finished up out here. Please thank the archpriest for his thoughtfulness."

The archpriest welcomed the old woman with open arms. "Well? What news?"

"Bad news. That boy is as cheerful as a canary."

"You just wait," said the archpriest. "He'll change his tune."

Pírolo returned and they sat down to supper. All through the meal he joked with the two servants while the archpriest sat there and shuddered.

"What work do you have lined up for me tomorrow?" asked Pírolo.

"Listen," said the archpriest, "I have a hundred pigs for you to drive to market and sell."

The next morning Pírolo drove the hundred pigs to market and sold them to the first merchant he met, all except for a sow as big as a cow. But before selling them, he cut off each one's tail and thus went away with ninety-nine pigtails. With money in his pocket now, he headed for home. He stopped in a field along the way, dug countless holes with a trowel, and planted the tails, leaving only their curls showing aboveground. Next he dug a vast hole and buried the sow, leaving only the curl of its tail showing. Then he cried at the top of his voice:

"Hurry, hurry, Don Raimondo,
Pigs you own are going to Inferno!
Downward do they rush to darkest dales;
Left to see are only curly tails!"

The archpriest looked out the window, and Pírolo frantically motioned for him to come outside. The archpriest came running.

"Who has ever suffered worse luck? I was here with the herd when I suddenly noticed them going under, right before my eyes. As you can see, they've disappeared all but for their tails! No doubt about it, they're tumbling straight down to Hell! Let's see if we can rescue a few, at least!"

The archpriest began tugging, but ended up with only a handful of tails. Pírolo, though, grabbed hold of the sow's tail and after tugging and tugging brought her out alive and in one piece and squealing like one possessed.

The archpriest was all ready to jump up and down in rage, but remembered the money and checked his anger. "Well, what more can we do but accept it," he said, feigning unconcern. "Accidents will happen." But he walked back to the house wringing his hands.

That night Pírolo asked as usual, "What do I have to do tomorrow?"

"I have a hundred sheep to go to market," replied the archpriest, "but I wouldn't want the same thing to happen that occurred today."

"Goodness, no!" said Pírolo. "We won't ever be that unlucky again!"

The next day he went to market and sold the sheep to a certain merchant, all except one that limped. He pocketed the money and headed for home. When he came to the field of the day before, he picked up a long, long ladder lying there on the ground, propped it against a poplar tree, and carried the lame sheep to the treetop and tied her up. Then he came back down, removed the ladder, and cried at the top of his voice:

"Hurry, hurry, Don Carmelo!
Lambs you own are bound for the rainbow!
Left behind in poplar's top
Is the lamb that limps and flops."

The archpriest came running, and Pírolo explained. "I was here with my sheep when all of a sudden I see them leap into the air as if summoned to Paradise. Only that poor crippled one there didn't make it and remained in the treetop."

The archpriest was as red as a beet, but again managed to feign unconcern. "What can you do but accept it. Those things will happen . . ."

At supper Pírolo asked what his next task would be, and the archpriest said, "My son, I have no more tasks for you. Tomorrow morning I shall say Mass in a neighboring parish. You can come along and serve Mass."

The next morning Pírolo rose early, shined the archpriest's shoes, put on a white shirt, washed his face, and went to wake up his employer. They left the house together, but as soon as they got out on the road it began to rain and the archpriest said, "Go back and get my wooden shoes. I don't want to muddy my nice shoes I say Mass in. I'll wait for you under this tree with the umbrella."

Pírolo ran home and said to the servants, "Quick, where are you? The archpriest said for me to give you both a kiss!"

"Kiss *us*? Have you lost your mind? We can just hear the archpriest saying such a thing!"

"Upon my word, he said to kiss you both! If you don't believe it, I'll let him tell you so himself!" He called out the window to the priest waiting outside, "One, father, or two?"

"Why, both of them, of course!" cried the archpriest. "Both of them!"

"You see?" said Pírolo, who gave them each a kiss. Then he picked up the wooden shoes and ran back to the archpriest, who asked, "What good would just one shoe have done me?"

When he got back home, the archpriest found the servant women sulking. "What's the matter?" he asked.

"What's the matter? You ask us that? What do you mean by giving the boy such orders? If we'd not heard with our own ears, we'd never have believed it!" And they told him about the kiss.

"That's the last straw," said the archpriest. "I must dismiss him at once."

"But you can't send field hands away," replied the servants, "until the cuckoo has sung."

"We'll just make believe the cuckoo is singing, then." He called Pírolo and said, "Listen, I have no more work for you, so Godspeed!"

"What!" replied Pírolo. "You know very well that you can't dismiss me before the cuckoo has sung."

"Very well, to be perfectly fair we'll wait for the cuckoo to sing."

The old servant killed and plucked a few hens, sewing all the feathers onto a waistcoat and a pair of breeches belonging to the archpriest. She then dressed up in all those feathers and went to the roof that night and sang, "Cuckoo! Cuckoo!"

Pírolo was at the supper table with the archpriest. "Well, bless my soul!" exclaimed the priest. "I do believe I hear the cuckoo singing."

"Oh, no," answered Pírolo. "March has scarcely begun, and the cuckoo never sings before May."

Yet there was no denying it was singing: "Cuckoo! Cuckoo!" Pírolo ran and got the shotgun hanging behind the archpriest's bed, opened the window, and took aim at that big bird singing on the rooftop. "Don't shoot! Don't shoot!" shouted the archpriest, but Pírolo fired away.

Down tumbled the feather-clad servant, riddled with shot.

This time the archpriest was blind with rage. "Pírolo, get out, and don't ever let me see you again!"

"Why? Are you angry, Father?"

"I certainly am!"

"Well, give me the three bags of money, and I'll go."

So Pírolo went home with four bags of money, in addition to all the proceeds from the sale of the pigs and sheep. He gave his brothers back their shares, opened up a haberdashery with his own, got married, and lived happily ever after.

(Bologna)

57

The Feathered Ogre

A king fell ill and was told by his doctors, "Majesty, if you want to get well, you'll have to obtain one of the ogre's feathers. That will not be easy, since the ogre eats every human he sees."

The king passed the word on to everybody, but no one was willing to go to the ogre. Then he asked one of his most loyal and courageous attendants, who said, "I will go."

The man was shown the road and told, "On a mountaintop are seven caves, in one of which lives the ogre."

The man set out and walked until dark, when he stopped at an inn. When the innkeeper learned of his mission, he said, "How about bringing me a feather too on your way back, since they are so beneficial."

"I'll be glad to," replied the king's man.

"And should you talk to the ogre, try and find out something about my daughter. She disappeared years ago and is now goodness knows where."

In the morning the man continued on his way. He came to a river and called the ferryman to row him to the other side. During the crossing, they got into conversation.

"Will you bring me a feather too?" asked the ferryman. "I know they bring luck."

"Yes, of course I'll bring you one."

"And if you have the chance, ask the ogre how come I've been at this job for so many years and can't get off the ferry."

"I'll certainly ask him."

The king's man disembarked and continued his journey on foot. At a fountain he sat down to eat a bite of lunch. Two well-dressed noblemen came by and also sat down, and the three of them got to talking.

"Why don't you bring us a feather too," they said.

"I certainly will."

"Also, would you ask the ogre something? In our garden is a fountain that once spewed gold and silver, but it has since dried up."

"I'll ask him why, without fail."

He moved on and walked until dark, when he knocked at a monastery. Friars answered the door, and he requested shelter.

"Come in, come in."

After hearing his story, the friars inquired, "But do you know what you are getting into?"

"I was told there are seven caves. At the back of one of them is a door I'm to knock on and be greeted by the ogre."

"My poor man," said the prior, "if you are unmindful of all the danger, you'll certainly lose your life. This is no laughing matter. I'll tell you about the ogre, in hopes you'll do us a favor."

"Of course I will."

"Listen to me, then. When you get to the mountaintop, you'll see seven caves. The seventh is the ogre's. Go down into that one, all the way to the end, where it will be pitch-dark. We'll give you a candle and matches to light your way. But be sure to go in right at noon, when the ogre is out. You'll find his wife there, a bright girl who will tell you exactly what to do. Beware of the ogre, who would eat you up in a minute."

"How good of you to tell me all these things I didn't know."

"Now here's what you are to find out for us. We lived here in peace for no telling how many years. But for the last ten, we've done nothing but wrangle. Some want one thing, others another, there is bickering, and things are always in turmoil. What is the meaning of it?"

The next morning the man scaled the mountain. He was at the top by eleven o'clock and sat down to rest. At the stroke of noon he slipped into the seventh cave. It was pitch-dark, but he lit the candle and discovered a door. The minute he knocked, a beautiful girl opened and asked, "Who are you? What brings you here? You don't know my husband! He eats every human being he sees!"

"I came for some feathers. Since I'm already here, I'll stay and try my luck. If I get eaten, that's that."

"Listen, I've been here for years and years and can't stand it any longer. Be very careful, and we'll both flee. He must under no circumstances see you, or he'll eat you. I'll hide you under the bed and when he retires for the night, I'll pull out the feathers. How many do you want?"

"Four." And he told her about king, innkeeper, ferryman, noblemen, friars, and the queries of each.

They talked as they ate their dinner. As it had grown late in the meantime, the young lady began getting the ogre's meal ready. "When he's hungry, he smells humans right away. After eating he no longer notices, luckily for you!"

At six o'clock a great clatter was heard at the door, and the man disappeared under the bed in a flash. In stormed the ogre sniffing and saying:

"Here, here,
There're stinking humans here.

186

There were, there are, they're hiding;
My nose informs me they are near!"

"Nonsense!" replied his wife. "Your hunger is making you imagine things. Sit down and eat."

The ogre ate, but he could still smell a man and went all through the house after dinner looking for him. It was at last bedtime, so they undressed, got under the covers, and the ogre went to sleep at once.

The man under the bed held his breath. "Listen closely," whispered the woman. "I'm going to pretend to be dreaming and pull out one of his feathers." She plucked a feather and slipped it under the bed to him.

"Ouch! What do you mean by plucking me?" yelled the ogre.

"Oh, dear, I was dreaming . . ."

"What were you dreaming?"

"I was dreaming about the monastery down below us. For the last ten years the friars have been so much at odds with one another that it's pure torture to be under one roof together."

"That's no dream but a fact," answered the ogre. "The friars are ill-tempered because ten years ago the Devil got into their monastery dressed as a priest."

"How could they get rid of him?"

"The real friars would have to start doing good deeds. Then they'd spot the Devil in their midst." At that, the ogre went back to sleep.

A quarter of an hour later, his wife pulled out another feather and passed it to the man under the bed.

"Ouch! That hurt!"

"I was dreaming."

"Again? What were you dreaming this time?"

"You know the fountain down below us in the garden of those two noblemen, which used to spew gold and silver? I dreamt it had gone dry. What on earth could that mean?"

"All of your dreams are true tonight. The fountain is stopped up and can't spew any more gold and silver. They would have to dig gently down to the mouth of the fountain, where they'd find a ball entwined with a sleeping snake. They would have to crush the snake's head beneath the ball before the snake awakened, and the fountain would spew gold and silver anew."

Again in a quarter of an hour she plucked another feather. "Ouch! I believe you've made up your mind to pluck me clean tonight."

"I'm sorry, I was dreaming."

"What now?"

"A ferryman down there on the river hasn't been able to leave his ferry for years."

"True. He doesn't realize that he should ferry a man across the river, collect his fare, and disembark before his passenger can. The traveler will then have to remain on the ferry."

The wife pulled out the fourth feather. "Confound it! What are you about?"

"I'm sorry. I keep on dreaming. I was dreaming of an innkeeper still looking for his daughter, years after her disappearance."

"You mean your father, because you are that innkeeper's daughter."

In the morning at six o'clock, the ogre rose, bid his wife goodbye, and went off. The man came out from under the bed with the four feathers wrapped in a package, took the young lady by the arm, and together they fled.

They stopped at the monastery to tell the friars, "The ogre said that one of you is the Devil. You must start doing all the good you can, and he will flee."

The friars all did one good deed after another until the Devil finally fled.

The couple next stopped by the garden to give the two noblemen a feather and explain to them about the snake. And it wasn't long before the fountain was again spewing gold and silver.

They came to the ferryman. "Here's your feather!"

"Thank you. And what did the ogre say concerning me?"

"I'll have to wait until I'm on the other bank to tell you."

Once the couple was safely on the opposite shore, they told the ferryman what to do.

Upon arriving at the inn, the king's man cried, "Innkeeper, here I am with your feather and your daughter!" Right away the innkeeper wanted to give his daughter to the man in marriage.

"Let me first take the king his feather and ask his permission."

He carried the feather to the king, who got well and rewarded him. The man said, "Now if Your Majesty permits, I'll be off to my wedding." The king doubled the reward, and the man took leave of him and returned to the inn.

What about the ogre? Discovering his wife gone, he set out in pursuit, fully intending to devour her and whoever was involved in her escape. He came to the river and jumped on the ferry. "Pay your fare," said the ferryman. The ogre paid, never dreaming the ferryman knew the secret. Before landing on the opposite shore, off jumped the ferryman, and the ogre could no longer leave the boat.

(Garfagnana Estense)

58

The Dragon with Seven Heads

There was once a fisherman whose wife bore him no children, even though they had been married for some time. One fine day the fisherman took his nets to the nearby lake to fish and caught a big, beautiful fish. The minute it was pulled out of water the fish began begging the man to let it go, promising in return to tell him about a pond in the region where he would make a much finer haul and in no time at all. Hearing a fish talk frightened the fisherman, and he didn't hesitate to free the fish, which immediately disappeared in the water. The fisherman went to the pond and caught so many fish in two or three hauls that he returned home more loaded down than a donkey.

His wife insisted on knowing how he had ever caught that many fish, so he told her in detail what had occurred. At that, the woman was furious with her husband. "Simpleton! How could you let such a fine fish get away? Be sure to catch it tomorrow and bring it home. I intend to prepare it in a stew that will really satisfy our craving for fish."

To please his wife, the fisherman returned to the lake the next day, cast his net, and again pulled up the talking fish. But this time also he yielded to the fish's begging and pleading and spared it, then made a splendid haul in the same pond as yesterday. When he came home and told his wife, she flew off the handle, put her hands on her hips, and blessed him out. "You dumb ox! Blockhead! Can't you see that you're cursed with luck? How can you turn your back on it? Either you bring me that fish tomorrow, or you'll be sorry you didn't. Is that clear?"

Dawn found the fisherman back at the lake. He cast his nets, pulled them up, and there again was the big fish, whose words and entreaties this time fell on deaf ears. The fisherman ran straight home, and his wife took the fish, which was still alive, and threw it into a tub of fresh water. Then they both stood by the tub admiring the fish and discussing the best way to cook it. At that, the fish poked its head above water and said, "Since I can't get out of dying, let me at least make my testament."

The fisherman and his wife consented, and the fish said, "When I'm dead, cooked, and halved, let the woman eat my meat, the mare my broth, the dog my head, and plant the three biggest fishbones in the garden. Hang my gall bladder from a beam in the kitchen. You will have children; should any of them come to grief some day, blood will ooze from my gall bladder."

After killing and cooking the fish, the two people followed its instructions to the letter. Then it came to pass that the woman, the mare, and the dog all three gave birth on the same night. The dog had three puppies, the mare three colts, and the woman three baby boys. The fisherman said, "How about that! Nine creatures born in one night!" The triplets were so much alike that it was impossible to tell them apart without a different emblem around each one's neck. As for the fishbones planted in the garden, they sprouted into three splendid swords.

When the children became big boys, their father gave them each a horse, a dog, and a sword and, as a present from himself, a shotgun apiece. In no time the firstborn grew weary of living at home in poverty and decided to go out and seek his fortune. He mounted his horse, took up his dog, sword, and shotgun and bid everyone farewell. To his brothers, he added, "Should the gall bladder hanging from the beam ever ooze blood, come in search of me, for I'll either be dead or in serious trouble. Farewell." And away he galloped.

After riding for days and days through unfamiliar territory he came to the gate of a big city draped in mourning. He entered and found all the inhabitants grief-stricken and dressed in black. At an inn where he went for dinner, he asked the reason for all the black, and the innkeeper explained. "There's a dragon with seven heads who comes down to the bridge every day at noon. If he is not given a maiden to eat, he will enter the city and devour everyone in his path. Lots are drawn daily. Today it's the turn of the king's daughter, who must be on the bridge at noon for the dragon to devour. The king has posted a proclamation that the man who rescues her will wed her."

The youth said, "There must be some way to save the king's daughter and free the city from such a scourge. I have a powerful sword, dog, and horse, and would like to be taken to the king."

Led to His Majesty at once, the youth asked permission to confront and slay the dragon.

"Young man full of zeal," replied the king, "note that many men before you have tried and lost their life, poor wretches. But if you feel like risking your life and conquer the dragon, you will have my daughter in marriage and inherit the kingdom at my death."

Undaunted, the youth took his dog and horse and went to sit on the parapet of the bridge.

At the stroke of twelve, here came the king's daughter, dressed from head to toe in black silk, with her retinue. When they were halfway across the bridge, her attendants turned back in tears, leaving her there by herself. She looked around and saw a man sitting on the bridge with a dog.

"Noble sir," she said, "what are you doing here? Didn't you know that a dragon is coming any minute to devour me and that he will eat you too if he finds you here?"

"I'm well aware of that, and I've come to set you free and marry you."

"My poor man," answered the princess, "flee, or the dragon will have two souls to devour today instead of just me. He's a dragon full of wiles. How can you expect to slay him?"

From just looking at the princess, the youth had fallen in love with her, and he said, "For the sake of your love I will risk my life, and what will be, will be."

They had just finished this discussion, when the palace clock struck noon. The earth began quaking, a chasm yawned, and out sprang the dragon with seven heads amid smoke and flames. He made straight for the princess, his seven mouths open and whistling for joy, since he had noticed he would feast on two humans this day instead of one. In a flash the youth was on his horse and charging the dragon as well as sicking his dog on the monster. Brandishing his sword, he swept off six of the seven heads, one after the other. Then the dragon asked to rest awhile and the youth, who was also out of breath, said, "Let us both rest a moment."

But the dragon rubbed his one remaining head on the ground and came back up with the other six heads reattached. Seeing that, the youth realized he had to sweep off all seven heads at once. He therefore rushed upon the dragon, swinging around his sword with all his might until every single head was off and rolling on the ground. Then he took his sword and cut out the seven tongues, asking the king's daughter, "Do you have a handkerchief with you?"

The princess gave him her handkerchief, in which he wrapped the seven tongues. He mounted his horse again and rode to an inn to wash and dress for his visit to the king.

As luck would have it, in a hovel near the bridge lived a very sly and wicked coalman, who had witnessed the combat from afar. He thought to himself, Let's outsmart this ninny who leaves the dragon heads lying around and wastes time getting all spruced up. He gathered the amputated heads into a bag and ran to the king brandishing a huge knife smeared with the dragon's blood.

"Sacred Crown!" he exclaimed. "Here before you stands the dragon slayer, and these are his seven heads which I cut off one by one with the knife you see here. Therefore, Sacred Crown, keep your royal promise and give me your daughter's hand in marriage!"

The king was quite taken aback at the sight of that ugly and sinister face. He was not convinced of the truth of the man's story, suspecting strongly that the zealous youth had been devoured and that the coalman had shown up at the last minute, when the dragon was already done for, and dealt only the finishing stroke. In any event, the royal promise could not be altered, and the king was obliged to reply, "If that's how it really happened, then my daughter is yours, so take her."

At that, the princess, who had been in the audience hall listening to the conversation, began screaming that the coalman was a liar, that it wasn't he who had slain the dragon but the young man who would arrive any minute. A heated quarrel followed, but the coalman stuck by his story, producing the heads in the bag as evidence. The king could not dispute it and had no choice but to order his daughter to calm down and get ready to marry the coalman.

Right away the king ordered the announcement made public. Three days were devoted to festivities, with a grand banquet on each day, at the end of which period the wedding would be celebrated. In the meantime the real dragon slayer arrived at the royal palace. But the guards at the front door refused to admit him under any circumstances; in the same instant he heard the town crier going through the city squares announcing the forthcoming wedding of the princess and the coalman. The youth argued in vain to be taken to the king; the guards were not to be moved. Finally the coalman appeared and ordered the young man thrown out at once. The youth therefore had no choice but return to the inn, seething with rage, and think of a way to prevent that marriage, expose the coalman's lie, and establish himself as the slayer of the dragon.

At court the table was laid and all the nobility invited. Seated next to the princess was the coalman dressed in velvet; since he was short in stature, seven cushions were placed under him to make him look a little taller.

After racking his brains back at the inn, the young man woke his dog sleeping at his feet and said, "Listen, Faithful, run to the palace to the king's daughter, make a fuss over her alone, no one else, and when they are all ready to sit down and eat, upset the table and flee. But be careful not to get caught."

The dog, who understood everything his master said to him, ran off, found the princess, put his front paws in her lap, whined, and licked her hands and face. She recognized him and was quite glad to see him; stroking him, she whispered in his ear and asked where her rescuer was. But the coalman was suspicious of all those caresses and ordered the dog driven out of the banquet hall. They were just serving the soup, so the dog caught hold of a corner of the tablecloth and pulled it clean off the table with everything on it, thus littering the floor with broken dishes. Then he flew down the stairs so fast that no one could catch him or even see which way he went. The confusion of the guests was too much for words. The banquet had to be called off, which caused something of a scandal.

When the second banquet came up, the youth said to his dog, "Go back, Faithful, and do the same thing over." Seeing the dog back, the princess laughed for joy, but the coalman, fearful and suspicious, insisted that the dog be driven out with the whip. The princess, however, stood up for the dog, and the coalman, in spite of his meanness, dared not defy her. This time too, as soon as the soup was served, the dog grabbed hold of the tablecloth, pulled everything off onto the floor, and fled like lightning. Guards and servants tore after him, but he was out of sight before they came anywhere near him.

Just before the third banquet, the young man said, "Go back, Faithful, and do the same thing once more, but this time let them follow you home to me."

The dog did just what he was supposed to, and here came the guards on his heels right into the room of the young man, whom they seized and carried to the king. The king recognized him. "But aren't you the man who wanted to rescue my daughter from the dragon?"

"I certainly am, Majesty, and rescue her I did."

At those words, the coalman shouted, "It's not so! I killed the dragon myself with my own two hands. To prove it I brought along the seven heads!" He ordered the heads laid at the king's feet.

Without losing countenance, the youth turned to the king, saying, "Maybe he brought the seven heads. They were so heavy I brought only the tongues. Let's look in those seven mouths and see if there's a tongue in each one."

The seven tongues were missing. Then the youth pulled out of his pocket the handkerchief in which he had wrapped them and described the combat in detail. But the coalman refused to concede defeat, claiming the tongues would have to be put back in place to be sure they fit. Every time a tongue went in exactly right, he flung one of the cushions off his chair in anger; when they got to the seventh tongue he disappeared under the table and fled. But he was caught at once and hanged by order of the king in the town square.

Now in the highest of spirits, king, bride and guests sat down to feast and conclude the marriage. Then night fell and everyone went to bed. At dawn the youth rose, opened the window and, seeing a forest full of birds before him, felt the urge to go hunting. His wife begged him not to go, as the forest was enchanted and whoever entered it never came back home. But the more the youth heard, the more he was tempted by the danger, so he took horse, dog, sword, gun, and departed. He had already shot many birds, when a violent storm arose, with thunder, lightning, and rain by the barrels. Soaking wet, the young man who had already strayed in the darkness then enveloping the forest, spied a cave and took shelter in it. It was full of white marble statues in various postures, but the youth was too tired and wet to pay much attention to them. He raked up some dry wood and, with the aid of his flintlock, lit a small fire to dry out his clothes and cook the birds.

In a little while, an old woman entered the cave seeking shelter. She was drenched through and through, her teeth chattered, and she begged the youth to let her warm up at his fire.

"By all means, ma'am," he replied. "You can keep me company."

The old woman sat down and offered the youth salt for the roasted birds, bran for the horse, a bone for the dog, and grease to grease the sword. But the minute the youth, the horse, and the dog ate, and the sword was greased, they all froze into statues.

Waiting in vain for her husband to return, the princess gave him up for dead, and the griefstricken king ordered the city draped in mourning.

Meanwhile back at the fisherman's house, from the time the firstborn son had left, his father and brothers looked daily at the gall bladder hanging from the beam. One day they found the kitchen inundated with blood, which was pouring from the gall bladder. At that, the second-born son said, "My big brother is either dead, or something terrible has happened to him. I'm going out and look for him. Farewell." He mounted his horse, and with dog, sword, and shotgun, galloped off.

All along the way he stopped and asked people if they had seen his brother. "Have you seen a man who looks exactly like me?"

Everybody would laugh. "That's a fine joke! Aren't you the same one who rode through here some time ago?"

"So the youth realized his big brother too had come this way, and he continued in the same direction. He came to the royal city, and when the people dressed in black saw him, they marveled. "Here he is, here he is! He's not dead after all! Hurrah! Long live our prince!"

They led him before the king, and the whole court, including the princess, took him for the firstborn. The king scolded him at great length for going off, and the second-born, without seeming puzzled, apologized, making up with the princess as well. So cleverly did he handle questions and answers that he learned all about his brother, his marriage, and his disappearance.

That night on going to bed, the second-born took off his sword and placed it blade upward in the middle of the bed, telling the princess they would sleep one on one side, the other on the other. The princess didn't understand why, but they went to bed and fell asleep.

He too rose at dawn and right away opened the window. Seeing the forest before him, he said, "I shall go hunting there."

"Isn't one narrow escape enough for you?" replied the princess. "Must I suffer more anxious moments?"

Her words fell on deaf ears, and he left with horse, dog, sword, and gun. He met with the same fate as the firstborn and thus remained in the cave as a statue. Waiting in vain for him to return, the princess felt certain he was dead this time, and once more the city put on mourning by order of the king.

Back at the fisherman's meanwhile, the kitchen was newly flooded with blood trickling from the gall bladder. The third-born set out at once in search of his brothers, taking horse, dog, sword, shotgun, and galloping off. He too inquired along the way: "Did you see two young men who each looked exactly like me riding through here?"

"What a clown you are!" exclaimed the people he stopped. "Are you going to continue to come by asking the same thing?"

So the third-born knew he was on the right road and kept on until he reached the city, where he was as joyously welcomed as if he had just risen from the dead. He too was taken for the firstborn by king, princess, and court. Like his brother, he went to bed at night with the princess, putting the sword in the middle of the bed and sleeping on one side of it, with her on the other. Seeing the forest from the window in the morning, he announced, "I am going hunting."

Again the princess was thrown into a state of dismay. "Are you bent on going to your doom? Do you love me no better than that? Every time you go hunting I'm worried to death about you."

But the third-born was dying to be off in search of his brothers and left immediately. Taking shelter in the cave out of the storm, he examined the statues one by one and recognized his two brothers. "There's mischief here, for sure," he said to himself, "so I shall watch my step."

He had just lit the fire and put on the birds to roast when the old woman appeared and, bowing and scraping, asked to warm herself. But the youth scowled and said, "Out of my way, you ugly witch, I want you nowhere near me."

Appearing hurt by such a welcome, the old woman whimpered, "Have you no love for a fellow human being? I would still like to give you a few little things to improve your supper: salt for the roast birds, bran for the horse, a bone for the dog, and even grease to keep your weapons from rusting."

"Horrid old hag, you're not going to catch me too!" he cried, and pounced on her, throwing her to the ground and holding her down with his knee. Gripping her throat with his left hand, he unsheathed his sword with his right and pressed the point to her neck, snarling, "Awful old witch! Give me back my brothers, or I'll slit your throat this very instant!"

The old woman protested she had never harmed a soul, but menaced by the youth's sword touching her windpipe, she finally confessed her witchcraft and promised to obey him if he would spare her life. She immediately pulled a jar of salve from her pocket to restore the statues to life. The youth wasn't about to let her go, and with his sword against her back, forced her to daub the statues. So one by one all the statues turned back into living persons, and the cave was full of people. When the brothers saw one another they joyfully embraced, while all the other men were speechless with gratitude toward the third-born. In all the turmoil the witch was slipping away, when the brothers saw her and ran up and cut her to bits. Now the spell of the forest was completely broken, and the firstborn carefully pocketed the jar of salve which brought the dead back to life.

Returning to the royal city in a body, the men got to talking, and the three brothers told one another what had happened to them. At the news his brothers had slept with the princess, the firstborn was seized with jealous rage and unsheathed his sword and slew them.

No sooner had he committed this crime than he repented and pointed the sword at his own throat. The other noblemen restrained him, and then he remembered the jar of salve. He anointed his dead brothers' wounds, and up they stood again as hale and hearty as ever. Overjoyed, the firstborn begged their forgiveness, which they granted, mentioning the sword in the middle of the bed of which they'd not had a chance to speak earlier. The three of them continued on until they reached the palace of the king.

They called the princess, who had nearly cried her eyes out. Seeing the triplets, she couldn't for the life of her say which one was her husband. The firstborn then identified himself and introduced his two brothers. The king married them to two daughters of noblemen they had freed, named them courtiers, and even invited the old fisherman and his wife to the palace.

(*Montale Pistoiese*)

59

Bellinda and the Monster

Once upon a time in Leghorn there was a merchant who had three daughters: Assunta, Carolina, and Bellinda. He was rich, and had brought his girls up in the lap of luxury. They were all three beautiful, but the youngest was so bewitchingly lovely that they had given her the name of Bellinda. Not only was she beautiful, but also kind, modest, and wise—every bit as much as her sisters were haughty, stubborn, spiteful, and always full of envy to boot.

When the girls were older, the richest merchants in town went and proposed to them, but Assunta and Carolina scornfully dismissed them. "Never will we marry a merchant!"

Bellinda, however, always had a courteous reply for her suitors. "I can't marry just now, for I'm still too young. We'll speak further of the matter when I get older."

As the saying goes, life is full of surprises. The father lost a ship with its entire cargo, and in no time he was ruined. Of all his former possessions, the only thing left was a cottage in the country. The only choice he now had was to move there with his daughters and till the soil as a farmer. Just imagine the faces the two older girls made upon hearing that. "No, indeed, Father," they said, "we're not about to move to the country. We're staying right here in town. Certain gentlemen of consequence have proposed to us."

But just let them seek out the gentlemen now! On hearing that the young ladies were left without a cent to their name, the sometime suitors all stole away, saying, "It serves them right! That will teach them a lesson. Now they'll get off their high horse." But equal to the men's delight over Assunta and Carolina's misfortune was everyone's sympathy for poor Bellinda, who had never turned up her nose at anyone. Two or three youths even asked her to marry them just as she was, beautiful and penniless. She wouldn't hear of it, however, for her heart was set on helping her father, whom she couldn't think of abandoning now. As things stood, she was the one who rose early in the country, did the housework, got dinner for her sisters and her father. Her sisters, however, always rose at ten o'clock and didn't lift a hand all day long. They were forever out of sorts with Bellinda, and called her "country wench," for taking such a wretched life in her stride from the start.

One day the father got a letter saying that his ship, which he had given up for lost, had reached Leghorn with part of its cargo intact. The older sisters, imagining they'd be back in town in no time and rich again, went wild with joy. Their father said, "I'm going to Leghorn now to see about recovering what is due me. What shall I bring you as a present?"

Assunta said, "I want a beautiful silk gown the color of air."

Then Carolina said, "Bring me, instead, a peach-colored gown."

Bellinda, however, remained silent and asked for nothing. Her father repeated his question, and she said, "Now is no time to be spending so much money. Just bring me a rose, and I'll be happy." Her sisters poked fun at her, but she paid no attention.

The father went to Leghorn, but just as he was about to claim his cargo, up rushed other merchants to prove he owed them money and that these goods were therefore not his. After much wrangling, the poor old man was left empty-handed. But not wanting to disappoint his daughters, he drew out the little money remaining to him and bought the air-colored gown for Assunta and the peach-colored one for Carolina. Then he hadn't a cent left. The rose for Bellinda was such a little thing, he decided, that it really made no difference whether he bought it or not.

Thus he headed back to the country. He walked and walked until nightfall; entering a forest, he soon lost his way. To make matters worse, snow began falling and a strong wind arose. The merchant took refuge under a tree, expecting to be torn to bits any moment by the wolves whose howling came from all directions. While he stood there glancing around, he caught sight of a light in the distance. He made his way toward it and at length saw a handsome palace all lit up inside. The merchant went in, but not a soul was anywhere to be seen; no matter where he looked, there was absolutely no one. A fire burned brightly in the fireplace, and the merchant, who was soaking wet, paused to warm himself. Somebody will surely come in now, he thought. He waited and waited, but not a living soul appeared. The merchant saw a table laden with delicacies of every variety, so he sat down and dined. Then he took up the lamp and passed into another room, where a fine bed had been carefully made; after undressing, he climbed into it and went to sleep.

When he woke up next morning, he couldn't believe his eyes: there on the chair beside the bed lay a brand-new suit of clothes. He dressed, went downstairs and out into the garden. A magnificent rosebush was blooming in the middle of a flowerbed. The merchant remembered his daughter Bellinda's wish and decided he could now fulfill this one too. He selected the most beautiful rose and plucked it. At that moment a roar came from behind the rosebush, and in the midst of the roses appeared a monster, so ugly that the mere sight of it was enough to reduce a person to ashes. It exclaimed, "How dare you steal my roses after I've lodged you, fed you, and clothed you! You shall pay for that rose with your life!"

The poor merchant fell to his knees and explained that the flower had been intended for his daughter Bellinda, who wanted no present but a rose. Hearing the story, the monster calmed down and said, "If you have such a daughter, bring her to me. I will keep her here with me, and she will live like a queen. But if you don't send her, I will pursue you and your family wherever you happen to be."

Quaking in his boots, the little old man could hardly believe it when he was told he was free to go. But first the monster had him go back inside the palace and pick out all the jewels, gold objects, and brocades that captured his fancy. These things filled a chest, which the monster would send to the merchant's house.

As soon as the merchant got back to the country, his daughters ran out to meet him. Simpering, the two older girls asked him for their presents. Bellinda, though, was truly happy over his return and as gracious as ever. He gave one of the dresses to Assunta, the other to Carolina. Then he looked at Bellinda and burst into tears as he handed her the rose and told her exactly what had happened.

The older sisters were quick to speak out. "We said so! Bellinda and her crazy ideas! A rose, mind you! Now we'll all have to suffer the consequences!"

Calm as usual, Bellinda said to her father, "The monster promised to harm none of you if I go to him? In that case I'll go, since it's better for me to sacrifice myself than for all of us to suffer."

Her father said that never, never would he take her there, and her sisters insisted she was crazy. Bellinda, though, would hear no more. She put her foot down and declared she was going.

The following morning, then, father and daughter set out at dawn. Earlier, however, upon arising, the father had found at the foot of his bed the chest with all the treasures he had selected at the monster's palace. Making no mention of it to the two older girls, he hid it under the bed.

They arrived at the monster's palace in the evening and found it all lit up. They went inside. On the first floor was a table laid for two, full of heavenly delights. Although Bellinda and her father had little appetite for these things, they nevertheless sat down to taste a few dishes. When they had finished eating, a great roar was heard, and in came the monster. Bellinda was speechless: he was far uglier than she had dared imagine. But little by little she took heart, and when the monster asked if she'd come of her own will, she answered quite frankly that she had.

The monster seemed pleased. He turned to the father, handed him a traveling bag full of gold, and ordered him to leave the palace at once and never set foot there again; the monster would see to it that the family had everything they needed. Heartbroken, the poor father kissed his daughter goodbye and returned home, pitifully weeping.

Left by herself (since the monster had bid her good night right after her father's departure), Bellinda undressed and got into bed and slept peacefully the whole night long knowing she had saved her father from no telling what catastrophes.

Next morning she arose refreshed and confident, and decided to look around the palace. On the door of her room was written *Bellinda's Room.* On the door of her wardrobe was written *Bellinda's Wardrobe.* In each of the beautiful frocks was embroidered *Bellinda's Frock.* And all around were placards that read:

> Queen art thou here,
> Thy every wish to us is dear.

In the evening when Bellinda sat down to dine, the customary roar was heard, and in walked the monster. "May I join you?" he asked.

Naturally polite, Bellinda replied, "You are the master."

"No," he said, "you are in charge here. The whole palace and everything in it are yours." He was silent for a while, as though lost in thought. Then he asked, "Am I really so ugly?"

Bellinda answered, "Ugly you are, but you have a kind heart which makes you almost handsome."

Then he asked, all of a sudden, "Bellinda, would you marry me?"

She trembled all over, not knowing what to reply. She thought, If I turn him down, goodness knows how he will feel. Then she took heart and said, "To tell the truth, I'm not really interested in marrying you."

The monster made no comment, but bid her good night and went away sighing.

Three months passed. And every evening during that time, the monster came and asked Bellinda the same thing, if she would marry him, and then went away sighing. The girl was now so used to it that she would have been hurt if he had missed one evening.

Every day Bellinda strolled in the garden, and the monster told her about the magic of the plants. Among the trees was a leafy one known as the tree of weeping and laughter. "Whenever its leaves turn upward," explained the monster, "that means there's joy in your family; when they droop, there is weeping at home."

One day Bellinda noticed the tree of weeping and laughter with all its leaves pointed upward. She asked the monster, "Why is it so jubilant?"

"Your sister Assunta is going to get married."

"Could I go to the wedding?"

"Of course," answered the monster. "But come back in a week, or else you'll surely find me dead. Take this ring. Whenever the stone clouds up, that means I'm sick, and you must rush back to me at once. Now gather together whatever things in the palace you'd like to take along as wedding presents, and put them in a trunk this evening at the foot of your bed."

Bellinda thanked him and took a trunk and filled it with silk gowns, fine lingerie, jewels, and gold coins. She put the trunk at the foot of her bed and went to sleep. In the morning she woke up in her father's house, and there with her was the trunk she had packed the night before. Everybody gave her a hearty welcome, even her sisters. But when they learned she was so happy and rich and the monster so kind, they were again green with envy, since they were far from wealthy themselves, in spite of the monster's presents; and to make matters worse, Assunta was marrying a mere carpenter. As spiteful as ever, they got Bellinda's ring away from her under the pretext of wearing it themselves a little while; then they hid it. Bellinda was quite upset over not being able to see the stone, and at the end of a week she wept and pleaded so with her sisters that her father ordered them to return the ring at once. As soon as she got it back she noticed the stone had become somewhat cloudy, so she left immediately for the palace.

The monster failed to appear at mealtime, and Bellinda grew worried; she looked all over for him and called and called. Only at dinner did he turn up, with a somewhat pained expression. "I was ill," he said, "and if you'd come any later, you wouldn't have found me alive. Don't you love me any more?"

"Of course I love you," she replied.

"And you would marry me?"

"That, no!" exclaimed Bellinda.

Two more months went by and the leaves again pointed upward on the tree of weeping and laughter, since this time Carolina was getting married. Bellinda went home once more with the ring and another trunk of treasures. Her sisters pretended they were glad to see her. Assunta was now meaner than ever, since her carpenter husband beat her every day. Bellinda told her sisters what a risk she had run by staying too long on her last visit, and said she couldn't tarry this time. But once more the sisters stole the ring. When they finally returned it, the stone had completely clouded over. Bellinda rushed home in alarm, but the monster showed up for neither lunch nor dinner. He came in next morning looking quite weak, and said, "I was ready to die. If you are late another time, it will be the end of me."

A few more months went by. One day, the leaves of the tree of weeping and laughter were drooping, and the tree appeared completely withered. "What's the matter at home?" Bellinda cried.

"Your father is dying," answered the monster.

"Let me go to him! I promise I'll come back on time!"

The joy of having his youngest daughter at his bedside put the poor merchant on the road to recovery. Bellinda stayed by him day and night, but one day while washing her hands she left the ring lying on the washstand and then couldn't find it when she went to put it back on. Frantic, she looked everywhere for it, and pleaded with her sisters to return it. When she finally recovered it, the stone was all black, except for a tiny dot on the edge.

She hastened to the palace, but it was pitch-dark and looked as though it had been vacant for the last hundred years. Screaming and crying, she called and called the monster, but there was no answer. She looked everywhere for him; as she was running through the garden she suddenly saw him lying under the rosebush and breathing what seemed to be his last. She got down on her knees and listened to his heart: it was still beating but very feebly. Then she kissed him and sobbed, "Monster, if you die, I'll be lost without you! If only . . . if only you could go on living, I'd marry you at once to make you happy!"

She had not finished speaking, when all at once the whole palace lit up and music and song poured from every window. Bellinda turned around, amazed. When she faced the rosebush again, the monster had vanished, and in his place, among the roses, stood a handsome knight. He bowed and said, "Thank you, dear Bellinda, for freeing me."

Bellinda was dumbfounded. "But I want the monster," she said.

The knight knelt at her feet and said, "Here is the monster. I was under a spell and obliged to remain a monster until a beautiful maiden promised to marry me the way I was."

Bellinda gave her hand to the youth, who was a king, and together they walked to the palace. At the door stood her father, who embraced her, and her two sisters. The sisters, out of spite, remained outside and became statues on each side of the door.

The young king made Bellinda his wife and queen, and they lived happily ever afterward.

(Montale Pistoiese)

60

The Shepherd at Court

A boy was tending the flock, when a lamb fell into a ravine and perished. The shepherd went home, and his parents, who had little love for him to begin with, screamed at him and beat him, then turned him out of the house into the night. Weeping, he wandered about over the mountain and found a hollow rock, which he lined with dry leaves and nestled in the best he could, stiff from the cold air. But he was unable to sleep.

Through the darkness, a man made his way to the rock and said, "You had the nerve to take my bed! What are you doing here at this time of night?"

Shaking with fright, the boy told how he had been turned out of the house, and begged the man to let him stay there the rest of the night.

The man said, "You were very clever to bring in dry leaves. The idea never occurred to me. Go on and stay here." And he lay down beside him.

The lad made himself as small as possible so as not to disturb him, keeping perfectly still to give the impression he was sleeping; but he couldn't shut his eyes for watching the man. Nor was the man sleeping, but mumbling to himself under the illusion the boy was asleep. "What present can I make this boy who lined the stone for me with leaves and who's thoughtful enough to stay on his side and not disturb me? I can give him a linen napkin which, unfolded, produces dinner for everybody present. I can give him a little box which, opened, produces a gold coin. I can give him a harmonica which, played, sets everyone within earshot to dancing."

This mumbling slowly put the boy to sleep. He awakened at dawn, thinking he had been dreaming. But there beside him on the bed of leaves lay the napkin, the little box, and the harmonica. The man was gone, and the boy had not even seen his face.

After walking some distance he came to a crowded city that was getting ready for a big tournament. The king of that city had staked his daughter's hand, together with the entire treasure of the state. The lad thought, Now I can test the little box. If it gives me the money needed, I too can line up to joust. He began opening and closing the box and, every time, it produced a shiny new gold piece. He took all the money and purchased horses, armor, princely clothes, engaged squires and servants, and passed himself off as the son of the king of Portugal. He won every match, and the king was bound to declare him his daughter's bridegroom.

But at court, the lad, having been raised with sheep, was as uncouth as could be: all his food he picked up in his hands, then wiped them on the curtains, and he was constantly slapping the ladies on the back. The king became suspicious. He dispatched ambassadors to Portugal and found out that the king's son, having dropsy, had never set foot outside the palace. So he ordered the lying lad imprisoned at once.

The palace prison was right under the banquet hall. When the boy walked in, the nineteen prisoners already there greeted him with a chorus of jeers, knowing he'd had the impudence to become the king's son-in-law. He let them jeer all they liked. At noon, the jailer brought the prisoners the usual pot of beans. The lad rushed up and kicked the pot over on the floor.

"Have you lost your mind? What will we now eat? You'll pay for this!"

"Shhhhhhh! Just wait," he replied. Pulling the napkin out of his pocket, he said, "For twenty," and unfolded it. Dinner for twenty appeared, including soup, many tasty dishes, and excellent wine. At that, they all hailed the lad as a hero.

Every day the jailer found the pot of beans overturned on the floor and the prisoners better fed and livelier than ever. So he went and told the king. Curious, the king went down into the prison and asked for an explanation. The lad stepped forward. "Listen, Majesty, I am the one providing my companions with food and drink far better than what's on the royal table. So if you'll accept, I invite you to dine with us and promise you'll go away happy."

"I accept," said the king.

The lad unfolded the napkin and said, "For twenty-one, and fit for a king." Out came the most wonderful dinner you ever saw and the king, delighted with the sight, took a seat in the midst of the prisoners and ate and ate.

When dinner was over, the king said, "Will you sell me the napkin?"

"Why not, Majesty? But on condition you let me sleep one whole night with your daughter, my rightful betrothed."

"Why not, prisoner?" replied the king. "But on condition you keep perfectly still and quiet on the edge of the bed, with the windows open, a lamp lit, and eight guards in the room. If that suits you, well and good. Otherwise you get nothing at all."

"Why not, Majesty? That's settled."

So the king got the napkin, and the boy slept an entire night with the princess, but with no possibility of talking to her or touching her. And in the morning he was taken back to prison.

Seeing him back, the prisoners all raised their voices in mockery.

"Hey, stupid! What a blockhead you are! Now we'll be back on our daily beans! A fine bargain you made with the king!"

But the lad didn't lose countenance. "Why can't we buy our dinner from now on with perfectly good money?"

"Who has any of that?"

"Take heart," he said, and started pulling gold pieces out of his purse. So they had grand dinners sent in from the inn next door, and continued to kick over the pot of beans on the floor.

The jailer went to the king again, and the king came down to investigate. As soon as he found out about the box, he asked, "Will you sell it to me?"

"Why not, Majesty?" he replied, making the same bargain as before. He gave the king the box, and slept with the princess another time without being able to touch her or talk to her.

Seeing him back, the prisoners resumed their taunts. "Well, here we are on beans once more, hurrah!"

"Joy is a good thing indeed. Whether we eat or not, we will dance."

"What!"

The lad pulled out the harmonica and began to play. The prisoners started dancing around him, with their ankle-chains clanking loudly. They broke into minuets, gavottes, and waltzes, and couldn't stop. The jailer rushed in, and he too started dancing, with all his keys jingling at his side.

In the meantime the king had just sat down to a banquet with his guests. Hearing the notes of the harmonica float up from the prison, they all jumped to their feet and began dancing. They looked like so many bewitched souls, and nobody knew what was going on: the ladies danced with the butlers, and the gentlemen with the cooks. Even the furniture danced. The crockery and crystal were smashed to smithereens; the roasted chickens flew off; and people butted the walls and ceiling beams. The king himself danced while yelling for everyone to stop. All of a sudden the lad stopped playing, and everyone fell to the floor at once, with heads spinning and legs collapsing.

Out of breath, the king went down to the prison. "Just who is being so funny?" he began.

"It's me, Majesty," answered the lad, stepping forward. "Would you like to see?" He blew a note, and the king took a dance step.

"Stop! Stop this instant!" he said, frightened, then asked, "Will you sell it to me?"

"Why not, Majesty? But under what conditions this time?"

"The same as before."

"Well, Majesty, here we're going to have to make a new bargain, or I'll play more music."

"No, no, please! Tell me your terms."

"Tonight I'll be satisfied with talking to the princess and having her answer me."

The king thought it over and ended up agreeing. "But I'm doubling the number of guards, and there'll be two lamps lit."

"As you like."

Then the king called his daughter to him in secret and said to her, "Listen carefully: you are to say no, and only no, to every question which that rascal asks you tonight." The princess promised she would.

Night fell, and the lad went to the bedchamber—which was brightly lit and full of guards—and stretched out on the edge of the bed at some distance from the princess. Then he said, "My bride, do you think that in this chilly night air we ought to keep the windows open?"

"No."

"Did you hear that, guards?" cried the lad. "By express orders of the princess, the windows are to be closed." The guards obeyed.

A quarter of an hour passed, and the lad said, "My bride, do you think it is quite right for us to be in bed and have all these guards around us?"

"No."

"Guards!" cried the lad. "Did you hear? By express orders of the princess, be gone and don't show your faces here any more." So the guards went off to bed, which struck them as almost too good to be true.

Letting another quarter of an hour pass, he said, "My bride, do you think it right to be in bed with two lamps lit?"

"No."

So he put out the lamps, making the room pitch-dark.

He came back and took his place on the edge of the bed, then said, "Dear, we are lawfully married, and yet we are as far apart as if we had a thornbush hedge between us. Do you like that?"

"No."

At that, he took her in his arms and kissed her.

When day dawned and the king appeared in his daughter's room, she said to him, "I obeyed your orders. Let bygones be bygones. This young man is my lawful husband. Pardon us."

Having no alternative, the king ordered sumptuous wedding festivities, balls, and tournaments. The lad became the king's son-in-law and then king himself, and there you have the tale of a shepherd boy lucky enough to plop down on a royal throne for life.

(Montale Pistoiese)

61

The Sleeping Queen

Spain was once ruled by the good and just King Maximilian. He had three sons: William, John, and little Andrew—the youngest and his father's favorite. Following an illness, the king lost his eyesight. Though all the doctors in the kingdom were summoned, none knew of any remedy. One of the oldest doctors suggested, "Since medical knowledge is limited in this case, send for a soothsayer." So, soothsayers from everywhere were called in. They pored over their books, but proved no wiser than the doctors in the end. With the soothsayers a wizard had slipped in, a stranger to everyone. After the others had all had their say, the wizard came forward and spoke. "I am familiar with cases of blindness like yours, King Maximilian. The cure is nowhere to be found but in the Sleeping Queen's city: it is the water in her well." People's amazement at those words had not yet died down before the wizard vanished and was never heard of again.

The king was eager to find out who he was, but no one had ever laid eyes on the man before. One of the soothsayers thought he might be a wizard from the vicinity of Armenia, come to Spain by means of magic. The king asked, "Could the Sleeping Queen's city also be thereabouts?" An old courtier replied, "We won't know where it is until we look for it. If I were younger, I would go in search of it myself, without delay."

William, the eldest son, stepped forward. "If anyone is to set out in search of the city, I am the one to go. It is only fitting that the firstborn put his father's health above all other concerns."

"Dear son," replied the king, "you have my blessing. Take money and horses and everything else you need. I will be expecting you back victorious in three months."

William went to the kingdom's port and boarded a vessel sailing for the Isle of Buda, where it was to anchor for three hours before continuing on to Armenia. At Buda he went ashore to see the island. As he strolled about, he met a charming lady and became so engrossed in talking to her that the three hours went by before he knew it. At the appointed time the ship unfurled its sails and departed without William. He was sorry at first, but the lady's company made him soon forget all about his father's illness and the original purpose of the voyage.

When the three months were up, with still no sign of William, the king began fearing the boy was dead, and to the pain of going blind was added that of losing a son. To console him, John, the middle boy, volunteered to go in search of his brother as well as the water. The king consented, although fearful that something would happen to this son too.

On the boat, John soon came in sight of the Isle of Buda. This time the ship was to anchor there for a day. John went ashore to look around the island. He strolled into a park of myrtles, cypresses, and laurels, which shaded lagoons of limpid water stocked with fish of every color of the rainbow. From there he proceeded through the town's beautiful avenues and streets to a square with a white marble fountain in its center. Encircling the square were monuments and buildings, and in their midst rose a majestic palace with gold and silver columns and crystal walls that sparkled in the sunlight. John spied his brother moving about on the other side of those crystal walls.

"William!" he cried. "What are you doing here? Why did you not come home? We thought you were dead!" And they embraced.

William told how, once he'd set foot on the island, he'd been unable to tear himself away and how he'd been received by the beautiful lady who owned everything in sight. "This lady is Lugistella," he added, "and she has a very lovely little sister named Isabel. If you like her, she is yours."

In short, the twelve hours went by, and the ship sailed without John. After a brief spell of remorse, he too forgot all about his father and the miraculous water and became a guest, like his brother, in the crystal palace.

When the three months were up, with no sign of the second son, King Maximilian was alarmed and, with the entire court, feared the worst. Then little Andrew boldly declared he would go in search of his brothers and the Sleeping Queen's magic water. "So you intend to leave me too?" said the king. "Blind and crushed as I am, I must give up my last son as well?" But Andrew revived his hope of seeing the three boys back safe and sound in addition to obtaining the wonderful water, so his father consented at last to his departure.

The ship dropped anchor at the Isle of Buda, where it would remain for two days. "You may disembark," the captain told Andrew, "but be back on time if you don't want to be left behind like two other young men who went ashore and have not been heard of since." Andrew realized he was talking of his brothers, who must be somewhere on the island. So he began looking around and found them in the crystal palace. They embraced, and the brothers told Andrew about the spell that kept them on Buda. "We are in a real paradise here," they told him. "We each have a beautiful lady. The mistress of the island is mine, her sister is John's. If you'll join us, I believe our ladies still have a cousin . . ."

But Andrew cut them off. "You've obviously lost your mind if you don't remember your duty to Father! I intend to find the Sleeping

Queen's magic water, and nothing can turn me from that resolution—neither riches, nor amusements, nor beautiful ladies!"

At those words, the brothers became silent and walked away in a huff. Andrew returned at once to the ship. The sails were unfurled, and favorable winds carried the vessel straight to Armenia.

As soon as he was on Armenian soil, Andrew asked everyone he met where the Sleeping Queen's city was, but apparently no one had ever heard of it. After weeks of vain search, he was directed to an old man living on a mountaintop. "He's an old, old man, as old as the world itself, by the name of Farfanello. If he doesn't know where this city is, nobody knows."

Andrew climbed the mountain. He found the bearded, decrepit old man in his hut and told him what he was seeking. "Dear youth," said Farfanello, "I have indeed heard of this place, but it is quite far away. First you have to cross an ocean, and that will take almost a month, to say nothing of the perils of sailing those waters. But even if you do get across them safe and sound, still greater dangers lie in store for you on the Sleeping Queen's isle, the very name of which suggests misfortune, since they call it the Isle of Tears."

Glad to have definite information at last, Andrew embarked at the port of Brindisse. The ocean crossing was hazardous because huge polar bears, capable of wrecking even big ships, swam in those waters. But Andrew, a courageous hunter, was not afraid, and the vessel steered clear of the polar bears' claws and arrived at the Isle of Tears. The port looked abandoned, and not a sound was to be heard. Andrew disembarked and saw a sentinel holding a gun, but the man was completely motionless. Even though Andrew asked him for directions, he continued to stand as still and silent as a statue. Next, Andrew approached the porters about his baggage, but they didn't move a muscle; some held heavy trunks on their backs, with one foot forward and raised. Andrew entered the city. On one side of the street he saw a cobbler, still and silent, halted in the midst of drawing thread through a shoe. On the other side of the street a coffeehouse keeper held a pot in position to pour a lady a cup of coffee, but both he and she were mute and motionless. Streets, windows, and shops were full of people, but they all looked like figures of wax in the strangest of postures. Even the horses, dogs, cats, and other creatures were standing dead still in their tracks. Moving through this thick silence, Andrew came to a splendid palace adorned with statues and tablets commemorating the island's past kings; on the façade was a frieze full of figures, with an inscription in radiant letters of gold: TO HER MAJESTY THE QUEEN OF LUMINOUS SOULS, WHO REIGNS OVER THIS ISLE OF PARIMUS.

"Where could this queen be?" wondered Andrew. "Could she be one and the same as the Sleeping Queen?" He went up a grand alabaster staircase and crossed several halls decorated with bas-reliefs and guarded at the doors by the customary men of arms, over whom a spell had been cast. In one hall marble steps led up to a dais, on which stood the throne surmounted by a canopy and displaying a diamond-studded coat-of-arms. A grapevine growing in a gold pot had trailed clear across the room and twined around the throne and canopy, adorning them with clusters of ripe grapes and vine leaves. That wasn't all: fruit trees of every kind in the garden had grown quite out of bounds, thrusting their branches through the windows into the hall. Andrew, who was hungry after so much walking, pulled an apple from one of those branches and bit into it. He'd no sooner done so than his eyesight dimmed, then left him altogether. "Woe is me!" he cried. "How will I now get about in this strange country peopled with nothing but statues?" He began groping his way out of the palace; but moving along, he suddenly stepped into a hole and plummeted through empty space, landing in water over his head. With a few strokes he came to the surface, and the minute his head was out of water, he realized he had regained his eyesight. He was at the bottom of a deep well, and high above him was the sky. "So this is the well," he said to himself, "the wizard was referring to. This is the water that will cure my father, if I ever manage to get out of here and carry some back to him." He spied a rope hanging in the well, took hold of it, and climbed out.

It was nighttime, so Andrew looked about for a bed to sleep in. He found a bedchamber royally decorated annd containing a large bed, in which a maiden of angelic beauty lay sleeping. The maiden's eyes were closed and her face was peaceful, so Andrew knew she had been put under a spell while she slept. After a little reflection, he undressed and slipped into bed beside her passing a delightful night without her giving any sign she knew he was there. At daybreak he jumped out of bed and wrote her a note, which he left on her bedside table: "To his great joy, Andrew, son of King Maximilian of Spain, slept in this bed on the 21st of March, in the year 203." He filled a bottle with the water that restored eyesight and plucked one of the apples that caused blindness, and set out for home.

The ship again called at the port of Buda, where Andrew stopped to visit his brothers. He told them of the wonders of the Isle of Tears and the night he had spent with the lovely maiden. Then he showed them the apple which took away one's eyesight and the water which restored it. Possessed with sudden envy, the two brothers hatched a plot against Andrew. They stole the bottle of magic water, leaving in its place one exactly like it. Then they informed him they would accompany him home in order to present their wives to their father.

No words can describe the joy of King Maximilian at the safe return of all three of his sons to Spain. After many hearty embraces, the king asked, "Which one of you was the luckiest?" William and John held their tongues, but Andrew spoke up. "Father, I make bold to say I was, for I found and brought back my lost brothers. I reached the Sleeping Queen's city and got the water that will restore your eyesight. I also got something else amazing, and I'm going to show you right this minute how it works."

He pulled out the apple and handed it to his mother to eat. The queen bit into it, went blind on the instant, and let out a scream. "Don't get upset, Mamma," said Andrew, taking out the bottle, "for a drop of this water will restore your sight and also that of Papa, who's been blind for so long."

But the water came from the bottle substituted by the older brothers, so she did not regain her sight. The queen wept, the king raged, and Andrew trembled in his boots. Then the two brothers came forward and said, "This has happened because he didn't find the Sleeping Queen's magic water. We found it ourselves, and here it is." Once the stolen water had touched their eyes, the two old people could see again as well as ever.

A big row followed. Andrew called his brothers thieves and traitors, and they turned around and called him a little liar. The king could make neither head nor tail of the dispute, but finally he sided with William and John and their wives, and said to Andrew, "Silence, you shameless wretch! You not only had no intention of curing me, but you meant to blind your mother as well! Guards, away with this ungrateful creature! Take him to the woods and slay him! And bring me back his heart, or more heads will roll!"

The soldiers dragged off Andrew, screaming and protesting, to a thicket outside the city. But Andrew managed to tell his story and convince them. So as to avoid staining their hands with innocent blood, the soldiers made him promise never to come back to town, then set him free. They returned to the king with the heart of a pig purchased from a farmer and slaughtered on the spot.

On the Isle of Tears nine months went by, and the sleeping maiden gave birth to a fine baby boy, as she brought him forth, she awakened. With the queen now awake, the spell was broken which Morgan le Fay had cast over her out of envy, and the whole city awakened and came back to life. The soldiers frozen at attention relaxed, those at ease jumped to attention, the cobbler finished drawing the thread through the shoe, the coffeehouse keeper overfilled the lady's cup, and the porters at the port shifted their loads to the other shoulder, since the first shoulder was a bit weary by now.

The queen rubbed her eyes and said, "I wonder who on earth was so bold as to make his way to the island and sleep in this room and thereby free me and my dear subjects from the spell we were under."

One of her maids of honor then handed her the note from the night table, so the queen knew her visitor had been Andrew, son of King Maximilian. Right away she wrote the king to send Andrew to her without delay, or else she would make war on Spain.

When King Maximilian got that letter, he called in William and John to read it and give their opinion. Neither one of them knew what to say. At last, William spoke. "We'll never know what this is all about unless somebody goes to the queen for an explanation. I'll go myself and see what I can find out."

William's trip was easier, since Morgan le Fay's spell had been broken and all the polar bears had disappeared. He went before the queen, saying he was Prince Andrew.

The queen, who was naturally distrustful, began to question him. "What day did you come here the first time? How did you find the city? Where was I? What happened to you in the palace? What do you see now that you didn't see before?" And on and on. William soon got all confused and started stammering, so the queen knew right away he was lying. She had his head chopped off and stuck on a spike atop the city gate, with the inscription: IF YOU LIE, THIS IS HOW YOU WILL DIE.

King Maximilian got a second letter from the sometime Sleeping Queen saying if he didn't send Andrew to her, troops were ready to move against him and reduce his kingdom, his people, his family, and himself to ashes. Long regretful of having ordered Andrew slain, the king wailed to John, "Now what do we do? How will we tell her that Andrew is dead? And why doesn't William come home?" John volunteered to go to the sometime Sleeping Queen himself. He reached the island, but the sight of William's head on the city gate told him all he needed to know, and he returned home at full speed. "Father!" he exclaimed, "we are done for! William is dead, and his head is on a spike atop the city gate. If I had gone in, there would have been another head next to it."

The king was beside himself with grief. "William dead? Also William! Now I know for sure Andrew was innocent, and all this has happened to punish me. Tell me the truth, John; confess your treachery before I die."

"Our wives are to blame!" said John. "We never went to the Sleeping

Queen ourselves, and we put a bottle of ordinary water in the place of Andrew's magic liquid."

Railing, weeping, and pulling out his hair, the king summoned the soldiers to take him to the spot where Andrew was buried. Among the soldiers this order caused great alarm. The king noticed it and was filled with new hope. "Out with it! I want the truth. Whatever it is you're guilty of, I give you my royal word that you are pardoned."

Trembling in their boots, the soldiers admitted that they had flatly disobeyed the order to slay Andrew. To their great surprise, the king began madly hugging and kissing them. Posted at every street corner was an announcement that whoever found Andrew would be richly rewarded for the rest of his life.

Andrew returned, to the infinite joy of his old father and the court, and set out at once for the Isle of Tears where he was given a hero's welcome.

"Andrew, who freed me and my people," said the queen, "you will be my husband and king forever!" For months afterward, all you heard on the island were songs of joy, so they called it the Isle of Happiness.

(Montale Pistoiese)

◈ 62 ◈

The Son of the Merchant from Milan

There was once a Milan merchant who had a wife and two sons. Of the two he favored the older, who was now big enough to help in the business. It wasn't that the merchant disliked the younger boy, but he still looked on him as a child and paid him little mind. The merchant had grown rich and now entered into only those transactions yielding immense profit. That was why he was going to France for the manufacture of certain goods, an undertaking he figured would be very, very profitable. The elder son was to accompany him as a matter of course, but his little brother, Menichino, begged to go with them. "Let me come too, Papa. I promise I'll be a good boy. I'll even be a big help to you. I won't stay here in Milan by myself." The merchant, though, didn't want to be bothered and threatened to slap Menichino if he didn't keep quiet.

The hour of departure arrived. The merchant and his son had their trunks brought out, and climbed into the carriage. It was nighttime, and what with the darkness and the commotion of getting away, the postilions failed to notice Menichino crouching on the footboard at the back of the carriage.

When the carriage stopped at the first relay station to change horses, the little boy jumped off so as not to be seen. He waited until the vehicle was again in motion before regaining his place on the footboard. By the time they reached the second post house, it was already daylight, so Menichino ran off and hid at a bend in the road until the carriage should come by and he could jump back on. But it went by so fast he failed to leap on in time and remained stranded there in the middle of the road.

Finding himself in a strange place, alone, penniless, and hungry, the lad felt like crying. But he took heart and proceeded to look around. Seated by the roadside was an old woman. "Where are you going all by yourself?" she asked. "Are you lost?"

"I am indeed, madam," said Menichino. "I was traveling with my father and my big brother, and the carriage took off from a relay station without me, so here I am, and I don't even know the way home to my mother. But there's nothing for me to do at home anyway. I really prefer to go out in the world and seek my fortune, since my father has left me stranded in the road."

He was thoughtful a while, then added, "Well, the truth of the matter is, he didn't know I was with him; I hid on the back footboard, so I could go to France too."

The old woman said, "Good for you! You have been truthful, and it's a good thing, for I am a fairy and knew the whole story anyway. If you are seeking your fortune, I'll tell you where to find it, if you are clever and respectful."

"I am young, I admit," answered Menichino, "but I don't believe I'm dumb for a boy of fourteen. So you can be sure, madam, I'll do everything you tell me, if you're so kind as to help me."

"Good boy!" replied the fairy. "Listen to this: the king of Portugal has a gifted daughter who can solve any riddle. The king has promised her in marriage to the man who gives her a riddle she can't solve. You're a bright boy, so find a riddle, and your fortune is made."

Menichino said, "Oh, yes, but how do you expect me to compose a riddle that baffles a young lady so brilliant. Only learned people could do that, not ignorant boys like me."

"Oh," said the old woman, "I'm only putting you on the right track. You'll manage the rest by yourself. I'm making you a present of this dog.

His name is Bello, and your riddle will originate with him. Take him and go in peace."

"Very well, madam, if you say so, I believe it. And thank you very much. You have been very kind, and that is what counts." He bid her farewell and left, but he set little store by what she had told him. Still, he took the dog by the leash and continued on his way.

Toward evening he came to a farmhouse and asked for food and shelter for the night. The woman who answered the door inquired, "What are you doing out at night by yourself with just a dog for company? Have you no mamma and papa?"

Menichino replied, "I wanted to go to France, so I hid at the back of the carriage, and Papa took off without me. Now I'm going to the daughter of the king of Portugal with a riddle, and this dog, a present from a fairy, will furnish me the riddle, and that way I'll get to marry the king's daughter."

The woman, a wicked soul, thought to herself, If this dog gives out riddles, I could steal him and send my own son to the princess. So she decided to kill Menichino.

She prepared a poisonous pastry for him and said, "We call this pizza, and I have prepared it for you. However, I can't put you up for the night here, since my husband forbids me to let strangers in. But you can go and sleep in a cabin of ours, which you will see as soon as you enter the woods. Take this pizza along and eat it in the cabin. I'll come and awaken you tomorrow morning and bring you some milk."

Menichino thanked her and went off to the cabin. But the dog was hungrier than the boy, and jumped up for the pizza, so Menichino broke off a piece and threw it to him. Bello caught it, and no sooner had he swallowed it than he started trembling, rolled over on his back with his paws up in the air, and died. Menichino looked on, open-mouthed, then threw away the rest of the pizza. At length, he gave a sudden start and exclaimed, "But here is the beginning of the riddle:

> 'Pizza slays Bello
> While Bello saves Menichino . . .

All I have to do now is find the rest of it!"

At that very instant, three crows flying overhead and spying the dead dog swooped down and started pecking on the carcass. A minute later, there lay three dead crows.

"There's the next line," said Menichino:

> " 'One dead soul slays three . . .' "

He tied the three crows together with the dog leash and slung them over his shoulder. Suddenly out of the woods rushed a band of armed robbers, gaunt from hunger. "What do you have there?" they asked. Menichino, who had only those three crows, was not frightened. "Three birds to be roasted," he answered.

"Hand them over," commanded the robbers, and off they went with the catch. Menichino hid in a tree to see what would happen. The robbers roasted the crows, and all six of them died.

At that Menichino added another line to his riddle:

> " 'Pizza slays Bello
> While Bello saves Menichino.
> One dead soul slays three,
> And three then kill six.' "

But seeing the robbers eat reminded him of his own hunger, and there was the spit all ready for roasting game. He took the gun of one of the dead robbers and fired on a bird in the tree. It happened to be a bird on a nest, and instead of hitting the bird the bullet struck the nest, which fell to the ground. Out of the broken eggs hopped a few baby birds with still no feathers. He placed them on the same hearth where the crows had roasted, and to light the fire, he tore out the pages of a book found among the robbers' spoils. Then he climbed back up the tree and fell asleep. He now had the complete riddle in his head.

When he reached Portugal's royal city, he hurried at once to the princess, dirty and ragged as he was from his long trip. The princess burst out laughing. "The impudence of this ragamuffin even to think of outwitting me and becoming my husband!"

"Wait until you've heard the riddle, Princess, before judging me," said Menichino. "Your royal father's decree applies to all men alike and makes no distinction among them."

"That's true enough and well put," said the princess. "However, there's still time for you to withdraw if you like and avoid a beating."

Menichino thought for a moment, then remembered the fairy's words and took heart. "My riddle goes like this," he said.

> " 'Pizza slays Bello
> While Bello saves Menichino.
> One dead soul slays three,
> And three then kill six.
> I fired on what I saw,
> But hit what I saw not.
> Flesh unborn was my feast,

Cooked with words and words and words.
Neither on earth nor in heaven have I slept,
So guess now, guess I pray thee, my queenlet.' "

As soon as Menichino had finished, the princess exclaimed, "Of course, that one is very easy! Pizza is one of your brothers or friends, and he has slain Bello your enemy. Bello saves you by dying, since that way he can't harm you any more. So far, so good? But before dying, Bello killed three others . . . uh . . . and those three, in turn . . . let's see . . ."

She propped her elbows on her knees, then her chin on her hands. Next she began scratching the nape of her neck and assuming some very unbecoming positions for a princess, in her effort to concentrate. "Flesh unborn. . . . Really! Cooked with words. . . . That means . . . If I could just unravel all this . . ." Finally she gave up. "You've got me. It's an impossible riddle. You'll have to explain it to me."

So Menichino told his story from start to finish and asked her to keep the royal promise. The princess replied, "You are right, I can't refuse. But I have no desire to be your bride. If you would come to some other agreement with my father, I would be very happy."

Menichino said, "It all depends. If the new terms suit me, I'll agree to them. But remember, I'm out in the world to better my lot, and if I can't marry the princess, then I must have some other prize of equal value."

"Don't worry," said the princess. "You'll have far more. You'll be a millionaire and in a position to realize your every wish. Just compare that with marrying a princess who wants no part of you and who would always be unhappy and irritated with you. Can you guess what I am giving you? The secret of the Sorcerer of Flower Mountain. With that secret in your possession, all your worries are gone forever."

"And where is that famous secret?"

"You must go and pick it up in person, from the Sorcerer of Flower Mountain. Go in my name, and he will give it to you."

Menichino wondered whether he should give up the certain for the uncertain, but the thought of being the princess's husband was more frightening than pleasant, so he asked for directions to Flower Mountain.

Flower Mountain was a great big mountain all but impossible to climb, and Menichino wore himself out getting to the top, where there was an immense castle surrounded by gardens. How they had ever managed to build it way up there on all those rocks was a complete mystery. Menichino knocked, and was met at the door by several gigantic beings who were neither men nor women and hideous enough to frighten Fright herself. Expecting still worse to come, Menichino looked at them calmly and asked to be taken before the Sorcerer. The Sorcerer's steward, a monstrous giant, stepped forward and said, "Boy, you don't lack courage, that's for sure. But you'd better not try to meet my master, since he has a weakness for good Christian folk and eats them alive and raw."

Menichino answered, "Be that as it may, I still have to speak with the Sorcerer in person. Please be so kind as to announce me."

The Sorcerer was still lounging on rich carpets and cushions, and learning that a boy was there to see him, he thought to himself, Here is a Christian morsel, good and fresh, for breakfast! In walked Menichino, and the Sorcerer asked, "Who are you? What do you want?"

"Put your mind at rest, sir," replied Menichino. "I've come on no evil mission. I'm just a poor boy seeking his fortune, and they directed me to you, since you are so charitable toward poor, unfortunate souls."

Hearing that, the Sorcerer broke into laughter that shook the whole palace. "May I ask who sent you?" So Menichino told the whole story. The Sorcerer propped himself on an elbow, the better to see the boy. "You are courageous and you're also truthful. You deserve this prize. My secret is this magic wand. I'm giving it to you, but heaven help you if it is ever mislaid or stolen! Each time you tap the ground with the wand and state your wish, it will be granted at once. Here you are. Now go in peace."

Going back down Flower Mountain, Menichino thought things over and decided he'd best return home now, dressed as a gentleman, and see if his parents and his brother were alive and still remembered him. "This will be the first test I put the magic wand to," he told himself. He tapped the ground with it and heard a tiny voice say, "Command!" Menichino replied, "I command a four-horse carriage, lackeys, grooms, and a noble lord's wardrobe." And there before him stood a carriage with magnificent horses, while servants came forward to dress him in clothing of the latest style. The horses were magic creatures and galloped straight to Milan without once stopping along the way.

When he got to Milan, Menichino discovered that his parents had moved out of their original palace. That business in France, instead of bringing his father still more wealth, had resulted in a loss of everything he owned. Now the family were tenants of a hovel on the outskirts of the city. When Menichino drew up to their door with horses and servants, his parents were dumbfounded, as you can well imagine. He didn't mention the wand, but said he had made a fortune in business and that he would provide for them all from now on. So, using the wand, he produced a large palace, explaining that it had been built by the fastest and most experienced workers in his service. The family moved in, to enjoy the greatest plenty of furnishings, clothes, horses, servants, and chamberlains, to say nothing of money itself.

They were all quite happy, except Menichino's brother, who was consumed with envy. Here he was the older son and his father's favorite, and he had to defer to and depend on his younger brother for everything! He worried himself sick to learn where Menichino actually got his money. He began spying on him through the keyhole, and seeing him always daydreaming with that wand, he resolved to steal it. Menichino kept the wand in his chest of drawers, so one day while he was away from home his brother entered the bedroom and stole it. Back in his own room he tried tapping the floor with the wand, but nothing happened, since in his hands the wand was powerless. He said, "I've made a mistake; this is no magic wand." So he returned to put it back and rummage through the drawers for another one. But he'd no sooner gone into the room than he heard Menichino coming upstairs. Afraid of being found out, he broke the wand in two and threw it out the window overlooking the garden.

Now Menichino did not use the wand every day, but only when he needed something. That's why he did not miss it right away. But the first time he went to look for it and found it gone, he was frantic. He imagined himself done for and all his wealth gone in a flash. In despair he went outside and was pacing up and down in the garden, when he happened to look up and see a broken wand hanging on a tree limb. His heart skipped a beat. He shook the tree and down fell the two halves of the wand. The instant they hit the ground a voice said, "Command!" Menichino's despair gave way to great joy: that was his wand, and even when broken, it still worked! He put it back together and resolved to be more careful in the future.

At that time, the king of Spain sent a dispatch to all lands: his only daughter had reached marriageable age, and the bravest knights of every nation were invited to joust for three days; the victor would marry the princess and inherit the kingdom. Menichino decided this would be a good time for him to become crown prince and then king. Tapping the wand, he ordered radiant armor, horses, and shields; then he left for Spain. To keep his name and origin a secret, he took lodgings at an out-of-the-way inn to await the day of the joust.

The joust took place in a busy clearing thronged on all four sides by lords and ladies of every description. On a canopied dais sat the king and the princess, conversing with the most important barons. All of a sudden trumpets sounded, and the crowd looked up to see knights riding onto the field with lances poised. The combat began at once, blows rained fast and furious upon the armor, but no one was unseated, since all were equally matched in strength and valor. Onto the field galloped a new knight, his visor down over his face and his coat-of-arms known to none. One by one he challenged the others to joust with him. One knight rode up to him, clashed, and fell to the ground. Another tried, and was disemboweled. A third had his lance broken, a fourth his helmet knocked off. That went on until they were all disabled by the unknown knight. But instead of parading victoriously around the field, he rode through the opening in the fence and galloped off. All the spectators were bewildered, and everybody from the king on down speculated on who this knight might be, but they got absolutely nowhere.

"We'll find out who he is if he comes back tomorrow," they said. As a matter of fact, the unknown knight returned, unseated everyone as he had done the day before, and escaped with his identity still a secret.

The king, both intrigued and offended, ordered his men on the third day to stop him, and the number of guards around the fence was doubled for that very purpose. The knight appeared and carried off the final victory, then went to bow before the royal dais. When the princess, aglow with admiration, threw him her embroidered handkerchief, he caught it and galloped toward the exit. The guards barred his way with their weapons, but with a few flourishes of his sword he opened up a path and fled, even though a lance had penetrated one of his legs.

The king ordered the entire city searched, indoors and out, and finally they found Menichino in that wretched inn, in bed because of a wound in his leg. They were not sure at first whether he was the unknown knight, since great knights wouldn't have taken such modest lodgings; but then they saw his wound bandaged with the princess's embroidered handkerchief, and there could be no more doubt. Led before the king, he was asked to reveal his identity, since he would become prince and heir to the throne, provided his character was spotless.

"My character is spotless," answered Menichino; "however, I am not a knight by birth, but the son of a merchant from Milan."

Princes and barons began clearing their throats and shuffling their feet, and the noise grew louder and louder as Menichino went on with his story.

When he had finished, the king spoke. "You are not a knight, but a merchant's son, and all your wealth is the fruit of magic. If this magic were lost, what would become of you and yours?"

The princess chimed in. "You see what peril I'm in, Father, as a result of the joust you ordered."

The barons commented, "Can we accept as our sovereign a man of lower birth than we are?"

The king shouted, "Silence, silence! What is this hubbub? Beyond all doubt this young man, according to what I promised as king, is entitled to marry my daughter and inherit the kingdom. But if he would agree to it—since you do not accept him and there's no telling how the people would take the matter—I propose that he give up my daughter for other prizes."

"What do you propose in exchange, Majesty?" asked Menichino. "If I profit from it, then I agree."

"I had in mind," stated the king, "paying you a pension of one thousand lire a year for the rest of your life."

"I accept," said Menichino, and a notary was summoned at once to draw up the contract. Immediately afterward Menichino left for Milan.

Back home he found the old merchant ailing, and not long afterward the old man gave up the ghost. The two brothers were left with their aging mother. The elder boy was more envious all the time, especially now that they were so wealthy and had no need whatever of the magic wand to stay rich. So he planned to have Menichino killed by two hired assassins. Menichino was accustomed to visit friends at their villa outside Milan, so the hired assassins lay in wait for him along the road. But he always carried the wand with him, and the instant he was out of the city, he tapped the ground with the magic stick.

"Command!"

"I command that the horse run like lightning." The assassins heard something rush past them, but it went so fast they couldn't make out what it was.

"We'll get him on his return, at nightfall."

But Menichino also galloped home at the same speed, and the assassins heard only a whir like a whistle.

The elder brother admitted the killers to the palace and took them to Menichino's bedroom. But Menichino had smelled a rat and commanded the wand to make his door impossible to open, so the assassins struggled in vain throughout the night to get in, until dawn at last put them to flight.

Then one day Menichino made a fatal mistake. He thought, If I keep going out with the wand, I'll end up being treacherously attacked and robbed. I'd better hide the wand in my bedroom. That he did, and went out hunting with his friends. But his brother, who always kept his eyes open, began turning drawers and wardrobes inside out until he came across that stick broken in half. "So this is really it!" he exclaimed. "Otherwise my brother wouldn't have bothered picking it up in the garden where I threw it! But this is going to be the end of it." At that, he rushed into the kitchen and flung it into the fire. The wand was reduced at once to ashes, and in the same instant, palace, money, horses, clothing, and everything else obtained through its magic turned to ashes as well.

At the time, Menichino was deep in the woods, where the shotgun he held, the horse he rode, and the hounds chasing the hare all turned to ashes and were swept away by the wind. He realized that his entire fortune was gone for good this time, all because of his folly, and he burst into tears.

It was now futile to go back to Milan. He decided he had best set out for Spain, where he still had that pension of one thousand lire a year awarded him by the king. So, full of sighs, he started off on foot.

On a ferryboat he met a man who sold oxen. As fellow travelers often do, they greeted one another and began to talk about themselves, continuing their conversation along the road after leaving the boat. Touched by Menichino's misfortune, the man asked the boy if he would like to help drive a herd to market wherever oxen were in demand. Menichino accepted the salary offered him, and the two of them made the rounds of the fairs. He had even saved up a little money, when one evening he was attacked by a band of assassins at an inn where he was passing the night with his companion. Together with the innkeeper and the oxen-dealer, Menichino took up arms to resist, but the assassins, who were quite numerous, overpowered and killed them. Thus ended the adventures and misfortunes of Menichino.

Nor did his brother fare any better. Poor again, he tried his hand at trade, but to no avail, and going from bad to worse, he ended up in a band of professional robbers. It is a well-known fact that robbers get away nine times out of ten, but the tenth time they get caught. That's what happened to him: the constables set a trap and watched him walk into it. He was shackled and thrown into prison, where he remained up to the day he was led out with the priest at his side to shrive him; then he was handed over to the executioner and beheaded.

(Montale Pistoiese)

◆ 63 ◆

Monkey Palace

Once there was a king who had twin sons, John and Anthony. As he was not quite sure which of the two was born first, no more than the court itself was of one mind on this subject, the king couldn't say who was crown prince. "To be completely fair to you both," he told the boys, "I want you each to go out into the world and seek a wife. The one whose bride presents me with the rarer and finer gift will be named crown prince."

The twins mounted their horses and galloped off, one in one direction, the other in the opposite.

Two days later John came to a large city. There he made friends with a marquis's daughter and told her about his father's proposal. As her gift, she gave him a tiny, sealed box to take to the king, and the young people announced their engagement. The king accepted the tiny box, but put off opening it until he should also have Anthony's bride's present.

Anthony rode and rode, but never reached any city. He found himself in a dense, pathless, seemingly endless forest, through which he had to cut his way with his sword. All at once he came to a clearing, at whose end rose a solid marble palace with glittering windowpanes. Anthony knocked, and who should answer the door but a monkey in steward's livery! It bowed to Anthony and motioned him into the palace. Two other monkeys helped him dismount and led his horse to the stable. He proceeded up a carpeted, marble staircase past monkeys galore sitting on the balustrade and silently bowing to him. Anthony walked into a hall where a table had been set up for a game of bridge. One monkey offered him a chair at the table as three other monkeys took the remaining places, and the game began. After a while they asked by signs if he was ready to dine. They led him to the dining room where monkeys in plumed hats sat at a sumptuously laid table waited on by monkeys in aprons. After dinner they picked up torches and lighted him to his bedchamber, where they left him for the night.

Although puzzled and uneasy, Anthony was so tired he fell asleep at once. But just when he was sleeping his soundest, a voice in the dark awakened him. "Anthony!"

"Who's calling me?" he asked, huddling beneath the covers.

"Anthony, what have you come here seeking?"

"I'm seeking a bride who will give the king a finer present than John's bride will. That way I'll be named crown prince."

"If you agree to marry me, Anthony," said the voice in the dark, "you'll get the finer gift and the crown."

"Let's get married, then," said Anthony scarcely above a whisper.

"Very well, send a letter to your father tomorrow."

The next day Anthony wrote his father that he was in good health and would soon be home with his bride. He entrusted the letter to a monkey, which jumped from tree to tree and wound up in the royal city. Although surprised by the unusual messenger, the king was very happy over the good tidings and offered the monkey hospitality at the palace.

The next night Anthony was again awakened by a voice in the dark. "Anthony, are you still of the same mind?"

"I certainly am!"

"Very well. Send your father another letter tomorrow."

The next day Anthony again wrote his father he was in good health and sent the letter off by a monkey. The king took this monkey also into the palace.

Thus, every night the voice asked Anthony if he'd changed his mind, and told him to write his father, while every morning a monkey left with a letter for the king. That went on for a month, and the royal city was now full of monkeys. They swarmed in the trees, on the rooftops, and over all the monuments. Shoemakers hammered heels onto shoes while monkeys sat on their shoulders and mimicked them. Surgeons performed operations as monkeys looked on and made off with scalpels and thread for cutting open and stitching up patients. Ladies went out walking, with monkeys perched on their parasols. The king no longer knew what to do.

At the end of the month, the voice in the dark at last said, "Tomorrow, we will go to the king together and get married."

In the morning when Anthony went downstairs, a handsome carriage was waiting at the door with a monkey coachman on the box and two monkey footmen hanging onto the back. And who should be sitting inside the coach on velvet cushions and wearing many jewels and an elaborate coiffure of sweeping ostrich plumes but a monkey! Anthony seated himself at her side, and off they drove.

When the couple arrived in the royal city, all the people pressed against the strange carriage and, to their utter amazement, discovered that Prince Anthony's bride was a monkey. All eyes turned to the king, who awaited his son on the palace doorstep, to see how *he* would react. But the king lacked no backbone: he didn't bat an eye, as though marrying a monkey were the most normal thing in the world. He merely said, "He chose her, so he has to marry her. Never does a king go back on his word." From the monkey's hands the king took a tiny, sealed box exactly like her sister-in-law's. The boxes were to be opened tomorrow, the day of the wedding. The monkey was shown to her room, where she requested to be left alone.

The next day Anthony went for his bride. When he entered her room, the monkey was standing before the mirror trying on her wedding dress. "How do I look?" she asked, turning around. Anthony was speechless: she was no longer a monkey but a tall and graceful maiden with blond hair and rosy cheeks, whom it was a real joy to behold. Anthony rubbed his eyes, unable to believe what he saw, but she said, "You are looking at none other than your bride." At that, they fell into each other's arms.

Outside the palace all the people had gathered to see Prince Anthony and his monkey bride. When he emerged instead with that beautiful maiden, their mouths flew open and they could only gape with admiration. Beyond the palace monkeys flanked the street and filled all trees, rooftops, and windowsills. When the royal couple passed, every monkey wheeled around and instantly changed into a human being. Some became ladies in long, trailing cloaks; others cavaliers with plumed hats and sabers; and the rest friars, farmers, or pages. They all moved in procession behind the couple going to be married.

The king opened the tiny boxes containing the presents. Out of the box from John's bride flew a live baby bird, and everybody marveled that it had stayed alive closed up so long. In its beak the bird clutched a walnut, inside which lay a tassel of gold.

The king opened the box from Anthony's wife. It too contained a live baby bird which, of all things, was holding a lizard. In the lizard's mouth was a hazelnut. The king cracked it open and beheld, carefully folded, a lace tablecloth embroidered by one hundred hands.

To John's dismay, the king was about to proclaim Anthony crown prince, when Anthony's bride spoke. "Anthony doesn't need his father's kingdom, since I am giving him mine as my dowry. By marrying me, he has freed us from the spell that made monkeys of us all!" Then all the monkeys newly changed into human beings hailed Anthony as their king.

John inherited his father's kingdom, and everybody lived in peace and harmony.

> Life they did indeed enjoy,
> But me they never would employ.

(Montale Pistoiese)

◈ 64 ◈

Rosina in the Oven

The young wife of a poor man died giving birth to a beautiful baby girl named Rosina. Since the man had to work and couldn't look after the baby, he took another wife who bore him a second daughter by the name of Assunta, who was as homely as the other child was beautiful. The two girls grew up together, went to school together, and were always together wherever they went. But every time Assunta would come home full of resentment and say to her mother, "Mamma, I won't go out with Rosina any more. People we meet always pay her so many compliments, telling her how rosy and beautiful and well-mannered she is. But they say I'm as black as coal."

"What difference does it make if you're dark-skinned?" replied her mother. "My complexion is somewhat dark, so you get your color from me. Your beauty is very special, don't you see?"

"Say what you like, Mamma," answered Assunta. "I'll not go out with Rosina any more no matter what."

Seeing the girl consumed with envy, her mother, who would have done anything to make her happy, said, "But what can I do?"

"Give her a pound of hemp to spin while she pastures the cows. If she comes home at night with hungry cows and messy spinning, beat her. Beat her day in day out, and she'll be a sight."

With some reluctance, the stepmother gave in to her daughter's whims, called Rosina, and said, "You will not go out with Assunta any more. You are to tend the cows and find grass for them, and while you're doing that, you will spin this pound of hemp. If you bring the cows home still hungry or the hemp not all spun, you will hear from me. But do your part, and I'll do mine."

Rosina, who wasn't used to being ordered around like that, was dumbfounded. But since her stepmother had already picked up a stick, she had no choice but to obey. She took the cows to the pasture, with her distaff full of hemp, repeating along the way, "Dear little cows, how will I ever mow grass for you if I have to spin this whole big distaff of hemp? Somebody is bound to lose!"

At those words one of the oldest cows wheeled around, saying, "Don't you worry, Rosina. Just cut the grass for us, and we'll spin all the hemp for you and wind it into a skein. All you have to do is say:

'Cow, little cow,
Spin, spin, use your mouth,
Wind, wind, use your horn,
Make my skein this very morn.' "

That night when Rosina returned, the cows went into the barn well fed; on her head she carried a nice bundle of hay, and under her arm a one-pound skein of spun hemp. Seeing that, Assunta was furious, and said to her mother, "Send her out again tomorrow with the cows, but give her two pounds of hemp, and if she doesn't spin every bit of it, beat her."

But once more all Rosina needed to do was say:

"Cow, little cow,
Spin, spin, use your mouth,

Wind, wind, use your horn,
Make my skein this very morn"

and by night the cows were fed, the bundle of hay had been mowed, and the two pounds of hemp were all spun and wound into a skein.

"But how on earth did you do it in one day?" asked Assunta, just seething inside.

"As you might expect," said Rosina, "there are always good souls around to lend a hand. My little cows came to my rescue."

Assunta went running to her mother. "Mamma, keep Rosina home tomorrow to do the housework. I'll take the cows out, and I want some hemp to spin while I'm tending them."

Her mother gave her the hemp, and Assunta led the cows out. She kept them stepping by whacking them on the rump and the tail until they reached the pasture. Then she put the hemp on their horns, but they didn't budge.

"Get busy! Why aren't you spinning?" screamed Assunta, whacking them with the stick. The cows began twisting their horns until the hemp became so stringy that it was nothing but a wad of tow.

That wasn't enough for Assunta, who said to her mother one day, "Mamma, I feel like eating some lamb's lettuce. Send Rosina out tonight to get some from that farmer's field."

To make her happy, her mother ordered Rosina to go and pick the farmer's lettuce. "What!" exclaimed Rosina. "You want me to go out and steal? But that's something I've never done before. What if the farmer saw someone trespassing in his field at night? He'd be sure to shoot me from his window!"

That's just what Assunta wanted to happen, and as she too had now taken to bossing Rosina about, she said, "Go on, you have to go, or we'll thrash you good!"

So Rosina went out in the dark and climbed over the hedge into the farmer's field; but instead of lamb's lettuce, she found a turnip. She caught hold of the turnip and pulled and pulled. Finally it came up, revealing a toad's nest underneath containing five tiny toads. "Oh, how darling!" she exclaimed, holding them to her bosom and making a fuss over them. But one fell on the ground and broke a leg. "Excuse me, little toad," said Rosina, "I didn't drop you on purpose."

The four little toads which she still held, finding her so kind, said, "Lovely maiden, you are kind, and we want to reward you. You shall become the most beautiful girl in the world and shine like the sun, even when it is cloudy. So be it."

But the lamed toad grumbled. "I certainly don't think she's kind. She has crippled me for life! She could have been more careful! The minute she sees a ray of sun, she shall change into a snake and never turn back into a woman without first passing through a fiery oven."

Rosina went home half cheerful and half terrified, and her beauty was so radiant that the night around her was like daylight. Seeing her walk in shining like the sun left her stepmother and sister speechless. Right away she told them how it had come about, concluding, "But I'm not to blame for it. Please don't send me into the sunlight, or I'll become a snake."

From then on, Rosina never went out of doors when the sun was shining, but only after sunset or when the sky was cloudy. She spent her days at the window working and singing. A great light shone from the window and was visible for miles around.

One day the king's son passed by, noticed the glow, and saw the maiden. "What is such a beauty doing in a hovel of peasants?" he wondered, and entered the house. Thus they came to know one another, and Rosina told him her whole story and about the curse placed on her.

The king's son said, "Come what may, you are too beautiful to remain in this hovel, you will be my bride."

At that point the mother stepped in. "Majesty, be careful, you're asking for trouble. Just realize that the first time a ray of sun touches her she will turn into a snake."

"This does not concern you," said the king's son. "You obviously hate this girl, but I order you to send her to my palace. I shall dispatch a sealed carriage for her, so that the sun will not fall on her along the way. From now on, you will have all the money you need. We agree, and goodbye."

Gritting their teeth, the stepmother and Assunta, unable to disobey the king's son, got everything ready for Rosina's departure. The carriage finally arrived, one of those old-fashioned carriages entirely closed, with only a round vent in the roof, and a groom behind decked in tassels, plumed hat, and sword. Rosina got into the carriage, and her stepmother climbed in after her to keep her company along the way. But the woman had first taken the groom aside and said, "Sir, if you would like a reward of ten crowns, open the vent in the roof when the sun strikes it."

"Yes, madam, as you say."

The carriage sped off, and as soon as it was noon and the sun shone straight down on the roof, the groom opened the vent, and a ray of sun fell upon Rosina's head. In a flash she changed into a snake and wriggled off into the woods, hissing.

When the king's son opened the carriage and found Rosina gone, he knew exactly what had occurred. Dismayed and tearful, he was on the verge of slaying the wicked stepmother. But everybody told him that the thing had been bound to happen to Rosina, if not then, at some other time. Finally he calmed down, although he remained sad and disconsolate.

Meanwhile the cooks had everything in the ovens and on the stoves and spits for the wedding feast, and the guests were all at the table. Learning that the bride had disappeared, they said, "Since we are already here, let's have the banquet just the same." So the cooks were ordered to fire up the oven. One cook was putting into the lighted oven a bundle of brushwood just brought in from the woods, when he spied a snake concealed in the bundle. There was no way to draw the brushwood back out, since it had already caught fire. He peered into the oven to get a look at the snake, and there out of the flames leapt a maiden stark naked, fresh as a rose, and more radiant than fire or sun! Petrified, the cook yelled, "Come here quick! I've just seen a maiden in the oven!"

At that cry the king's son rushed into the kitchen, with the whole court behind him. He recognized Rosina, took her into his arms, and then the wedding was celebrated, and from that time on Rosina lived happily, and no one begrudged her anything more.

(Montale Pistoiese)

◈ 65 ◈

The Salamanna Grapes

There was once a king who had a very beautiful daughter of marriageable age. A neighboring king had three grown sons, who all fell in love with the princess. The princess's father said, "As far as I am concerned, you are all three equal, and I couldn't for the life of me give any one of you preference over the other two. But I wouldn't want to be the cause of any strife among you, so why not travel about the world for six months, and the one who returns with the finest present will be my son-in-law."

The three brothers set out together, and when the road branched off in three different directions each went his separate way.

The oldest brother traveled for three, four, and five months without finding a thing worth taking home as a present. Then one morning of the sixth month in a faraway city, he heard a hawker under his window: "Carpets for sale! Fine carpets for sale!"

He leaned out the window, and the carpet seller asked, "How about a nice carpet?"

"That's the last thing I need," he replied. "There are carpets all over my palace, even in the kitchen!"

"But," insisted the carpet seller, "I'm sure you have no magic carpet like this one."

"What's so special about it?"

"When you set foot on it, it takes you great distances through the air."

The prince snapped his fingers. "There's the perfect gift to take back. How much are you asking for it, my good man?"

"One hundred crowns even."

"Agreed!" exclaimed the prince, counting out the hundred crowns.

As soon as he stepped onto it, the carpet went soaring through the air over mountains and valleys and landed at the inn where the brothers had agreed to meet at the end of the six months. The other two, though, had not yet arrived.

The middle brother also had traveled far and wide up to the last days without finding any suitable present. And then he met a peddler crying, "Telescopes! Perfect telescopes! How about a telescope, young man?"

"What would I do with another telescope?" asked the prince. "My house is full of telescopes, and the very best, mind you."

"I bet you've never seen magic telescopes like mine," said the telescope seller.

"What's so special about them?"

"With these telescopes you can see a hundred miles away and through walls as well."

The prince exclaimed, "Wonderful! How much are they?"

"One hundred crowns apiece."

"Here are one hundred crowns. Give me a telescope."

He took the telescope to the inn, found his big brother, and the two of them sat and waited for their little brother.

The youngest boy, up to the very last day, found nothing and gave up all hope. He was on his way home when he met a fruit vendor crying, "Salamanna grapes! Salamanna grapes for sale! Come buy nice Salamanna grapes!"

The prince, who'd never heard of Salamanna grapes, since they didn't grow in his country, asked, "Just what are these grapes you're selling?"

"They are called Salamanna grapes," said the fruit vendor, "and there're no finer grapes in the world. They also work a special wonder."

"What do they do?"

"Put a grape in the mouth of someone breathing their last, and they will get well instantly."

"You don't say!" exclaimed the prince. "I'll buy some in that case. How much are they?"

"They are sold by the grape. But I'll make you a special price: one hundred crowns per grape."

As the prince had three hundred crowns in his pocket, he could only buy three grapes. He put them in a little box with cotton around them and went to join his brothers.

When they were all three together at the inn, they asked each other what they had bought.

"Me? Oh, just a little carpet . . ." said the oldest boy.

"Well, I picked up a little telescope . . ." replied the middle boy.

"Only a little fruit, nothing more," said the third.

"I wonder what's going on at home right now. And at the princess's palace," one of the boys said.

The middle boy casually pointed his telescope toward their capital city. Everything was as usual. Then he looked toward the neighboring kingdom, where their beloved's palace was, and let out a cry.

"What's the matter?" asked the brothers.

"I see our beloved's palace, a stream of carriages, people weeping and tearing their hair. And inside . . . inside I see a doctor and a priest at somebody's bedside, yes, the princess's bedside. She lies there as still and pale as a dead girl. Quick, brothers, let's hurry to her before it's too late. . . . She's dying!"

"We'll never make it. That's more than fifty miles away."

"Don't worry," said the oldest brother, "we'll get there in time. Quick, everybody step onto my carpet."

The carpet flew straight to the princess's room, passed through the open window, and landed by the bed, where it lay like the most ordinary bedside rug, with the three brothers standing on it.

The youngest brother had already taken the cotton from around the three Salamanna grapes, and he put one into the princess's pale mouth. She swallowed it and immediately opened her eyes. Right away the prince put another grape into her mouth, which regained its color at once. He gave her the last grape, and she breathed and raised her arms. She was well. She sat up in bed and asked the maids to dress her in her most beautiful clothes.

Everybody was rejoicing, when all of a sudden the youngest brother said, "So I'm the winner, and the princess will be my bride. Without the Salamanna grapes she'd now be dead."

"No, brother," objected the middle boy, "if I'd not had the telescope and told you the princess was dying, your grapes would have done no good. For that reason I will marry the princess myself."

"I'm sorry, brother," put in the oldest boy. "The princess is mine, and nobody will take her away from me. Your contributions are nothing compared with mine. Only my carpet brought us here in time."

So the quarrel the king had wanted to avoid became ever more heated, and the king decided to put an end to it by marrying his daughter to a fourth suitor who had come to her empty-handed.

(Montale Pistoiese)

◈ 66 ◈

The Enchanted Palace

A king of long ago had a son named Fiordinando who never took his nose out of his books. He was always shut up in his room reading. From time to time he would close the book and gaze out the window at the garden and the woods beyond, then resume his reading and musing. Never did he leave his room except for lunch or dinner, or maybe for a rare stroll in the garden.

One day the king's hunter, a bright young man who as a child had played with the prince, said to the king, "May I call on Fiordinando, Majesty? I've not seen him for quite some time."

The king replied, "By all means. Your visit will be a pleasant diversion for my fine son."

So the hunter entered the room of Fiordinando, who looked him over and asked, "What brings you to the court in those hobnailed boots?"

"I am the king's hunter," explained the young man, who went on to describe the many kinds of game, the ways of birds and hares, and the different parts of the woods.

Fiordinando's imagination was kindled. "Listen," he said to the youth, "I too shall try my luck at hunting. But don't say anything to my father, so he won't think it was your idea. I'll simply ask him to let me go hunting with you one morning."

"At your service, as always," replied the young man.

The next day at breakfast, Fiordinando said to the king, "Yesterday I read a book on hunting which was so interesting I'm dying to go out and try my luck. May I?"

"Hunting is a dangerous sport," replied the king, "for someone who is new to it. But I won't keep you from something you think you might like. For a companion I'll let you have my hunter, who is unequaled as a hunting dog. Don't ever let him out of your sight."

Next morning at sunrise Fiordinando and the hunter mounted their horses with their guns on shoulder straps and off to the woods they galloped. The hunter aimed at every bird or hare he saw and laid it low. Fiordinando tried his best to keep pace, but missed everything he shot at. At the end of the day the hunter's game bag was bulging, whereas Fiordinando hadn't brought down so much as one feather. At dusk Fiordinando spied a small hare hiding under a bush and took aim. But it was so small and frightened he decided he would simply run up and grab it. Just as he reached the bush, the hare darted off, with Fiordinando close behind. Every time he was right upon it, the hare would run far ahead, then stop, as though it were waiting for Fiordinando to catch up, only to elude him again. In the meantime Fiordinando had strayed so far from the hunter that he could no longer find the way back. Again and again he called out, but no one answered. By now it was completely dark, and the hare had disappeared.

Weary and distressed, Fiordinando sat down under a tree to rest. It was not long before he saw what seemed to be a light shining through the trees. He therefore got up, made his way through the underbrush, and emerged in a vast clearing, at the end of which stood the most ornate of palaces.

The front door was open, and Fiordinando called out, "Hello! Is anyone at home?" He was answered with dead silence; not even an echo came back to him. Entering, he found a large hall with a fire burning in the fireplace and, nearby, wine and glasses. Fiordinando took a seat to rest and warm up and drink a little wine. Then he rose and passed into another room where a table was set for two persons. The cutlery, plates, and goblets were gold and silver; the curtains, tablecloth, and napkins were pure silk embroidered with pearls and diamonds; from the ceiling hung lamps of solid gold the size of baskets. Since no one was there and he was hungry, Fiordinando sat down to the table.

He had scarcely eaten his first mouthful when he heard a rustle of dresses coming down the steps, and in walked a queen followed by twelve maids of honor. The queen was young and extremely beautiful of figure, but her face was hidden by a heavy veil. Neither she nor the twelve maids of honor said one word during the entire meal. She sat across the table in silence from Fiordinando while the maids quietly served them and poured their wine. The meal thus passed in silence, and the queen carried her food to her mouth under that thick veil. When they had finished, the queen rose, and the maids of honor accompanied her back upstairs. Fiordinando also rose and continued his tour of the palace.

Coming to a master bedchamber with a bed all turned down for the night, he undressed and jumped under the covers. Behind the canopy was a secret door: it opened, and in walked the queen, still mute, veiled, and followed by her twelve maids of honor. With Fiordinando leaning on his elbow and gaping, the maids of honor undressed the queen all but for her veil, put her in bed beside Fiordinando, and left the room. Fiordinando was sure she would say something now or unveil her face. But she had already fallen asleep. He watched the veil rising and falling with her breath, thought about it a minute, then he too fell asleep.

At dawn the maids of honor returned, put the queen's clothes back on her, and led her away. Fiordinando also got up, ate the hearty breakfast he found waiting for him, and went down to the stables.

His horse was there eating oats. Fiordinando climbed into the saddle and galloped off to the woods. The whole day long he looked for a road that would take him back home, or for some trace of his hunting companion, but he only got lost anew, and when night fell, there stood the clearing and palace once more.

He went inside, and the same things happened as the evening before. But the next day as he was galloping through the woods he met the hunter, who'd been looking for him for the last three days, and together they returned to the city. When the hunter questioned him, Fiordinando made up a tale about a lot of complicated mishaps, but said nothing about what had really happened.

Back at the royal palace Fiordinando was like a changed person. His eyes wandered constantly from the pages of his book to the woods beyond the garden. Seeing him so moody, listless, and absorbed, his mother began pestering him to tell her what he was brooding over. She kept nagging until Fiordinando finally told her from beginning to end what had happened to him in the woods. He made no bones about being in love with the beautiful queen and wondering how to marry her when she neither spoke nor showed her face.

"I'll tell you what to do," replied his mother. "Sup with her one more time. When the two of you are seated, accidentally knock her fork off the table. When she bends over to pick it up, pull off her veil. You can be sure she'll say something then."

No sooner had he received that advice than Fiordinando saddled his horse and raced off to the palace in the woods, where he was welcomed in the usual manner. At supper he knocked the queen's fork off the table with his elbow. She bent over, and he tore off her veil. At that, the queen rose, as beautiful as a moonbeam and as fiery as a ray of sun. "Rash youth!" she screamed. "You have betrayed me. Had I been able to sleep one more night beside you without speaking or unveiling my face, I would have been free from the spell and you would have become my husband. Now I'll have to go off to Paris for a week and from there to Peterborough, where I'll be given in prize at a tournament, and heaven knows who will win me. Farewell! And note that I am the queen of Portugal!"

In the same instant she vanished, along with the entire palace, and Fiordinando found himself alone and abandoned in the thickest part of the underbrush. It was no easy task to find his way home, but once he got there, he didn't waste a minute. He filled a purse with money, summoned his faithful hunter, and departed on horseback for Paris. They wore themselves out riding, but didn't dismount until they reached an inn in that famous city.

Nor did he spend long resting up, for he wished to learn if the queen of Portugal really was there in Paris. He began pumping the innkeeper. "What's the news around here?"

The innkeeper replied, "None to speak of. What sort of news do you expect?"

"There's all kind of news," replied Fiordinando. "News about wars, feast days, famous people passing through the city . . ."

"Oh!" exclaimed the innkeeper, "come to think about it, there is a piece of interesting news: five days ago the queen of Portugal arrived in Paris. In three more days she'll leave for Peterborough. She's a very beautiful lady and highly educated. She enjoys exploring unusual spots, and strolls outside the city gate near here every afternoon with twelve maids of honor."

"And it's possible to get a look at her?" asked Fiordinando.

"Why not? When she walks in public, any passer-by can see her."

"Wonderful!" said Fiordinando. "In the meantime get dinner for us and serve it with a bottle of red wine."

Now the innkeeper had a daughter who rejected all wooers, mind you, because none of them suited her. But the instant she laid eyes on Fiordinando getting out of his saddle, she told herself he would be the only one she would ever consider. She went to her father at once to say she had fallen in love and to ask him to find a way for her to marry the stranger. So the innkeeper said to Fiordinando, "I hope you'll like Paris and have the good fortune to find yourself a lovely bride here."

"My bride," replied Fiordinando, "is the most beautiful queen in the world, and I am trailing her all over the globe."

The innkeeper's daughter, who was eavesdropping, was seized with rage. When her father sent her to the cellar after the wine, she thrust a handful of opium into the bottle. Fiordinando and the hunter went outside the city after dinner to await the queen of Portugal, but suddenly they became so drowsy that they sank to the ground and slept like logs. Shortly thereafter the queen came by, recognized Fiordinando, bent over him, called his name, caressed him, shook him, and rolled him over and over; but there was no waking him. Then she slipped a diamond ring from her finger and placed it on his brow.

Now in a cave nearby lived a hermit who had witnessed the whole scene from behind a tree. As soon as the queen left, he tiptoed out, picked up the ring from Fiordinando's brow, and retreated with it to his cave.

When Fiordinando awakened, it was already dark, and it took him a while to recall where he was. He shook the hunter awake, and together they cursed the red wine for being too strong and lamented over missing the queen.

The second day they said to the innkeeper, "Give us white wine, but make sure it's not too strong." The daughter, however, drugged the white wine too, and the young men went back only to end up snoring in the middle of the meadow.

At a loss to awaken Fiordinando, the queen of Portugal placed a lock of her hair on his brow and fled. The hermit emerged from the grove of trees and made off with the lock. When Fiordinando and the hunter awakened in the middle of the night, they had no idea what had taken place.

Fiordinando became suspicious of the sleep that came over him every afternoon. It was now the last day before the queen would be leaving for Peterborough, and he intended to see her at all costs. He thus told the innkeeper to bring him no more wine. But the daughter now drugged the soup. So, upon arriving in the meadow, Fiordinando felt his head drooping already. He pulled out two pistols and showed them to the hunter. "I know you're loyal," he said, "but I warn you that if you don't stay awake today and keep me awake, you are going to get it. I'll unload both of these into your head, and I don't mean maybe."

At that, Fiordinando stretched out and began to snore. To stay awake, the hunter tried pinching himself repeatedly, but between one pinch and the next his eyes would close, and the pinches became rarer and rarer, until he too was snoring.

The queen arrived. With cries, embraces, slaps in the face, kisses, and shakes, she did her best to awaken Fiordinando. But realizing she would not succeed, she began weeping so violently that instead of tears a few drops of blood trickled down her cheeks. She wiped the blood off with her handkerchief, which she placed over Fiordinando's face. Then she got back into her carriage and sped straight to Peterborough. Meanwhile the hermit came out of the cave, picked up the handkerchief, and stood by to see exactly what would happen.

When Fiordinando woke up at night and realized he'd missed his last chance to see the queen, he was fit to be tied. He pulled out the pistols and was about to carry out his threat of unloading them in the sleeping hunter's head, when the hermit grabbed him by the wrists and said, "That poor fellow is blameless. The culprit is the innkeeper's daughter who drugged the red wine, the white wine, and the soup."

"Why would she do a thing like that?" asked Fiordinando. "And how do you know so much about it?"

"She's in love with you and gave you opium. I know all about it from peeping through the trees at everything that goes on here. For the last three days the queen of Portugal has come by and tried to awaken you, leaving on your brow a diamond, a lock of her hair, and a handkerchief moist with tears of blood."

"And where are these things now?"

"I took them away for safekeeping, since there are many thieves around here who would have stolen them before you ever got to see them. Here they are. Look after them, because if you act sensibly, they will bring you luck."

"What am I to do?"

"The queen of Portugal," explained the hermit, "has gone to Peterborough where she will be given in prize at a tournament. The knight who jousts with this ring, this lock of hair, and this handkerchief on the tip of his lance will be invincible and wed the queen."

Fiordinando didn't have to be told twice. He sped from Paris to Peterborough, where he arrived in time to enter the list of jousters, but under a false name. Illustrious warriors had arrived from all over the world with wagonloads of luggage, servants, and arms as shiny as the sun. In the heart of the city a large arena had been surrounded with viewing stands, and there the knights were to contend on horseback for the queen of Portugal.

With his visor lowered, Fiordinando won the first day, thanks to the diamond on the tip of his lance. He won the second day with the lock of hair. He won the third with the handkerchief. Horses and men fell by the dozens until not a one was left standing. Fiordinando was proclaimed victor and the queen's bridegroom. Only then did he open his helmet. The queen recognized him and swooned for joy.

There was a grand wedding, and Fiordinando sent for his mother and father, who had already given him up for dead and gone into mourning. He introduced his bride to them, saying, "This is none other than the little hare I pursued, the veiled lady, and the queen of Portugal whom I have freed from an awful spell."

(Montale Pistoiese)

◈ 67 ◈

Buffalo Head

A farmer was angrily hoeing the wretched soil of his field, when his hoe struck something hard. He dug gently around it, unearthing a buffalo head twice as big as any other buffalo head. Its horns stood up, its fur gleamed, and its eyes were open and bright, so that it really looked as if it were alive. It was alive, in fact, for when the farmer made ready to bring his hoe down on the ugly thing with all his might, the head opened its mouth and spoke. "Stop! Don't kill me. I will be the making of one of your daughters if you spare me."

Suspecting that magic was in play here, the farmer carefully picked up the head and carried it to the edge of the field, where he put it down and covered it with his coat. When his elder daughter brought him his midday meal, he said to her, "Go look at what's under my coat."

The girl lifted the coat and let out a scream. "Oh, what a hideous monster!" She went flying back home, screaming all the way.

Seeing her return so frightened her mother thought something might have happened to her husband, so she said to her middle daughter, "Go to your father and find out if he needs anything."

She too was directed by her father to look under the coat, and she too fled like lightning, screaming at the top of her voice. "What a dreadful snout!"

The mother then called her youngest daughter, who was also the smartest and most courageous of the three, and sent her to the field. When her father told her to look under the coat, the little girl obeyed. A smile spread over her face, and she reached out and petted the buffalo head. "My, what a pretty little head! What fine horns! What fine whiskers! Papa, where did you find this wonderful buffalo head?"

At those compliments, the buffalo head looked up and whined happily. "Would you come and live with me, you lovely child?"

"If Papa lets me, I'll come right now."

The farmer didn't have the heart to refuse. The buffalo head led the way, capering on its horns, while the child followed dancing and clapping for joy.

In a clearing in the heart of the woods was a trapdoor. Buffalo Head opened it with one of her horns and sprang through it. Reaching the bottom, she called to the little girl, "Remove your wooden shoes and come on down. Be careful, though, for the steps are glass." The child went down the glass steps and found herself in a princely drawing room, and there in an armchair sat Buffalo Head.

The little girl was quite happy in this underground home. Buffalo Head was even better to her than any real mother would have been, teaching her how to keep house, cook and iron, along with many other things. The child was good at everything, even reading and writing. She grew by leaps and bounds, and before you knew it she had become a beautiful maiden and so fond of Buffalo Head that she called her "Mamma."

The girl had not been in this hole too many years before she began saying, "Mamma, let me go up to the clearing for a little fresh air."

Buffalo Head didn't think much of the idea, but the girl kept on begging until Buffalo Head finally gave her a silver dress and a stool and told her she could sit in the clearing and knit.

While she was knitting in the clearing, a hunter who had lost his way came by and saw her. He was the son of the king of that territory. He got to talking to her, and in no time he was in love with the maiden.

"Beautiful maiden," he said, "you delight me in every way imaginable. If you have no objections, I would like to marry you."

"I have no objections myself," she replied, "but I must first see what my mother has to say." She walked back to the trapdoor and disappeared down the glass staircase.

Buffalo Head did not say no, but, "Do as you like. If you wish to leave me, then leave. But remember to be grateful. I'm responsible for all the good that has come your way, even the king's son who just asked you to marry him."

The king's son promised to return in a week with maids of honor and knights and royal carriages to take his bride away. Meanwhile, with Buffalo Head's help, the bride-to-be got her trousseau ready, and it was a trousseau fit for a queen. "Remember, now," Buffalo Head told her repeatedly, "when you leave this home, see that you leave nothing behind you. If you forget something, great misfortune could overtake you."

But when the king's son arrived with his train, the bride was so excited and distracted that not only did she leave her comb behind, she also forgot to bid Buffalo Head farewell and ran off without even closing the trapdoor behind her.

The bridal party had already traveled a great distance, when the bride suddenly clapped her hand to her forehead. "We must go back, we must go back, Majesty! I have forgotten my comb."

The king's son replied, "Are you afraid there'll be no combs in my palace? Or that the shops in town will have none?"

"I'm afraid something awful will happen to me," she said in a tearful voice, "since Mamma told me to leave none of my things behind if I didn't want to have bad luck. Please, Majesty, let's go back." So the prince ordered the horses turned about, and they rode back into the woods.

The trapdoor was still wide open. The bride ran downstairs and started looking for her comb.

"Oh, you'd already gone off?" asked Buffalo Head.

"Yes, Mamma, and in the excitement of getting away I forgot my comb, and now I can't find it."

"You forgot your comb, did you?" said Buffalo Head. "Just your comb? Go look in your dresser drawer."

Quite upset by now, the bride pulled out a drawer and bent over to rummage through it. When she stood back up and saw herself in the mirror, she let out a scream: her head had changed into a large buffalo head. "Mamma, Mamma! Something terrible has happened to me! Come here quick! Help me!"

Buffalo Head said, "There is nothing I can do. This is what you get for being ungrateful. You went off without even telling me goodbye."

"What will my bridegroom now say?"

"He will have to take you as you are, for he promised to marry you."

In short, all the maiden could do was wind a thick veil around her head and return to the prince. "Why on earth are you all muffled up?" he asked her. The girl explained that her eyes had suddenly become sore and swollen.

At court, the prince's mother and all the noble ladies could hardly wait to see this great beauty. But, under the pretext of sore eyes, she arrived heavily veiled and showed her face to no one. At last came the hour when she was alone with the prince and had to lift the veil. Imagine his

240

surprise on discovering that his bride had become a monster! He pressed his hands tightly over his eyes and refused to take a second look. His first thought was to have her burned at the stake, but then he consulted his mother, who persuaded him to close up the poor thing in the palace attic instead. The rumor spread through the court that he kept her locked up out of jealousy. Only his mother knew his secret and sympathized with him the sadder he became. One day she said to him, "Dear son, you will have to get rid of that buffalo head and start looking around for a bride who is worthy of you."

"How can I get rid of her when I've already promised to marry her?"

"There's a way, I assure you," said his mother. "At the court, there are two graceful young ladies whose sole dream is to marry you. Let us have a contest between them and the buffalo head to see who spins a pound of flax the best, in only one week. The winner will be your bride."

The prince followed the suggestion. The noble ladies shut themselves up in separate rooms and went to work at once painstakingly spinning their pound of flax. The poor bride, however, did nothing but sit and weep over her misfortune. Saturday evening she slid down a rope and ran to Buffalo Head's trapdoor in the woods. "Mamma," she whined, "please help me. Get me out of this mess, please. I know you can! After showing me so much kindness, you have made me the most miserable girl alive."

Buffalo Head replied, "You think ingratitude is nothing? I'm sorry but I can't help you. I can only give you this walnut. Offer it to the king's son tomorrow and tell him to eat the kernel in exchange for the pound of flax he gave you to spin."

On Sunday the noble ladies took their finely spun yarn to the queen to judge. "Not bad!" she commented. "But it's not perfect, either; it's not completely uniform. Let's now have a look at this other one's work."

The bride came forward with the walnut. "Do you want to make a fool of me too?" said the king's son. He nevertheless cracked open the nut and found a skein of thread perfectly spun from a pound of flax and finer than any you ever saw.

But the queen remarked, "The thread is fine indeed; that, we cannot deny. But would you marry a monster solely because of the wonder she's worked with a pound of flax? There must be a second test. This time we'll give these ladies a week to make a linen shirt, and the best seamstress will be your bride."

Once more the ladies were in their rooms and bent over their work: stitch by stitch, minute by minute, they worked. But the bride wept without cease and didn't even go near her material. Saturday evening she slipped down the rope and back to Buffalo Head. "Mamma, please help me! Forgive me for going off without saying goodbye. Have you really ceased to love your daughter?"

"Can't you do anything but weep and complain?" said Buffalo Head. "It's certainly not my fault you're in this predicament. You can't say I didn't warn you in time. All I can give you is this hazelnut. Take it to the king's son to crack open and eat. If he doesn't like it, let him spit it out!"

When the king's son cracked the hazelnut, out came a shirt embroidered entirely in gold, with certain stitches so close and fine that no human eye could see them.

The queen spoke. "We shall now have the final test. One week from today a grand ball will take place. Order these three ladies to get ready for it, and the most beautiful will be your bride."

As soon as they got back to their rooms, the two noble ladies each began exerting all their skill to become the most beautiful. They were so busy rubbing down their bodies with aromatic oils, making up their faces, setting their hair in every way imaginable, and trying on dress after dress, that they had no time to sleep. And could looking glasses be worn out from use, nothing of theirs would have remained at the end of that week. As for the bride, what more would you expect her to do, with that bunglesome buffalo head on her shoulders, than cry all week long and then slip back on Saturday night to the trapdoor in the woods.

"You've come back to bawl?" asked Buffalo Head.

"Mamma, what on earth will I do now? If you don't forgive me right away, I will lose my bridegroom for good."

"You brought this all on yourself. You ran off like a dog, and after all the nice things I'd done for you."

"I didn't do it on purpose, Mamma, don't you see? I was too happy and excited to realize what I was doing that day."

"And now, if you had that day to live over, what would you do?"

"Oh, Mamma, I would hug and kiss you goodbye, and I wouldn't leave any of my belongings behind, and I would close the trapdoor securely when I went out."

"All right, then, I forgive you," said Buffalo Head. "Go look for your comb."

The bride went to the dresser, opened her drawer, rummaged through it, and found the comb. How great was her amazement, as she stood up again, to see her original head in the mirror, but twice as lovely now as before! Jumping and screaming for joy, she ran to Buffalo Head and hugged and kissed her, thanking her over and over for her kindness.

On Sunday the whole court gathered in the royal ballroom before the king and queen seated on the high throne and their son standing at the foot of the steps. The three ladies came forward, heavily veiled from head to foot. The prince lifted the first one's veil and muttered, "What's this? She's nothing but padding from top to bottom!"

The second lady moved up, and the prince lifted her veil. "My word! This one is all ribbons and paint!" He dared not lift his bride's veil, but when he did lift it at last, he was dumbfounded. "That's my wife! That's how she looked the first time I ever saw her, knitting out in the heart of the woods. Only, she's far lovelier now than then! Mother dear, I've made my choice, I will marry this kind and beautiful maiden."

He took her by the hand and drew her onto the throne beside him, while the whole court hailed her as their future queen. From that day on they were together and as happy as happy could be.

(Montale Pistoiese)

◈ 68 ◈

The King of Portugal's Son

The king of Portugal had a son named Peter, who was dying to get married if only he could find a girl that suited him. On his way home from hunting one day, Peter spied on a shoemaker's doorstep a very beautiful girl with thick golden hair, sparkling brown eyes, and rosy cheeks. "She is certainly beautiful enough to be my wife," said Peter to himself. He got to the palace, put his gun up, changed into clothing appropriate to his rank, then went back out. "Come what may, I'm going to have a little chat with her," he told himself. "It's a shame she's only a shoemaker's daughter!" Thinking such thoughts, he reached the shop, struck up a conversation, and found the girl to be not only beautiful but also quite refined. In short, he fell head over heels in love with her and asked: "Will you have me for your spouse?"

"What!" she laughed. "You're joking! You are a king's son, and I'm the daughter of a poor shoemaker. We've nothing in common."

"I'm serious," replied Peter. "I want you, and I don't care who your father is. If you like me, I will marry you."

To make a long story short, they became engaged and, walking on air, Peter returned to the palace, as it was now dinner time.

At table he passed up the soup, then the main dish, and when they came to dessert, he said, "Father, I've decided to get married, and I've found my bride."

The king was overjoyed at first, but upon learning who the girl was, he exclaimed, "What! A shoemaker's daughter? That's no bride fit for a king. What would the nobility say? What would all the people say when they saw a shoemaker's daughter on the throne of Portugal? No, a thousand times no, this wedding cannot take place."

"Father," answered Peter, "I'm sorry you're unhappy about it, but I gave the girl my word, my royal word. So you see I've no choice now but to wed her."

"That being the case," said the king in dismay, "keep your promise, by all means. But outside this palace and this kingdom. Here, I want to see neither you nor her."

The ceremony took place in a few days, but without pomp, and then the newlyweds and one maid climbed into a coach and headed for Paris. When it was night, Peter, his bride, and the maid, worn out from the day's journey, went sound asleep in the coach while the drivers whipped the horses onward. It was so dark that, upon arriving at a crossroads, the drivers made a slip and, instead of going to the right, took the road to the left into a dense forest and immediately lost their way. All of a sudden, out rushed a herd of wild beasts, which attacked horses and drivers and devoured them all in a flash. At the uproar, Peter awakened and called the drivers, but there was, naturally, no answer, as they were dead. He climbed out of the coach, and there on the ground lay only the boots of those unfortunate men and the hoofs of the horses. Frightened, the women also left the coach and, in an attempt to get out of the forest, the three of them ran until they came to a clearing, where they dropped from exhaustion. Peter threw up a shelter of branches, in which they rested for the remainder of the night, half dead from fear and running.

At dawn Peter got up before the others and went outside. Some distance away was a fountain, so he picked up his gun, which he always carried with him, and went to wash. Reaching the fountain, he removed his hat and placed his diamond ring on it, so he could wash his hands and face. But as he rinsed, a little bird swooped down, picked up the ring, and flew into a tree with it. Peter grabbed his gun and ran after the bird; but when he took aim, the bird flew to another tree, with the king's son running after it. Peter thus spent the entire day running from tree to tree, without managing to shoot the bird. Night fell, and the little bird went to roost, but it was now too dark for Peter to see a thing. As he hated to lose his ring, he decided to spend the night under the tree and shoot the bird at daybreak. He was actually up before dawn, with his gun aimed at the roost, but the little bird outsmarted him and got away again. What with the bird flying and Peter running, they went quite far away and came to a very high wall, over which the bird disappeared.

It was a thick wall without doors or windows. Peter decided to skirt it and, before too long, found himself in the heart of the woods. There he saw a tree so tall that one of its limbs extended over the wall, so Peter climbed to the top and took a look. On the other side of the wall was a beautiful garden, in which he saw the bird calmly pecking. Peter slid from the branch to the top of the wall, then jumped safely into the garden. With his gun aimed, he crept up on the bird, but this time too it got away, flying over the wall and disappearing into the woods. Peter was now a prisoner in the garden. He tried to scale the wall, but there was no possible way to escape.

In the thick of Peter's struggle, a sorcerer appeared. His eyes blazed as he yelled, "Rogue! Thief! I've caught you at last. Now I know who's been pulling up my plants!"

"No, indeed, sir," answered Peter. "There's surely a mistake. I slipped in here for an entirely different reason, and nothing was further from my mind than destroying or stealing anything of yours."

But the sorcerer refused to listen to reason, and his eyes gleamed with rage: he was bent on putting Peter to death. Seeing himself in a hopeless predicament, Peter fell to his knees and begged the sorcerer not to kill him, telling in detail what had befallen him.

"Very well," replied the sorcerer, "in time I'll see whether or not you're telling the truth. Meanwhile, come with me into my palace."

They went into the palace, where they found the sorceress, wife of the sorcerer. "What's new, my husband?" she asked.

"I found this young man tearing up our beautiful garden. What shall we do with him?"

After hearing the whole story, the sorceress said, "Well, if he's told the truth, we must spare him. Let's test him, husband, to see whether he's a liar or not, and whether he's good for something or good for nothing. Afterward we'll decide what to do with him."

So Peter was put to work in the large garden looking after the flowers and vegetables. He took pains to satisfy the two sorcerers and obey them at all times. The sorcerers were delighted over how beautifully he kept the garden and, all in all, looked on him as their very own son.

Several months went by, and one day the sorcerer said, "Listen, Peter, you are now to dig up this little field here, because I shall sow it in a particular manner of my own." Peter set to work digging, and what should he see as he bent over but the little bird with the ring, which flew right down to the worked ground and began scratching around in it! Peter didn't hesitate a minute, but ran for his gun, aimed, fired and, this time, brought the bird down dead. He touched its crop and felt the ring still there.

At the blast, the sorcerer had come running. "What happened? What happened?" he cried.

"Look, Uncle"—as he now called the sorcerer—"here's the clear proof I'm a nobleman, and that I was telling the truth the first time I set foot in your splendid garden. You remember my telling you about the little bird and the ring? Well, I've killed the bird at last, and the ring is still in its crop."

"This means," replied the sorcerer, "that you can consider yourself as my true son and just as much the owner of everything here as I am."

So Peter lived there as the son of the sorcerer and the sorceress, but he disliked being forever cooped up in that garden and constantly hinted he would like nothing better than to leave. The sorcerer, who truly loved him like a son, realizing what he wanted, said, "Listen, to go outside this wall is quite dangerous, for the surrounding woods are full of wild animals. I'll never understand how you got here without being eaten alive. But if you wait for the day when there's a storm at sea, you will see the water rise to the top of the wall and ships arrive and moor to those spires up there on the roof. If you are patient, you will be able to sail away on one of those ships."

Several months went by before the sorcerer finally announced, "Tomorrow there will be a storm at sea, Peter. If you still want to leave, get ready. I hate to see you go, but do as you will. However, first go into the treasure storeroom and take as much money as you like."

Peter didn't have to be begged. He went down into the treasury and filled his pockets with beautiful coins.

The next morning when he got up, he saw that the sorcerer had spoken the truth: the sea was on a level with the top of the wall, with the ships moored up at the battlements. Peter went to one of the ships and asked, "Captain, what is your destination?"

"I'm bound for the port of Spain."

"Fine! I too will sail for Spain if you'll take me aboard."

He bid the sorcerer and sorceress farewell, thanking them for the kindness they had shown him, went aboard the ship, and landed in Spain a few days later and went to an inn. He had no idea what he had come to the port of Spain to do, so he asked the servant at the inn, "Is there any way to get work here in the city?"

"Why not? There's a man whose job is precisely to find jobs for people, and he comes by here every morning."

When the man showed up, Peter went to him and was told, "If you're interested, the governor is looking for a footman."

Peter said he was interested, the man took him to the governor, and Peter became his confidential servant. Every day he took his master's children to school. Now the master used to give the children a pocketful of coins so that they would learn to practice charity in the street. To whoever asked in God's name, they gave a penny. And everybody who got a penny from the children would then receive five pence from Peter, who had all that money given him by the sorcerer.

Word of this instantly spread all over the city, and the people began grumbling about the governor, saying, "The footman would make a far better governor than the old skinflint who presently governs." A great tumult arose: the people went and shouted under the governor's windows, "Down with him! Down with the governor! We want Peter the footman for our governor!"

But Peter went to the window and signaled for the people to behave, at which they grew calm and left.

Now the governor had a marriageable daughter who was in love with Peter. When she saw that the people wanted him in her father's place, she made such a fuss that the governor had to let her marry Peter. In the meantime Peter continued his almsgiving, only now, instead of five pence, he gave ten. A still greater tumult resulted than before, and the governor thought it wiser to withdraw to one of his country villas. Peter took his place and governed so well that everyone without exception was delighted.

Let's go back a bit, to the wife and maid Peter had left in the shelter of branches when the bird flew off with his ring. Finding Peter gone, the two women went everywhere looking for him and thus passed through many cities and towns. After months and months of travel, they too ended up in the port of Spain. They took lodgings in an inn, had a hairdresser cut their hair short and a tailor made them men's clothing, and asked the servant at the inn if there was work to be had in some house or other.

"There's a man," explained the servant, "who looks for servants in particular for the rich. Speak to him of the matter."

The man arrived, spoke with the two women, and said, "It just so happens our new governor needs a cook and a footman. I'll take you both to him."

They came to terms, and the shoemaker's daughter took the job of cook, while her maid became the footman. But so much time had passed that Peter did not recognize them, nor did they recognize Peter.

Not many days later Peter said to his wife, the governor's daughter, "I won't be home for dinner today. Certain noblemen have invited me out, so I'll leave you by yourself."

"By all means, go," replied his wife. "Rather than stay here by myself and be bored, I'll go to Papa's villa and keep him company awhile. I'll even stay for a few days." So they each went their own way.

Remaining behind at the palace were the cook and the footman—that is, the two women in disguise. The cook said, "I'm going to give the kitchen a thorough cleaning while the master and mistress are away. Hold on to this ring my husband gave me when we became engaged, as I don't want to damage it."

The footman took the ring and slipped it on, so as not to lose it. Then he went to put the master bedchamber in order, and to avoid scratching the ring, he removed it and placed it on the chest of drawers. But once he had finished, he forgot to put the ring back on his finger.

In the evening Peter returned, dined in high spirits, then went to bed. The next morning as soon as he opened his eyes he saw the ring sparkling on top of the chest of drawers. "Whose ring is this?" he wondered, picking it up and examining it from every angle. He had a strong feeling he had seen it before. He rang the bell and asked the footman who put the ring there.

"Oh, please excuse me, sir," replied the footman, "I'm all to blame. I forgot and left the ring there myself. It's not mine, though; it belongs to the cook."

"Send the cook to me, then," said Peter, and the cook came up too.

To make a long story short, what with questions, answers, and explanations, they all ended up recognizing each other. But if the women were excited, Peter was far less so, for he was thinking of his other wife there in Spain, and had no idea how to get out of such a mess. When the governor's daughter returned from the country, Peter took heart and told her of his past and how his first wife had turned up there at the palace. "Tell me now how to get out of this," he concluded, "for I frankly don't know what to do."

His second wife answered, as though nothing were amiss. "Is that all that's bothering you? Do you think I'm jealous? Even if you have two wives instead of one, so what? The Turks have as many as twelve."

Peter couldn't believe his ears. Was it possible to live with two wives who were not at one another's throats?

At nightfall, Peter said, "Well, who is going to sleep with me tonight?"

The governor's daughter replied, "It's only fair for your first wife to, after all this time apart."

So Peter went off to bed with his first wife. But not an hour had gone by when the door opened and the governor's daughter entered with a pistol in each hand. One bullet in Peter's head, one bullet in his wife's head, and the jealous and false-hearted woman was revenged.

At the noise the whole palace woke up and, running into Peter's room, beheld that awful scene. The guards immediately seized the governor's daughter, who was led to the square next morning into the throng of outraged citizens, fastened to a pyre, coated with pitch, and burned alive for the crime she had committed.

(Montale Pistoiese)

◈ 69 ◈

Fanta-Ghirò the Beautiful

In olden times there was a king who had no sons, but only three beautiful daughters. The oldest was Carolina, the next Assuntina, and the youngest was called Fanta-Ghirò the Beautiful, since she was the loveliest of the three.

The king, who was always sick and irritable, stayed shut up in his room the whole day long. He had three chairs—a sky-blue one, a black one, and a red one. Every morning upon going in to greet him, his daughters were quick to note in which chair he sat. If it was sky-blue, that meant high spirits. But the black one spelled death, and the red one war.

One day the girls found their father in the red chair. "Father!" exclaimed the eldest. "What's happened?"

The king replied, "I've just received a declaration of war from the king next door to our land. What will I do? I'm ailing, as usual, and there's no one to take command of the army for me. Where can I get a good general at a moment's notice?"

"If you'll allow me," said the oldest girl, "I'll be your general myself. Do you think I couldn't command the soldiers?"

"Don't be silly! That's no task for a woman!" said the king.

"Do let me try," begged the girl.

"Try. Very well, we shall try it," said the king, "but understand that if, along the way, you get to talking about women's work, you march straight back home."

She agreed to that condition, and the king ordered his trusted squire, Tonino, to mount his horse and ride with the princess to war, but to bring her straight home to the palace the first time she mentioned women's work.

So the princess and squire rode off to war, with the whole army behind them. They had already gone a good way when they came to a cane field and started through it. The princess exclaimed, "What magnificent canes! If we had them at home, we could make any number of distaffs for our spinning!"

"Whoa, princess!" cried Tonino. "I'm under orders to take you back to the palace. You've brought up women's work." They wheeled their horses around, and the whole army about-faced and followed them home.

Then the second girl went to the king. "Majesty, I will take command of the army myself."

"Under the same conditions as your sister?"

"The very same."

They set out on horseback, she and the squire side by side, with the army right behind them. On and on they galloped. They went through the cane field, and the princess said nothing. They passed by a pile of vine stakes, and the princess said, "Look at these fine stakes, Tonino. So straight and thin! If we had them at home, there's no telling how many spindles we could make."

"Whoa, princess!" cried Tonino the squire, stopping her horse. "Back home you go! You brought up women's work."

So the whole army, bag and baggage, took the road back to town.

The king no longer knew which way to turn, when Fanta-Ghirò came to him.

"No, a thousand times NO!" he replied. "You're too young. How could you command an army if neither of your sisters could?"

"Is there any harm in letting me try, Papa? I promise I won't let you down or disgrace you. Let me try."

So it was agreed that Fanta-Ghirò would go to war. She dressed as a warrior, with helmet, armor, sword, and two pistols, and galloped off with Tonino at her side. They passed the cane field without comment; they passed the pile of vine stakes, also without comment. Thus they reached the border. "Before going into battle," said Fanta-Ghirò, "I'd like a word with the enemy king."

The enemy king was a handsome young man. The minute he laid eyes on Fanta-Ghirò he suspected she was a maiden rather than a general, and invited her to his palace to agree on the reasons for the war before going into battle.

They arrived at the palace, and the king ran to his mother. "Mamma, Mamma," he said, "listen! I've brought home with me the general in command of the enemy forces, but just wait until you see him!

> Beautiful Fanta-Ghirò
> With eyes so black and speech so low:
> She's a maiden, I know, I know!"

His mother replied, "Take him into the armory. If the general is really a girl, arms won't interest her at all, and she won't even look at them."

The king led Fanta-Ghirò into the armory. Fanta-Ghirò took down the swords hanging on the walls carefully noting how you gripped them and how heavy they were. Then she moved on to the guns and pistols, breaking them to see how they were loaded. The king ran back to his mother. "Mamma, the general handles weapons like a man. But the more I look at him, the more I'm convinced of what I say.

> Beautiful Fanta-Ghirò
> With eyes so black and speech so low:
> She's a maiden, I know, I know!"

His mother said, "Take him into the garden. If the general is a girl, she will pick a rose or a violet and pin it on her bosom. If he is a man, he will choose the Catalonian jasmine, sniff it, and then stick it behind his ear."

So the king and Fanta-Ghirò went for a stroll in the garden. She reached for the Catalonian jasmine, plucked a blossom, sniffed it, then stuck it behind her ear. In great distress, the king returned to his mother. "The general did what a man would do, but I stick to what I've said all along.

> Beautiful Fanta-Ghirò
> With eyes so black and speech so low:
> She's a maiden, I know, I know!"

Realizing that her son was head over heels in love, the queen said, "Invite him to dinner. If the general holds the bread against his chest when he cuts it, then the general is a girl. But if he holds it in the air and cuts it, he is a man for sure, and you have fallen in love for nothing."

But the results of this test were no better. Fanta-Ghirò cut her bread like a man. The king, however, continued to say to his mother:

> "Beautiful Fanta-Ghirò
> With eyes so black and speech so low:
> She's a maiden, I know, I know!"

"Well, put him to the final test," proposed the queen. "Invite him to swim with you in the fishpond in the garden. If the general is a girl, she will certainly refuse."

He extended the invitation, and Fanta-Ghirò replied, "Of course! I would love to go swimming; not now, though, but tomorrow morning." She took Tonino the squire aside and said, "Leave the palace and return tomorrow morning with a letter bearing my father's seal. The letter should say: 'Dear Son, Fanta-Ghirò, I am deathly ill and wish to see you before I die.'"

The next day they went to the fishpond. The king undressed and dived in first, then invited Fanta-Ghirò to do the same.

"Please wait a little longer, for I'm wet with perspiration," she said, listening for approaching hoofbeats of the squire's horse.

The king insisted that she get undressed. Fanta-Ghirò replied, "I don't know what it is, but I suddenly feel quite uneasy, as though something terrible were about to happen somewhere."

"Nonsense! Nothing is going to happen," answered the king. "Get undressed and jump in! The water is fine. What could go wrong?"

At that moment hoofbeats were heard, and up rode the squire and handed Fanta-Ghirò a letter with the royal seal.

Fanta-Ghirò turned pale. "I'm terribly sorry, Majesty, but this is bad news. My father lies on his deathbed and is asking for me. I must depart at once. All you and I can do is make peace, and if any matters remain to be settled, you will find me at home in my kingdom. Farewell. I will go swimming with you some other time."

The king stayed in the fishpond, alone and naked. The water was cold, and he gave way to despair: Fanta-Ghirò was surely a girl, but she had left before he could prove it.

Before leaving, Fanta-Ghirò stopped by her room to get her things. On the bed, she placed this note:

> Woman came and woman went,
> But of her presence gave the king no hint.

After the king found and read the note, he continued to stand there like a fool, half angry and half jubilant. He ran to his mother. "Mamma, Mamma, I guessed it, the general was a girl after all!" And without giving his mother time to reply, he jumped into his carriage and sped off in the tracks of Fanta-Ghirò.

When Fanta-Ghirò got home, she embraced her father and told him how she had won the war and made the enemy king abandon his plans for an invasion of their kingdom. At that moment the clatter of wheels was heard in the courtyard. It was the enemy king arriving, head over heels in love. As soon as he saw Fanta-Ghirò, he asked, "General, will you marry me?"

The nuptials were celebrated, the two kings made peace, and when Fanta-Ghirò's father died, he left everything to his son-in-law, and Fanta-Ghirò the Beautiful became queen of two kingdoms.

(Montale Pistoiese)

◈ 70 ◈

The Old Woman's Hide

There was a king with three daughters. He was going to the fair and, before leaving home, asked his daughters what they wanted as a present. The oldest said a kerchief, the next a pair of high-top shoes, while the third asked for a box of salt. The two older girls, who were jealous of their little sister, said to their father, "Do you know why that awful girl wants salt? For no other reason than to pickle you."

"I see!" said the father. "She intends to pickle me, does she? Well, I'll turn her out." And that he did.

Turned out of her home with only her nursemaid and a purse of gold coins, the poor girl had no idea where to go. All the young men she met were bothersome, so the nursemaid had an idea. A woman a hundred years old had just died and was being buried, and the nursemaid asked the gravedigger, "Would you sell us the old soul's hide?" After much haggling, the gravedigger picked up a knife, skinned the old woman wrinkle by wrinkle, and sold her whole hide together with face, white hair, fingers, and nails. The nursemaid then tanned the hide, stitched it onto cambric, and clothed the girl in it. People looked at the old centenarian and couldn't get over her spry gait and her voice as clear as a bell.

Whom should they meet but the king's son, who asked the nursemaid, "Just how old is that old soul?"

"Ask her yourself," replied the nursemaid.

"Granny, can you hear me? How old are you?"

"Me? I'm a hundred and fifteen," laughed the girl.

"Good heavens! And where do you come from?"

"From my town."

"Who were your parents?"

"I'm my own mother and father."

"What is your occupation?"

"Having a good time!"

Amused, the king's son said to the king and queen, "Let's take this old soul to the palace. She'll entertain us as long as she lives."

So the nursemaid left the girl at the royal palace, where they gave her a room on the mezzanine, and whenever the king's son had nothing else to do, he'd go in to talk to the old woman and laugh at her droll remarks.

One day the queen said to Rotten Eyes (they called her that because of the blear eyes of the old woman's hide), "What a shame you can't do any more work with those eyes!"

"But I sure could spin as a girl!" replied Rotten Eyes.

"Try spinning this little bit of flax," said the queen, "just to be doing something."

As soon as the old woman was alone, she locked her door, removed the hide, and spun thread that was a marvel to behold. The king's son, the queen, and the whole court were amazed that a decrepit, shaky, half-blind old woman had been able to turn out work like that.

The queen next had her make a blouse. As soon as she was alone, she cut it out, sewed it up, and embroidered the front with the daintiest gold flowers you ever saw. People no longer knew what to think. But the king's son was suspicious and peeped through the keyhole the next time the old woman locked her door. Just what did he see? The old woman removed the hide, and there stood a maiden as beautiful and radiant as the sun. Without thinking, the king's son broke the door down and embraced the girl, who was quite embarrassed and tried to cover herself. "Who are you?" asked the king's son. "Why did you disguise yourself like that?"

The girl confessed that she too was the child of a king, who had cursed her and turned her out of his palace.

The king's son went at once to his parents and said, "I've found a king's daughter, mind you, to marry."

The wedding festivities were proclaimed, and all kings and queens from near and far were invited. Among them was the bride's father, but he didn't recognize her beneath her veil and diadem. The bride had her father's food cooked separately and without salt, except the roast. The soup was brought in and the guests ate it, but after one spoonful her father ate no more of his. Next came boiled meat, but the father scarcely tasted his. Then came fish, which he left completely untouched. "I'm not hungry," he explained. But when the roast was served, he liked it so much that he helped himself to it three times. Then his daughter asked why he'd not touched the other dishes but had relished the roast. The king said he would never understand it, but the roast had been so tasty, and everything else so tasteless.

"So now you see," replied his daughter, "how awful food is without any salt in it? That's why your daughter asked for salt when you went to the fair and those wicked sisters of mine said it was to pickle you . . ."

At that, the father recognized his daughter, embraced her, begged her forgiveness, and punished the envious sisters.

(Montale Pistoiese)

◈ 71 ◈

Olive

Once upon a time a rich Jew lost his wife in childbirth and had to leave his newborn daughter to be raised by a farmer who was a Christian.

In the beginning, the farmer was reluctant to take on the burden. "I have children of my own," he explained, "and can't bring up your little girl in the Hebrew faith. She'd always be with my children and become accustomed to our Christian ways."

"It doesn't matter," replied the Jew. "Please do me a favor and keep her, and you will be repaid. If I've not come back for her by the time she's ten, then you are free to do as you please, for that will mean I'll never return and the child will be with you for good."

The Jew and the farmer came to an agreement, and the Jew left on a journey to distant lands to look after his business. The baby was nursed by the farmer's wife who, finding her so gentle and pretty, became as attached to her as if she'd been her own daughter. The child learned to walk in no time, played with the other children, and did everything children of her age do; but no one ever spoke to her about the Christian commandments. She heard everybody else saying their prayers, but she didn't know what religion was and remained in ignorance of it up to her tenth year.

When she reached ten, the farmer and his wife looked for the Jew to show up any day and reclaim her. But her eleventh year also passed, then the twelfth, thirteenth, and fourteenth, without any sign of the Jew. So they concluded he had died. "We've now waited long enough," they said. "It's high time to have this daughter baptized."

They had her instructed in the faith, then baptized in pomp, with the whole town looking on. They named her Olive and sent her to school to learn women's skills as well as reading and writing. So by the time she was eighteen, Olive was a truly fine girl, well-mannered, loving, beautiful, and cherished by all.

The farmer and his family were now happy and their minds at rest, when one morning a knock was heard on the door. They opened up. It was the Jew. "I've come for my daughter."

"What!" exclaimed the mother. "You said if you'd not returned by the time she was ten for us to do as we pleased, since she'd then be our daughter. Eighteen years have passed. What right can you possibly have to her? We had her baptized, and now Olive is a Christian girl."

"I don't care," replied the Jew. "I didn't show up sooner, because I couldn't. But the girl is my daughter, and I'm taking her back."

"We're not giving her to you, that's for sure!" screamed the whole family in unison.

A bitter quarrel ensued. The Jew took the matter to court, and the court granted him custody of the girl, since she was his own daughter. The poor farmer and his family therefore had no choice but comply with the law. They all wept, the most heartbroken of all being Olive herself as her father was a total stranger to her. With tears streaming down her face, she broke away from those good people who'd been her mother and father for so long.

Bidding her goodbye, the woman slipped Olive a copy of the Office of the Blessed Virgin and urged her never to forget she was a Christian. With that, those two gentle souls separated.

When they got home, the first thing the Jew said to Olive was, "Here, we are Jews, and you are too. You will believe what we believe. Heaven help you if I ever catch you reading the book the woman gave you. The first time I'll throw it into the fire and beat you, and the next time I'll cut off your hands and turn you out of the house. Watch your step, because I mean what I say."

Under those threats, poor Olive had no choice but publicly pretend she was Jewish. Locked in her room, though, she said the Office and Litanies of the Blessed Virgin while her faithful maid kept watch, in case her father should unexpectedly appear. All precaution was useless in the end, for the Jew caught her by surprise one day kneeling and reading from the book. Seized with rage, he flung the book into the fire and beat her unmercifully.

That did not discourage Olive. She had her maid buy her a second book like the first and continued to read in it. But the Jew was suspicious and, without seeming to, watched her constantly. Finally he burst into her room and caught her again. This time, without a word, he led her to a workbench, motioned for her to stretch out her hands, and, with a sharp knife, cut them clean off. Then he ordered her taken into the woods and abandoned.

The unfortunate girl remained there more dead than alive, and with no hands what could she now do? She set out and walked and walked until she came to a large palace. She thought of going in and asking for alms, but the palace was surrounded by a high doorless wall, which enclosed a beautiful garden. Jutting out over the top of the wall were branches of a pear tree laden with ripe, yellow pears. "Oh, if only I had one of those pears!" exclaimed Olive. "Is there any way of reaching them?"

The words were no sooner out of her mouth than the wall opened and the pear tree bent down its branches, so that Olive, who had no hands, could reach the pears with her teeth and eat them while they were still on the tree. When she had eaten her fill, the tree raised its branches once more, the wall closed back together, and Olive returned to the woods. She now knew the secret and went and stood under the pear tree every day at eleven o'clock and made a meal off the fruit. Then she would return to the densest part of the woods and get through the night the best she could.

These were very fine pears, and one morning the king who lived in the palace decided to sample them, so he sent his servant out to pick a few. The servant came back quite dismayed. "Majesty, some animal has been climbing the tree and gnawing the pears down to the core!"

"We'll catch him," said the king. He built a hut out of branches and lay in wait every night, but he lost sleep that way while the pears continued to be nibbled. He therefore decided to watch in the daytime; at eleven o'clock he saw the wall open, the pear tree bend down its branches, and Olive bite into first one pear and then another. The king who had been ready to shoot, dropped his gun in amazement. All he could do was stare at the beautiful maiden as she ate and then disappeared through the wall, which closed behind her.

He called his servant at once, and they scoured the woods for the thief. Suddenly they came upon her sleeping under a bush.

"Who are you? What are you doing here?" asked the king. "How dare you steal my pears? I was about to shoot you down with my shotgun!"

By way of reply, Olive showed him her stumps.

"You poor girl!" exclaimed the king. "What villain mutilated you so cruelly?" After hearing her story, he said, "I don't care about the pears. Come and live in my palace. My mother the queen will indeed keep you with her and look after you."

So Olive was presented to the queen, but the son mentioned neither the pear tree's bending down nor the wall's opening by itself, lest his mother think the girl a witch and detest her. The queen did not actually refuse to take Olive in, but she had no love for her and gave her little to eat, for the simple reason that the king was too charmed by the handless maiden's beauty. To rid him of any idea he might have, she said, "My son, it's time you looked about for a wife. Any number of princesses could be yours for the asking. Take servants, horses, and money, and travel around until you have found her."

The king obediently departed and was away visiting courts in many lands. But six months later he came home and said, "Don't be angry with me, Mamma. There's no shortage of princesses in this world. But I met none so kind and beautiful as Olive. So I've decided Olive is the maiden I'll marry."

"What!" exclaimed the queen. "A handless girl from the woods? You know nothing about her! Would you disgrace yourself like that?"

But the queen mother's words fell on deaf ears, and the king married Olive without further delay.

Having a daughter-in-law of unknown origins was more than the old queen could bear, and she lost no opportunity to be rude and mean to Olive, taking care on the other hand not to displease the king. Wisely, Olive never made any protest.

In the meantime Olive expected a baby, to the great joy of the king; but certain neighboring kings suddenly declared war on him, obliging him to lead his soldiers to the defense of the kingdom. Before leaving, he wanted to entrust Olive to his mother, but the old queen said, "No, I can't assume such responsibility. I too am leaving the palace and shutting myself up in a convent."

So Olive had to stay at the palace by herself, and the king urged her to write him a letter every day. Thus the king left for the battlefield and the old queen for the convent, while Olive remained at the court with all the servants. Every day a messenger left the court with a letter from Olive to the king, but at the same time an aunt of the old queen plied between court and convent to inform her of everything that went on. Upon learning that Olive had given birth to two fine babies, the old queen left the convent and returned to the palace under the pretext of coming home to help her daughter-in-law. She called the guards, forced Olive out of bed, thrust a baby under each of their mother's arms, and told the guards to take the young queen back to the woods where the king had found her.

"Leave her there to starve to death," she said to the guards. "Your heads will roll if you disobey my orders and if you ever breathe a word of this!"

Then the old queen wrote her son that his wife had died in childbirth along with her babies; so that he would believe the lie, she had three wax figures made, then held a grand funeral and burial in the royal chapel. For the ceremony she wore mourning and wept many tears.

Off at war, the king couldn't get over the unfortunate event, nor did he suspect foul play on the part of his mother.

But let's go back to Olive, handless in the middle of the woods and dying of hunger and thirst, with those two babies in her arms. She walked onward until she came to a pool of water where a little old woman was washing clothes.

"My good woman," said Olive, "please squeeze the water out of one of your cloths into my mouth. I'm dying of thirst."

"No," replied the old woman, "do as I say: kneel down and drink right from the pool."

"But can't you see I have no hands and must hold my babies in my arms?"

"That doesn't matter. Go on and try."

Olive knelt down, but as she bent over the pool, both babies slipped out of her arms, one after the other, and disappeared under the water. "Oh, my babies! My babies! Help! They're drowning! Help me!"

The old woman didn't budge.

"Don't be afraid, they won't drown. Fish them out."

"How can I? Don't you see I have no hands?"

"Plunge in your stumps."

Olive immersed her stumps in the water and felt her hands growing back. With her hands she then grabbed hold of the babies and pulled them up safe and sound.

"Be on your way now," said the old woman. "You no longer lack hands to do for yourself. Farewell." She was out of sight before Olive could even thank her for her fine deed.

Wandering about the woods in search of a refuge, Olive came to a brand-new villa with the door wide open. She went in to ask for shelter, but no one was there. A kettle of porridge was boiling on the hearth next to some heavier foods. Olive fed her children, ate something herself, then went into a room where there was a bed and two cradles; she put the two children to bed, then lay down herself. She thus lived in the villa without ever needing a thing or seeing a soul anywhere around the place.

But let's leave her and go back to the king, who went home when the war was over and found the town in mourning. His mother tried to comfort him, but he was more and more unhappy as time went on. In an effort to cheer up, he decided to go hunting. In the woods he was overtaken by a storm, and it looked as though the earth would yawn under all the thunder and lightning. "If only I might die!" said the king to himself. "What reason do I have to go on living without Olive?" Through the trees he spied a faint light and moved toward it in search of shelter. He knocked, and Olive opened. He did not recognize her, and she said nothing, but welcomed him cordially and invited him up to the fire to warm himself while she and the children bustled about to make him comfortable.

The king watched her, thinking how much like Olive she was; but noticing her perfectly normal hands, he shook his head. As the children jumped and played around him, he said, "I might have been blessed with children like that, but they died, alas, with their mother, and here I am all alone and miserable!"

Meanwhile Olive had gone to turn down the guest's bed and called the children to her. "Listen," she whispered to them, "when we go back in the other room, ask me to tell you a story. I'll refuse and even threaten to slap you, but you keep on begging me."

"Yes, Mamma, we'll do that."

So when they got back to the fireside, they began saying, "Mamma, tell us one of your stories!"

"What are you thinking of! It's late, and the gentleman is tired and doesn't want to hear any story!"

"Come on, Mamma, please!"

"If you're not quiet, I'll slap you!"

"Poor little things!" said the king. "How could you slap them? Go on and make them happy. I'm not at all sleepy and would like to hear a story too."

With that encouragement, Olive sat down and began her tale. The king gradually became serious, listened more and more anxiously, asking repeatedly, "And then? And then?" because it was the life story exactly of his poor wife. But he didn't dare get his hopes up, for the mystery of the hands was still unexplained. Finally he broke down and asked, "And about her hands that were cut off, how did that turn out in the end?" Olive therefore told about the old washerwoman.

"Then it's you!" cried the king, and they hugged and kissed. But after they had given vent to their joy, the king's face darkened. "I must return to the palace at once and punish my mother as she deserves!"

"No, not that!" said Olive. "If you really love me, you must promise not to lay a hand on your mother. She will be sorry enough as it is. The poor old soul believed she was acting in the interest of the kingdom. Spare her life, since I forgive her for all she has done to me."

So the king returned to the palace and said nothing to his mother.

"I was uneasy about you," she said to him. "How did you get through the night out in that storm?"

"I passed a good night, Mamma."

"What!" said the queen, growing suspicious.

"Yes, at the home of kind-hearted people who kept my spirits up. It was the first time since Olive's death I've felt cheerful. By the way, Mamma, is Olive really dead?"

"What do you mean? The whole town was at the funeral."

"I'd like to put some flowers on her grave, and see with my own eyes . . ."

"Why all the suspicion?" asked the queen, flushed with anger. "Is that any attitude for a son to have toward his mother, doubting her word?"

"Go on, Mamma, enough of these lies! Olive, come in!"

In walked Olive leading their children. The queen, who had been crimson with rage, now turned white with fear. But Olive said, "Don't be afraid, we'll do you no harm. Our joy over finding one another again is too great to feel anything else."

The queen returned to the convent, and the king and Olive lived in peace for the rest of their life.

(Montale Pistoiese)

❖ 72 ❖

Catherine, Sly Country Lass

One day a farmer hoeing his vineyard struck something hard. He bent over and saw that he had unearthed a fine mortar. He picked it up, rubbed the dirt off, and found the object to be solid gold.

"Only a king could own something like this," he said. "I'll take it to my king, who will most likely give me a handsome present in return!"

At home he found his daughter Catherine waiting for him, and he showed her the mortar, announcing he would present it to the king.

Catherine said, "Beyond all doubt, it's as lovely as lovely can be. But if you take it to the king he'll find fault with it, since something is missing, and you'll even end up paying for it."

"And just what is missing? What could even a king find wrong with it, simpleton?"

"You just wait; the king will say:

'The mortar is big and beautiful,
But where, you dummy, is the pestle?' "

The farmer shrugged his shoulders. "The idea of a king talking like that! Do you think he's an ignoramus like you?"

He tucked the mortar under his arm and marched straight to the king's palace. The guards weren't going to let him in, but he told them he was bringing a wonderful gift, so they took him to His Majesty. "Sacred Crown," began the farmer, "in my vineyard I found this solid gold mortar, and I said to myself that the only place fit to display it was your palace. Therefore I am giving it to you, if you will have it."

The king took the mortar and turned it round and round, running his eye over every inch of it. Then he shook his head and spoke:

"The mortar is big and beautiful,
But missing is its pestle."

Catherine's words exactly, except that the king didn't call him a dummy, since kings are well-bred persons. The farmer slapped his brow and couldn't help but exclaim, "Word for word! She guessed it!"

"Who guessed what?" asked the king.

"I beg your pardon," said the farmer. "My daughter told me the king would say just those words, and I refused to believe her."

"This daughter of yours," said the king, "must be a very clever girl. Let's see just how clever. Take her this flax and tell her to make me shirts for a whole regiment of soldiers. But tell her to do it quickly, since I need the shirts right now."

The farmer was stunned. But you don't argue with a king, so he picked up the bundle (which contained only a few measly strands of flax), bowed to the king, and set out for home, leaving the mortar without receiving a word of thanks, much less anything else.

"My daughter," he said to Catherine, "you are really in for it now." And he told her what the king had ordered.

"You get upset over nothing," replied Catherine. "Give me that bundle." She took the flax and shook it. As you know, there are always scalings in flax, even if it has been carded by an expert. A few scalings dropped on the floor, so tiny you could scarcely see them. Catherine gathered them up and said to her father, "Here. Go right back to the king and tell him for me that I will make him the shirts. But since I have no loom to weave the cloth, tell him to have one made for me out of this handful of scalings, and his order will be carried out to the letter."

The farmer didn't have the nerve to go back to the king, especially with that message; but Catherine nagged him until he finally agreed.

Learning how cunning Catherine was, the king was now eager to see her with his own eyes. He said, "That daughter of yours is a clever girl! Send her to the palace, so that I'll have the pleasure of chatting with her. But mind that she comes to me neither naked nor clothed, on a stomach neither full nor empty, neither in the daytime nor at night, neither on foot nor on horseback. She is to obey me in every single detail, or both your head and hers will roll."

The farmer arrived home in the lowest of spirits. But his daughter merrily said, "I know how, Daddy. Just bring me a fishing net."

In the morning before daybreak, Catherine rose and draped herself with the fishing net (that way she was neither naked nor clothed), ate a lupin (that way her stomach was neither empty nor full), led out the nanny goat and straddled it, with one foot dragging the ground and the other in the air (that way she was neither on foot nor on horseback), and reached the palace just as the sky grew lighter (it was neither day nor night). Taking her for a madwoman in that outlandish get-up, the guards barred the way; but on learning that she was just carrying out the sovereign's order, they escorted her to the royal chambers.

"Majesty, I am here in compliance with your order."

The king split his sides laughing, and said, "Clever Catherine! You're just the girl I was looking for. I am now going to marry you and make you queen. But in one condition, remember: you must never, never poke your nose into my business." (The king had realized that Catherine was smarter than he was.)

When the farmer heard about it, he said, "If the king wants you for his wife, you have no choice but to marry him. But watch your step, for if the king quickly decides what he wants, he can decide just as quickly what he no longer wants. Be sure to leave your workclothes hanging up here on a hook. In case you ever have to come home, you'll find them all ready to put back on."

But Catherine was so happy and excited that she paid little attention to her father's words, and a few days later the wedding was celebrated. There were festivities throughout the kingdom, with a big fair in the capital. The inns were filled to overflow, and many farmers had to sleep in the town squares, which were crowded all the way up to the king's palace.

One farmer, who had brought to town a pregnant cow to sell, found no barn to put the animal in, so an innkeeper told him he could put it under a shed at the inn and tether it to another farmer's cart. Lo and behold, in the night, the cow gave birth to a calf. In the morning the proud owner of the cow was preparing to lead his two animals away when out rushed the owner of the cart, shouting, "That's all right about the cow, she's yours. But hands off the calf, it's mine."

"What do you mean, it's yours? Didn't my cow have it last night?"

"Why wouldn't it be mine?" answered the other farmer. "The cow was tied to the cart, the cart's mine, so the calf belongs to the owner of the cart."

A heated quarrel arose, and in no time they were fighting. They grabbed props from under the cart and struck in blind fury at one another. At the noise, a large crowd gathered around them; then the constables ran up, separated the two men, and marched them straight into the king's court of justice.

It was once the custom in the royal city, mind you, for the king's wife also to express her opinion. But now with Catherine as queen, it happened that every time the king delivered a judgment, she opposed it. Weary of that in no time, the king said to her, "I warned you not to meddle in state business. From now on you'll stay out of the court of justice." And so she did. The farmers therefore appeared before the king alone.

After hearing both sides, the king rendered this decision: "The calf goes with the cart."

The owner of the cow found the decision too unjust for words, but what could he do? The king's judgment was final. Seeing the farmer so upset, the innkeeper advised him to go to the queen, who might find a way out.

The farmer went to the palace and asked a servant, "Could you tell me, my good man, if I might have a word with the queen?"

"That is impossible," replied the servant, "since the king has forbidden her to hear people's cases."

The farmer then went up to the garden wall. Spying the queen, he jumped over the wall and burst into tears as he told how unjust her husband had been to him. The queen said, "My advice is this. The king is going hunting tomorrow in the vicinity of a lake that is always bone-dry at this time of year. Do the following: hang a fish-dipper on your belt, take a net, and go through the motions of fishing. At the sight of someone fishing in that dry lake, the king will laugh and then ask why you're fishing where there's no water. You must answer: 'Majesty, if a cart can give birth to a calf, maybe I can catch a fish in a dry lake.'"

The next morning, with dipper dangling at his side and net in hand, the farmer went off to the dry lake, sat down on the shore, lowered his net, then raised it as though it were full of fish. The king came up with his retinue and saw him. Laughing, he asked the farmer if he had lost his mind. The farmer answered him exactly as the queen had suggested.

At that reply, the king exclaimed, "My good man, somebody else had a finger in this pie. You've been talking to the queen."

The farmer did not deny it, and the king pronounced a new judgment awarding him the calf.

Then he sent for Catherine and said, "You've been meddling again, and you know I forbade that. So now you can go back to your father. Take the thing you like most of all in the palace and go home this very evening and be a farm girl once more."

Humbly, Catherine replied, "I will do as Your Majesty wills. Only, I would ask one favor: let me leave tomorrow. Tonight it would be too embarrassing for you and for me, and your subjects would gossip."

"Very well," said the king. "We'll dine together for the last time, and you will go away tomorrow."

So what did sly Catherine turn around and do but have the cooks prepare roasts and hams and other heavy food that would make a person drowsy and thirsty. She also ordered the best wines brought up from the cellar. At dinner the king ate and ate and ate, while Catherine emptied bottle after bottle into his glass. Soon his vision clouded up; he started stuttering and at last fell asleep in his armchair, like a pig.

Then Catherine said to the servants, "Pick up the armchair with its contents and follow me. And not a word out of you, or else!" She left the palace, passed through the city gate, and didn't stop until she reached her house, late in the night.

"Open up, Daddy, it's me," she cried.

At the sound of his daughter's voice, the old farmer ran to the window. "Back at this hour of the night? I told you so! I was wise to hold on to your workclothes. They're still here hanging on the hook in your room."

"Come on, let me in," said Catherine, "and don't talk so much!"

The farmer opened the door and saw the servants bearing the armchair with the king in it. Catherine had him carried into her room, undressed, and put into her bed. Then she dismissed the servants and lay down beside the king.

Around midnight the king awakened. The mattress seemed harder than usual, and the sheets rougher. He turned over and felt his wife there beside him. He said, "Catherine, didn't I tell you to go home?"

"Yes, Majesty," she replied, "but it's not day yet. Go back to sleep."

The king went back to sleep. In the morning he woke up to the braying

of the donkey and the bleating of the sheep, and saw the sunshine streaming through the window. He shook himself, for he no longer recognized
the royal bedchamber. He turned to his wife. "Catherine, where on earth
are we?"

She answered, "Didn't you tell me, Majesty, to return home with the
thing I liked best of all? I took *you*, and I'm keeping you."

The king laughed, and they made up. They went back to the royal
palace, where they still live, and from that day on, the king has never
appeared in the court of justice without his wife.

(Montale Pistoiese)

◈ 73 ◈

The Traveler from Turin

There once lived in the city of Turin a well-to-do gentleman with three
sons. The oldest was Joseph, a clever youth who dreamed constantly of
taking a trip: he was eager to see the city of Constantinople. His father,
who wanted him to marry, have children, and become his heir, was
reluctant to let him go; but Joseph thought of nothing but journeys.
Finally the middle son got married, and the father then looked to him as
the one who would take his place in the business and continue his name.
He therefore consented to the departure of Joseph, who embarked for
Constantinople with a trunk full of personal belongings, sundries, and
money.

A storm came up on the high seas, the boat began pitching, and the
sailors lost all control of it. Now off course, it crashed on a reef. All the
passengers were swept under the waves and drowned. Joseph, having
jumped from the sinking ship, straddled his precious trunk and spent a
whole night buffeted by the storm until the wind, at last, brought him
ashore on an island. The sun rose, and the sea became calm once more.
The island appeared deserted, although abounding in fruit trees.

But while Joseph was exploring the surroundings, a band of savages
dressed in animal skins popped up. Joseph went over to them and asked
for hospitality and also if they would carry his trunk, but he couldn't
make them understand him. He then pulled out a gold coin and offered

it to them. They looked at it as though they didn't know what to do with it. He showed them his watch, and he might as well have shown them the heel of a shoe. He produced a knife and cut off the branch of a tree. When they saw that, their interest perked up, and many reached for the knife. In sign language, Joseph explained he would give it to none but their leader, so they took up his trunk on their shoulders and led him to the cave where their king lived.

Joseph and the king became good friends in no time. The youth stayed in the royal cave and learned the savages' language. And he taught them many things they didn't know, such as how to bake bricks and build houses from the island's rich supply of clay and limestone. The king named him viceroy and ended up offering him his daughter in marriage. Joseph did not welcome this last honor, either because he was already in love with a beautiful savage, or because the king's daughter was the ugliest girl he had ever seen. But he was alone in the midst of this uncivilized people, on an island from which there was no escape, and woe to him should he displease the king! He had no choice but consent to the marriage. He and his beloved separated in tears, but their bond of love remained as strong as ever. Joseph married the king's daughter, while his beloved, so as to arouse no suspicion, also got married, becoming the wife of an old fisherman. The youth, from a material standpoint, couldn't have been better off. He wasn't king, but he wasn't far from it. Just one thing was lacking: happiness. He felt cooped up on the island like a slave and was sorry now he hadn't listened to his father.

Unexpectedly, the king's daughter became ill and died. All the kingdom went into deep mourning, while the king was inconsolable, weeping and grieving without cease. In an effort to console him, Joseph said, "But, Majesty, you must accept this loss. True, you no longer have your daughter, but I'm still here to keep you company."

"Alas!" said the king. "I'm weeping not only over losing my daughter but also over losing you."

"Losing me?" exclaimed Joseph. "What do you mean, Majesty?"

"Don't you know the laws of these parts?" asked the king. "If a husband or a wife dies, the survivor must be buried with the deceased. Our laws and customs require it. You have to comply."

In vain did Joseph protest and wail. The funeral procession formed. The pallbearers carried the coffin in which his wife was laid out as a queen; Joseph followed, half frozen with fear, then came all the people moaning and weeping. The tomb was a vast underground cavern sealed off by a boulder. Whenever anyone died, the boulder was rolled aside and he was entombed there with all his wealth. It was Joseph's wish to have his trunk of treasures lowered into the cavern with him. The people thus obliged him, also giving him food enough for five days, and a lamp. When the ceremony was over, they rolled the boulder back into place and left Joseph alone with the corpse.

Using his lamp, he decided to explore the cavern. It was full of deceased people, some who had died recently and others who were now nothing but skeletons; with the dead were treasures of gold, silver, and precious stones. He thought how worthless all this wealth was to him, condemned by that savage practice to end his days there. Tired and heavyhearted, he sat down on his trunk, pulling out his watch every now and then, checking the time, and grappling with the idea of dying. Shortly after midnight he heard the sound of trampling. He glanced around and saw an animal something like a huge ox enter the cavern. It approached a body, caught it by the hair, jerked it around onto his back, and disappeared into the dark. The next night at the same hour, the animal returned and carried off another body. This time Joseph followed him. The cavern ended in a passageway that ran downhill, and from the roar of water that reached his ears from the bottom, he realized that the passage led into the sea. The discovery filled him with joy; he was now sure he would get out of the cavern alive, but he didn't want to flee empty-handed when all that wealth lay within such easy reach. He therefore put off his escape until the next day, since it was almost daylight and the islanders might see him leaving.

He spent the day getting together the treasure he would carry off, when suddenly he heard the customary dirge sung in funeral processions and saw the door of the cavern rolled back. The body of a man was lowered, followed by a living woman with a lamp and a basket of food. Joseph hid behind a boulder, waiting for the cavern to be closed before showing himself to his companion-in-misfortune. When she spied his trunk at the back of the cavern, she went up to it and wept. "My poor Joseph! He is most certainly dead by now, and the same awful fate awaits me." At that, Joseph recognized the woman: she was his beloved, who had married an old fisherman just now deceased. He came out of hiding and embraced her, saying, "No, I've not died yet, and I don't intend to die. You and I will flee this tomb together."

Once the woman was over her initial shock and convinced that Joseph was not a ghost, she said, "No one ever came out of here alive. How can you be so hopeful in spite of that?" Joseph told her about his discovery, and they ate the new provisions she had brought and waited for the arrival of the ox.

As soon as the ox had come and disappeared with a body, Joseph trailed him until he saw the gleam of moonlight on the sea at the bottom of the cavern and the ox swimming away with the corpse on his back.

Then Joseph, too, jumped into the water, swam around the island, crawled through the dark up to the mouth of the cavern and, with great effort, moved aside the boulder. He lowered the rope he had brought along, and his friend, who waited uneasily in the cavern, tied it around as much treasure as he could draw up at a time. He pulled up bag after bag of gold, silver, and jewels filched from the dead. The last item to come up was Joseph's trunk, and it was followed by his friend.

Safely out of the cavern with the treasure, they headed for another kingdom on the same island, arriving before daybreak. They introduced themselves to the king, told him their story, and were hospitably welcomed into the king's own dwelling.

Joseph spent many years in that kingdom, and had three sons. Although he lived comfortably and had become the king's chief minister, he constantly yearned to go back to Turin, his birthplace. He built a boat, under the pretext of using it for pleasure, and sailed out to sea and back in the same day with his wife, so as not to arouse the king's suspicions. Then one calm night he embarked with his wife and sons and the trunk and all the treasures, and rowed out of sight of the island. Upon spotting a distant ship in the moonlight, he sounded a call for help through the waterspout. The ship happened to be going to Constantinople. Thus, Joseph saw the dream of his youth come true. He went to Constantinople and opened up a jewelry store with the treasures from the cavern of the dead. Then, rich and happy, he went back to Turin, where his old father was still waiting for him.

(Montale Pistoiese)

◈ 74 ◈

The Daughter of the Sun

A king and a queen who had waited for ages were at last about to have a child. They called in the astrologers to learn if it would be a boy or a girl and under what planet it would be born. After looking at the stars, the astrologers said the baby would be a girl; she was destined, they added, to be loved by the Sun before she was twenty and to bear his daughter. The king and queen were quite outdone to learn their daughter would have a child by the Sun who stays in the sky and can't marry. To ward off such a fate, they had a tower built with windows so high up that not even the Sun himself could reach to the bottom of it. The baby girl was shut up inside with her nurse, to remain there until she turned twenty, without once seeing the Sun or being glimpsed by him.

The nurse had a daughter the same age as the king's, and the two little girls grew up together in the tower. One day when they were almost twenty and musing on the wonderful things that must be in the world outside the tower, the nurse's daughter said, "Let's try climbing up to those windows by placing one chair on top of the other. That way we'll get an idea of what's outside."

No sooner said than done! They piled up chairs all the way to the windows, looked out, and beheld trees, river, soaring herons and, high in the sky, clouds and the Sun. The Sun saw the king's daughter, fell in love with her, and sent her one of his rays. From the instant that ray touched her, the girl began expecting a baby—the daughter of the Sun.

The Sun's daughter was born in the tower, and the nurse, fearful of the king's anger, carefully wrapped it in royal swaddling clothes and carried it to a patch of broadbeans, where she abandoned it. In no time afterward the king's daughter turned twenty and her father let her out of the tower, thinking the danger had passed. He had no idea that everything had already happened and that the baby girl born to his daughter and the Sun lay weeping in a bean patch.

Now through that field passed another king on his way hunting. He heard the wails and, pitying the beautiful little baby left among the beans, took her home to his wife. They found a wet nurse for her, and the child was brought up at the palace just as though she were the king and queen's own daughter along with their son, who was only a little older than she.

Growing up together, the boy and girl eventually fell in love with each other. The king's son was eager to marry her, but his father was unwilling for him to wed a foundling and sent her away from the palace to a distant and isolated house, in hopes the boy would forget her. The king never dreamed that the girl was the daughter of the Sun and endowed with all the magic skills lacking in men.

As soon as she was out of the way, the king betrothed his son to a girl of royal birth. The day of the wedding, sugared almonds were sent to all the relatives and friends of the bride and groom, including the girl found in the bean patch.

When the king's messengers knocked on her door, she came down to open it, but without her head. "Oh, I'm sorry," she said, "but I was combing my hair and left my head on the dresser. Let me fetch it." She showed the messengers into the house, replaced her head, and smiled.

"Now tell me what I should give you to take back as a wedding present." She led the messengers into the kitchen. "Open up, oven!" she commanded, and the oven door opened. The Sun's daughter looked at the messengers and smiled. "Into the oven with you, wood!" and the wood flew into the oven. The Sun's daughter again smiled and commanded, "Light up, oven, and call me when you're hot!" She turned to the messengers and asked, "Well, what's the good news?"

Deathly pale, with their hair standing straight up, the messengers groped for words, when the oven cried, "My lady!"

The Sun's daughter said, "Excuse me," plunged headlong into the fiery oven, turned around, and stepped back out holding a beautiful pie all ready to serve. "Take this to the king for the wedding banquet."

When the messengers returned wild-eyed and, speaking scarcely above a whisper, told all the things they had seen, no one would believe them. But the bride, who was jealous of the girl everyone knew as the prince's first sweetheart, said, "That's nothing! I used to do those things all the time when I lived at home."

"All right," answered the bridegroom, "let's see you do them here for us."

"Indeed I will, but . . ." she began as he pulled her into the kitchen.

"Wood, into the oven," said the bride, but the wood didn't budge. "Fire, light up," but the oven remained cold. The servants lit it themselves, and as soon as it was hot, the boastful bride insisted on getting into it. She wasn't all the way in before she had burned to death.

After a short time, the king's son was persuaded to take another wife. The day of the wedding, messengers went back to the Sun's daughter with sugared almonds. Instead of answering the door when they knocked, she came through the wall and greeted them. "Excuse me, but the door doesn't open from the inside. I always have to come through the wall and open it from the outside. There, it's open now. Please walk in."

Leading them to the kitchen, she asked, "So, what should I prepare this time for the wedding of the king's son? Wood, into the fire! Fire, light up!" All this took place in a split second right before the eyes of the messengers, who broke out in a cold sweat.

"Skillet, onto the burner! Oil, into the skillet! And call me when you are hot!"

In a little while the oil called, "My lady, I'm ready!"

"Here we go," said the Sun's daughter, smiling, and thrust her fingers into the boiling oil. At once the ten fingers turned into the most beautiful fried fish you ever saw. The Sun's daughter wrapped them up herself, since her fingers had grown back in the meantime, and handed them to the messengers with a smile.

When the new bride, who was as jealous and boastful as the first, heard the dumbfounded messengers' tale, she said, "You should see the fish *I* fry!"

The bridegroom took her at her word and had oil boiled in a skillet. The vain soul thrust in her fingers, and the pain from her scald killed her.

The queen mother took the messengers to task. "With your tales you're the death of all the brides!"

However, the king and queen found their son a third bride, and messengers went out with sugared almonds on the day of the wedding.

"Hello, I'm up here!" said the Sun's daughter when they knocked. Looking all about them, they spied her up in the air. "I was just taking a little stroll on a spiderweb. I'll be right down." She climbed down the spiderweb and took the almonds.

"This time I truthfully don't know what to do about a present," she said. After thinking it over, she called, "Knife, come here!" The knife came forward, she caught hold of it, and cut off one of her ears. Attached to her ear was a strip of gold lace which came out of her head as though unwinding from her brain. She pulled and pulled until the lace came to an end. Then she put her ear back in place, tapped it gently with her finger, and all was just as before

It was such beautiful lace that the whole court wanted to know where it came from, so the messengers, despite the seal placed on their lips by the queen mother, ended up telling about the episode with the ear.

"Oh!" exclaimed the new bride, "I've trimmed all my gowns with lace I obtained in that very manner."

"Take the knife and let's see you do it!" directed the bridegroom.

The idiot therefore cut off one of her ears. Instead of lace, out flowed so much blood that she died.

The king's son went on losing wives and was now more in love than ever with that maiden. He eventually got sick, and no one knew how to cure him, for he neither ate nor laughed.

They sent for an old sorceress, who advised, "You must feed him barley pap made from barley that is sown, grown, reaped, and made into pap all within the hour."

The king was frantic, for barley like that had never been seen. At last they thought of the maiden who could work so many wonders and sent for her.

"Yes, indeed, I'm familiar with barley like that." In a flash she had it sown, grown, reaped, and made into pap, well before the end of the hour.

She insisted on taking the pap in person to the king's son, who lay in bed with his eyes closed. But it was vile pap, and he took one taste and spat it out, and some flew into the maiden's eye.

"How dare you spit pap into the eye of the daughter of the Sun and granddaughter of the king!"

"You're the daughter of the Sun?" asked the king, who was standing nearby.

"I am."

"And a king's granddaughter?"

"I am."

"And here we thought you were a foundling! In that case you can marry our son!"

"Of course I can!"

The king's son got well that very instant and married the Sun's daughter, who from that day onward became like all other women and did no more strange things.

(Pisa)

◈ 75 ◈

The Dragon and the Enchanted Filly

There was once a king and queen who had no children. The royal couple constantly prayed for a baby and gave generously to the poor. At last the queen found herself with child, and the king sent for astrologers to find out if it would be a boy or a girl and under what star it would be born. The astrologers replied, "You will have a boy, who the minute he turns twenty will take a wife, and in the same instant he will slay her. Otherwise he would turn into a dragon." The king and queen were all smiles when the astrologers informed them they would have a son who would marry at the outset of his twentieth year. But upon hearing the rest of the prophecy they burst into tears.

The son was born and grew into a fine young man. That was no little comfort to his parents, who nevertheless shuddered at the thought of his terrible fate. As his twentieth birthday approached, they sought a wife for him and asked for the hand of the queen of England.

Now the queen of England had a talking filly who told her owner everything and was, beyond all doubt, her best friend. As soon as the queen became engaged, she spread the word to the filly. "You have no cause to rejoice," replied the filly, a bewitched animal who knew everything. "The truth of the matter is . . ." and she revealed the prince's strange destiny. The queen was horrified and wondered what she should now do. "Listen carefully," said the filly. "Tell your bridegroom's father the queen of England will not ride to the wedding in a carriage but on horseback. Come wedding day, you will mount me and proceed to the church. The instant I paw the ground, throw your arms around my neck and leave everything to me."

In the wedding procession the filly, draped in gala trappings, stood beside the bridegroom's carriage, while on her sat the queen of England in her wedding dress. Every now and then the queen peered through the carriage window at the bridegroom with a sword on his lap and at her in-laws holding watches and awaiting the exact time when he had been born. Suddenly the filly pawed the ground with all her might, then sped off like lightning, with the bride holding on for dear life. The fatal hour had struck, and the bridegroom's parents dropped their watches. Right before their eyes the king's son had turned into a dragon, sending king, queen, and courtiers fleeing for their lives from the overturned carriage.

The filly reached a farmhouse and drew to a halt. "Dismount," she told the queen, "and go in and tell the farmer to give you his clothes in exchange for your royal ones." The farmer could hardly believe he was getting a real queen's dress, and a wedding dress at that. In exchange, he gave her his coarse shirt and breeches. The queen came back out dressed as a farmer, jumped into the saddle, and continued on her way

They came to the palace of a second king, and the filly said, "Go to the stables and see if they'll engage you as a stableboy." She did, impressing the people as a bright boy who also had that fine filly, so they said, "We'll hire you with your filly to work here with us."

Now the king had a son the girl's age exactly. The boy no sooner saw the new groom than a certain thought struck him, which he confided to his mother. "Mamma, I may be wrong, but I believe that new stableboy is a girl and one that appeals to me."

"No, no," replied his mother, "you're all wrong. If you want to find out for sure, take him to the garden and show him the flowers. If he makes a bouquet, then you'll know your stableboy is actually a girl. If he pulls a flower and sticks it in his mouth, he's a man."

The prince called the stableboy into the garden and said, "Would you like to make a bouquet of flowers?"

But the filly who knew everything had already warned the false stableboy, who replied, "No, thanks, I don't care much for flowers," and pulled a blossom and stuck it in her mouth.

"What did I tell you? He's a man for sure," replied the queen when the prince related the incident.

"I don't care what you say, Mamma. I'm more convinced than ever that the stableboy is really a girl."

"Try something else, then: invite him to the table to cut the bread. If he holds it up to his chest, your stableboy is actually a girl. If he holds it away from him to cut it, then he's a man."

This time too the filly warned her mistress, who held the bread away from her like a man. The prince was still not convinced, though.

"The only thing left," said his mother, "is to see him fence. Arrange a match with him."

The filly taught the girl all the subtleties of fencing, but concluded, "This time, my dear, you will be found out."

No fault could ever be found with her fencing, but in the end she fainted from exhaustion. And that way they finally discovered she was a girl. The prince was so deeply in love by now that he resolved at all costs to marry her.

"Marry her without any idea who she is?" exclaimed his mother.

They asked her to tell her story, and upon learning that she was the queen of England, the prince's mother made no more objections to the wedding, which was celebrated with great pomp.

A little later, the new wife found herself with child, when the king was summoned to war. But being an old man, he sent his son in his place. The prince urged his parents to write him as soon as the baby was born, mounted his wife's filly, and rode off to battle. Before leaving, however, the filly gave her mistress three hairs from her mane, saying, "Hide these in your bosom. Break them in an emergency, and I'll come to your aid."

In due time the princess gave birth to twins, a boy and a girl, who were the most beautiful children you ever saw. The king and queen wrote their son the good news at once. Now as the messenger was riding to the prince with the letter, a dragon lay in wait for him midway to the battlefield. It was none other than the other king's son who had turned into a monster on his wedding day. Seeing the messenger approach, it blew its noxious breath down the road, and the man fell from his saddle in a deep sleep. Then the dragon pulled the letter from the messenger's pocket, read it, and forged a new one saying the princess had borne two dogs, a male and a female, thus turning the whole town against her. This letter went into the pocket of the messenger, who, finding nothing amiss upon awaking, got back onto his horse and rode to the prince.

When the prince read the letter, he turned as white as a sheet, but said nothing. He sat down immediately and wrote a reply: "Be they male dogs or females, keep them for me and take good care of my wife."

On the way back, the messenger was again spotted by the dragon and put to sleep by its breath as he came down the road. The dragon replaced the letter the messenger bore with one that read: "Burn wife and children at the stake in the town square. If king and queen do not comply, they too shall go up in flames."

Such a reply threw the town into alarm. What was the meaning of the prince's fury? But instead of burning those innocent souls, the king and queen put the princess into a boat along with the children, two nurses, food, water, and four oarsmen, and secretly launched them on the sea. Then they carried to the town square three dummies resembling the princess and her babies and set them afire. The citizens, who had grown to love the princess, were outraged and vowed revenge.

The princess sailed across the sea and was put ashore with her babies. She was walking along the deserted strand, when before her loomed the dragon. She had already given herself and the children up for doomed, when she remembered the three hairs from her filly's mane. She pulled them from her bosom, broke one, and saw an impenetrable thicket instantly spring up around her. But the dragon plunged into it and twisted his way right on through. She broke another hair, and out gushed a river wide and deep. The dragon had a time in the swift current, but he finally got across the river as well. She frantically broke the last hair as the dragon was about to seize her, and a tongue of fire shot up and expanded into a mighty fire. But the dragon passed through the fire, too, and had her in his hands, when onto the shore galloped the filly.

They faced one another, filly and dragon, and then began to fight. The dragon was taller, but the filly reared and kicked and bit so furiously that she laid him low and crushed him to bits. The princess rushed up to embrace the filly, but her joy was shortlived, for the filly closed her eyes, hung her head, and dropped lifeless to the ground. The princess wept as though she had lost her own sister, recalling all that the filly had done for her.

She was there weeping with her children, when she happened to look up and see a large palace she didn't remember seeing before. Moving closer, she noticed a beautiful lady at a window motioning to her to come inside. She entered with her children, and the lady embraced her, saying, "You don't recognize me, but I'm the filly. I was under a spell and couldn't change back into a woman until I'd slain a dragon. When you broke the hairs of my mane, I left your husband on the battlefield and ran to you. Killing the dragon, I broke the spell."

Let's leave them for the time being and turn to the husband when he saw the filly flee from the battlefield. He thought to himself, Something must have happened to my wife! and hurried to win the war so he could go home.

When he got back to town, all the citizens rose up against him. "Tyrant! Monster!" they screamed. "What crime had that poor woman and her children committed?" He couldn't for the life of him understand what the people were talking about. When his father and mother, wrathful and griefstricken, produced the letter received from him, he said, "This is not from me!" He showed them the letter he had received, and they realized then that both letters had been forged by no telling whom.

After rounding up the mariners who had rowed his wife to that deserted shore, the prince put out to sea with them immediately. He came to the spot where they had disembarked, saw the dead dragon and then the dead filly, and lost all hope. But while he was weeping, he heard his name called: it was the beautiful lady at the palace window. He went inside, and the lady announced she was the filly and led him into a room, where he found his wife and children. They hugged and kissed, wept and cried. Then, together with the beautiful lady who'd been a filly, they departed. Everyone was overjoyed to have them back, and from that time on they were always together and as happy as happy could be.

(Pisa)

◈ 76 ◈

The Florentine

There was once a Florentine who went out every evening into society and listened to the talk of people who had traveled and seen something of the world. He never had anything to tell them, however, for he'd never been away from Florence a day in his life and therefore felt like a perfect blockhead.

He was thus filled with the urge to travel and knew no peace until he had sold all his belongings, packed his bags, and set out. He walked all day long and at dusk requested shelter for the night at the house of a priest. The priest invited him to supper and inquired, as they ate, the reason for the man's journey. Upon learning that the Florentine was traveling solely to come back to Florence with something to talk about, he said, "I, too, have often had such a desire. If you like, we can travel together."

"Fine!" said the Florentine. "I never dreamed I'd be so lucky as to have a traveling companion."

The next morning they set out together, the Florentine and the priest.

At nightfall they came to a farm and requested shelter. The farmer asked, "What is the purpose of your journey?" Hearing their account he, too, felt the urge to travel and left at daybreak with them.

The three had gone quite some distance, when they came to the palace of a giant. "Let's knock on the door," proposed the Florentine, "so when we go back home we can tell about a giant."

The giant answered the door himself and invited them in. "If you'll stay with me," he told them, once they were inside, "I can certainly use a priest in the parish and a farmer on the farm. Although there's no particular need for a Florentine, a place can be found for him too."

The three men talked the matter over. "No doubt about it, working for a giant, we'd see some pretty unusual things. Goodness knows how much we'd have to talk about afterward!" So they accepted his offer. He led them off to bed, with the understanding that all the details would be worked out the next morning.

The next morning the giant said to the priest, "Come and let me show you the parish records." They went into a room and closed the door. The Florentine, who had a great deal of curiosity and feared that he might miss something interesting, put his eye to the keyhole. As the priest bent over the records, the giant raised his sword, cut off his head, and threw the remains through a trapdoor.

This will certainly be something to tell back in Florence! thought the Florentine. The only trouble is, people won't believe me when I tell them.

"I have put the priest where he belongs," announced the giant. "Now I'll look after the farmer. Come along and let me show you the records of the farm."

Unsuspecting, the farmer followed the giant into that same room.

Through the keyhole the Florentine saw him bend over the records and the giant's sword come down on his neck. Then the farmer's remains went through the trapdoor.

The Florentine was gloating over how many extraordinary things he would now be able to tell back home, when it suddenly occurred to him his turn would be next; in that case he wouldn't get to tell a single thing. More and more he felt like running away, when the giant emerged from the room and said he would have lunch before looking after the Florentine. They sat down to the table, but the Florentine was thinking so hard about how to escape that he couldn't swallow a thing.

Now the giant had one eye that squinted. At the end of lunch, the Florentine spoke. "What a pity! You are so handsome, but that eye there..."

The giant was ill at ease whenever anyone noticed that eye, and he began blinking, frowning, and squirming in his chair.

"I know of an herb," continued the Florentine, "which cures every eye disorder. I think I even saw some growing on the lawn of your park."

"Really?" replied the giant. "There's some right here at the palace? Let's go get it at once!"

He led him through the palace and onto the lawn, while the Florentine carefully noted where the keys were kept and how to get out when the time came to flee. On the lawn he picked a common weed. They went back inside, and he put it on to boil in a pot of oil.

"This is going to hurt very badly," he told the giant. "Can you stand the pain without moving a muscle?"

"Of course I can!" answered the giant.

"To make sure you keep perfectly still I'd better tie you to this marble table. If I don't, you'll be sure to move, and the operation will be a failure."

As he was anxious to get that eye corrected, the giant consented to being tied to the marble table. When he was all bound up like a sausage, the Florentine poured the pot of boiling oil into his eyes, blinding him. Then in a flash he was down the steps, rejoicing to himself. "This, too, will I relate!"

The giant let out a howl that shook the whole house, jumped up with the marble table bound to his back, and ran after the man as best he could. But realizing he would never catch him now, he fell back on a trick. "Florentine!" he yelled. "O Florentine! Why are you running away from me? Don't you want to finish the operation? How much will you take to finish it? Would you like this ring?" He threw him a ring. It was an enchanted ring.

"How about that!" said the Florentine. "I'll take this back to Florence and show it to anybody who doesn't believe me!" But he'd no sooner picked it up and slipped it on than his finger turned to marble and weighted him completely to the ground. There he lay motionless, for the finger weighed tons and tons. He vainly tried to pull his finger out of the ring, but it stuck fast to him. The giant was almost upon him. At his wit's end, the Florentine pulled out his pocketknife and cut off the finger. That way he escaped and the giant caught him no more.

He reached Florence with his tongue hanging out, and gone forever was his urge not only to travel far and wide but also to talk about his journeys. As for the finger, he said he had cut it off mowing the grass.

(Pisa)

◈ 77 ◈

Ill-Fated Royalty

There was a king in Naples who had three sons. Being an old man, he decided to give his eldest son the queen of Scotland as a wife. Shortly after his son's marriage the old king died, and the son inherited the throne. But the other two brothers could not bear being ruled by him, and their hatred grew to the point that they resolved to kill him. They thought of all the ways to do so, and finally one of them suggested, "Let's set fire to the palace. Everyone inside, including the king, will perish. Then we'll build ourselves another palace." So the two of them, in league with other rogues in the city, formed a plot to burn down the royal palace. But one of the conspirators had a change of heart and turned the king's spy. Seeing that, the other conspirators realized they hadn't a minute to lose, so they surrounded the palace and set fire to it.

When the queen, who happened to be downstairs, saw the flames, she leaped from the window into the garden with Elizabeth, her maid of honor. At the end of the garden was a door, through which they passed into safety. From a distance the palace was seen going up in smoke and flames along with all those who were trapped inside.

The two women entered a forest. They walked the whole day, but the queen, who was expecting a baby, grew very tired. Toward evening they ran into twelve murderers. "You women there! What are you doing in this forest?"

"Misfortune has driven us here," answered the queen.

The murderers seized the maid of honor, bound her to a tree, and left her there, while they took the queen home with them to keep house and cook for them. They also showed her their medicine cabinet, in case they should ever come in wounded. One day when the queen was by herself and looking in the cabinet, she found a bottle labeled POISON: CAUSES DEATH WITHIN TWELVE HOURS. She took the bottle and sprinkled all the food with its contents. When the murders came home to dinner, she put everything on the table and fled. Off their guard, the murderers ate and were poisoned to death.

The queen went through the forest looking for her maid of honor, but found no trace of her, near or far. She was tired and in pain. In the heart of the forest she came upon a hollow tree, went in to rest, and all at once labor pains started, and she gave birth to a boy. She stayed inside the tree all night nursing him. In the morning two shepherds passing by heard the baby crying and went up to the tree. Seeing the woman with the newborn baby, they gave her their assistance, carried her home with them, and said, "Here you will be the lady of the house, and we will provide for you." So the queen, with her little son, dwelled in the shepherds' house, while back in Naples her two treacherous brothers-in-law lived in their newly built palace and reigned undisturbed.

The shepherds were rich and had a big house. One day while they were out, the queen decided to look around.

She opened a door and saw a very long flight of stairs. Climbing it, she came to a door already ajar and pushed it open. There sat a pensive youth, with his face in his hands. The queen made a move to withdraw, but he raised his head and told her to come in. They asked each other how they happened to be in that house, and told each other their stories.

"I am the son of the king of Portugal," said the youth. "My father and his chamberlain both got married the same day. The day I was born, the chamberlain's wife gave birth to a baby girl. From the beginning, we children were together, and as we grew up we fell in love with each other. No one knew about our love, but I swore I would never go to the altar with any girl but my lovely Adelaide. Meanwhile my father had grown old and decided to marry me off. He sent ambassadors to the queen of England to conclude the marriage. I didn't have the courage to say I loved Adelaide and let the negotiations continue. One day I was finally forced to tell her I was about to marry another girl. No memory is more bitter to me than the pain and wrath of my beloved on hearing my confession. Hopeless and indignant, she dismissed me, and I deserved it. Meanwhile my father went on with lavish preparations for the wedding. He had three entrances made into the banquet hall: one for princes, one for maids of honor, and one for pages. During the ceremony the bride observed my distress.

" 'Listen,' she said to me, 'if you are not marrying me of your own free will, I'll gladly go back home.' "

"With courtesy, I told her I was gloomy by nature. And I married her. After the wedding we took our places on the throne. The whole court filed by, with all the princes advancing through one door, maids of honor through another, and pages through the third. The last page was a lad dressed in white, with a large bouquet of flowers. He came to the foot of the throne, bowed, and mounted the steps to offer the flowers to the bride. As she reached out for them, the little page drew a dagger from the flowers and stabbed her right there on the throne beside me. The page was seized by the guards and carried before my father. No sooner did he stand before the king than he drew out a second dagger and planted it in his own heart. Upon trying to revive him, they discovered the page was a woman; I bent over her and immediately recognized Adelaide, who was already dead.

"Upset as I was at the time, I told my father the reason for her vengeance. When he heard my account, my father, the sternest of men, ordered me locked up in a tower and the keys thrown into the sea. There I remained until I managed to slide down a long rope and flee into the forest. After walking for miles I came to a hollow tree trunk and fell asleep inside it. The next morning I was awakened by two shepherds who took pity on me and brought me here, where they treat me like a son. But tell me how you ended up in this house."

The queen therefore told her story, and they both realized they were persecuted by fate. "Listen," said the prince, "since death claimed your husband and my wife and chance brought us together, let's get married. We'll ask the shepherds for two horses and go to Scotland, your home-land."

The queen agreed, and they asked the shepherds, who had come home in the meantime, for two horses, promising to knight the men for all their kindness. The shepherds let out two horses, and the prince took the queen's child into his saddle and they departed.

They had to go over the mountains, and the trail was full of perils. The queen's horse suddenly shied, took a false step, and went plunging into the ravines below. The poor prince saw this as one more instance of the misfortune that plagued him. Broken-hearted, he continued on to Scotland with the baby and tidings of what had happened to the queen.

But she was not dead. She had gone over the cliff, but landed on soft ground. Although bruised, she was alive. Regaining her senses, she looked about her and, in the midst of the precipices, spied a cottage. She approached it and knocked, but no one answered. She later returned when it grew dark, and knocked again; but there was still no answer. She settled down to wait. Around midnight a man covered with hair arrived carrying on his back a load of dead animals.

"What are you doing here?" he asked the queen.

"I'm seeking shelter for tonight."

The hairy man knocked, and this time his wife opened the door. The queen stepped into the dark house, and they gave her a place to sleep. In the morning the hairy man went out hunting, and his wife brought the queen a cup of broth. Seeing the woman in the light, the queen exclaimed, "Elizabeth!" And the hairy man's wife said, "Yes, I'm Elizabeth."

She was the queen's own maid of honor. The hairy man had found her tied to the tree where the murderers had left her, and brought her home with him. Every night he came in from hunting with animals for her to skin. But he was unkind to her because she would not love him.

The two women embraced and made a big to-do over each other, telling their stories in turn. "But how are we going to get away from here?" asked the queen. "Have you no opium?"

"Yes, I can get some," replied Elizabeth.

They drugged the hairy man's wine and, once he was sleeping, killed and buried him.

At the bottom of the mountain was a door that led directly into Scotland, and the key to the door was in the possession of the hairy man. Elizabeth got it, opened the door, and she and the queen walked through it into Scotland. On seeing the queen, who had twice been reported dead, all of Scotland was overjoyed. The queen's father was glad to give his daughter to the prince of Portugal, but before celebrating the wedding, he wanted to make war on the two usurpers reigning in Naples. He dispatched ambassadors to ask the prince's father to be his ally, and the king of Portugal provided him with a good-sized army.

Before leaving for war, all the generals gathered round the throne to swear allegiance. Beside the king stood the two betrothed as the generals each stepped forward to present their swords. When it was the general of Portugal's turn, the queen suddenly dashed from the throne and hugged and kissed the general, and they both fainted. She had recognized him as her husband, the king of Naples, who had not died in the fire, but fled from his brothers and enlisted under a false name in the service of the king of Portugal.

Then the queen said to the prince, "I can't marry you, since this is my husband right here. But Elizabeth, my maid of honor, is the daughter of the king of Spain and would be fully worthy of you."

The prince agreed, and they sent for the king of Portugal, under the pretext that his general was ailing. The king came, and seeing his son alive whom he had given up for lost, he was greatly moved and apologized for having shut him up in the tower.

They attacked, vanquished, and put to death the usurpers of Naples, returning the throne to the king and his queen and little son. Elizabeth married the prince of Portugal, and from then on there was no more misfortune.

(Pisa)

◈ 78 ◈

The Golden Ball

There was once a king who had three daughters. The oldest was in love with the baker who brought bread to the palace, and the baker was in love with her. But how was he to ask for her hand? He went to the king's secretary and explained the situation to him.

"Have you lost your mind?" exclaimed the secretary. "The idea of a baker falling in love with the king's daughter! Heaven help you if the king should find out!"

"That's just why I came to see you," replied the baker, "so you could prepare him for the news."

"Heaven help us both!" sighed the secretary, putting his hands to his head. "I wouldn't dream of telling the king. He would be furious and punish me as well as you."

But the baker was a persistent youth and kept after the secretary day in day out to talk to the king. To get rid of him, the secretary at last said, "Very well, I'll speak to His Majesty; but mark my words, no good will come of it!"

Availing himself of a day when the king was in a good humor, he said, "Majesty, if you permit, I would like to tell you something, provided you promise not to become angry with me."

"Go right ahead," replied the king, and the secretary repeated the baker's speech.

The king turned pale. "What! How can you be so bold? Guards! Here! Seize him!" They were already on the way out with the poor secretary when the king recalled his promise not to be angry with him, so he ordered his three daughters seized instead and kept on bread and water.

He confined them for six months. Then he decided he ought to let them get a little air and sent them out in a closed carriage with servants for a drive along the highway. When they had gone halfway, the road was suddenly enveloped in heavy fog, out of which stepped a sorcerer. He opened the carriage door, pulled the three girls out, and carried them off with him.

When the fog lifted, the servants found the carriage empty. They called and called and searched and searched, but to no avail. They returned to the palace empty-handed, and all the king could do was issue a proclamation: whoever found the three girls would be allowed to choose one of them for his wife.

The baker, who had been driven from the palace at once, joined two friends to travel about the world. One evening, in a forest, they found a house all lit up. They knocked, and the door swung open by itself. They went in and up the stairs, but no one was in sight. There was, however, a table laid for three, with supper ready and waiting. They sat down and ate, then found three bedrooms with beds turned down for the night, so they got into them and went to sleep.

In the morning they found three shotguns next to their beds. There was food in the kitchen, but it had not yet been cooked. They decided, therefore, that two of them would go hunting while the third friend stayed behind to cook. The baker was one of the two to go hunting. The friend to stay behind went to light the fire. While he was putting in the coal, out of the fireplace bounded a golden ball and went capering around his feet. No matter what he did, the man couldn't get out of its way. He tried kicking it off, only to have it bounce right back between his feet. The more he kicked, the harder it was to dodge the ball. Finally, at the hardest kick yet, the ball burst open, and out jumped a little hunchback brandishing a cudgel. He was a tiny hunchback who could just reach up to the man's knees with his cudgel, but he dealt such furious blows that the man soon fell to the ground, his legs one solid bruise. Then the hunchback reentered the ball, which closed and disappeared up the chimney.

Half dead, the man dragged himself on his hands and knees to his room and fell onto his bed. Cooking was the furthest thing from his mind. And being a wicked-hearted man, he thought to himself, Since I took a beating, my friends ought to take a beating. So when the other two returned and found him in bed and no dinner on the table, he explained, "Nothing's wrong with me, the bad coal of this region just made me dizzy."

The next morning he felt better and went hunting with the baker. The other friend who stayed behind to cook went to light the fire, and out of the fireplace bounced the golden ball. This friend also tried kicking it out of the way until the little hunchback finally jumped out and beat him so badly that he, too, had to take to his bed half dead.

"The coal made me dizzy too," he explained when his companions came home and found no dinner waiting for them.

"I don't understand this," said the baker. "I'll try tomorrow myself and see how I feel."

"By all means, do! You'll certainly be dizzy too!" said the two who had already taken the lickings.

The next day the baker remained behind. He went to light the fire, and the ball began rolling about between his feet. He walked forward and backward, with the ball constantly bouncing around him. He stood up on a chair and the ball hopped up on the chair. He stood on top of a table, and the ball hopped up on the table. He put the chair on top of the table and then, standing on the chair, calmly plucked a chicken, allowing the ball to bounce freely in and out of the chair legs.

In no time the ball grew tired of bouncing. It opened, and out stepped the little hunchback. "Young man," he said, "I like you. Your friends kicked me, and you didn't. I beat them, but I will help you."

"Wonderful," said the baker. "Help me cook, then, since you've already made me waste enough time. Go get the wood for me and hold it while I chop it up."

The hunchback bent over to steady a log and the baker raised his ax, but instead of bringing it down on the log, he struck the hunchback as hard as he could on the neck and cut off his head. Then he threw the remains into the well.

The friends returned, and the baker said, "You poor things. It was a far cry from coal! The blows are what hurt you!"

"What! You didn't receive any?"

"Not a one; and what's more, I beheaded the hunchback and threw him into the well."

"Go on, you don't mean it!"

"If you don't believe it, let me down into the well, and I'll bring him up to you."

The friends tied a rope around his waist and lowered him into the well. Halfway down the well was a large window all lit up, through which the baker saw the king's three daughters sewing and embroidering in a locked room. Just imagine the joy of the lovers over this reunion. But the king's daughter said to him, "Flee for your life! It's time for the sorcerer to return. Come back tonight while he's asleep and free us!"

Overjoyed, the baker continued on down to the bottom of the well, picked up the hunchback's body, and carried it up to show to his friends.

That same night, he had them lower him into the well with a saber to free the king's daughters. He entered by the big window, and there was the sorcerer asleep on a sofa, with the king's three daughters fanning him. Should they stop fanning for one instant, he would awaken. "Let's try fanning him with the saber," proposed the baker. The sorcerer awakened, but he was already dead, with his head cut off by a sweep of the saber and sent flying to the bottom of the well.

The king's daughters opened the dresser drawers, which were full of sapphires, diamonds, and rubies. The baker filled a basket with them, tied it to the rope, and had his friends draw it up. Then he had the girls pulled up one by one.

"Here, take this walnut," said the first girl as he tied the rope around her.

"Here, take this almond," said the second girl when her turn came.

The third girl was his beloved and, being the last, she was able to give him a nice kiss in addition to a hazelnut.

It was now the baker's turn to be pulled up, but he distrusted his two friends who had already proved untrue to him by their silence about the hunchback, so instead of tying the rope around himself, he attached it to the beheaded sorcerer. Up, up went the body, then all of a sudden crashed to the bottom of the well, since the friends had let go of the rope in order to take the king's daughters home and tell the king they had freed them.

Noticing that the friends had ceased pulling, the king's daughters began shouting. "What! You're going to leave him down there after he set us free?"

"Hush, dears!" replied the two rogues, "if you know what's good for you! You'll go back to the palace with us like good girls and say yes to everything we say."

Thinking the men had actually rescued his daughters, the king embraced them and made a big to-do over them; although they were scarcely to his liking, he promised each of them one of his daughters. But the daughters found all kinds of excuses for putting off the wedding and waited day after day for the poor baker to return.

Abandoned down in the well, the baker remembered the three gifts from the king's daughters. He broke open the walnut and found the glittering clothes of a prince. He cracked the almond, and out rolled a carriage drawn by six horses. He broke open the hazelnut, and out marched a regiment of soldiers.

So, dressed as a prince, riding one of the horses in the team, and followed by the regiment of soldiers, he journeyed from the world below ground to the world above ground and came to the city of the king.

Learning of the arrival of such an important nobleman, the king sent ambassadors to him. "Do you come in the name of peace, or of war?"

"Peace to those who love me, war to those who have betrayed me," replied the baker.

"There he is, there he is, that's our deliverer!" shouted the king's three daughters, who had climbed to the top of the tower to look through the telescope.

"That's my bridegroom, that's my bridegroom!" exclaimed the eldest girl.

"What does that peasant in disguise want?" said the two friends, who armed themselves and went onto the field.

The entire regiment fired, and the two traitors fell to the ground, dead.

The king welcomed the newcomer as the victor and the rescuer of his daughters.

"I am the baker you dismissed, Majesty!" said the young man.

The king was so taken aback that he abdicated, and the baker reigned happily from then on with his bride.

(Pisa)

❖ 79 ❖

Fioravante and Beautiful Isolina

A king getting on in years, with a grown son unwilling to learn a thing, grew worried and sent for the boy. "Fioravante," he said, "I'm taking great pains to instruct you in important matters, but it's like beating my head against a wall. How can I leave the crown to you?"

"Dear Father," replied Fioravante, "I'm in love with a certain maiden, and all you tell me goes in one ear and out the other."

Now the maiden was Sandrina, a poor weaver, which displeased the king. "But what will people say? A king's son in love with a weaver? Have you no sense of decorum?" He decided to write his brother, the king of Paris, to let Fioravante come to his court for a while and that way forget the weaver.

Fioravante chose a sleek horse and set out on the journey. He was going through a dark and dense forest inhabited by wolves when the sky suddenly filled with clouds, followed by distant thunder, then lightning, and a heavy downpour. Fioravante dismounted and took shelter under his horse; he stretched out, lit his pipe, and waited for the rain to stop. When the storm was finally over he made out a small light in the distance, which led him to a cottage; he knocked and was met at the door by an old woman.

"You have come to a place, sir, where . . ." she began in a wheezy voice, when behind her appeared a huge man with a beard down to his chest.

He happened to be a murderer, and asked the youth, "Who are you?"

Fioravante replied, "I am the son of the king of London, and I'm going to my uncle, the king of Paris, to get an education."

"If you want me to spare you, change clothes with me and pretend I am you and that you are my servant, so that the king of Paris will take me for his nephew. But give me away, and I'll kill you. Is that clear?"

"Quite!"

They went to Paris. The uncle welcomed the murderer, thinking him to be the nephew he was expecting. Fioravante was put in the stable to curry the horses and eat fodder with them.

One day the murderer said to the king, "All kings have fine teams for their carriages, and you don't. How come?"

"I could have the finest teams of all the kings in the world," explained the king, "but the horses are wild and graze in a herd in the meadows. No one has ever managed to capture a team for my carriage."

"My servant," answered the murderer, "can capture all the teams you want, or at least he boasts he can."

"Let him try," said the king, "but if he fails, his head will roll!"

The murderer ran to Fioravante and said, "The king has decided you are to get him a team of horses from his herd grazing in the meadows. And if you fail, you will be beheaded."

Fioravante, who was tired of staying in the stable all the time, saddled a horse and rode to the meadows. On the way he passed through a garden full of all kinds of flowers and plants; riding near an oak he heard a voice that sounded like a woman's saying, "Fioravante! Fioravante!"

He was quite surprised, for this was the first time he had ever been in these parts, and he didn't see how anybody could know his name. "Who is it?" he asked, and out of the oak's trunk walked a beautiful filly, who said, "Don't be afraid! If you want to carry off a team from the herd, leave your horse and get on my back."

Fioravante tied his horse to a tree, mounted the filly, and rode bare-back to the meadows. He went up to the herd, threw out his lasso, caught two horses, and put a halter on them as easily as you please.

When the king saw the team, which was the most beautiful anybody had ever beheld, he said to the murderer, "That's a superior man you have there for a servant. But let's see now if he can break them."

Following the filly's advice, Fioravante by means of a beater used in threshing grain taught the horses to be quiet and obedient.

The king said, "I'm going to have that servant of yours to dinner."

But the murderer objected. "Sovereign uncle, it is better not to do that, since he's accustomed to eating fodder like the horses, and if he tastes something different, there's no telling what ideas he might get." He thus persuaded the king not to invite Fioravante, and every chance he got he would say, "Sovereign uncle, you are getting on in years, unfortunately. Who will inherit your crown? You have no sons, and I would hate to end up with it myself . . ."

"Sons have I none," replied the king. "But I had a daughter as fair as day, who died when she was fourteen. I've never even visited her tomb, as she's buried in a convent far, far away, in India. And I still mourn her."

"Don't cry, Majesty," said the false nephew. "My servant says he can bring your daughter back safe and sound."

"But who on earth is this servant to raise the dead?"

"I'm just going by what he says," replied the murderer.

"Your servant boasts too much. Tell him he'd better make good his boast, for his head will surely roll if he fails to bring me Isolina."

Fioravante had his filly, who said to him, "Don't be discouraged. Go to your master and ask him for a goblet of pure crystal; a solid gold cage with gold sticks, bars, and drinking trough; and a ship that doesn't leak the least bit."

Fioravante went to the murderer and explained it all to him, and the king ordered everything readied.

Fioravante set sail for India, with his faithful filly aboard the ship. In the middle of the sea he saw a fish jumping up out of the water. "Catch it!" said the filly, and Fioravante reached out and grabbed the fish at its next leap. "Put it in the crystal goblet now," said the filly.

After disembarking in India, they set out for the convent. A bird swooped down, and the filly said, "Catch it!" Fioravante seized it and was then told: "Put it in the gold cage."

They reached the convent, and Fioravante asked the abbess where Isolina, the daughter of the king of Paris, was buried. The abbess lit a taper, led him into church, pointed to the tomb, and left him there.

Fioravante began digging. He dug and dug, and under the earth appeared the king's daughter adorned in gold and diamonds and as fresh as a sleeping maiden. He went to lift her from the tomb, but she stuck fast, as though she had become part of the stone. He then sought advice from the filly waiting outside the church. The filly said, "Isolina is missing a blond tress from around her head. Without it she cannot be torn from the tomb. Ask the nuns where they put it. When she gets it back, she will come off the stone as easily as a rose petal."

Fioravante went and knocked on the abbess's door and asked about the tress. "The tress was lost at sea," explained the abbess, "during the journey to the convent for burial."

At that, the filly said, "What you now must do is throw the fish you have in the goblet back into the sea, telling him to bring you in exchange for his freedom Isolina's tress."

The tress was at the bottom of the sea, where two dolphins were engaged in a tug of war with it. Set free, the fish swam swiftly between them, seized the tress in his mouth, and made off with it. The dolphins both hoisted their tails, wondering where in the world the tress had gone. In revenge they began eating right and left all the little fish they met. In the meantime the fish brought the tress to Fioravante waiting on shore. He thanked him and let him go.

Fioravante put the tress around Isolina's head, bent down to raise her, and she came up as light as a feather. He carried her aboard the ship, but she was dead, and Fioravante wondered if seeing her come home like that wouldn't heighten her father's present grief. But the filly said, "Go ask the abbess where Isolina's soul is. When she has it back, the maiden will breathe again."

The abbess said, "Isolina's soul is too far away for anyone to get. It is at the top of a mountain as steep as a tower and so high up that, to reach it, you would have to go through red air, green air, and black air, and contend with wild beasts of every species and nation."

"What am I to do?" Fioravante asked the filly.

"Let the bird out of the gold cage."

The bird soared upward all the way to the red air, turning solid red; then penetrated the green air, turning solid green; and from there passed into the black air, turning solid black. Above the black air rose the mountain peak, where Isolina's soul rested enclosed in a little phial. The bird took the phial in its beak, swooped back down through the air, again turning black, then green, then red, and landed on Fioravante's deck with the phial. Fioravante set the bird free and emptied the phial into Isolina's mouth. Isolina drew a deep breath, the color flowed back to her cheeks, and she spoke. "Oh, how I have slept!"

The anchor was drawn up, and the ship set sail.

The king was at the port of Paris waiting for them. As soon as he saw that Fioravante had brought his daughter back alive, he was beside himself with joy, and said to the false nephew, "There's no finer man on earth than your servant!"

He gave orders for a grand banquet, with kings and queens as guests, and he also wanted the servant to be present. This time the false nephew failed to sway him with the fodder argument and had no choice but to extend the invitation. But he told Fioravante, "You are not to open your mouth at the table. You are used to living with the horses, so you will neigh. Is that clear?" The look he gave Fioravante plainly warned that he would kill him at the first word the youth said.

When Fioravante told the filly, she advised: "Do as you like. But tell the king. 'I'll come to the banquet only if you invite my filly, too.' "

Upon hearing that strange condition, the king was of a mind to refuse, but the man had brought his daughter back to life, so he wasn't in a position to quibble and consented. Finding themselves at the same table with the servant and his filly, all the kings and queens were quite puzzled. Once they finished eating and drinking they began talking, with everybody joining in. Only Fioravante remained silent.

"How can you be so quiet," they asked him, "after all the experiences you've had?" He smiled, but kept his mouth shut.

At that, the filly reared and, placing her front hoofs on the table, spoke: "With your permission, ladies and gentlemen, I shall speak for him."

Hearing a filly talk frightened everyone to death. But they were still more astounded by Fioravante's story, which she told from beginning to end. The false nephew made a move to flee, but was seized by the guards at once.

The king said, "Fioravante, my nephew, take this murderer and punish him as he deserves."

"All Fioravante has to do," said the filly, "is mount me, with the murderer tied to my tail. We'll go for a romp through the city. If the man comes back alive, fine for him."

They broke into a trot, dragging the murderer right along with them, kicking him right and left, splashing him with mud, and banging him against every stone in the path. When they returned, the murderer was no longer breathing.

The king said, "Fioravante, you brought my daughter back to life. It is right for you to marry her."

"If my father permits me to marry my cousin," replied Fioravante, "I will do so."

The king of Paris wrote to London to his brother, who was overjoyed that his son had forgotten about the weaver.

After the wedding Fioravante went to the filly and said, "Guess what: I've married the king's daughter."

"I know," said the filly. "There's no more hope for me now . . ."

"What do you mean, filly? I'll always cherish you, I'll never abandon you . . ."

"Yes, you will, you'll forget all about me . . ."

That night, before retiring with his bride, Fioravante wanted to go and say good night to the filly, as he did every evening. Isolina objected. "It's dark now, you can go see her tomorrow." Fioravante agreed. In the morning when he went down to the stable, the filly said, "You see? You are forgetting me . . ."

"But, filly . . ."

"All right, here's the last test to see if you love me: take your sword and cut off my head."

"No! Never!"

"That means you don't love me."

With a heavy heart, Fioravante raised his sword and whack! cut off her head in one sweep. And what did he then see? From the filly's neck emerged, fully dressed, a lovely maiden, white and rosy like an apple. It was Sandrina the weaver, his first love. "You see, Fioravante," she said, "to save your life I had a sorceress cast a spell over me, and I took the form of a filly. And now you have abandoned me . . .'

"O Sandrina! Had I known it was you, I would never have married Isolina! If only I could go back in time!"

But that was impossible, so to make amends, he gave Sandrina a rich dowry and married her to a merchant of Fiesole, while he remained king of London and Paris and the spouse of beautiful Isolina with the blond tress.

(Pisa)

Fearless Simpleton

A man had a nephew who was as stupid as could be. The boy understood nothing; on the other hand, nothing frightened him. Now the man left home, instructing his nephew to watch out for robbers and not let them steal any belongings from the house. The boy began wondering. "What are robbers? What are belongings? I'm afraid of nothing."

The robbers appeared and said, "What are you doing here, boy? We have come to rob you."

"So what are you waiting for? Go on and rob me. Is anybody stopping you? Do you think I'm afraid?" And he let them steal everything in the house.

The uncle returned and found the house ransacked. He asked his nephew, "Did you send for the robbers?"

"Me? I was here on the doorstep. The robbers came. They said, 'What are you doing here? We have come to rob you.' 'And who's stopping you?' I said to them. 'How dumb can you be!' So they went ahead and robbed us. I had nothing to do with it."

The man thought of his priest brother, who could perhaps teach the boy something. "You are going to your priest uncle," he told him.

"What is a priest uncle? I don't know of any priest uncles or anybody else. If we must go to my priest uncle, let's go!"

The first evening the priest uncle said to him, "Tonight you will go and put out the lights in church."

The nephew replied, "What are lights? What is a church? I don't know of any lights or any church. I'll go wherever you say, I'm afraid of nothing."

The uncle had given instructions that while his nephew was putting out the lights, the sacristan was to lower a basket of flaming candles and say, "Get into the basket, whoever wants to see the kingdom of heaven."

The nephew saw the basket, heard the voice, and said, "What heaven? What heaven? I don't know of any heaven. Wait, let me get in."

He took a knife and cut the rope. The sacristan went to pull up the basket and ended up with nothing but rope.

The next evening the priest uncle ordered the sacristan to get into a coffin and pretend to be dead, in order to frighten the nephew. "Tonight," he told the boy, "you are going to wake a dead man."

"What is a dead man? What is wake? I'll go anywhere." And he went into church to wake the dead man. A small candle flickered near the corpse, while the rest of the church was pitch-dark. The corpse slowly raised one leg. The boy watched and didn't move a muscle.

The dead man raised his head, and the boy yawned. Then the dead man spoke: "You, there! I'm still alive!"

The boy replied, "If you are alive, you are going to die now." He picked up a candle-snuffer, struck him on the head, and killed him. Then he went back and told his priest uncle, "That dead man hadn't finished dying, so I finished him off myself."

(Leghorn)

◈ 81 ◈

The Milkmaid Queen

There was once a king and queen who had no children. A little old woman told their fortune. "You may choose between having a son who will leave home and never be seen again, or a daughter whom you'll manage to keep until she's eighteen, provided you watch her closely."

The king and queen settled for the daughter, who in due time came into the world. The king had a magnificent underground palace built, and there the little girl was reared without the slightest notion of what was aboveground.

Upon reaching her eighteenth year, she begged her governess to open the door for her, and the governess finally obeyed. The girl walked through the door and found herself in the garden. She was charmed at the sight of the sun, which she saw for the first time, and with the various hues of the sky and the flowers. But a bird with large wings swooped down from the sky, took her in his claws, and flew off with her.

On and on he flew, finally landing on a farmhouse and leaving the girl on the roof. Two farmers, a father and his son, were in the field at the time and saw something gleaming on the roof of their house. They climbed a ladder to the roof and found a maiden wearing a crown of sparkling diamonds. Now the farmer had five daughters who were milkmaids, and he kept the maiden at home with them. Every month they sold a diamond from the crown, and that supported the whole family.

When the diamonds were all gone, the maiden said, "I don't want to live off of you, Mamma [as she now called the farmer's wife]. Go to the queen of this country and have her give you something to be embroidered."

The woman went, but the queen said, "How do you expect one of your daughters who's always been a milkmaid to embroider?" She scornfully gave her a bit of canvas, which the girl embroidered so exquisitely that the queen was speechless when the farmer's wife returned it. The woman went home with two gold coins and a dustcloth which the girl was to embroider next. A few days later the farmer's wife carried a masterpiece of embroidery to the queen, who gave her three gold coins and an old torn skirt. When the skirt was returned, it looked like part of an evening dress.

"But where on earth did your daughter learn to embroider so beautifully?" asked the queen.

"The nuns taught her."

"That may well be, but this is no country girl's work. No matter, I want her to embroider all my son's wedding clothes."

Upon learning that this milkmaid was embroidering his wedding clothes, the king's son just had to meet her and went to her while she was working. Being a rather mischievous young man, he got on her nerves. One day he took her quite by surprise and kissed her. At that, the milkmaid aimed her embroidery needle at his chest, thrust it into his heart, and killed him.

The maiden was led before the tribunal, which was made up of the king's four daughters. The oldest proposed a death sentence; the second life imprisoinment; the third twenty years; whereas the youngest, who was the most kind-hearted and understood that her brother had brought it on himself, recommended imprisonment for eight years in a tower with the prince's body, the constant sight of which should make the girl repent. The youngest daughter's counsel prevailed, and the maiden was led to the tower. As she passed, the king's youngest daughter whispered, "Don't be afraid. I will help you."

True to her word, she sent the prisoner in the tower the choicest dishes from the court table every day.

The prisoner had been closed up in the tower three years, when there appeared in the sky the bird with the large wings that had kidnapped her. It lit on top of the tower, built a nest, and laid eggs which hatched into ten baby birds. "Bird, O bird," said the prisoner daily, "take me away from here the same as you took me away from home."

Next to the tower stood the palace of the king's three older daughters. One day while they were at the window they overheard the maiden's words and reported them to the king.

"Knock that bird from the tower," ordered the king. The guards shoved the nest off the tower with their lances, and the ten baby birds fell out and were crushed to death upon hitting the ground.

That evening the prisoner saw the big bird hovering over the dead baby birds, with a tuft of special grass in its beak. It stroked them with the grass, and the birds revived. "Bird, O bird," said the prisoner, "bring me some of that miraculous grass!"

The bird flew off and returned with a tuft of grass in its talons. The maiden took it and ran to stroke the body of the king's son with it. Little by little, the king's son revived. I don't know which of the two was the happier: they embraced, kissed, and made much fuss over each other.

They didn't tell a soul the good news, except the king's youngest daughter who, in her elation over the wonderful surprise, sent them all kinds of delicacies every day; since her brother requested a guitar, she sent that, too.

Now, the two lovers locked in the tower spent their time playing the guitar and singing. In the palace next door the three older daughters of the king heard this guitar music and singing, and decided to go and see what was up. But the king's son stretched back out in his coffin, and the maiden feigned total bewilderment. The sisters went home none the wiser, but that night they once more heard music and singing coming from the tower.

They kept after the king until he ordered the prisoner moved to another prison. When the guards went for her, whom did they see emerge but the maiden arm in arm with the king's son, who was alive and as sound as a bell.

The whole royal family, looking out the window, was dumbfounded.

"Papa, Mamma, sisters," said the king's son, "I would like to introduce my bride." His little sister clapped for joy.

But the idea of having a milkmaid for their sister-in-law was more than the three older sisters could bear, and they lost no opportunity to mock and humiliate her.

"Before the wedding," said the bride, "I must go home and see my parents. Tell me what gifts you want me to bring you."

"Uh . . . a bottle of milk!" said the first sister-in-law.

"Ah . . . bring me some ricotta!" said the second sister-in-law.

"Eh . . . I'll take a basket of garlic!" snapped the third sister-in-law.

The milkmaid departed, but she didn't return to the farm; she went home to her real father, the king who had kept her closed up so long in the palace underground. One week later she returned to her bridegroom in a handsome carriage drawn by white horses. "What! The milkmaid in a carriage?" exclaimed the sisters-in-law, upon seeing her drive up.

Out stepped the milkmaid holding the presents. To her first sister-in-law she handed the milk; it was in a silver bottle wrapped in gold cloth. The second sister-in-law got the ricotta, silver ricotta in a gold basket. To the third sister-in-law she handed the basket of garlic—diamond cloves and emerald leaves.

"And you brought nothing to me who've always loved you dearly?" asked the youngest daughter.

The milkmaid opened the carriage door, and out stepped a handsome youth. "This is my little brother who was born while I was away from the court. He will be your husband."

(Leghorn)

◈ 82 ◈

The Story of Campriano

There was once a man, a tiller of the soil, named Campriano. He had a wife and a mule. Yokels from backward Ciciorana sometimes passed through the field he was working and called to him, "Hey, Campriano, what are you doing?" They would ask him if he was ready to go home, and frequently he and his mule would walk a little way with them.

One morning Campriano slipped a few gold pieces he had saved into his mule's rear end. When the yokels from Ciciorana came by, Campriano said, "Wait for me, I'm going home, too." He loaded his things onto the mule and joined the group in conversation. It was springtime and the fresh grass relished by animals abounded, so the mule, which had eaten his fill, soon cut loose and dropped the money his owner had hidden in him.

The yokels from Ciciorana exclaimed, "Why, Campriano, your mule makes droppings of money!"

"That's right," replied Campriano. "Without him, I'd never manage. He's my fortune."

Right off the bat they said, "Campriano, you must sell him to us! You must!"

"I'm not selling him."

"But if you did, what would you ask for him? A whole lot?"

"I wouldn't sell him for all the money in the world. You'd have to offer me . . . no less than three hundred crowns."

The yokels of Ciciorana dug into their pockets and all together came up with three hundred crowns. They led the mule away, and the minute they got home they told their wives to spread sheets in the stable to catch all the gold that would be dropped during the night.

In the morning they ran to the stable and found the sheets loaded with manure. "Campriano has cheated us! We'll kill him!" With that, they grabbed up pitchforks and shovels and marched off to Campriano's house.

His wife answered the door. "Campriano isn't here, he's out in the vineyard!"

"We'll get him out of the vineyard!" they shouted, and marched on. At the vineyard, they called to him. "Come out, Campriano! We are going to kill you!"

Campriano emerged from the rows of vines. "Why?"

"You sold us the mule, and he doesn't turn out any money!"

"Let me ask how you treated him," said Campriano.

"We treated him fine. He had sweet broth to drink and fresh grass to eat!"

"Poor animal! If he's not dead by now, he will be shortly! He's accustomed to eating roughage that shapes into durable coins, don't you see? Wait a minute, and I'll come and look at him. If he's still all right, I'll take him back. If not, you'll keep him and hold your peace. But first, I have to stop by my house a minute."

"All right! Go ahead, but come straight back. We'll wait here."

Campriano ran to his wife and said, "Put on a pot of beans to boil. But when we return, pretend to pull it out of the cupboard while they boil. Is that clear?"

Campriano accompanied the Ciciorana yokels to the stable and found the mule standing in the middle of the dung-laden sheets. "It's a wonder he's still alive," he said. "This animal is no good for work any more. But how could you! If I'd only known you'd break him down that way! Poor thing!"

The yokels were puzzled. "What do we do now?"

"What do you do now? I have nothing more to say, and you shouldn't either!"

"You have a point!"

"It was just one of those things. Come to my house to dinner, and let's forget the whole business once and for all."

They got to Campriano's and found the door closed. Campriano knocked, and his wife emerged from the barn, pretending to finish her chores and enter the house only at that moment.

The fire was out in the kitchen. Campriano said, "What! You've not cooked my dinner yet?"

"I just got back from the field," she replied. "But I'll scrape up something right away."

She set the table for everybody, then opened the cupboard, where the pot of beans boiled.

"What!" exclaimed the yokels of Ciciorana. "A pot that boils all by itself in the cupboard? How does it do that without any fire underneath it?"

"Goodness knows what we'd do without that pot!" replied Campriano. "How could my wife and I go out together to work if we weren't sure of finding the soup ready and waiting when we got back?"

"Campriano," said the yokels, "you must sell it to us."

"Not for all the money in the world!"

"Campriano, things didn't work out with the mule. To make up for it, you have to sell us the pot. We'll give you three hundred crowns." Campriano sold the pot for three hundred crowns, and they left.

His wife said to him, "They were ready to kill you over the mule. How will you get out of this one?"

"Leave everything to me," said Campriano. He went to a butcher, bought an ox-bladder, and filled it with raw blood. He said to his wife, "Here, put this bladder in your bosom, and don't be afraid when I throw a knife at you."

The yokels of Ciciorana arrived carrying clubs and stakes. "We want your head! Give us back our money, or we'll kill you!"

"Now, now, calm down! Let's hear what it is this time."

"You told us that pot boiled without fire. We went out to work with our wives, and when we came back, the beans were as raw as ever!"

"Easy, now, easy! It must be the fault of that confounded wife of mine. I'm going to ask her if she didn't switch pots on me . . ."

He called his wife and asked, "Honestly, did you switch pots on these men?"

"Of course I did. You go and give things away without asking me anything. Then I have to do the work! I don't want to part with that pot!"

Campriano let out a yell. "You wretch!" He grabbed a knife, flung it at her, striking the bladder hidden in her bosom, and blood squirted all over the place. Down fell the woman in a whole pool of it.

The two yokels of Ciciorana turned as pale as ghosts. "You mean you'd kill a woman, Campriano, over a pot?"

Glancing at his wife all covered with blood, Campriano pretended to be sorry. "Poor thing, we'll just have to revive her!" He pulled a straw from his pocket, placed it in the woman's mouth, blew three times into it, and the woman rose as sound and fresh as ever.

The two yokels were wide-eyed. "Campriano," they said, "you must give us that straw."

"No, indeed," replied Campriano. "I'm often overcome with the urge to kill my wife. If I didn't have that straw, I couldn't revive her afterward."

They begged and pleaded with him and ended up giving him another three hundred crowns, so Campriano let them have the straw. They went home, picked a fight with their wives, and knifed them. They were apprehended while still blowing into the straw, and imprisoned for life.

(Lucchesia)

◈ 83 ◈

The North Wind's Gift

A farmer by the name of Geppone lived on a prior's farm up on a hillside where the north wind always destroyed his crops. As a result, poor Geppone and his family often went hungry. One day he made a decision. "I shall go in search of this wind that persecutes me." He said goodbye to wife and children and headed for the mountains.

As soon as he got to Ginevrino Castle he knocked on the door. The North Wind's wife peeped out the window. "Who's knocking?"

"It's Geppone. Is your husband there?"

"He went out to blow through the beech trees awhile and will be back shortly. Come inside and wait for him."

An hour later, the North Wind returned.

"Good day, Wind."

"Who are you?"

"I'm Geppone."

"What do you want?"

"Every year you ruin my crops, as you well know. All because of you, my family and I are starving to death."

"What did you come to me for?"

"To ask you to make up for all the suffering you've caused me."

"What can I do?"

"I leave that up to you."

The North Wind's heart went out to Geppone, to whom he said, "Take this box and open it whenever you get hungry. Order whatever you wish and you will get it. But tell no one about the box, or you'll lose it and end up with nothing at all."

Geppone thanked him and departed. Halfway home, as he went through the woods, he got hungry and thirsty. He opened the box and said, "Bring out wine, bread, and something to eat with it," whereupon the box produced a hearty loaf, a bottle, and a ham. Geppone had a fine feast right there in the woods and then continued on his way.

Just before he reached his house, he met his wife and children, who had walked down the road to meet him. "How did you fare? Did things go well?"

"Quite nicely," he replied, leading them all inside. "Everybody sit down to the table." He then said to the box, "Wine, bread, and all the rest for everybody here." So they all had a fine dinner. When the meal was over, Geppone said to his wife, "Don't tell the prior I brought this box back, or he'll want it and take it away from me."

"I wouldn't dream of it!"

The prior sent for Geppone's wife.

"Your husband is back, is he? And how did everything go? Fine? I'm glad to hear it. And what did he bring back worth a mention?" Thus, one thing led to another, and before you knew it the cat was out of the bag.

The prior sent for Geppone at once. "Geppone, my good man, I hear you have a very valuable box. May I see it?" Geppone was inclined to deny the whole story, but now that his wife had blabbed, what could he do but show the priest the box and how it worked.

"Geppone, you just have to give it to me."

"Then what will I do?" replied Geppone. "You know I lost all my crops and have nothing to eat."

"If you give me the box, I'll give you all the grain you want, all the wine you can drink, and whatever else you ask for."

Geppone, poor soul, gave in. And what did he get in return? The prior let him have a few sacks of wretched seed, and that was that. Geppone was again as badly off as ever, and all because of his wife, mind you.

"You caused me to lose the box," he said, "and to think that the North Wind advised me not to mention that box to a soul! Now I'd never have the nerve to go back to him."

In the end, though, he took heart and set out for the castle. He knocked, and the Wind's wife looked out. "Who's there?"

"Geppone."

Then the Wind also looked out. "What do you want, Geppone?"

"You remember the box you gave me? Well, my landlord took it away from me and won't give it back, and now I live in hunger and poverty."

"I told you to tell no one about the box. So go away, since I'm giving you nothing more."

"Please, you alone can make good this loss."

A second time the Wind's heart went out to Geppone, and he pulled out a gold box and gave it to him. "Don't open this one unless you are famished. Otherwise it won't obey you."

Geppone thanked the Wind and headed home through the valleys with the box. Hunger soon got the better of him, and he opened the box and said, "Provide!"

Out of the box jumped a big, strapping man holding a club and began thrashing poor Geppone for dear life.

As soon as he could, Geppone shut the box and continued on his way, all stiff and bruised. To his wife and children who'd come down the road to meet him and find out how things went, he said, "All right. I brought back a finer box than the other one." He sat them around the table and opened the gold box. This time, out came not just one, but two big, strapping men with clubs and set upon the family. The wife and children screamed for mercy, but the men didn't let up until Geppone got the box closed.

"Now go to the prior," he instructed his wife, "and tell him I brought back a much nicer box this time than the last."

The wife went, and the prior asked her the usual questions. "So Geppone's back? What did he bring home this time?"

"Just imagine, Prior, he brought a nicer box than the other one. It's solid gold, and the beautifully cooked dinners it serves are a dream. But Geppone wouldn't part with this box for the world."

The priest sent for Geppone at once. "Geppone, Geppone, you don't know how glad I am you're back. And with another box. Show it to me."

"If I do, you'll take this one away from me, too."

"No, I won't, I promise."

Geppone showed him a corner of the glittering box. The priest couldn't contain himself a second longer. "Geppone, give it to me, and I'll give you back the other one. What do you need with a gold box? I'll give you the other one for it and then something."

"All right, return the other one, and I'll give you this one."

"Agreed."

"But beware of opening this box, Prior, unless you are famished."

"This box couldn't come my way at a better time," said the prior. "I'm expecting the bishop tomorrow and many other priests. I'll keep them all fasting till noon, then open the box and offer them a big dinner."

In the morning, after saying their Masses, all the priests started milling around the prior's kitchen. "He refuses us breakfast this morning," they said. "Just look, the fire's out, and the larder's empty."

But those in the know said, "Just you wait! At dinner time he's going to open a box and serve a meal finer than any we could imagine."

In marched the prior and seated everybody around the table, in the center of which gleamed the gold box, with all eyes now upon it. The prior opened the box, and out leaped six strapping men with clubs and began clubbing the priests for all they were worth. Under that onslaught the prior dropped the box, which lay open on the floor, so the men went on pounding the life out of the dinner guests. Geppone, who was hiding nearby, noticed the box and shut it. Otherwise the men would have beaten the priests to death. So that was the meal they got, and it appears that they were unable to say their office in the evening. Geppone kept both boxes, never lent them out again, and from that day on lived a life of ease.

(Mugello)

◈ 84 ◈

The Sorceress's Head

There was once a king who had no children. He was always imploring heaven to send him a child, but all his prayers were in vain. One day he had gone to pray as usual, when he heard a voice. "Do you want a boy who will die, or a girl who will flee?"

He didn't know what to say and kept silent. He went home, summoned all his subjects, and asked what reply he should make. They answered, "If the boy is to die, that's the same as having no child. Ask for the girl. You can keep her under lock and key and she can't flee."

The king went back to his prayers and heard the voice. "Do you want a boy who will die, or a girl who will flee?"

"A girl who will flee," he replied.

So nine months later the queen gave birth to a beautiful baby girl. Many miles outside the city the king had a large park with a palace in the middle of it. He took the baby girl there and shut her up with a nurse. Her father and mother rarely visited her, so that she wouldn't think about the city and decide to run away.

When the maiden was sixteen, the son of King Giona came by there. Seeing her, he fell in love with her and bribed the nurse with a great deal of money to let him into the palace. Overcome with love for one another, the two young people got married without their parents' knowing a thing about it.

Nine months later the princess gave birth to a fine baby boy. The next time the king called on her, he was met by the nurse, whom he asked how his daughter was getting along. "Beautifully, Your Highness," replied the nurse. "Would you believe she's just had a baby?" The king refused to have anything more to do with the girl.

She continued to live in her palace with her husband and their son. When the boy reached fifteen without ever having seen his grandfather, he said to his mother, "Mamma, I would like to meet my grandfather."

"Go to his palace, then, and meet him," answered his mother.

He rose bright and early, took a horse and a goodly supply of money, and departed.

His grandfather made no fuss over him; he hardly looked at him and said nothing. Cut to the quick by such a cold welcome, the young man said three or four months later, "What do you have against me, Grandfather? Why won't you even talk to me? For you, I'd go and cut off the sorceress's head."

"That's just what I wanted," replied the grandfather, "for you to go and cut off the sorceress's head."

Now this sorceress was so hideous that all who laid eyes on her turned to stone, and the old king was certain that would be his grandson's fate. The youth chose a fine horse, a goodly supply of money, and departed.

Along the way he met a little old man, who asked, "Where are you going, my boy?"

"To the sorceress, to cut off her head."

"Oh, goodness me! For that you'll need a horse that can fly, since you'll have to go over a mountain swarming with lions and tigers that would devour you and your horse in a flash."

"But where can I find a horse that flies?"

"Just a minute, and I'll get you one," replied the old man. He disappeared and returned with a magnificent flying horse.

"Now listen to me," said the old man. "You cannot look directly at the sorceress, or you'll turn to stone. You must watch her in a mirror, which I'll now explain how to get. Walk down the road a little way and you'll come to a marble palace and a garden of flowering peach trees. There you will see two blind women, who have only one eye between the two of them. Those women have the mirror you need. The sorceress spends her time in a meadow full of flowers, whose scent alone is enough to cast a spell over you. Beware of it. And look at the sorceress only in the mirror, or you'll turn to stone."

With the flying horse he hurdled the mountain infested with bears, tigers, and snakes, which all lunged after him. But he soared high and escaped them.

With the mountain behind him, he traveled and traveled and finally saw a marble palace in the distance. "That must be the blind women's palace," he said to himself. These blind women had only one eye between the two of them, and they passed it back and forth to one another. The young man didn't dare knock, but went for a stroll in the garden while the women ate their dinner. When they'd finished, they too strolled into the garden, and he climbed a tree so they wouldn't see him. They were in conversation, and the one who had the eye at the moment held it up to glance about her. "Oh, you should see these fine new mansions the king has built!" she exclaimed.

"Give me the eye," replied her sister, "and let me look too."

The woman held out the eye, and the young man reached down from the tree and took it.

"So you're not giving it to me?" said the other sister. "You want to see everything all by yourself?"

"But I gave it to you!"

"No, you did not!"

"I put it right into your hand."

They argued and argued until it dawned on them that neither sister had the eye. Then they said very loudly, "That means somebody's in the garden and has taken our eye. If this person is here, please give us back our eye, since we have only one between the two of us. Name what you want in return, and we'll reward you with it."

The youth then came down the tree and said, "I took the eye. You must give me your mirror in exchange for it, since I have to kill the sorceress."

"Gladly," replied the blind women, "but you must first return the eye so we can find the mirror." He courteously returned it, and the blind women went into the palace and came back out with the mirror, for which he thanked them and continued on his way.

On and on he traveled until the air grew sweet with flowers, and the nearer he got to them, the stronger the scent became. He reached a handsome palace in the middle of a meadow full of flowers. The sorceress was strolling in the meadow. He had meanwhile mounted his horse backward and looked at her only in the mirror, with his back to her. The sorceress, who was confident of her power to turn people to stone, did not run or make any effort to protect herself. Facing the other way and looking into the mirror, he rode right up to her, swung his sword around behind him, and cut off her head. Then he put the head into a bag out of sight. It had dripped a little blood, though, which changed into serpents on contact with the ground. Thanks to the flying horse, he got safely away.

He took a different road home, passing through a seaport along the way. Beside the sea was a chapel, which the youth entered and found a beautiful maiden dressed in mourning and weeping. At the sight of the young man, she cried, "Be gone! Be gone! If the dragon comes, he will eat you, too! I'm here waiting for him, since today it's my turn to be eaten. He eats one person alive every day."

"No, no, beautiful maiden, I will free you."

"It's impossible to kill a dragon like this one!" she said.

"Don't be afraid. Jump up on my horse," said the youth, and helped her into the saddle.

In that instant a great din and splashing was heard. The youth, after telling the maiden to close her eyes, pulled the sorceress's head from the bag. Just as the dragon stuck his head out of the water, he saw the sorceress's head, turned to stone, and sank to the bottom of the sea.

The maiden was the king's daughter, and the king gave her in marriage to the young man, promising to make him his heir if he stayed there. But the youth thanked him and said he already had his own kingdom to which he had to return. He took the princess with him and went first to his grandfather, who was surprised and dismayed to see him come back alive.

"Grandfather," said the youth, "didn't you want me to go and cut off the sorceress's head? I went, and I've brought it back to you. If you don't believe me, just look!"

He pulled it from the bag, and his grandfather turned to stone. Then the young man went to his parents, and they all returned to the grandfather's kingdom.

> And there they lived a life happy and long,
> But nothing did they ever give me for my song.

(Upper Val d'Arno)

◈ 85 ◈

Apple Girl

There was once a king and a queen who were very sad because they had no children. The queen kept asking, "Why can't I bear children the same as the apple tree bears apples?"

Now it happened that instead of bearing a son, the queen gave birth to an apple, but an apple redder and more beautiful than any you ever saw. The king placed it on a gold tray on his balcony.

Across the street from the king lived a second king, who happened to be standing at his window one day and saw, on his neighbor's balcony, a beautiful maiden as fair and rosy as an apple bathing and combing her hair in the sun. Open-mouthed, he stood staring at her, never having seen so lovely a maiden. But the minute the girl realized she was being observed, she ran back to the tray and disappeared inside the apple. The king had fallen madly in love with her.

He racked his brains and ended up crossing the street and knocking on the door, which the queen answered. "Majesty," he said to her, "I have a favor to ask of you."

"By all means, Majesty," replied the queen. "Any way neighbors can help one another out . . ."

"I would like to have that magnificent apple on your balcony."

"Do you know what you're asking, Majesty? I'm that apple's mother, mind you, and I had to wait a long time before I had her."

But the king wouldn't take no for an answer, so the other king and queen had to grant his wish, in order for them all to remain good neighbors. Thus he went home with the apple, which he took straight to his own room. He put out everything necessary for her toilette, and the maiden would emerge every morning to bathe and arrange her hair while he looked on. That was all she did. She neither ate nor talked; she only bathed and arranged her hair, then went back inside the apple.

The king lived with his stepmother, whose suspicions were aroused by her stepson's constant seclusion in his room. "I'd give anything to know what my son is up to!"

War broke out, and the king had to go off and fight. It broke his heart to leave his apple. He called his most trusted servant to him and said, "I'm leaving the key to my room with you. See that nobody goes in. Put out water and a comb every day for the apple girl, and make sure she has everything she needs. And don't forget, she tells me everything." (That wasn't so, the girl never said a word, but the king thought it wise to tell his servant the contrary.) "If a hair of her head is harmed during my absence, you'll pay with your life."

"Have no fear, Majesty, I will look after her to the very best of my ability."

As soon as the king was gone, the stepmother queen went to all lengths to get into his room. She put opium into the wine of his servant and stole the key from him when he fell asleep. She unlocked the door and turned the room upside down in search of clues to her stepson's strange behavior; but the more she searched, the less she found. The only thing out of the ordinary in the room was that splendid apple in a golden fruit bowl. "It must be this apple that is always on his mind!"

Queens, as you well know, always have a small dagger concealed in their sashes. She took out her dagger and began pricking the apple all over. Out of every wound flowed a rivulet of blood. The stepmother queen grew frightened, ran away, and replaced the key in the sleeping servant's pocket.

When the servant awakened, he had no idea what had happened to him. He ran into the king's room and found blood all over the place. "Oh, my goodness, what will I do now?" he exclaimed and fled.

He went to an aunt of his who was a fairy and possessed all the magic powders. The aunt took a powder suitable for apples under spells and another for bewitched maidens, and blended them.

The servant returned to the apple and sprinkled all the wounds with the mixture. The apple burst open, and out stepped the maiden in bandages and plaster casts.

The king came home, and for the first time, the maiden spoke. "Would you believe that your stepmother stabbed me all over with her dagger? But your servant has nursed me back to health. I am eighteen and was under a spell. If you like, I will be your bride."

"If I like! Indeed I do!"

The wedding was celebrated, to the great joy of both palaces. The only person missing was the stepmother, who fled and was never heard of again.

> Merrily through life they went,
> But were only content
> To give me one cent
> I never spent.

(Florence)

◈ 86 ◈

Prezzemolina

There was once a husband and a wife who lived in a pretty little house. And this little house had a window overlooking the fairies' garden.

The woman was expecting a baby and had a craving for parsley. She looked out the window and saw in the fairies' garden a large patch of parsley. Waiting until the fairies were out, she descended into the garden by means of a rope ladder, gorged herself with parsley, then climbed back up the ladder and closed the window behind her.

The next day, and the next, and the next, she did the same thing. Not a day went by that she didn't eat her fill of parsley. At last the fairies, out walking in the garden, noticed that their parsley was almost all gone.

"Know what we should do?" said one of the fairies. "Pretend we've all gone out, while one of us stays behind and hides. That way we'll find out who's stealing the parsley."

When the woman descended into the garden, out jumped a fairy. "Ah-ha! I've caught you at last, you villain!"

"Please don't be angry with me," said the woman. "I crave parsley because I'm expecting a baby . . ."

"We will forgive you," answered the fairy. "But if the baby is a boy you must name him Prezzemolino—that is, Little Parsley-Boy. If it is a girl, you are to call her Prezzemolina—that is, Little Parsley-Girl. And when the child is older, no matter whether it is a boy or a girl, we will take it away from you!"

The women returned home in tears. When her husband learned of her pact with the fairies, he was furious. "You awful glutton! Don't you see you've brought this on yourself?"

The baby was a girl, Prezzemolina. In time her parents forgot all about the pact with the fairies.

When Prezzemolina was a big girl, she started to school. And every day on her way home she met the fairies, who would say, "Child, tell Mother to remember what she owes us."

"Mamma," said little Prezzemolina as she walked in, "the fairies say for you to remember what you owe them." At those words her mother would wince, but say nothing.

One day her mother had other things on her mind. Prezzemolina came in from school and said, "The fairies say for you to remember what you have to give them." Without thinking, her mother replied, "Of course. Tell them to go ahead and take it."

The next day the child went to school. "Well, does Mother remember?" asked the fairies.

"Yes, she said you can go ahead and take what she owes you."

The fairies didn't have to be told twice. They grabbed Prezzemolina and away they flew.

When her mother didn't see her come home, she began to worry more and more. All of a sudden she remembered what she had told the little girl, and exclaimed, "Oh, dear! What have I done! And there's no way to make up for it . . ."

The fairies took Prezzemolina to their house and showed her the grimy room where they kept their coal. "See this room, Prezzemolina? When we come home tonight, the walls must be milk-white and have all the birds of the air painted on them. Or else we'll eat you alive." They went off and left Prezzemolina crying her heart out.

A knock was heard at the door. Prezzemolina went to open it, sure the fairies were back and that her time had come. But in walked Memé, the fairies' cousin. "Why are you crying, Prezzemolina?" he asked.

"You would cry too," replied Prezzemolina, "if you had to make these filthy black walls milk-white and then paint all the birds of the air on them before the fairies get home! If I don't do it, they will eat me alive!"

"Give me a kiss," said Memé, "and I'll do every bit of the work myself."

But Prezzemolina replied:

> "Rather let the fairies eat me,
> Than allow a man to kiss me."

"The answer is so charming," said Memé, "that I'll do the work all the same."

He waved his magic wand, and the walls became as white as white could be, with all the birds on them, the way the fairies had directed.

Memé went away, and the fairies came back. "Well, Prezzemolina, did you get your work done?"

"I certainly did, ladies. Come see."

The fairies looked at one another. "Tell the truth, Prezzemolina; our cousin Memé came by, didn't he?"

Prezzemolina replied:

> "I saw neither your cousin Memé
> Nor my dear mother who brought me to the light of day."

The next morning the fairies put their heads together. "How can we get to eat her? There must be some way . . . Prezzemolina!"

"What is it, ladies?"

"Tomorrow morning you are to go to Morgan le Fay and tell her to give you Handsome Clown's box."

"All right," replied Prezzemolina, and set out the next morning. After walking and walking she met Memé, the fairies' cousin, who asked, "Where are you going?"

"To Morgan le Fay, to get Handsome Clown's box."

"But don't you know she will eat you alive?"

"So much the better for me. That way it will be all over."

"Here," said Memé. "Take with you these two pots of grease. You'll come to a gate that slams in your face. Grease it, and it will let you through. Also take along these two loaves of bread. You'll come upon two dogs in a fight. Throw them the bread, and they will let you pass. Also take with you this string and awl. You will meet a cobbler pulling out his beard and hair with which to stitch shoes. Give him these things, and he will let you pass. Then take these brooms. You will meet a woman baking bread and sweeping out the oven with her bare hands. Give her the brooms, and she will let you pass. And be sure you act quickly."

Prezzemolina took along grease, loaves, string, and brooms, which she gave to the gate, the dogs, the cobbler, and the woman baking bread. They all thanked her. Then she came to a town square and went up to knock on the door of Morgan le Fay's palace.

"Just a minute, child," said Morgan le Fay, "just a minute." But Prezzemolina knew she had to act quickly, so she flew up two flights of steps, found Handsome Clown's box, and made off with it as fast as her legs would carry her.

Hearing her flee, Morgan le Fay yelled out the window, "Woman baker sweeping out the oven with your bare hands! Stop that child, stop her!"

"Would I be such a fool? She was thoughtful enough to bring me brooms, so I won't have to sweep out the oven any longer with my hands."

"Cobbler stitching shoes with your beard and hair! Stop that child, stop her!"

"Would I be such a fool? She was thoughtful enough to bring me string and awl, so I won't have to pull out my beard and hair any longer to stitch up shoes."

"Dogs in a fight! Stop that child!"

"Would we be such fools? She gave us a loaf of bread apiece!"

"Slamming gate! Stop that child!"

"Would I be such a fool? She greased me from head to foot!"

So Prezzemolina got safely through. No sooner was she out of all danger than she began wondering. "Just what could be in Handsome Clown's box?" At last she gave in to the temptation to open it.

Out jumped a whole regiment of tiny, tiny men, who went marching off to the sound of their band, and there was no stopping them now at any cost. Prezzemolina tried her best to get them back into the box; but for every one she caught, ten others slipped away from her. She burst into tears, and just at that moment Memé arrived.

"Curiosity killed the cat!" he said. "Now you see what a fine mess you're in!"

"I was just going to take a peep."

"There's no way to make up for this. But if you give me a kiss, I'll get the men back into the box."

She replied:

> "Rather let the fairies eat me,
> Than allow a man to kiss me."

"You put that so nicely that I'm going to help you all the same." He waved his magic wand, and all the tiny men retreated into Handsome Clown's box.

The fairies were anything but pleased to hear Prezzemolina knock at the door. "How on earth did she ever get away from Morgan le Fay alive?"

"A pleasant day to each of you," she said. "Here is the box."

"Ah, you clever girl. . . . And what did Morgan le Fay have to say?"

"She told me to give you her best wishes."

"So that's it!" whispered the fairies to each other. "We are to eat her ourselves." That evening Memé called on them. "Know what, Memé? Morgan le Fay didn't eat Prezzemolina. We are to do so ourselves."

"Fine!" exclaimed Memé. "Wonderful!"

"Tomorrow when she's finished all the housework, we'll have her put on a large laundry tub of water to boil. When the water boils we'll grab her and throw her in."

"Yes, yes," he said. "I agree, it's a good idea."

When the fairies left the house, Memé went to Prezzemolina. "Listen, Prezzemolina. They intend to boil you alive in the laundry tub. But you must say there's no wood and go to the cellar to fetch some. Then I will come in."

So the fairies told Prezzemolina she had to do the wash and put the tub of water on to boil. She lit the fire, then announced: "But there's hardly any wood left."

"Go to the cellar and get it."

When Prezzemolina went down to the cellar, she heard Memé say, "I am here, Prezzemolina."

He took her by the hand and let her to the back of the cellar where many candles were burning. "These are the fairies' souls. Blow them out!" She started blowing, and for each candle that went out, a fairy died.

Finally only one candle was still burning, the biggest of them all. "This is Morgan le Fay's soul!" They both blew with all their might until they had put it out too and become the sole heirs to everything belonging to the fairies.

"Now you will be my bride," said Memé, and Prezzemolina gave him the kiss he had been waiting for.

They took up residence in Morgan le Fay's palace, making the cobbler a duke and the baker a marquise. They kept the dogs with them at the palace and left the gate right where it was, carefully greasing it every now and then.

> Thus they lived happily ever afterward,
> But nothing did they give to me their bard.

(Florence)

87

The Fine Greenbird

There was once a nosy king who went prowling in the evenings under the windows of his subjects to hear what they said about him in private. It was a time of unrest, and the king feared the people were hatching some plot against him. Thus, lurking near a humble country dwelling at dusk, he overheard three sisters on their porch in a spirited discussion.

The eldest said, "If I could marry the king's baker, I would make as much bread in a single day as the court eats in a whole year, so taken am I with that handsome young baker!"

The middle girl stated: "For my husband, I would like the king's vintner, and you would see me intoxicate the whole court with one glass of wine, so much does that vintner delight me!"

Then they asked the youngest girl, who held her tongue. "And whom would you marry?"

The youngest, who was also the loveliest of the three, answered, "I would take the king himself, and I would give him two rosy-faced, golden-haired sons, and a rosy-faced, golden-haired daughter with a star on her brow."

Her sisters made fun of her. "Poor little thing! You ask for so little!"

The nosy king, who had heard every word, went home, and the next day he sent for the three sisters. The girls were very frightened, for these were dangerous times when everyone was viewed with suspicion, and anything could happen. They got to the palace, quite upset, but the king said, "Don't be afraid. Just tell me what you were saying last night on your porch."

More taken aback than ever, they stammered, "Uh . . . we were . . . uh, we weren't saying anything."

"Weren't you saying you wanted to get married?" prompted the king. And he kept on until the eldest finally repeated what she'd said about wanting to marry the baker. "Very well, you shall have him," said the king. So the eldest girl got the baker for her husband.

The middle girl admitted she wanted the vintner. "Your wish is granted," said the king, and he gave her the vintner.

"And you?" he asked the youngest. Blushing from head to toe, she told him what she had said last night.

"If your wish to marry me came true, would you keep your promise?"

"I would do my best," said the girl.

"In that case you shall become my wife, and we'll see which of you girls is the most faithful to her word."

It galled the elder sisters, the baker's and vintner's wives, to be now so much lower in station than their lucky little sister-turned-queen-over-night, and their envy deepened when they learned that the queen was with child.

Meanwhile, the king had to go to war against his cousin. "Remember your promise," he said to his wife as he departed, leaving her in the care of his sisters-in-law.

While he was at war, his wife gave birth to a rosy-cheeked, golden-haired boy. How do you think her sisters reacted to that? They took the baby away and put a monkey in its place. They gave the baby to an old woman to drown. The old woman took the baby to the river in a basket. Reaching the bridge, she heaved her burden over the railing, basket and all.

The basket floated downstream and was soon seen by a boatman, who rowed after it. He caught hold of it, saw that beautiful child, and took him home to his wife to nurse.

To the king on the battlefield the sisters sent word that his wife had given birth to a monkey rather than a rosy-cheeked, golden-haired baby boy, and they wanted to know what they should do. "No matter whether it is a monkey or a baby boy," replied the king, "take care of my wife."

When the war was over he came home, but he no longer felt the same toward his wife. He still loved her, of course; but he was disappointed she hadn't kept her promise. Meanwhile the wife found herself expecting another child, and the king hoped things would go better this time.

But to get back to the first baby, the boatman happened to notice the little boy's hair one day. He said to his wife, "Just look at it! Doesn't it look like gold?"

The wife agreed. "It certainly does. It is gold!" They cut off a lock and went out and sold it. The goldsmith weighed it on his scales and paid them a gold sequin for it. From then on, the boatman and his wife would cut off a lock of the boy's hair every day and sell it. In no time they were rich.

Meanwhile the king's cousin started another war, and the king went off and left his wife awaiting their second child. "Remember your promise!" he told her as he departed.

This time too, while the king was away, the queen gave birth to a rosy-cheeked, golden-haired baby boy. Her sisters took the baby away and put a dog in its place. The baby was given to the same old woman, who threw him into the river in a basket, like his brother.

"What's going on?" asked the boatman upon seeing a second baby land in the river. Then he realized that this boy's hair would double their fortune.

Still at war, the king heard from his sisters-in-law. "This time, Majesty, your wife was delivered of a dog. Write us what to do with her." By way of reply, the king wrote: "No matter whether the dog is male or female, take good care of my wife." At last he came back to town very long-faced. But he truly loved his wife and still hoped that things would go well the third time.

As luck would have it, the cousin declared a third war, again while the queen was with child. The king had no choice but to go. He said to his wife, "Farewell, and remember your promise. You failed to give me the two golden-haired boys. Try to give me the little girl with the star on her brow."

She bore the baby girl, a beautiful rosy-cheeked, golden-haired child with a star on her brow. The old woman got her little basket ready and threw the baby into the river. The sisters put a small tiger cub in bed in its place. They wrote the king about the tiger that had been born and asked what he wanted done with his wife. He wrote back: "Whatever you like, just so I never see her in the palace again upon my return."

The sisters pulled her out of bed and carried her down to the cellar. There they walled her up, leaving only an opening for her head. Every day they took her a morsel of bread and a glass of water, then each of them gave her a slap in the face: that was her daily meal. Her rooms were walled up, and no trace at all was left of her. When the war was over and the king came home, he never mentioned her, nor did anyone else. He was now sad all the time.

The boatman, who had also found the little basket containing the baby girl, now had three fine children, who grew by leaps and bounds. With their golden hair, he amassed quite a fortune. One day he said, "We must now think about their future, poor dears, and build them a palace, for they are growing up." So, right across from the king's, he had an even larger palace built, with a garden that included all the wonders of the world.

Meanwhile the boys had become young men, and the girl a graceful young lady. The boatman and his wife had died, and the children lived together in this handsome palace, rich beyond belief. As they always wore their hats, no one knew they had hair of gold.

From the windows of the king's palace, the baker's and vintner's wives would gaze at them, never dreaming they were the young people's aunts. One morning these aunts saw the brothers and their little sister without their hats on, seated on a balcony, and cutting each other's hair. It was a sunny morning, and the golden hair gleamed so brightly that it blinded you. The thought suddenly occurred to the aunts that these might be their sister's children who had been thrown into the river. They began spying on them regularly, observing that they cut their hair every morning only to have it long again the next day. From then on, the two aunts were on pins and needles because of their crimes.

At the same time, the king, too, had taken to studying the neighboring garden and the children that lived there. He thought to himself, Those are just the kind of children I wanted my wife to give me. They look exactly like the ones she promised me. But he hadn't seen their golden hair, since they always kept their heads covered.

He got into conversation with them. "What a wonderful garden you have!"

"Majesty," replied the girl, "we have here in this garden all the beautiful things in existence. If you deem us worthy of the honor, you are welcome to walk here."

"With great pleasure. Since we are neighbors, why don't you come to my palace for dinner tomorrow?"

"Oh, Majesty," they said, "that would inconvenience the entire court too much."

"No," insisted the king, "your visit will make us very happy."

"In that case, we accept and will be there tomorrow."

When the sisters-in-law learned of the invitation, they flew to the old woman supposed to have murdered the poor little things. "Menga, what did you really do with those babies?"

"I threw them into the river, basket and all, but the basket was light and remained afloat. I didn't stay to see whether it ever sank or not."

"Wretch!" exclaimed the aunts. "The children are alive, and the king has seen them. If he learns who they are, we are done for. You must keep them from coming to the palace and do away with them once and for all."

"I will," replied the old woman.

Disguised as a beggar woman, she paused before the gate of their garden. Just then the girl was looking around her property and saying, as usual, "What does our garden lack? Nothing, for we have right here every beautiful thing in existence!"

"Ah, you say you have everything?" asked the old woman. "I know of one thing you lack, my child."

"What thing?"

"The dancing water."

"Where can you get . . ." began the child, but the old woman had disappeared. The girl burst into tears. "And here I thought we had everything in our garden, but . . . but we don't have the dancing water. The dancing water. . . . There's no telling how lovely it is!" And on and on she sobbed.

Coming home and finding her so upset, her brothers asked, "What's the matter? Why do you weep?"

"Please, leave me alone. I was here in the garden saying to myself that we had every beautiful thing in existence, when an old woman came to the gate and said, 'You think you have everything, but you have no dancing water.'"

"Is that all you're crying about?" asked the elder brother. "I'll go and get it myself, so you'll be happy." He removed the ring he was wearing and slipped it on his sister's finger. "If the stone changes color, that's a sign I am dead." He then mounted his horse and galloped off.

He had already gone a good way when he met a hermit, who asked, "Where are you going, my lad?"

"I am seeking the dancing water."

"My poor child!" answered the hermit. "They are sending you to your death. Are you unaware of the danger of the quest?"

"However dangerous it is, I must find the water."

"Listen to me, then," said the hermit. "Do you see that mountain? Scale it and you will come to a large plateau, in the middle of which rises a beautiful palace. Before the front door stand four giants holding swords. Watch out: if their eyes are closed you must not go past them. Is that clear? But when their eyes open, then you can go in. About the door: if it's open, don't go in; if it's closed, then push it open and walk in. You will come upon four lions: if their eyes are closed, don't go past them; pass only when their eyes open and you will come to the dancing water." The boy bid the hermit farewell, mounted his horse, and rode up the mountain.

Up there he found the palace with the front door open and the four giants with their eyes closed. "Yes, indeed, wait . . ." he told himself. The instant the giants opened their eyes and the front door closed, he went in. He waited for the lions also to open their eyes and moved past them. There was the dancing water. The boy filled the bottle he had brought along, and the minute the lions reopened their eyes, he took to his heels.

Just imagine the joy of the little sister, who'd spent all those days anxiously watching the ring, when her brother walked in with the dancing water. They hugged and kissed, then they placed two golden basins in the garden and poured into them the dancing water, which, to the little girl's great delight, leaped from one basin to the other. She was sure she now had every beautiful thing in existence right there in her garden.

The king passed by and wanted to know why they had not come to dinner; he had waited and waited for them. The little girl explained that their garden had lacked the dancing water, so her older brother had been obliged to fetch it. The king had much praise for the new addition and extended the three young people another invitation for the following day. The old woman, sent back by the aunts, saw the dancing water and felt her blood boil. "You have the dancing water now, but you still don't have the musical tree," she said to the little girl and vanished.

The brothers came home. "If you love me, dear brothers, you must bring me the musical tree."

This time it was the second brother's turn to say, "Why, of course, my little sister. I'll go and get it for you."

He gave his sister his ring, mounted his horse, and galloped all the way to the hermit who had helped his brother.

"Oh!" exclaimed the hermit. "The musical tree is a hard nut to crack. Here's what you have to do: scale the mountain, beware of the giants, the front door, and the lions, just as your brother did. You will then come to a little door with a pair of scissors over it. If the scissors are closed, don't go through the door. If they are open, go on through. You will then come upon a huge tree making music with its every leaf. Climb the tree and break off its highest branch. Plant it in your garden, and it will take root there."

The youth went up the mountain, found every sign favorable, and went in. He made his way up the tree through all the musical leaves and got the highest branch. Accompanied by its melody, he returned home.

When it was planted, the branch became the most beautiful tree in the garden, filling it with its music.

The king, who was rather outdone over the children's failure to show up the second time, was so delighted with that music that he reinvited all three of them for the next day.

The aunts at once dispatched the old woman. "You're satisfied with the advice I gave you? The dancing water, the musical tree! Now all you need is the fine Greenbird, and you will possess every beautiful thing in existence."

Here came the boys. "Little brothers, who is going after the fine Greenbird for me?"

"I am," replied the oldest, and he was off.

"This is truly unfortunate," said the hermit. "So many have gone after it, and no one has come back. Go to the same mountain, enter the same palace, and you will find a garden full of marble statues. They are noble knights who, like you, tried to capture the fine Greenbird. Flying through the trees in the garden are hundreds of birds. The fine Greenbird is the one that talks. He will speak to you, but don't dare answer, regardless of what he says."

The youth made it to the garden full of statues and birds. The fine Greenbird perched on his shoulder and said, "So you too have come, my good knight? And you think you can catch me? You are mistaken. Your aunts send you here to your death, and they keep your mother walled up alive . . ."

"My mother walled up alive!" exclaimed the youth, and as he spoke he immediately became a marble statue.

The sister never let a minute go by without looking at the ring. Seeing the stone turn blue, she screamed. "Help! He's dying!" And the other brother jumped into his saddle and galloped away.

He, too, reached the garden, and the fine Greenbird said to him, "Your mother is walled up alive."

"What! My mother walled up alive!" he cried, and turned to marble.

The sister was looking at the second brother's ring and saw it turn black. She did not go to pieces, but dressed up as a knight, took a phial of dancing water and a branch from the musical tree, saddled their fastest horse, and departed.

The hermit said to her, "Look out: if you answer the Greenbird when he speaks, you are done for. Rather, pull out one of his feathers, dip it into the dancing water, and touch each statue with it."

When the fine Greenbird saw the maiden dressed as a knight, he perched on her shoulder and said, "You're here, too? Now you will become like your brothers. Do you see them? One, two, and you will make three . . . Your father at war . . . Your mother walled up alive . . . And your aunts thumbing their noses at her . . ."

She let him run on, and the bird grew hoarse repeating his words in her ear. He was about to fly off when the maiden seized him, pulled a feather out of one of his wings, dipped it in the phial of dancing water, and passed it under the noses of her petrified brothers; they came back to life and embraced her. Next, the three of them stroked all the other statues and had a retinue of knights, barons, princes, and sons of noblemen. They made the giants sniff the feather, so the giants revived, too, and finally they brought the lions back to life. The fine Greenbird perched on the musical branch and allowed himself to be caged. In a grand procession, they all left the palace on the mountain, which magically vanished into thin air.

When the aunts looked out of the royal windows into the garden with the dancing water, the musical tree, and the fine Greenbird, and saw brothers and sister mingling with all those joyful princes and barons, they grew weak in the knees. The king decided to invite everybody to dinner.

They came, and the little sister brought along the fine Greenbird perched on her shoulder. As they were sitting down to the table, the fine Greenbird said, "One person is missing!" Everybody stood stockstill.

The king then counted everybody in his household to see who could be absent, but the fine Greenbird went on saying, "One person is missing!"

They had no idea whom else to bring in, when it suddenly dawned on the children. "Majesty! Could it be the queen walled up alive?" The king ordered her unwalled at once. The boys embraced her, and the little girl with the star on her brow helped her into a tub filled with dancing water and brought her out again as sound as ever.

Then they went back to their dinner, with the queen dressed as a queen at the head of the table and her two sisters green with envy.

Everybody was about to take the first bite of their food, when the fine Greenbird blurted out, "Only what I peck!" That was because the two aunts had poisoned the food. The guests ate only those portions the fine Greenbird pecked, and no one was poisoned.

"Now let us hear what the fine Greenbird has to tell us," proposed the king.

The fine Greenbird hopped onto the table before the king and said, "King, these are your children." They uncovered their heads, and everyone saw that they all three had golden hair, and the little sister a golden star on her brow. The fine Greenbird kept talking and told the whole story.

The king embraced his children and begged his wife to forgive him. Then he summoned his two sisters-in-law and the old woman, and said to the fine Greenbird, "Bird, now that you have disclosed everything, give out the sentence."

The bird said, "For the sisters-in-law, a gown of pitch and a greatcoat of fire. Throw the old woman out the window."

Thus was it done, and king, queen, and children lived happily ever after.

(Florence)

88

The King in the Basket

There was once a woodcutter at the king's court who had three daughters. One day the king ordered him to go and cut down a forest in a remote region of the country, something that would keep him away from home for years. The man couldn't very well tell the king, "No, I won't go," for he had to earn his bread; but at the same time he disliked the idea of going so far away and leaving his daughters at home by themselves.

He left the king and went home very much upset. "Girls, His Majesty has assigned work to me, and I must leave you. But before I go, I want you to allow one thing in particular."

"What, Papa?"

"Allow me to wall up the front door to keep everyone out and you safe inside."

"If that is your will, Papa, we consent."

The woodcutter had the front door walled up and supplied the girls with money and all else they would need, saying, "Take this nice big basket, tie the well-rope to it and let it down with money in it whenever any street peddlers come by selling something you want."

Weeping, they kissed each other goodbye, then he departed, and the masons immediately walled up the opening they had left in the entrance for the father to get out.

Now confined in earnest, the three girls were always leaning out of the window. The king saw them and thought they were the prettiest girls he had ever seen. He therefore dressed up as a peddler and went under their window crying, "Beautiful skeins of gold for sale! Beautiful skeins of gold for sale!"

The girls decided to buy some for their embroidery and called the peddler.

"What can I do for you, my ladies?" he asked.

"How much are those gold skeins?"

"Three crowns," he answered, and it was indeed a high price, since he was a king and didn't have too clear an idea about money. But the girls sent one crown down to him in the basket and told him to put in the skeins.

"Be careful, because they are heavy. Do you think you can pull them up?"

"Why not? There are three of us here!"

The girls pulled the basket up with great difficulty, and in it saw a man. They were going to let go of the rope, but the man grabbed hold of the window, saying, "Don't! I am the king! Knowing you were here by yourselves, I came to keep you company."

Clutching one another, the girls replied, "Majesty, we are poor girls. How can we possibly receive a person of your rank?"

"Don't give it a thought," said the king. "I've not come in search of luxury. I'm here to spend an hour with you, simply because you are beautiful and also kind, I'm sure." Then he added, "What a pity your father's away! I would so like to ask him to let you attend three grand balls I'm giving, starting tonight."

"That's much too kind of you," replied the girls, bowing. "Much too kind."

"When your father returns, I'll have other balls," said the king, "and you will come." He thus made pleasant conversation with them for an hour, then had the girls let him back down in the basket.

All the sisters could talk about afterward was the king's visit. The youngest said, "You can be sure you'll lower me in the basket tonight!"

"You? What on earth for?"

"Lower me and you'll see." She persuaded them, and the sisters let the basket down from the window.

The girl, whose name was Leonetta, went to the royal palace with the basket. She entered by the kitchen door, which was unguarded, as all the guards stood at the front door. The cooks, too, were out at the time peeping through the door at the arriving guests, so there was no one tending the stoves. Leonetta picked up items of food right and left and thrust them into her basket: roast chicken, skewered lamb, macaroni, almond cakes. Everything she was unable to carry off she doused with water and ashes, ruining it completely. Then she fled with the basket crammed full of the finest food in the world. Arriving home she whistled to her sisters to lower the rope and draw her and the basket up.

The next day when they heard, "Beautiful skeins of gold for sale!" they lowered the basket and drew up the king, whose face wore a grim expression.

"Majesty, what's wrong today?"

"Oh, my girls, you'd never imagine what happened to me last evening! When time came to sit down to the table, the servants went into the kitchen and found every bit of the food doused with ashes and water. Not a thing was fit to eat. They threw themselves at my feet and swore they were innocent, and I believed them. I'm either ill-starred, or else a traitor is trying to overthrow me. I've stationed guards everywhere for tonight's ball. When I find out who's scheming against me, I'll make mincemeat of him."

The girls sympathized with the king. "Oh, Majesty, how could anyone have done such things to you! You must be joking." The most indignant of the girls was Leonetta. "The idea of treating such a kind-hearted king like that! Whoever did it must be a madman!"

The king left, somewhat comforted by the three sisters' strong expression of sympathy.

In the evening, Leonetta said to the other two girls, "Come on, let me down and be quick about it!"

"Have you lost your mind?" asked her sisters. "Tonight you're staying home! After what you did last night, do you think we would let you out again? You heard what the king said, didn't you?"

Back and forth they argued, until the sisters finally lowered Leonetta to the ground. She went straight to the royal palace with the basket, but instead of entering through the kitchen, which was guarded, she went down into the cellar, where the best bottles and finest flasks were all hers. When the basket was full, she uncorked every one of the casks and fled.

The next day they pulled the king up, who was glummer than ever.

"Majesty, what on earth has happened to you?"

"It's too awful for words, my girls. They didn't bother the kitchen last night, but when the banquet was in full swing and I ordered wine brought to the guests, the servants went to the cellar and found it flooded with wine up to their knees. All the casks were uncorked and still spewing wine."

"Majesty, you don't mean it!"

"These are dangerous times, dear girls, I feel sure people are plotting to overthrow me. I shall double the number of guards tonight, and heaven help any traitor I lay my hands on! There'll be nothing left of him when I finish!"

"You're exactly right, Majesty," said Leonetta. "The idea of people abusing a kind-hearted soul like you!"

That evening the sisters were determined not to let Leonetta out, but she made such a fuss that they put her into the basket, saying, "All right, do as you please, but we're writing Papa at once and disclaiming all responsibility for your actions."

This time the kitchen and the cellar were packed with guards. Leonetta slipped into the cloakroom and filled her basket with all the mantles, furs, plumed hats, and boots it would hold. Then she set fire to what was left and fled.

At home the first thing the sisters did every morning was conceal all the loot, so the king wouldn't see it when he came. That day they really had a time hiding all those clothes, which they'd spent the whole night trying on. They dressed in their usual attire, but Leonetta forgot to take off a stolen pair of silver pumps.

When the king came up in the basket, his hair was all disheveled, and he had circles under his eyes. "Would you believe, my girls," he began, "they even tried to set fire to my palace! Luckily we caught it in time, but the cloakroom was heavily damaged. I will give no more balls from now on. I'm even of a mind to abdicate."

"Traitors!" exclaimed Leonetta, echoing the king's thoughts. "The idea of abusing such a kind-hearted soul!"

Time came for the king to leave, and the sisters were lowering him in the basket, when he noticed Leonetta's silver pumps and realized they were the very ones that had disappeared from his wardrobe. "You traitress!" he cried, reaching out to grab hold of the window. But the sisters all three let go of the rope, and the king plummeted to the ground in the basket. The girls were almost sure he'd killed himself, when they saw him get up and go hobbling off.

He reached the palace and began plotting his revenge at once. He wrote the woodcutter to return immediately, as he had to talk to him. The woodcutter, who had expected to be away no telling how long, was overjoyed to return, all the more so when he heard the king ask for the hand of one of his daughters.

"Any one of the three," said the king, "you want me to have."

The woodcutter went home and put the matter before the girls.

The oldest said, "No, Papa, I wouldn't want him for my husband."

"Neither would I, Papa," said the second girl, "since . . ."

Right off the bat, Leonetta said, "I'll take him."

The woodcutter went back and told the king, "Majesty, I informed my three daughters of the proposal. The first one answered: 'I wouldn't want him'; the second, 'Neither would I, since . . .' But the third said: 'I'll take him.'"

The king said to himself, "So she's actually the boldest and the one who did all the mischief." To the woodcutter he replied, "In that case I'll marry the third girl."

The wedding was set for a few days hence. The bride had at her beck and call many maids of honor from the royal household, to whom she said, "Listen, I intend to play a joke on the king."

"What will you do, my lady?"

"Don't breathe a word of it to a soul, under any circumstances. I'm going to make a life-size pastry woman, with sugar and honey for a heart.

326

She'll have strings tied to her that will permit her to say yes or no. I'll put her in bed in my place and see if the king notices it."

The maids of honor set to work and made the pastry woman. The girl had her put into the bridal bed dressed in her own nightgown and bonnet.

The wedding was followed by the banquet, supper, and finally it was time to go to bed. Leonetta asked to go upstairs before the king, and hid under the bed, holding the strings to make the pastry woman move.

The king came into the room, closed the door, and said, "It's now between the two of us, my dear! At last you're in my hands! Do you remember telling me, 'You're such a kind-hearted soul, Majesty'?"

"Yes, I remember," said Leonetta from under the bed, making the pastry woman nod her head.

"You do? And just who wrecked my kitchen?"

"I did, Majesty," said Leonetta, and the pastry woman lying in bed shook her head and moved her hands.

"What a hypocrite you are! And who wrecked the cellar?"

"I did, Majesty!"

"And the cloakroom?"

"I did that too, Majesty!"

"And you think I'll ignore those outrages?"

"I don't know, Majesty!"

She'd no sooner got the words out of her mouth than the king unsheathed his sword and, thinking the pastry woman was his bride, thrust it into her heart. Sugar and honey spurted all over him.

"There, I've killed you," he began shouting; "you asked for it!" Tasting sugar and honey on his lips, he exclaimed, "But you were made of sugar and honey! We could have been happy together! How I would love you, were you still alive!"

From under the bed, Leonetta replied in a faint voice, "Alas I'm dead . . ."

"Oh, what have I done!" sighed the king. "My Leonetta of sugar and honey . . . If you were here now, I would love you so!"

"I'm now dead," answered Leonetta.

"Since you are dead, I too will be better off dead!" said the king and prepared to impale himself on his sword.

"Don't, I'm alive, I'm alive!" cried Leonetta, leaping from under the bed and embracing him.

They fell into each other's arms, kissed, and from then on loved each other and were as happy as happy could be.

(Florence)

89

The One-Handed Murderer

There was once a miser king, so miserly that he kept his only daughter in the garret for fear someone would ask for her hand and thus oblige him to provide her with a dowry.

One day a murderer came to town and stopped at the inn across the street from the king's palace. Right away he wanted to know who lived over there. "That's the home of a king," he was told, "so miserly that he keeps his daughter in the garret."

So what does the murderer do at night but climb up on the king's roof and open the small garret window. Lying in bed, the princess saw the window open and a man on the ledge. "Help! Burglar!" she screamed. The murderer closed the window and fled over the rooftops. The servants came running, saw the window closed, and said, "Your Highness, you were dreaming. There's no one here."

The next morning she asked her father to let her out of the garret, but the king said, "Your fears are imaginary. No one in the world would ever think of coming up here."

The second night the murderer opened the window at the same hour. "Help! Burglar!" screamed the princess, but again he got away, and no one would believe her.

The third night she fastened the window with a strong chain and, with pounding heart, stood guard all by herself holding a knife. The murderer tried to open the window, but couldn't. He thrust in one hand, and the princess cut it clean off at the wrist. "You wretch!" cried the murderer. "You'll pay for that!" And he fled over the rooftops.

The princess showed the king and the court the amputated hand, and everybody finally believed her and complimented her courage. From that day on, she no longer slept in the garret.

Not too long after that, the king received a request for an audience from an elegant young stranger who wore gloves. He was so well-spoken that the king took an instant liking to him. Talking of this and that, the stranger mentioned that he was a bachelor in search of a genteel bride, whom he would marry without a dowry, being so wealthy himself. Hearing that, the king thought, This is just the husband for my daughter, and he sent for her. The minute the princess saw the man she shuddered, having the strong impression she already knew him. Once she was alone with her father, she said, "Majesty, I'm all but sure that's the burglar whose hand I cut off."

"Nonsense," replied the king. "Didn't you notice his beautiful hands and elegant gloves? He's a nobleman beyond any shadow of a doubt."

To make a long story short, the stranger asked for the princess's hand, and to obey her father and escape his tyranny, she said yes. The wedding was short and simple, since the bridegroom couldn't remain away from his business and the king was unwilling to spend any money. He gave his daughter, for a bridal present, a walnut necklace and a worn-out foxtail. Then the newlyweds drove off at once in a carriage.

The carriage entered a forest, but instead of following the main road it turned off onto a scarcely visible trail that led deeper and deeper into the underbrush. When they had gone some distance, the bridegroom said, "My dear, pull off my glove."

She did, and discovered a stump. "Help!" she cried, realizing she'd married the man whose hand she had cut off.

"You're in my power now," said the man. "I am a murderer by profession, mind you. I'll now get even with you for maiming me."

The murderer's house was at the edge of the forest, by the sea. "Here I've stored all the treasure of my victims," he said, pointing to the house, "and you will stay and guard it."

He chained her to a tree in front of the house and walked off. The princess remained by herself, tethered like a dog, and before her was the sea, over which a ship glided from time to time. She tried signaling to a passing ship. On board they saw her through their telescope and sailed closer to see what the matter was. The crew disembarked, and she told them her story. So they set her free and took her aboard, together with all the murderer's treasure.

It was a ship of cotton merchants, who thought it wise to conceal the princess and all the treasure underneath the bales of cotton. The murderer returned and found his wife gone and the house ransacked. She could have only escaped by the sea, he thought to himself, and then saw the ship disappearing into the distance. He got into his swift sailboat and caught up with the ship. "All that cotton overboard!" he ordered. "I must find my wife who has fled."

"Do you want to ruin us?" asked the merchants. "Why not run your sword through the bales to see if anyone is hiding in them?"

The murderer started piercing the cotton with his sword and, before long, wounded the girl hiding there. But as he drew his sword out, the cotton wiped the blood off, and the sword came out clean.

"Listen," said the sailors, "we saw another ship approach the coast, that one down there."

"I'll investigate at once," said the murderer. He left the ship carrying cotton and directed his sailboat toward the other ship.

The girl, who had received a mere scratch on her arm, was put ashore in a safe port. But she protested, saying, "Throw me into the sea! Throw me into the sea!"

The sailors talked the matter over, and one oldtimer in their midst whose wife had no children, offered to take the girl home with him, together with part of the murderer's jewels. The sailor's wife was a good old soul and gave her a mother's love. "Poor dear, you will be our daughter!"

"You are such good people," said the girl. "I'm going to ask just one favor: let me always stay inside and be seen by no man."

"Don't worry, dear, nobody ever comes to our house."

The old man sold a few jewels and bought embroidery silk, so the girl spent her time embroidering. She made an exquisite tablecloth, working into it every color and design under the sun, and the old woman took it to the nearby house of a king to sell.

"But who does this fine work?" asked the king.

"One of my daughters, Majesty," replied the old woman.

"Go on! That doesn't look like the work of a sailor's daughter," said the king, and bought the tablecloth.

The old woman used the money to buy more silk, and the girl embroidered a beautiful folding screen, which the old woman also took to the king.

"Is this really your daughter's work?" asked the king. He was still suspicious, and secretly followed her home.

Just as the old woman was closing the door, the king walked up and stuck his foot in it; the old woman let out a cry. Hearing the cry from her room, the girl thought the murderer had come after her and she fainted from fright. The old woman and the king came in and tried to revive her. She opened her eyes and, seeing that it was not the murderer, regained her senses.

"But what are you so afraid of?" asked the king, charmed with this girl.

"It's just my bad luck," she replied, and would say nothing more.

So the king started going to that house every day to keep the girl company and watch her embroider. He had fallen in love with her and finally asked for her hand in marriage. You can just imagine the old people's amazement. "Majesty, we are poor people," they began.

"No matter, I'm interested in the girl."

"I am willing," said the maiden, "but on one condition."

"What is that?"

"I refuse to see all men regardless of who they are, except you and my father." (She now called the old sailor her father.) "I will neither see them nor be seen by them."

The king consented to that. Jealous beyond measure, he was delighted she wanted to see no man but him.

Thus were they married in secret, so that no man would see her. The king's subjects were not at all happy over the matter, for when had a king ever married without showing the people his wife? The strangest of rumors began circulating. "He's married a monkey. He's married a hunchback. He's married a witch." Nor were the people the only ones to gossip; the highest dignitaries at the court also talked. So the king was forced to say to his wife, "You must appear in public for one hour and put an end to all those rumors."

The poor thing had no choice but obey. "Very well, tomorrow morning from eleven till noon I will appear on the terrace."

At eleven o'clock, the square was more packed than it had ever been. People had come from all over the country, even from the backwoods. The bride walked onto the terrace, and a murmur of admiration went up from the crowd. Never had they seen so beautiful a queen. She, however, scanned the crowd with uneasiness, and there in its midst stood a man cloaked in black. He brought his hand to his mouth and bit it in a threatening gesture, then held up his other arm, which ended in a stump. The queen sank to the ground in a swoon.

They carried her inside at once, and the old woman said over and over, "You would have to show her off! You would have to show her off against her will. Now just see what's happened!"

The queen was put to bed, and all the doctors were called in, but her illness baffled everyone. She insisted on remaining shut up and seeing no one, and she trembled all the time.

Meanwhile the king received a visit from a well-to-do foreign gentleman with a glib tongue and full of flattery. The king invited him to stay for dinner. The stranger, who was none other than the murderer, graciously accepted and ordered wine for everyone in the royal palace. Casks, barrels, and demijohns were brought in at once, but every drop of the wine had been drugged. That evening, guards, servants, ministers, and everybody else drank their fill and, by night, they were dead drunk and snoring, the king loudest of all.

The murderer went through the palace making sure that on the stairs, in the corridors and all the rooms there was no one who wasn't flat on his back and sleeping. Then he tiptoed into the queen's room and found her hunched up in a corner of her bed and wide-eyed, almost as though she expected him.

"The hour has come for my revenge," hissed the murderer. "Get out of bed and fetch me a basin of water to wash the blood from my hands when I've cut your throat."

The queen ran out of the room to her husband. "Wake up! For heaven's sake, wake up!" But he slept on. Everybody in the whole palace slept, and there was no way in the world to wake them up. She got the basin of water and returned to her room.

"Bring me some soap, too," ordered the murderer as he sharpened his knife.

She went out, tried once more to rouse her husband, but to no avail. She then returned with the soap.

"And the towel?" asked the murderer.

She went out, got the pistol off of her sleeping husband, wrapped it in the towel and, making a motion to hand the towel to the murderer, fired a shot point-blank into his heart.

At that shot, the drunk people all woke up at the same time and, with the king in the lead, ran into her room. They found the murderer slain and the queen freed at last from her terror.

(Florence)

❖ 90 ❖

The Two Hunchbacks

There were two hunchbacks who were brothers. The younger hunchback said, "I'm going out and make a fortune." He set out on foot. After walking for miles and miles he lost his way in the woods.

"What will I do now? What if assassins appeared . . . I'd better climb this tree." Once he was up the tree he heard a noise. "There they are, help!"

Instead of assassins, out of a hole in the ground climbed a little old woman, then another and another, followed by a whole line of little old women, one right behind the other, who all danced around the tree singing:

> "Saturday and Sunday!
> Saturday and Sunday!"

Round and round they went, singing over and over:

"Saturday and Sunday!"

From his perch in the treetop, the hunchback sang:

"And Monday!"

The little old women became dead silent, looked up, and one of them said, "Oh, the good soul that has given us that lovely line! We never would have thought of it by ourselves!"

Overjoyed, they resumed their dance around the tree, singing all the while:

"Saturday, Sunday
And Monday!
Saturday, Sunday,
And Monday!"

After a few rounds they spied the hunchback up in the tree. He trembled for his life. "For goodness' sakes, little old souls, don't kill me. That line just slipped out. I meant no harm, I swear."

"Well, come down and let us reward you. Ask any favor at all, and we will grant it."

The hunchback came down the tree.

"Go on, ask!"

"I'm a poor man. What do you expect me to ask? What I'd really like would be for this hump to come off my back, since the boys all tease me about it."

"All right, the hump will be removed."

The old women took a butter saw, sawed off the hump, and rubbed his back with salve, so that it was now sound and scarless. The hump they hung on the tree.

The hunchback who was no longer a hunchback went home, and nobody recognized him. "It can't be you!" said his brother.

"It most certainly is me. See how handsome I've become?"

"How did you do it?"

"Just listen." He told him about the tree, the little old women, and their song.

"I'm going to them, too," announced the brother.

So he set out, entered the same woods, and climbed the same tree. At the same time as last, here came the little old women out of their hole singing:

"Saturday, Sunday,
And Monday!

333

Saturday, Sunday,
And Monday!"

From the tree the hunchback sang:

"And Tuesday!"

The old women began singing:

"Saturday, Sunday,
And Monday!
And Tuesday!"

But the song no longer suited them, its rhythm had been marred.

They looked up, furious. "Who is this criminal, this assassin? We were singing so well and he had to come along and ruin everything! Now we've lost our song!" They finally saw him up in the tree. "Come down, come down!"

"I will not!" said the hunchback, scared to death. "You will kill me!"

"No, we won't. Come on down!"

The hunchback came down, and the little old women grabbed his brother's hump hanging on a tree limb and stuck it on his chest. "That's the punishment you deserve!"

So the poor hunchback went home with two humps instead of one.

(Florence)

❖ 91 ❖

Pete and the Ox

A woman was cooking some chickpeas. A needy girl passed by and begged for a bowl of them. "If I give them to you," replied the woman, "what will I then eat myself?" At that, the poor girl cursed her. "May all the peas in the pot become so many children for you!" Then she continued on her way.

The fire went out, and from the pot, like chickpeas boiling over, popped one hundred little boys as tiny as peas screaming, "Mamma, I'm hungry! Mamma, I'm thirsty! Mamma, pick me up!" They scattered into all the drawers, ovens, and pots. Frightened out of her wits, the woman scooped up these little creatures by the handful, thrust them into her mortar, and crushed them with the pestle as though she were making mashed peas. When she thought she'd finally slain them all, she began getting dinner for her husband. But reflecting on what she'd done, she burst into tears, saying, "If only I'd spared the life of at least one of them! He'd now be a help to me and take his father's dinner to the shop!"

Just then, she heard a tiny voice. "Don't cry, Mamma, I'm still here!" It was one of the little sons, who'd escaped death by hiding behind the handle of the jug.

The woman was overjoyed. "Come here, my dear! What is your name?"

"Pete," replied the child, sliding down the jug and landing on the table.

"Well done, my little Pete!" exclaimed the woman. "You are now to go to the shop with your father's dinner." She put everything into the basket and set it on Pete's head.

Pete left, and all you saw was the basket, which looked as though it were walking by itself. He asked a couple of people the way, scaring the life out of them, for they thought the basket itself was talking. Finally he reached the shop and called, "Papa, I've brought your dinner to you."

"Who's that calling me?" wondered his father. "I've never had any children!" He came out and saw the basket, under which a tiny voice was heard. "Papa, lift the basket and you will see me. I'm your son Pete, born this very morning."

The man lifted the basket, and there stood Pete. "Well done, Pete!" said his father, who was a locksmith. "You will now come round to the farmers' houses with me to see if they've anything that needs mending."

At that, the father put Pete into his pocket, and off they went. They talked without stopping along the way, and everybody thought the man had lost his mind to be talking to himself that way.

He asked around at the different houses, "Do you have anything to be repaired?"

"We do indeed," he was told, "but we wouldn't trust a crazy man like you to mend a thing."

"Crazy? What do you mean? I'm much smarter than you any time."

"Well, why do you constantly talk to yourself as you go from place to place?"

"That's not true. I was talking to my son."

"Just where is this son?"

"In my pocket."

"You see what we meant? You are crazy."

"Look here!" he said, reaching into his pocket and bringing his hand out with Pete straddling one of his fingers.

"Oh, what a fine little man! Please hire him out to us to keep watch over our ox."

"Would you like that, Pete?"

"Yes, I would."

"Well, I'll leave you here and stop by for you tonight."

Pete was placed on one of the ox's horns, and it looked as though the ox was in the field unguarded. Two thieves came by and decided to steal it. But Pete cried, "Farmer, farmer, come quick!"

The farmer came running, and the thieves asked, "My good man, where is that voice coming from?"

"Oh, that's Pete talking. Don't you see him perched up there on the ox's horn?"

The thieves spotted Pete and said to the farmer, "Let us have him a few days and we'll make you rich." The farmer sent Pete off with the thieves.

With Pete in their pocket, the thieves went to the king's stable to steal horses. The stable was locked, but Pete crawled through the keyhole, opened the door, untied the horses, and came outside with them, hidden in one of the horse's ears. The thieves, who were waiting for him, mounted the horses and galloped home, where they said to Pete, "We're tired and will go to bed. You give the horses their oats."

Pete started fastening feedbags onto the horses, but he was so sleepy that he fell into one of the bags and went fast asleep. Unaware he was in there, the horse ate him up along with the oats.

When he didn't come back, the thieves went down to the stable to look for him. "Pete, where are you?"

"Here I am," replied a tiny voice. "I'm in the belly of one of the horses."

"Which one?"

"This one right here!"

The thieves slit open a horse, but he wasn't in it. "No, it's not this one. Which horse are you in?"

"In this one!" So the thieves slit open another horse.

They went on slitting open horses until they had killed every last one of them, but they still didn't find Pete. They were tired by then and said, "That's a crying shame! He was so useful to us, and now we've gone and lost him! To make matters worse, we've lost all the horses too!" They dragged the carcasses into the field and went back to bed.

A hungry wolf came by, spied the butchered horses, and had quite a feast. Pete still happened to be hiding in the belly of one of the horses, and the wolf swallowed him with all the rest. There he was in the wolf's belly now, and when the wolf got hungry again and headed for a nanny goat tethered in a field, Pete began yelling inside the wolf's belly, "Wolf, wolf!" until the owner of the goat heard and put the wolf to flight.

The wolf said, "What's the matter with me to be making these sounds? I must have gas on my stomach!" and he began breaking wind.

There, it's all gone, he thought. Now I'll go and eat a sheep.

But when he neared the sheepfold, Pete started up again in his belly, crying "Wolf, wolf!" until he'd awakened the shepherd.

The wolf was worried. "I still have gas making all that noise inside me," and he went back to breaking wind. He broke wind once, then again, and the third time Pete too slipped out and hid behind a bush. Thus unburdened, the wolf returned to the sheepfold.

Three robbers came by and sat down to count the money they'd stolen. "One, two, three, four, five," said one. From his hiding place, Pete mimicked him. "One two three four five . . ."

The robber said to his companions, "Shut up, you're confusing me. One more word out of you, and I'll let you have it." Then he started over: "One, two, three, four, five . . ."

"One two three four five," piped Pete.

"You didn't hear what I said? Now you'll see!"

He killed him and turned to the other robber. "You'll get the same thing if you make a sound." Again he started over. "'One, two, three, four, five . . ."

Pete repeated: "One two three four five . . ."

"That wasn't me, I swear!" said the other robber.

"Don't try to fool me! Now it's your turn," and he killed him. "At last I can count the money in peace and keep every penny for myself. One, two, three, four, five . . ."

"One two three four five," piped Pete.

The robber's hair stood straight up. "There's somebody hiding around here. I'd better flee." He fled, leaving all the money right there.

Carrying the bag of money on his head, Pete went home and knocked on the door. His mother opened up and saw nothing but the bag of money. "Pete!" she exclaimed. She lifted the bag, and there stood her son, whom she embraced.

(Florence)

The King of the Peacocks

A king and a queen had two sons and a little girl who was the apple of their eye. They always gave her all their love, and even had a nursemaid just for her in the palace. Now the king was taken sick one day and died. The queen looked after the kingdom for a while afterward, but then she, too, fell ill. At death's door, she entrusted her two sons with the care of their little sister, after which she drew her last breath and died.

In the meantime the little girl had grown up without ever leaving the palace. Her sole pastimes were gazing out of the window at the countryside, singing softly, chatting with the nursemaid who was now her governess, and embroidering. One day as she stood at the window, a peacock emerged from the woods, flew up, and lit on the window ledge. The girl made a big to-do over him, serving him birdseed and inviting him inside. "How handsome he is!" she exclaimed. "Until I've found the king of the peacocks, I'll not marry!" She kept the peacock with her all the time, but shut him up in a wardrobe whenever anyone came in.

It wasn't long before the brothers remarked to one another, "Our dear little sister never wants to go out of the house. If that continues, she will be in a bad way. Let's see if she wishes a husband." They went to her and made known their thoughts. "As long as you are unmarried, we will not get married ourselves. Do you feel like taking a husband?"

"No, I don't."

"You just think you don't. Look at these portraits here of all the kings, pick out the one you like, and we'll ask him if he wants you."

"I tell you I want no husband . . ."

"For our sake, will you make a choice?"

"If you insist, I will. But the choice must be mine."

"Of course."

At that, the sister opened the wardrobe and brought out the peacock. "See this?"

"Yes, it is a handsome peacock."

"I'll not marry until I find the king of the peacocks."

"Where is he?"

"I've no idea, but I will marry only him."

"In that case we'll try to find him." They entrusted the girl to her governess, chose a reliable governor to look after the kingdom, and each went their separate way in search of the king of the peacocks.

They asked all around, but no one had ever heard of the king and took them for fools. But the two youths didn't give up hope and continued their search. One evening the older boy met an old man who was part-sorcerer. "Tell me, is there a king of the peacocks?"

"There certainly is."

"What is he like? Where does he live?"

"He's a handsome young man who dresses like a peacock. Peru is his kingdom, and to see him you have to go there."

The youth thanked and rewarded the man, then headed for Peru. After traveling a great distance, he entered a meadow planted with strange trees, and all around him voices said, "There he is! There he is! He's come to offer his sister in marriage to the king! Make way for him!"

The youth glanced about, but still saw nothing but a flutter of brightly colored feathers in the air.

"Where am I?" he asked.

"In Peru," answered the voices. "In the realm of the king of the peacocks."

"Would you please tell me where he is?"

"With great pleasure. Go to the right and you'll come to a beautiful palace. Say to the guards, 'Royal secret!' and they will let you in."

"Thank you!"

"Don't mention it!"

These trees are very courteous, thought the youth, but sorcery is surely afoot here. He walked on, coming to a palace adorned with blue, white, and violet peacock feathers that gleamed in the sun like gold. The front door was flanked by guards dressed as peacocks, and you really couldn't tell whether they were men or birds. "Royal secret!" said the youth, and they let him in. In the middle of a hall stood a throne of precious stones, with an aureole of peacock feathers whose eyes sparkled like stars. On the throne sat the king dressed from head to foot in feathers and indistinguishable from a bird. The youth bowed and, at a signal from the king, all the courtiers left the room. "Speak, I am listening," said the king.

"Lord, I am the king of Portugal," began the youth, "and I have come to ask if you will accept my little sister in marriage. Pardon my boldness, but my sister has resolved to wed no one but the king of the peacocks."

"Do you have her portrait with you?"

"Here it is, Majesty."

"She is lovely! I like her! I agree to this marriage!"

"Majesty, thank you! My sister will be delighted, as we all are." He bowed, and proceeded to leave.

"Stop!" ordered the king. "Where are you going?"

"To fetch her, Majesty."

"No, whoever enters the kingdom of the peacocks can no longer leave it. You are a stranger to me. For all I know, you could be a spy for an enemy king, or a thief aiming to rob me. Write home, send my portrait, and wait for an answer."

"I will do just that," replied the youth. "But where will I lodge while I'm waiting for the answer?"

At a signal from the king, guards entered and seized the young man by the arms. "You will wait in prison," said the king, "until your sister arrives."

In the meantime the second brother had returned home after a fruitless search. When the letter came from Peru, he ran to his sister with the portrait of the king of the peacocks. "That's my bridegroom," said the girl, "exactly the one I wanted! Let's make haste and leave, I can hardly wait to see him!" They started packing up the trousseau, prepared the luggage and horses, and ordered the finest ship of the whole fleet.

"To reach Peru we have to cross the sea," the brother explained to the nursemaid. "How can we protect my sister against wind, dampness, and bright sunlight?"

"That's simple," said the nursemaid. "Drive her to the shore in a carriage, bring the ship up close to land, and roll the carriage up the gangplank. That way she can make the crossing comfortably inside her carriage, without getting in the wind or spoiling her bridal gown." The plan was adopted.

Now the nursemaid had a daughter as ugly as a demon, who was envious and wicked to boot. Upon learning that the princess was going off to be married, she began whimpering to her mother. "She's getting married and I'm not, she's getting a king and I'm getting nothing; all eyes will be on her and none will be on me . . ."

"True," said the nursemaid, "that occurred to me, too." And she began dreaming of a scheme to marry her daughter instead of the princess to the king. After some hard thinking she came up with a plan. She ordered her daughter a carriage and wedding dress exactly like the princess's, and said to the captain of the ship, "Here are two million just for you, now listen to what you are to do. In the last carriage that comes aboard sits my daughter. At night when everybody is sleeping, take the princess, throw her into the sea, and put my daughter in her place."

The captain was afraid to consent, but two million was a fortune, and he thought, When I have it in my pocket, I can flee to a distant country and enjoy it. So he haggled a bit over the amount, then agreed to do the deed.

When time came to depart, all the carriages were lined up side by side on the ship, but at the last minute the princess started crying for her little dog. "He's been my companion for so long I just can't go off and leave him!" So her brother ran ashore, picked up the dog, and brought him to the carriage. The little dog cuddled up on the mattress and the ship set sail.

At nightfall, the nursemaid went to the bride's carriage. "The weather is good, a favorable wind is blowing, and we'll be in Peru tomorrow. Go to sleep and get your rest." The princess fell asleep dreaming of the king of the peacocks and the grand welcome she would receive upon her arrival.

At midnight, the captain eased open the carriage door, picked up the mattress with princess and dog on it, and threw it into the sea.

Nearby, the nursemaid's daughter waited in the shadows, and the captain put her into the bride's carriage.

Falling, the princess woke up and found herself in the middle of the sea, while the ship continued on its way out of sight. But the mattress, as light as a feather, floated instead of sinking. A gentle wind rose and swept it right to Peru with the girl on it in her wedding gown and the dog at her side.

Toward daybreak, a sailor of Peru, whose house was by the sea, heard a dog barking in the distance. "Do you hear that dog?" he asked his wife.

"Yes, there must be someone in danger."

"I thought so, too. It's almost day, so I'll go outside and take a look."

He dressed, picked up a harpoon, and went out on the shore. And there in the dawn he saw something floating along briskly and accompanied by barking. As it came closer, the sailor stepped into the water, reached out with his harpoon, and drew the raft to him. Imagine his surprise on beholding a girl asleep in a wedding dress and a little dog making such a fuss! He brought her in to shore very gently so as not to awaken her, but she stirred and asked, "Where am I?"

"You've landed at the house of a poor but kind-hearted sailor and his wife. Come and stay with us."

At that hour, the accursedly hideous girl was arriving in Peru, shut up in her carriage. When the procession reached the meadow of strange trees, a clamor went up on all sides:

"Cuckoo! Cuckoo!
What an eyesore, the queen of Peru!"

And in the air swirled thousands of peacock feathers. The brother who had accompanied her on the voyage, rode up on horseback and, at the sound of those cries coming from no telling where, his heart stood still. "This is a bad sign," he said to himself. "Something will surely happen to us!" He ran to the carriage, opened the door, and was stunned at the sight of the hideous girl. "How did you ever become so ugly? What happened? Was it the sea, the wind, the sun that did it? Tell me!"

"How should I know?"

"Here's the king! Now everybody's head will roll!"

In the middle of a band of feather-clad soldiers appeared the king of the peacocks. The soldiers raised long gold trumpets and blew a flourish. The trees cried:

> "Long live the king! Long live the king!
> The bride is a homely thing!"

and the air was so thick with swirling feathers that it looked like fog.

"Where is the bride?" asked the king.

"Here she is, Lord . . ."

"Is this the beautiful maiden whose praises were sung to me?"

"What do you expect, Majesty? It must be the fault of the wind, the sea breeze . . ."

"What a wind and what a sea! Shut up, impostors! You intended to cheat me, but you'll learn that the king of the peacocks is not to be deceived! Take them to prison and prepare gallows for each." At that, the king of the peacocks walked away scowling: he was out of sorts not only because of the insult, but especially because of his love for the beautiful maiden whose picture he wore around his neck and never tired of admiring.

Let's leave the king and those unfortunate prisoners and go back to the beautiful princess in the home of the poor sailor. The next morning she asked the sailor's wife, "Would you have a little basket?"

"Yes, madam."

"Give it to me, so I can take care of our dinner." She called the dog and gave him the basket, saying, "Go to the king's and get our dinner."

Holding the basket handle in his mouth, the little dog ran to the king's kitchen, picked up a roasted chicken, thrust it into the basket, and went flying back to his mistress. They ate a good dinner that day at the sailor's house, and even the dog had his share of bones to gnaw on.

The next day the dog returned to the king's kitchen with the basket and made off with a huge fish. This time the cook went and told the king, who ordered him to catch the dog the next time, or at least find out where he went.

So, when the dog ran off with a leg of lamb the next day, the cook followed and saw him enter the sailor's house. He returned and told the king.

"I'll go after him myself tomorrow," declared the king. "Or could I already be the laughingstock of everybody?"

The next morning as soon as the dog had left with the little basket, the princess put on her bridal gown and sat down in her room to wait. "If someone comes looking for the dog," she said to the sailor and his wife, "send him in to me."

It wasn't long before the dog was back home with dinner in the basket, and right behind him ran the king with two peacock soldiers. "Have you seen a dog?" they asked the sailor.

"Yes, Majesty."

"Why is he always stealing my dinner?"

"He does it of his own free will, to provide us with something to eat. We never taught him to do that ourselves."

"Where did you get him?"

"He's not ours. He belongs to a bride who is here with us."

"I want to see her."

"Come in, come in, Majesty. You will excuse us: this is the house of poor people." They brought him in, and there before him was the maiden of the picture, dressed as a bride. "I am the daughter of the king of Portugal, and you, my lord, have my brothers in prison."

"Can it be?" said the king of the peacocks.

"Look, here is the portrait you sent me. I have always worn it next to my heart."

"I can't make head or tail of all this," said the king. "Wait for me here, I'll be right back." Off he went like an arrow. He reached the palace and released the two brothers from prison. "Your sister has been found. You are back in my esteem, but tell me how everything happened."

"We are just as much in the dark as you. The more we think about it, the less we understand."

The king then called in the nursemaid and her daughter, threatened them, and was informed of their entire scheme. He had them thrown into prison where the two brothers had been, armed all the soldiers, donned his finest feathers and, accompanied by the band, marched off at the head of his army to the poor mariner's house to fetch his bride.

> "Yes indeed! Yes indeed!
> This is the queen we heed!"

cried the trees, and millions of feathers filled the air and blotted out the sunlight.

When they arrived at the palace, the wedding was celebrated and followed by a grand banquet. The nursemaid and her accursedly ugly daughter were hanged on the gallows that had been readied for the brothers. They never were able to catch the captain of the ship, since he had fled to the end of the earth to enjoy his two million.

(Siena)

◈ 93 ◈

The Palace of the Doomed Queen

In bygone days there lived an old widow who earned her bread by spinning. She had three daughters who also were spinners. Although they toiled day and night at their spinning wheels, the three spinners could never lay up a cent, as they earned barely enough for their daily needs. One day the old woman got sick and ran a high fever, and three days later she was near death. Calling her tearful daughters around her, she said, "Don't weep. Nobody lives forever. I've lived a long life, and now it's my turn to die. What really breaks my heart is to leave you so poor. But since you know how to earn your living, you will manage somehow, and I'll beg heaven to help you. All I have to leave you as a dowry are the three balls of spun hemp there in the cabinet." After those words, she drew her last breath and died.

A few days later the sisters got to talking. "This Sunday," they said, will be Easter Sunday, and here we are with nothing for a decent Easter-dinner."

Mary, the oldest sister, suggested: "I'll sell my ball of thread and we'll buy the dinner." So, on Easter morning, she took her thread to market. It was excellent thread and brought a goodly sum, with which Mary bought bread, a leg of lamb, and a bottle of wine. She was on her way home with them when a dog rushed up behind her, seized the leg of lamb and the bread, broke the bottle, and fled, nearly scaring the poor girl to death. When she got home, she told her sisters what had happened, and that day they had to be content with a few crusts of brown bread.

"I will go to market tomorrow," announced Rose, the middle girl, "and we'll just see if the dog dares to give me any trouble."

She went, sold her ball of thread, bought giblets, bread, and wine, then headed for home by a different road. Lo and behold, the dog ran after her too, grabbed the giblets and bread, broke the bottle, and fled. Rose, bolder than Mary by far, ran after him, but he was too fast for her and she went home all out of breath and told her sisters what had happened. So, for the second day in a row, they feasted on brown bread.

"Tomorrow it's my turn to go to market," announced Nina, the youngest, "'and we'll just see if the dog pulls the same thing on me."

Next morning she left the house much earlier than her sisters had on the preceding days, took her ball of thread to market, sold it, and bought provisions aplenty. As she walked home by another road, up rushed the dog, broke the bottle, and made off with everything else. Nina struck out after him and chased him all the way to a palace, into which he disappeared. She said to herself, "If I meet anyone inside, I'll tell them about the dog running away with our dinner for the last three days, and I'll make them pay me for all the food we've lost." Then she entered the palace.

She came to a fine kitchen with the fire burning brightly and things cooking over it in pots and pans; roasting on a spit was a leg of lamb. Lifting the lid of a pot, Nina saw meat stewing which she'd bought only a little while ago, and there in another pan were the giblets! She opened a cupboard and beheld there three loaves of bread. She moved on through the house without meeting a living soul, but the table in the dining room was set for three persons. They seem to have cooked dinner just for us, thought Nina, and with our own food! If my sisters were here, I'd sit down to the table right away!

At that moment she heard a cart going down the street. Looking out the window, she recognized its driver and asked him to tell her sisters she was waiting for them there, where a fine dinner was all ready to be served.

When the sisters arrived, Nina told them what had happened and said, "Let's sit down to the table. If the occupants of the house come in, we'll simply say we're only eating our own food."

The sisters were not so bold, but being quite hungry by this time, they finally took their places at the table. It had grown dark, and the three girls suddenly saw the windows close and the lamps light up. They were still marveling over it, when in came dinner and arranged itself before them. "We thank whoever's serving us and sparing us the trouble of getting dinner ourselves," said Nina. "And now, sisters, let's begin," she urged, and bit into the lamb.

Paralyzed with fear, the sisters could scarcely eat and spent the whole time glancing about them, expecting any minute to see some monster rush in. But Nina said, "If they didn't want us here for dinner, they shouldn't have cooked for us, lit the lamps, and served us at the table."

After dinner they were soon sleepy, so Nina led them through the house until they came to a bedchamber with three nice beds all turned down. "Let's go to bed now," she proposed.

"No," said the sisters, "let's go home. It's so frightening here."

"You ninnies!" snapped Nina. "We're comfortable here, and you want to leave! I'm going to bed, come what may!"

She'd no sooner persuaded them to remain there than a voice was heard at the bottom of the stairwell:

"Nina, come light my way."

The sisters were terrified. "Merciful heavens! Who can it be? Don't go, Nina!"

"I will go," said Nina, who picked up the lamp and went down the steps. She found herself in a room where a queen was chained and darting flames from her mouth, ears, and nose.

"Listen, Nina," said the queen, speaking amid the flames, "would you like a fortune?"

"Yes."

"You'll need the help of your sisters as well."

"I'll tell them."

"There'll be awful things to do, mind you, and if you get scared, you'll die."

"I'll persuade them to do what must be done."

"Very well. Open those three chests there. They're full of queens' robes, gilded and bejeweled. I was the queen of Spain, mind you. I fell in love with a young man of this town and because of him I am in Hell today. Now after all the wrong he's done me, he intends to marry another woman, but I want to see him suffer in Hell with me, which is only fair. Tomorrow, put on my robe, arrange your hair exactly like mine, and lean on the balustrade with a book in your hand. At a certain time the young man will come by and say, "Madam, may I call on you?" Say yes, invite him in to coffee, and give him this poisoned cup. When he drops dead, bring him down here, open this chest, throw him in, and light four candles around him. I was very rich. Here is a list of my assets which you can reclaim from my stewards, who're stealing everything I own."

Nina went back upstairs and related everything to her sisters. "Swear you'll help me, or heaven help you!" The next morning she dressed up to look exactly like the dead queen and went to the balustrade with a book. Hoofbeats were soon heard, and a young man rode up and stopped to look at her. Nina nodded in greeting.

"May I call on you, madam?"

"Please do."

The young man dismounted and climbed the steps to the palace.

"Let us have a cup of coffee together."

"With pleasure." He drank from the poisoned cup and dropped dead.

Nina called her sisters to help carry the body downstairs, but they refused and she said, "If you don't help me I'll kill you also!" She grabbed him by the head while the sisters caught hold of his feet, and they went downstairs to the closed chest surrounded by four candles. The sisters shuddered and wanted to drop the body and flee. "You just try to get away," said Nina, "and I'll show you a thing or two!" The sisters knew better than to defy her and remained right there.

Nina opened the chest: in it sat the queen on a throne of flames. They put in her beloved beside her, and she took him by the hand and said, "Come with me to Hell, you wicked soul. That way you won't leave me again."

With a great din the trunk slammed and sank out of sight.

Nina revived her sisters, who had fainted, and led them back upstairs to recover from the shock. Then they retrieved all the wealth in the hands of the stewards and became the richest girls in the world. A few years later the sisters got married, and Nina gave them each a dowry fit for a princess. Finally she, too, got married and ever after lived like a queen.

(Siena)

◈ 94 ◈

The Little Geese

Once upon a time a flock of little geese were on the way to the marshes to lay their eggs. Halfway there, one of them stopped. "My sisters, you'll have to go on without me. I must lay my eggs at once and I'll never make it to the marshes."

"Wait!"

"Hold it!"

"Don't leave us!"

But the little goose was not to be swayed. They embraced, said goodbye, promised to meet on the way back, and the goose took cover in the woods. Under an old oak tree, she made a nest of dry leaves and laid her first egg. Then she went in search of fresh grass and clear water for her lunch.

She returned to the nest at sunset, but the egg was gone. The little goose was frantic. The next day she decided to go up the oak tree and lay her second egg in the safety of the branches. Then she came back down the tree quite pleased with herself and went off in search of food as on the day before. When she got back, the egg was gone. The goose thought, There must be a fox in the woods feeding on my eggs.

She went to the town nearby and called at the blacksmith's shop.

"Sir, would you make me a little house of iron?"

"Yes, if you will lay me a hundred pairs of eggs."

"Very well, put a basket out here for me, and I'll lay the eggs while you're building my little house."

The goose squatted down, and at every blow of the smith's hammer on the iron house, she laid an egg. When the smith struck the two-hundredth blow, the goose laid the two-hundredth egg and jumped from the basket. "Sir, here are the hundred pairs of eggs I promised you."

"Mistress goose, here's your little house all finished."

The goose thanked him, took the house on her back, carried it to the woods, and set it down in a clearing. "This is the perfect place for my little ones. There's fresh grass here for them to eat, as well as a stream to swim in." Quite content, she shut herself up to lay the last of her eggs in peace.

The fox meanwhile had been back to the oak, but found no more eggs. He therefore snooped around the woods until he reached the clearing and saw the little iron house. I'll just bet the goose is inside, he thought, and rapped on the door.

"Who is it?"

"It's me, the fox."

"I can't answer the door, I'm sitting on the eggs."

"Open up, goose."

"No, because you'll eat me."

"No, I won't, little goose, open up!"

He waited.

"I'm warning you, goose, if you don't open up this minute,

> I'll climb to the rooftop,
> Dance a dance called the contradanse,
> Your house will topple,
> And you won't stand a chance."

The goose replied:

> "Climb to my rooftop,
> Dance your contradanse,
> House will stand,
> And on will play the band!"

The fox hopped to the roof and stomped every inch of it. But, would you believe it, the more he stomped, the more solid the iron house became. Exasperated, he jumped down and ran off, while the goose split her sides laughing.

For a while after that the fox kept under cover, but the goose kept a sharp eye open whenever she went outside. The eggs had hatched into many, many goslings.

One day a knock was heard on the door.

"Who is it?"

"It's me, the fox."

"What do you want?"

"I came to tell you there's a fair tomorrow. Shall we go together?"

"Gladly. When will you be by for me?"

"Whenever you like."

"Well, drop by about nine. I can't go earlier, I have to look after my little ones."

They said goodbye, the best of friends. The fox was already licking his lips, certain of gobbling up goose and goslings in two bites.

But next morning the goose rose at dawn, fed her little ones, kissed them goodbye, cautioned them to open up to no one, and went off to the fair.

No sooner had the clock struck eight than the fox was there knocking at the little iron house.

"Mamma's not home," said the goslings.

"Open up and let me in," ordered the fox.

"Mamma said not to."

The fox said to himself, "I'll eat you all later." Then, out loud, he asked, "How long has Mamma been gone?"

"She went out early this morning."

That was all the fox needed to hear, and he went running off as fast as his legs would carry him. The poor goose had finished her shopping and was on the way home, when she saw the fox coming down the road lickety-split with his tongue hanging out. "Where, oh, where can I hide?" wondered the goose. At the fair, she'd bought a huge soup tureen. She put the lid on the ground, squatted upon it, and pulled the pot over her.

The fox came to a dead stop. "My, my, what a pretty little altar! I think I'll say a prayer." He knelt before the tureen, said his prayer, left a gold coin as an offering, and ran on his way.

The little goose peeped out, picked up the gold piece, gathered together the soup tureen, and sped home to give her goslings another kiss.

Meanwhile the fox looked up and down at the fair for the goose, but found her nowhere around. "But she must still be here, since I didn't meet her on the road," he said, and resumed his search for her. The fair was over, the vendors put away their unsold goods, took down their stalls, but there was no trace of Mistress Goose. "She's given me the slip once more!"

Famished, he went to the iron house and knocked.

"Who is it?"

"It's me, the fox. Why didn't you wait for me?"

"It was hot. Also, I thought I'd meet you along the way."

"But what road did you take?"

"There's only one."

"Then how come we missed each other?"

"We didn't. I was inside the little altar . . ."

The fox was furious. "Open up, goose."

"No, because you'll eat me."

"I'm warning you, goose,

> I'll climb to your rooftop,
> Dance a good old contradanse,
> House will topple,
> And you won't stand a chance."

The goose replied:

> "Climb to my rooftop,
> Dance your dumb contradanse,
> House will stand,
> And on will play the band!"

Stomp, stomp, stomp! Up and down he stomped, and the iron house became stronger and stronger.

For days and days afterward, the fox didn't show his face. But one morning there was a knock.

"Who is it?"

"It's me, the fox. Open up."

"I can't, I'm busy."

"I just wanted to say that Saturday is market day. Will you go with me?"

"With pleasure. Stop by for me."

"Tell me exactly when to come, so the same thing won't happen that did last time."

"Let's say seven o'clock, I can't make it any earlier."

"All right," he agreed, and they parted, the best of friends.

Saturday morning, before daybreak, the goose tidied the little ones' feathers, fetched them fresh grass, cautioned them to open up to no one, and was off. It was hardly six o'clock when the fox arrived. The goslings told him Mamma was already gone, and he set out after her.

The goose was idling before a stall of melons, when the fox appeared in the distance. To run away now was out of the question. On the ground she spied a huge melon, pecked a hole in it, and slipped inside. The fox arrived and scoured the market for the goose. "Could be she's not here yet," he said, and sidled up to the melon stall to pick out the best melon for himself. He bit into one, tasted a second, but the rind of every one he tried was too bitter, and he brushed them all aside. His eyes finally fell on the huge melon lying on the ground. "This one has to be good!" he exclaimed and took the biggest bite yet. The goose, who happened to be looking in that very direction, saw a tiny window open up, and spat through it.

"Ugh! It tastes awful!" exclaimed the fox, and sent the melon rolling. It sped down a long slope, crashed into a rock, and the goose jumped out and flew home.

The fox searched the market until sundown, then went and knocked at the iron cottage. "Goose, you failed to keep your promise and come to market."

"I most certainly was there. I was inside that huge melon."

"Ah, you gave me the slip again! Now open up!"

"No, because you'll eat me."

"I'm warning you, goose,

> I'll climb to your rooftop,
> Dance a good old contradanse,
> House will topple,
> And you won't stand a chance."

The goose replied:

> "Climb to my rooftop,
> Dance your dumb contradanse,
> House will stand,
> And on will play the band!"

Stomp, stomp, stomp, but there was no jarring of the iron house.

More time went by. Then one day the fox was back knocking at the door: "Come on, goose, let's make up. To forget the past, we'll have a fine supper together."

"Gladly, but I have nothing to your liking to offer you."

"I'll bring the food, and all you'll have to do will be to cook and serve it." True to his word, he made one trip after another, bringing salami, mortadella, cheese, and chicken pilfered during his rounds. At last the iron cottage was chock-full of food.

The day of the supper arrived. The fox had been fasting for the past two days, so as to be as hungry as possible. He naturally gave no thought to cheese or mortadella, but to the tasty meal he would make off of goose and goslings. He went to the iron house and called, "Are you ready, goose?"

"Yes, whenever you are; everything is done. But I'm afraid you'll have to come through the window. Our big table with all the food on it reaches all the way to the door and keeps me from opening it."

"I don't care. The only problem is getting to the window."

"I'll throw down a rope. Slip your head through the noose, and I'll pull you up."

The fox, who was dying to gobble up the goose, stuck his head through the noose without noticing that it was a slipknot. The goose began pulling him up, and the knot tightened. The more the fox kicked, the more he choked, strangling, with his eyes and tongue popping out. The goose was still uneasy, though, so she let go of the rope and he went crashing to the ground, now dead for sure.

"Come, my little ones," she then said, opening the door, "come and eat the fresh grass and swim in the brook." So the little geese finally got to go outside and play.

One day the goose heard a clamor and a flapping of wings. It was the season for geese to return from the marshes. "If only that were my sisters!" She went out on the road and saw a flock arriving, followed by all the newborn goslings. Like all good sisters, they made a grand to-do over each other, and the goose told them about her close calls with the fox. The sisters liked the cottage so much that they all went to the blacksmith and had him make one for each of them. And to this day, in a clearing somewhere stands the town of the geese, who all live in little iron cottages, safe from the fox.

(Siena)

95

Water in the Basket

There was once a widowed mother who married a widowed father, and they each happened to have a daughter by their first marriage. The mother loved her own daughter, but not her husband's. She sent her own child for water with the jug, and her stepdaughter she sent with the basket. But the water would all run out of the basket, and the stepmother beat the poor girl every day.

One day as the stepchild was filling her basket, it slipped out of her hand and was swept off by the stream. She began running downstream asking everyone she met, "Did you see my basket go by?" but they all told her, "Go farther downstream and you'll find it."

She soon met an old woman sitting on a rock in the middle of the stream examining herself for fleas. "Have you seen my basket?" asked the girl.

"Come here," replied the old woman. "I have your basket. But first be so good as to look down my back and see what's biting me."

The girl killed vermin by the hundreds, but so as not to embarrass the old woman, she said, "Pearls and diamonds."

"You shall have pearls and diamonds yourself," replied the old woman. When all the fleas were off, she said, "Come with me," and they went to her house, which was one big rubbish heap. "Do me a favor, my girl, and make my bed. Do you see anything in it?" It too was crawling with vermin, but the girl politely replied, "Roses and jasmines."

"You shall have roses and jasmines yourself. Do me another favor now and sweep the house. What do you see to sweep out?"

"Rubies and cherubs," answered the girl.

"You shall have rubies and cherubs yourself." Then she opened a wardrobe containing all kinds of clothes and asked, "Do you want a silk dress or one of cotton?"

"I'm a poor girl, as you can tell, so give me a cotton dress."

"I'm giving you the silk one." She gave her a handsome gown of silk, then opened a jewel case. "Would you like gold or coral?"

"I'll take coral."

"But I'm giving you gold," and she slipped a gold necklace on her. "Do you want crystal earrings, or diamond earrings?"

"Crystal."

"But I'm giving you diamond ones," and she put them on her, adding, "You shall be beautiful, your hair shall be golden, and when you comb it, down one side shall pour roses and jasmines; down the other, pearls and rubies. Go home now, but don't turn around when the donkey brays. When the cock crows, turn around."

The girl set out for home. The donkey brayed, but she didn't turn around. The cock crowed, she turned around, and on her forehead appeared a star.

Her stepmother asked, "Who in the world gave you all those things?"

"An old woman, who'd found my basket, gave them to me for killing the fleas on her."

"Now I know I love you," said the stepmother. "Henceforth you'll go for water with the jug, while your sister takes the basket." To her own daughter she whispered, "Go for water with the basket, let it slip away from you in the stream, and go after it. And may you have the same luck as your sister!"

The stepsister marched off, threw the basket into the water, then ran after it. Farther downstream she met the old woman. "Did you see my basket go by?"

"Come here, I have it. Look down my back and see what's biting me." The girl began killing vermin, and the old woman asked, "What is it?"

"Fleas and the itch."

"You shall have fleas and the itch yourself."

She took the girl to make the bed. "What do you see there?"

"Bedbugs and lice."

"You shall have bedbugs and lice yourself."

She had her sweep the house. "What do you see?"

"Disgusting filth!"

"You shall have disgusting filth yourself."

Then she asked her if she wanted a dress of sackcloth or one of silk.

"A silk dress!"

"But I'm giving you sackcloth."

"A pearl necklace, or a necklace of rope?"

"Pearl!"

"But I'm giving you rope."

"Golden earrings or tinsel?"

"Golden!"

"But I'm giving you tinsel. Go home now and turn around when the donkey brays, but don't turn around when the cock crows."

She went home, turned around when the donkey brayed, and on her forehead sprouted a donkeytail. It was useless to cut it off, it only grew right back. The girl screamed and cried:

"Mamma, Mamma, this is how it goes:
My head is now a tail down past my nose;
The more of it I cut, the more it grows."

As for the girl with the star on her brow, the king's son asked for her hand in marriage. On the day he was supposed to fetch her in his carriage, her stepmother said to her: "Since you are marrying the king's son, do me one last favor before you leave: wash out the barrel for me. Climb into it and I'll come and help you in a minute."

The girl climbed into the barrel while her stepmother went off to get a kettle of boiling water to throw upon her and scald her to death. The woman intended to dress the ugly girl in the wedding dress and take her to the king all veiled so that he wouldn't know the difference until too late. Meanwhile the ugly girl walked by the barrel. "What are you doing in there?" she asked her half-sister.

"I'm here because I'm to marry the king's son."

"Let me get in, so I'll be the one to wed him."

As accommodating as ever, the beautiful girl climbed out, while the ugly one took her place. The mother returned with the boiling water and poured it into the barrel. She thought she'd killed the stepdaughter but, discovering it was her own child, she began screaming and crying at the top of her voice. Her husband came in about that time, having heard everything from his daughter, and gave the woman the beating of her life.

The beautiful daughter married the king's son and lived happily ever after.

Wide is the sheet, narrow is the street;
To tell your tale after mine is meet.

(Marche)

SVPERBIA · VMILTA

96

Fourteen

There was once a mamma and a papa with thirteen sons. One more was born, and they named him Fourteen. He grew by leaps and bounds, and when he got to be a big boy, his mamma said to him, "It's time you too were out digging in the field and helping your brothers. Take this basket with your lunch and theirs and join them."

She gave him a basket containing fourteen round loaves, fourteen cheeses, and fourteen liters of wine, and Fourteen set out. Halfway there, he got hungry and thirsty and ate all fourteen loaves and cheeses and drank all fourteen liters of wine.

His brothers, who had to go hungry, said to him, "Grab a hoe and get to work."

"I sure will," answered Fourteen, "but I need a hoe weighing fourteen pounds."

His brothers found him a hoe weighing fourteen pounds, and Fourteen said, "Shall we race one another digging to the end of the field?"

All fourteen of them began digging, and Fourteen got to the end of the field first.

From that time on, Fourteen worked with his brothers. He did the work of fourteen boys, but also ate the food of fourteen, and his brothers became as thin as rails.

His mother and father then said to him, "Go off for a while into the world!" That he did. He came across a wealthy farmer who needed fifteen laborers. "I do the work of fourteen and eat enough for fourteen, so I want the pay of fourteen," the boy told him. "If you hire me under those terms, I'll work for you."

The wealthy farmer decided to try him out and hired one other man, who together with Fourteen made fifteen. They began digging, and to every one thrust of the man's hoe Fourteen gave fourteen and in no time dug up the whole field.

Once the field was dug up, the wealthy farmer was unwilling to give him the pay and nourishment of fourteen men, so he thought up a way to get rid of Fourteen. "Listen," he said, "you must now perform another service for me. Go to Hell with seven mules and fourteen buckets and fill them with Lucibello's gold."

"I sure will," said Fourteen. "Just give me some tongs weighing fourteen pounds."

With the tongs in his possession, he drove the mules to Hell. When he got there, he said to the devils standing around the gate, "Bring Lucibello to me."

"What do you want with our chief?" asked the devils.

Fourteen handed them the letter from his employer asking for fourteen buckets of gold.

"Come on down," replied Lucibello.

As soon as he arrived underground, fourteen devils pounced on him to eat him alive. But Fourteen clamped his tongs to the tongue of each devil and tortured them all to death. Now only Lucibello the chief was left.

"How am I going to fill the buckets with gold after you've killed the fourteen devils who were to fill them?"

"I'll fill them myself," said Fourteen. He filled the buckets and said, "Thanks, and so long."

"Just a minute," replied Lucibello. "You don't expect to walk off like that, do you?"

The devil opened his mouth to eat the boy, but Fourteen clamped the tongs to his tongue, picked him up, threw him over his shoulder, and galloped off with the mules laden with gold.

He reached his employer's house and tied the devil to a leg of the kitchen table.

"What do you want me to do now?" asked Lucibello.

"Take my master and return to Hell with him."

The devil didn't have to be begged, and Fourteen became one of the richest farmers alive.

(Marche)

◇ 97 ◇

Jack Strong, Slayer of Five Hundred

Once upon a time in Rome there was a woodcutter named Jack. As he was cutting a limb from an oak tree one day, the limb fell on him and broke his leg, putting him in the hospital for three months. When he couldn't stand the hospital a minute longer, he ran away and came down here to Marca. One day he unbandaged the wound, and flies swarmed all over it. So what did Jack do but slap and kill them as fast as they lit.

When no more came buzzing around him, he counted the dead ones on the ground: there were a good five hundred. He made a sign and hung it around his neck: I AM JACK STRONG, SLAYER OF FIVE HUNDRED. He went into the city and took lodgings at an inn.

The next morning the governor sent for him. "Since you are so strong," said the governor, "go after the giant here in the vicinity who is robbing everyone."

Jack went into the brush and walked until he came upon a shepherd. "Where is the giant's cave?" asked Jack.

"What business have you there? The giant will gobble you up in one mouthful," replied the shepherd.

Jack said, "Sell me three or four white cheeses." And he went away with an armload of white cheeses. On reaching the giant's cave, he began stamping his feet to make a racket. Out came the giant. "Who's there?"

Jack picked up a cheese and said, "Shut up, or I'll crush you like this stone," and he squeezed the ricotta until it oozed through his fingers.

At that, the giant asked him if he wanted to be his partner. Jack said yes, threw away the other cheeses, and joined up with the giant.

The next morning, the giant was out of wood, so he took a long, long rope and went into the woods with Jack. He uprooted an oak with one hand, uprooted another with the other hand, and said to Jack, "Now you gather a few oaks yourself."

Jack answered, "Look here, giant, would you have a slightly longer rope? I'd like to put it all the way around the woods and pull up everything at once so as not to have to make a second trip."

The giant replied, "Never mind. I don't want you tearing up the whole place. What I've gathered here will be enough for now, so let well enough alone." He picked up all the uprooted oaks, and Jack didn't have to carry a thing.

One day the giant wanted to have a contest with the spinning top: whoever threw it the greatest distance would win one hundred crowns. For cord he took a windmill cable, for a spinning top a millstone. After making a throw of almost a mile, he walked to the top, pointed out how far it had gone, and said to Jack, "Now it's your turn."

Jack dared not touch the millstone, which he couldn't have budged an inch, but he began yelling, "Looook out! Looooooooook out, everybody!"

The giant squinted. "Who are you calling to? Who's down there? I don't see anybody."

"I'm talking to the people across the sea!"

"Well, never mind about throwing the top. You would send it so far we would never get it back," and he gave him the hundred crowns without making him throw the stone.

Jack then proposed a contest himself. "You clever man, let's see which one of us can thrust his finger further into an oak tree trunk."

The giant accepted. "We'll stake another hundred crowns on that!"

Earlier, Jack had taken a gimlet and a knife and made a hole in an oak, then re-covered it with bark, so you couldn't see it. The contest started, and the giant stuck his finger halfway into the trunk. Jack aimed for the hole he had made and shoved more than half his arm through it.

The giant gave him the hundred crowns, but was no longer at ease with such a strong man around him, so he sent him away. He waited until Jack got part way down the mountain; then he sent a lot of huge rocks rolling after him. But Jack, who distrusted the giant, had hidden in a cave. When he heard the rocks coming down, he yelled, "What's that falling out of the sky, flakes of plaster?"

The giant said to himself, "Heavens! I threw boulders down on him, and he calls them flakes of plaster. It's better to have a man like that for a friend than an enemy." So he called Jack back to the cave, but still went on thinking how he might get rid of him. One night while the strong man was sleeping, the giant tiptoed up to him and dealt him a blow on the head. But it so happened that every night Jack put a pumpkin on his pillow and slept with his head at the foot of the bed. As soon as the giant smashed the pumpkin, he heard Jack say, "Little do I care if you've beaten in my head. But you're going to pay for disturbing my sleep!"

The giant was now more fearful than ever. He thought, I'll take him into that wood and leave him; then the wolves will tear him to bits. He said to Jack, "Come on, we're going for a walk."

"All right," agreed Jack.

"Would you like to run a race?" asked the giant.

"Let's," answered Jack. "Just let me get a slight head start, since your legs are longer than mine."

"Fair enough! I'll give you ten minutes."

Jack struck out and ran until he met a shepherd with his sheep. "Will you sell me one?" he asked. He bought it, pulled out his knife, slit the sheep open and flung its intestines, liver, and innards into the road. "If a giant asks about me," he said to the shepherd, "tell him that, to run faster, I ripped out my intestines and then ran like the wind; and show him the intestines here on the ground."

Ten minutes later, here came the giant at top speed. "Did you see a man running this way?" he asked the shepherd.

The shepherd told him about the intestines and pointed to them. The giant said, "Give me a knife so I can do the same thing," and he ripped open his belly from top to bottom, fell to the ground, and gave up the ghost. Jack who had climbed a tree, jumped down, borrowed two buffalo, and dragged the giant into the city, where the governor had him burned in the middle of the square. And Jack was rewarded with food for the rest of his life.

(Marche)

◈ 98 ◈

Crystal Rooster

There was once a rooster that went strutting about the world. He found a letter lying in the road, picked it up with his beak, and read:

Crystal Rooster, Crystal Hen, Countess Goose, Abbess Duck, Goldfinch Birdie: Let's be off to Tom Thumb's wedding.

The rooster set out in that direction, and shortly met the hen.

"Where are you going, brother rooster?"

"I'm going to Tom Thumb's wedding."

"May I come, too?"

"If you're mentioned in the letter."

He unfolded the letter again and read: *Crystal Rooster, Crystal Hen . . .* "Here you are, here you are, so let's be on our way."

They continued onward together. Before long they met the goose.

"Oh, sister hen and brother rooster! Where are you going?"

"We are going to Tom Thumb's wedding."

"May I come, too?"

"If you're mentioned in the letter."

The rooster unfolded the letter again and read: *Crystal Rooster, Crystal Hen, Countess Goose . . .* "Here you are, so let's be on our way!"

The three of them walked and walked and soon met the duck.

"Where are you going, sister goose, sister hen, and brother rooster?"

"We are going to Tom Thumb's wedding."

"May I come, too?"

"Yes, indeed, if you are mentioned here." He read: *Crystal Rooster, Crystal Hen, Countess Duck, Abbess Duck . . .* "You're here all right, so join us!"

Before long they met the goldfinch birdie.

"Where are you going, sister duck, sister goose, sister hen, and brother rooster?"

"We are going to Tom Thumb's wedding."

"May I come, too?"

"Yes, indeed, if you're mentioned here!"

He unfolded the letter again: *Crystal Rooster, Crystal Hen, Countess Goose, Abbess Duck, Goldfinch Birdie* . . . "You are here too."

So all five of them walked on together.

Lo and behold, they met the wolf, who also asked where they were going.

"We are going to Tom Thumb's wedding," replied the rooster.

"May I come, too?"

"Yes, if you're mentioned here!"

The rooster reread the letter, but it made no mention of the wolf.

"But I want to come!" said the wolf.

Out of fear they all replied, "All right, let's all go."

They'd not gone far when the wolf suddenly said, "I'm hungry."

The rooster replied, "I've nothing to offer you . . ."

"I'll just eat you, then!" He opened his mouth wide and swallowed the rooster whole.

Further on he again said, "I'm hungry."

The hen gave him the same answer as the rooster had, and the wolf gobbled her up too. And the goose and the duck went the same way.

Now there was just the wolf and the birdie. The wolf said, "Birdie, I'm hungry!"

"And what do you expect me to give you?"

"I'll just eat you, then!"

He opened his mouth wide . . . and the bird perched on his head. The wolf tried his best to catch him, but the bird flitted all around, hopped from branch to branch, then back to the wolf's head and on to his tail, driving him to distraction. When the wolf was completely exhausted, he spied a woman coming down the road with the reapers' lunch in a basket on her head. The bird called to the wolf, "If you spare my life, I'll see that you get a hearty meal of noodles and meat which that woman is bringing the reapers. As soon as she sees me, she'll want to catch me. I'll fly off and hop from branch to branch. She'll put her basket down and come after me. Then you can go and eat everything up."

That's just what happened. The woman came up, spied the beautiful little bird, and immediately reached out to catch him. He then flew off a little way, and she put down her basket and ran after him. So the wolf approached the basket and started eating.

"Help! Help!" screamed the woman. The reapers came running with scythes and sticks, pounced upon the wolf, and killed him. Out of his belly, safe and sound, hopped crystal rooster, crystal hen, countess goose, abbess duck, and together with goldfinch birdie they all went to Tom Thumb's wedding.

(Marche)

◈ 99 ◈

A Boat for Land and Water

Once a king issued this decree:

> The man who builds a boat
> That glides o'er land and water
> Will surely wed my daughter.

Now in that country was a father with three sons, and all he had to his name were a horse, a donkey, and a piglet. When the oldest son heard the decree, he said to his father, "Papa, sell the horse and with the proceeds buy me tools for building boats, and I'll build a boat that glides over land and water and wed the king's daughter."

He kept after his father, who finally gave in for the sake of a little peace and sold the horse and bought the tools. The son rose bright and early and went off to the woods with the tools to cut the timber for the boat.

He was already halfway through building the boat, when a little old man came walking by. "What are you working on there, my lad?"

"Just what I please."

"And what pleases you, may I ask?"

"Barrel staves," replied the boy.

"You shall find barrel staves all cut out for you," said the old man and left him.

The next morning upon returning to the woods where he had left the boat half built, together with the timber and tools, the young man found only a pile of barrel staves. He went home crying as though his heart would break and told his father what had happened. You can just imagine the bad mood that put the man in, who had sold his horse to humor the boy whose neck he now could have wrung!

Less than a month later the middle son was itching to try his luck building such a boat. He went to work on his father and kept begging and pleading until the man finally had to sell his donkey and buy him the proper tools. The lad took them to the woods right away and cut his timber. He had the boat half finished, when the old man showed up asking, "What are you making, my lad?"

"I'm making what I please."

"And what pleases you, may I ask?"

"Broom handles!"

"You shall find broom handles all cut out for you!" said the old man and turned away.

The boy went home that night, dined, slept, and at dawn returned to the woods. His experience was exactly like his brother's: there lay only a pile of broom handles.

When he too came home heartbroken, his father shouted; "It serves you right! It serves you both right for having such foolish ideas! And it serves me right, too, for ever listening to you!"

At that point, the youngest son, who had been listening, spoke up; "We've gone this far, so we might as well go the rest of the way. Papa, I want to try too. Let's sell the piglet, replace the tools, and who knows but what I might succeed where they failed."

In short, the piglet was sold and the youngest son went off to the woods with the tools. He was already halfway through building the boat, when the same old man showed up. "My lad, what are you doing?"

"I'm building a boat to glide over land and water."

"You shall find a boat all built and ready to glide over land and water," said the old man and left him.

That night the boy went home, dined, slept, and at dawn returned to the woods. There stood the boat finished down to the smallest detail, with the sails unfurled. He went aboard and commanded, "Boat, glide over land," and the boat moved through the woods as smoothly as if it had been on water and came out before his house; his father and brothers were too amazed for words.

Then the boy repeated his command, "Boat, glide over land," and headed for the king's palace, skimming over mountains, plains and, naturally, any rivers that had to be crossed along the way.

He now had the boat but no crew. He came to a river fed by a creek, but water from the creek was not reaching the river, since just above the mouth of the river a huge man was kneeling on the bank drinking the creek dry.

"Good heavens, what a gullet!" exclaimed the boy. "How about coming along to the king's palace with me?"

The huge man took one more gulp, gurgled, and said, "Gladly, now that my thirst is somewhat quenched." Then he came aboard.

The boat sped over water and over land and came to where a huge man was turning a hefty buffalo on a spit.

"Hello!" called the boy from the boat. "Would you like to come to the king's palace with me?"

"Gladly," he answered. "Just give me time to eat this morsel here."

"By all means."

At that, the man popped the buffalo off the spit and into his mouth as if it were a roasted thrush. Then he came aboard and they moved onward.

The boat skimmed lakes and fields and came to another huge man leaning against a mountain.

"Hello!" called the captain of the boat. "Would you like to come along to the king's palace with me?"

"I can't move."

"Why can't you?"

"Because if I don't lean against the mountain, it will fall down."

"Let it fall."

The man moved away, holding the mountain up with one hand, and jumped into the boat. The boat had no sooner sailed off than a boom and a rumble were heard, and the mountain came crashing down.

Gliding over roads and hills, the boat finally drew up before the king's palace. The boy disembarked and said, "Sacred Crown, with my own two hands I built this boat to go over land and water. Please keep your promise now and give me your daughter in marriage."

The king, who wasn't expecting this, was dismayed and regretted his decree. Now he would have to give his daughter to some pauper he'd never laid eyes on.

"I'll give you my daughter," replied the king, "on condition you and your crew eat every mouthful of the banquet I shall offer you, without leaving so much as one chicken wing or raisin on your plates."

"Very well. When is this banquet to take place?"

"Tomorrow." And he ordered a banquet of one thousand dishes, thinking, This ragamuffin certainly won't have a crew capable of putting away a spread like that.

The captain of the boat showed up with only one member of the crew, the man who ate buffaloes like roast birds. He ate and ate, chewing ten dishes one after the other, then swallowing a hundred whole, and on and on until he'd polished off one thousand. The king, who looked on speechless, snapped out of his amazement to ask the servants, "Is there any-thing left in the kitchen?"

"There're still a few leftovers."

The leftovers were brought in, and the man ate everything down to the last crumb.

"Of course you'll marry my daughter, but first I want to offer your crew all the wine in my cellar, which you must drink to the last drop."

The drinker of rivers came in and drained a cask, then a barrel, then a demijohn. He even got his hands on the two flasks of malmsey the king had set aside for himself and emptied those with all the rest.

"Understand," said the king, "I'm not in the least opposed to giving you my daughter. But there's the matter of the dowry that comes with her: dresser, cupboards, bed, washstand, linen, treasure chests, and everything else in the house. You must take it all away in one trip, immediately, with my daughter seated on top of the load."

"Do you feel like a little work?" said the boy to the man who held up mountains.

"Do I!" he replied. "That's my weakness!"

They went up to the palace, and the boy said to the movers, "Are you ready? You can begin loading things onto his back."

They brought out wardrobes, tables, and trunks of jewels and piled them on the big man's back all the way up to the roof. To get on top of the pile, the king's daughter had to climb to the tower of the palace and step off from there. Once she was on, the huge man called out, "Hold on tight, princess." He ran all the way to the boat with everything and jumped aboard.

"Now fly, my boat," commanded the boy, and the boat sped through the streets and town squares out across the fields.

Looking from his balcony, the king shouted, "After them! Seize them and bring them back in chains!"

The army charged after them, but was checked by the cloud of dust the boat stirred up.

The boy's father was overjoyed to see his youngest son return with a whole boatload of treasure and the king's daughter in a wedding dress. The boy had the most beautiful palace in the world built, gave one floor of it to his father and brothers, one to each of his companions, and kept all the rest for himself and his bride, the king's daughter.

(Rome)

The Neapolitan Soldier

Three soldiers had deserted their regiment and taken to the open road. One was a Roman, one a Florentine, while the smallest was a Neapolitan. After traveling far and wide, they were overtaken by darkness in a forest. The Roman, who was the oldest of the three, said, "Boys, this is no time for us all three to go to sleep. We must take turns keeping watch an hour at a time."

He volunteered for the first watch, and the other two threw down their knapsacks, unrolled their blankets, and went fast asleep. The watch was almost up, when out of the forest rushed a giant.

"What are you doing here?" he asked the soldier.

"None of your business," replied the soldier, without even bothering to turn around.

The giant lunged at him, but the soldier proved the swifter of the two by drawing his sword and cutting off the giant's head. Then he picked up the head with one hand and the body with the other and threw them into a nearby well. He carefully cleaned his sword, resheathed it, and called his companion who was supposed to keep the next watch. Before awakening him, though, he thought, I'd better say nothing about the giant, or this Florentine will take fright and flee. So when the Florentine was awake and asking, "Did you see anything?" the Roman replied, "Nothing at all, everything was as calm as could be." Then he went to sleep.

The Florentine began his watch, and when it was just about up, here came another giant exactly like the first, who asked, "What are you doing here?"

"That's no business of yours or anybody else's," answered the Florentine.

The giant sprang at him, but in a flash the soldier drew his sword and lopped off his head, which he picked up along with the body and threw into the well. His watch was up, and he thought, I'd better say nothing of this to the lily-livered Neapolitan. If he knew that things like this went on around here, he'd take to his heels and we'd never see him again.

So, when the Neapolitan asked, "Did you see any action?" the Florentine replied, "None at all, you've nothing to worry about." Then he went to sleep.

The Neapolitan watched for almost an hour, and the forest was perfectly still. Suddenly the leaves rustled and out ran a giant. "What are you doing here?"

"What business is it of yours?" replied the Neapolitan.

The giant held up a hand that would have squashed the Neapolitan flatter than a pancake, had he not dodged it, brandished his sword, and swept off the giant's head, after which he threw the remains into the well.

It was the Roman's turn once more to keep watch, but the Neapolitan thought, I first want to see where the giant came from. He therefore plunged into the forest, spied a light, hastened toward it, and came to a cottage. Peeping through the keyhole, he saw three old women in conversation before the fireplace.

"It's already past midnight, and our husbands are not yet back," said one.

"Do you suppose something has happened to them?" asked another.

"It might not be a bad idea," said the third, "to go after them. What do you say?"

"Let's go right now," said the first. "I'll carry the lantern that enables you to see a hundred miles ahead."

"And I'll bring the sword," said the second, "which in every sweep wipes out an army."

"And I'll bring the shotgun that can kill the she-wolf at the king's palace," said the third.

"Let's be on our way." At that, they threw open the door.

Hiding behind the doorpost with sword in hand, the Neapolitan was all ready for them. Out came the first woman holding the lantern, and swish! her head flew off before she could say a single "Amen." Out came the second, and swish! her soul sped to kingdom come. Out came the third and went the way of her sisters.

The soldier now had the witches' lantern, sword, and shotgun and decided to try them out immediately. "We'll just see if those three dotards were telling the truth." He raised the lantern and saw an army a hundred miles away besieging a castle, and chained on the balcony was a she-wolf with flaming eyes. "Let's just see how the sword works." He picked it up and swung it around, then raised the lantern once more and peered into space: every last warrior lay lifeless on the ground beside his splintered lance and dead horse. Then the Neapolitan picked up the gun and shot the she-wolf.

"Now I'll go and see everything from close up," he said.

He walked and walked and finally reached the castle. His knocks and calls all went unanswered. He went inside and walked through all the rooms, but saw no one until he came to the most beautiful chamber of all, where a lovely maiden sat sleeping in a plush armchair.

The soldier went up to her, but she continued to sleep. One of her slippers had dropped off her foot, and the soldier picked it up and put it in his pocket. Then he kissed her and tiptoed away.

He was no sooner gone than the sleeping maiden awakened. She called her maids of honor, who were also sleeping, in the next room. They woke up and ran to the princess, exclaiming, "The spell is broken! The spell is broken! We have awakened! The princess has awakened! Who could the knight be who freed us?"

"Quick," said the princess, "look out the windows and see if you see anyone."

The maids looked out and saw the massacred army and the slain she-wolf. Then the princess said, "Hurry to His Majesty, my father, and tell him a brave knight came and defeated the army that held me prisoner, killed the she-wolf that stood guard over me, and broke the evil spell by kissing me." She glanced at her bare foot and added, "And then he went off with my left slipper."

Overjoyed, the king had notices posted all over town: WHOEVER COMES FORWARD AS MY DAUGHTER'S DELIVERER SHALL HAVE HER IN MARRIAGE, BE HE PRINCE OR PAUPER.

In the meantime the Neapolitan had gone back to his companions in broad daylight. When he awakened them, they asked immediately, "Why didn't you call us earlier? How many hours did you watch?"

But he wasn't about to tell them all that had happened and simply said, "I was so wide-awake I watched the rest of the night."

Time went by without bringing a soul to town to claim the princess as his rightful bride. "What can we do?" wondered the king.

The princess had an idea. "Papa, let's open a country inn and put up a sign that reads: HERE YOU CAN EAT, DRINK, AND SLEEP AT NO CHARGE FOR THREE DAYS. That will draw many people, and we'll surely hear something important."

They opened the inn, with the king's daughter acting as innkeeper. Who should then come by but our three soldiers as hungry as bears, and singing as usual, in spite of hard times. They read the sign, and the Neapolitan said, "Boys, here you can eat and sleep for nothing."

"Don't believe a word of it," replied his companions. "They just say that, the better to cheat people."

But the princess-innkeeper came out and invited them in, assuring them of the truth of every word of the sign. They entered the inn, and the princess served them a supper fit for a king. Then she took a seat at their table and said, "Well, what news do you bring from the world outside? Way off in the country like this, I never know what's going on elsewhere."

"We have very little of interest to report, madam," answered the

Roman who then smugly told of the time he was keeping watch when suddenly confronted by a giant whose head he cut off.

"Zounds!" exclaimed the Florentine. "I too had something similar happen to me," and he told about his giant.

"And you, sir?" said the princess to the Neapolitan. "Has nothing ever happened to you?"

His companions burst out laughing. "You don't think he would have anything to tell, do you? Our friend here is such a coward he'd run and hide for a whole week if he heard a leaf rustle in the dark."

"Don't belittle the poor boy like that," said the maiden, who insisted that he too tell something.

So the Neapolitan said, "If you really want the truth, I too was confronted by a giant while you two were sleeping and I killed him."

"Ha, ha, ha!" laughed his companions. "You'd die of fright if you so much as saw a giant! That's enough! We don't want to hear any more, we're going to bed." And they went off and left him with the princess.

She served him wine and coaxed him to go on with his story. Thus, little by little, he came out with everything—the three old women, the lantern, the shotgun, the sword, and the lovely maiden he had kissed as she slept, and her slipper he had carried off.

"Do you still have the slipper?"

"Here it is," replied the soldier drawing it from his pocket.

Overjoyed the princess kept filling his glass until he fell asleep, then said to her valet, "Take him to the bedchamber I prepared especially for him, remove his clothes, and put out kingly garb for him on the chair."

When the Neapolitan awakened next morning he was in a room decorated entirely in gold and brocade. He went to put on his clothes and found in their place robes for a king. He pinched himself to make sure he wasn't dreaming and, unable to make heads or tails of a thing, he rang the bell.

Four liveried servants entered and bowed down to him. "At Your Highness's service. Did Your Highness sleep well?"

The Neapolitan blinked. "Have you lost your mind? What highness are you talking about? Give me my things so I can get dressed, and be done with this comedy."

"Calm down, Highness. We are here to shave you and dress your hair."

"Where are my companions? Where did you put my things?"

"They are coming right away, you will have everything immediately, but allow us first to dress you, Highness."

Once he realized there was no getting around them, the soldier let the servants proceed: they shaved him, dressed his hair, and clothed him in a kingly outfit. Then they brought in his chocolate, cake, and sweets. After breakfast he said, "Am I going to see my companions or not?"

"Right away, Highness."

In came the Roman and the Florentine, whose mouths flew open when they saw him dressed in such finery. "What are you doing in that costume?"

"You tell me. Your guess is as good as mine."

"Goodness knows what you've cooked up!" replied his companions. "You must have told the lady some pretty tall tales last night!"

"For your information, I told no tall tales to anyone."

"So how do you account for what's happening now?"

"I'll explain," said the king, coming in just then with the princess in her finest robe. "My daughter was under a spell, and this young man set her free."

By questions and answers, they got the entire story.

"I am therefore making him my daughter's husband," said the king, "and my heir. As for yourselves, have no fears. You will become dukes, since had you not slain the other two giants, my daughter would not be free today."

The wedding was celebrated to the great joy of all, and followed by a grand feast.

> On the menu was chicken à la king:
> Long live the queen!
> Long live the king!

(Rome)

◈ 101 ◈

Belmiele and Belsole

There was once a father with two such fair and beautiful children that the boy was called Belmiele, and the girl Belsole. The man's duties as chief steward at the royal court took him away from his children, as the king lived in another town. Hearing their looks so highly praised, the king, who had never seen these children, said to his chief steward, "Since you have such a handsome son, bring him to the court, and I will make him a page."

The father fetched the boy and left the girl with her nurse. Belmiele became page to the king, who liked the boy so much that he kept him on at the palace as page after the chief steward's sudden death and even entrusted him with the special privilege of dusting the paintings in his art gallery. Belmiele could never dust the portraits without pausing to admire one in particular, that of a lady. Time after time the king caught him there, entranced, with his feather duster at rest.

"What's so fascinating about that portrait?"

"Majesty, this portrait is the perfect likeness of my sister, Belsole."

"I don't believe you, Belmiele. I looked the world over for a lady like the one in that portrait, but I didn't find her. If your sister resembles that lady, bring her here, and she will be my bride."

Belmiele wrote the nurse to bring Belsole to him immediately, as the king wanted to marry her. Now the nurse, mind you, had a daughter uglier than sin and as envious as could be of Belsole's beauty. After receiving Belmiele's order, she set out with Belsole and her own ugly daughter. As they were to travel by sea, they all three boarded a boat.

On board, Belsole fell asleep, and the nurse said, "So that's how it is! She has all the luck and gets to marry the king! Wouldn't it be better if he married my daughter?"

"Indeed it would!" replied the daughter.

"Leave everything to me. I have absolutely no use for that silly creature."

Meanwhile Belsole awoke, saying, "Nurse, I'm hungry."

"I have bread and sardines, but not even enough for me."

"Please, nurse, give me a tiny little bit."

So that dreadful woman gave her a small piece of bread and some sardines, but mostly sardines and practically no bread, and in no time the girl was dying of thirst. "Oh, nurse, I'm so thirsty."

"I have scarcely any water, but if you like, I'll give you some salt water."

Feeling her throat utterly parched, Belsole said, "I'll even drink salt water." After one sip, though, she was thirstier than ever.

"Nurse, I'm thirstier than ever."

"All right," replied the cruel woman, "I'm really going to give you a drink of water now." She grabbed her around the waist and flung her into the sea.

A whale happening by just then saw Belsole in the water and swallowed her whole.

The nurse came into the king's port, and there stood Belmiele on the pier with arms outstretched to embrace his sister. Instead, he saw that ugly face wearing a bridal veil. His arms dropped. "What's happened? Is this my sister? My starry-eyed sister? My sister with the gracious smile?"

"Ah, my son," said the nurse, "if you only knew how sick she was! In a short time she turned into this."

The king stepped forward. "So this is the beauty you were boasting about! This is the maiden as fair as day? To me she looks like something the cat dragged in! What a fool I was to believe you and promise to marry her! A king can't go back on his word, so I've no choice but to wed her, But you, my page, will change your calling this very day and go out to tend the ducklings."

So the king married the nurse's daughter, but treated her more like his servant than his wife.

Meanwhile Belmiele took the ducklings down to the seashore. He sat on the beach to watch them swim, and he thought of his misfortunes and of Belsole the way she was before joining him at the king's court. Suddenly he heard a voice at the bottom of the sea:

> "Whale, my whale,
> Stretch out thy tail
> To the shore of the sea,
> Where Belmiele would speak with me."

Belmiele couldn't imagine what the voice was talking about, when lo and behold, out of the water stepped the loveliest of maidens dragging a chain around one ankle. She looked exactly like Belsole, so much so that it could only have been Belsole, who was now more beautiful than ever.

"Sister, what on earth are you doing here?"

"I am here because of the nurse's treachery, brother." And she told him her story while tossing seed of gold and pearl to the ducklings.

"You cannot mean it, sister!" gasped poor Belmiele.

"The nurse threw me overboard and put her own daughter in my place," replied Belsole, and hung colored tassels on the ducklings.

Night fell and the sea turned black. "Farewell, brother," said Belsole, and gradually sank back underwater to where the chain was attached.

Belmiele assembled the ducklings all decked in tassels and headed homeward along the shore, with ducklings chanting:

> "Crò! Crò! We come from brine,
> On gold and pearls we dine.
> Belsole's fair, as fair as day,
> Our lord and king would love her, yea!"

People stopped in their tracks to listen and were flabbergasted: never had they heard ducklings chant like that. That night, back in the royal roost, instead of going to sleep, the ducklings chanted straight through the night:

> "Crò! Crò! We come from brine,
> On gold and pearls we dine.
> Belsole's fair, as fair as day,
> Our lord and king would love her, yea!"

A scullery boy heard them and told the king the next morning what the ducklings Belmiele tended had chanted throughout the night. The king listened only with half an ear at first, then curiosity got the better of him, and he decided to trail Belmiele when he took the ducklings out.

He hid in the reeds and heard the voice at the bottom of the sea:

> "Whale, my whale,
> Stretch out thy tail
> To the shore of the sea,
> Where Belmiele would speak with me."

Up came the maiden with the chain around her ankle and swam to shore. Beholding her in all her beauty, the king emerged from the reeds, saying, "You are my bride beyond any shadow of a doubt!" They introduced themselves and, together with Belmiele, considered how to free her from the whale that kept her chained. The king and Belmiele took a rock equal to Belsole's weight, cut the chain, and fastened the rock to it in the maiden's place. Then arm in arm, the king and Belsole walked back to the royal palace, followed by Belmiele with his flock of ducklings chanting:

> "Crò! Crò! We come from brine,
> On gold and pearls we dine.
> Belsole's fair, as fair as day,
> Our king has wed her with no delay!"

When the nurse and her daughter heard this chant and saw the procession arriving, they fled from the palace, never to be seen again.

(Rome)

The Haughty Prince

There was once a merchant who had a daughter, and in the evenings he took her out into society. One evening while the girl was out in society, she saw a gentleman pull a snuffbox from his pocket and take snuff. On the cover of the snuffbox was a portrait. It was the portrait of the king of Persia's son with seven veils over his face, and the maiden fell in love with him.

She went home and said to her father, "Papa, I've fallen in love with the king of Persia's son. Please go and propose to him for me, and take him my portrait."

Now the king of Persia's son was well known for two things: for being unusually handsome and unusually haughty. He was too handsome for human eyes; in fact, lest someone see him, he wore seven veils over his face and kept to the throne room, where he never addressed a living soul, with the exception of his mother.

After his daughter had spoken, the merchant replied, "Dear daughter, you had better forget this son of the king of Persia."

But the girl by now was so smitten she could think of nothing else. She began pestering her father and went on so that the merchant decided to satisfy her by going to the king of Persia's son who wore seven veils over his face and telling him about the girl's love.

The queen received him, and took the maiden's portrait off to show to her son.

"Do you wish to see the portrait, my son?"

"Tell him to throw it in the toilet."

The queen relayed the message, and the poor father argued, "But my poor daughter is crying her eyes out!"

The haughty prince's mother went back to her son, "My son, the man says his daughter is crying her eyes out."

"Then give him these seven handkerchiefs for her!"

"But my daughter will kill herself!" objected the wretched father, when the queen brought him the handkerchiefs.

"He said she will kill herself," the queen repeated to her son.

"Then give him this knife for her to kill herself with."

The old man returned to his daughter with those cruel answers. After a few minutes of silence, she said, "Father, in this matter we must be firm. Give me a horse and a purse of money, and let me be off."

"Have you lost your mind?"

"Crazy or not, I intend to go out into the world."

She left and went out into the world. Night overtook her in the country. She spied a light and, approaching it, came to a house where a woman watched over her dying son. The girl said, "Go and get some sleep while I watch over your son."

As she kept watch, the lamp went out, and everything was pitch-black. She groped about for a taper, but none was to be found. "I must see if there's someone in the neighborhood who can give me a light." She ran out, circled around, and in the distance, saw a ray of light. She approached it and found an old woman putting wood under a caldron filled with oil.

"Will you help me light my lamp, ma'am?"

"If you help me," answered the old woman.

"Do what?"

"Work a spell on a young man, the son of those country people who live up there"—and she pointed to the house where the dying boy lay. "When this oil has boiled away, the young man will be dead."

"I'll help you," said the maiden. "I'll pile on the wood, and you look to see if the caldron is boiling."

The old woman bent over to see if the caldron was boiling, and the girl grabbed her by the legs and thrust her headfirst into the boiling oil, holding her there until she was good and dead. Then she lit her lamp, put out the fire, and hurried back to the farmhouse, where she found the youth completely recovered and getting out of bed. There was, naturally, great rejoicing in that humble house. "You are going to be my wife!" declared the young man. "No, I must move on," was her reply, and the next morning, loaded down with gifts, she continued on her way.

She came to a town and went to work for a man and his wife. The husband, poor soul, had been sick in bed for years, of a mysterious ailment that left all the doctors baffled. While working for them, the girl grew suspicious of the wife. She started spying on her and, one evening, hid behind a curtain to see just what the woman did at night. In came, the wife, awakened her husband, gave him a cup of opium to drink and, the minute he fell back asleep, opened a casket, saying, "Out with you, dear daughters, now is the time."

Out of the casket crept a brood of vipers. They fell on the sleeping man and began sucking his blood. When their thirst was quenched, the wife pulled the snakes off, brought out a little pot she kept hidden behind a picture, and made them spit out all the blood they had sucked. Then she carefully anointed her hair with the blood, replaced the reptiles in the casket, and said:

Over wind and over sea,
Take me to old Benevento's walnut tree.

At that, she disappeared.

So what did the girl do but carefully anoint her own hair with blood from the pot, repeat the woman's words, and end up all of a sudden inside a large cask full of witches who danced and worked magic spells. When day dawned, the girl, so as to be back home before her mistress, thought, I must now find the magic counter-formula. So she tried saying:

Under wind and under sea,
Take me from old Benevento's walnut tree.

Instantly she found herself back home. Her mistress returned and saw her sleeping as though nothing at all had happened.

But the next morning the girl said to her master, "Tonight, pretend to drink the cup your wife brings you, but don't swallow one drop of it."

He heeded her and thus remained awake. When his wife went to pull the snakes off, he jumped up and slew her. She'd not drawn her last breath before the husband was completely well. "How can I thank you?" he asked the maiden. "Don't leave. I want you to stay here with me forever."

But she wouldn't hear of it. She took all the money the master gave her and set out again.

After some distance she came to another town and took lodgings at an inn. The innkeeper had a young son, who for quite some time had been in bed and taken no food or drink, sleeping night and day. The girl said, "Allow me, and I will cure him."

That night she kept watch. Ten o'clock struck: nothing. Eleven: nothing. At the stroke of midnight, bang! Two holes appeared in the ceiling, and through them dropped two bundles, a white one and black one. Upon hitting the ground, the white bundle became a beautiful lady, and the black one a maidservant carrying a tray with dinner on it. The lady slapped the sleeper and he woke up; then they set the table and dined as though nothing at all were amiss. At the first crow of the cock, the beautiful lady gave the young man another slap and he fell back asleep at once. The two women rolled themselves into new bundles, one white and one black, and vanished through the holes in the ceiling.

When it was day, the girl said to the sick boy's parents, "If you want that poor soul to get well, take heed of what I say. You must do five things: first, have all the cocks in the country killed; secondly, muffle all the bells; thirdly, have a black coverlet embroidered with all the stars and hang it outside the window; fourthly, light a bonfire under the window; lastly, have a mason stand by, on the roof, to plug up two holes."

The next night the two women bundles descended into the room and proceeded to dine with the young man. From time to time they glanced at the window to see if it was getting light, but the stars shone as brightly as ever. Hour after hour passed, but it remained dark outside; not a cock was to be heard, much less a hen. The two women bundles went to the window to see why in the world the night was lasting so long. They reached out and felt, not the air, but a coverlet, which suddenly fell, revealing the sun already high in the sky. At that, they turned themselves frantically into bundles again and rushed for the ceiling. But the mason had already carefully replaced tiles, beams, and plaster, thus cutting off their escape. They then dashed to the window, but saw the bonfire below it. Like it or not, they had to jump; scorched, they fled for dear life. In all their haste they had forgotten to slap the young man, who therefore remained awake and free from the evil spell.

His parents were wild with joy and ran up to embrace him. His first words were: "What a girl! I will marry her!" But she had other plans and, with an armful of presents from the innkeeper, was again on her way.

She met a little old woman, who asked, "Where are you going?"

"I am seeking the haughty prince."

"Listen," said the old woman. "I am well aware you have had your share of suffering. Here's a little magic wand for you. Command what you will, and it will be granted. The haughty prince happens to be in this town." At that, the old woman vanished. The merchant's daughter then went and stood before the haughty prince's palace, tapped the ground with the wand, and said, "I do hereby command that a palace equal in size to the haughty prince's spring up instantly, with seven windows exactly like his. Let my palace be placed so that the windows at one end are close to the prince's, and those at the other end are far from his."

Right away, opposite the king's palace, sprang up a second palace exactly as specified. It was morning, and the haughty prince rose and saw a brand-new palace across from his. He looked out, and opposite his window was the most distant window of the other palace. At it stood a maiden so beautiful that, to get a better look at her, the haughty prince removed the first of his seven veils. He told his servants at once, "Pick out the two most beautiful bracelets in the treasure house and take them straight to the maiden, asking for her hand in my name."

Carrying the bracelets on velvet cushions, the servants went to relay the message. But when the maiden saw the gifts, she said, "Put those bracelets on the front door to serve as door knockers, which are just what I needed." With that, she dismissed the men.

The next day the girl appeared at the second window, and the haughty prince removed another veil and looked from his second window. Then he sent the servants to her with a diamond necklace. "This necklace," she said, "will make a nice chain for the dog you now have tied up with a rope."

On the third day, the maiden was at the third window, and the haughty prince, without his third veil, stood at his third window too. He sent the servants to her with two pearl earrings. "Take these earrings," she ordered, "and put them in the dog's harness bell for bell-clappers."

On the fourth day, from the fourth window, she told the servants who brought her a precious embroidered shawl to put it down for a doormat, and on the fifth day, since the prince had taken off his fifth veil as well and sent her an engagement ring set with a walnut-size diamond, she told them to give the diamond to the porter's children to play with.

The sixth day they brought her the queen's crown. "Put it under the pot as a tripod."

But meanwhile they had come to the seventh window, where they stood face to face. The haughty prince removed his last veil and so delighted the merchant's daughter that she said, "Very well, I will marry you."

> A feast of chicken, bread, and pride . . .
> Long live the bride! Long live the bride!

(Rome)

◈ 103 ◈

Wooden Maria

There was once a king and queen, who had a very beautiful daughter named Maria. When Maria was fifteen, her mother became deathly ill. The king, weeping at his wife's bedside, swore he would never remarry, whereupon the queen said, "My husband, you're still young and have a daughter to raise. I am leaving you this ring: you must take as your wife the lady on whose finger the ring fits."

When the period of mourning was over, the king started looking around for a new wife. Numbers of ladies came forward, but they all withdrew after trying on the ring; it was too loose on half of them and too tight on the other half. "That means I'm not supposed to remarry right now," said the king, "so we'll just drop the matter for the time being." And he put the ring away.

One day while poking around the house, the daughter came across the ring in a drawer. She slipped it on and could not get it off. "What will Papa now say?" she wondered. She took a piece of black cloth and bandaged her finger. When her father noticed the bandage, he asked, "What's the matter, daughter?"

"Nothing, Papa, I just got a little scratch on my finger."

But a few days later her father, deciding to take a look at her finger, undid the bandage and saw the ring. "Oh, dear daughter," he exclaimed, "you are to be my wife!"

Shocked by such talk, Maria ran off to hide and tell her nursemaid. "If he brings the matter up again," said the nursemaid, "agree to it, but say that you want a betrothal gown the color of meadows and printed with all the flowers in the world. There's no such dress on earth, and you'll have a good excuse not to do what he wants."

When he heard the condition, the king immediately called in a faithful servant, supplied him with a bag of gold coins and a good horse, and sent him out into the world to look for a gown the color of meadows and printed with every flower. He traveled about for six months, but there was no finding the dress. At last he entered a city full of Jews and asked a textile merchant, "Would you have such and such a silk material?"

"What do you mean, would I have it?" replied the Jew. "I have something far finer than that."

So the king presented his daughter with just the gown she had requested. Maria flew to her nursemaid, in tears. "Be not dismayed, my child," said the nursemaid, "but request another gown for the announcement of the banns: one the color of the sea, with all the fish embroidered in gold."

In a few months' time the servant also obtained that dress down in the city of the Jews. The nursemaid then suggested that Maria ask for a wedding dress far more splendid than the other two gowns: it was to be the color of air and display sun, planets, and all the stars. The servant made a third trip, and six months later the dress was ready.

"Now," said the king, "there's no more time to lose, my daughter. In one week we will get married."

The ceremony was set, but in the meantime the nursemaid had made the girl a wooden outfit which covered her from head to toe and could float in the sea.

The day of the wedding Maria told her father she was going to take a bath. Then in the tub of water she secured a dove, to which was tied a second dove outside the tub. As the second bird pulled against the first in an attempt to fly away, the first vigorously beat its wings, thus splashing water loudly like someone taking a bath. In the meantime Maria slipped into her wooden outfit, concealing the three gowns the color of meadows, water, and air in the wooden skirt, and fled. As the splashing in the other room continued, her father suspected nothing.

Maria went down to the sea and started walking on the water in the dress that floated. On and on she went over the waves and came to a place where a king's son was fishing with some fishermen. At the sight of the wooden lady walking on the sea, he said, "I've never seen a fish like that. Let's catch it and see what it is." He had the nets spread, and they drew her in to shore.

"Who are you? Where do you come from?" they asked.

And Maria answered:

> "I am Wooden Maria,
> Fashioned far from here;
> Made with much skill,
> I go where I will."

"And what can you do?"

"Everything and nothing."

So the king's son carried her to the royal palace and had her tend the geese. The news that the court had as a solid wooden maiden as a goose girl created a sensation, and people came from far and wide to watch her following the geese around over meadows and over ponds, walking or floating as she pleased.

But on Sundays, when no one was looking, Wooden Maria removed her floating garb, let down her beautiful black locks over her bare back, and climbed a tree. There she would comb her hair, while the geese gathered round the tree and sang:

> "Pa-paparapà,
> Lovely girl there up high
> Like the Moon, like the Sun in the sky,
> King's child or emperor's, that's no lie!"

Every evening Wooden Maria went back to the palace with a basket of eggs, and one evening she found the king's son getting ready to go to a ball and started joking with him.

> "Where do you go, son of the king?"
> "I don't have to tell you a thing."

"Take me with you to the dance!"
"I'll kick you in the seat of your pants!"

And he gave her a kick. Maria returned to her roost, slipped on the gown the color of meadows and printed with all the flowers in the world, and went off to the ball herself.

There the unknown lady was the belle of the ball in her gown, the likes of which had never been seen. The king's son invited her to dance with him, and asked her where she came from and what her name was. Maria said, "I am the Countess Thwartboot." The king found it hard to believe, never having heard that name; not a soul knew the lady, and she said nothing other than "Thwartboot." The king's son had truly fallen in love with her, and made her a present of a gold pin, which she put in her hair, then flew away from the ball laughing. The king's son ordered his men to pursue her and see where she went, but she scattered a handful of gold coins behind her. The men stopped to pick them up and got to squabbling among themselves.

The next evening, torn between hope and despair, the king was getting ready for the ball, when Wooden Maria came in with her eggs and said, "Majesty, you're going dancing again tonight?"

"Don't bother me. I've other things to think about!"

"Can't you take me with you?"

Losing patience, the king's son grabbed up a small shovel from the hearth and struck her.

Wooden Maria returned to the poultry house, donned the sea-colored gown displaying all the fish in the sea, and went off to the ball. The king's son was overjoyed to dance with her again. "Tell me who you are this time."

"I am the Marquise Thwartshovel," answered Maria, and not one more word would she say.

The king's son gave her a diamond ring, and she ran away just as she had done the evening before, getting rid of the servants once more by means of gold coins. The king was now more in love than ever.

The next evening he was in no mood for Wooden Maria's wit. As soon as she came asking him to take her to the ball, he hit her on the back with the reins, as he was just harnessing his horse. At the ball he met the lady wearing a dress still lovelier than those of the evenings before, this one the color of air and displaying sun, planets, and stars. The lady told him she was the Princess Thwartreins. The king's son gave her a medallion bearing his picture.

And not even that evening were the servants able to follow her.

The king's son was overcome with love sickness, and the doctors were completely baffled. He refused to eat and wouldn't touch even one spoonful of soup. One day he said to his mother, who was constantly after him to eat something, "Yes, I feel like a pizza. Make it yourself, Mamma, with your own hands."

The queen went into the kitchen, and there was Wooden Maria. "Leave everything to me, Majesty, I'll gladly help you," she said, and set to work kneading dough and baking the pizza.

The king's son took a bite, liked it, and was complimenting his mother, when he suddenly bit into something hard—the pin he had given the beautiful stranger. "Mamma, just who made this pizza?"

"I did. Why?"

"No, you didn't. Tell me the truth, who made it?"

So the queen had to admit Wooden Maria had helped her. Right away the son said to have her make him another one.

Wooden Maria's second pizza arrived, and in it the king's son found his diamond ring. "Wooden Maria must know something about the beautiful stranger," said the young man to himself, and ordered that she make him a third pizza. When he found the medallion bearing his picture, he jumped out of bed and ran to the poultry yard. There he found all the geese gathered round the tree singing:

> "Pa-paparapà,
> Lovely girl there up high
> Like the Moon, like the Sun in the sky,
> King's child or emperor's, that's no lie!"

And looking up in the branches, he saw the beautiful stranger coming out of the wooden hide and combing her locks. Maria told him her story, and in less than no time they were man and wife and as happy as happy could be.

(Rome)

Louse Hide

There was once a king who, during a leisurely stroll one day, found a louse on him. A king's louse, he thought, is to be respected. So instead of delousing himself, he took it home to the royal palace and cared for it. The louse grew fat, as fat as a cat, and spent the whole day in a chair. Then it got as fat as a pig and had to be moved to an easy chair. When it became as fat as a calf, it had to be put in a barn. But the louse continued to fatten and soon outgrew the barn, so the king had it slaughtered. Once it was slaughtered, he had it skinned and the hide nailed to the palace door. Then he issued a proclamation: whoever guessed which animal's hide it was would have his daughter in marriage; but whoever guessed wrong would be condemned to death.

No sooner was the proclamation out than a long line of men formed before the royal palace. They guessed and lost. The hangman worked day and night. Now the king's daughter, unbeknownst to her father, had a lover, and was on pins and needles until she found out from certain servants in the know that the hide belonged to a louse. In the evening when the lover showed up as usual under her window, she said in a low voice, "Tomorrow, go to my father and say the hide is that of a louse."

But he didn't catch her words. "Of a mouse, you say? A giant mouse?"

"No, louse!" answered the king's daughter, raising her voice.

"Grouse?"

"Louse! Louse!" she yelled.

"Oh, I get it! I'll see you tomorrow." With that, he left.

But under the window of the king's daughter a hunchback cobbler had his workbench, and overheard the whole conversation. "We'll just see now who marries you," he said to himself, "me or that man there." And in a flash, without even removing his smock, he jumped up and ran to the king. "Sacred Crown, I have the honor to come before you and guess what hide you have here."

"Be careful," said the king. "Ever so many men have already lost their lives guessing."

"We'll just see if I lose mine too," said the hunchback. The king showed him the hide. The hunchback took a good look, sniffed it, pretended to rack his brains, and said, "Sacred Crown, I have the honor to inform you that it doesn't take so much effort to recognize what animal this hide is from: it is from a louse."

The king was quite put out by the hunchback's cleverness, but without a word, since a king's promise is sacred, he sent for his daughter and right away declared her the rightful bride of the hunchback. The poor girl, who had been sure of marrying her lover on the morrow, was now beside herself with woe.

The little hunchback became king and she was his queen. But having to live with him took all the joy out of life. In her service was an old chambermaid who would have given anything to see the queen laugh again and said one morning: "Sacred Majesty, I saw three hunchback buffoons passing through town dancing and singing and playing, and everybody just dying laughing. How about me bringing them to the royal palace to entertain you too a bit?"

"Are you out of your mind?" said the queen. "What would the hunchback king say if he came in and found them here? He'd think we'd brought them in to mock him!"

"Don't worry," replied the chambermaid. "If the king comes in, we'll hide them in the trunk."

So the three hunchback musicians went to the queen and entertained her royally, and the queen split her sides laughing. Right in the middle of their act the doorbell rang loudly: the hunchback king was back.

The chambermaid grabbed the three hunchbacks by the neck, thrust them into the big cupboard, and locked the door. "All right, all right, I'm coming!" she said, and went and let the king in. They ate supper and then went out for a walk.

The next day was the day the king and queen received visitors, and the hunchbacks were completely forgotten. The third day the queen said to the maid, "By the way, what became of the hunchbacks?"

"Oh, my goodness!" exclaimed the maid, clapping her hand to her forehead. "I forgot all about them! They're still there in the cupboard!"

They opened the cupboard at once, and what should they find but three dead hunchbacks. They had died from lack of food and air, and they looked quite sulky.

"Now what will we do?" asked the frightened queen.

"Don't worry, I'll think of a way out of this," replied the maid, and she took one of the hunchbacks and stuffed him in a sack. She then called a porter. "Listen, in this sack is a thief I killed with a slap as he was stealing the crown jewels." She opened the sack and showed him the hump. "Now take him on your back and, without letting anyone see you, throw him into the river. I'll pay you when you return."

The porter flung the sack over his back and went to the river. Meanwhile that devil of a chambermaid stuffed the second hunchback into another sack and placed it beside the door. The porter returned to be paid, and the maid said, "How do you expect to be paid when the hunchback is still here?"

"But what game are we playing?" asked the porter. "I just now threw him into the river."

"Here's proof you didn't do it very well. Otherwise he wouldn't still be here."

Shaking his head and grumbling, the porter once more loaded the sack onto his back and trudged off. When he returned to the royal palace a second time what should he find but the sack with the hunchback, and the chambermaid as angry as a hornet. "Am I not right, you don't know how to throw him into the river? Can't you see he's back again?"

"But this time I tied a stone to him before throwing him in!"

"Tie on two! Just let that sack come back here again, and not only will I not pay you, I'll beat the daylights out of you!"

Once more the porter took up the sack, walked to the river, tied two boulders to it, and threw the third hunchback into the water. He watched carefully to see that it didn't reappear, then returned to the royal palace.

As he was climbing the steps, he met the hunchback king on his way out and thought to himself, Damnation! The hunchback has escaped again, and that old witch will now beat me for sure! In a blind rage, he grabbed the hunchback by the neck and shouted, "You hangman of a hunchback, how many times do I have to throw you into the river? I tied one stone to you, and you came back up. I tied on two, and here you are back again! How can you be so ornery? I'm going to fix you for good!" At that, he put his hands around the hunchback's throat and strangled him. Then he caught him by the neck and dragged him straight to the river, where he tied four rocks to his feet and hurled him into the water.

When the queen learned that her husband had gone the way of the other three hunchbacks, she showered the porter with presents: gold, precious stones, hams, cheese, and wine. Without hesitation she then married her first love and from then on was happy as happy could be.

> "Wide is the leaf, narrow is the way,
> Tell yours now, as I have had my say."

(Rome)

105

Cicco Petrillo

There was once a couple who had a daughter, for whom they had found a husband. All the relatives were invited to her wedding, after which everybody sat down to the table. Right in the middle of dinner, the wine gave out, so the father said to his newlywed daughter, "Go down to the cellar and fetch more wine."

The bride went to the cellar, placed the bottle under the cask, opened the tap, and waited for the bottle to fill up. While waiting, she got to thinking, Today I got married. In nine months I'll have a son and name him Cicco Petrillo, dress him, put shoes on him. He'll become a big lad . . . but suppose Cicco Petrillo dies? Oh, my poor son!" At that she burst out crying as no one has ever cried before.

The tap was still open and the wine meanwhile ran all over the cellar. The diners sat and waited for the bride to come back, but there was no sign of her. Her father said to his wife, "Go down to the cellar and see if she's fallen asleep, by chance."

The mother went down to the cellar and found her daughter in a flood of tears. "What on earth's the matter, daughter? What happened?"

"Ah, Mamma, I was thinking that today I got married and in nine months I'll have a son and name him Cicco Petrillo, and what if Cicco Petrillo then dies?"

"Oh, my poor, poor grandson!"

"Oh, my poor, poor son!"

And the two women burst out weeping together.

The cellar meanwhile was filling up with wine. The diners upstairs continued to wait for wine, but no wine came. The father said, "They could have both had a stroke. I must go and see."

He went down to the cellar and found the two women bawling like two newborn babies. "What in God's name has happened?" he asked.

"Oh, my husband, if you only knew! We got to thinking that this daughter of ours is now married and in practically no time she'll be having a son we'll name Cicco Petrillo, and what if Cicco Petrillo goes and dies?"

"Ah!" screamed the father. "Our poor darling Cicco Petrillo!"

And they all three broke down and wept right in the middle of the wine.

When nobody returned, the bridegroom said, "But what kind of stroke could they have had down in the cellar? Let me go and see." And he went downstairs.

Hearing all that wailing, he asked, "What the deuce has come over you to wail like that?"

"Ah, dear husband!" answered the bride. "We were thinking that now we are married and will have a son we'll name Cicco Petrillo, and what if our Cicco Petrillo should die?"

At first the bridegroom thought they were joking, but realizing they were serious, he blew up. "I always figured you were all a trifle stupid, but never to such a degree as this. It would be my luck to get mixed up with such simpletons! But I'm not staying! I'm leaving, and you, my dear, can be sure you'll never see me again unless, on my travels, I should meet three people crazier than you!" At that, he left the house and never once looked back.

He walked until he came to a river, where a man was trying to unload a boatful of hazelnuts with a pitchfork.

"What are you doing, my good man, with that pitchfork?"

"For some time I've tried, but I can't pick up a single one."

"Naturally! Why not try the shovel?"

"The shovel? Bless my soul, I never thought of it!"

"That's one!" said the bridegroom. "He's even dumber than all my wife's family put together."

He continued on his way until he reached another river. There a farmer was working himself to death watering two oxen with a spoon.

"What on earth are you doing?"

"I've been here for three hours and can't seem to quench these animals' thirst."

"Why not let them put their muzzles in the water?"

"Their muzzles in the water? What a dandy idea! I never thought of that."

"He's number two," said the bridegroom, and moved on.

After going some distance he spied a woman in the top of a mulberry tree holding out a pair of breeches.

"What are you doing up there, my good woman?"

"Oh, let me tell you!" she replied. "My husband died, and the priest told me he went up to Paradise. I'm waiting for him to return and step back into his breeches."

That makes three! thought the bridegroom. It seems that *everybody* is dumber than my wife, so I'd better go back home!

And so he did and was glad of it, since they say that if you look far enough you can always find something worse.

(Rome)

106

Nero and Bertha

This particular Bertha was a poor woman who did nothing but spin, being a skillful spinner.

One day as she was going along she met Nero, the Roman emperor, to whom she said, "May God grant you health so good you'll live a thousand years!"

Nero, whom not a soul could abide because he was so mean, was astounded to hear someone wishing him a thousand years of life, and he replied, "Why do you say that to me, my good woman?"

"Because a bad one is always followed by one still worse."

Nero then said, "Very well, bring to my palace all you spin between now and tomorrow morning." At that, he left her.

As she spun, Bertha said to herself, "What will he do with the thread I'm spinning? I wouldn't put it past him to hang me with it! That hangman is capable of everything!"

Next morning, right on time, here she was at Nero's palace. He invited her in, received the thread she had spun, and said, "Tie the end of the ball to the palace door and walk away as far as you can go with the thread." Then he called his chief steward and said, "For the length of the thread, the land on both sides of the road belongs to this woman."

Bertha thanked him and walked away very happy. From that day on she no longer needed to spin, for she had become a lady.

When word of the event got around Rome, all the poor women went to Nero in hopes of a present such as he had given Bertha.

But Nero replied, "The good old times when Bertha spun are no more."

(Rome)

The Love of the Three Pomegranates

A king's son was eating at the dinner table. While slicing the ricotta, he cut his finger, and a drop of blood fell on the white cheese. He said to his mother, "Mamma, I would like a wife white like milk and red like blood."

"Why, my son, whoever is white is certainly not red, and whoever is red is by no means white. But go out all the same and see if you can find such a girl."

The son set out. After some distance he met a woman, who asked, "Where are you going, young man?"

"How can I confide my secret to a woman? The very idea!"

On and on he went, and met a little old man, who asked, "Where are you going, young man?"

"You I'll tell, respected sir, who will certainly hear further of me. I'm seeking a girl both milk-white and blood-red."

"My son, whoever is white is not red, and whoever is red is not white. Take these three pomegranates, however. Open them and see what comes out. But do so only beside the fountain."

The youth opened a pomegranate, and out jumped a very beautiful girl white like milk and red like blood, who immediately cried:

> "Dear young man, bring me some water,
> Otherwise I'm Mother's dead daughter!"

The young man dipped up water in the hollow of his hand and offered it to her, but he was too late: the beautiful creature was dead.

He opened another pomegranate, and out jumped another beautiful girl saying:

> "Dear young man, bring me some water,
> Otherwise I'm Mother's dead daughter!"

He brought her water, but she was already dead.

He opened the third pomegranate, and out jumped a girl still more beautiful than the other two. The young man threw water in her face, and she lived.

She was as naked as the day her mother gave birth to her, so the young man threw his own cloak over her, saying, "Climb this tree while I go for clothes to dress you in and a carriage to take you to the palace."

The girl remained in the tree beside the fountain. Now every day, this fountain was visited by the ugly Saracen woman, who came there for water. As she went to dip up water with her earthen pot, she saw the maiden's face reflected on the surface of the fountain from the tree, and sighed:

> "Why must I, who am so beautiful,
> Trudge home with water by the potful?"

At that, she slammed the pot down, smashing it to smithereens. When she got home, her mistress said, "Ugly Saracen, how dare you return with no water and no crock!" She therefore picked up another earthen pot and returned to the fountain, where she again saw that image in the water. "Ah, I am truly beautiful!" she said to herself, adding:

> "Why must I, who am so beautiful,
> Trudge home with water by the potful?"

Again she slammed down the crock. Again her mistress scolded her. Again she went to the fountain and smashed still another pot. Up to then the maiden had merely looked on from the tree, but now she had to laugh.

Ugly Saracen looked up and saw her. "Oh, it's you? You are the one who made me smash three pots to smithereens? But you are truly beautiful! Just a minute, I want to do your hair for you."

The maiden was reluctant to come down the tree, but Ugly Saracen insisted. "Let me dress your hair, so that you will be still more beautiful."

Helping her down, Ugly Saracen undid the maiden's hair and found a hairpin, which she thrust into the poor girl's ear. A drop of blood fell from the maiden, then she died. But when the drop of blood hit the ground, it changed into a wood pigeon, which flew away.

Ugly Saracen went and settled in the tree. The king's son returned in the carriage and, seeing her, said, "You were milk-white and blood-red. How on earth did you become so dark?"

Ugly Saracen replied:

> "Out came the sun
> And made me dun."

"But how could your voice have changed so?" asked the king's son. She replied:

> "The wind came up,
> My voice came down."

"But you were so beautiful, and now you are so ugly!" said the king's son.

She replied:

> "Also rose the breeze
> And caused my face to freeze."

That was that. He took her into the carriage and carried her home.

From the moment Ugly Saracen settled down in the palace as the wife of the king's son, the wood pigeon would alight on the kitchen window ledge every morning and say to the cook:

> "Cook, O cook of the cursèd kitchen,
> Tell me, tell me
> What the king is doing with old Ugly Saracen."

"He eats, drinks, and sleeps," replied the cook.
The wood pigeon said:

> "Please, a bit of soup for me,
> And plumes of gold I will give thee."

The cook served her a plate of soup, and the wood pigeon gave a little shake and shed a few feathers of gold. Then she flew off.

The next morning she was back:

> "Cook, O cook of the cursèd kitchen,
> Tell me, tell me
> What the king is doing with old Ugly Saracen."

"He eats, drinks, and sleeps," replied the cook.

> "Please, a bit of soup for me,
> And plumes of gold I will give thee."

She ate her soup, and the cook took the golden feathers.

A little later, the cook decided to go to the king with the whole story. The king listened carefully, and replied, "Tomorrow when the wood pigeon returns, catch it and bring it to me. I shall keep it."

Ugly Saracen, who had eavesdropped and heard everything, knew only too well that the wood pigeon would be her undoing, so next morning she beat the cook to the window when the pigeon lit, pierced it through with a spit and killed it.

The wood pigeon died, but a drop of blood fell in the garden and right there a pomegranate tree sprang up at once.

This tree had the magic property that whoever was dying and ate one of its pomegranates got well. And there was always a long line of people begging Ugly Saracen for a pomegranate.

Finally only one pomegranate remained on the tree, the biggest one of all, and Ugly Saracen announced: "I will keep this one for myself."

An old woman came to her, asking, "Will you give me that pome-granate? My husband is dying."

"I have only one left, and I am keeping it for decoration," replied Ugly Saracen, but the king's son objected. "Poor old thing, her husband is dying, you can't refuse her."

So the old woman went back home with the pomegranate. She got home and found her husband already dead. "That means I keep the pomegranate for decoration," she told herself.

Every morning the old woman went to Mass. And while she was at Mass, the girl would come out of the pomegranate, light the fire, sweep the house, do the cooking, and set the table. Then she would go back inside the pomegranate. Finding everything in order upon her return, the old woman was baffled.

One morning she went to confession and told her confessor all about it. He replied, "Know what you should do? Tomorrow morning pretend to go out to Mass, but hide somewhere at home instead. That way you'll see who's doing all your housekeeping."

The next morning the old woman pretended to leave the house, but stopped outside the door. The maiden emerged from the pomegranate and started on the housework and the cooking. The old woman came back in and caught the girl before she could reenter the pomegranate.

"Where do you come from?" asked the old woman.

"Peace to you, ma'am, don't kill me, don't kill me!"

"I'm not going to kill you, but I want to know where you come from."

"I live inside the pomegranate . . ." And she related her story.

The old woman dressed her in peasant garb like her own, since the maiden was still as naked as the day she was born, and on Sunday took her to Mass with her. The king's son was also at Mass and saw her. "My heavens!" he exclaimed. "I do believe that's the maiden I met at the fountain!" So he lay in wait for the old woman on the road.

"Tell me where that maiden came from!"

"Don't kill me!" whimpered the old woman.

"Don't worry, I only want to know where she comes from."

"She comes from the pomegranate you gave me."

"She was in a pomegranate too?" exclaimed the king's son, who turned to the maiden and asked, "How on earth did you get into a pomegran-ate?" And she told him everything.

He returned to the palace with the girl, and had her tell the whole story once more in front of Ugly Saracen.

"Did you hear that?" the king's son asked Ugly Saracen when the girl had finished her tale. "I don't want to be the one to condemn you to death. Condemn yourself."

As there was now no way out, Ugly Saracen said, "Coat me with pitch and burn me to death in the center of the town square."

So was it done, and the king's son married the maiden.

(Abruzzo)

◈ 108 ◈

Joseph Ciufolo, Tiller-Flutist

There was once a youth named Joseph Ciufolo, who played the flute when he wasn't tilling the soil. One day he was dancing through the fields and playing his flute to relax awhile from all his digging, when he suddenly spied a corpse lying on the ground beneath a swarm of flies. He put down his flute, walked up to the body, shooed the flies away, and covered the dead man with green boughs. Returning to the spot where he had left his hoe, he saw that the hoe had gone to work by itself and already dug up half the field for him. From that day on, Joseph Ciufolo was the happiest tiller alive: he would dig until he got tired, then take his flute out of his pocket, while the hoe went on digging by itself.

But Joseph Ciufolo worked for a stepfather who bore him no love and wished to turn him out of the house. In the beginning the man said Joseph was a good worker but lazy; next he said Joseph dug a whole lot but badly. Joseph Ciufolo therefore took his flute and left home.

He went around to all the landowners, but none of them would give him any work. Finally he met an old beggar, and asked him for work to keep body and soul together.

"Come along with me," said the beggar, "and we will share alms."

So Joseph Ciufolo started going around with the beggar and singing:

> "Succor us, please, please succor us,
> In the name of Jesus and all His Saints!"

Everybody gave alms to the old man, but to Joseph Ciufolo they all said, "What's a young man like you doing out begging? Why don't you work for a living?"

"Nobody will hire me," replied Joseph Ciufolo.

"That's what you say. There's the king with so many untilled fields that he's offering good wages to anyone willing to cultivate them."

Joseph Ciufolo went to the king's fields and took the old man whose alms he had been sharing. The fields had never been worked by anyone. Joseph Ciufolo dug them up, sowed them, weeded them, then harvested the crops. Whenever he wearied of reaping he would play his flute; and once he was weary of playing, he would sing:

> "Sickle so brisk, sickle so gay,
> Master's child with me, away!"

Hearing the singing, the princess looked out the window. She saw Joseph Ciufolo and fell in love with him. But she was a princess, and he a tiller; the king would never consent to their marriage. So they decided to run away together.

They fled at night in a boat. They were already on the high seas, when Joseph Ciufolo remembered the beggar. He said to his beloved, "We must fetch the old man, since he shared his alms with me. I can't go off and leave him like that." At that very moment they saw the old soul coming up behind the boat, walking on the water as if it had been dry land. Reaching the boat, he said, "We agreed to divide everything we had, and I always shared with you everything I own. Now you have the king's daughter and must give half of her to me." At that he handed Joseph Ciufolo a knife to cut his bride in half.

Joseph Ciufolo took the knife with a trembling hand. "You are right," he said, "you are perfectly right." He was on the point of cutting his bride in two, when the old man stopped him.

"Stop! I knew you were a just man. I am the dead man, mind you, whom you covered with green boughs. Go in peace, and may the two of you always be happy."

The old man walked away on the waves. The boat came to an island rich in all good things, with a princely palace awaiting the newlyweds.

(Abruzzo)

Bella Venezia

A mother and a daughter kept an elegant inn, where kings and princes passing through town would stop. The innkeeper's name was Bella Venezia, and while travelers sat at the table, she would strike up a conversation. "What town do you come from?"

"From Milan."

"Did you ever see any woman in Milan lovelier than I am?"

"No, I've never seen a soul lovelier than you."

When it came time to settle the accounts, Bella Venezia would say, "Normally that would be ten crowns, but you need give me only five"—for she charged anyone only half-price when he told her he'd never seen a lovelier woman than herself.

"Where are you from?"

"Turin."

"And is there anyone in Turin lovelier than I am?"

"No, a woman lovelier than you I have never seen."

Then, at reckoning time, Bella Venezia said, "Normally I charge six crowns, but you need give me only three."

One day the innkeeper was asking a traveler the usual question, "And did you ever see a lovelier woman than myself?" when her daughter went through the room. And the traveler replied, "Indeed I have."

"Who?"

"Your daughter."

This time in making up the bill, Bella Venezia said, "It's normally eight crowns, but I'm asking you for sixteen."

That evening the mistress called in the kitchen boy. "Go to the seashore, build a hut with just one tiny little window, and close up my daughter in it."

Thus Bella Venezia's daughter was imprisoned day and night in that hut by the sea. She heard the breaking of the waves, but was able to see no one other than the kitchen boy, who came to her daily with bread and water. But in spite of being shut up there, the maiden grew lovelier by the day.

A stranger riding along the beach on horseback saw that hut all boarded up and drew closer. He peeped through the tiny window and made out in the dimness the most beautiful maidenly face he had ever laid eyes on. A bit frightened, he spurred his horse and galloped off.

That night he stopped at Bella Venezia's inn.

"What town are you from?" asked the innkeeper.

"Rome."

"Did you ever see anyone lovelier than myself?"

"I certainly have," replied the stranger.

"Where?"

"Closed up in a hut by the sea."

"Here's your bill. It's only ten crowns, but I want thirty from you."

In the evening Bella Venezia asked the kitchen boy, "Listen, would you like to marry me?"

The kitchen boy couldn't believe his ears.

"If you want to marry me," continued Bella Venezia, "you must take my daughter into the woods and kill her. Bring me back her eyes and a bottle of her blood, and I'll marry you."

The kitchen boy was eager to marry the mistress, but he didn't have the heart to kill her daughter, who was all beauty and goodness. So he took the girl to the woods and left her. To get eyes and blood to carry back to Bella Venezia, he killed a lamb, which is innocent blood. And the mistress married him.

Alone in the woods, the girl screamed and cried, but no one heard her. Toward nightfall she spied a light in the distance. Drawing near, she heard many people talking and, frightened to death, hid behind a tree. It was a rocky, desolate place, and twelve robbers had come to a halt before a white boulder. One of them said, "Open up, desert!" and the boulder swung outward like a door. The inside was all lit up like a large palace. The twelve robbers went in, and the last one said, "Close up, desert!" and the boulder swung to. The girl hidden behind the tree bided her time. In a little while, a voice inside said, "Open up, desert!" The door opened, and out filed the twelve robbers, the twelfth ordering, "Close up, desert!"

Once the robbers were out of sight, the girl went to the white boulder and said, "Open up, desert!" and the door swung open for her. She stepped into the lighted interior and commanded, "Close up, desert!"

Inside was a table laid for twelve, with twelve plates, twelve loaves of bread, and twelve bottles of wine. In the kitchen twelve chickens were on a spit ready to roast. The girl tidied up the place, made the twelve beds, and roasted the twelve chickens. Hungry by then, she ate a wing from every chicken, took a bite of every loaf of bread and a sip of wine from every bottle. When she heard the robbers coming back, she hid under a bed. The twelve bandits didn't know what to think when they saw the house so tidy, the beds made, and the chickens roasted. Then they noticed a wing missing from every chicken, a bite from every loaf, and a sip from every bottle, and said, "Somebody must have come in here." Tomorrow it was agreed one of them would remain behind to stand guard.

The smallest of the robbers stayed, but he went outside to watch while the girl meanwhile came out from under the bed, put everything in order, ate the twelve chicken wings, the twelve chunks of bread, and drank the twelve swallows of wine.

"You're good for nothing!" said the ringleader, when he returned and saw that the house had been visited again. He assigned someone else to stand guard the next day. But this man also remained outside, while the girl was indoors. So for eleven days straight every robber tried keeping watch, but failed to discover the girl and was bawled out by all the others for being so stupid.

On the twelfth day, the chief decided to stand guard. Instead of watching outside, he remained inside and thus saw the girl come out from under the bed. Grabbing her by the arm, he said, "Don't be afraid. Now that you are here you can stay, and we will treat you as our little sister."

So the girl remained with the robbers, keeping house for them, and every evening they brought her jewels, gold pieces, rings, and earrings.

The youngest robber delighted in dressing up as a grand nobleman to commit his robberies, and he would stop at the best inns. He thus went to Bella Venezia's one evening for dinner.

"Where do you come from?" asked the innkeeper.

"From the heart of the forest."

"Have you ever seen any woman lovelier than myself?"

"I certainly have," replied the robber.

"Who is she?"

"A girl we have with us."

So Bella Venezia knew her daughter was still alive.

Every day an old woman would come to the inn asking for alms, and this old woman was a witch. Bella Venezia promised one-half of her fortune to the witch if she could track down and kill that daughter.

One day while the robbers were out, the girl was standing at the window singing, when an old woman came by and said, "Brooches for sale! Brooches for sale! Lovely maiden, may I come in? I'll show you a pin that's a real gem, for your hair."

The maiden invited her in, and the old woman, going through the motion of showing her how nice a pin would look in her hair, thrust it into her scalp. The girl died.

When the robbers came home and found her dead, they all burst into tears, tough as they were. They chose a tall tree with a hollow trunk and buried her inside it.

Now the king's son was out hunting. He heard the dogs barking and, moving closer, saw them all scratching on the trunk of a tree. The king's son looked inside and found a very beautiful maiden who was dead.

"If you were alive, I would marry you," he said. "Even though you are dead, I can't tear myself away from you." He sounded his hunting horn, assembled his hunters, and had her taken to the royal palace. Without his mother the queen's knowledge, he had the beautiful maiden put in one of the rooms and would stay there the whole day admiring her.

Suspicious, the mother burst into the room. "So that's why you didn't want to come out! But she's dead! How could you possibly be interested in her?"

"Dead or not, I can't live apart from her!"

"You can at least have her hair fixed!" said the queen, and sent for the royal hairdresser. He came and, combing her hair, broke his comb. He picked up another comb and broke that one too. Thus, one right after the other, he broke seven combs. "What on earth does this girl have in her head?" asked the royal hairdresser. "I shall take a look." And he touched the head of a pin. He pulled ever so gently and, as the pin came out, the maiden regained her color, opened her eyes, drew her breath, said, "Oh," and stood up.

The wedding was celebrated. Tables were also set up in the streets. Whoever wanted to eat, ate; whoever didn't want to, didn't.

> O Lord!
> A hen for every sinner!
> And for me, sinner of sinners,
> A hen and several roosters!

(Abruzzo)

◈ 110 ◈

The Mangy One

A king had no sons, and that distressed him. In the grip of this distress he was riding through a forest, when he met a knight on a white horse.

"Why so sad, Majesty?" inquired the knight.

"I have no sons," explained the king, "and my kingdom will be lost."

"If you want a son," said the knight, "sign a pact with me. When this son turns fifteen, you will bring him to this spot in the forest and give him to me."

"I'd sign any agreement whatever just to have him," answered the king. Thus the pact was signed, and the son was born.

He was a little boy with golden hair, and he wore a gold cross on his chest. Day by day he grew in stature and in wisdom. Before fifteen years had rolled around, he had already completed all his schooling and become an expert at handling weapons. Just three days before the boy's fifteenth birthday, the king locked himself in his apartments and wept. The queen didn't know what to make of this weeping, until the king informed her of the pact that was about to come due, and then she too wept and wept. The boy saw his parents in tears and was puzzled. His father said, "Son, I shall now take you to the forest and entrust you to your godfather, who sanctioned your birth by a pact."

So father and son rode to the forest in silence. Other hoofbeats were suddenly heard: it was the knight on the white horse. The youth rode up alongside him and, without a word, the father, with tears streaming down his face, wheeled his horse around and rode back whence he had come. The youth continued onward alongside the mysterious knight, through parts of the forest never before traversed. At last they came to a large palace, and the knight said, "Godson, you will live here and be the master of the house. Three things only I forbid you: to open this little window, to open this cupboard, and to go down into the stables."

At midnight, the godfather rode away on his white horse and didn't return until dawn. After three nights, as soon as he was by himself, the godson was overcome with curiosity about the forbidden little window. He opened it and found himself peering into smoke and flames, since the window opened onto Hell. The youth stared into Hell to see if he recognized anyone there, and whom should he see but his own grandmother. She saw him too and cried out from the depths, "Grandson, dear grandson, who brought you here?"

"My godfather!" replied the young man.

"No, no, no, dear grandson. That man is not your godfather, he's the Devil. Flee for your life, grandson. Open the cupboard and take with you a sieve, a cake of soap and a comb. Then go down into the stables, where you will find your horse. Flee, and when the Devil comes after you, throw down those three objects in his path. You'll cross the Jordan River, and then be out of his reach for good."

The next minute the youth was already galloping away on his horse named Horseradish. When the godfather returned and found him gone, together with horse and objects from the cupboard, he launched out against the damned souls and raked them over the coals. Then he set out after the fugitive. Godfather's white horse galloped a hundred times faster than Horseradish, and would certainly have caught up with him, had the godson not thrown down in his path the comb, which changed into a forest so dense that the godfather had to struggle for quite some time to get through it. When he was finally out of the forest and galloping again, the godson let him almost catch up, then threw down the sieve. The sieve changed into a marsh, which godfather had a time crossing, after no little wallowing. He'd almost caught up with the boy for the third time, when the godson threw down the cake of soap. The soap changed into a slippery mountain, and no matter where the godfather's horse set foot, he took many more steps backward than forward. In the meantime the godson had come to the bank of the Jordan River, and spurred Horseradish forward into the current. Horseradish swam across to the other side, while the godfather, who had finally got over the mountain, gave vent to his anger over being unable to pursue him beyond the Jordan River, by unleashing thunder, lightning, wind, rain, and hail. But the youth was already on the opposite shore and galloping off to the royal city of Portugal.

In Portugal, so as not to be recognized, the youth decided to hide his golden hair and therefore bought an ox bladder from a butcher. He put it on his head, and thus looked as if he had the mange. He tethered Horseradish in a meadow, and nobody could steal him, for during his stay in the Devil's stables, the horse had learned to eat humans.

Wearing the bladder over his head, the youth strolled by the king's palace. The gardener saw him and, learning that he was seeking work, engaged him as his helper. The gardener's wife began grumbling when her husband brought him home, for she wanted no mangy man in her house. So, to please her, the husband sent him to a wood hut nearby, telling him he was not to set foot in their house again.

At night the youth stole softly from the hut and went off and untied Horseradish. He dressed up in a king's red suit, removed the bladder from his head, and his golden hair gleamed in the moonlight. He rode Horseradish through various maneuvers in the royal garden, jumping over hedges and ponds, and engaged in feats of skill, such as tossing into the air three shiny rings, gifts from his mother, that he wore on his three middle fingers, and catching them on the tip of his sword.

At the same time, the daughter of the king of Portugal happened to be at her window gazing at the garden in the moonlight, and saw the young rider with gold hair and dressed in red going through all those maneuvers. "Who can he be? How did he get into the garden?" she wondered. "I'll watch where he goes when he leaves." She therefore saw him leave, just before dawn, by a gate that led into the meadow where he kept his horse tethered. She still had her eyes on the gate, when a few minutes later the mangy one who helped the gardener came through it, and she closed her window so as not to be seen.

The next night she sat at the window and waited. At last she saw the mangy one come out of the hut and go through the gate. In a few minutes the rider with golden hair came back through it; this time he was dressed from head to toe in white, and resumed his maneuvers. Just before dawn he left, and in no time the mangy one returned. The princess began to suspect some connection between the mangy one and the rider.

The third night the same things took place; only the rider was dressed in black. The princess said to herself, "The mangy one and the rider are one and the same."

The next day she went down into the garden and told the mangy one to bring her some flowers. He made three nosegays: a big one, a middle-sized one, and a little one; he put them in a basket and carried them to her. The larger nosegay was fitted into the ring from his middle finger, the middle-sized one into the ring from his ring finger, and the smaller one into his pinkie ring. The princess recognized the rings and returned the basket full of gold doubloons.

The mangy one carried the basket back to the gardener, doubloons and all. The gardener began scolding his wife. "Just look at that!" he said. "You won't allow him inside our house, but the princess calls him into her rooms and fills his basket with gold doubloons!"

The next day the princess wanted the mangy one to bring her some oranges. He brought her three: one ripe, one half-ripe, one green. The princess put them on the table, and the king asked, "Why are you bringing green oranges to the table?"

"They are what the mangy one brought in," answered the princess.

"Let's see what this mangy one has to say for himself; bring him in," said the king. When the mangy one stood before him, the king asked why he'd picked three oranges of varying degrees of ripeness.

"Majesty," replied the mangy one, "you have three daughters: one is marriageable, a second is only halfway ready for marriage, while the third still has a few years to wait."

"True," replied the king, and issued a proclamation: LET ALL MY OLDEST DAUGHTER'S SUITORS LINE UP. HE TO WHOM SHE GIVES HER HANDKERCHIEF WILL BE CHOSEN.

There was a grand parade under the royal windows. First came all the sons from reigning families, then all the barons, next all the knights, then the artillery men, and finally the foot soldiers. Bringing up the rear was the mangy one, and the princess gave her handkerchief to him.

Upon learning that his daughter had chosen the mangy one, the king turned her out of the house. She went off to live in the mangy one's hut. He gave her his own bed and made do with a couch by the fire, saying a mangy one may not come close to the daughter of the king. So he's really and truly mangy, thought the princess to herself. Good heavens, what have I gone and done! She already regretted her choice.

War broke out between the king of Portugal and the king of Spain, and all the men went off to fight. They said to the mangy one, "All the men are going to war; are you who have taken in the king's daughter staying here?" They had already made plans to give him a lame horse, so that he would be killed in battle. The mangy one took the lame horse to the meadow where Horseradish was tethered, dressed himself in red from head to toe, donned a breastplate his father had given him, and rode off to war on Horseradish. The king of Portugal found himself hemmed in by the enemy: up galloped the red knight, put the enemy to flight, and saved the king's life. No enemy soldier on the field could come anywhere near the king, for the knight dealt cutting blows right and left while his horse filled the enemy horses with terror. That was how the first day's battle was won.

Each evening the king's daughter went to the palace to hear the latest news from the battlefield. When they told her about the golden-haired knight in red, who'd saved the king's life and brought victory to his army, she couldn't help thinking, That's my knight, the rider I used to see at night in the garden. And here I've gone and chosen the mangy one! With a heavy heart she returned to the hut and found the mangy one asleep by the fire, huddled up under his old cloak. The princess could no longer keep back her tears.

At dawn the mangy one rose, took the lame horse, and went off to battle. But he first stopped by the meadow as usual, exchanging the lame horse for Horseradish and his rags for a white suit. He donned the breastplate and removed the ox bladder from his head of golden hair. That day too the battle was won, thanks to the knight in white.

Upon hearing this latest piece of news in the evening and then going home and finding the mangy one sleeping by the fire, the king's daughter was more woeful than ever.

The third day the golden-haired knight showed up on the field dressed entirely in black. This time the king of Spain himself was there, together with his seven sons. So what did the golden-haired knight do but single-handedly confront all seven of them at once. He slew them one by one until they were all vanquished. But the seventh son, before dying, wounded him on the right arm with a sword. At the end of the battle the king of Portugal wanted to have the wound dressed, but the knight had already disappeared, as on the other evenings.

Hearing that the golden-haired knight had been wounded, the king's daughter was deeply grieved, as she was still in love with that stranger. She went home feeling more bitter than ever toward the mangy one and stared at him with contempt as he slept curled up next to the fire. But as she looked at him, she caught a glimpse, through his unbuttoned cloak, of a bandage around his arm. Then she noticed that, under this cloak, he wore a costly black-velvet outfit. Nor was that all she observed: sticking out from under the ox bladder was a lock of golden hair.

Wounded, the youth had been unable to change as on the other evenings; dead tired, he'd dropped down on his couch and fallen asleep.

The king's daughter stifled a cry of surprise and joy and uneasiness all in one and tiptoed out of the hut, so as not to awaken him, and went flying to her father. "Come see who won your battles! Come see!"

Followed by the whole court, the king went to the wood hut. "Yes, it is he all right!" said the king, recognizing the knight under his disguise. They woke him up and would have carried him out on their shoulders, but the king's daughter had called in the surgeon to dress his wound. The king wanted to celebrate the wedding right then and there, but the young man said, "First I must go and inform my father and mother, for I too am the son of a king."

The father and mother came to meet their son, whom they had given up for dead, and everybody sat down together to the wedding banquet.

(Abruzzo)

◈ 111 ◈

The Wildwood King

A king had three daughters. Two of them were neither beautiful nor ugly, but the youngest was beautiful beyond words. Whenever any man came asking for the hand of the eldest girl, he would fall in love with the youngest. None of the girls, therefore, managed to get married. The two older daughters formed a plot against the youngest; they told their father they had both had a dream: their sister would run away from home with a common soldier. Lest the dream come true and his youngest daughter disgrace the royal house, the king called in a general and ordered him to take the girl walking in the woods of the wildwood king and there kill her with his sword.

So they went into the wildwood king's woods, the maiden and the general. "All right," said the girl after a time, "let's go back home now."

"No, Your Highness," replied the general. "I'm sorry, but I have orders to kill you right here."

"And why would you want to kill an innocent girl like me?"

"King's orders," answered the general, unsheathing his sword. But the sight of the poor girl so frightened moved him to pity, and he took away her clothes to dip them in a lamb's blood and present them to the king as proof of her death.

The girl remained in the woods weeping, terrified at the thought of the wildwood king who lived in those woods and ate everyone who crossed his path. After crying for a while, she dried her eyes and fell asleep in a hollow tree trunk.

In the morning, the old wildwood king was out hunting and in pursuit of a wounded stag. But instead of finding the stag, he came upon the sleeping maiden. Seeing how beautiful she was, he awakened her and asked, "Would you like to come with me? Don't be afraid." The girl accepted, and followed the wildwood king to his house in the depths of the woods, where he led a sad, solitary life, going out hunting and never seeing a living soul. The girl began keeping house for him, and the old hermit loved her like a daughter.

In the morning as soon as she got up, she braided her hair at the window, and a parrot lit on the ledge and said:

> "In vain are you pretty and neat,
> You will become the forest king's meat."

Hearing those words, the girl started crying. The wildwood king came in from hunting and, seeing her upset, asked, "What's the matter?" The girl told him what the parrot had said.

"Do you know what you should tell him?" asked the wildwood king.

> "Parrot, parrot, hear this ban:
> Your feathers for my fan,
> Your meat is for my pan,
> Your master will become my man!"

When she repeated those words the next day the parrot, out of spite, shook himself so vigorously that he flew off minus half his feathers. The parrot belonged to a king in the vicinity, who seeing his bird come back half-plucked, asked the servants, "Who is pulling out the parrot's feathers?"

The servants replied, "Every morning he flies off in the direction of the wildwood king's house and comes home plucked."

"I shall follow him tomorrow morning," said the king, "and find out what is going on."

So next morning, riding through the woods, he followed the parrot's flight and came up to the window where the most beautiful maiden he had ever seen was arranging her hair. The parrot lit on the window ledge and sang:

> "In vain are you pretty and neat,
> You will become the forest king's meat."

And the maiden replied:

> "Parrot, parrot, hear this ban:
> Your feathers for my fan,
> Your meat is for my pan,
> Your master will become my man!"

And the parrot shook out all his feathers.

Then that king went to the wildwood king and asked for the girl's hand in marriage. The wildwood king gladly gave her to him, although it grieved the old king to be separated from her. She thanked him and bid him farewell, leaving him by himself in the depths of the woods.

The wedding banquet was also attended by the bride's royal father, who asked her forgiveness for all the suffering he had caused her at the prompting of the wicked sisters.

And the parrot? He flew off and was never seen again.

(Abruzzo)

◈ 112 ◈

Mandorlinfiore

There was a wife and a husband, and a baby was about to be born to them. The father went to the door to see who was passing by, for his son would become exactly like that person.

Some loose women came down the street, and the father cried to his wife, "Don't have him just now, for heaven's sake!"

Some thieves went by, and again the father cried out, "Not yet, not yet, please!"

Then the king came by, and in that instant the baby was born, and it was a boy. So father, mother, grandmother, and aunts shouted, "The king has just been born! The king has just been born!"

The king heard them and entered the house. He asked for an explanation, and they told him the whole story. Then the king said he would take the baby home with him and raise him. Father and mother blessed the child and handed him over to the king.

Along the way the king got to thinking, Why should I raise a child that will only be my undoing in the end? He pulled out a knife, planted it in the baby's throat, and left him lying in the middle of an orchard of almond trees in bloom.

The next day two merchants came that way. They found the baby still alive, so they bandaged up his wound, and one of them took him home to his wife. They were a rich couple with no children of their own, and they came to love him dearly. They named him Mandorlinfiore, which means "almond blossom."

Mandorlinfiore grew to be a handsome and clever youth. Then, quite unexpectedly, the merchant had a son of his own. One day when this second son was already a big boy, he got into an argument while playing with Mandorlinfiore and called him a bastard. Mandorlinfiore went and told his mother, and thus learned how he had been found in an orchard. At that he decided to leave home and nothing the merchant and his wife said could make him change his mind. After going some distance he reached the city of the king who had stabbed and then abandoned him. Finding Mandorlinfiore such an intelligent youth, the king engaged him as his secretary, for he naturally did not recognize him.

Now the king had a daughter as fair as day, named Belfiore, and the youth fell in love with her. When the father realized that his daughter was being courted by his secretary, rather than deprive himself of a secretary he really needed, he chose to send his daughter to live in the house of another king, his brother. In his unhappiness Mandorlinfiore fell sick, and it was then that the king, visiting his bedside, discovered the scar on his neck. He remembered the baby he had knifed, and asked the secretary where he was born. "I was found in a field of almond trees in bloom," replied the youth.

So the king resolved to put him to death. He told him to carry a letter to the other king, his brother, and Mandorlinfiore departed. The letter said the young man was to be hanged immediately. But Belfiore, who had got wind of her lover's arrival, was waiting for him and brought him in on the sly through a small secret entrance. Once they were alone, Belfiore asked to see the letter her father was sending her uncle, but Mandorlinfiore said no, he'd promised to give it to no one except its addressee. But when the youth went to sleep, Belfiore opened the letter and read it. That way she discovered the trap set by her father, and she and Mandorlinfiore together sought a way to get the better of the king. In place of the letter, they put another that said the youth was to marry Belfiore immediately, and Mandorlinfiore went back through the secret entrance, purchased princely garb and a gilded carriage and then returned with the letter. Uncle called in niece and told her that, by order of her father, he was to give her in marriage, and Belfiore pretended to be thunderstruck. The marriage was celebrated, and when the king learned of it, he was so mad he died.

(Abruzzo)

❖ 113 ❖

The Three Blind Queens

There were three sons of a king, but the king was dead, like the queen. The nursemaid was the one who ran the house. The three king's sons wanted to get married, and had three portraits of three girls they liked.

They said to the ambassadors, "Go all over the world. If you find three girls like the portraits, bring them back for us to marry." The ambassadors searched the world over without finding anyone. Finally they saw a fisherman's three daughters, who alone looked like the portraits. The ambassadors had them dressed as queens and presented to the three sons of the king. They liked one another and got married.

War broke out. The three noble sons departed, leaving the nursemaid in charge of the house. But the nursemaid, with those three queens at home, couldn't have her own way as in the past. So she told a minister to kill them and, as proof, to bring her back three pairs of eyes. The minister said to the queens, "The weather is truly delightful today, let's go for a ride." They got into a carriage, and the carriage kept right on going until it came to the foot of a mountain. The three queens got out, and then the minister. The minister drew a sword and sighed. "I have the honor to announce I shall kill you and carry your eyes back to the nursemaid." The three queens replied, "No, listen, don't kill us. Leave us here on the mountain. As for the eyes, we will give them to you ourselves."

They gouged out their eyes and gave them to the minister, who wept. When the three royal sons returned and asked about the three queens, the nursemaid said they had died in an accident. The three widowers swore they would never take any other wives.

Inside a cave, the three blind queens lived on herbs and roots. They were each expecting a baby, and one night, each of them gave birth at the same time to a fine baby boy. The mothers and their babes lived on herbs and roots. When the herbs and roots ran out, to keep from starving to death the mothers drew lots to see whose baby they would eat. It fell to the oldest, so her baby was eaten. Next it was the turn of the second sister's baby. The youngest sister, whose turn had now come, picked up her baby and groped her way hastily out of the cave.

She found another cave and a spot with many herbs. So the baby grew to be a big boy, went hunting with a gun made from a reed, and brought his mother back something to eat. Then he found the other two blind women and took them to his mother's cave.

One of the royal sons, who happened to be his father, was out hunting one day and met the boy in a forest. "Come with me," said the king's son.

"I'll go and tell my mother," replied the boy. His mother said all right, and the boy left her.

The nursemaid pretended to be glad to see him, but behind his back she scowled. When it came to handling weapons, the boy was the best and bravest man in the whole kingdom. The nursemaid decided to impose on him a task that would get him out of the way for good. A long time ago a princess of the family had been kidnapped by the fairies. The nursemaid said to the three royal sons, "This youth could go and find the girl." At that, the three sons instructed him to go and look for her.

First the young man went to the blind women's cave for advice, after which he set out. In a desert stood a black and white palace. He drew near, and a plaintive voice called to him, "Do you see where I am? Do you see? Turn around!" The youth replied, "No, if I turn around, I'll change into a tree." He entered the black and white palace. Three yellow candles were burning in one of the rooms. The young man blew them out in one puff. That broke the evil spell, and he suddenly found himself back at the palace of the three royal sons, together with the very lovely princess and his mother and aunts, who had all recovered their eyes. The youth married the princess. At dinner everyone told a story. The three queens told theirs, and the nursemaid got to shivering so much that, to warm her up, they coated her with pitch and roasted her alive.

(Abruzzo)

◈ 114 ◈

Hunchback Wryneck Hobbler

A king was out strolling. He looked at the people, the swallows, the houses, and was content. A little old woman passed, minding her own business. She was a very well bred old soul, but she limped a little, and was also a trifle hunchbacked and, in addition, had a wryneck. The king stared at her and said, "Hunchback wryneck hobbler! Ha, ha, ha!" And he laughed heartily in her face.

Now this old woman was a fairy. She looked the king in the eye and said, "Go on, laugh your fill. We'll just see who's laughing tomorrow."

At that, the king went into another peal of laughter. "Ha, ha, ha!"

This king had three daughters who were beautiful girls indeed. The next day he called them to go out walking with him. The oldest girl showed up with a hump on her back. "A hump?" asked the king. "How on earth did you get that?"

"Well," explained the daughter, "the maid made up my bed so badly that I got a big hump last night."

The king began pacing the floor; he felt uneasy.

He sent for his second daughter, who showed up with a wryneck. "What's the meaning of coming in now with a wryneck?" asked the king.

"Here's what happened," replied the second daughter. "While the maid was combing my hair, she pulled out a hair . . . and here I am now with a wryneck."

"And this girl?" said the king, noticing his youngest daughter limp into the room. "Just why is she limping now?"

"I went out into the garden," explained the third daughter, "and the

409

maid picked a jasmine blossom and flung it at me. It fell on my foot and lamed me."

"But who is this maid?" screamed the king. "Have her come before me at once!"

The maid was called and had to be dragged before the king by the guards because, in her words, she was ashamed to be seen: she was hunchbacked, wrynecked, and hobbling—the very same old woman as the day before! The king recognized her instantly and yelled, "Coat her with pitch and burn her to death!"

The old woman shrank and shrank until her head was the size of a nail and just as pointed. There was a tiny hole in the wall, and she squeezed through it and disappeared from sight, leaving behind only her hump, wryneck, and lame foot.

(Abruzzo)

◈ 115 ◈

One-Eye

There were two friars who went out begging. Darkness fell on the mountains. From a cave shone a little light.

"Lord of the house," they called, "will you give us shelter for the night?"

"Come in," thundered a voice that echoed on the mountain.

The friars entered, and there before the fire was a giant with one eye in his forehead, who said, "Welcome, you will be comfortable here."

He stepped behind the friars, who were shaking like leaves, and closed up the entrance with a boulder that one hundred persons all together couldn't have budged.

"I have one hundred sheep," said One-Eye, "but the year is long and I must save them. So which one of you two should I eat first, Little Friar or Big Friar? You decide, by drawing lots."

Lots were drawn, and it fell to Big Friar. One-Eye ran him onto a spit, and put him over the fire to roast. As he turned the spit, he sang, "Fatty tonight, Shorty tomorrow, Fatty tonight, Shorty tomorrow!"

Little Friar was torn between grief over losing his companion and eagerness to avoid a like fate. When Big Friar was done, One-Eye began eating him and also gave Little Friar a leg to sample. Little Friar pretended to eat, but threw the meat over his shoulder.

Once he'd picked Big Friar's bones clean, One-Eye threw himself down on the straw to sleep. Little Friar curled up by the fire and also pretended to go to sleep. When he heard One-Eye snoring like a pig rooting in the earth, he took the spit, heated the point to a glow, and—zing!—thrust it into the giant's single eye.

The blinded giant jumped to his feet howling and waving his hands in all directions in an effort to grab Little Friar. But Little Friar darted into the flock of sheep. One-Eye began feeling the sheep, one by one, but Little Friar got out of his way every time. Then the giant said, "Just wait until daylight!"

At that, Little Friar quietly took the ram, skinned it, and wrapped himself up in the fleece. When it was day, One-Eye lifted the boulder away from the mouth of the cave and planted himself at the entrance, one leg on one side, the other on the other, so as to be able to feel everything that came out and to let the sheep pass, but not Little Friar. He called the ram first of all, and Little Friar came forward on all fours ringing the bell around his neck. One-Eye stroked him on the back and said, "Go on." Then he felt the sheep as they went out one by one. So Little Friar was free and took to his heels, overjoyed to be out of the cave.

But once all the sheep were out, One-Eye began feeling around the cave, and his hands came upon the flayed ram. He then realized that the ram he had felt a few minutes ago was none other than Little Friar in disguise, and he ran out of the cave after him. He groped his way along, sniffing the air and, smelling the friar near at hand, he yelled, "Little Friar, you gave me the slip! You're smarter than I am! Here's a ring for a souvenir of your victory!" He threw him a ring, which Little Friar caught and put on. But it was a magic ring: once it was on his finger, Little Friar tried to run away from One-Eye but ran to him instead. The harder he tried to flee, the closer he came to the giant. He tried to remove the ring, but it would no longer come off. When he was almost in the giant's reach, he cut off the finger with the ring and flung it in the giant's face: right away he was free and able to flee.

One-Eye opened his mouth and swallowed Little Friar's finger, saying, "I at least got a taste of you!"

(*Abruzzo*)

116

The False Grandmother

A mother had to sift flour, and told her little girl to go to her grandmother's and borrow the sifter. The child packed a snack—ring-shaped cakes and bread with oil—and set out.

She came to the Jordan River.

"Jordan River, will you let me pass?"

"Yes, if you give me your ring-shaped cakes."

The Jordan River had a weakness for ring-shaped cakes, which he enjoyed twirling in his whirlpools.

The child tossed the ring-shaped cakes into the river, and the river lowered its waters and let her through.

The little girl came to the Rake Gate.

"Rake Gate, will you let me pass?"

"Yes, if you give me your bread with oil."

The Rake Gate had a weakness for bread with oil, since her hinges were rusty, and bread with oil oiled them for her.

The little girl gave the gate her bread with oil, and the gate opened and let her through.

She reached her grandmother's house, but the door was shut tight.

"Grandmother, Grandmother, come let me in."

"I'm in bed sick. Come through the window."

"I can't make it."

"Come through the cat door."

"I can't squeeze through."

"Well, wait a minute," she said, and lowered a rope, by which she pulled the little girl up through the window. The room was dark. In bed was the ogress, not the grandmother, for the ogress had gobbled up Grandmother all in one piece from head to toe, all except her teeth, which she had put on to stew in a small stew pan, and her ears, which she had put on to fry in a frying pan.

"Grandmother, Mamma wants the sifter."

"It's late now. I'll give it to you tomorrow. Come to bed."

"Grandmother, I'm hungry, I want my supper first."

"Eat the beans boiling in the boiler."

In the pot were the teeth. The child stirred them around and said, "Grandmother, they're too hard."

"Well, eat the fritters in the frying pan."

In the frying pan were the ears. The child felt them with the fork and said, "Grandmother, they're not crisp."

"Well, come to bed. You can eat tomorrow."

The little girl got into bed beside Grandmother. She felt one of her hands and said, "Why are your hands so hairy, Grandmother?"

"From wearing too many rings on my fingers."

She felt her chest. "Why is your chest so hairy, Grandmother?"

"From wearing too many necklaces around my neck."

She felt her hips. "Why are your hips so hairy, Grandmother?"

"Because I wore my corset too tight."

She felt her tail and reasoned that, hairy or not, Grandmother had never had a tail. That had to be the ogress and nobody else. So she said, "Grandmother, I can't go to sleep unless I first go and take care of a little business."

Grandmother replied, "Go do it in the barn below. I'll let you down through the trapdoor and then draw you back up."

She tied a rope around her and lowered her into the barn. The minute the little girl was down she untied the rope and in her place attached a nanny goat. "Are you through?" asked Grandmother.

"Just a minute." She finished tying the rope around the nanny goat. "There, I've finished. Pull me back up."

The ogress pulled and pulled, and the little girl began yelling, "Hairy ogress! Hairy ogress!" She threw open the barn and fled. The ogress kept pulling, and up came the nanny goat. She jumped out of bed and ran after the little girl.

When the child reached the Rake Gate, the ogress yelled from a distance; "Rake Gate, don't let her pass!"

But the Rake Gate replied, "Of course I'll let her pass; she gave me her bread with oil."

When the child reached the Jordan River, the ogress shouted, "Jordan River, don't you let her pass!"

But the Jordan River answered, "Of course I'll let her pass; she gave me her ring-shaped cakes."

When the ogress tried to get through, the Jordan River did not lower his waters, and the ogress was swept away in the current. From the bank the little girl made faces at her.

(Abruzzo)

117

Frankie-Boy's Trade

A woman had an only child, Frankie-Boy, and was anxious for him to learn a trade. The son replied, "Find me a master, and I will learn the trade." So his mother found him a blacksmith as a master.

Frankie-Boy went to work at the smithy, where he accidentally brought down the hammer on his hand. Back home he went to his mother. "Mamma, find me another master, I'm not cut out to be a blacksmith."

Mamma looked around for another master, and this time found a cobbler. Frankie-Boy worked at the cobbler's and accidentally ran the awl through his hand. Home to his mother he went. "Mamma, find me another master; I'm not cut out to be a cobbler either."

His mother replied, "Son, I have only ten ducats left. If you learn the trade, well and good! If not, I don't know of anything else to do for you."

"If that's the case, Mamma," said Frankie-Boy, "you'd better give me the ten ducats and let me go out into the world and see if I can't learn a trade on my own."

His mother gave him the ten ducats, and Frankie-Boy set out. Along the way, in the heart of a forest, four robbers sprang out and cried, "Face to the ground!"

"Face to the ground?" repeated Frankie-Boy. "How do you mean?"

"Face to the ground!"

"Show me how I'm supposed to do," replied Frankie-Boy.

The robbers' ringleader thought to himself, This fellow is even more persistent than we are. What if we took him into our band? So he asked him, "Young man, would you like to be one of us?"

"What trade will you teach me?" said Frankie-Boy.

"Our own respectable trade," replied the ringleader, "We accost people, and if they refuse to hand over their money, we assassinate them. Then we feast, drink, and go for a stroll."

So Frankie-Boy began roving the highways with the band. A year later the leader died, and Frankie-Boy took his place. One day he ordered the whole band to go out and prowl while he stayed behind to guard their booty. An idea occurred to him. "I could take all this money here, load it onto a mule, and slip away without leaving a trace." And that's just what he did.

He reached his mother's house and knocked. "Mamma, I'm home, open up!" His mother opened the door and found herself face to face with her son, who held a mule by the halter. He immediately began unloading sacks of money.

"But what trade did you learn?"

"The *respectable trade*, Mamma, where one eats, drinks, and goes for a stroll."

His mother, who didn't know what he was talking about, concluded it must be a good trade and asked no more questions. Now his mother happened to be very chummy with the archpriest. The next morning she went to the priest and said, "Father, did you know your old friend was back?"

"He is? Has he learned a trade?"

"Yes, he has. He's learned the *respectable trade*. He eats, drinks, and goes for a stroll. And he's earned a muleload of money."

"He has, has he?" said the archpriest, who knew better. "Send him around to see me. I want to talk to him . . ."

Frankie-Boy went to see him. "Well, old friend," asked the priest, "have you really and truly learned a good trade?"

"Yes, indeed."

"If that is so, we must make a bet."

"What are we betting?"

"I have twelve shepherds and twenty dogs. If you can steal a ram from my flock, I'll give you one hundred ducats."

"My friend," replied Frankie-Boy, "if you have twelve shepherds and twenty dogs, how can I possibly succeed? I don't know what to say. Let us try all the same."

He dressed up as a monk and went to the shepherds. "O shepherds, hold your dogs, I'm only a poor man of God."

The shepherds tied up their dogs. "Come closer, father, come up to the fire and warm yourself along with us."

Frankie-Boy sat by the fire with the shepherds, drew out of his pocket a piece of bread, and began eating. Then he took a flask from his pack and pretended to drink—just pretending, mind you, for the wine was drugged. A shepherd spoke up. "That's the spirit, father! Eat and drink, offering no one a taste or sip!"

"Sir," replied Frankie-Boy, "one swallow of this is sufficient for me." And he offered him the flask. The shepherd drank, as did the others, and when the wine was all gone they felt very drowsy. In no time they were asleep. "It would happen that just when we wanted to talk to the monk a bit, all of you had to fall asleep!" said the only shepherd to stay awake. But the words were scarcely out of his mouth before he too yawned and went sound asleep.

When Frankie-Boy was sure they were all fast asleep he undressed them one by one and reclothed them in the garb of monks. Then he took the fattest ram and left. Back home, he killed and roasted it, then sent a leg to the archpriest.

When the shepherds awakened and saw themselves dressed as monks, they knew at once they had been robbed. "What will we now tell our master?" they wondered.

"You go and explain," said one. "No, you go," said the other. But no one was willing, so they decided to go to him in a body. They knocked. When the priest's servant saw them, she said, "Father, the porch is full of monks who wish to come inside!"

"I have to say Mass this morning," replied the priest. "Tell them to go away."

"Open up, open up!" cried the shepherds, who finally all burst into the house.

Seeing his shepherds dressed as monks, the archpriest knew it was none other than Frankie-Boy's doing, and muttered, "So he really did learn the trade!" He sent for him and gave him the hundred ducats.

"Now, my friend," said the priest, "let's play the return match. We will stake two hundred ducats this time. There's a church in the country which belongs to our parish. If you succeed in taking anything at all from that church, you win. I'm giving you eight days to do it."

"Very well," replied Frankie-Boy.

The archpriest sent for the hermit who stayed at the church and said, "Be on your guard; a man will attempt to steal something out of the church. Be on the lookout night and day."

"Have no fear, father! Just arm me well, and leave the rest to me."

Frankie-Boy let seven days and seven nights go by. On the last evening, he went up close to the church and hid by one of the corners. Now the hermit, poor thing, who'd not slept a wink for seven days and seven nights, came to the door and said to himself, "For seven nights he's not showed up. Tonight is the last one. Six o'clock has already struck, and he hasn't come. That's a sign he's afraid to. But who knows? I shall go and relieve myself and then get some sleep."

Out he went to the privy while Frankie-Boy, who had heard every word, darted into church like a cat and hid. The hermit returned, bolted the doors, then threw himself down, dead tired, on the church floor and fell asleep. At that, Frankie-Boy collected all the statues in church and put them around the hermit, at whose feet a sack was placed. Then he dressed up as a priest, went up the altar steps, and started preaching. "Hermit of this church, the time of thy salvation is nigh!"

The hermit slept on.

"Hermit of this church, the time of thy salvation is now!"

The hermit awakened and, seeing all those saints around him, said, "Lord, pray tell what I must do!"

"Get into the sack," replied Frankie-Boy, "since the time of thy salvation is now!"

The poor hermit squeezed into the sack. Frankie-Boy came down the altar steps, slung the sack over his shoulder, and off he went. He reached the archpriest's house and threw the sack into the middle of the room. "Ugh!" groaned the hermit inside the bag.

"Here you are, my friend. Just take a look at what I brought away from the church."

The archpriest opened the sack and found himself nose to nose with the hermit.

"Frankie-Boy, my friend," said the archpriest, "here are your two hundred ducats. I see you have really mastered the trade. Let us be friends, or you'll have me in the sack too."

(Abruzzo)

◈ 118 ◈

Shining Fish

There was a good old man whose sons had died, and he had no idea how he and his wife would now survive, for she too was old and ailing. Every day he went to the woods to gather firewood, and he would sell the bundle to buy bread and thus keep body and soul together.

One day as he was making his way through the woods and groaning, he met a gentleman with a long beard, who said, "I'm aware of all your troubles, and I will help you. Here is a purse containing a hundred ducats."

The old man took the purse and fainted. When he came to, the gentleman had disappeared. The old man went home and hid the hundred ducats under a heap of manure, without breathing a word to his wife. "If I gave her the money, it would be gone in no time . . ." Next day he returned to the woods, as usual.

That evening he found the table spread with a feast. "How did you manage to buy all this?" he asked, already alarmed.

"I sold the manure," said the wife.

"Wretch! Hidden in it were one hundred ducats!"

The next day the old man went through the woods sighing louder than ever. Again he met the gentleman with the long beard. "I am aware of your bad luck," said the gentleman. "Calm down. Here are one hundred ducats more."

This time the old man hid them in an ash pile. The next day his wife sold the ashes and fixed another hearty meal. When the old man came in and saw it, he couldn't eat a single bite and went off to bed tearing out his hair.

He was weeping in the woods the next morning, when back came the gentleman. "This time I shall give you no money. Take these twenty-four frogs out and sell them and with the proceeds buy yourself a fish—the biggest one to be had."

The old man sold the frogs and bought a fish. At night he realized that it gleamed; it put out an intense light that shone all around. Holding it was like carrying a lantern. In the evening he hung it outside his window to keep it fresh. It was a dark and stormy night. The fishermen out at sea couldn't find their way in over the waves. Seeing the light at the window, they rowed toward it and were saved. They gave the old man half their haul and made an agreement with him that if he hung up the fish at the window every night, they would always divide their night's catch with him. That they did, and the good old man knew no more hardship.

(Abruzzo)

◈ 119 ◈

Miss North Wind and Mr. Zephyr

Once upon a time Miss North Wind felt the urge to get married. She went to Mr. Zephyr and said, "Sir Zephyr, how would you like to be my husband?"

Mr. Zephyr was a fellow attached to money, and didn't care for women. So, without beating around the bush, he replied, "No, Lady North, because you haven't a penny to your name for a dowry."

Cut to the quick, Miss North Wind began blowing with all her might, without a minute's respite, at the risk of bursting her lungs. For three days and three nights straight, she blew; and for three days and three nights, it snowed up a storm. All the fields, hills, and villages were blanketed in snow.

When Miss North Wind had spread her silver everywhere, she said to Mr. Zephyr, "Here's my dowry that you said I didn't have! Will that do?" Then she went off to rest up from her labor of the last three days.

Mr. Zephyr didn't bat an eye. He shrugged his shoulders and then *he* started blowing. He blew for three days and three nights, and for three days and three nights, fields, hills, and villages sweltered under intense heat that melted away every single bit of the snow.

When Miss North Wind was thoroughly rested, she awakened and saw that her dowry was all gone. She ran to Mr. Zephyr, who mockingly asked, "Where did all your dowry go, Lady North? Do you still want me for your husband?"

Miss North Wind turned her back on him. "No, Sir Zephyr, I'd never want to be your wife, since you're capable of squandering my entire dowry in a day."

(Molise)

◆ 120 ◆

The Palace Mouse and the Garden Mouse

While he was gnawing on a cheese in the pantry, a mouse was given such a scare by the house cat that he ended up somehow or other out in the middle of the garden.

He hid under a head of lettuce and began thinking. After much thought, he remembered that his father, God rest his soul, had once told him about a fieldmouse friend of his who lived in the garden under a fig tree. So he went round and round, found the burrow, and entered it.

His father's friend had also died, but his son was there. They introduced themselves, and the fieldmouse was so hospitable that, for two days, the palace mouse forgot all about pantry, cheese, and cat.

But by the third day he'd had his fill of turnips and hated the mere smell of them, so he said, "My friend, I must not impose on you any longer."

"Why must you leave so soon, my friend? Stay at least one more day."

"No, my friend, they are waiting for me back home."

"Who is waiting for you?"

"An uncle . . . Listen, I have an idea. Walk me home. We'll have lunch together, and you'll come back here."

The field mouse, who was dying to see the house of a palace mouse, accepted, and they headed for the palace.

Once they were out of the garden, they climbed up a trellis and went through the little window of the pantry.

"What a charming house!" exclaimed the field mouse. "And what a delightful smell!"

"Go on down, my friend, don't be bashful. Make yourself at home."

"No, thank you, my friend. I'm inexperienced and might not be able to find my way back. I'd better stay here on the windowsill . . ."

"Well, wait a minute," said the palace mouse, and went down into the pantry by himself.

As he made his way to a piece of bacon, the cat lurking nearby jumped out and grabbed him.

"Eeeeeeeek! Eeeeeeeeeeek!" squealed the poor little victim.

The field mouse's heart pounded, and he thought to himself, What's he saying? Unkkkkkkk? Unkkkkkkkkkkk? So that's his uncle! A fine reception indeed! If that's how he receives his nephew, just imagine what he would do to me, a total stranger!

And in one bound he was back in the garden.

(Molise)

◆ 121 ◆

The Moor's Bones

A widower king with one son remarried and then died. The son remained with his stepmother, who paid him no mind whatever, since she was in love with a Moor and had eyes only for him. The king's son, faithful to his father's memory, began to detest the Moor. They went hunting together, and the prince killed and buried him in the heart of the forest.

When the Moor failed to return, the queen became worried and went out looking for him with her dog. Drawing near to the grave, the dog could smell the Moor and began barking and digging. He dug and dug until he came to the body. The queen finished uncovering it, and returned to the palace with the skull and the bones from the arms and legs. She had the skull made into a cup decorated with gold and precious stones. The leg bones went into a chair, and the arm bones became the frame of a mirror.

Then to get even with her son, she said to him, "You killed the Moor, so I'm condemning you to death. I will spare you only if, in three months' time, you can explain the meaning of this riddle:

'I drink Moor, I sit Moor,
I look up and see Moor.' "

The youth went out into the world to find the answer. He asked everyone he met, but no one could solve the riddle. When the time was up but for one day, he stopped at a haystack that housed a father, a mother, and a daughter. He asked for something to eat, and the father and mother replied, "We have nothing; we are so poor that we live in a haystack."

"We have only one hen," said the daughter. "Let's wring its neck and feed our visitor."

The father and mother didn't like the idea of killing their only hen, but the daughter said, "Let's wring its neck; this is certainly a king's son!"

She cooked the hen, put it on the table, and invited the king's son to carve it. He served the father the legs, the mother the breast, the daughter the wings, and kept the head for himself.

At night he was given a bed on the haystack. He and the father slept on one side, mother and daughter on the other. In the night he woke up and heard the daughter saying to her mother, "Did you notice how the king's son carved the hen? He gave Papa the legs, since he goes out and gets food for us. He gave you the breast, since you are the mother and nursed me as a baby. He gave me the wings, since I am beautiful like an angel of paradise. And he ate the head himself, since he will be his subjects' head."

Hearing that, the king's son thought, I'm sure this girl would understand my mother's riddle. And when it was day, he asked her.

"That's easy," she replied. " 'I drink Moor' refers to the queen's drinking cup. 'I sit Moor' refers to her chair. 'I look up and see Moor' refers to her mirror."

The youth left her a purse of gold coins, and promised to return and marry her. He went back to his stepmother, but instead of giving her the answer, said, "I didn't find the solution; I am ready to die."

The stepmother had the gallows erected at once.

The whole town gathered in the square around the youth, who already had his neck in the noose, and cried, "Spare him! Spare him!"

"To be spared," answered the queen, "he must explain the riddle."

"Well, for the last time," said the judge to the king's son, "can you explain the meaning of 'I drink Moor'?"

Only then did the youth say, "Yes, it means the queen had herself a cup made out of the Moor's skull."

"And 'I sit Moor,' what does that mean?" asked the judge.

"It means that the chair where the queen sits is made out of the Moor's leg bones."

"And 'I look up and see Moor'?"

"That means that the queen's mirror is framed by the Moor's arm bones."

So the judge also went to see the mirror frame.

The king's son then said, "And the riddle in its entirety means that the queen must hang for thinking of the Moor, both living and dead, and forgetting my late father."

At that, the judge condemned the queen to death.

The king's son returned to the haystack and wed the wise maiden.

(Benevento)

◈ 122 ◈

The Chicken Laundress

There was once a washerwoman who had no children. One day while she was hanging out clothes, she saw a mother hen with seven chicks running along behind her. "Holy Mother," she said, "even if you helped me have a hen for a daughter, I would be happy."

Thus she actually gave birth to a chicken. The washerwoman was happy, and loved her, and before long, this daughter became a big hen, the likes of which had never been seen.

One day the hen went about the house saying, "Co, co, co, give me the clothes and I'll go and wash them!" And she sang that song the whole day long.

The washerwoman at first turned a deaf ear to her, then lost patience and threw her an old rag. The hen took it in her beak and began flapping her wings, continuing to flutter until she reached a deserted terrain. There she put the cloth down on the ground, and in its place rose a palace. The hen climbed the palace steps, walked through the front door and, in that moment, turned into a beautiful young lady.

From the palace came numerous fairies, who dressed her like a queen and prepared a fine feast for her. After eating she went out on her balcony for a while. The king's son, who was hunting in the vicinity, saw her and fell in love with her. He hid nearby and waited for her to come out. He finally saw her emerge and turn into a hen.

The hen gave the palace one peck, the palace turned back into a rag and, holding the rag in her beak, the hen flew off. The king's son ran after her.

"How much will you take for this hen?" he asked the washerwoman.

"I wouldn't sell her for all the gold in the world!" said the poor woman.

But the king's son went on so, that the washerwoman was unable to say no and therefore parted with her hen daughter.

The king's son carried her to his palace and made her a nest in a basket beside his bed. In the evening he went off to dance. The hen waited until he was gone, then shook her feathers, turned back into a young lady, and ran off to the ball herself.

When she entered the ballroom, the king's son recognized her and hurried away at once; he ran home, looked in the basket and, seeing the chicken feathers, threw them into the fire. Then he returned to the ball and danced with the young lady, pretending he had not recognized her.

He went home late, but the hen wasn't there. The king's son went to bed and pretended to go to sleep. Then, in stole the young lady on tiptoe and, thinking no one saw her, went to don her chicken feathers again. She approached the basket, but the feathers were gone. Terrified, she was glancing about her, when the king's son rose and took her in his arms, saying, "You will be my bride!"

(Irpinia)

123

Crack, Crook, and Hook

Once upon a time there were three rogues—Crack, Crook, and Hook. They made a bet to see who was the craftiest of the three. They set out walking. Crack walked ahead of the others and saw a magpie sitting on her nest in a treetop. He said, "Do you want to see me take the eggs out from under that magpie without her noticing it?"

"Yes, let's see you do that!"

Crack climbed the tree to steal the eggs and, while he was taking them, Crook cut the heels from his shoes and hid them in his hat. But before he'd put his hat back on his head, Hook had filched them from him. Crack came down the tree and said, "I am the craftiest rogue, since I stole the eggs from under the magpie."

"No, I'm the craftiest," said Crook, "since I cut off the soles from under your shoes without your noticing it." And he removed his hat to show him the heels, but they were gone.

Then Hook spoke up. "I'm the craftiest; I stole the heels from out of your hat. And since I am the craftiest, I intend to separate from you two. I'll do far better by myself."

He went his own way, accumulating so much that he became quite rich. He changed cities, got married, and opened up a pork-butcher's shop. The other two, in the course of roaming and thieving, came to this city and saw the shop. "Let's go inside," they said to one another. "It might be very worth our while!"

They went in and found only the wife there. "Fine lady, will you give us something to eat?"

"What do you want?"

"A slice of cheese."

While she was cutting the cheese, the two glanced all around to see what there was to snitch. They spied a quartered pig hanging up and signaled each other they'd fetch it at night. Hook's wife noticed, but said nothing. When her husband came home, she told him everything. Master thief that he was, he caught on right away. "That must be Crack and Crook, who mean to steal the pig. Fine! Just you wait!" He took the pig and put it in the oven. In the evening he went to bed. When it was night, Crack and Crook came to steal the pig. They looked everywhere, but couldn't find it. So what did Crook decide to do but steal up to the bed, to the side on which Hook's wife was sleeping, and say, "Listen, I don't see the pig any more. Where did you put it?"

Thinking it was her husband, the wife answered, "Go back to sleep! Don't you remember putting it in the oven?" Then she went back to sleep.

The two rogues went to the oven, removed the pig, and left. Crook went out first, then Crack with the pig on his back. Passing through the pork-butcher's garden, Crack noticed some soup herbs growing there. He caught up with Crook and said, "Go back to Hook's garden and pick us a few herbs which we'll boil together with a pig leg when we get home."

Crook went back to the garden, while Crack continued on his way.

Meanwhile Hook woke up, went to look in the oven and, finding the pig gone, glanced into the garden and saw Crook picking herbs for soup. "Now I'll let him have it!" he said to himself. He picked up a thick bunch of herbs he had in the house and ran outside without letting Crook see him.

He caught up with Crack, who walked bent over under the weight of the pig on his back, and signaled he would carry the pig a while. Thinking it was Crook returning with the herbs, Crack took the bunch and passed him the pig. Once the pig was on his back, Hook turned around and ran back home.

In a little while Crook caught up with Crack carrying the herbs and asked, "What did you do with the pig?"

"You have it!"

"I do? I have nothing at all!"

"But you changed with me and gave me the herbs to carry just a minute ago."

"When did I do that? You sent me off to see about the soup!"

They finally realized they had been outwitted by Hook, truly the craftiest rogue of them all.

(Irpinia)

124

First Sword and Last Broom

Once there were two merchants who lived directly opposite one another. One had seven sons, the other seven daughters. Every morning the one with the seven sons would throw open his window and greet the one with the seven daughters, saying, "Good morning, merchant with the seven brooms." And the other one never failed to take offense; he withdrew into his house and wept for anger. To see him in such a state upset his wife, who would ask him every time what the matter was, but the husband never answered and went on weeping.

The youngest of the seven daughters was seventeen, lovely as a picture, and her father's pride and joy. "If you love me as much as you claim, dear Father," she said one day, "tell me your trouble."

"Dear daughter, the merchant across from us greets me every morning with 'Good morning, merchant with the seven brooms,' and every morning I stand there with no idea how to answer him."

"Oh, Papa, is that all that's bothering you?" replied the daughter. "Listen to me. When he says that to you, you answer back, 'Good morning, merchant with the seven swords. Let's make a bet: let's take my last broom and your first sword and see which one gets the scepter and crown of the king of France first and brings them back here. If my daughter wins, you will give me all your goods; and if your son wins, I lose all my goods.' That's what you must tell him. And if he agrees, make him sign a written contract at once that spells out the terms."

Open-mouthed, the father listened to this speech from beginning to end. When it was over, he said, "But, daughter, do you realize what you're advising? Do you want me to lose everything I own?"

"Papa, have no fear, leave it all to me. Just make the bet, and I'll see to the rest."

That night the father couldn't sleep a wink and waited impatiently for day to dawn. He appeared on his balcony earlier than usual, and the window across the street was still closed. It opened all of a sudden, revealing the father of the seven sons, who came out with his usual "Good morning, merchant with the seven brooms!"

The other merchant was all ready for him. "Good morning, merchant with the seven swords. Let's make a bet: I'll take my last broom, and you your first sword; we'll supply them with a horse and a purse of money apiece, and then just see which one makes it back with the crown and scepter of the king of France. We'll stake all our wares; if my daughter wins, all your goods will be mine; if your son wins, all my goods will be yours."

The other merchant stared at him a moment, then burst out laughing and shook his head as if to say the father of the daughters was crazy.

"What, you're afraid? You have no confidence?" said the father with the seven daughters.

Cut to the quick, the other man replied, "For my part, I agree; let's sign the contract and send them off." And he went to tell his oldest son everything immediately. Thinking he would be traveling in the company of that beautiful daughter, the oldest son was all smiles. But when it was time to leave and she came out dressed as a man and seated in the saddle on a white filly, he realized this was no laughing matter. In fact, once their parents had signed the contract and said "Ready, set, GO!" the filly took off at full speed, and his own sturdy horse had a hard time indeed following.

To reach France, it was necessary to cross a dense, dark, and pathless forest. The filly sprang right through it, as though on home ground, winding to the right of an oak, to the left of a pine, leaping a holly hedge, and constantly advancing. In contrast, the merchant's son was at a loss to steer his sturdy horse: first he rammed his chin into a low tree branch and fell from the saddle, then the horse's hoofs sank into a mire concealed beneath dead leaves, and the animal landed flat on its belly. Next they got all tangled up in a briar patch and couldn't for the life of them get free. The girl with her filly had already made it through the forest and was galloping miles ahead.

To reach France, it was necessary to go over a mountain full of crags and gorges. She had come to its slopes, when she heard the hoofbeats of the sturdy horse of the merchant's son behind her. The filly galloped straight up the mountain, as though on home ground, winding her way around the boulders by leaps and bounds and continuing right on to the top, whence she descended to the flatlands. But the youth maneuvered his horse upward by jerking on the reins and, in no time, a landslide carried him back to the bottom and left him crippled.

The girl was now far ahead on the road to France. But to reach France it was necessary to cross a river. As though on home ground, the filly knew just where a ford was and jumped into the water, galloping through it as fast as on a beaten track. When they emerged from the water onto the other bank, they looked around and saw the youth approaching the river and spurring his horse into it after her. But he didn't know where the ford was and, when the horse's hoofs no longer touched solid ground, the current swept away both rider and steed.

In Paris and dressed as a man, the girl went to a merchant who hired her as his helper. He was a supplier to the royal palace and began sending goods to the king by the youth of such handsome appearance. When the king saw the merchant's helper, he asked, "Who are you? You look like a foreigner to me. What brought you here?"

"Majesty," replied the helper, "my name is Temperino—Penknife—and I was carver to the king of Naples. A series of mishaps has brought me here."

"What if I found you a position as carver in the royal house of France?" asked the king. "Would you like that?"

"Majesty, it would be a godsend."

"Very well, I'll speak to your master."

Reluctantly, the merchant released his helper to the king, who arranged for the youth to become carver. But the more the king looked at him, the more certain he was of something, and finally he told his mother about it one day.

"Mamma, there's something about this Temperino that's puzzling. He has beautiful hands, a slim waist; and he plays and sings, reads and writes. Temperino is the girl I've lost my heart to!"

"My son, you have lost your mind," replied the queen mother.

"I assure you, Mamma, Temperino is a girl. How can I prove it?"

"Here's the way," said the queen mother. "Take him hunting. If he hunts only quail, then Temperino is a girl with a mind only for roast bird. If he hunts goldfinches, then he's a man who delights in the chase."

So the king gave Temperino a gun and took him hunting. Temperino mounted the filly, which he had insisted on bringing along. To trick him, the king shot only quail. But every time a quail appeared, the filly turned away, and Temperino realized he was not supposed to shoot quail. "Majesty," Temperino then said, "may I be so bold as to ask if you think shooting quail is a test of skill? You already have enough to roast. Shoot some goldfinches as well, which is more difficult."

When the king got home, he said to his mother, "True, he went for goldfinches rather than quail, but I'm still not convinced. He has beautiful hands and a slim waist, plays and sings, reads and writes. Temperino is the girl I've lost my heart to!"

"My son, put him to another test," answered the queen. "Take him to the garden to pick salad. If he carefully picks just the tips, then Temperino is a girl, since we women are more patient than men. If he pulls up the whole plant, roots and all, then he is a man."

The king went into the garden with Temperino and began plucking salad, taking only the tips of the plants. The carver was about to do the same thing, when the filly, who had come along too, began pulling up plants by the roots; Temperino understood he was to do that. He hurriedly filled a basket with uprooted salad plants, to which the dirt still clung.

The king took the carver past the flowerbeds. "See the beautiful roses, Temperino?" he said. But the filly directed her muzzle at another flowerbed.

"Roses stick your hands," said Temperino. "Pick yourself some carnations and jasmines, not roses."

The king was disappointed, but he did not give up hope. "She has beautiful hands and a slim waist," he repeated to his mother. "She sings and plays, reads and writes. Temperino is the girl I've lost my heart to."

"At this stage, my son, the only thing left for you to do is take her swimming with you."

So the king said to Temperino, "Come along, let's go swimming in the river."

At the river, Temperino said, "Majesty, you get undressed first." The king undressed and slipped into the water.

"Now you come in too!" he said to Temperino.

At that instant a great neighing was heard, and the filly came galloping up excited and foaming at the mouth. "My filly!" cried Temperino. "Wait, Majesty, I must go after my excited filly!" And she ran off.

She ran to the royal palace and said to the queen, "Majesty, the king is in the river without his clothes and some guards, not recognizing him, want to seize him. He sent me to fetch his scepter and his crown to identify him."

The queen picked up scepter and crown and handed them to Temperino. When she had them, Temperino got on the filly and galloped away, singing:

> "As a maiden I came, as a maiden I return,
> So the scepter and the crown do I earn."

She crossed river, mountain, forest, and arrived home, and her father won the bet.

(Naples)

125

Mrs. Fox and Mr. Wolf

There was once a wolf and a fox who called each other brother and sister, and made a pact to share everything they were each lucky enough to catch.

The wolf, going about sniffing the air, caught a whiff of sheep and said to the fox, "Sister, I'm going to take a look in these pastures to see if a flock is grazing there."

He went, and landed right in the middle of a flock. He'd no sooner sunk his teeth into a lamb than he had to flee for his life, carrying the animal in his mouth. But he wasn't quick enough, and received a thrashing that put him in bed for a week.

"Since it cost me so many blows," reasoned the wolf, "I shall keep this lamb all for myself." He hung it up inside the fireplace hood and said nothing about it to the fox.

"How about those sheep? Did you catch them?" asked the fox.

"Sister, it's dangerous to go after them. Leave them alone, that's my advice."

The fox, who didn't believe him, said to herself, "I'll fix you now!"

She had discovered a hiding place full of honey, which smugglers had buried. "Brother," she said to the wolf, "I found a place full of honey, something too good to be true! One of these days we'll go and see it!"

She departed, instead, by herself, without a word to the wolf, found the honey, tasted it, and licked her lips. "Ah, what a delicious thing!"

Still aching from those blows, the wolf would ask her every time he met her, "Sister, when are we going to see that honey?"

"Oh! What do you expect from me, brother? I traveled quite a distance!"

"But, sister, where did you go to be away so long?"

"Brother," replied the fox, "I was in a town called Taste-It."

The next day the wolf had finished eating the lamb, and asked the fox, "Well, sister, shall we go?"

"Oh, dear, brother, it's so far away!"

"But you were gone a long time . . . Where did you go?"

"Brother, I'm exhausted. Just imagine, I went to a town called Pilfer-It."

The poor wolf returned the day after. "Shall we go take a look, sister?"

And the fox finally said, "Tomorrow we will go."

But no sooner had she left the wolf than she departed alone. She went straight to the hiding place and ate the rest of the honey. She was licking the bottom of the pot, when the smugglers arrived, but the fox ran away as fast as her legs would carry her.

The next day they set out, she and the wolf. "Brother, we have to go to a town quite far from here. If you want to come along, follow me. It's a town called Finish-It!" Still limping from all those blows, the wolf followed as best he could.

When they reached the top of a hill, the fox said, "Here we are in Finish-It. You go on ahead while I stay behind and watch, so no smugglers will come up and beat us."

The poor wolf went, but the smugglers who had discovered their honey gone, were also keeping watch. The wolf got there, but all he found were potsherds smeared with honey. Hungry as he was, he began licking the potsherds, when all of a sudden the smugglers pounced on him and beat him black and blue.

From her lookout, the fox feasted her eyes on the dancing wolf. When he finally managed to flee and come back to her, groaning every step of the way, she said, "Goodness, brother, what happened?"

"Sister!" he moaned, "can't you see they've beaten me to death? Let's run away fast if we don't want to catch any more!"

"Run away? How can I ever, since I've turned my ankle? No, I can't run!"

So with the wolf all beat up and impatient to flee, and the fox pretending to limp, they headed home.

"Oh, brother," groaned the fox, "how will I ever make it with this ankle? Carry me some way on your back."

The wolf had no choice but to take her on his back. And thus they moved along, the hale fox astride the half-dead wolf, while she sang:

> "Look, look, get a kick,
> The dead one bears the quick!"

"Why are you singing that, sister?" asked the wolf.

"Why, brother, they are the words to the song I'm singing to cheer you along the way."

They got home. So bruised was the wolf from all the blows, and so exhausted from lugging the fox on his back, that he fell lifeless to the ground and never revived. And that was how the fox got even with him for eating the lamb all by himself.

> Cock-a-doodle-doo,
> The wolf has left you.

(Naples)

126

The Five Scapegraces

In Maglie there was a mother and father who had one son, and this son was a devil if there ever was one. He was always pawning something or other, or else selling it outright. He stayed out all night and, in short, was a hard cross for the two old people to bear. One evening his mother said, "Husband, that boy will be the death of us. Let's make whatever sacrifice necessary and send him away from home."

The next day his father bought him a horse, and borrowed one hundred ducats to give him. When the son came in at noon, his father said, "My son, you can't go on like this. Here are one hundred ducats and a horse. Get out and start earning your own living."

"Very well," replied the son, "I'll go to Naples." He set out, riding this way and that and, in the middle of a field, spotted a man on all fours. "Handsome youth," called the boy from Maglie, "what are you doing there? What is your name?"

"Lightning."

"And your last name?"

"Streak."

"Why that name?"

"Because my specialty is chasing hares." He'd no sooner spoke than one darted by. In four bounds, he caught it.

"Not bad! I have an idea," said the boy from Maglie. "Come along with me to Naples. I have a hundred ducats." Lightning didn't have to be begged, and the two of them departed, one on horseback and the other on foot.

Soon they met another. "And what is *your* name?"

"Blindstraight."

"What kind of a name is that?" The words weren't out of his mouth before a flock of crows flew overhead, pursued by a falcon. "Let's see what you can do."

"I shall put out the left eye of the falcon and bring him down." With that he drew his bow, and the bird dropped to the ground with an arrow in his left eye.

"What do you say, friend, to coming along with us?"

"Certainly I'll come. Let's be off."

They reached Brindisi. In port a hundred stevedores were working, but there was one in particular who bore a heavier burden than a mule, as though it were nothing at all.

432

"Look at that!" exclaimed all three travelers. "Let's ask him his name."

"What's your name?" asked the youth from Maglie.

"Strongback."

"Well, guess what: we want you to come along with us. I have one hundred ducats and enough to eat for us all. When I run out of food and money, then you will all provide for me."

Imagine the dismay of the other stevedores over the departure of Strongback, who was such a help to them all! They began crying, "We'll give you another four pence, we'll give you another four pence, if you stay with us!"

"No, no!" said Strongback. "Leisure is better—eating, drinking, and going for a stroll."

All four of them moved onward, stopped off at a tavern where they ate like pigs and drank all the wine they could hold. Then they were again on their way. They'd not gone five or six miles before they ran into a youth with his ear to the ground.

"What are you doing down there? What's your name?"

"Rabbitears," he replied. "I hear all the conversations in the world, be they kings', ministers', or lovers'."

"Let's see if you're telling the truth," said the youth from Maglie. "Cock your ears and listen to what they're saying in Maglie, in that house in front of the column."

"Just a minute," he replied. He put his ear to the ground. "I hear two old people talking by the fireside, and the old woman says to the old man: 'Thank God you went into debt, husband. It was worth it to get that devil out of our house and have a little peace at last.' "

"You've not made that up," said the youth from Maglie. "Only my mother and father could say those things."

They resumed their journey and came to a place where many bricklayers were working and sweating under a hot morning sun.

"How do you poor souls manage to work at this hour?"

"How do we manage it? We have somebody who cools us off." They looked and saw a youth fanning the workers with his breath. "Puffffffff. Pufffffffff."

"What's your name?" they asked him.

"Puffarello," he replied. "I can imitate all the winds. Foooooooooooo! That's the north wind. Pooooooooooooo! That's the southeast wind. Ffffffffff! That's the east wind." And he went on imitating winds, blowing with all his might. "If you order a hurricane, I can even produce a hurricane." He blew, and trees began crashing to the ground and rocks flying through the air with all the fury of the gods.

433

"That will do!" they told him, and he calmed down.

"Friend," said the youth from Maglie, "I have one hundred ducats. Will you come along with me?"

"Let's go," he answered. They made a rollicking band all together. Telling one tall tale after another, they came to Naples. The first thing they did was go and eat, naturally. Next they went to a barber, then dressed up and went for a stroll, to lord it over everybody. In three days' time, the hundred ducats were running low, and the youth from Maglie said, "Friends, the air of Naples doesn't suit me. Let's go off to Paris, which is better."

After a long distance they arrived in Paris. On the city gate was written:

> The man who defeats the king's daughter in a foot-race
> Will have her as his wife.
> But whoever loses, loses his life.

The youth from Maglie said, "Lightning, here's where you come into the picture." He went up to the royal palace and spoke to a steward. "Sir, I am traveling for my own pleasure. This morning as I entered the city I read the challenge issued by the king's daughter, and I want to try my luck."

"My son," replied the steward, "just between the two of us, she is a madwoman. She does not wish to get married, and is constantly thinking up all these tricks to send many, many fine men to their death. It grieves me to see you join them."

"Nonsense! Go and tell her to pick the day; I am ready any time."

Everything was set for Sunday. The youth from Maglie went to tell his companions. "Guess what! The big day is Sunday!" They went off to the inn to eat a hearty meal and plan what to do. Lightning Streak said, "You know what you should do? Send me to her Saturday evening with a note saying you have a fever and can't race, but that you're sending me to run in your place. If I win, she'll still marry you. If I lose, you're still under the obligation of going to your death."

That's what they did, and Sunday morning the people lined both sides of the street that had been swept free of every speck of dust. At the appointed hour, out came the princess dressed as a ballerina and took her place beside Lightning Streak. Everyone looked on, wide-eyed. The signal sounded, and the princess was off like a hare. But in four bounds Lightning Streak passed and left her one hundred feet behind. Just imagine the applause and cheers! Everyone shouted, "Hurrah, Italian youth! She's finally met her match, that madwoman! That will sober her!"

She went home quite long-faced, and the king said, "My daughter, such a contest was your idea, and now it is your turn to be angry, whatever good that will do you."

But let's leave the princess and turn to Lightning Streak. He went back to the inn and sat down to a feast with his companions. Right in the middle of it, Rabbitears said "Shhh!" and put his ear to the ground the way he always did. "We're in trouble. The princess says she won't have you for a husband at any cost. She says the race won't count, that another one must be run. She's now asking a sorceress to find a way to make you lose. And the sorceress tells her she'll cast a spell over a precious stone and have it set in a ring. The princess is to give you the ring before the race, and once you have it on your finger you'll no longer be able to remove it, and your legs will give way beneath you."

"That is where I come in," said Blindstraight. "Before the start of the race, hold out your hand, and I will shoot the stone out of the ring with an arrow. Then we'll see what our princess can do!"

"Wonderful! Wonderful!" they all shouted, and worried no more about it.

The next morning a note came to the sick youth from the princess congratulating him on his friend's skill; but if he didn't mind, she wanted to run another race next Sunday.

Sunday even more people lined the street than the first time. At the appointed hour, she came out with her legs bared like an acrobat's. She approached the Italian and offered him a ring. "Good youth, since you are the only one ever to defeat me in a race, I am presenting you with this ring as a remembrance from your friend's bride." She slipped the ring on his finger, and his legs started trembling and gave way beneath him. Blindstraight, who was looking straight at him, cried, "Hold out your hand!" Slowly and with great difficulty he stretched out his hand, and right at that moment the trumpet sounded. The princess had already run past him. Blindstraight drew his bow, the arrow knocked the ring off, and Lightning Streak in four bounds was right on the heels of the princess. He leaped over her as in a game of leapfrog, causing her to fall on her face, and ran on ahead.

But the real show was the people! Cheers went up and hats were tossed into the air. Rejoicing over the defeat of the haughty princess, they picked him up and carried him in triumph all over town on their shoulders.

When the five scapegraces were at last alone, they began hugging and slapping one another on the back. "We are rich!" said the youth from Maglie. "Tomorrow I'll be king, and I'd just like to see anyone try to turn you out of the royal palace! Tell me what you want me to name each of you."

"Chamberlain," replied one.

"Minister," said another.

"General," put in a third.

But Rabbitears motioned to them to be silent. "A message is coming through!" And he threw himself to the ground to listen. At the royal palace they were talking about offering a large sum of money as a settlement and refusing him the princess's hand.

"Here's where I come into the picture," announced Strongback. "I'll make them pay, down to their very souls."

The next morning, the youth from Maglie dressed up and went to the palace. Outside the throne room, he met a councilor. "My son, will you take advice from someone older than you? If you marry that madwoman, you're doing nothing but taking the devil into your home. Instead, ask for whatever sum you wish, and go in peace."

"Thank you for your advice," replied the youth, "but I don't like naming a round sum. Let's do it this way: I'll send a friend of mine to you, and you load onto his back all you can."

So Strongback showed up with fifty hundred-pound sacks and said, "My friend sent me here for you to load me down."

All the people at court looked at one another, certain that this young man was mad. "I'm serious," he said, "hurry up!" They entered the treasury and proceeded to fill one of the sacks. Twenty persons were then needed to lift it. When they finally got it on his back, they asked, "Will that do?"

"Are you joking?" he asked. "To me that's like a tiny straw."

They went on filling sacks and exhausted the pile of gold. Then they started on the pile of silver, and all their silver ended up on Strongback. Next they took up copper, and not even that sufficed. They crammed in all the candlesticks and crockery, and Strongback still did not stoop under the weight.

"How do you feel?" they asked.

"Shall we bet I can even take on the palace?"

His companions came along and saw a mountain advancing all by itself on two little feet, and they all left the city, in gay spirits.

They had gone five or six miles when Rabbitears, who bent over to listen from time to time, said, "Friends, at the royal palace, they are in council. Can you imagine what the councilor is saying? 'Majesty, is it possible that four good-for-nothings have left us stark naked, that we can't even buy a penny's worth of bread? They took everything we owned! Quick, let's send a regiment after them and blow them to bits!'"

"If that's the case," said the youth from Maglie, "we are done for. We got out of all the other difficulties, but now what can we do against shotguns?"

"Silly youth!" exclaimed Puffarello. "Have you forgotten that I can whip up a hurricane and knock every one of them down? You go on ahead, and I'll show you what I can do!"

Hoofbeats were heard just then. As soon as they came within range, Puffarello began blowing, gently at first—ff, ff—then stronger—ffffff!—blinding them with clouds of dust; then with all his might—ffffffffffffff-ffffffff!—and the soldiers fell beneath their horses, trees were uprooted, walls crumbled, cannons went whirling through the air!

When he was certain of having dashed them all to bits, Puffarello rejoined his companions and said, "The king of France was not expecting that! Let him remember it and tell his sons."

So they returned to Maglie by the grace of God, divided up the fortune, each taking four million, and whenever they were all together after that, they would say, "Down with the king of France and that mad daughter of his!"

(Terra d'Otranto)

◈ 127 ◈

Ari-Ari, Donkey, Donkey,
Money, Money!

There was once a mother and a son. The mother sent her son to a monk to be instructed in godly matters, but the boy was in no mood to learn a thing. The neighbors advised her to send him to the village school, where Schoolmaster Squall kept them hopping. Master Squall tried his best, but he couldn't even drum into the boy his A-B-C-'s, so he finally kicked him out of school. The boy went home jumping for joy. Seeing him back, his mother grabbed the broom and thrashed him. "Get out of my house, you rascal! Don't ever let me see you again!"

He left home and set out on the road. After some distance, he came to a garden with no wall around it. As he was hungry, he climbed a pear tree and started eating pears.

Right in the middle of his meal, he heard, "H'm, h'm! I smell human flesh around here!" Under the pear tree appeared Pappy Ogre, the owner of the garden, sniffing the air.

"I am indeed human flesh," said the boy in the pear tree. "I'm a poor lad kicked out of the house by his mother."

"Come down, then," said Pappy Ogre, "and I'll take you to my house."

He took him home, dressed him in other clothes, and let him stay there. "You will live with me now, and no one will beat you any more." Every morning Pappy went out to work and carried the boy along. That continued for two years. Then one day the boy was very long-faced.

"Why do you look so sad?" asked Pappy.

"I want to see my mamma. Goodness knows how many tears she's shed since I left."

"You're really worrying about your mamma? I'll let you go see her, then. I'll give you a donkey to take her as a present. When you get home, take him inside and say: "Ari-ari, donkey, donkey, money, money!" And the donkey will drop money from his rear end. But watch out along the way that nobody steals him from you!"

The boy departed with the donkey. After going half a mile, he said to himself, "I just want to see if this donkey really drops money." He looked about him to be sure no one was around, dismounted, and said, "Ari-ari, donkey, donkey, money, money!" The donkey went "Prrrr-rrrrrr!" raised its tail, and dropped numbers and numbers of coins.

Pappy Ogre, who had climbed up in the tower of his house to spy on the boy's movements, said, "There, he's gone and done it!"

The boy stuffed his pockets with coins and got back on the donkey. He came to an inn and asked for the best room in the house for his donkey. The innkeeper wanted to know why.

"Because my donkey drops money."

"What do you mean, he drops money?"

"You have only to say, 'Ari-ari, money, money!'"

"Oh, no, my boy," replied the innkeeper, "we'll put him in the stable and cover him up with a sack so he won't sweat. Don't worry, no one will touch him."

With all that money, the boy ordered his fill of food and drink, then went off to bed. The innkeeper went down into the stable, took away the boy's donkey, and left in its place one that looked just like it. The boy got up in the morning and asked, "You didn't say a word to my donkey, did you?"

"No, what should I have said to him?"

"All right, all right," he replied. He then climbed on the donkey and rode home to his mother. "Open up, Mamma, your Tony's home!"

"Merciful heavens! So you're finally back! I thought you'd fallen off the face of the earth!"

The son walked in. "How are you doing, Mamma?"

"I'm worn out! I washed a tubful of stuff and, for all my work, earned a few peas!"

"Is that so? You're eating this mess?" He picked up the pot and threw it out the door. Just imagine how the poor woman screamed and wailed when she saw her peas go sailing through the air!

"Don't cry, Mamma, I'll make you rich!" He pulled the blanket off the bed and spread it on the floor, then led in the donkey and said, "Ari-ari, money, money!"

Yes, he really expected the donkey to drop gold! "Ari-ari, money, money!" he continued to say, but nothing dropped. Then he grabbed a stick and—bam, bam, bam!—thrashed him so hard that the donkey at last let out everything he had inside him. When the mother saw the blanket full of manure, she jerked the stick out of his hand and began pounding him.

Long-faced, the son made his way back to Pappy Ogre's. When Pappy saw him, he said, "So you've come back, have you? Very well, you'll settle down now and not cry for Mamma any more."

A little time went by, and the boy began whining to go see his mamma. Pappy gave him a table napkin and said, "Don't do anything foolish. When you get to Mamma's, say, 'My table napkin, make ready the table!'"

The boy left. When he came to the place where he'd tested the donkey, he pulled out the napkin and said, "My table napkin, make ready the table!" Out came all kinds of good things—macaroni, meatballs, sausage, blood pudding, tasty wine.

"What a feast!" he sighed. "Now Mamma need weep no more over spilled peas!"

He ate his fill and more besides, then said, "My table napkin, clear the table!" and was on his way once more. He came to the same inn. The minute they saw him, they all asked, "Well, Tony, how is everything?"

"Fine. What's for dinner?"

"A few turnips and Neapolitan kidney beans, my son, since this is an inn for carters!"

"Pooh! I'm not eating that disgusting stuff. I'll now show you what a real meal is." He pulled out the napkin and said, "My table napkin, make ready the table!" Out came poached fish, baked fish, veal cutlet, wine, and all kinds of other good things. When he'd eaten his fill and more besides, he stuffed the napkin into his vest pocket and said, "I'd just like to see you make off with this the way you did with the donkey! Look where I'm putting it!" But right at that moment, from all he'd eaten and drunk, he fell fast asleep and had to be carried off to bed. They took the napkin away and left him one that looked just like it. The next morning he got up, saying, "So you didn't take this away from me!" Then he continued his journey.

He reached his mother's and knocked at the door. "Who is it?"

"It's me, Mamma."

"Oh, dear, you're back again? Away with you! Get away from this house."

"No, Mamma, let me in. This time I have something for you that will make you happy for life!"

When his mother let him in, he asked, "What's for supper tonight?"

"What am I having? A few mustard greens I picked behind the statue of Our Lady of Sorrows in the master's garden."

The son grabbed the frying pan and emptied it out the window.

"You murderer! You wretch! You're forcing me to go hungry again. God knows how Vito Borgia abused me when he caught me picking the greens, and now you come, you murderer, and pitch them out the window!"

"No, no, dear Mamma!" he replied. "Take this rag of a table napkin and just see what comes out of it. My table napkin, make ready the table! My table napkin, make ready the table!"

But no matter how many times he repeated "My table napkin, make ready the table!" absolutely nothing happened. He yanked it this way and that, reducing it to tatters good for nothing but a dishrag. His mother gave him a mighty whack and once more turned him out of the house.

So he went back to Pappy once more. "What happened to you this time, stupid boy? Didn't I tell you that you'd get into more trouble?" The boy had no choice, then, but return to his former routine, digging in the field.

After a while, though, he was again yearning for his mother. Pappy said, "All right, my son, this is the last time. Take this club and, when you get to your mother's, say, 'My club, let me have it, let me have it!'"

Weeping, the boy left Pappy and was on his way. Nosy as ever, when he came to the usual place, he had to try it out, and said, "My club, let me have it, let me have it!" Once in motion, there was no stopping the club. It thrashed him right and left, whirling round like a lathe.

Up in his tower, Pappy Ogre doubled up with laughter. "That should put some sense into that head of his!"

The boy screamed, "My club, be still! My club, you have killed me!"

"Give it to him, give it to him!" cried Pappy from his tower top. When he saw that the boy had had enough, he said, "Now be still," and the club stopped.

The boy reached the inn in the lowest of spirits. "Back again, Tony? How is everything, lad? How come you're all bandaged up?"

"I don't want to talk about it. I'm going to bed. Keep this stick for me, but beware of ever saying, 'My club, let me have it, let me have it!'"

When it was night, the innkeeper picked up the club and tested it, saying, "My club, let me have it, let me have it!" The club began thrashing the daylights out of him and all his family, flying around like a woolwinder. "Help! Help! Christians to the rescue, it is killing us!"

The boy ran in. "Give me back the donkey and the table napkin, or else I won't take back the club."

They gave him back the donkey and the table napkin. When he had made sure they were really his, he took back the club and left. He reached his mother's house with club, donkey, and table napkin.

Hearing the knocking on the door, his mother opened a peephole and saw him there with another donkey. "You bandit! You rogue! Away with you, away with you, and may they catch you and skin you alive!"

He said, "Club, give her a couple of whacks, but go easy."

The club went flying through the peephole and—bam! bam!—let her feel a couple of blows.

"You monster! You turncoat! Would you beat your own dear mother?"

"Open up wide if you want the club to stop."

His mother flung open the door, and he rode in on the donkey. "No, not the donkey, for heaven's sake! You're not going to dirty my house again, are you?" she shouted over and over.

"Well, my club," he said, "give her two more."

She therefore quieted down immediately. The son pulled the blanket off the bed and made the donkey drop a pile of gold pieces. Then he took out the table napkin and ordered it to make ready the table. They sat down and ate and drank their fill, while here we are, dying of thirst.

(Terra d'Otranto)

The School of Salamanca

There was once a father who had an only son. To this son, who showed he was shrewd, the father said, "My son, by being thrifty, I have managed to save up a hundred ducats, and I would like to double the sum. But I'm wary of investing it, lest I lose every bit of the money, for in one way or another, men are all rogues. I worry day and night over what to do. Tell me your thoughts on the matter. What does that brain of yours advise?"

The son was silent awhile, as though lost in thought, and when he had carefully reflected, he said, "Papa, I've heard of the school of Salamanca where one may learn any number of things. If I can enter it with our hundred ducats, you can be sure I'll know what to do when I come out and rake in the money for you with little effort."

This idea appealed to the father, and early the next day they set out for the mountain. After some distance they came to a hermitage. "Hello in there!"

"Hello, hello, who comes hither?"

"A good Christian soul exactly like yourself!"

"Here the cock crows not, the moon shines not; what brings you, solitary soul? Do you bring clippers to clip my eyelashes? Do you bring shears to shear my hedges?"

"I bring clippers to clip your eyelashes, and shears to shear your hedges." No sooner was that said than the door of the hermitage flew open, and father and son stepped inside. They trimmed the big old man's long eyelashes with the scissors, and once he was able to look out and see them, they asked his advice.

The hermit approved of their decision, gave the boy much advice, and said in conclusion, "When you reach the top of that mountain way over there, strike the ground with the wand I am giving you, and out will come an old man far older than I am: he is the Master of Salamanca."

They talked on a bit, then separated. For two days and two nights, father and son walked and, reaching the mountaintop, they did what the hermit had told them. The mountain opened, and there stood the Master.

At that, the poor father fell to his knees and, with tears in his eyes, told why he had come. The Master, totally impassive and hard-hearted like all masters, took the hundred ducats and invited father and son into his dwelling. He led them through rooms and rooms and rooms packed with animals of all species. As he passed them, he whistled, and all the animals turned into dazzling young men. The Master said to the father, "You need worry no more about your son. He'll be treated even better than a nobleman. I will instruct him in the mysteries of science and, if at the end of the year, you are able to distinguish him from all these animals, you'll take him back home together with the hundred ducats you have given me. But if you're unable to recognize him, he will remain with me forever."

At those woeful words, the poor father began weeping. But then he took heart, embraced his son, kissed him goodbye over and over, and made his way back home alone.

Morning and evening the Master instructed the youth, who caught on at once and made enormous progress. In almost no time he was so clever that he could figure out things by himself. In sum, when the year rolled around, the pupil knew everything the Master knew, good and bad.

The father, meanwhile, was on the way to get his son, and the poor old man was worried, having no idea how he would recognize his son in the midst of all those animals. He was climbing the mountain, when he heard the wind blow, and a voice in the wind spoke. "Wind I am, and a man will I become." And there before him stood his son.

"Papa," said the youth, "listen to me: the Master will take you into a room full of pigeons. You will hear a pigeon cooing. That will be me." Then he said, "Man I am, wind will I become." At that, he turned back into wind and flew away.

Overjoyed, the father pushed on to Salamanca. When he reached the mountaintop, he struck the ground with the wand, and—bang!—there stood the Master! "I've come for my boy," explained the father, "and may God help me recognize him!"

"Fine, fine!" replied the Master. "But you'll certainly fail. Come with me."

He took him from one end of the house to the other, upstairs, downstairs and all around to confuse him. When they entered the room containing pigeons, he said, "It's up to you now: tell me if your son is in there; if not we will move on."

In the midst of those pigeons, a magnificent white and black one began strutting around and cooing. "Coo, cooo, COO . . ." Right off the bat the father said, "This one is my son, I just know it is he, my blood tells me so . . ."

The Master was mortified, but what could he do? He had to abide by the pact and hand over the son as well as the hundred ducats, which he hated even more to lose.

Overjoyed, father and son went home and, as soon as they arrived, invited relatives and friends to a big banquet, and everyone joyfully ate and drank. After a month of merrymaking, the son said to his father, "Papa, the hundred ducats are still here, we've not yet doubled the sum. If we built ourselves a cottage, the money wouldn't even pay for the bricks. What did I go to school for? Wasn't it to learn how to rake in money? Listen to me: the fair of Saint Vitus takes place tomorrow in Spongano. I will turn into a horse with a star on my head, and you will take me to the fair to sell. Watch out, for the Master will surely come and recognize me. But sell me for no less than one hundred ducats and *without the halter*. Remember that; my life depends on it."

The next day, right under his father's nose, he changed into a fine horse with a star on his head, and off they went to the fair. Everybody flocked open-mouthed around the beautiful animal, they all wanted it but, hearing the owner ask one hundred ducats for it, they all backed away. The fair was almost over, when an old man came sidling up, looked the horse over, and said, "How much are you asking for it?"

"One hundred ducats, halter not included."

Hearing that figure, the old man grumbled a little. Then he balked, saying it was too much. But seeing that the owner would not come down on his price, he began counting out the money. The father was pocketing the money and hadn't yet removed the halter from the horse, when that cursed old man, quick as lightning, leaped onto the horse's back and fled like the wind. "Stop! Stop!" frantically cried the father. "I have to get the halter. The halter doesn't go with the horse!" But he'd vanished without a trace.

Astride the horse, the Master whipped him to top speed. The blows fell so fast and thick that the animal bled all over and would have soon dropped, had luck not brought them to a tavern. The Master dismounted, led the wounded horse into the stable, tied him to an empty manger, and left him still wearing the halter and with neither fodder nor water.

Working as a servant at the tavern was a girl who was a marvel to behold and, while the Master was upstairs dining, she chanced to walk through the stable. "Ah, poor horse!" she exclaimed. "Your owner must really be base to leave you here like that without fodder or water and all bloody! But I'll look after you." The first thing she did was lead him to the fountain to drink and, so that he could do so with ease, she removed the halter.

"A horse I am, and an eel will I become!" said the horse, once out of the halter, and transformed into an eel, he jumped into the fountain.

Hearing him, the Master pushed aside the plate of macaroni he was eating and flew downstairs, livid with rage. "Man I am, and a conger will I become!" he screamed and jumped into the water, turning into a conger and pursuing the eel.

The disciple, though, did not lose heart, but said, "Eel I am, and a dove will I become!" And swish! out of the water he flew, now a beautiful dove. The sorcerer then said, "Conger I am, and a falcon will I become!" Now a falcon, he flew after the dove. They flew and flew, with the Master always on the verge of overtaking pupil and, at length, they came to Naples. Outside in the king's garden sat the princess under a tree. She happened to be looking up at the sky and suddenly saw the poor dove pursued by the falcon, and the sight moved her to pity. Seeing her, the disciple said; "Dove I am, and a ring will I become." He became a gold ring and dropped into the princess's bosom. The falcon swooped down and lit on the roof of the house across the way.

At night, when the princess undressed and removed her corset, the ring fell into her hands. Bringing it closer to the candlestick to examine it, she heard these words: "My princess, forgive me for coming to you like this without your leave, but it's a matter of life and death. Allow me to appear in my true form, and I will tell you my whole story."

Hearing that voice, the princess almost died of fright, but curiosity then got the better of her and she granted him permission to show himself. "Ring I am, and a man will I become!" The ring gleamed brighter, and there stood a dazzling young man. The princess was fascinated and couldn't take her eyes off of him. Then when she heard of all his accomplishments and the misfortunes he was enduring, she fell in love with him and insisted that he remain with her. In the daytime the youth turned back into the ring, which she wore on her finger. At night when they were alone, he took back his human form.

But the Master didn't stand idly by. One morning the king woke up in terrible pain. All the doctors were called, and they made him take every medicine known to man, but his suffering did not lessen. The princess was grieved, and the youth still more so because he knew all this was the Master's doing. As a matter of fact, here came a foreign doctor to the palace, from a country at the end of the earth, and he claimed that if they let him into the king's room, he would cure him. They showed him in at once, but the princess saw the ring gleaming more intensely and realized that the youth wanted a word with her. She shut herself up in her chamber, and the young man said, "What a mistake you have made! That doctor is the Master! He will cure your father but, for his pay, he will demand the ring! Refuse to give it up, but if the king orders you to, then throw it on the floor as hard as you can!"

Things happened that way: the king got well and told the doctor, "Name whatever you want, and I will give it to you." At first the doctor pretended to want nothing, but at the king's insistence, he asked for the ring on the princess's finger. She screamed, cried, and finally fainted; but feeling the king grab her hand to take the ring by force, she suddenly jumped up, slipped it from her finger, and threw it to the floor.

As soon as she hurled it, a voice was heard. "Ring I am, and a pomegranate will I become!" The pomegranate broke open on the floor, and seeds scattered all over the room.

"Doctor I am, and a cock will I become!" said the Master, turning into a cock and proceeding to eat the seeds one by one. But one seed landed under the long skirt of the princess, who kept it hidden there.

"Pomegranate I am, and a fox will I become!" said the seed, and out from under the princess's skirt jumped a fox and ate the cock in one gulp.

The pupil had outwitted the Master! The fox turned back into a young man, told the king his story, and the next day all the cannons were fired in honor of the princess's marriage.

(Terra d'Otranto)

◈ 129 ◈

The Tale of the Cats

A woman had a daughter and a stepdaughter, and she treated the stepdaughter like a servant. One day she sent her out to pick chicory. The girl walked and walked, but instead of chicory, she found a cauliflower, a nice big cauliflower. She tugged and tugged, and when the plant finally came up, it left a hole the size of a well in the earth. There was a ladder, and she climbed down it.

She found a house full of cats, all very busy. One of them was doing the wash, another drawing water from a well, another sewing, another cleaning house, another baking bread. The girl took a broom from one cat and helped with the sweeping, from another she took soiled linen and helped with the washing; then she helped draw water from the well, and also helped a cat put loaves of bread into the oven.

At noon, out came a large kitty, the mamma of all the cats, and rang the bell. "Ding-a-ling! Ding-a-ling! Whoever has worked, come and eat! Whoever hasn't worked, come and look on!"

The cats replied, "Mamma, every one of us worked, but this maiden worked more than we did."

"Good girl!" said the cat. "Come and eat with us." The two sat down to the table, the girl in the middle of the cats, and Mamma Cat served her meat, macaroni, and roast chicken; but she offered her children only beans. It made the maiden unhappy, however, to be the only one eating and, noticing the cats were hungry, she shared with them everything Mamma Cat gave her. When they got up, the girl cleared the table, washed the cats' plates, swept the room, and put everything in order. Then she said to Mamma Cat, "Dear cat, I must now be on my way, or my mother will scold me."

"One moment, my daughter," replied the cat. "I want to give you something." Downstairs was a large storeroom, stacked on one side with silk goods, from dresses to pumps, and on the other side with homemade things like skirts, blouses, aprons, cotton handkerchiefs, and cowhide shoes. The cat said, "Pick out what you want."

The poor girl, who was barefooted and dressed in rags, replied, "Give me a homemade dress, a pair of cowhide shoes, and a neckerchief."

"No," answered the cat, "you were good to my little ones, and I shall give you a nice present." She picked out the finest silk gown, a large and delicately worked handkerchief, and a pair of satin slippers. She dressed her and said, "Now when you go out, you will see a few little holes in the wall. Push your fingers into them, then look up."

When she went out, the girl thrust her fingers into those holes and drew them out ringed with the most beautiful rings you ever saw. She lifted her head, and a star fell on her brow. Then she went home adorned like a bride.

Her stepmother asked, "And who gave you all this finery?"

"Mamma, I met up with some little cats that I helped with their chores, and they gave me a few presents." She told how it had all come about. Mother could hardly wait to send her own idle daughter out next day, saying to her, "Go, daughter dear, so you too will be blessed like your sister."

"I don't want to," she replied, ill-mannered girl that she was. "I don't feel like walking. It's cold, and I'm going to stay by the fire."

But her mother took a stick and drove her out. A good way away the lazy creature found the cauliflower, pulled it up, and went down to the cats' dwelling. The first one she saw got its tail pulled, the second one its ears, the third one had its whiskers snatched out, the one sewing had its needle unthreaded, the one drawing water had its bucket overturned. In short, she worried the life out of them all morning, and how they did meow!

At noon, out came Mamma Cat with the bell. "Ding-a-ling! Ding-a-ling! Whoever has worked, come and eat! Whoever hasn't worked, come and look on!"

"Mamma," said the cats, "we wanted to work, but this girl pulled us by the tail and tormented the life out of us, so we got nothing done!"

"All right," replied Mamma Cat, "let's move up to the table." She offered the girl a barley cake soaked in vinegar, and her little ones macaroni and meat. But throughout the meal the girl filched food from the cats. When they got up from the table, heedless of clearing away the dishes or cleaning up, she said to Mamma Cat, "Give me the stuff now you gave my sister."

So Mamma Cat showed her into the storeroom and asked her what she wanted. "That dress there, the nicest! Those pumps with the highest heels!"

"All right," replied the cat, "undress and put on these greasy woolen togs and these hobnailed shoes worn down completely at the heels." She tied a ragged neckerchief around her and dismissed her, saying, "Off with you, and when you go out, stick your fingers in the holes and look up."

The girl went out, thrust her fingers into the holes, and countless worms wrapped around them. The harder she tried to free her fingers, the tighter the worms gripped them. She looked up, and a blood sausage fell on her face and hung over her mouth, and she had to nibble it constantly so it would get no longer. When she arrived home in that attire, uglier than a witch, her mother was so angry she died. And from eating blood sausage day in, day out, the girl died too. But the good and industrious stepsister married a handsome youth.

> A pair so handsome and happy
> We are ever happy to see;
> Listen, and more will I tell to thee.

(Terra d'Otranto)

130

Chick

A husband and a wife had seven children. The father was a farmer and, as a great famine raged, they were starving to death. At night, while the children slept, their father and mother lay awake worrying. "My wife, this life is unbearable," said the man. "It breaks my heart to see our little ones starving."

"It is truly sad," replied his wife, "but what can we do?"

"Tomorrow, when I go to the woods, I'll take the children along and leave them there. It's better to lose them all at once than watch them waste away like candles."

"Sh!" cautioned the wife. "Don't let them hear us talking."

"Don't worry, they're all sound asleep."

But the smallest of the seven children, a hunchback they called Chick, wasn't asleep and heard every word.

In the morning when they got up, their mother called them, got them ready, kissed them with tears in her eyes, and said, "Off with you, good children, you're going with Papa today."

They set out, and along the roadside Chick picked up as many white pebbles as he could find and put them in his pockets. Once they left the road and entered the woods Chick, aware of what his father had in mind, dropped a pebble every step of the way to mark the path they took. In the middle of the woods their father went off and left them. Night fell, and the children screamed and cried. Chick spoke up. "What are you afraid of, silly children? I'm going to lead the way, and we'll go back home."

"Yes, yes, little brother," they all chimed in, "what are we to do?"

"Just come with me." And he followed the white pebbles out of the woods. Day was breaking when they got back home, dead tired.

"My dears!" exclaimed their mother, overjoyed to see them again. "How did you find the way back?"

"Chick showed us the way," answered the older brothers.

The children remained at home, but it was not long before their father decided to take them back to the woods, since there was no letup of the famine. Their mother sold everything they had left in the house, in order to buy seven long loaves of bread. Next morning she gave each child a loaf, kissed them goodbye, and sent them all off to the woods with their father.

This time the father walked behind Chick, to make sure he didn't strew white pebbles. But instead of eating his loaf of bread, Chick crumbled it all up in his pocket and dropped crumbs every step of the way through the woods. Finding themselves alone once more at nightfall, the brothers began to bawl, but Chick said, "Have no fear, we're going back home this time too." And he started looking for the bread crumbs he had dropped. But what crumbs the ants had not carried off, the birds had eaten, so Chick could no longer find the way. His brothers shed more tears. "Wait a minute," said Chick and, like a squirrel, scampered to the top of the tallest tree around. He saw a light in the distance. "That's the direction we have to take."

After walking a long way they came to a house. They knocked, and out came Mammy Ogress. With her long stringy hair and teeth like corkscrews and eyes like lanterns, she seemed more of an ogress than she really was. She said, "My, my! Where in the world are you children going at this hour of night?"

"Madam," answered Chick, "we have lost our way. We saw your light and came here."

"Dear me, children, I must hide you, because when Pappy Ogre comes in, he'd eat you in one bite. I have roasted him a sheep to satisfy his appetite. If you don't make a sound I'll put you to bed with my own children. I have seven, the same number as you."

Pappy Ogre came home, and began saying, "Mm! Mmm! I smell human flesh around here."

"Always the same old tune!" replied his wife. "Sit down here and eat the nice mutton I roasted for you. Mind your own business and keep your hands off the poor dears. Seven little brothers ended up here after losing their way, and I took them in, since we too have seven children we wouldn't want to see harmed."

"Give me that mutton, then," answered Pappy Ogre. "I'm tired and want to get to bed early."

When Pappy Ogre's seven children went to bed, they each wore a crown of flowers on their head. They slept in a big bed, at the foot of which Mammy Ogre placed Chick and his brothers. As soon as she left the room, Chick began wondering, "Why do her children wear those crowns? There's something behind all that." And he took the crowns off of Pappy Ogre's sleeping children and put them on his brothers and himself.

No sooner had he finished than Pappy Ogre tiptoed in, bent over the bed and, since the room was dark, began feeling the children. He touched Chick and his brothers on the head and, feeling the crowns of flowers, let the boys be. Then he touched his own sons one by one and, finding no crowns, ate them. There in the dark, Chick trembled like a leaf. Pappy

Ogre gulped down his last son, licked his lips, and said, "Now that I've eaten them my wife can preach all the charity she wants to." At that, he left. Chick woke his brothers up at once. "Let's get out of here quick." They eased the window open and dropped to the ground. Through the woods they ran and ran until they came to a cave, in which they hid.

When Mammy Ogress got up the next morning, she found neither her seven sons nor the seven stray children and, from the marks on the bed, she realized what had happened. She began pulling out her hair and screaming. "Monster! Murderer! Come see what you have done!" Pappy Ogre ran in, stunned. "What! Our own sons were not wearing the crowns of flowers? How could that have happened? Give me my great boots that travel a hundred miles an hour. I will hunt down those rogues and eat them raw." He put on his boots and combed the earth, but he didn't find the children, because they were hidden in the cave.

Dead tired from his search, Pappy Ogre sank to the ground and fell asleep just a stone's throw from the cave where the seven brothers were in hiding. Chick, who was always out looking for food, found him stretched out there. He called his brothers. "Quick! Let him have it, all of you!" They each took their knife they cut bread with and stabbed him all over, until he looked like a strainer. When they were sure he was dead, they pulled off his boots, into which all seven of them climbed and went speeding to Mammy Ogre's house.

"Mammy," they said to her, "Pappy sent us to tell you he's fallen into the clutches of robbers, and if you don't give them all his money, they will kill him. So you'd believe us, he lent us his boots."

Mammy Ogress got all her husband's money, diamonds, and gold and gave them to the seven brothers. "Of course, my boys, go and free him."

In one step of the great boots, the seven brothers reached the house of their mother and father and made them rich. Chick went off to Naples and, with those boots that traveled a hundred miles an hour, became a courier, because in those days there were no locomotives or steamboats. Thus the little hunchback made his whole family rich, and lived happily ever after.

(Terra d'Otranto)

131

The Slave Mother

There was once a husband and a wife, well-off tenant farmers, who managed the farm of the leading nobleman of the province, in the vicinity of Otranto. They had five sons, and the farmer's wife, after finishing her chores and putting supper on to cook for the men coming in from their work, would sit on the doorstep of their house each evening and say her rosary.

One evening as she was about to make the sign of the cross, she heard the owl call, "Farmer's wife, farmer's wife! When do you want wealth, in youth or in old age?"

"Good heavens!" exclaimed the farmer's wife, crossing herself in haste.

It was the hour when the men came in from the fields. They sat down to the table and ate quite heartily. That poor woman was a bit disturbed. "What's the matter?" asked her husband and sons. She replied that she didn't feel well.

The next evening when she began saying her rosary anew, she again heard the owl. "When do you want wealth, in youth or in old age?"

"Mother of God!" exclaimed the farmer's wife. "This is serious!" She went straight to her husband about it.

"Wife," said the farmer, "if the same thing happens again, tell the owl that you want wealth in old age, since one always gets through youth somehow, but in old age a person can't have too many comforts."

So when the owl called the third evening, the farmer's wife said, "You're back again? In old age I want it, is that clear?"

Time went by. One evening, fed up with eating the same vegetables all the time, the husband and sons said: "Mamma, tomorrow, God willing, make us a salad of mixed greens."

In the morning the farmer's wife took her apron with the deep pocket and a knife, and went out for salad greens. The farm was on a promontory overlooking the sea, and the farther out she went, the nicer became the greens. "What fine greens!" she said. "This evening my sons and husband will have something to feast on!" She picked greens here, there, and yonder and wound up right on the seashore. And while she was bending over gathering a particular kind of chicory, some Turks sneaked up, seized her, dragged her off to a boat, and sped away over the sea. In vain did she scream and beg for mercy; not for the life of them would they let her go.

But let's leave her screaming her head off and turn to the poor husband and sons when they came in that evening. Instead of seeing the house open, as always, and supper ready, they found the door closed. They called, they knocked, and finally they broke the door down. When they had made sure she was nowhere in the house, they went around and asked the neighbors if any of them had seen her. "Yes," said the farmers in the vicinity, "we saw her go out with her apron, but we didn't see her return."

Just imagine the grief of those men! Night fell, and they lit lanterns and went out into the fields crying, "Mamma! Mamma!" They also peered down into the wells. Finally they gave up hope of finding her and returned home in tears.

Then they dressed in mourning and received callers for three days straight. But since everything in this world passes, they again returned to work in the fields as before.

Two years later it happened they had to plow a large field for sowing it in grain. The sons and the old man each took a team of oxen and started plowing. As he plowed, the old man's plow caught on something in the earth. Unable to get it loose by himself, he called his oldest son. They pulled and pulled and finally saw that it was caught in an iron ring. They tugged on the ring, and up came a large stone slab. Underneath it was a room.

"Oh, Papa!" said the son, "What do I see down there? May I go down?"

"No," answered the old man. "Let's leave everything as it is. Tonight we'll come back and see what this is all about." Thus they separated.

In the evening they took the farmhands aside and got them good and drunk. Once the men were snoring, the old man and his five sons took the lantern and returned to the stone slab. They raised it, descended to the underground vault, and found seven pots full of gold pieces. They stared at one another open-mouthed, at a loss for words, and with no idea what to do. "My sons," said the old man, "don't just stand there like blockheads. Run get the cart and hurry back."

The sons raced back with the cart, loaded on all the treasure, and took it off and hid it.

The next day—exactly two years and one month from the time of the disappearance of the poor farmer's wife—they went to the owner of the farm and said that they wouldn't stay at the farm any longer, their hearts were no longer in it. They turned the property back over to him, offered the farmhands a fine feast, and set out for Naples. Arriving there, they took off their country clothes and donned fine new ones. They bought a palace and called in schoolmasters and language teachers to teach them the ways of gentlemen. Then they went to the theater and other similar things. The old man grew a pigtail, as was the custom in those days. They took up speaking like the Neapolitans, and even changed their names; it was no longer Renzo, or Cola, but *Don* Pietrino, *Don* Saveruccio; every nice name they heard, they latched onto it. Nobody they used to know would have recognized them any more.

One day all five brothers happened to be together in the square of the Immacolatella, where a slave market was being held, with both dark Moorish girls and white girls for sale. Among the white girls stood out one who was especially beautiful. The minute the boys got back home they cried, "Father, Father!" (They no longer said "Papa.")

"What is it, my sons?"

"We saw any number of beautiful slave girls. Shall we buy one?"

"What!" said the father. "You want to bring a slut into the house? Indeed you won't! If there's an old woman among them, we'll take her."

He went to the square himself, looked the slaves over, and spotted an old one among them—rather, she looked old just then from all the knocks she had taken and all the work forced on her, poor soul. "How much do you want for her?" the father asked the slave dealer.

"One hundred ducats."

He paid, and they took her home. It wrung one's heart watching that poor soul moving about in rags, so they bought her new clothes and put her in charge of the house.

In the evening, as always, the sons attended the theater. But the old man never went out. When the poor woman saw the five brothers leave, she would sigh and weep. One evening after lighting the young gentlemen down the stairs, she came back up weeping, and the old gentleman closed the book he was reading and called to her.

"Why do you always sigh and weep when you see my sons?"

"Sir," replied the slave woman, "if you knew what was in my heart you wouldn't ask me!"

"Sit down and tell me," said the old man.

"Well, I have never been the slave you bought me for. I was married to a farmer and had five sons like Your Honor's . . ." and she went on with her story. When she came to the part about going out for salad greens and being kidnapped by the Turks, the old man rose, embraced her, and covered her with kisses. "My wife, my wife, I am that very farmer, and the five boys are your own sons. One day, after years of suffering, since we thought you were dead, we came upon a fortune while plowing the field. So what the owl told you has come true."

Just imagine the good woman's joy over miraculously finding her husband and sons after seventeen years of slavery. As they clasped one another relating past woes, their sons returned from the theater. Seeing the two old people lavishing so many caresses on one another, they said, "And he didn't want us to buy a young woman!"

"No, my sons," said the father, "this is your mother, whom we mourned for so many years."

Just imagine the sons! Over and over they embraced and kissed her, saying, "Mamma, you've worked and suffered quite enough. From now on you will command and enjoy every luxury."

Maids and servants came and dressed her as the great lady she actually was, with a muff for winter and a fan for summer.

Thus they lived in peace and contentment, spending their old age in the lap of luxury.

(Terra d'Otranto)

◈ 132 ◈

The Siren Wife

There was once a beautiful woman married to a mariner. This mariner used to sail off and stay for years at a time and once, while he was away, a king of the region fell in love with his wife and finally persuaded her to run away with him. When the mariner got back, he found the house empty. Time passed, and the king tired of the woman and dismissed her. Repentant, she returned to her husband and begged his forgiveness on bended knee.

Although he still loved her as much as ever, the mariner was so offended by her faithlessness that he turned his back on her, saying, "I'll never forgive you. You'll get the punishment you deserve. Prepare to die."

Tearing her hair, the woman begged and pleaded with him, but all to no avail. The mariner had the faithless wife loaded onto his ship as though she were a sack of grain, weighed anchor, and sailed off.

Reaching the high seas, he said, "Your time has come." At that, he picked her up by the hair and threw her into the waves. "I am now avenged," he said, changing course and sailing back into port.

The wife sank to the floor of the sea, right where the Sirens congregated.

"Look what a beautiful young woman they've thrown into the sea," said the Sirens. "The idea of such a lovely creature being eaten up by the fish! Let's rescue her and take her in with us!"

So they took the wife by the hand and led her to their underwater palace which was all lit up and glowing. And one Siren combed her black hair, another perfumed her arms and bosom, another put a coral necklace around her neck, still another slipped emerald rings on her fingers. The wife was too amazed for words. "Froth!" she heard them call her. "Froth, come along with us!" She realized that would be her name among the Sirens. She entered the grand hall of their palace and found it full of women and handsome youths who were dancing, and she too began dancing.

What with so many comforts and celebrations, the wife's days flew by in joy. But the memory of her husband would often return to haunt her and make her sad.

"Aren't you happy here with us, Froth?" asked the Sirens. "Why are you so quiet and downcast?"

"Nothing at all is wrong, I assure you," she would answer, but she was unable to force a smile.

"Come, we'll teach you to sing." They taught her those songs of theirs which make sailors dive into the sea when they hear them. So Froth took her place in the Sirens' choir, which rose to the surface to sing on moonlight nights.

One night the Sirens saw a vessel approaching full sail. "Come on, Froth, we're going up to sing!" said the Sirens and began their song:

> "This is the song of the full moon,
> Of the moon so round and here so soon;
> Is it the comely Siren you wish to see?
> Then, jump, O sailor, into the sea!"

At that, a man was seen leaning over the railing of the ship, bewitched by the music, and next thing you knew, he flung himself into the waves. By the light of the moon, Froth had recognized him: it was her husband.

"This one we'll turn into coral!" the Sirens were already saying.

"Or into white crystal! Or else shell!"

"Wait! Please wait!" exclaimed Froth. "Don't kill him! Don't work any more magic on him!"

"But why are you showing so much pity for him?" asked her companions.

"I don't know . . . I'd like to work a spell over him myself . . . in my own way, you'll see. . . . Please, let him live for twenty-four hours more."

After seeing her so sad all the time, the Sirens didn't have the heart to say no, and shut the mariner up in a white palace on the floor of the sea. It was now day, and the Sirens went off to sleep. Froth approached the white palace and sang a song that went like this:

> "This is the song of the moon when it's full,
> I knew you in life and you were ungrateful,
> Now I've become a Siren
> You I will save and me they will condemn."

The mariner pricked up his ears and realized that the one singing could be none other than his wife. He grew hopeful, realizing deep down that he had already forgiven her and regretted drowning her.

Now the Sirens slept in the daytime and went about the sea at night spreading their nets for sailors. Froth waited until night, opened the white palace and was reunited with her husband. "Be quiet," she told him. "The Sirens have just gone off and can still hear us! Hold on to me and let me carry you." Like that she swam and swam for hours until they came in sight of a large ship.

"Cry to the sailors for help!" said Froth.

"Here, down here! Help! Help!"

A rowboat was lowered from the ship. They rowed toward the survivor and pulled him on board.

"The Siren . . ." he said. "The Siren . . . The Siren, my wife . . ."

"He's gone crazy in the water," said the rescuers. "Now, now, calm down, friend, you're safe. There's no siren around here!"

The mariner made it back to his town, but all he could think of from then on was his siren wife, and he was unhappy. "I drowned her and now she has saved my life," he thought. "I will go sailing until I find her. I will save her, or else drown myself."

And thinking those thoughts, he penetrated a forest up to a walnut tree where the fairies were said to gather.

"My good lad, why are you so sad?" said a voice next to him. He turned around and there stood an old woman.

"I'm sad because my wife is a Siren and I don't know how to bring her back."

"You seem like a good lad to me," said the old woman, "and I will help you get your wife back. But on one condition. Do you agree?"

"I'll do whatever you say."

"There's a flower that grows only in Sirens' palaces and which is called 'the loveliest.' You must get this flower and bring it back here at night and leave it under this walnut tree. Then you shall have your wife back."

"But how can I do it? Get a flower from the floor of the sea?"

"If you would have your wife back, you must find the way."

"I'll try," said the mariner. He went to the port at once, boarded his ship, and weighed anchor. When he reached the high seas, he started crying his wife's name. He heard water splashing and saw her swimming in the wake of the ship. "My wife," said the mariner, "I want to save you, but to do so I must get a flower which grows only in Sirens' palaces and which is called 'the loveliest.' "

"That is impossible," said the wife. "The flower is there and gives off a heavenly scent, but it is a flower the Sirens stole from the fairies, and the day it goes back to the fairies, all the Sirens will die. I'm a Siren too, so I would die along with the rest of them."

"You won't die," said the mariner, "because the fairies will save you."

"Come back tomorrow and I'll have an answer for you."

The next day the mariner went back. His wife reappeared in the sea. "Well?" he asked.

She answered: "In order for me to bring you the flower called 'the loveliest,' you must sell everything you own and, with the proceeds buy the finest jewels there are in the strongboxes of the goldsmiths of every city in the kingdom. At the sight of the jewels, the Sirens will stray from the palace and I'll be able to pick the flower."

In no time the mariner had sold all he owned and bought the most splendid jewels in the kingdom. He loaded the ship with jewels, hanging them in clusters from all the yards where they gleamed in the sun. Like that, he sailed over the sea.

Thirstier for jewels than everything else, the Sirens began to surface on the waves and follow the boat, singing:

> "This is the song of the noonday sun
> Your boat overflows with gems that stun;
> Good sailor, pause here a while,
> Give us rings and chains and pins in style."

But the mariner kept on going, and the Sirens followed along behind, getting farther and farther from their palace.

All of a sudden a great rumbling came from under the sea. The waters billowed higher than ever before and all the Sirens were swept under and drowned. Out of the water flew an eagle, with the old fairy and the mariner's wife astride, and disappeared into the distance.

When the mariner got home, his wife was already there waiting for him.

(Taranto)

The Princesses Wed to the
First Passers-By

There was once a king with four children—three girls and one boy, who was the crown prince. On his deathbed, the king sent for the prince and said, "Son, I'm dying. You must do as I order: when your sisters are old enough to marry, have them go out on the balcony, and the first man who comes down the street is to be their husband, no matter whether he is an ignorant peasant, a learned master, or a nobleman."

When the oldest girl reached a marriageable age, she went out on the balcony. A barefooted man came by.

"Friend, stop here for a minute."

"What is it, Majesty?" asked the man. "Don't delay me, for my pigs are penned up and I have to take them out to pasture."

"Sit down. We have to have a word in private. I must give you my oldest sister in marriage."

"Your Majesty is joking. I am only a poor swineherd."

"And you will marry my sister, in accordance with my father's will."

So the princess and the swineherd were married and left the palace.

Now came time to marry off the second sister. He put her on the balcony, and the first man that passed was called into the house.

"Your Majesty, don't delay me. I've set snares and have to go see if there are any birds in them."

"That makes no difference. Come in for a moment, I've got to talk to you."

And he offered him his sister's hand. "Majesty, how can that be?" asked the man. "I'm a poor fowler I can't marry into a royal family."

"My father has so decreed," replied the young king, and the second sister was wed to the fowler and departed with him.

When the third sister went out on the balcony, who should pass by but a gravedigger, and however much it grieved the brother (since he adored his little sister), he sent her off as the gravedigger's wife.

Left alone in the palace, with all his sisters gone, the young king thought, What if I should do as my sisters? Whom would it be my lot to marry? He went out on the balcony. An old washerwoman came scurrying by, and he called to her, "Friend, O friend, wait a minute . . ."

"Just what do you want?"

"Come inside a minute, I have to speak to you. It's urgent!"

"What is so urgent? It's urgent for me to get to the river and wash these clothes."

"Come in here, will you? I order you to!"

"Go on, try and bully old women." She looked him squarely in the eye and let out a curse. "Go look for lovely Floret!" At that, she turned and walked off.

The king grew weak in the knees and had to lean on the railing of the balcony. He was overcome with longing, which he at first thought was for the sisters he had lost. Instead it was the name, lovely Floret, which had gone to his head. He said to himself, "I must leave this house and travel the world over until I find lovely Floret."

He combed half the world, but no one knew anything about lovely Floret. He'd been journeying for three years, when he found himself in a field one day and ran into a herd of pigs, then another, and still another. He was swept along in the herd and making his way forward, he soon came to a large palace. He knocked and said, "Hello, anybody home? Give me shelter for the night!"

The palace door opened, revealing a great lady. She saw the king and threw her arms around his neck. "Dear brother!" she exclaimed. And the king recognized his oldest sister, who had married a swineherd. "Dear sister!" he exclaimed.

And here came the swineherd brother-in-law dressed as a great lord, and they showed him around their magnificent palace, telling him the other two sisters had homes every bit as fine.

"I'm out seeking lovely Floret," explained the King.

"We know nothing about her," she said. "But go to our sisters; they might be able to help you."

"And should you ever find yourself in danger," said the brother-in-law who had been a swineherd, "take these three pig bristles, throw one on the ground, and you'll get out of every difficulty."

The king continued on his way and after going a great distance found himself in a forest. On every tree branch in the forest, birds had lit. They flew from tree to tree, and the sky was no longer visible for all the birds that fluttered in the air. They all chirped together, in a deafening chorus. In the heart of the forest rose the palace of the second sister, who was even better off than the first one with her husband, once a poor fowler and now a great lord. Neither did they know anything about lovely Floret, and directed the king to the third sister. But before bidding him farewell, his brother-in-law gave him three bird feathers. In case of danger, all he had to.do was drop one of them, and his safety would be assured.

The king continued on his way and, at a certain point, began to see graves on both sides of the road, graves that became ever more numerous as he advanced, until the whole countryside revealed nothing but graves. Thus he reached the palace of the third sister whom he loved best of all, and his brother-in-law who had been a gravedigger gave him a small bone from a corpse, instructing him to drop it in case of danger. And his sister told him yes, she knew the city where lovely Floret lived. She directed him to an old woman whom she had helped and who would certainly help him.

The youth reached the town of lovely Floret, who was the king's daughter. Opposite the king's palace stood the house of that old woman, who gratefully welcomed the brother of her benefactress. From the window of the old woman's house the young king could see lovely Floret looking out at dawn, covered with a veil, a flower of loveliness at the sight of which he would have fallen out of the window, had the old woman not been holding on to him.

"But don't attempt to ask for her hand, Majesty," cautioned the old woman. "The king of this town is cruel and imposes impossible tasks on the suitors. He beheads all those who fail."

But the young man was unafraid and went to the father of lovely Floret and asked for her hand. The king had him shut up in an immense storage room with bins and bins of apples and pears, telling him that unless he ate all the fruit in a single day, his head would roll. The youth remembered the pig bristles from his swineherd brother-in-law and threw them on the floor. At once a chorus of grunts arose, and pigs poured in from every direction—pigs, pigs, pigs, an ocean of grunting, rooting pigs that ate up everything in sight, overturning all the bins and gobbling up every apple and pear without leaving a single core.

"Hurrah," said the king. "You will marry my daughter. But there's a second test. The first night you spend with her, you must put her to sleep with the song of most beautiful and musical birds ever seen and heard. Otherwise your head will roll tomorrow."

The bridegroom recalled the three feathers from his fowler brother-in-law and threw them down. At that, the sky was darkened by a cloud of birds with wings and tails of every color. They lit in the trees, on spires and rooftops, and began singing such soothing music that the princess fell asleep with a smile on her lips.

"Yes, indeed," said the father-in-law, "you have won my daughter. But since you are man and wife, by tomorrow morning you must have a baby that can say Papa and Mamma. Or else I'll behead you and her too."

"There's time between now and tomorrow morning," replied the bridegroom and, taking leave of the king, he remained with lovely Floret.

In the morning he remembered the little bone from the gravedigger brother-in-law. He threw it on the floor, and lo and behold, the bone changed into a beautiful baby boy holding a golden apple and calling Papa and Mamma.

The father-in-law king came in, and the baby went to him and insisted on placing the golden apple on the tip of his crown. The king then kissed the baby, blessed the newlyweds and, removing his crown, placed it on the head of his son-in-law, who now had two crowns.

There was a grand celebration attended by the three brothers-in-law—swineherd, fowler, and gravedigger—and their wives.

(Basilicata)

❖ 134 ❖

Liombruno

There was once a fisherman who had no luck at all. For three years he'd not caught so much as an anchovy. To survive, he and his wife and four children had sold everything to their name and were now living on charity. But each day he still put his boat into the water and rowed out to where he would lower his nets. Then he pulled them up without so much as a crab or a mussel in them, and let out awful curses.

One day, precisely while he was cursing over an empty net, who should appear in the middle of the sea but the Evil One and ask, "Why are you so angry, mariner?"

"Who wouldn't be with bad luck like mine? I fish nothing from this sea, not even a piece of rope to hang myself with!"

"Listen, mariner," replied the Evil One. "Make a pact with me, and you'll have fish every day and become a rich man."

"On what condition?" asked the fisherman.

"I want your son," answered the Evil One.

The fisherman began trembling. "Which one?"

"The one who is not yet born, but who will arrive shortly."

The fisherman reasoned that for some years now no sons had been born to him, nor was it very likely he would ever have any more. So he said, "Very well, I agree."

"In that case," said the Evil One, "when your son is thirteen, you will hand him over to me. And starting this very day, your hauls will be abundant."

"But what if this son should not come into the world?"

"Don't worry, your nets will still be full of fish, and you will owe me nothing."

"I just wanted to be sure." Then he put his name to the agreement.

Once the pact was concluded and the Evil One had disappeared over the sea, the fisherman pulled his nets up full of giltheads, tuna, mullet, and squid. And it was the same way the next day, and the next. The fisherman grew rich and was already saying, "I pulled a fast one!" when lo and behold a son was born to him, as fair as fair could be, and who would surely become the handsomest and strongest of all his sons. He named him Liombruno.

While the fisherman was in the middle of the sea one day, the Evil One turned up again ."Hello, mariner."

"What can I do for you?"

"Remember your promise and what you owe me. Liombruno is mine."

"Yes, indeed, but not before he turns thirteen."

"See you again in thirteen years." And he vanished.

Liombruno grew, and his father grieved when he saw him becoming ever handsomer and stronger, for the fatal day was approaching.

The thirteen years were all but up, and the fisherman began hoping that the Evil One had forgotten the pact, when lo and behold, here he came while the fisherman was rowing upon the sea one day. "Well, mariner."

"Woe is me!" said the mariner. "Yes, I know, the time has come. Tell me what I am to do."

"Bring him to me tomorrow."

"Tomorrow," repeated the father, weeping.

And the next morning he told Liombruno to bring him his lunchbasket at noon to a deserted spot on the shore, where the fisherman would fetch it and then go right back to his fishing without first coming home. The boy went, but saw no one. His father had gone way out to sea, so as not to be around when the Evil One showed up. Not finding his father, the boy sat down on the shore to wait for him. To pass the time, he made some little crosses out of the driftwood that had washed ashore and placed them around him in a circle, humming to himself as he did so. While he was humming there in the middle of the circle and holding a cross in his hand, who should arrive by sea but the Evil One and ask, "What are you doing there, boy?"

"I'm waiting for my father."

"You must come with me," said the Evil One, but he drew no closer, since the boy was encircled by those crosses.

"Undo those crosses this minute!" he ordered.

"I will not!"

The Evil One's eyes, mouth, and nose then began flashing fire and so frightened Liombruno that he quickly undid the crosses, but there was still the one he held in his hand.

"Undo that one too, and quickly!"

"I will not!" wept the boy as he faced the Evil One, who continued to flash fire. Just then an eagle was seen in the sky. It swooped down, seized hold of Liombruno's shoulders with its claws, and soared off into the sky with him, right under the nose of the Evil One, who was furious.

The eagle carried Liombruno to a high mountain top, then changed into a very beautiful fairy. "I am Fata Aquilina," she said, "and you will live with me and be my spouse."

A princely life began for Liombruno. He was fed and reared by the fairies, who instructed him in the arts and in the use of weapons. But after several years up there, he grew homesick and asked permission of Fata Aquilina to visit his father and mother.

"Go ahead, and carry riches to your old parents," said the fairy, "but you must return to me at the end of the year. Take this ruby; whatever you ask of it will be granted. But beware of revealing that I am your wife."

When the people back in Liombruno's village saw a knight so richly arrayed arrive, they made way for him and watched him dismount at the door of the old fisherman. "What business do you have with those poor people?" they asked, but Liombruno made no reply.

His mother answered the door, and Liombruno, without revealing who he was, asking for lodging. Great was the embarrassment of those poor old people over having to put up a lord so noble and rich in bearing. "Ever since we lost our youngest and most beloved son," they explained to him, "nothing else in the world has mattered to us, and we have let this house go to wrack and ruin."

But Liombruno proved he could adapt to anything and, that night, fell asleep on a couch, as though he were right at home.

When everybody was finally asleep in the house, Liombruno said to the ruby, "Dear ruby, transform this poor hovel into a palace with noble furnishings, and also make our beds as soft and comfortable as possible." And the ruby turned all those wishes into realities.

Next morning the fisherman and his wife awakened in a bed so soft that they sank way down. "Where are we?" asked the old woman, frightened. "Husband, where are we?"

"How would I know, wife?" answered the fisherman. "But I've never been more comfortable!"

And their amazement increased when they opened the window and sunlight streamed into a princely bedroom. In place of the ragged clothes they had left on the chair lay clothing embroidered with gold and silver. "Where on earth have we ended up?"

"In your own house," replied the knight as he entered their room, "and my house too, since I am your son Liombruno you thought you had lost forever."

So began for the old fisherman and his wife, reunited with their son, a life of joy and luxury. Then one day the boy informed them that he had to go away. He gave them chests of jewels and precious stones and took his leave, promising to return for a visit every year.

Riding back to Fata Aquilina's castle, he came to a city where a tournament was being announced. Whoever won for three days straight would receive the king's daughter in marriage. With the magic ring on his finger, Liombruno felt like showing off, so he entered the tournament the first day, defeated everybody, and fled without disclosing his name. The second day he went back, carried off another victory, and again slipped away. The third day the king stationed more guards around the tournament, and the victor was stopped and led before the royal dais.

"Unknown knight," said the king. "You entered the contest and won. Why do you refuse to reveal who you are?"

"Pardon me, Majesty, I dared not come into your presence."

"You were victorious, knight, and now you must wed my daughter."

"It grieves me to be unable to do so, Majesty!"

"And why can you not wed her?"

"Majesty, your daughter is the most graceful of maidens, but I already have a wife a thousand times more beautiful than your daughter."

At those words a great hubbub arose in the court. The princess turned crimson, and all the noblemen began whispering to one another. Solemn and impassive, the king spoke. "Knight, in order to allow your boast, you must at least show us this consort of yours."

"Yessirree," chimed in the noblemen all together, "we too wish to behold this beauty."

Liombruno fell back on the ruby. "Ruby, dear ruby, bring Fata Aquilina here."

But the ruby, although able to fulfill every request, could not produce Fata Aquilina, who was the source of its magic. And the fairy, full of indignation because Liombruno had bragged about her, responded to the ruby's summons by sending him the least of her serving women.

But even the least of Fata Aquilina's attendants was so comely and so richly clad that the king and his entire court could do nothing but gape at her.

"Your wife is indeed beautiful, knight!" they said.

"But she is not my wife!" said Liombruno. "She is but the least of my wife's attendants."

"Well, what are you waiting for to show us your wife herself?" said the king.

Liombruno repeated to the ruby, "Ruby, I want Fata Aquilina here."

This time Fata Aquilina sent her first serving woman.

"Ah, that is truly beauty herself!" they all said. "Surely she is your wife!"

"No," answered Liombruno. "She is only her first attendant."

"Let's be done with this comedy!" said the king. "I order you to send for your true wife."

Liombruno had hardly looked at the ruby a third time, when in a splendor like the sun's appeared Fata Aquilina. Dazzled, all the noblemen of the court stood stock-still, the king bowed his head, and the princess burst into tears and fled. But Fata Aquilina approached Liombruno, as though she meant to take his hand, and took away the ruby, exclaiming, "Traitor! You have lost me and will find me no more unless you use up seven pairs of iron shoes looking for me!" At that she vanished.

The king pointed at Liombruno. "I see now. You won by no power of your own, but thanks to the ruby. Servants, thrash him!" And the knight was thrown out and beaten and then abandoned in the middle of the street, black and blue, in tatters, and without a horse.

As soon as he was strong enough to get up, he headed dejectedly for the city gate. Hearing a great pounding of hammers, he realized he had come to a blacksmith's shop, which he entered. "Sire," he said, "I need seven pairs of iron shoes."

"What for? Did you make a bargain with the Eternal Father to live hundreds of years and use up all those shoes? As far as I'm concerned, I can make you ten pairs or as many as you say."

"What business of yours is it if I use them? All I need do is pay you, right? Make me the shoes and keep quiet!"

Receiving the shoes, he paid for them, slipped on a pair, and put three in one side and three in the other side of a knapsack and continued on his way. Night overtook him in the middle of the forest. He heard voices arguing; three thieves were arguing over how to divide up their booty.

466

"You, there, good fellow! Come and be our judge. We shall let you decide what goes to each of us."

"What's to be divided among you?"

"A purse that produces one hundred ducats every time it's opened. A pair of boots that carry their wearer faster than wind. And a cloak that makes you invisible."

"Let me try out these things first, if I am to be your judge. Yes, the purse does what you say. The boots: they are certainly comfortable. Now the cloak: let me button this button. Can you see me?"

"Yes."

"Now can you?"

"Yes, we still see you."

"Now?"

"No, now we can't."

"And you won't any more, either!" said Liombruno. Invisible in the cloak, racing faster than wind in the magic boots, and clutching the one-hundred ducat purse, he skimmed valleys and forests.

He saw smoke and came in sight of a cottage covered with brambles, in a deep and somber gorge. He knocked, and an old woman's voice called out, "Who's knocking?"

"A poor Christian soul seeking shelter."

The cottage door opened, and a decrepit old woman said, "Oh my poor boy, what on earth possessed you to stray into these parts?"

"Ma'am," said Liombruno, "I'm looking for my wife, Fata Aquilina, and won't rest until I've found her."

"What will we do now when my sons come home? They will eat you alive."

"Why? Who are your sons?"

"You don't know? This house is the domicile of the Winds, and I am old Voria, their mother. My sons will be back any minute now."

Voria hid Liombruno in a trunk. From out of the distance came a whirring, like a fierce swaying of trees and snapping of branches, with howls echoing through the mountain ravines. Leading was North Wind, freezing cold with icicles hanging from his clothing. Next came Northwest Wind, Northeast Wind, and Southwest Wind. They had already sat down to supper when Voria's last son arrived, Southeast Wind, the one who was always late, and the minute he came in, the house grew very warm.

The first thing all these Winds said to their mother upon entering was, "What a strong smell of human flesh! Some man is in this house!"

"You're dreaming, my sons! What human being could ever penetrate these wild-goat haunts?"

467

But the Winds went on sniffing every so often and talking about the smell of humans. Voria meanwhile set before them a steaming polenta, which they greedily ate. When they had eaten their fill, Voria said, "You imagined you smelled humans because you were hungry, didn't you?"

"Now that we are full," replied Northwest Wind, "even if we had a human right here within reach, we wouldn't touch him."

"You're sure you wouldn't?"

"Absolutely. We wouldn't harm a hair of his head."

"Well, then, if you swear by St. John to do him no harm, I will produce a real live human being."

"What's that, Mamma? A man here? But how did he ever make it? We certainly will swear by St. John to do him no harm if you'll let us see him."

So amid the Winds' gusts, which almost blew him over, Liombruno came out and, at their questions, told his story.

When they learned of his search for Fata Aquilina, each one thought hard and then admitted, one by one, that in all their travels about the world, they had never come across her. Only Southeast Wind had not spoken. "Southeast Wind," said Voria, "do you know anything about her?"

"I certainly do," replied Southeast Wind. "I'm not half asleep like my brothers, who can never find anything. Fata Aquilina is sick from love. She weeps constantly, saying her husband betrayed her, and now she's at death's door, out of grief. Gallows bird that I am, I delight in cutting up around her palace, tearing open windows and doors and messing up everything, down to the bedclothes."

"You wonderful Southeast Wind! You must help me!" said Liombruno. "You must show me the way to this palace. I am Fata Aquilina's husband, and it's not true at all I'm a traitor. I too will die of grief if I don't find her."

"I don't know what to do," said Southeast Wind, "since the way there is too complicated to explain to you. You would have to come along with me, but I go with such speed that no one can keep up with me. I'd have to carry you on my back, but how could I do that? I'm all air, and you would slip off, for sure."

"Don't worry," said Liombruno. "Just go ahead, and I'll keep up."

"But you have no idea how I fly! Yet, if you want to try, we'll set out tomorrow at dawn."

Next morning, Liombruno with purse, boots, and cloak left with Southeast Wind. Every few minutes Southeast Wind wheeled around and called, "Liombruno! O Liombruno!"

"Yes, what is it?" Liombruno was way ahead of him. Southeast Wind couldn't get over it.

"We are here," Southeast Wind at length announced. "That is your beloved's balcony." With a sudden gust, Southeast Wind tore open the window. Liombruno lost no time in leaping through it, wrapped in his invisible cloak.

Fata Aquilina was in bed, and one of her serving women asked, "How do you feel my lady? A little better?"

"Better? With this infernal wind rising once more? I'm half dead."

"Can't I bring you something—a little coffee, maybe, some chocolate, a cup of broth?"

"Nothing. I care for nothing at all."

But the serving woman kept on until she persuaded her to drink a spot of coffee. She brought in a demitasse and left it on the bedside table. Invisible, Liombruno picked it up and drank the coffee. Thinking the fairy had finished it, the serving woman then brought her a cup of chocolate, which Liombruno also drank. The woman returned with a cup of broth and a breast of pigeon. "My lady, since you drank your coffee and your chocolate, I believe your appetite is coming back. Try this broth and breast of pigeon and you'll get stronger."

"What coffee are you talking about? And what chocolate?" said the fairy. "I haven't touched a thing."

The serving women exchanged glances that said, She's losing her mind.

But as soon as they were alone, Liombruno took off his cloak. "Dear wife, do you recognize me?"

The fairy threw her arms around her neck and forgave him. They swore their love for each other, declaring how they had suffered during their separation. And they gave a big banquet at the palace, inviting all the Winds to whirl around the windows in celebration.

(Basilicata)

Cannelora

Once a king whose wife bore him no children issued a decree, stating:

> Whoever can advise the king and queen
> How to have children
> Will become, after the king,
> The richest man in the land.
> But whoever proves wrong
> Will be beheaded out of hand.

This decree spurred many persons to try, and they advised all sorts of things, but every adviser ended up headless.

At last a poor old man came forward, bearded and dressed in rags. "Majesty," he said, "order a sea dragon fished up, and its heart cooked by a maiden. Just from smelling the aroma of the frying dragon, she will begin to expect a baby. After the maiden has cooked the dragon, the queen is to eat it, and she too will begin expecting a baby, and both babies will be born at exactly the same time."

Although skeptical, the king followed all the old man's instructions: he had the dragon caught, gave it to a beautiful country lass to cook, and the minute she inhaled those cooking fumes, she felt herself with child. The queen's son and the cook's were born on the same day and looked as much alike as twins. And the same day even the bed gave birth to a little bed, the wardrobe to a little wardrobe, the coffer to a little coffer, and the table to a little table.

The queen's boy was named Emile, and the cook's Cannelora. They grew up as brothers, loving each other dearly, and in the beginning the queen also loved them both. But as they grew, it annoyed her more and more to see no difference between her son and the other boy, who she feared might prove more intelligent and luckier than the little prince. So she explained to Emile that Cannelora was not his brother but a cook's son, and forbade them to treat each other as equals. But the two boys were so devoted to one another that they paid no attention. Then the queen took to mistreating Cannelora. But Emile protected him and became ever fonder of him, while the queen seethed with rage.

One day as the two boys were enjoying themselves casting bullets for hunting, Emile stepped outside for a moment, and the queen drew near the fireplace. Finding Cannelora alone, she threw a red-hot bullet in his face with the aim of killing him. But the bullet merely grazed him above his eyebrows, leaving a deep burn on his forehead. The queen was about to pick up another bullet with the tongs, when Emile returned, so she pretended that all was well and left the room.

Although the burn pained him, Cannelora pulled his hair down over his forehead and gave Emile no clue of what had happened, but went on casting bullets, gritting his teeth. A little later he said, "Dear brother, I've decided to leave this house forever and go out and seek my fortune."

Emile couldn't understand. "But why, brother? Aren't you happy here?"

With tears in his eyes and his hat pulled down over his forehead, Cannelora replied, "Brother, fate doesn't want us living together. I must leave you." All Emile's protests were in vain. Cannelora picked up his double-barrel shotgun—the offspring of another gun and born at the time the dragon's heart was cooked—and went out into the yard with Emile. "Dear brother, it grieves me to leave you, but I am giving you this remembrance." He thrust his sword into the ground, out of which spurted a fountain of clear water. He stuck the sword into the ground a second time, and next to the water sprang up a myrtle tree. "When you see this water grow muddy and this myrtle tree drying up," said Cannelora, "it will be a sign some grave misfortune has befallen me."

After those words they embraced, with tears in their eyes. Then Cannelora mounted his horse and rode off, leading his dog by a leash. After some distance he came to a crossroads. One road led into a vast forest, the other into other parts of the world. Right there at the junction was a garden, in which two gardeners were quarreling and about to come to blows. Cannelora entered the garden and asked why they were quarreling.

"I found two piasters," said one man, "and my friend here wants one of them because he was standing next to me when I found it."

"I saw the money first," said the other man, "rather, we both saw it at the same time."

Cannelora drew four piasters from his pocket and gave each of the men two of them. The gardeners couldn't thank him enough, and kissed his hand. He moved on, taking the way that led into the forest. Then the gardener who had ended up with four piasters shouted to him, "Young man, if you go that way, you'll never get out of the forest. Take the other road instead."

Cannelora thanked him and took the other road. Much farther on, he ran into a group of young ruffians hitting and torturing a snake. They had already cut off the tip of its tail to watch it wriggle all by itself. "Let the poor creature go!" cried Cannelora, and the serpent slithered off with its tail mutilated.

Cannelora entered a large forest, and night fell. It was icy cold. From every direction came the howls of wild animals, and Cannelora feared for his life. All of a sudden, amid the cries, appeared a beautiful maiden holding a light, and took Cannelora by the hand.

"Poor youth!" she said. "Come warm up and rest at my house." Cannelora thought he must be dreaming. Unable to utter a sound, he followed the maiden. Leading him inside, she said, "Do you remember the snake you rescued from the ruffians? I am that snake. If you look at my little finger, you will see that the end has been cut off. That came about when they cut off the tip of the snake's tail. And now I'll save your life, just as you saved mine."

Cannelora was overjoyed. The fairy built him a fire, brought out food, and they dined together. Then they went off to rest for the night, each one in a separate room. In the morning the fairy embraced and kissed him, saying, "Go now, dear youth. You will meet with still more suffering, but the day will come when we'll be reunited and happy."

Cannelora didn't know what she meant, but he kissed her one more time and departed, with tears in his eyes. He came to a forest and saw among the trees a doe with golden horns. He aimed his double-barreled gun, but the doe ran off, with him in pursuit. He thus came to a cave in the heart of the forest. At that instant, a great storm broke: hailstones the size of eggs fell, so Cannelora took refuge in the cave. While he waited there, he heard a tiny voice outside in the rain. "Will you let me come in, good youth, out of the storm?"

Cannelora looked out and saw a snake. He knew that helping snakes brought him good luck, so he said, "Come in, make yourself at home."

"But," said the snake, "I'm afraid the dog will bite me. You couldn't tie him up, could you?"

Cannelora tied him up.

"But," said the snake, "the horse could stomp me with his hoofs."

Cannelora fettered the horse. "Now," said the snake, "I'm uneasy because your gun is loaded. What if it went off somehow and killed me? I'm scared."

Cannelora humored the snake by unloading his gun, then said, "All right, now you can come in without fear."

The snake entered and immediately changed into a giant. With the dog and horse tied up and the gun unloaded, Cannelora was defenseless. With one hand the giant grabbed him by the hair and with the other uncovered a tomb there in the cave and buried him alive.

Meanwhile at the king's house, young Emile knew no peace. Every day he went into the garden and looked at the fountain and myrtle tree, and one day he found the water muddy and the myrtle tree dried up.

"Woe is me!" he exclaimed. "Some misfortune has befallen my brother Cannelora. I will go out in the world and find him and see what I can do for him."

Neither the king nor the queen could stop him. He took up his gun, sent his dog running on ahead of him, mounted his horse, and rode off. At the crossroads he saw the two men's garden and, yes, met the one who had ended up with four piasters. "Welcome back, young man!" said the gardener, doffing his cap. "Do you recall the four piasters you gave me the other time? And I told you one road was dangerous and advised you to take the other, remember?"

"I certainly do," said Emile, who was so happy to learn which way Cannelora had gone that he gave the man four more piasters and continued on his way. He too reached the forest where Cannelora had met the beautiful fairy with the end of her little finger missing.

"Welcome, friend of my husband!" said the fairy as she appeared before Emile.

Amazed, Emile asked, "But who are you, madam?"

"I am the fairy betrothed to your Cannelora."

"Tell me, then, is Cannelora alive? If so, please tell me where he is, for I'm anxious to go to him."

Tears came into the fairy's eyes. "Do hurry, for our dear boy is suffering in an underground tomb. But beware of the false snake." At that, she disappeared.

Emile took heart and pressed on. He reached the forest, where he too pursued the dog with the golden horns, then was caught in the storm and took refuge in the cave. The tiny snake came crawling up asking if it could come in and get warm, and he said yes. Just as it wished, he tied up the dog and then the horse, but when it asked about the gun, Emile remembered what the fairy had told him, and said, "Oh, yes, you want me to unload it, do you?" He took aim and fired two shots into the snake, and what should he then see at his feet but a dead giant with two bullet wounds in his head gushing blood by the buckets. At the same time many voices underground cried, "Help, help, good soul! You've come at last to rescue us!"

Emile opened the tomb and out came Cannelora, followed by a long line of princes, barons, and knights, buried there for years and years and living on bread and water. Emile and Cannelora fell into each other's arms. Then the brothers and all the noblemen rode out of the forest in a grand cavalcade.

They set out to find the fairy without the tip of her little finger, whom they soon saw coming to meet them and followed by other very beautiful fairies, but none so beautiful as she was. Taking Cannelora by the hand, she helped him from his horse, embraced him, and said, "My dearest, our worries are over. You saved my life, and I'm going to make you the happiest man alive. You shall be my husband."

Then she called another fairy, the most beautiful after herself, and said, "Belle, kiss Emile, my husband's dearest friend and also a prince. Be his wife, and you both will be happy."

Then she addressed the other fairies. "Each of you pick the nobleman you like best, give him a kiss, and become his bride."

So there was a grand wedding of fairies, and lucky were the men who were there! Then everyone, including Emile and Cannelora, took his bride home, and there was great rejoicing throughout the kingdom. Poor girls were given the means for a fine wedding, but poor little me was not present, so here I am empty-handed!

(Basilicata)

◈ 136 ◈

Filo d'Oro and Filomena

There was once a shoemaker's daughter named Filomena (Nightingale), whose father and mother were very old. One day her mother said to her, "Filomena, go to the market-gardener and buy a cabbage for soup. If you don't find the gardener, pick the cabbage yourself and leave the money on the ground for him. But be careful not to pick any savoys instead of cabbage."

The girl went to the market-garden, but the farmer was not there. She went to pick a cabbage but accidentally pulled up a savoy. In its place she left a crown. She'd no sooner put the money on the ground than it disappeared and a little crystal window opened. A handsome youth appeared at the window and said, "Come to me, lovely maiden. I'm madly in love with you!"

The next thing she knew—as though drawn by a magnet—Filomena found herself underground with him, in a room fit for a queen. The youth kissed her and said, "I am Filo d'Oro [Gold Thread], and you will be my bride." Then he gave her a bag of money and said, "Go home to your parents, but come back to see me every day. You will always find the savoy you pulled up this morning in the same place. Pull it up again and throw in a crown where the root was. That way you will see me again. But make sure you are the only one who sees me. No one else must look upon me."

Filomena went home overjoyed and told her parents everything that had befallen her. The two old people couldn't get over it, and days of plenty began for them. Every day the girl would go to the garden and come home with a bag of money. But her mother was dying to see this bridegroom of her daughter's. "Let me see him just once," she begged. "I am your mother!"

"No, Mother, for if you look upon him, my luck will leave me."

"But you can at least let me see the place where he appears. That much you can do!"

So the girl ended up taking her mother there.

"This is the market-garden, and this is the savoy. Now goodbye, Mother, you must go."

The old woman pretended to leave, but hid behind a walnut tree instead. Filomena pulled up the savoy, threw in the crown, saw the little window; but this time there was no Filo d'Oro peering through the crystal panes. The old woman who was dying to see what her son-in-law looked like, threw a walnut at the window. The panes shattered, and the youth's face appeared, flushed with rage, then immediately vanished along with the window and everything else. Back came the savoy where it had been, but it could no longer be uprooted as before.

Filo d'Oro, mind you, was the son of an ogress, who wanted to marry him to a princess; but the fairies had destined him to wed a shoemaker's daughter. So the ogress had said, "May you see only one woman in the world, and should you look upon a second, may you die!" And to keep him from seeing any women except the one he would wed, she had shut him up in that underground dwelling.

The fairies, who wanted to rescue him from his mother's curse, had brought it about that the first woman seen by Filo d'Oro was Filomena, and they made him fall in love with her. But the minute he saw her mother, the curse took effect, and he died in the ogress's arms.

Finding herself with a dead son on her hands, and all because of her curse, the ogress began tearing out her hair. As Filo d'Oro had formerly been put under a spell by the fairies, his dead body did not decay. His mother buried him up to the waist and went to look on his beautiful face every day and weep.

Meanwhile Filomena, grieved over the disappearance of her bridegroom, had left home and gone out into the world looking for Filo d'Oro. One night she stopped under an oak tree to sleep. In this oak a pair of doves had alighted, and Filomena heard them singing:

"Dead is Filo d'Oro,
Coo-coo-roo-COO,
But on lives lovely Filomena . . .
Coo-coo-roo-COO,
Let her kill us,
Coo-coo-roo-COO,
Then burn us,
Coo-coo-roo-COO,
Then smear him with our ashes,
Coo-coo-roo-COO,
Thus Filo d'Oro would she save,
And he would rise up from his grave,
Coo-coo-roo-COO."

After the song was over, Filomena waited for the two doves to fall asleep; then without a sound, she climbed the oak, grabbed hold of them, and killed them. From her post high in the tree, she saw a little light off in the forest. She came down and made her way toward the light. It came from a hut, and the girl went in and asked for fire to burn up the doves. The hut was occupied by a fairy baker, who put the doves in the fire and, after hearing Filomena's tale, said, "My daughter, keep the dove ashes in this pot, and also carry along this basket of figs with you. Then go up to the ogress's windows. You will find her spinning at the window and, to stretch the thread, she lets the spindle down to the ground from the window. You are to pick up the spindle and stick a fig on it. The ogress will eat the fig, thank you, and invite you in. But be careful, for she can gobble you up. Don't go in until she has sworn by Filo d'Oro's soul not to eat you. Then you'll show her you have the ashes to revive her son, and leave everything else to fate."

Tickled pink, the maiden thanked the fairy baker and went to the house of the ogress, who was spinning at the window. The girl stuck a fig on the spindle, the ogress drew the spindle up, saw the fig, and ate it. "Excellent!" she exclaimed. "Let the good soul who stuck a fig on my spindle come inside so I can kiss you."

"No, because you'll eat me!" replied Filomena.

The ogress threw the spindle back out, and Filomena stuck on another fig. "Come in and let me kiss you! I won't eat you, I promise!" said the ogress, after eating the fig.

"I don't trust your promises," said Filomena, putting another fig on the spindle.

"Come on, I swear by Filo d'Oro's soul I won't eat you."

So Filomena went inside the house. But when the ogress discovered she was Filo d'Oro's wife bringing ashes to revive him, she took the pot away from her at once and brought her son back to life herself. Then she shut him up underground again so he wouldn't see Filomena, and hatched quick plans to wed him to that princess.

"Really?" commented the fairies, who had decided the shoemaker's daughter would get Filo d'Oro. "In that case, we'll just put a curse on the princess: in one month the earth shall yawn under her feet, and she shall fall into Hell."

In the meantime the ogress kept Filomena there as her servant, racking her brains for an excuse to gobble her up.

"You must make five down mattresses for Filo d'Oro, who's getting married," she told her. "Here are the ticks: fill them with down in twenty-four hours' time, or I'll eat you."

Filomena wrung her hands and wept. But Filo d'Oro, mind you, under the fairies' spell, could change his form and, thus transformed, he came out of the underground palace. He became a man with a beard and went to Filomena. "Lovely maiden," he said, "kiss me, and I'll get you all the down you need in a flash."

But Filomena replied, "Were you Filo d'Oro, not one kiss but a thousand would you get. But I refuse to kiss you, even though my life depends on it."

The bearded man smiled and disappeared. Then through the windows into the room flew thousands of birds of every species. In and out they flew, flapping their wings and strewing the room with feathers galore of every color. The carpet of feathers became thicker and thicker, so in twenty-four hours' time Filomena was able to pack five mattresses as the ogress had ordered.

The ogress said to herself, "This is surely my son's doing. But we'll just see who wins in the long run." She said to Filomena, "You must go to my ogress sister, who lives on the Mountain of Entertainment, and get her to give you the music box."

Now to reach the Mountain of Entertainment, it was necessary to cross the River of Serpents, the River of Blood, and the River of Bile, and once you did and got to the ogress's house, you then ran the risk of being eaten alive. All the poor maiden could do was weep.

But lo and behold, here came a man with whiskers, who was none other than Filo d'Oro in disguise. "Give me a kiss," he said, "and I'll show you how to make off with the box and get back here safe and sound."

"Were you Filo d'Oro, I'd give you a thousand kisses," replied Filomena. "But I would rather end up in the mouths of both ogresses than give you a single kiss."

Deeply moved by her loyalty, Filo d'Oro said, "Even if you won't kiss me, I'll still help you. When you reach the River of the Serpents, say, 'Oh, what macaroni! It looks so good I could eat three bowls of it!' When you reach the River of Blood, say, 'What wine! I'd gladly drink three glasses of it!' At the River of Bile, say, 'What milk! It looks so good I could drink three cups with pleasure!' That way, you will get to the ogress's house. Take this shovel, which will come in handy. Farewell." And the man with whiskers disappeared.

Filomena set out, and repeated what the man with whiskers had told her to say. When the serpents heard themselves called macaroni, they separated and let her through. The blood hearing itself called wine also separated, as did the bile upon hearing itself called milk.

She scaled the Mountain of Entertainment and came to the ogress's house, which she entered, scared to death. In the kitchen, a servant was baking bread. This servant girl was a poor maiden like Filomena, and had fallen into the ogress's clutches through misfortune. Three times a week she had to rake the embers out of the oven with her bare hands and put in the bread. The poor thing would suffer terribly from all her burns. But once the bread was baked and removed from the oven, the ogress magically healed the burns, so the girl didn't die. But when time came to bake again, she had to suffer all over.

Seeing Filomena enter, the girl shouted, "Be gone, for heaven's sake! What are you doing here anyway? Don't you know the ogress will eat you?"

"If she doesn't, then her sister will," said Filomena. "So I might as well get what I came for."

"What's that?"

"The music box."

"Listen, we must help each other. I see you are carrying a shovel. Give it to me, so I can put things into the oven and remove them without burning myself, and I'll get you the music box, which I alone can find."

Filomena gladly gave her the shovel and left with the music box. Meanwhile the ogress returned, missed the box, and screamed, "I've been robbed! Serpents, eat her alive!"

"No," said the serpents. "She called us macaroni!" At that, they let her through.

"Blood! Drown her!"

"No," answered the blood, "she called me wine!" And it let her through. "Bile! Sweep her under!"

"No," replied the bile. "She called me milk!" And it let her through.

But once the three rivers were behind her, Filomena was overcome with curiosity to know what music and song were in the box. She opened it, heard a "Zing!" and an "Ooh!" and that was it. The box was empty, music and song had escaped together. Filomena burst into tears.

Lo and behold, here came a man with sideburns, who was none other than Filo d'Oro. "Will you give me a kiss? I'll get music and song back into the box for you."

As usual, she answered, "If you were Filo d'Oro, one thousand kisses. But for you, nothing."

"But I am Filo d'Oro!" The man with sideburns disappeared, and in his place stood Filomena's husband. Trembling with emotion, she fell into his arms and kissed him a thousand times, while music and song came back into the box and were heard throughout the countryside.

"Go home in good spirits, Filomena," said Filo d'Oro, "and we'll be man and wife in three days."

Overjoyed, the maiden returned to the ogress. Sure that the girl had been swallowed by the rivers or devoured by her sister, the ogress had already set her son's wedding to the princess cursed by the fairies for three days hence. Seeing Filomena walk in with the music box, she turned livid with rage and said, "My son's wedding will take place in three days. You will hold the candlestick during the ceremony."

Filo d'Oro seemed willing to go through with the ceremony. But he wanted it to take place at midnight. All the guests were waiting, but there was no sign of the wedding procession. Filomena held the candlestick and became more and more uneasy as the minutes ticked by. Lo and behold, here came the procession, with Filo d'Oro holding onto the princess's arm. At that moment the church bell sounded—dong, dong, dong, twelve strokes. The earth yawned beneath the princess, and she disappeared into the flames.

Filo d'Oro took Filomena by the hand. "This is my bride," he said, and heavenly music wafted from the music box.

Filo d'Oro and Filomena got married. At that, the ogress let out a shriek, put her hands to her forehead, and pronounced this curse: "You who have charmed my son can have a baby without dying in childbirth only when I put my hands to my forehead like this!"

At that threat, Filomena grew weak in the knees. But Filo d'Oro squeezed her hand and gave her courage.

Sometime later, Filomena began expecting a baby. "When you see the time has come to be delivered of the child," Filo d'Oro told her, "dress in mourning and go to my mother. She'll ask you the reason for the mourning and you will say, 'Because Filo d'Oro is dead.'"

Filomena did just that. When the ogress heard "Filo d'Oro is dead," she put her hands to her forehead, very upset and crying, "My poor son," and Filomena gave birth right away, with no risk, to a fine baby boy.

Filo d'Oro then came in and, seeing him alive, the ogress forgave him and his wife and blessed the baby. So they lived in peace all the rest of their days.

(Basilicata)

◈ 137 ◈

The Thirteen Bandits

There were once two brothers, it is said—one a rich cobbler, the other a poor farmer. One day the farmer was in the country and saw thirteen men under an oak tree, each with a wicked-looking knife that would scare anyone to death. Bandits! thought the farmer, and hid. He watched them go up to the oak tree and heard their chief say, "Open up, oak!" The trunk yawned, and one by one the bandits went in. The farmer continued to watch from his hiding place. In a little while the bandits came out, one by one, with the leader bringing up the rear. "Close up, oak!" he said, and the oak went back together.

When the bandits were gone, the farmer decided to try it himself. He went up to the tree and said, "Open up, oak!" The tree opened, and he went in. There were stairs that led underground. He went down and found himself in a cave containing thirteen piles of treasure from floor to ceiling; there were several heaps of gold, several of diamonds, and several of napoleons. The farmer stared and stared, feasting his eyes on all the glitter, and once his eyes had got their fill he proceeded to fill his pockets, beginning with those in his coat, then those in his pants; finally he pulled his pants up tight against his seat and crammed all the empty space with gold pieces and went slowly jingling home.

"What happened to you?" asked his wife, when he came walking in like that. He emptied pockets and pants and told her everything. He thought he could best count the money by using one of those bottles in which wine is measured; but having nothing like that on hand, he had to send to his brother's to borrow one. The cobbler wondered, "What on earth could my brother be measuring? He never has anything to his name. I shall find out." So what did he do but stick a fishbone on the bottom of the measuring bottle.

When the bottle was returned to him, he checked at once to see what had caught on the fishbone. Just imagine the expression on his face when he beheld a napoleon!

He went to his brother right away. "Tell me who gave you that money!" And the farmer told him. The cobbler then said to him, "Brother, you just have to take me to that place too! I have children to support, and need money worse than you do!"

So the two brothers took two pack animals and four sacks to the tree and said, "Open up, oak!" They filled up their sacks and left. Back home, they divided up the gold, diamonds, and napoleons, and now had ample means to live comfortably. Therefore they said to each other, "We are well fixed now. Let's not even think of returning to that place, if we don't want to lose our life!"

Though the cobbler had agreed, he couldn't resist deceiving his brother and going back one more time by himself to plunder, since he was the kind of man who never got enough. He went and waited for the bandits to emerge from the oak, but he failed to count them when they went off. He paid for his folly: instead of thirteen, only twelve came out; one stayed behind to keep watch, since they had realized someone was coming into the cave and robbing them. The bandit leaped out, taking the cobbler by surprise, butchered him like a pig, and strung him up on two branches.

When he failed to come home, his wife went to the farmer. "Dear brother-in-law, I know something bad has happened! Your brother went to that oak tree again and hasn't come back!"

The farmer waited until nighttime and went to the oak. Strung up on the branches was the quartered body of his brother. He untied it, loaded it onto his donkey, and carried it home to a wailing wife and children. So as not to bury him quartered, they called in one of his fellow cobblers to sew him back together.

The cobbler's widow, with all the money left her, bought a tavern and became a tavern-keeper.

Meanwhile the bandits were going about town in search of the person who had inherited the money. One went to the cobbler who had sewn the body back together and said to him, "Friend, could you stitch up this shoe a bit?"

"Are you joking?" he asked. "I stitched up a shoe-mender. Do you think I couldn't sew up a shoe?"

"Who was this mender of shoes?"

"A fellow cobbler who had been completely quartered. The husband of the tavern-keeper."

So the bandits learned that the tavern-keeper was the one to profit from the stolen riches. They got a large cask, and eleven of them hid inside. The cask was put on a cart, and the other two drew it along the street. They stopped at the tavern and said, "Good lady, will you let us leave this cask here for the time being? And will you feed us?"

"Make yourselves comfortable," replied the tavern-keeper, and put on macaroni for the two carters. Meanwhile the daughter, who was playing nearby, heard noise inside the cask. She listened closely and heard, "Now we'll take care of this woman!" The girl jumped up and ran to tell her mother. In a split second the woman grabbed up a kettle of boiling water and dashed it into the cask, scalding the bandits to death. Then she went and served the other two macaroni. She poured out drugged wine for them and, when they fell asleep, cut off their heads. "Now go for the judge," she told her daughter.

The judge arrived, recognized the thirteen bandits, and rewarded the tavern-keeper for crushing such lawlessness.

(Basilicata)

◆ 138 ◆

The Three Orphans

A man with three sons died of illness. The three sons became three orphans. One day the oldest announced: "Brothers, I am leaving home and going out to seek my fortune." He came to a city and began crying out in the streets:

> "Whoever would have me as his helper,
> Him do I want for a master!"

An important gentleman appeared on a balcony. "If we can reach an agreement," he said, "I'll take you on as a helper."

482

"Fine, offer me whatever you wish."

"But I expect obedience."

"In all things will I obey you."

Next morning the gentleman called the boy and said, "Take this letter, mount this horse, and away. But never touch the reins, for if you do, the horse will turn back. You have only to let him gallop, for he knows the way to the place where you are to deliver the letter."

He mounted and rode off. On and on he galloped, coming at length to the edge of a deep ravine. I'll surely go plunging down there! thought the orphan and pulled on the reins. The horse wheeled about and was back at the palace in a flash.

Seeing him back, the master said, "So you didn't go where I sent you! You are dismissed. Go over to that pile of money, take as much as you like, and get out."

The orphan filled his pockets and was off. As he stepped outside, he fell straight down into Hell.

As for the other two orphans, when their big brother failed to return, the second oldest also decided to leave home. He took the same road, came to the same city, and he too proceeded to cry:

> "Whoever would have me as his helper,
> Him do I want for a master!"

The gentleman came out and called to him. They made a bargain, and next morning the boy was given the same instructions as his brother and sent off with the letter. He too pulled on the reins as soon as he came to the edge of the ravine, and the horse turned back. "Now," said the master, "take as much money as you like and get out!" He filled his pockets and left. He went out and straight down to Hell.

When neither one brother nor the other returned, the little brother also left home. He traveled the same road, came to the same city, cried, "Whoever would have me as his helper, him do I want for a master." The gentleman appeared, invited him in, and said, "I offer you money, food, and whatever you want, on condition that you obey me."

The orphan consented, and next morning the master gave him the letter and all the instructions. When he got to that drop-off of the road, the boy looked straight down the rocky precipice and felt his flesh creep, but he thought to himself, God help me!, closed his eyes, and when he reopened them, he was already on the other side.

On and on he galloped and came to a river as wide as a sea. He though, What choice do I have if I really must drown? All the same, God help me! At that, the water divided, and he crossed the river.

On and on he galloped and came to a swollen stream of blood-red water. He thought, Here's where I drown for sure. All the same, God help me! He plunged into the stream, and the water divided before the horse.

On and on he galloped and came to a forest so thick that not even a little bird might fly through it. Here I'm doomed, thought the orphan, but so is the horse. God help me! and he galloped onward into the forest.

In the forest he came upon an old man sawing on a tree with a blade of wheat. "What on earth are you doing?" he asked him. "Do you expect to cut down a tree with a blade of wheat?"

"One more word out of you, and I'll also cut your head off."

The orphan galloped away.

He rode and rode and came to an arch of fire, with a lion on each side. "I'll surely get burned going through there, but so will the horse. Forward, with God's help!"

On and on he galloped and came upon a woman kneeling on a stone and praying. At that point the horse drew to a sudden halt. The orphan realized that the letter was for the lady and gave it to her. She opened and read it, then scooped up a handful of sand and threw it into the air. The orphan remounted his horse and took the way back.

Upon his return, the master, who was the Lord, said to him, "The ravine, mind you, is the chute into Hell; the water, the tears of my mother; the blood, the blood of my five wounds; the forest, the thorns of my crown; the man sawing the tree with the blade of wheat, Death; the fiery arch, Hell; the two lions are your brothers, and the kneeling lady is my mother. You obeyed me. Take all the money you want from that pile of gold."

The orphan wanted nothing, but ended up taking a single gold coin, and thus left the Lord.

The next day he went shopping and spent the coin, but he always found it in his pocket and lived happily ever afterward.

(Calabria)

Sleeping Beauty and Her Children

There was once a king and queen who had no children, which made the whole court as sad as if it had been in mourning. The queen prayed night and day, but no longer knew which saint to turn to, for they all turned a deaf ear to her. Finally one day she prayed this prayer: "Blessed Mother, help me to have a daughter even if she should have to die at fifteen from pricking her finger on a spindle!"

Lo and behold, she began expecting a child, and the loveliest of baby girls was born to her. They had a grand christening and named her Carol. Nobody on earth was happier than the king and queen over the blessing they had received.

The child grew by leaps and bounds and became ever more graceful. When she was almost fifteen, the queen recalled the vow she had made. She told the king, whose grief was indescribable. Right away he issued an order for all the spindles throughout the kingdom to be destroyed. Anyone in whose house a spindle was found would be beheaded without fail. Those persons who earned their living with a spindle were to go to the king, and he would then support them. Not satisfied with his proclamation, the king had his daughter locked up in her room for greater security, and ordered that she was to see no one at all.

Alone in her room, Carol entertained herself by looking out the window. Now there was an old woman living across the street. One knows how some old people are: they get so wrapped up in themselves they can't for the life of them think of anything else. This old woman had kept a spindle and a wad of cotton wool, and whenever the mood struck her, she would spin a while on the sly.

Spinning at the window to get a little sun, the old woman caught the attention of the king's daughter. Carol had never before seen anybody make such strange movements with their hands, and her curiosity was aroused. "Ma'am! Oh, ma'am!" she called. "What are you doing there?"

"I'm spinning this teensy wad of cotton wool, but don't you tell a soul!"

"May I try to spin a little myself?"

"Of course, dear. Just don't let anyone see you doing it!"

"All right, ma'am, I'll let a little basket down into the street, and you put those things in the basket, where you'll find a present for yourself."

So she let a purse of money down to the old woman and pulled up the spindle and cotton wool. As happy as happy could be, she tried to spin.

She spun the first thread, then the second, but the third time the spindle slipped and the point stuck in her right thumb under the nail. The maiden fell to the floor, dead.

When the king knocked at his daughter's door and got no answer, he tried the door and found it bolted from the inside (she'd locked it so she wouldn't be caught spinning). Then he had it broken down and saw Carol lying lifeless on the floor next to the spindle.

There are no words to describe the grief of the king and queen. Poor dear, as beautiful as the girl had always been, she looked as though she were only sleeping; nor did her face even grow cold. She just didn't breathe any more, nor did her heart beat any longer, as though a spell had been cast over her.

Her poor father and mother stood at her bedside for weeks on end, hoping she would come back to life.

They couldn't believe she was dead, and so they refused to bury her. They had a castle built on a mountaintop with no door, but only a window high up from the ground. Inside they laid their daughter on a wide bed surmounted by a canopy embroidered in gold and full of flowers, and they dressed her in her bridal dress, which had seven skirts with silver bells. After placing one last kiss on that face as fresh as a rose, they left the castle by a door that was immediately walled up.

One day long after that, another king, who was young and had been left an orphan with his queen mother, was out hunting in those parts, and chance led him right up to that castle. "What can it be?" he wondered. "A castle with no doors and only one window? What on earth is it?" The dogs ran around the castle and wouldn't stop barking, while the young man was dying of curiosity to know what was inside. But how was he to get in? The next day he returned with a rope ladder, which he threw up to the window and thus managed to climb inside.

At the sight of the maiden lying among the flowers with her face as fresh and beautiful as a rose, he almost swooned away. He got hold of himself, eased up to the bed, reached out and touched her forehead, discovering it was still warm. So she's not dead! he thought, unable to take his eyes off her. He stayed there until night, expecting her to awaken any minute, but she didn't awaken. He returned the next day too, and the next; by then he couldn't bear to be away from her so much as an hour. He kissed her repeatedly and all but devoured her with his eyes. In short, he was in love with her, and the queen mother couldn't imagine what was eating her son and keeping him away from home all the time.

The young king's love was so intense that the sleeping maiden gave birth to twins, a boy and a girl, and you never saw two more beautiful children in your life. They came into the world hungry, but who was to nurse them if their mamma lay there like a dead woman? They cried and cried, but their mother didn't hear them. With their tiny mouths they began seeking something to suck on, and that way the boy child happened to find his mother's hand and began sucking on the thumb. With all that sucking, the spindle tip lodged under the nail came out, and the sleeper awakened.

"Oh, me, how I've slept!" she said, rubbing her eyes. "But . . . where am I? In a tower? And who are these two babies?" She was growing more and more puzzled, when the young king, having climbed up to the window as usual, jumped into the room.

"Who are you? What do you want with me?"

"Oh! You're alive! Speak to me, my love!"

After their initial amazement was over, they began talking, learned of each other's royal origins, rejoiced, and embraced one another as man and wife. They named the baby boy Sun, and the baby girl Moon.

The king returned to his court, promising to come back for his bride and shower her with fine gifts and arrange for their wedding. But the poor girl was born under a truly unlucky star: the minute the king reached his palace, he fell sick. His illness was so serious that he was practically unconscious and refused all food. All he did was repeat:

> "O Sun, O Moon, O Carol,
> If only I had you at my table!"

Hearing those words, his mother suspected her son had been bewitched, and she had the woods combed in search of the place where he used to go every day. When she learned that in an isolated castle lived an unknown young woman with whom her son was madly in love and who had borne him two babies, she was seized with fierce hatred, awful woman that she was. She sent two soldiers to the castle to order the young woman to give them Sun, since the king was sick and wanted to see him. Though Carol didn't want to, she had no choice but to obey, and weeping, handed him over to the soldiers.

The soldiers returned to the palace, where the queen awaited them on the stairs. She took the baby in to the cook. "This," she said, "you are to roast for the king."

But the cook was a good man and didn't have the heart to slay the baby. He gave him to his wife, who kept him hidden and nursed him. In the baby's stead, he roasted a lamb and took it in to the sick king. Seeing the food, the king as usual sighed and said:

> "O Sun, O Moon, O Carol,
> If only I had you at my table!"

Passing him the dish, his mother said, "Eat, my dear, you're feasting on your very own!"

The young man heard those words and looked up at her, but he didn't understand what she meant.

The next morning the cruel woman sent the same soldiers back to the castle to fetch the girl child. This time too the baby was saved by the cook and given over to his wife to nurse; in her stead another lamb was roasted. Again the king's mother said, "Eat, you're feasting on your very own!" In a whisper he asked her what she meant, but his mother made him no answer.

The third day the soldiers were ordered to fetch the young woman. The poor thing, frightened to death, followed the two men, dressed in her bridal gown of seven skirts with silver bells on them.

The queen was on the stairs waiting for her and began slapping her right and left as soon as the girl came up to her. "Why do you beat me?" asked the poor soul.

"Why? You cast a spell over my son, you ugly witch, and now he's dying! But do you see where you will end up yourself?" asked the queen, pointing to a kettle of boiling pitch.

In the meantime the king heard nothing, since in his room was a band of musicians his mother had sent for, saying the doctors had prescribed that to cheer him up.

Words can't describe the terror of the poor girl when she saw the kettle prepared for her and learned that she was to die. "Take off those skirts," ordered the queen, "and then I'll throw you into the pitch."

Trembling, the young woman obeyed. She removed the first skirt, and the silver bells rang. Indistinctly the prince heard the music of the bells, and it struck him as a familiar sound. He opened his eyes, but the drummer meanwhile beat the bass drum, so he decided he must not have heard right after all.

The young woman removed her second skirt, and the bells rang louder. The prince raised his head and was almost sure he heard Carol's skirts, but about that time the cymbalist played the cymbals, and he couldn't hear anything else. Then he imagined he heard still more jingling, clearer yet, and he strained his ears to listen. The young woman thus removed one skirt after the other, and each time the bells rang louder, until at last they resounded through the whole palace.

"Carol!" cried the king, and jumped out of bed, weak and shaky as he was. He made his way downstairs and saw his beloved about to be thrown into the kettle.

"Stop!" he shouted. He grabbed his sword, thrust it against the queen, and said, "Confess your sins!"

When he learned that the children had been served up as a meal, he ran in to kill the cook, but in the kitchen they told him at once that the babies were safe and sound, and he was so thrilled that he laughed and danced like a madman.

In the meantime the queen had been thrown into the kettle, which was just the place for her. The cook received a handsome present. And the king with Carol, Sun, and Moon, lived happily ever afterward.

> Whether a long tale or a short tale,
> Let's hear yours, now you've heard mine.

(Calabria)

❖ 140 ❖

The Handmade King

Once there was a king whose wife had died and left him with a daughter on his hands. The daughter had reached marriageable age, and there came asking for her hand sons of kings, marquis, and counts, but she rejected them all.

Her father sent for her and asked, "Daughter, why is it you do not wish to marry?"

"Papa," she replied, "if you would have me marry, give me one hundred and seventy-six pounds of flour and the same measure of sugar, for I want to create my betrothed with my own two hands."

The king shrugged his shoulders and said, "All right, you shall have them." He gave her the sugar and the flour, and the daughter shut herself up in her chamber with a kneading trough and a sieve and began sifting flour. Six months she devoted to the sifting and refining, and six months to the kneading and shaping. When the kneading and molding was done, she disliked the way it had turned out, so she undid it and started over. The second time it finally came out the way she wanted it, and she stuck in a pepper for a nose. Standing him in a niche, she called her father and said, "Papa, Papa, look at my betrothed. His name is King Pepper."

Her father looked him over and found him to his liking. "He is handsome, but he doesn't talk!"

489

"Just you wait! In time he will speak."

Every day the king's daughter went before King Pepper in the niche and said:

> "O King Pepper, made by hand,
> But pen to paper put I not;
> Six months to refine thee,
> Six months to fashion thee,
> Six months to undo thee,
> Six months to redo thee,
> Six months in the niche,
> And thou shall speak our speech!"

And for six months the girl continued to sing him that little song. At the end of six months, King Pepper started talking.

"I can't talk to you," he said. "I must first speak to your father."

The girl ran to her father. "Come here quick, Father, my betrothed is talking!"

The king came and began talking with King Pepper of this and that, and in the end King Pepper asked for the hand of his daughter. Overjoyed, the king gave orders for a grand banquet and invited King Pepper to dinner. Preparations began for the wedding, which took place two days hence in the presence of all the reigning monarchs from nearby and faraway.

Among these sovereigns was also a queen named Turk-Dog. The minute Turk-Dog laid eyes on King Pepper, she was infatuated with him and secretly resolved to steal the handmade king from his bride.

Following the nuptials, the newlyweds began to lead a very happy life, but King Pepper never went outdoors. At last the king commented on it to his daughter. "My daughter, why is it you and your husband never go out? An excursion every now and then would do you so much good!"

"Yes, Papa, you are right. As a matter of fact, I feel like a drive in our carriage this very day."

They had the horses harnessed, and the princess went out for a drive with King Pepper. Turk-Dog, who was forever on the lookout for a chance to kidnap King Pepper, began following them in her carriage. When they reached the country, King Pepper decided to get out and walk around a bit. Suddenly a strong gust of wind arose and swept King Pepper away. He was blown right up to the carriage of Turk-Dog, who put out her cloak and caught him. His wife and the coachman looked everywhere for him, but nowhere was he to be found. Griefstricken, the princess returned to the palace. "And your husband?" inquired her father.

"A gust of wind swept him away! I shall shut myself up in my room with my misery, and I don't want to hear another thing."

But she didn't stay shut up for long. Unable to stand her grief any longer, she took a horse and a purse of money, asked for her father's blessing, and rode off in search of King Pepper.

One night in a forest, she was listening to the cry of animals, when she saw a light and knocked at a house. "Who goes there?"

"A good Christian soul. Please give me shelter this night, so the animals won't eat me."

"To these parts come no Christians, only animals and serpents. If you are a Christian, make the sign of the cross."

"In the name of the Father and of the Son and of the Holy Ghost."

The door opened, and there stood an old man with a long beard, who said, "Princess, what are you doing roaming around this countryside infested with wild animals?"

"I'm out seeking my fortune. I molded myself a husband with my own two hands"—and she told him her story.

"Princess," said the old man, "you must first find your husband. Meanwhile take this chestnut and don't lose it. Tomorrow morning, continue on your way until you come to another house; there you will find my brother, and you will ask him!"

The next day the princess found the other hermit, who gave her a walnut to keep with the chestnut, and he told her the way to their third brother's house. The third hermit, who was older than the other two put together, gave her a hazelnut and said, "Go this way until you come to a large palace. Adjoining this palace, which belongs to Turk-Dog, is an uglier one, which is the prison. When you are under the palace windows, break open the chestnut and out will come treasure which you are to hawk. At the sound of your voice, Turk-Dog's maid will appear and invite you inside. Turk-Dog will ask how much you want for the item you are selling. Don't ask for money; just say you want to be left alone with Turk-Dog's husband. He is none other than King Pepper. If you don't manage to speak to King Pepper tonight, break open the walnut and proceed to sell its contents. If you don't succeed the second night either, then break open the hazelnut."

When she got to the palace, the princess broke open the chestnut. Out of it came a golden loom, with a maiden seated at it and weaving pure gold. The princess began crying. "Hallo! Who wants to buy a lovely loom, together with a maiden weaving pure GOLD?"

The palace maid looked out the window and said to Turk-Dog, "Majesty, Majesty, look what wonderful things are for sale! Do buy them to go in your gallery, for they are rare indeed."

The princess was invited inside and upstairs. Turk-Dog asked her, "What do you want for these things?"

"I want no money, but only to spend one night in a room by myself with Your Majesty's husband."

Turk-Dog was unwilling to allow that, but her maid talked her into it, so Turk-Dog served King Pepper drugged wine, put him to bed, and then told the seller of the loom, "You may go in now."

The princess couldn't for the life of her awaken King Pepper. She sang to him:

> "O King Pepper, made by hand,
> But pen to paper put I not;
> Six months to refine thee,
> Six months to fashion thee,
> Six months to undo thee,
> Six months to redo thee,
> But now Turk-Dog possesses thee;
> Awake, my king, and let us flee!"

But King Pepper heard nothing. Thus, daylight overtook her singing and weeping.

She had already departed in despair, when she remembered the hermit's advice and cracked open the walnut. Out of it came a golden tambour and a maiden embroidering with pure gold. The princess began crying. "Hallo! Who wants to buy a fine golden tambour and a maiden embroidering with pure GOLD?" The palace maid looked out and called her inside.

"What do you want for it?" asked Turk-Dog.

"I want no money, but only to spend tonight as well with your husband."

But this night, too, Turk-Dog put opium in King Pepper's wine. So one more night the princess sang and wept all in vain.

For a second night the prisoners in the building next to the palace were kept awake by those songs and laments, and they decided that if King Pepper came out tomorrow, they would call to him from their barred windows and tell him about all the wailing which inspired their pity as well as kept them awake.

Thus, when King Pepper emerged from the palace into the sunlight, the prisoners poked their hands through the bars and motioned to him, saying, "Majesty, how can you sleep so soundly at night? We hear someone weeping and calling, 'King Pepper, I am your wife!' Then we hear her singing that she fashioned you with her own two hands, spending six months molding you, and six more unmolding you. How could you be deaf to all that?"

King Pepper thought to himself, If I hear nothing, Turk-Dog must be drugging my wine. Tonight I won't drink any.

Meanwhile, the poor young lady was in the depths of despair, because only the hazelnut was left. She cracked it open, and out came a little golden basket and a maiden sewing with pure gold. The princess cried out, "Hallo! Who wants to buy a pretty little golden basket and a maiden sewing with pure GOLD?" She was brought inside and made the same bargain as for the previous nights.

When she was at last alone with a sleeping King Pepper, she was all ready to resume her song, when King Pepper (who had pretended to drink and was now pretending to sleep) opened his eyes and said, "Shhhhh, my wife. We will flee tonight. How did you ever find me?"

"King Pepper, I never gave up the search!" And she told him of all her trials.

He explained that Turk-Dog had always held him under a spell and prevented his escape, but that now the spell was weaker while Turk-Dog thought he was drugged.

They opened the door, made sure Turk-Dog was sleeping her soundest, mounted the princess's horse one behind the other, and off they sped.

Finding them gone the next morning, Turk-Dog pulled out her hair, strand by strand, and when no hair was left she tore off her head and perished.

The couple on horseback reached the palace of the princess's father. He was standing on his balcony, saw them ride up, and cried, "My daughter! My daughter!"

> They danced and sang and ate and ate,
> But us they gave not e'en a blessèd date.

(Calabria)

The Turkey Hen

Once there was a king and queen. In giving birth to a baby boy, the queen died, leaving the king with two motherless children, the boy and a little girl not much older. The poor father was so griefstricken that he spent his days weeping. He went on like that for a year, and then he too died.

He had a brother and, before dying, commended the poor orphans to him. Their uncle solemnly promised to look after them, but the minute the king was dead, all his brother could think of was getting the crown for himself and ruling over the kingdom. He was a tyrannical king; he kept his niece and nephew closed up underground, and when the boy reached ten years of age, he began sending him to the fields every day to oversee the men tilling the soil.

The boy grew up under that daily routine and, up to his seventeenth year, was unaware that he and his sister were royal children. They didn't even know that the present king was their uncle, but thought he had only taken them in out of charity.

With Christmas coming on, an old woman who raised geese and turkeys and knew the children's real rank, took pity on them. She said to herself, "Tomorrow is Christmas Eve, and those poor children are all alone. If their father, God rest him, were alive, they would have a feast and all kinds of entertainment! The late king would have done fine things for them! Everybody celebrates Christmas, including me, who keeps geese! But those poor dears have nothing. I shall make them a present of one of my turkeys, so they too can celebrate Christmas. But how will I get it to them? I can't go through the main door, since the guard always stands there. . . . I'll call the girl to the window."

So, on the morning of Christmas Eve, the old woman rose, took the fattest turkey, and began calling through the window, "Young lady, oh, miss! Today is Christmas Eve, and I want to make you a present of this turkey. You and your brother eat it, for me!"

The girl looked out. "Thank you, thank you very much, kind lady. But what can I give you in return? I have nothing to my name . . ." She didn't want to take the gift, but the good old woman was so insistent that the girl finally accepted it.

That morning being a holiday, the brother didn't go to the fields, but took the accounts to the king. Waiting for him to come back, the sister put the turkey in a dark room so no one would see it, and locked the door. Left by itself, the turkey began scratching about on the ground and digging. It dug and dug, and at length uncovered a trapdoor. Toward evening the brother came back with something to eat. They sat down to the table, brother and sister, and as they ate she said, "Guess what, brother. This morning a good-hearted old woman made me a present of a turkey hen."

"Where did you put it?" asked the brother.

"I hid it in the dark room, and I'm going to feed it now."

When the brother, who was weary, had gone to bed, the girl took up the candle and went to look after the turkey. She saw where it had been digging, spied the trapdoor, and said, "Look what the turkey has found!" She opened the trapdoor, and there was a staircase. "I shall go down," said the girl. She went down the steps and saw a king's outfit, including helmet, sword, and armor; only the crown was missing. "Whose things are these?" wondered the girl. "No matter, I'm taking them." And she carried them up to her room.

Next morning when he woke up, the brother saw helmet, sword, and armor by his bed. "Where did these things come from?"

"Would you believe it?" answered the sister. "The turkey hen went to digging, and down underneath was a trapdoor and staircase. I went below and found all these things."

"But they are for a king!" said the brother.

"Yes, they are! Isn't that nice! Put them on, brother, and let me see how you look in them. Go on, put them on!" And she helped her brother into them and clapped her hands for joy.

In that instant trumpets and drums were heard: it was Christmas Eve, and musicians had come to play under the windows of the royal palace.

The girl threw open the window, and before all the people in the square appeared the boy dressed as king, in helmet, sword, and armor.

Everybody began crying, "This is our king! This is our king!"

Hearing these cheers, the palace guards gave the alert. A great tumult arose in the crowd. The whole court began crying, "What is it? What's going on?"

Taking note of the turmoil inside the palace, the people on the outside proceeded to shout, "Down with him!" or else "Hurrah!" Meanwhile people crowded into the square from all over the city, and the greater the uproar grew, the more people it drew.

Quite pale, the king appeared on the steps, and moved forward to address the people, but they had had quite enough of his tyranny and rushed upon him with stones and clenched fists. So furiously did they beat him that he finally died, as he deserved. Then they took the royal crown and put it on his nephew's head, amid cheers and fireworks.

As king, the youth ruled justly, and everyone was happy and loved him. For his part, he was so pleased with the way things had turned out that he made a vow: every Friday all the poor people in the realm were to come to the palace, and he would distribute alms to them in person. Poor people from all over came and received alms from him, Friday after Friday. Once after a long and tiring day, he was about to leave when he saw an old blind woman come forward with a maiden about twelve years old. In a voice that inspired pity, the maiden said, "Royal Majesty, be kind to this poor blind soul, that God may reward you."

The king gave the old woman alms, but as he did so he eyed the girl, who was very beautiful, and said, "Good lady, come back every Friday, but stay apart from everybody else, don't mingle with the other poor people, so that I can see you."

The two women went off blessing him, and the king remained quite wistful; it seemed like a hundred years before next Friday when he would see if the old woman and the maiden returned. Friday finally came again, and the king looked at every single person until he spotted the two women a little bit apart from the others, according to his instructions. He waved to them, gave them a little more money than usual, then said to the maiden, "Throw away those rags you're wearing and make yourself a new dress. Wear it next Friday when you return."

Next Friday the maiden came in a cotton dress and new shoes, and the king gave her still more money. And so each week she showed up better dressed than the week before, and at last wore a muslin dress that made her look like a rose.

The king said to her, "Next Friday, you be the first one to come up."

The king had fallen in love and was always very wistful at home. His sister noticed it and asked, "What's the matter, brother?"

"Nothing . . . I've a headache . . ." At last he couldn't keep his love a secret any longer and said, "There's a poor girl I've fallen in love with, and I would like to marry her."

He never dreamed his sister would consent to his marrying a poor girl. But she was a good soul and loved her brother; besides, she too had known poverty. The only thing she asked was to see the girl.

Therefore next Friday the king's sister accompanied her brother when he went to distribute alms. The first in line was the beautiful beggar girl, who was so lovely that the sister said to the king, "Do as your heart commands." Thus the king married the beggar girl.

The day of the wedding the king said to his sister, "I'm now getting married, but nothing else will change and you will still be the mistress of the house."

The bride, however, now so rich after being so poor, became arrogant. She envied her sister-in-law who was mistress of the house and kept all the keys. Thus, bit by bit, she proceeded to turn her husband against his sister. She had him take the keys away from her, and forced him to scold her, even though she didn't deserve it in the least. Yet, the poor girl was ever kinder. But the new wife went on so about her sister-in-law to the king that he finally said to her, "Wife, what would you have me do?"

She answered, "When it's night, have her taken to the woods and slain. And to make certain they've killed her, order her heart brought back along with her amputated hands and bloody gown." The husband couldn't refuse. He ordered the executioner to take his sister into the depth of the woods at midnight and kill her, and to bring him back as proof of the deed her heart, hands, and gown.

Thus was it done. At midnight the poor girl was awakened and seized by two hired assassins. "What do you want of me?"

"By order of your brother the king, you are to come with us!"

They packed her into a carriage, and off they went, out into the country. When they reached the woods, the assassins said to each other, "So we must now kill her, for no reason. She's never done us any harm, poor thing!"

"I'm not killing her, that's for sure," said the other man. "You do it!"

"Well, what are we going to do? We have to present the king with heart, hands, and gown smeared with blood. We've no choice but to kill her."

At that instant a bleating was heard: it was a little lamb that had strayed and remained behind in the woods for the night. They seized it and said to the princess, "Take off your gown, and we'll slay the lamb and take its heart. But as much as we hate to do it, we have to cut off your hands; that is an order. You just have to bear it!" They did what they said they would and carried off the lamb's heart and the bloody hands wrapped in the gown.

The princess remained there with blood spurting from her wrists. At the sight of those poor remains, the king couldn't keep back his tears, and said, "My sister, you were so happy over my marriage, and now you are dead through my wife's doing!" Thus, thinking back to the past, he regretted what he'd done and called out, weeping, "My sister! My sister!"

While he grieved, his sister was in the woods with the blood running out of her veins. It so happened that an English lord came riding through the woods right then in his small carriage. Hearing moans, he drew near, saw her, and asked who had wounded her. The princess answered that wild animals had eaten her hands, and the Englishman remembered that he had some cloth in his carriage and gave it to her to stanch the blood. Then he invited her to get in, and took her off with him. The lord was married, but had no children. You can just imagine the happy life the girl led at his house. So she wouldn't be without hands, the lord had some wax ones made for her.

In spite of all her sorrows, the princess, who was twenty, was as beautiful and fresh as a rose. She was on the balcony when a foreign king came down the street and saw her. He liked her and asked the Englishman for her hand in marriage. The lord consented to the match, but in all honesty told him the girl had wax hands. The king said that made no difference, so he married her and took her off with him to the palace.

In a few months' time the princess began expecting a baby, when war was declared against her husband, who led his army to fight the enemy.

While he was away, the princess gave birth to two beautiful children, a boy and a girl. But the ministers, who disliked being ruled by a woman, especially one whose origin was a mystery to them, decided to take advantage of the circumstances and get rid of her.

So what did they do but write the king and tell him his wife had brought forth two little dogs, because of which they were awaiting orders from him as to what they should do with the queen.

The king nearly died from the shock and wrote back for them to await his return when he himself would see what was to be done. But the ministers, who were bent on getting rid of the queen at all costs, woke her up in the middle of the night, strapped a knapsack to her, putting a baby in each side of it, and abandoned her on a deserted shore.

The poor soul started weeping. Alone, hungry and thirsty, with those two stubs for hands, she had no idea how she would manage. She came to a pool of water and bent over to drink from it. While she was bending over, one of the babies slipped out of the rucksack and disappeared under the water. Just imagine her grief: there she was handless and unable to fish him up.

In that instant a handsome old man appeared before her, saying:

> "Plunge in your stub
> And get back hand and babe."

The princess immersed her mutilated arm in the water and felt her hand grow back. She grabbed the child at once and took him back in her arm. With that movement, the other baby slipped out of the knapsack and disappeared under the water.

Again the old man said:

"Plunge in your stub
And get back hand and babe."

So she got her other hand back and fished out the child and then she was able to nurse them both. Next, the good old man led her to a hilltop where a beautiful house stood. He invited her in, saying, "Remain here, and you will want for nothing. I will not abandon you."

Let's leave the princess and return to the king who was her husband. At the end of the war he came home, and how great was his grief on finding his wife gone! He asked for an explanation, but the ministers told him they were as much in the dark as he was: she had left in the night with the two little dogs she had brought forth. The king knew no more peace without his wife, and began combing the countryside in search of her.

Meanwhile, the queen's brother, from the moment he regretted his deed, kept to the house and let his beard grow down to his knees, out of grief over killing his innocent sister. And he imprisoned his wife who had been the cause of his injustice. His ministers kept after him until they finally got him to go out hunting one day for the sake of a little exercise. Once in the country, absorbed as he was in thought, he strayed from the ministers and lost his way. All of a sudden it began to rain, and the king took shelter under an oak tree.

It so happened that the other king as well, the husband out looking for his wife, was going through those same woods, and took refuge under the oak. Thus they met for the first time, for although they were both kings, they'd never seen each other before in their whole life. They spied a light and headed toward it in the rain. That light came from the good old man's house where their sister and wife lived.

They knocked. The old man answered the door and immediately offered them shelter. They went inside, and there was the queen. She recognized them, but they did not recognize her.

"Since it is raining," said the old man to her, "these two gentlemen here need shelter and ask hospitality of us."

"We are honored to have them here," she replied. "I was just getting supper for my children."

"So we'll all eat together," said the old man.

They were almost at the end of supper when the old man said to the two children, "Dear children, tell us a nice story now, so we'll hear you too."

The little girl, who was the more eager, then started talking. She told the story of her mother, from the time she had been taken to the woods by the hired assassins up to the moment of her marriage. Hearing those details, the brother said to himself, "But in that case she's none other than my sister!"

When the little girl stopped talking, the little boy took up from there and told the rest of the story, from the time his mother had married the king up to the moment the good old man had brought them to the hilltop, to the house they were in at that very moment. Listening to this account, the king said to himself, "So this lady is my wife, and these beautiful children are my very own? Why was I ever informed she had given birth to dogs?"

When the children had finished the tale, the old man said, "Gentlemen, this is your story." The two men embraced the lady. One of them asked her forgiveness, while the other kissed the children with tears in his eyes. The old man looked on joyfully; he was none other than St. Joseph, and as a sign of his good deed, his staff blossomed all over. "Now that I've done my part," he said, "I'll give you my blessing," after which he disappeared.

(Calabria)

◈ 142 ◈

The Three Chicory Gatherers

There was a poor mother who had three daughters. When chicory was in season, the three girls would go out with their mother to gather chicory. One day the mother and two of her daughters walked on ahead, while the oldest daughter lagged behind, having spied an enormous chicory plant which she did her best to uproot. She tugged and tugged, but the plant wouldn't budge. Then she pulled with all her might and the plant came up with so much dirt around the roots that a big hole remained, at the bottom of which was a trapdoor. The girl opened it to find an underground room where a dragon sitting in a chair said, "Mmmm! I smell human flesh! Mmm!"

"Please don't eat me," begged Teresa, "we are poor folks. I'm the daughter of a chicory vendor, and I came here to pick chicory. Poverty drives us to it."

"Well, stay here," replied the dragon, "and look after my house while I go hunting. I'm leaving your dinner here—the hand of a man. If you eat it, I'll marry you when I come back. But if I find you've not eaten it, I'll cut off your head."

Trembling all over, Teresa replied, "Yes indeed, Sir Dragon, I'll surely eat it!"

The dragon went hunting, and the poor girl from time to time would go look at that human hand in the pot and draw back in horror. How can I? she thought. How could I ever eat a human hand? It was almost time for the dragon to return, so she threw the hand into the lavatory and poured a bucket of water over it. "Now it's gone," she said to herself, "and the dragon will think I ate it."

The dragon returned and asked, "Did you eat the hand?"

"Yessirree, I did . . . It wasn't bad."

"Now we'll just see," said the dragon, and shouted, "Hand, where are you?"

"I'm in the lavatory!"

"You rascal! You threw it into the lavatory!" He grabbed the girl by the arms, carried her into a room full of beheaded dead persons, and cut off her head too.

In the evening when the mother came in from gathering chicory and didn't see Teresa, she asked the other girls, "Where is Teresa?"

"She stayed right with us," said the sisters, "up to a certain spot. Then she disappeared."

So they all went through the field calling, "Teresa! Teresa!" but there was no answer. They returned home weeping, but even though they had gathered a lot of chicory to sell and then buy food, every mouthful of food bought with that chicory was like poison to them, since they had paid with the loss of Teresa.

When Teresa still didn't return, her sister Concetta said, "Mamma, I shall go back to the same fields for chicory, in hopes of finding some trace of Teresa."

That she did, and right in the spot where Teresa had last been seen, she spied a big head of chicory. She tugged and tugged and finally uprooted it. There beneath the roots was the trapdoor. Concetta went down and found the dragon sitting in a chair. "Mmmm!" he said. "I smell human flesh!"

"Please don't eat me! I'm just a poor girl, and I've already lost a sister!"

"Your sister is here, with her head cut off because she refused to eat a human hand. Now you stay and look after my house. And for dinner you are to eat this human arm. If you do, I'll marry you. If you don't, I'll kill you like your sister."

"Yessirree, Sir Dragon, anything you say!"

The dragon went hunting, and Concetta, utterly horrified, had no idea what to do with the arm all ready for her on a plate and garnished with radishes. After racking her brains, she dug a hole and buried it.

The dragon returned and asked, "Did you eat the arm?"

"Yessirree, Sir Dragon, I really had a feast!"

"We'll just see now. Arm, where are you?"

"Underground!" cried the arm.

So the dragon cut off Concetta's head as well.

At home, when Concetta didn't return, they all went to pieces. "Now two of them are gone," they wailed.

The third daughter, Mariuzza, said, "Mamma, we can't lose two girls like that. I'm going out and look for them."

She too found the big chicory plant and uprooted it. She too met the dragon, who told her, "Your sisters are closed up in that room with their heads chopped off. You will come to the same end if you fail to eat this human foot I'm leaving here for you in the soup tureen."

Mariuzza humbly replied, "Yessirree, Sir Dragon, I'll do just as you order."

The dragon went hunting. Mariuzza racked her brains for a way out. Then an idea occurred to her: she took the bronze mortar and ground the foot to a powder with the pestle, then poured the powder into a stocking and hid it beneath her clothes on her stomach.

The dragon returned and asked, "Did you eat the foot?"

Mariuzza smacked her lips. "I can't tell you how good it was! I'm still licking my lips!"

"We'll just see now. Foot, where are you?"

"On Mariuzza's stomach."

"Hurrah! Hurrah!" exclaimed the dragon. "You will be my wife!" And he entrusted her with all the keys except the one to the room of those people he had murdered.

To celebrate their betrothal, Mariuzza served him wine. He emptied one bottle after another until Mariuzza had served him half the wine cellar, and he continued to drink. When she saw he was good and drunk, she said, "Now will you give me that key?"

"No, that one, no."

"Why? Why won't you give it to me?"

"Because . . . the dead souls are in there."

"If they are dead, you certainly don't expect them to come back to life, do you?"

"I can revive them . . ."

"Go on! You can?"

"Of course. I have the salve . . ."

"Where do you keep it, then, you fibber?"

"Uh . . . in the cabinet . . ."

"So you'll never die yourself?"

"Me, yes . . . the dove in the cage . . ."

"What does the dove have to do with it?"

"If you cut off the dove's head, you'll find an egg in its brain . . . and if you break the egg over my forehead . . . I'm done for . . ." Raving on, he put his head down on the table, dead drunk.

Mariuzza rummaged through the whole house until she found the dove. She cut off its head, got the egg, and—"crack!"—broke it over the sleeping dragon's forehead. He started, jerked a few times, then died for good.

The girl found the salve, opened up the room, and proceeded to anoint the dead people. The first one was a king, who gave a start like someone waking up suddenly. "How I've slept! Where am I? Who woke me up?" But Mariuzza paid him no mind and went on anointing the others, her sisters first of all, and then kings, princes, counts, and knights, of whom there seemed to be no end.

Countless kings and other noblemen wanted to marry the three sisters. Mariuzza said, "Here's what you must do: Play a round of morra, and the winner will choose the girl he wants."

They played morra, and a king won, choosing the oldest girl for himself. Next it was the turn of a prince, who took the second girl. Finally another king won, and chose Mariuzza.

Meanwhile, one of the barons nervously repeated, "Quick! Quick! Why are you wasting so much time? The dragon will return any minute and kill us all again!"

"Have no fear," said Mariuzza. "I slew the dragon myself."

"Hurrah! Hurrah!" they all shouted. "So we've nothing more to fear!" Each of them took a horse, divided up the dragon's treasure, and rode with the three betrothed to the city. They had a grand wedding celebration, and everyone was happy, especially the three girls' mother, who didn't have to go out any more to pick chicory.

(Calabria)

Beauty-with-the-Seven-Dresses

Once there was a father of two boys. Sensing his last hour approach, he called in his older son and said, "Son, I'm about to die. There's no more hope for me. Tell me which you prefer, my solemn blessing or a sum of money?"

Without beating around the bush, the son replied, "Give me the money, for with just the blessing I'd go hungry."

Then the father called his younger son and put the same question to him.

"Money matters little to me," said the younger boy. "I prefer your solemn blessing."

The father died and they carried him to the cemetery. The little boy, who'd received only the solemn blessing, wept heartily, while the big boy, who'd inherited all the property, was thinking of the best way to use it. He ended up opening a café and taking his place behind the counter, while the little brother, whose name was Francesco, went out into the world to seek his fortune.

One evening after walking quite a distance, he saw a little light far ahead of him and said, "If the Lord so wills, I must get to that place." He thus came to a house and knocked. Accompanied by seven ladies, Beauty-with-the-Seven-Dresses came down and offered him food and shelter. In the morning Beauty got into conversation with the young man, and was so taken with his good looks and manners, that she ended up saying she wanted him for her husband. She was a very beautiful and gracious maiden, and a few days later they got married.

One day while they were looking out the window at the garden, Beauty said to her husband, "Ciccillo, do you see that fine seven-part frock there?" (She spoke of it that way, since it included seven dresses, one inside the other.) "Do you see that seven-part frock hanging on the tree?"

"I certainly do!" he answered. "Why do you ask me?"

"I'm going to tell you. If a bird should light on the frock and you caught it, you wouldn't see me any more. If you shot the bird, the frock would fly away, and I would go through fire and water. Should worse come to worst, dress in a red outfit, which has already been laid out in this room, and leave home in search of me. I'll see to it that you find me again."

It happened one day that while the husband was out hunting and shooting birds, a bird lit right on the seven-part frock. So wrapped up in the hunt was Ciccillo that, without thinking, he fired at the bird. The seven-part frock immediately soared into the air and vanished from sight. Ciccillo then remembered his wife's warning. Frantic, he ran back to the palace at once, fearing the worst. When Beauty saw him, she asked, "What's the matter?" but he dared not tell her. Then she looked up at the tree and found the seven-part frock gone. At that, she began pulling out her hair and saying, "I've been betrayed! Betrayed! Now they'll come and take me away. Remember, if that happens, husband, to dress in red and don't abandon me."

Let's leave them and follow the seven-part frock which had taken flight at the shot. On and on it flew until it reached a palace, went through the window, and came to rest before the table of a king who was in the process of writing. The king scrutinized the seven-part frock and wondered whose it was. He asked all around, but no one knew a thing about it. Then an old woman, aware of the king's inquiries about the owner of the seven-part frock, went to the palace, announcing, "Majesty and lords, I can find the owner of this dress."

"What will it take to do so?" asked the king.

"Here's what I need. Fix me a bottle of drugged rosolio and a pound of sweets that have also been drugged. Leave everything else to me. Then I'll need a carriage with a good driver; I'll ride in it with a dagger concealed in my bosom."

The king provided her with all those things, and the old woman rode off in style.

When they had gone a certain distance, she said to the coachman, "Wait for me here and be sure to come when I call you." It was raining, but the old woman walked straight up to the palace of Beauty-with-the-Seven-Dresses. She knocked at the front door, and the husband came down with the seven ladies to let her in. The old woman asked for shelter for the night because it was raining, and he gladly welcomed her and invited her to table with them. At the table the old woman pulled out the rosolio and sweets, which were all drugged, and said, "These are not fit for important people like you, but do eat them for my sake. My daughter has just married, and I brought along this little bit so that you can celebrate the occasion too."

Once the sweets were eaten, the couple and all the other guests dropped to the floor like pears. The old woman then pulled out the dagger and thrust it all the way through the husband. Then she called the coachman, who was waiting outside, and the two of them together picked up Beauty, one by the head, the other by the feet, and carried her into the carriage as she slept. Once they had her inside, they galloped off to the king.

The king was anxiously awaiting them, and when the old woman arrived, he had Beauty-with-the-Seven-Dresses put in a room by herself until she should awaken. In the morning he went to her and found her awake and weeping over her misfortune. He tried to comfort her a bit, then all of a sudden asked, "When shall we get married?" At that proposition, Beauty began screaming at the top of her voice. Since there was no way to quiet her, the king took to his heels. A month later, he returned and repeated his proposal. Beauty replied, "When you find a man dressed entirely in red." The king drew a sigh of relief, and telegraphed at once throughout the world. But Ciccillo was dead, stabbed by that old woman, and the man dressed in red was not to be found.

One day the big brother who had opened up a café went broke; reduced to poverty, he decided to change countries and try his luck elsewhere. He happened to take the same road as his brother Ciccillo, and when the seven ladies answered the door, they thought he was the dead man, so much did the two resemble one another.

"You've risen from the dead?" they asked.

"What!" he replied, amazed.

"Or maybe you had a brother who looked like you?"

"Yes, I did," he said. "But why do you ask me?"

"Come with us and you will see," answered the ladies, drawing him into a room where there was a dead man. This dead man was his brother, and the minute he saw him he began weeping and wailing. "Oh, my brother! My brother!" The ladies comforted him, telling him how Ciccillo had been treacherously slain, and they invited him to remain there with them.

While this youth was standing on the doorstep one morning, he saw two lizards, a little one and a big one. The big one killed the little one, then went and pulled up a herb, with which it rubbed the dead little lizard until it revived. Seeing that, the young man thought, Who knows but what my brother might revive if rubbed with that same herb. It certainly won't hurt to try. He pulled the herb, rubbed his brother's entire body with it, and he too came back to life. He asked about his wife at once and, remembering her warning, he dressed in red and left home to look the world over for her.

Now that very day Beauty was to marry the king: they'd not been able to find the man in red, whom she had finally given up for dead. He came into the city where the marriage was to be celebrated, and the inhabitants, at the sight of a man dressed in red after so much fruitless searching on their part, stopped him and carried him to the king. The king hastened to tell Beauty that the man in red had been found, thus fulfilling her one condition and clearing the way for the wedding. Beauty replied that she first had to talk to the man in red, alone and behind closed doors, so he was brought into her room, where they spent the night relating their misfortunes and making plans for the future. Beauty had all the keys of the palace, and once the king was fast asleep, they got up, loaded two donkeys with sacks of money, and fled.

After traveling all day long it grew dark again and they saw a stable. They made their bed the best they could on the hay under a loft. In the loft above, a drunkard snored and tossed in his sleep. Tossing and turning, he fell from the loft and ended up between the husband and wife, sinking down into the hay without even waking up or ceasing to snore. In the morning Beauty was the first one awake, and called her husband, "Ciccillo, get up, it's late. Let's get on our donkeys with the money and be off." Her husband was still fast asleep, though, and didn't hear her. But the drunkard, at the mention of money, answered right off the bat, "By all means, let's be on our way!" It was still dark, and the two of them groped their way to the donkeys laden with money and left. When it was day, Beauty realized her companion was not her husband and began to protest. His only response was to reach out and knock her to the ground, leaving her there weeping as he made off with the two donkeys. She had no idea how to find her husband again, for she had already gone a good way with the drunkard. She went back until she came to a haystack and saw a farm boy. She begged and pleaded with him until he gave her his clothes, and thus dressed as a man, she was able to continue her journey in less peril.

Not a trace of her husband was to be found. In order to support herself, she therefore decided to work for a miller, who happened to be miller to the king's notary. She kept the miller's accounts and wrote such a beautiful hand that the notary, who had never seen such fine writing, asked the miller who kept his accounts. Learning that a farm boy did, the notary took Beauty into his service, and she kept the notary's accounts, which were presented to the king. The king too was impressed with the beautiful writing and just had to have the talented farm boy to work for him.

In the meantime, the other king who wanted to marry Beauty-with-the-Seven-Dresses had died. He had taken his own life, butting his head against the wall upon finding his bride missing the morning of the wedding. Who should inherit his kingdom but the king who kept the farm boy in his service! This king called in the farm boy and ordered him to go to the dead king's city and announce throughout the realm that he would govern them on behalf of their new king. The farm boy replied that if he was to govern, he needed absolute authority over every citizen, which the king granted him.

On his arrival at the dead king's, he had the news published in all the towns, inviting everybody marked by any unusual event to come before the new governor, who would give them each a purse of money.

The news spread, and the first person to show up to tell her story was the old woman who had stabbed the husband and kidnapped his wife. "You old wretch!" exclaimed the governor. "You have the nerve to come and tell me that?"

He ordered her seized and thrown into a caldron of boiling water.

The old woman was followed by the drunkard, who told his story. "You thief!" exclaimed the governor. "You robbed a woman and have the nerve to admit it?" He had him dragged to the gallows and hanged as a dangerous thief. After those two were taken care of, here came the husband to tell his story.

They recognized one another and fell into each other's arms. Then the governor went and changed clothes, reappearing in the seven-part frock and every bit as lovely as a rosebud. They had a fine dinner and were reunited with the big brother and the seven ladies. Ciccillo was named king, and thus ended his misfortunes.

(Calabria)

◈ 144 ◈

Serpent King

A king and a queen had no children. The queen would pray and fast, but still no children were born to her. She happened to be walking in the fields one day and saw animals of all kinds—lizards, birds, snakes—together with their offspring, and said, "All the animals have young ones—just see the little lizards, the baby birds, the tiny snakes—while I bear nothing at all!" A serpent passed by, with his brood crawling along behind. "I would be satisfied with just a serpent child!" exclaimed the queen.

Now it happened that she too found herself with child, and the whole court rejoiced with her. At last came the day, and what should she be delivered of but a serpent! The court was dumbfounded, but the queen recalled her remark, realizing her prayer had been answered exactly, and she loved that serpent son as much as if it were a baby boy. She put it in an iron cage and fed it the same thing they ate—soup and meat—at midday and in the evening.

The serpent ate enough for two and grew bigger every day. When he was enormous, he said to the maid one day when she went into his cage to make up his bed:

> "Tell Papa dear
> That a wife I want in here
> With beauty and money!"

The maid took fright and didn't want to enter his cage any more. The queen, however, required her to take his meals in to him, and the serpent repeated:

> "Tell Papa dear
> That a wife I want in here
> With beauty and money!"

The maid told the queen, who wondered, "What can we do?"

She called in one of their tenant farmers and said, "I'll pay whatever you ask if you'll bring me your daughter."

The wedding was celebrated. The serpent seated himself at the banquet table. In the evening the newlyweds went off to bed. At a certain hour the serpent awakened and asked his bride, "What time is it?"

It was around four in the morning, and the bride replied, "It's the hour when my father rises, takes his hoe, and goes off to the fields."

"So you're a farmer's daughter!" exclaimed the serpent, and he bit her on the throat and killed her.

When the maid came in with breakfast in the morning, she found the bride dead. The serpent said:

> "Tell Papa dear
> That a wife I want in here
> With beauty and money,
> BEAUTY AND MONEY."

The queen then called in a cobbler who lived across the street and who also had a daughter. They agreed on the price, and the wedding was celebrated.

Around five o'clock in the morning the serpent awakened and asked his bride what time it was. "It's the hour," she replied, "when my father gets up and starts hammering at his cobbler's bench."

"So you're a cobbler's daughter!" said the serpent, and killed her with one bite on the throat.

Next time, the mother asked for the daughter of an emperor. The emperor was reluctant to give his daughter in marriage to a serpent. He took counsel with his wife who, as the girl's stepmother, was dying to get rid of her and persuaded her husband to marry his daughter to the serpent king. The emperor's daughter went to her dead mother's grave and asked, "Dear Mother, what am I to do?"

From the grave her mother answered, "Go ahead and marry the serpent, my daughter. But on your wedding day, put on seven dresses, one over the other. When bedtime comes around, say you want no maid, that you will undress by yourself. When you are finally alone with the serpent, say to him, 'I'll take off one piece of clothing, and you'll take off one piece of clothing.' Then you'll take off your first dress, and he his first skin. Again say, 'I'll take off one piece of clothing, and you'll take off one piece of clothing.' He will remove his second skin, and so forth."

Everything happened just as the dead mother said it would: for each dress she removed, the serpent removed a skin. Once his seventh skin was off, there stood the handsomest youth in the world. They went to bed. Around two o'clock in the morning the bridegroom asked, "What time is it?"

"The hour when my father gets home from the theater."

A little later he again asked, "What time is it?"

"The hour when my father sits down to supper."

When it was broad daylight he asked once more, "What time is it?"

"The hour when my father calls for his coffee."

At that, the little king kissed her and said, "You are my true bride, but you must tell no one that I am a human being at night, or you will lose me." At that, he changed back into a serpent.

One night the serpent said to her, "If you would have me become a human being in the daytime as well, you must follow my instructions."

"I'll do exactly what you say, my husband."

"Every night there is music and dancing at the court; you must attend. Everybody will invite you to dance, but you are to dance with no one. When you see a knight come in dressed in red, who will be none other than myself, rise from your seat and dance with me."

The hour struck when the court gathered together. The princess entered the ballroom and took a seat. At once princes and marquis came up and invited her to dance, but she replied she was so comfortable there just looking on that she would not get up. The king and queen found that a trifle rude, but assuming that she refused out of regard for her husband who couldn't attend the ball, they said nothing.

Suddenly a knight dressed in red entered the ballroom. The princess got up and began dancing with him, and danced with him throughout the evening.

The ball ended, and as soon as the king and queen were alone with their daughter-in-law, they seized her by the hair, saying. "What do you mean by refusing everyone's invitation and then dancing the whole evening with that stranger? How dare you embarrass us like that!"

Upon retiring, the young wife told the serpent how his parents had abused her. "Pay no attention to it," said her husband. "You must endure this for three nights in a row, and at the end of the third night I will become a man forevermore. Tomorrow night I will be dressed in black. Dance only with me, and if they beat you for it afterward, endure it for my sake."

That evening the princess again refused every invitation. But when the knight in black entered, she danced with him.

"Do you intend to disgrace us like that every evening?" thundered the king and queen. "Either you do what we say, or else!" And they took a stick and beat her black and blue.

Aching all over and weeping, she told her husband, who said, "Dear wife, just one more night. I will come dressed as a monk."

So the third evening, after rejecting all the notables of the court, the princess proceeded to dance with the monk. That was the last straw! Right then and there, in front of everybody, king and queen each snatched up a stick, and blows rained on the girl and on the monk.

The monk tried to dodge the blows, but unable to avoid them, he suddenly changed into a bird, an enormous bird that crashed through the windowpanes and flew off. "Now see what you've done!" exclaimed his wife. "That was your son!"

When they heard that their drubbing had prevented their son from breaking the spell and turning back into a man for good, the king and queen began tearing out their hair and embracing their daughter-in-law and asking her pardon.

But the princess replied, "There's not a minute to lose." She picked up two bags of money and was off in the direction the bird had flown. She met a glazier weeping over all his broken panes and asked, "What's the matter, sir?"

"A furious bird flew through here and wrecked my whole shop."

"And what would all that glass have cost, since the bird happens to be mine?"

"My master told me it was worth fifty crowns."

The princess opened one of the bags of money and paid him. "Now please tell me which way he went."

"That way, in a perfectly straight line!"

After going some distance she came to a goldsmith's shop. The master was not in. His apprentice was keeping shop and weeping. "What's the matter, young man?" asked the princess.

"A furious bird flew through here and wrecked everything. Now my master will come back and beat me to death."

"What are all these gold objects worth?"

"Let him kill me, I can't think any more!"

"No, I must pay you, for the bird was mine."

The boy told her all the prices, drawing up an endless list. "Damages come to six thousand crowns," he finally said.

"Here you are. Which way did the bird go?"

"Straight ahead."

The princess moved on, and the apprentice paid his master three thousand crowns, keeping the rest for himself and opening up a shop of his own.

After some distance the princess came to a tree in which birds of all kinds roosted. Among them she spotted her husband. "Dear husband, do come back home with me!" But the bird didn't budge.

The princess climbed the tree. "Please come home, husband!" She wept and begged in a manner that would have drawn tears from a stone. All the other birds in the tree were deeply touched by her pleas and said, "Go on, go on home with your wife. Why don't you want to?"

But the bird's only response was to peck out one of her eyes. The wife went on begging and pleading with him, though, and shedding tears with her other eye. What did the bird then do but peck out that eye too. "Now I'm blind," wailed the poor thing. "Guide me, dear husband!" But the bird brought down his beak two times, lopping off both her hands.

Then he flew off, lit on the roof of his parents' palace, and turned back into a human being. There was great rejoicing at the court, and his mother whispered to him, "You did the right thing in killing that ugly woman!"

Meanwhile the princess groped her way along the road, saying, "What will become of me now with no hands and no eyes?" A little old woman happened by and asked, "What is the matter, pretty maiden?"

The princess told her story, and the little old woman, who was the Madonna, said, "Put your arms into this fountain." She plunged the stumps into the water, and her hands grew back.

"Now wash your face," instructed the Madonna. She washed her face, and her eyes reappeared.

"And now take this magic wand. It will give you everything you ask for."

The princess commanded a handsome palace across from the king's, and at once had a palace studded with diamonds inside and out, with a golden mother hen and a brood of chicks strutting through the rooms. Many golden birds also flew through the high-ceilinged halls, through which servants and doormen in golden livery passed. She sat on a throne herself, closed off from view by a canopy and muslin curtains.

Upon looking out the window in the morning, the king's son beheld the palace. "Papa, Papa!" he exclaimed. "Look at that wonderful palace!" No matter which way he turned, all he saw were golden animals strutting and flying. "What important people they must be to have a palace like that put up in a single night!"

At that moment the princess stood up and put her head through the curtains, and the king's son saw her. "Papa, Papa! What an extraordinary young woman! I want her for my wife!"

"Go on, there's no telling who she is! Do you think she would even look at you? You're wasting your time even thinking about her."

But the king's son had made up his mind, and sent her a cloth embroidered in gold. The beautiful neighbor took it and threw it to the mother hen and her brood. The maid went home and told the prince, and the king and queen said to him: "What did we tell you? She's not the least bit interested in you."

"But *I* like her!" he said, and sent her a ring. She gave it to the birds to peck on. The maid who had delivered it was ashamed to go back to the lady's palace any more.

After much thought, the prince had a coffin built, stretched out inside, and had himself carried by his neighbor's palace. When she saw him in the coffin, she came outside and leaned over the box to look at him. He raised up and recognized her. "My wife! How happy I am to find you again! Come on back to our palace!"

She looked at him with eyes of steel. "Have you forgotten what you did to me?"

"I was under a spell, dear wife."

"But to free you, I danced with you for three evenings, and your parents beat me."

"Had you not done so, I would have remained a serpent."

"And when you were a bird, were you still a serpent? You pecked out my eyes and lopped off my hands with your beak!"

"Had I not done so, I would have remained a bird, dear wife."

She thought it over, then said, "In that case, you were right. Let us go back together as man and wife."

When the king and queen heard the whole story, they asked her forgiveness, sent for her father who was an emperor, and made music and danced for an entire month.

(Calabria)

◆ 145 ◆

The Widow and the Brigand

Once there was a poor widow who had one son, and together they went out looking for work. As they walked along, the boy threw stones at the birds and brought a few down. In the evening, when it grew dark in the mountains, the boy lit a fire and said to his mother, "You stay here and cook these birds while I go and see if I can get something else."

He went off into the country and came to a place where a statue stood. In the statue's hand was a rope, and the pedestal bore this inscription: WHOEVER TAKES THIS ROPE AND GIRDS HIMSELF THEREWITH WILL HAVE INVINCIBLE STRENGTH.

The boy took the rope and tied it about his waist. At once he felt new strength flow through him, and he caught hold of a tree and pulled it up, roots and all.

But let's leave the son and go back to the mother. Near the fire passed a brigand on horseback. He saw the widow, went up to her, and extended her an invitation to go off on horseback with him, if she would.

"Let me alone," replied the widow. "My son will be back any minute now and kill you."

But this woman, who was still young and beautiful, had captured the brigand's fancy, and he refused to leave her. "Go on!" he told her. "Do you think I'm afraid of your son?"

In that very instant the boy arrived, girt with the magic rope. The brigand said, "Your son is no bigger than that?"

"Who are you? What are you doing here?" said the boy.

"What am I doing? I'm going to kill you, that's what!"

"You just watch out," replied the boy, and dealt the brigand a back-handed slap, knocking him from his horse. Then he cut off his head, dug a grave, and buried head and body. That was the end of the brigand. The boy then mounted his victim's horse. "I'm off again," he told his mother. "Wait here for me." And away he galloped.

After riding a good while he came to a field in the middle of which stood a very tall palace. He rode all around it, but found no doors. He circled it one more time and found a door wide open. After tying up his horse, he went inside and climbed the stairs to the floor above. There was a table set for seven, with seven loaves of bread. The plates were laden with food and the bottles were full of wine, so he ate and drank a little bit from every one of them, and took a nibble of each loaf of bread. Then he went looking for a place where he could hide and see who came in, and found a room where they stored men they had murdered and pickled.

While he was there, in rushed six brigands and took their places at the table. One of them said, "Somebody has been eating out of my plate!"

Another echoed, "Mine too!"

And everybody else chimed in, "Mine too!"

They were baffled, but began eating as usual. When the meal was over, they noticed one empty place at the table. They counted themselves and discovered only six men present.

"Shouldn't there be seven of us?" they asked. "One of our number didn't return."

"There's something very strange about that," said one of them. "He must be dead."

Another one spoke. "I'm going to see what they're saying in the exhibit hall." The exhibit hall was where they kept the men they had murdered and pickled. The boy was hiding behind the door, and grabbed the brigand and slit his throat the minute he stepped into the room. Then he pitched the remains down into the stable.

When their companion failed to return, the other five grew worried, and another brigand went to look for him. The boy slit his throat too. So, one by one, all six of them were slain.

The boy went and brought his mother to the palace, which became their home. Every day he would go hunting while his mother stayed behind in the palace, and they always had plenty to eat and drink.

One day while the boy was out hunting, a brigand came by, entered the palace, caught sight of the widow, and said, "What! Here by yourself?"

"By myself?" answered the woman. "I'm waiting for my son to come back from hunting."

They struck up a conversation, one thing led to another, and they fell in love.

From that day on, when the son went out to hunt, the brigand would join the mother. But the widow was constantly telling him, "Please be careful, for if my son should find out about this, that would be the end of you and me."

The brigand began saying, "Why don't we simply kill him, get rid of him once and for all?"

"But he's my son!" protested the widow. "I brought him into the world myself!"

"Is that what's holding you back? If you gave him life, couldn't you just as easily take it away from him?"

Saddened, the woman answered, "You tell me what's to be done."

"Here's what you should do: pretend to be sick and tell him you need a little of the lioness's milk. He'll go to the lioness for it, the lion will eat him alive, and then there'll be just the two of us and we'll have peace at last."

That's what the widow did. Feigning sickness, she said to her son, "I'll die for sure if I don't get a little of the lioness's milk."

"All right, Mamma," he said. "I'll go for some and bring it back to you."

He went into the forest and found the lion. "Brother," said the lion, "what are you seeking around here?"

"Brother Lion," the boy answered, "I came to get a little of Sister Lioness's milk, which I need for my sick mother."

"Why, of course," replied the lion, and filled him a bottle of milk. Then he said, "Brother, I'm also giving you this lion cub. Take good care of him, for he will be a great help to you."

The boy returned to his mother with the bottle of milk and the lion cub. Deep fear came over the mother; then she drank the milk and said she was well. The next day the son went hunting with the lion cub, and the brigand came inside. "Would you believe it?" began the widow. "My son came home with the lioness's milk and a lion cub."

The brigand said, "Pretend to be sick a second time and tell him you need the she-bear's milk. He'll go out for it, the bear will devour him, and then we'll have peace and quiet."

The son went out for the she-bear's milk. When he reached the bear's den, the bear said, "Brother, what business brings you?"

"Brother Bear," replied the boy, "I've come to you, since my mother is sick and needs a little of Sister Bear's milk to cure her."

"You're more than welcome to it," answered the bear, who filled him a bottle of milk and also handed him a bear cub. "Take this bear cub home with you, and you'll see how helpful he will be to you."

When he came in with the she-bear's milk and a bear cub, his mother almost fainted.

She told the brigand about it the next morning, and he said, "This son of yours must be some kind of devil. Know what you should do? Pretend to be sick again and say you need the tigress's milk. This time he won't get away."

The son, who never dreamed he was being deceived, went off for the tigress's milk. On his arrival, the tigress asked, "Brother, what have you come for?"

"Sister Tigress, I came because my mother is sick and needs a bit of your milk."

"Yes, indeed, brother," replied the tigress, and filled him a bottle. "Take this tiger cub too, who will be a help to you one day."

When his mother saw him come back with the tigress's milk and a tiger cub, she too said to herself, "This son of mine must be a devil!"

The brigand didn't know what to think. "Know what you should do?" he said to the woman. "Tell him to take you down to see the stable. In the stable is a heavy chain. Begin playing around with it and, as though you're joking, chain up your son. He will let you do it. Then chain him up tight. I'll be hiding there and, once you have him securely chained, I'll jump out and kill him."

The mother got her son chained up in the stable. The brigand rushed out with his knife, but when the boy saw him he cried, "Lion cub! Bear cub! Tiger cub! Eat this brigand!"

Out rushed lion cub, bear cub, and tiger cub and devoured him. With one thrust of his arms the boy burst the chain. His mother had already fled and taken refuge under the bed.

"Lion cub! Bear cub! Tiger cub!" ordered the boy. "Eat that traitress in three mouthfuls!" And that was the end of the deceitful mother.

The boy mounted his horse and rode off with lion cub, bear cub, and tiger cub, to seek his fortune.

(*Greci di Calabria*)

The Crab with the Golden Eggs

There was once a bricklayer who was married and had two sons. He fell sick and was no longer able to work. After spending all his savings, he began selling everything he owned, down to the tiles off his roof. One day when the family cupboard was bare, he said, "I'm going hunting to see if I can't get a few birds."

Not a single bird was to be seen that day, but on the way back home he spied a crab poised on a rock. He took it alive and put it into his game-pouch. "I'll carry it home to my children to play with," he told himself.

The children shut it up in a small cage. The next morning they saw that it had laid an egg. They took it to their father, who exclaimed, "Why, it's a golden egg!"

He went out and sold the egg, coming home with six ducats. The crab laid an egg every night, and in no time the bricklayer was a rich man with his daily income of six ducats.

Next door to the bricklayer lived a tailor, who began to say, "I just can't understand how this bricklayer has become so rich!" After a little spying, the tailor soon realized that the source of his neighbor's wealth was the crab. Now the tailor had three children—two boys and a girl. I could marry my daughter to the bricklayer's son, he thought.

The nuptials were set. "I'm furnishing my daughter's dowry," announced the tailor to the bricklayer, "but you must put up the crab for your son's dowry."

"So long as it is my son's," replied the bricklayer.

When the tailor had the crab in his possession, he carefully examined it and found writing on its belly. The tailor knew how to read, and this is what he read:

WHOEVER EATS CRAB AND SHELL WILL ONE DAY BE KING. WHOEVER EATS CRAB AND CLAWS WILL FIND A PURSE OF MONEY EVERY MORNING UNDER HIS PILLOW.

Seeing that, the tailor thought, All I have to do is give this crab to my two sons to eat.

He killed it and put it on the grill to cook, then went off to get his sons. No sooner had he gone out than the bricklayers's sons came in. At the sight of the crab on the grill, their mouths began to water.

"Let's eat it," they said. "You take the shell, and I'll take the claws." That they did. The tailor returned and found the crab gone. A great dispute arose, and the wedding was called off. Horrified at the misfortune they'd caused, the bricklayer's two boys said, "Let's leave home and go out into the world." At that, they departed.

In the first town they came to they stopped for the night at an inn. In the morning upon awakening, the younger brother found a purse full of money beneath his pillow. "Brother," he said, "here in this place they have taken us for thieves. To tempt us, the lady-innkeeper placed this purse of money under my pillow." Then he went to give it back to her, saying, "We are no thieves, as this will prove."

The woman was thunderstruck but, a sly one from way back, she concealed her surprise and took the purse. "Oh, yes," she replied, "I have a habit of leaving money lying around."

The next day there was another purse like the first under the younger brother's pillow. "They continue to be suspicious of us here," he said, "so we'd better move on." He gave this purse as well to the innkeeper.

"I assure you I didn't leave it there on purpose," said the woman as she took the purse. "I'm so absent-minded."

The brothers asked for their bill, paid it, and departed. Night overtook them in a forest, where they slept on the ground, with a rock for a pillow. Next morning, beside the rock, lay the customary purse of money.

"That confounded innkeeper!" exclaimed the younger brother. "She made it all the way here! This time we'll not return the money. That will teach her!"

But since he found the purse every morning wherever he slept, he finally realized he got it by luck and not from the innkeeper.

Coming to a crossroads, the brothers decided to go their separate ways. "Take this knife," said the younger to the older. "As long as it gleams, you need have no fear for my safety. When it tarnishes, though, you can mourn for me."

"Take this bottle of water," said the older. "As long as it's clear, I am all right. When it clouds up, you can mourn for me."

They divided the money and bid one another farewell.

The older brother came to a city where the king had died. The council of ministers made an announcement: "Let us proceed as follows: let us loose a pigeon, and the person on whose head it alights we will crown king."

It lit on the older brother's head, and he found himself surrounded by carriages, an army, and a band. They conducted him to the palace, dressed him as a king, crowned him, and he began to reign.

The younger brother came to another city and took rooms at an inn opposite the palace of a princess. She was unmarried and spent her days on her balcony. She spied the younger brother on his balcony at the inn, and they entered into conversation. One topic led to the next, and the princess at length said, "If you will so honor me, I'll expect you at my house for a bit of entertainment."

"The honor is all mine," replied the youth.

When he arrived, the princess proposed, "Let's play cards awhile to pass the time."

They began playing, but the princess won every round. The youth lost thousands upon thousands of ducats; still he never ran out of money, thanks to the full purse every morning beneath his pillow. The princess was at a loss to understand how he could be so rich. She consulted a sorceress, who explained, "That stranger carries a charm within him, wherefore his money never runs out. In his body is half a crab, and that's why he finds a full purse under his pillow every morning."

"How can I get that charm for myself?" asked the princess.

"Follow these directions of mine," replied the sorceress. "Pour him a glass of wine with this medicine in it. The medicine will bring up everything he has on his stomach, including the crab-half. Carefully wash the morsel and swallow it. Then, in the morning, the purse will appear under your pillow instead of his."

The princess did as she was told and now she was the one to find all the money. The young man was reduced to poverty and had no choice but to sell everything he owned and resume his travels through the world. He walked and walked until he was too weak from hunger to go any farther. He dropped down in a meadow and, so as to have something in his mouth at least, he reached out and pulled a handful of grass, which he ate. It happened to be a species of chicory, and no sooner did he bite into it than he changed into a donkey. "At least I won't be hungry any longer," he reasoned, "since I'll feed on grass." He next nibbled a plant resembling cabbage. As soon as he bit into it, lo and behold, he changed back into a man. "These plants will be the making of me!" he concluded, and picked a specimen of the grass that changed one into a donkey and a specimen that changed one back into a man. Then he dressed as a gardener and went to hawk these herbs under the princess's window. "Chicory, chicory, fine chicory for sale!"

The princess called him in, saw the tender white chicory, tasted it at once, and changed into a donkey. The youth immediately put a halter on her and led her down the palace steps, without anyone suspecting it was the princess.

He rode the donkey to a place where a large group of men were working for the king. He got them to hire him with his donkey and made the animal carry double loads of rock, beating it to keep it stepping.

"Why are you so hard on that poor donkey?" the men would ask.

"That's my business," he would answer. The men went and told the king.

The king sent for him. "Why do you want to break that animal down?"

"That's what it deserves," said the youth, who noticed that the king wore a knife on his belt, the knife given him by none other than his little brother.

"Give me the money I gave you at the crossroads," said the younger brother.

"How dare you speak like that to a sovereign!" said the king.

"And how should I speak to you? I recognize you. You are my brother! Here is the bottle you gave me!"

The brothers recognized one another and embraced. The younger one told about the donkey which was a princess. "If she returns your half of the crab," advised the royal brother, "turn her back into the lady she was."

They gave the donkey the particular medicine that made everything come up, and she spit out the crab-half. Then they fed her the plant that looked like cabbage, and she became a lady once more.

The king named his brother general, and I have remained just what I always was.

(Greci di Calabria)

◈ 147 ◈

Nick Fish

Once upon a time in Messina there was a mother with a son named Nick, who spent all his time, day and night, swimming in the sea. His mother was constantly calling to him from the shore, "Nick! Oh, Nickie! Come out of the water, will you? You're no fish, are you?"

But he would always swim farther out. From so much yelling, the poor mother got a kink in her intestines. One day when he'd made her scream herself hoarse, she pronounced a curse on him. "Nick, may you turn into a fish!"

Obviously heaven was listening that day, for the curse took effect: in a flash, Nick became half fish and half man, with webfeet like a duck and a frog's throat. Never more did Nick set foot on land, which upset his mother so much that soon after she died.

The rumor reached the king that there was a creature in the sea of Messina who was half man and half fish. The king therefore ordered all his sailors that if any of them saw Nick to tell him the king wanted to talk to him.

One day as a sailor headed his boat toward the open sea, he saw Nick swimming nearby. "Nick!" he said. "The king of Messina wishes a word with you!"

So Nick Fish immediately swam up to the king's palace.

The king smiled as he approached. "Nick Fish," he said, "you are such an expert swimmer, I would like for you to swim around the whole island of Sicily and tell me where the sea is deepest and what is to be seen in that spot!"

Following orders, Nick Fish swam around the entire coast of Sicily. In a short time he was back. He related that on the floor of the sea he had seen mountains, valleys, caves, and all kinds of fish. The only time he'd been frightened was when he passed by the lighthouse, since he'd been unable to find the bottom at that point.

"Well," asked the king, "what is Messina built on? You must go down and see."

Nick dived in and remained under water a whole day. Then he came back up and said to the king, "Messina is built on a rock, which rests on three columns: one of them is sound, another is splintered, and the third is broken.

> O Messina, Messina,
> One day you will be leaner!"

The king was amazed, and decided to carry Nick Fish to Naples to see the floor of the volcanoes. Nick went down and afterward related that first he'd found cold water, then hot water and, in certain places, springs of fresh water. The king was skeptical, so Nick asked for two bottles, filling one with hot water and the other with fresh water.

But the king was tormented in the back of his mind by the notion that at Lighthouse Point the sea was bottomless. He took Nick Fish back to Messina and said, "Nick, tell me approximately how deep the water is here at the lighthouse."

Nick went down and stayed for two days. When he returned to the surface he informed the king he had not seen the bottom, since a column of smoke was pouring from beneath a rock and clouding the water.

The king, who could no longer contain his curiosity, said, "Dive from the roof of the lighthouse."

The lighthouse stood at the very tip of the promontory and in bygone times had lodged a sentinel who would signal the tides with a trumpet and hoist a flag to warn vessels to keep to the deep. Nick Fish leaped from that lookout. The king waited one day, then a second, then a third, but there was still no sign of Nick. Finally he emerged as pale as a ghost.

"What's the matter, Nick?" asked the king.

"I nearly died of fright," he explained. "I saw a fish in whose mouth alone a large ship would fit! So he wouldn't swallow me, I hid behind one of the three columns that hold up Messina!"

The king listened open-mouthed, and was as anxious as ever to know how deep the water was at Lighthouse Point. But Nick said, "No, Majesty, I'm too frightened to dive one more time."

Unable to persuade him, the king removed his crown studded with dazzling gems and threw it into the sea. "Go after it, Nick!"

"Majesty, the idea! The crown of the kingdom!"

"The only crown of its kind in the universe," said the king. "Nick, you must fetch it!"

"If you order it, Majesty," replied Nick, "I shall go down. But my heart tells me I'll never come up again. Give me a handful of lentils. If I escape, you'll see me emerge. But if the lentils come to the surface, that's a sign I'll never return."

They gave him the lentils, and Nick plunged into the sea.

The king waited and waited. After an interminable wait, the lentils floated up. To this day one still awaits the return of Nick Fish.

(Palermo)

◇ 148 ◇

Gràttula-Beddàttula

Once there was a merchant with three grown-up daughters. The oldest was Rosa, the second Joanna, and the third Ninetta, the most beautiful of the three.

One day a splendid opportunity for gain came the merchant's way, and he returned home lost in thought. "What's the matter, Papa?" asked the girls.

"Nothing, my daughters. A golden opportunity has just turned up, but I can't go off and leave you here by yourselves."

"Is that all that's stopping you?" asked the oldest girl. "All you need do is get in provisions enough to last us for the time you'll be away, seal up the doors with us inside the house, and we'll see each other again when it's God's will to bring you back to us."

That's what the merchant did: he bought a large supply of food and instructed one of his servants to call up to his oldest daughter every morning from the street to see if she had any errands for him to run. Bidding them goodbye, he asked, "Rosa, what do you want me to bring you?"

"A gown the hue of the sky."

"And you, Joanna?"

"A gown the color of diamonds."

"And you, Ninetta?"

"Please bring me, Father, a beautiful date-palm branch in a silver vase. If you don't, may your ship move neither forward nor backward."

"You wicked girl!" exclaimed her sisters. "Don't you realize you might cast a spell over your father by such talk?"

"Not at all," replied the merchant. "Let her alone. She's little and can say what she pleases."

The merchant departed, and disembarked at just the right place. He made the important deal and then decided to buy the dress Rosa had asked for and the one Joanna had requested, but he forgot all about Ninetta's date-palm branch. He boarded his ship and gained the open sea, but a frightful storm arose with thunder, lightning, and angry waves, and there the ship sat, moving neither forward nor backward.

The captain was at his wit's end. "Where on earth did this storm come from?" At that, the merchant recalled his daughter's spell and spoke up. "Captain, I forgot to make a certain purchase. If we don't want to be shipwrecked, we must turn around and go back into port."

The instant they turned the helm, the weather changed and, with the wind behind them, they glided back into port. The merchant went ashore, bought the date-palm branch, stuck it in a silver vase, and went back on board. The mariners hoisted the sails and, after three days of smooth sailing, the vessel reached its destination.

In the meantime, while the merchant was away, the three girls stayed in the house with the doors sealed. They had everything they needed, even a well in the courtyard where they could always get fresh water. One day the oldest sister accidentally dropped her thimble into the well, and Ninetta said, "Don't worry, sisters; just lower me into the well, and I'll fetch the thimble."

"Go down into the well? You must be joking!" said the oldest.

"Of course I'll go down and get the thimble." So the sisters lowered her into the well.

Ninetta found the thimble floating on the surface of the water, and she picked it up. But when she raised her head, she noticed a hole in the wall of the well, with light coming through it. She removed a brick and beheld on the other side of the wall a beautiful garden with all kinds of flowers, trees, and fruits. Dislodging more bricks, she made an opening and slipped into the garden, where the finest flowers and fruits were all hers. She filled her apron with them, slipped back into the well, replaced the bricks, and called up to her sisters, "Pull me up!" She returned aboveground as fresh as a rose.

Seeing her emerge from the well with an apronful of jasmine and cherries, her sisters asked, "Where did you get all those fine things?"

"What difference does it make? Let me down again tomorrow, and we'll get the rest."

Now that garden belonged to the crown prince of Portugal. Finding his flowerbeds stripped, he took his poor gardener severely to task.

"I'm completely in the dark. How could such a thing possibly happen?" the gardener was careful to answer. But the prince ordered him to keep a sharper lookout from then on, if he knew what was good for him.

The next day Ninetta was all ready to go down into the garden. She said to her sisters, "Girls, let me down!"

"Are you drunk or out of your mind?"

"I'm neither drunk nor crazy. Let me down." And they had to let her down.

She pulled out the bricks and stepped into the garden. After gathering a good apronful of flowers and fruit, she cried, "Pull me back up!" But while she was leaving, the prince came to the window and saw her hop away like a hare. He ran into the garden, but she was already gone. He called the gardener, "Which way did that girl go?"

"What girl, Majesty?"

"The one who's picking the flowers and fruit in my garden."

"I saw nothing at all, Majesty, I swear."

"Very well, tomorrow I will take your place."

So the next day, hidden behind a hedge, he saw the girl slip through the bricks into the garden and fill her apron with flowers and fruit up to her chin. Out he jumped and tried to grab her, but with the speed of a cat she jumped back through the hole in the wall and closed it up with the bricks. The prince examined the entire wall, but found no spot where the bricks were loose. He waited for her the next day, and the next, but

Ninetta had received such a scare upon being discovered that she stopped going down into the well. The prince, who had found her as beautiful as a fairy, was so upset that he fell sick. But none of the doctors in the kingdom could say what his trouble was. The king consulted all the physicians, wise men, and philosophers. First one and then another spoke, and finally the floor was given to a certain Wisebeard. "Majesty," said this Wisebeard, "ask your son if he likes a certain young lady. That would explain everything."

The king sent for his son and asked him. The boy told him everything, saying he'd have no peace until he married this girl. Wisebeard said, "Majesty, have three days of social affairs at the palace, and issue a decree for fathers and mothers of every station in life to bring their daughters, under pain of death." The king was in agreement, and issued the decree.

Meanwhile the merchant had returned from his trip, had the doors unsealed, and given the dresses to Rosa and Joanna, and the date-palm branch in the silver vase to Ninetta. Rosa and Joanna were dying for a ball to be given somewhere and began working on their outfits. But Ninetta stayed shut up in her room with her date-palm branch and thought of neither parties nor balls. Her father and sisters said she was crazy.

When the decree was announced, the merchant went home and told his daughters. "How wonderful! How simply wonderful!" exclaimed Rosa and Joanna. But Ninetta shrugged her shoulders and said, "You two go, I have no desire to."

"Oh, no, my daughter," said her father. "You must go, under pain of death; death is nothing to play with."

"What difference does it make whether I go? Do you expect the whole world to know you have three daughters? Just say you have two."

They argued back and forth, and the evening of the first ball Ninetta stayed home.

No sooner had her sisters left than Ninetta turned to her date-palm branch:

> "Lovely date-palm, Gràttula-Beddàttula,
> Come forth and dress up Nina,
> Make her more beautiful than ever."

At those words, out of the date-palm branch came one fairy, then another, then many, many more, all carrying gowns and jewels without equal. They gathered round Nina, and some bathed her, some plaited her hair, some dressed her. In no time they had her fully clothed and decked with necklaces, diamonds, and other precious stones. When she was one dazzling jewel from head to toe, she got into a carriage, rode to the palace, climbed the stairs, and left everyone open-mouthed with admiration.

The prince recognized her and ran immediately to tell the king. Then he approached her, bowing and asking, "How are you, madam?"

"As well in winter as in summer."

"What is your name?"

"Ah, my name . . ."

"Where do you live?"

"In a house with a door."

"On what street?"

"On Whirlwind Lane."

"Madam, you will be the death of me."

"As you will!"

And so, genteelly conversing, they danced away the whole evening, leaving the prince quite out of breath, while she was still as fresh as a rose. When the ball was over, the king, who was concerned about his son, inconspicuously instructed his servants to follow the lady and find out where she lived. She got into her carriage, but noticing she was being trailed, she undid her hair, and pearls and precious stones fell onto the road. The servants were upon them at once, like chickens going after feed, and the lady was completely forgotten. She had the horses whipped to a gallop and vanished.

Arriving home before her sisters, she said:

"Lovely date-palm, Gràttula-Beddàttula,
Come down and undress Nina,
Make her just the same as ever."

At that, she found herself stripped of her finery and dressed in her usual housedress.

Her sisters came home. "Ninetta, Ninetta!" they exclaimed, "you don't know what a lovely ball you missed. There was a beautiful lady there who looked a little like you. Had we not known you were here at home, we would have mistaken her for you."

"Yes, I was here all the time with my date-palm branch."

"But tomorrow you just have to come with us."

Meanwhile, the king's servants returned to the palace empty-handed. "You good-for-nothing creatures!" said the king. "The idea of neglecting my orders for a few trifles! Heaven help you if you don't follow the lady all the way home tomorrow evening!"

The next evening as well Ninetta refused to accompany her sisters to the ball. "She's lost her mind," they said, "over her date-palm branch!

We're going, though!" And they were off. Ninetta turned at once to the branch.

"Lovely date-palm, Gràttula-Beddàttula,
Come forth and dress up Nina,
Make her more beautiful than ever."

And the fairies plaited her hair, dressed her in gala robes, and covered her with jewels.

At the palace everyone stared with admiration, especially her sisters and her father. The prince was by her side at once. "Madam, how are you?"

"As well in winter as in summer."

"What is your name?"

"Ah, my name . . ." And so on.

The prince let matters be, and invited her to dance. They danced the whole evening long.

"Goodness me!" said one sister to the other. "That lady is the spitting image of Ninetta!"

While the prince accompanied her to her carriage, the king signaled to the servants. Seeing herself followed, Ninetta pulled out a handful of gold pieces. But this time she aimed at the faces of the servants, hitting some on the nose and others in the eyes. Thus they lost sight of the carriage and went crawling back to the palace, looking so much like whipped dogs that even the king felt sorry for them. But he said, "The final ball is tomorrow evening. You must find out something by hook or by crook."

Meanwhile Ninetta was saying to her branch:

"Lovely date-palm, Gràttula-Beddàttula,
Come down and undress Nina,
Make her just the same as ever."

In the twinkling of an eye she was changed back to her usual self, and her sisters arrived and told her once more how much that elegant and bejeweled lady resembled her.

The third evening was like the previous ones. Nina went to the palace lovelier and more radiant than she had ever been. The prince danced with her even longer than before and melted with love, like a candle.

At a certain hour as Ninetta was preparing to leave, she was called before the king. Shaking like a leaf, she went up and bowed.

"Maiden," said the king, "you have made sport of me for the past two nights, but the third night you won't get away with it."

"But what on earth have I done, Majesty?"

"What have you done? You have made my son fall madly in love with you. Don't expect to escape."

"What sentence awaits me?"

"You are sentenced to become the prince's wife."

"Majesty, I am not free. I have a father and two older sisters."

"Have the father brought to me."

When the poor merchant heard he was wanted by the king, he thought, A royal summons bodes ill. Having several frauds on his conscience, he got goose pimples. But the king pardoned him on every count, and asked him for Ninetta's hand for his son. The next day they opened up the royal chapel for the marriage of the prince and Ninetta.

> They were as happy as happy could be,
> While here we sit, tap-tapping our teeth.

(Palermo)

◈ 149 ◈

Misfortune

Once, so the story goes, there were seven children, all of them girls and daughters of a king and queen. War was declared on their father. He was captured and dethroned, while his wife and children were left to shift for themselves. To make ends meet, the queen gave up the palace, and they all squeezed into a hovel. Times were hard, and it was a miracle if they got anything to eat. One day a fruit vendor came by. The queen stopped him to buy a few figs. While she was making her purchase, an old woman passed, asking for alms. "Goodness me!" said the queen. "I wish I could help you, but I can't. I am poor too."

"How do you happen to be poor?" asked the old woman.

"You don't know? I am the queen of Spain, humbled by the war waged against my husband."

"You poor thing. But do you know why everything is going badly for you now? You have under your roof a daughter who is truly ill-starred. You'll never prosper again as long as she stays at home."

"You don't mean I should send one of my daughters away?"

"Alas, my good lady, that's the only solution."

"Who is this ill-starred daughter?"

"The one who sleeps with her hands crossed. Tonight while your daughters are sleeping, take a candle and go and look at them. The one you find with her hands crossed must be sent away. Only in that way will you recover your lost domains."

At midnight the queen took the candle and filed past the beds of her seven daughters. They were all asleep, some with hands folded, others with their hands under their cheeks or pillows. She came to the last girl, who happened to be the youngest, and found her sleeping with her hands crossed. "Oh, my poor daughter! I really am obliged to send you away."

As she said that, the young lady awakened and saw her mother holding a candle and weeping. "What's wrong, Mother?"

"Nothing, my daughter. An old beggar-woman happened by and explained that I'll prosper only after sending away that daughter of mine who sleeps with her hands crossed. The unfortunate girl turns out to be you!"

"That's all you're weeping over?" replied the daughter. "I'll dress and leave at once." She put her clothes on, tied her personal effects up in a bundle, and was off.

After going a great distance she came to a desolate moor where only one house stood. She approached, heard the sound of a loom, and saw some women weaving.

"Won't you come in?" said one of the weavers.

"Thank you."

"What is your name?"

"Misfortune."

"Would you like to work for us?"

"I certainly would."

She set to work sweeping and doing the housework. In the evening, the women said to her, "Listen, Misfortune, we are going out tonight. After we've locked the door on the outside, you are to lock it on the inside. When we return in the morning, we'll unlock it on the outside, and you'll unlock it on the inside. You must see that no one steals the silk, braiding, or cloth we have woven." With that, they left.

When midnight struck, Misfortune heard a snipping of scissors. Candle in hand, she rushed to the loom and beheld a woman with a pair of scissors cutting all the gold cloth from the loom, and she realized her Evil Fate had followed her here. In the morning her mistresses returned; they unlocked the door from the outside, and she unlocked it from the inside. As soon as they came in, their eyes fell on the shreds littering the floor.

"You shameless wretch! Is this how you repay us for taking you in? Begone with you!" And they dismissed her with a kick.

Misfortune walked on through the countryside. Before entering a certain town, she stopped before a shop where they sold bread, vegetables, wine, and other things, and asked for alms. The shopkeeper's wife gave her a bit of bread and a glass of wine. When the shopkeeper returned, he took pity on her and told his wife to let her stay and sleep in the shop that night on the sacks. The shopkeeper and his wife slept upstairs, and in the middle of the night they heard a commotion below. Rushing downstairs to see what was going on, they found the casks uncorked and wine running all over the house. At that, the husband went looking for the girl and found her atop the sacks groaning as though caught in a nightmare. "Shameless wretch! Only you could be responsible for all this mess!" He took a stick and beat her, then put her out of the shop.

Not knowing which way to turn, Misfortune ran off, weeping. At daybreak she met a woman doing her laundry.

"What are you looking at?"

"I'm lost."

"Can you wash and iron?"

"Yes, indeed."

"Well, stay and help me. I'll do the lathering and you'll do the rinsing."

Misfortune began rinsing the clothes and hanging them up to dry. As soon as they dried, she gathered them up to mend, starch, and press.

Now these clothes were the prince's. When he saw them, he was struck by how beautifully they had been done. "Signora Francisca," he said, "you've never done such a good piece of work. I really must reward you for it." And he gave her ten gold pieces.

Signora Francisca used the money to dress Misfortune up and buy a sack of flour to bake bread. Two of the loaves were ring-shaped and seasoned with anise and sesame seed. "Take these two ring-shaped loaves to the seashore," she told Misfortune, "and call my Fate, like this— 'Hallooooo! Fate of Signora Franciscaaaaa!'—three times. At the third call my Fate will appear, and you will give her a ring-shaped loaf and my regards. Then ask her where your own Fate is and do the same with her."

Misfortune walked slowly to the seashore.

"Hallooooo! Fate of Signora Franciscaaaa! Halloooo! Fate of Signora Franciscaaaaa! Halloobooo! Fate of Signora Franciscaaaa!" Signora Francisca's Fate came out. Misfortune delivered the message, gave her the ring-shaped loaf, and then asked, "Fate of Signora Francisca, would you be so gracious as to inform me of the whereabouts of my own Fate?"

"Hear me through: follow this mule trail a piece until you come to an oven. Beside the pit of oven-sweepings sits an old witch. Approach her gently and give her the ring-shaped loaf, for she is your Fate. She will refuse it and insult you. But leave the bread for her and come away."

At the oven Misfortune found the old woman, who was so foul, blear-eyed, and smelly that the girl was almost nauseated. "Dear Fate of mine, will you do me the honor of accepting—" she began, offering her the bread.

"Away with you! Be gone! Who asked you for bread?" And she turned her back on the girl. Misfortune put the loaf down and returned to Signora Francisca's.

The next day was Monday, washday. Signora Francisca put the clothes in to soak, then lathered them. Misfortune scrubbed and rinsed them; when they were dry, she mended and ironed them. When the ironing was finished, Signora Francisca put everything in a basket and carried it to the palace. Seeing the clothes, the king said, "Signora Francisca, you won't pretend you've ever washed and ironed that nicely before!" For her pains, he gave her ten more gold pieces.

Signora Francisca bought more flour, made two ring-shaped loaves again, and sent Misfortune off with them to their Fates.

The next time she did his wash, the prince, who was getting married and anxious to have his clothes perfectly laundered for the event, rewarded Signora Francisca with twenty gold pieces. This time Signora Francisca bought not only flour for two loaves; for Misfortune's Fate she purchased an elegant dress with a hoop skirt, a petticoat, dainty handkerchiefs, and a comb and pomade for her hair, not to mention other odds and ends.

Misfortune walked to the oven. "Dear Fate of mine, here is your ring-shaped loaf."

The Fate, who was growing tamer, came forward grumbling to take the bread. Then Misfortune reached out and grabbed her and proceeded to wash her with soap and water. Next she did her hair and dressed her up from head to foot in her new finery. The Fate at first writhed like a snake, but seeing herself all spick-and-span she became a different person entirely. "Listen to me, Misfortune," she said. "For all your kindness to me, I'm making you a present of this little box," and she handed her a box as tiny as those which contain wax matches.

Misfortune went flying home to Signora Francisca and opened the little box. In it lay a piece of braid. They were both somewhat disappointed. "What a piece of nothing!" they said, and stuffed the braid away in the bottom of a drawer.

The following week when Signora Francisca took clean wash back to the palace, she found the prince quite depressed. Being on familiar terms with him, the washerwoman asked, "What's the matter, my prince?"

"What's the matter? Here I am all ready to get married, and now we find out that my betrothed's bridal gown lacks a piece of braid which cannot be matched anywhere in the kingdom."

"Wait a minute, Majesty," said Signora Francisca, and ran home, rummaged through the drawer, and came back to the prince with that special piece of braid. They compared it with what was on the bridal dress: it was identical.

The prince said, "You have saved the day for me, and I intend to pay you the weight of this piece of braid in gold."

He took a pair of scales, placing the braid in one plate and gold in the other. But no amount of gold made the scales balance. He then tried measuring the braid's weight with a steelyard, but this too was unsuccessful.

"Signora Francisca," he said, "be honest. How can a little piece of braid possibly weigh so much? Where did you get it?

Signora Francisca had no alternative but to tell the whole story, and the prince then wanted to see Misfortune. The washerwoman dressed her up (they had gradually accumulated a little finery) and took her to the palace. Misfortune entered the throne room and gave a royal curtsy; she was a monarch's daughter and by no means ignorant of courtly decorum. The prince welcomed her, offered her a seat, then asked, "But who are you?"

"I am the youngest daughter of the king of Spain, who was dethroned and imprisoned. My bad luck forced me out into the world where I have endured insults, contempt, and many beatings"—and she told him all.

The first thing the prince did was send for the weavers whose silk and braid the Evil Fate had cut up. "How much did this damage cost you?"

"Two hundred gold crowns."

"Here are your two hundred gold crowns. Let me tell you that this poor maiden you cast out is the daughter of a king and queen. That is all, be gone!"

Next he summoned the owners of the shop where the Evil Fate had tapped the casks. "And how much damage did you suffer?"

"Three hundred crowns' worth."

"Here are your three hundred crowns. But think twice next time before thrashing a poor king's daughter. Now out of my sight!"

He dismissed his original betrothed and married Misfortune. For matron of honor he gave her Signora Francisca.

Let us leave the happy couple and turn out attention to Misfortune's mother. After her daughter's departure, fortune's wheel began to turn in her favor: one day her brother and nephews arrived at the head of a mighty army and reconquered the kingdom. The queen and her children moved back into their old palace and all the comforts and luxuries they had formerly enjoyed. But in the back of their minds they thought of the youngest daughter, of whom they had heard absolutely nothing in all the time she had been gone. Meanwhile the prince, upon learning that Misfortune's mother had regained her kingdom, sent messengers to inform her of his marriage to her daughter. Ever so pleased, the mother set out for her daughter's with knights and ladies-in-waiting. Likewise with knights and ladies-in-waiting, the daughter rode to meet her mother. They met at the border, embraced over and over, with the seven sisters standing around, every bit as moved as their mother, while there was great rejoicing in one kingdom and the other.

(Palermo)

◈ 150 ◈

Pippina the Serpent

There was once a merchant with five children—four little girls and a boy. The boy was the oldest, a handsome youth by the name of Baldellone. The luck of the merchant shifted, and he went from rich to poor. The only way he could get along now was on charity, and to make matters even worse his wife began expecting another baby. Seeing his parents in such dire straits, Baldellone kissed them goodbye and left for France. He was an educated youth, and when he got to Paris he entered the service of the royal palace and was finally promoted to captain.

Back home, meanwhile, the merchant's wife said to her husband, "The baby is about to be born, and we have no baby things. Let's sell the dining room table, the only thing we have left, so we can get baby things."

They called in secondhand dealers passing through the street and sold them the table. That way the merchant was able to buy all the necessaries for the baby. The baby was born, a girl of matchless beauty, and father and mother were so moved that they burst into tears. "Dear daughter, it breaks our heart to see you born into such poverty!"

The infant grew by leaps and bounds and, when she was about fifteen months old, she began walking all by herself and playing in the straw where her father and mother slept. One day while playing in the straw, she called out, "Mamma, Mamma! Look, look!" and held out hands full of gold pieces.

Her mother couldn't believe her eyes. She took the coins, slipped them into her blouse, called in a baby-sitter, and ran to market. She bought this and that, shopping to her heart's content, and by noon they were finally able to have a real meal, for a change.

"Do tell me, Pippina, where did you get those nice shiny little things?" prodded the baby's father. And she answered, "Right here, Papa," pointing to a hole in the straw. In it was a jar full of coins. All you had to do was thrust in your hand for it to fill up with money.

So the family was able to hold up its head once more and resume its former way of living. When the child was four, her father said to his wife, "Wife, I think it's time to have a charm put on Pippina. We certainly have the money, so why not have her charmed?"

To have children charmed in those days, people would go halfway to Monreale, to a place where four fairy sisters lived. They took Pippina there in a coach and presented her to the four sisters. The fairies explained what to prepare, agreeing to come to the merchant's house on Sunday for their ceremony.

So on Sunday, right on time, the four sisters arrived in Palermo, where they found everything ready for them. They washed their hands, mixed up a bit of Majorcan flour, made four fine pies, and sent them off to be baked.

In a little while the baker's wife smelled a delightful aroma coming from the oven. Unable to check her gluttony, she pulled out one of the pies and ate it. Then she made another one exactly like it, only with regular flour and water drawn from the trough in which she washed the oven broom. But it rivaled the others in shape, and no one could distinguish it from the original three.

When the pies were back in the merchant's house, the first fairy cut one of them, saying, "I charm you, lovely maiden, so that every time you brush your hair, pearls and other precious stones will come pouring forth."

"And I," said the second fairy, cutting another pie, "I charm you to become more lovely yet than you already are."

The third fairy stood up. "And I charm you so that every fruit out of season you might desire will instantly be there."

"I charm you," began the fourth fairy, cutting into the pie filled with oven sweepings, when a cinder flew out of it and landed in her eye. "Ouch! That hurt!" exclaimed the fairy. "Now I'm going to put you under a monstrous spell. When you see the sun, you shall become a black serpent!" At that, the four sisters vanished.

The father and mother burst into tears: their baby girl wouldn't be able to see the sun any more!

But let's leave them and turn to Baldellone, who was bragging in France about his father's vast wealth while, for all he knew, his parents didn't have a penny to their name. But with his constant big talk he impressed everyone; as the proverb says:

> He who goes abroad
> Presents himself as count, duke, or lord.

The king of France was curious as to whether there was any truth to all this wealth of Baldellone, so he dispatched a squire to Palermo with instructions on what to observe and report back. The squire went to Palermo, asked for Baldellone's father, and was directed to a handsome palace with countless liveried doormen. He entered and beheld rooms with walls of gold, and valets and servants galore. The merchant gave the squire a royal welcome, invited him to the table and, when the sun had set, brought in Pippina. The squire was charmed at the sight of her; never before had he seen such a lovely maiden. He returned to France and told the king.

The king sent for Baldellone. "Baldellone, go to Palermo, run to your house and fetch me your sister Pippina, whom I wish to marry."

Baldellone, who didn't even know he had a sister, could make little sense out of all this talk, but he obeyed the king and departed for Palermo. Now in Paris Baldellone had a girl friend, who insisted that he take her with him to Palermo.

Upon his arrival in Palermo, Baldellone found his family prosperous once more. He renewed his old ties with them, met his sister, and announced that the king of France wanted to marry her. That delighted everyone. But when the girl who had come from France with Baldellone saw Pippina, she was consumed with envy and began plotting to undo her and become queen herself.

In a few days, Baldellone had to depart with Pippina. "Goodbye, Papa." "Goodbye, dear son." "Farewell, Pippina." "Farewell, Mamma, so long little sisters." Then they were off. To reach Paris, one travels first by sea, then overland. Baldellone closed Pippina up in the ship, and never let her see a single ray of sunshine, while his girl friend kept her company. When the ship pulled into port, he had his sister and his friend taken off board in a large sedan chair sealed against the sun. Baldellone's friend was furious at the thought they were nearing Paris where Pippina would soon become queen, while she herself would be only a captain's wife.

"Pippina," she began, complaining, "it's stifling in here; let's open a curtain!"

"Please, my sister, you will be my undoing!"

After a while she started up again. "Pippina, I'm burning up in here!"

"No you're not, be calm . . ."

"Pippina, I'm suffocating."

"Even so, you know good and well I can't open this thing!"

"Really?" At that, the woman snatched a penknife and rent the leather ceiling of the sedan chair. A ray of sunlight shone straight down upon Pippina, and she changed into a black serpent that went wriggling down into the dusty road and disappeared under a nearby hedge of the king's garden.

Seeing the chair arrive empty, Baldellone let out a cry. "My poor sister! And poor me! How will I ever tell the king, who is expecting her?"

"What are you worrying about?" said his friend. "Tell him I am your sister, and all will be well." Baldellone ended up doing exactly that.

When the king saw her, he turned up his nose slightly. "Is this the beauty without compare? No matter; a king's promise is a king's promise. I have no choice but to marry her."

He married her, and they lived together. Baldellone was fit to be tied: not satisfied to deprive him of his sister, that traitress had then abandoned him for the king! The new queen was well aware Baldellone would never forgive her for those two things, so she began scheming to get him out of the way as well.

"Majesty," she said, "I'm sick and need figs."

Figs were out of season, and the king replied, "Just where do you expect to find figs this time of year?"

"They are to be had. Tell Baldellone, and he will go after them."

"Baldellone!"

"Yes, Majesty?"

"Go pick a few figs for the queen."

"Figs at this time of year, Majesty?"

"In season, out of season, that's all the same to me. I said figs, and figs it must be. Otherwise your head will roll."

Sad and downcast, Baldellone went to the garden and burst into tears. Lo and behold, out of the flowerbed crawled a black serpent, who asked, "What's the matter?"

"My sister!" exclaimed Baldellone. "Now I too am in great difficulty!" and he informed her of the king's command.

"Oh, that's nothing to fret about. I have special power to bring forth fruit out of season. You want figs, you say? All right!" A beautiful basket of ripe figs appeared.

Baldellone ran to the king at once with them. The queen ate every last one of them, and it's a shame they didn't poison her! Three days later she was hankering for apricots. Pippina the serpent brought forth apricots.

Her next craving was for cherries, so Pippina produced cherries. Then came a call for pears. But we forgot to say that the charm worked for figs, for apricots, and for cherries, but not for pears.

Baldellone was sentenced to die. He asked one last favor: that his grave be dug in the royal garden. "Granted," replied the king. Baldellone was hanged and buried, and the queen drew a sigh of relief.

One night the gardener's wife awakened and heard a voice in the garden saying:

> "Baldellone, O dear brother,
> Buried here amid dark verdure,
> While the author of your fate
> Now plays queen to my intended mate."

The woman woke her husband up. They tiptoed outside and saw a dark shadow wriggling away from the captain's tomb.

In the morning, when the gardener went out as usual to make a bouquet of flowers for the king, he found the flowerbeds strewn with pearls and precious stones. He carried them to the king, who was greatly amazed.

The next night the gardener stood watch with his gun. At midnight a shadow loomed beside the tomb, saying:

> "Baldellone, O dear brother,
> Buried here amid dark verdure,
> While the author of your fate
> Now plays queen to my intended mate."

The gardener took aim and was about to fire, when the shadow said, "Put down your gun! I was baptized and confirmed the same as you were. Come closer and look at me." So saying, she lifted her veil, showing a face of matchless beauty. Then she undid her braids, and out of her hair poured pearls and precious stones. "Tell the king this," said the maiden, "and tell him I'll meet him here tomorrow night." The sky grew light, and the maiden changed into a serpent and wriggled away.

The next night at the usual time the shadow had scarcely appeared and said,

Baldellone, O dear brother,

when the king went up to her. The maiden lifted her veil and told an amazed king her story.

"Tell me, how can I free you?" said the king.

"Here's what you can do: leave tomorrow on a horse that runs like the wind and go all the way to the Jordan River. Dismount on its bank, and you will see four fairies bathing in the water—one with a green ribbon around her tress, another with a red one, a third with a blue one, and the last with a white one. Take away their clothes lying there on the riverbank. They will want them again, but don't dare give them back! Then the first fairy will throw you her green ribbon, the second her red ribbon, the third her blue ribbon; but only when the fourth fairy has thrown you her white ribbon, and then her tress, shall you return their clothes, for my evil spell by then will be lifted."

The king needed to hear no more. He left the next morning at dawn and put his kingdom behind him. After traveling a great distance, thirty days and thirty nights later, he reached the Jordan River, found the fairies, and did everything prescribed by Baldellone's sister. When he had the white ribbon and the tress in his hand, he said, "I'm now leaving you, but you can be sure I'll repay you."

Back in his kingdom, he ran at once to the garden, called the serpent, and stroked her with the tress. Pippina immediately changed back into the most beautiful maiden ever seen. She attached the tress to her hair, and from then on had nothing more to fear.

The king called the gardener and said, "Now listen to what you must do. Take a large ship, put Baldellone's sister on board, and sail off in the night. Return to port a few days later under a foreign flag and leave everything else to me."

The gardener carried out the plan down to the tiniest detail, and three days later turned the ship around and hoisted the English flag. From the royal palace you could see the sea. The king looked out and said to the queen, "What ship is this? Look! It's one of my relatives arriving. Let's go and meet him."

The queen, who was always ready to show off, dressed in the twinkling of an eye. She went on board and found herself face to face with Pippina. If I weren't certain Baldellone's sister became a black serpent, she thought, I'd swear this is she ...

After much fanfare, they disembarked with the newcomer, heaping praise on her beauty. "Tell me," said the king to the queen, "what punishment would a person deserve for harming a creature like this?"

"Oh," answered the queen, "who could be so wicked as to hurt this jewel?"

"But supposing there were someone, what would he deserve?"

"He would deserve to be thrown through this window and then burned alive!"

"And that's just what we are going to do!" snapped the king. "This lady is Baldellone's sister whom I was supposed to marry, and you, envious soul, came along and made her turn into a serpent, so you could take her place. You are now going to pay for deceiving me and for making this poor dear suffer. You have already pronounced your sentence with your own lips. Guards! Seize this wicked woman, throw her through the window, and burn her to death at once!"

No sooner said than done. The liar was dashed through the window and burned right there on the ground next to the palace. The king asked Baldellone's sister to forgive him for hanging her innocent brother. She replied, "Let's let bygones be bygones and go see what can be done in the garden."

They went to the tomb and raised the stone cover. Baldellone's body was almost intact. With a small brush, Pippina applied a certain salve to his neck, and Baldellone began breathing again, then moving, then rubbing his eyes like someone awakening, and finally he stood up. The scene was indescribable. They hugged and kissed, and the king, giving orders for grand festivities, sent for the merchant and his wife and married Pippina in great pomp.

(Palermo)

◈ 151 ◈

Catherine the Wise

Here in Palermo they tell, ladies and gentlemen, that once upon a time there was a very important shopkeeper in the city. He had a daughter who, from the time she was weaned, proved so wise that she was given her say on every single matter in the household. Recognizing the talent of his daughter, her father called her Catherine the Wise. When it came to studying all sorts of languages and reading every kind of book, no one could hold a candle to her.

When the girl was sixteen, her mother died. Catherine was so grief-stricken that she shut herself up in her room and refused to come out. There she ate and slept, shunning all thought of strolls, theaters, and entertainment of any kind.

Her father, whose life centered on this only child of his, thought it advisable to hold a council on the matter. He called together all the lords (for, even though a shopkeeper, he was on familiar terms with the best people) and said, "Gentlemen, you are aware I have a daughter who is the apple of my eye. But ever since her mother's death, she's been keeping to the house like a cat and won't for the life of her stick her head outside."

The council replied, "Your daughter is known the world over for her vast wisdom. Open up a big school for her, so that as she directs others in their studies, she will get this grief out of her system."

"That's a splendid idea," said the father, and called his daughter. "Listen, my daughter, since you refuse every diversion, I have decided to open a school and put you in charge of it. How does that suit you?"

Catherine was instantly charmed. She took charge of the teachers herself, and they got the school all ready. Outside they put up a sign: WHO-EVER WISHES TO STUDY AT CATHERINE THE WISE'S IS WELCOME, FREE OF CHARGE.

Numbers of children, both boys and girls, flocked in at once, and she seated them at the desks, side by side, without distinction. Someone piped up, "But that boy there is the son of a coal merchant!" "That makes no difference: the coal merchant's son must sit beside the prince's daughter. First come, first served." And school began. Catherine had a cat-o'-nine-tails. She taught everyone alike, but woe to those that didn't do their lessons! The reputation of this school even reached the palace, and the prince himself decided to attend. He dressed up in his regal clothes, came in, found an empty place, and Catherine invited him to sit down. When it was his turn, Catherine asked him a question. The prince didn't know the answer. She dealt him a back-handed blow, from which his cheek still smarts.

Crimson with rage, the prince rose, ran back to the palace, and sought out his father. "A favor I beg, Majesty: I wish to get married! For a wife, I want Catherine the Wise."

The king sent for Catherine's father, who went at once, saying, "Your humble servant, Majesty!"

"Rise! My son has taken a fancy to your daughter. What are we to do but join them in matrimony?"

"As you will, Majesty. But I am a shopkeeper, whereas your son is of royal blood."

"That makes no difference. My son himself wants her."

The shopkeeper returned home. "Catherine, the prince wants to wed you. What do you have to say about that?"

"I accept."

The wool for the mattresses was not wanting, no more than the chests of drawers; in a week's time everything needed had been prepared. The prince assembled a retinue of twelve bridesmaids. The royal chapel was opened, and the couple got married.

Following the ceremony the queen told the bridesmaids to go and undress the princess for bed. But the prince said, "There's no need of people to undress or dress her, or of guards at the door." Once he was alone with his bride, he said, "Catherine, do you remember the slap you gave me? Are you sorry for it?"

"Sorry for it? If you ask for it, I'll do it again!"

"What! You're not sorry?"

"Not in the least."

"And you don't intend to be?"

"Who would?"

"So that's your attitude? Well, I'll now teach *you* a thing or two." He started unwinding a rope with which to lower her through a trapdoor into a pit. "Catherine," he said when the rope was ready, "either you repent, or I'll let you down into the pit!"

"I'll be cooler there," replied Catherine.

So the prince tied the rope around her and lowered her into the pit, where all she found was a little table, a chair, a pitcher of water, and a piece of bread.

The next morning, according to custom, the father and mother came to greet the new wife.

"You can't come in," said the prince. "Catherine isn't feeling well."

Then he went and opened the trapdoor. "What kind of night did you spend?"

"Pleasant and refreshing," replied Catherine.

"Are you considering the slap you gave me?"

"I'm thinking of the one I owe you now."

Two days went by, and hunger began to gnaw at her stomach. Not knowing what else to do, she pulled a stay out of her corset and started making a hole in the wall. She dug and dug, and twenty-four hours later saw a tiny ray of daylight, at which she took heart. She made the hole bigger and peered through it. Who should be passing at that moment but her father's clerk. "Don Tommaso! Don Tommaso!" Don Tommaso couldn't imagine what this voice was, coming out of the wall like that. "It's me, Catherine the Wise. Tell my father I have to talk to him right away."

Don Tommaso returned with Catherine's father, showing him the tiny opening in the wall. "Father, as luck would have it, I'm at the bottom of a pit. You must have a passageway dug underground from our palace all the way here, with an arch and a light every twenty feet. Leave everything else to me."

The shopkeeper agreed to that and in the meantime he brought her food regularly—roast chicken and other nourishing dishes—and passed it through the opening in the wall.

Three times a day the prince peered through the trapdoor. "Are you sorry yet, Catherine, for the slap you gave me?"

"Sorry for what? Just imagine the slap you are going to get from me now!"

The workers finally got the underground passage dug, with an arch and a lantern every twenty feet. Catherine would pass through it to her father's house after the prince had looked in on her and reclosed the trapdoor.

It wasn't long before the prince was fed up with trying to get Catherine to apologize. He opened the trapdoor. "Catherine, I'm going to Naples. Have you nothing to tell me?"

"Have a good time, enjoy yourself, and write me upon your arrival in Naples."

"So I should go?"

"What? Are you still there?"

So the prince departed.

As soon as he shut the trapdoor, Catherine ran off to her father. "Papa, now is the time to help me. Get me a brigantine ready to sail, with housekeeper, servants, festive gowns—all to go to Naples. There let them rent me a palace across from the royal palace and await my arrival."

The shopkeeper sent the brigantine off. Meanwhile the prince had a frigate readied, and he too set sail. She stood on her father's balcony and watched him leave, then she went aboard another brigantine and was in Naples ahead of him. Little vessels, you know, make better time than big ones.

In Naples Catherine would come out on the balcony of her palace each day in a lovelier gown than the day before. The prince saw her and exclaimed, "How much like Catherine the Wise she is!" He fell in love with her and sent a messenger to her palace. "My lady, the prince would like very much to pay you a visit, if that won't inconvenience you."

"By all means!" she replied.

The king came regally dressed, made a big fuss over her, then sat down to talk. "Tell me, my lady, are you married?"

"Not yet. Are you?"

"Neither am I, isn't it obvious? You resemble a maiden, my lady, who captured my fancy in Palermo. I should like you to be my wife."

"With pleasure, Prince." And a week later they got married.

At the end of nine months Catherine gave birth to a baby boy that was a marvel to behold. "Princess," asked the prince, "what shall we call him?"

"Naples," said Catherine. So they named him Naples.

Two years went by, and the prince decided to leave town. The princess didn't like it, but he had made up his mind and couldn't be swayed. He drafted a document for Catherine saying the baby was his firstborn and in time would be king. Then he left for Genoa.

As soon as the prince had gone, Catherine wrote her father to send a brigantine to Genoa immediately with furniture, housekeeper, servants, and all the rest, and have them rent her a palace opposite the royal palace of Genoa and await her arrival. The shopkeeper loaded a ship and sent it off to Genoa.

Catherine also took a brigantine and reached Genoa before the prince. She settled down in her new palace, and when the prince saw this beautiful young lady with her royal coiffure, jewels, and wealth, he exclaimed, "How much like Catherine the Wise she is, and also my wife in Naples!" He dispatched a messenger to her, and she sent back word she would be happy to receive the prince.

They began talking. "Are you single?" asked the prince.

"A widow," answered Catherine. "And you?"

"I'm a widower, with one son. By the way, you look just like a lady I used to know in Palermo, not to mention one I knew in Naples."

"Really? We all have seven doubles in the world, so they say."

Thus, to make a long story short, they became man and wife in one week's time.

Nine months later, Catherine gave birth to another boy, even handsomer than the first. The prince was happy. "Princess, what shall we call him?"

"Genoa!" And so they named him Genoa.

Two years went by, and the king grew restless once more.

"You're going off like that and leaving me with a child on my hands?" asked the princess.

"I am drawing up a document for you," the prince reassured her, "stating that this is my son and little prince." While he made preparations to leave for Venice, Catherine wrote her father in Palermo for another brigantine with servants, housekeeper, furniture, new clothes and all. The brigantine sailed off to Venice. The prince departed on the frigate. The princess left on another brigantine and arrived before he did.

"Heavens!" exclaimed the prince when he beheld the beautiful lady at her casement. "She too looks exactly like my wife in Genoa, who looked exactly like my wife in Naples, who looked exactly like Catherine the Wise! But how can this be? Catherine is in Palermo shut up in the pit, the Neapolitan is in Naples, the Genoese in Genoa, while this one is in Venice!" He sent a messenger to her and then went to meet her.

"Would you believe, my lady, that you look like several other ladies I know—one in Palermo, one in Naples, one in Genoa—"

"Indeed! We are supposed to have seven doubles in this life."

And thus they continued their customary talk. "Are you married?" "No, I'm a widow. And you?" "I am a widower, with two sons." In a week's time they were married.

This time Catherine had a little girl, radiant like the sun and moon. "What shall we call her?" asked the prince.

"Venice." So they baptized her Venice.

Two more years went by. "Listen, princess, I have to go back to Palermo. But first, I'm drawing up a document that spells out that this is my daughter and royal princess."

He departed, but Catherine reached Palermo first. She went to her father's house, walked through the underground passage and back into the pit. As soon as the prince arrived, he ran and pulled up the trapdoor. "Catherine, how are you?"

"Me? I'm fine!"

"Are you sorry for that slap you gave me?"

"Have you thought about the slap I owe you?"

"Come, Catherine, say you're sorry! Otherwise I'll take another wife."

"Go right ahead! No one is stopping you!"

"But if you say you're sorry, I'll take you back."

"No."

The prince then formally declared that his wife was dead and that he intended to remarry. He wrote all the kings for portraits of their daughters. The portraits arrived, and the most striking was of the king of England's daughter. The prince summoned mother and daughter to conclude the marriage.

The entire royal family of England arrived in Palermo, and the wedding was set for the morrow. What did Catherine do in the meantime but have three fine royal outfits readied for her three children—Naples, Genoa, and Venice. She dressed up like the queen she actually was, took the hand of Naples, clothed as crown prince, climbed into a ceremonial carriage, followed by Prince Genoa and Princess Venice, and they drove off to the palace.

The wedding procession with the prince and the daughter of the king of England was approaching, and Catherine said to her children, "Naples, Genoa, Venice, go and kiss your father's hand!" And the children ran up to kiss the prince's hand.

At the sight of them, the prince could only admit defeat. "This is the slap you were to give me!" he exclaimed, and embraced the children. The princess of England was dumbfounded; she turned her back on everybody and stalked off.

Catherine explained all the mystery to her husband about the ladies who looked so much alike, and the prince couldn't apologize enough for what he had done.

> They lived happily ever after,
> While here we sit grinding our teeth.

(Palermo)

◈ 152 ◈

The Ismailian Merchant

A king went hunting with his men. The sky clouded up, and it began to pour down rain. The men ran off in every direction, and the king lost his way and took refuge in an isolated cottage.

In the cottage lived an old man, whom the king asked, "Will you give me shelter?"

"Come up to the fire and dry yourself, Majesty."

The king hung up his wet clothes and stretched out on a couch to sleep. In the night he awakened and heard the old man talking. Finding him nowhere in the house, he went to the door. The sky was clear once more, and the stars had come out. There sat the old man on the doorstep. "To whom are you talking, my good man?" asked the king.

"I was talking to the planets, Majesty," answered the old man.

"What were you telling them?"

"I was thanking them for the luck they have brought me."

"What luck, good man?"

"They favored me by giving my wife a son tonight, and you too they have favored this same night by giving your wife a daughter. When my son grows up, he will become your daughter's husband."

"You ill-bred old man! How dare you tell me such nonsense! You shall certainly pay for it!" He got dressed again, and at dawn took the road that led back to the palace.

Along the way he was met by knights and valets who had come in search of him. "We have good tidings, Majesty! The queen gave birth last night to a fine baby girl!"

The king rode to the royal palace, and the minute he dismounted, in the midst of all the court welcoming him and the nurses showing him the baby, he issued a decree: let all the baby boys born in the city last night be found and slain. Soldiers went through the city, which in an hour's time they had thoroughly searched. Only one baby boy had been born during the night. They tore it away from its mother, by order of the king, and carried it off to the forest.

The soldiers were two in number, and when they raised their sword over the baby, they were moved to pity. "Must we really put this innocent creature to death? There's a dog; let's kill it and smear its blood on the swaddling clothes, which we'll take back to the king. We'll leave the baby here, to the mercy of God." They did just that, and the baby remained in the forest crying.

An Ismailian merchant named Giumento happened along, on his way to trade his wares. He heard the child wailing, found him in the bushes, quieted him down, and took him home to his wife. "Wife, this time I bring back something I did not buy—a little baby I found in the middle of the forest. We have no children ourselves and now heaven has given us one."

They reared him and kept him with them up to his twentieth year, in all of which time he thought he was the merchant's own son. On his twentieth birthday the merchant said, "My son, I am growing old and here you are a man now: take charge of my accounts, books, and coffers. You will look after all my trade abroad."

The youth packed boxes and suitcases and, accompanied by his servants, left home with the blessing of the merchant and his wife. He arrived in Spain, where news of such a rich merchant reached the royal palace. The king sent for him to come to the palace and show his jewels. Now the king of Spain was the very king who had ordered the baby killed. He called in the princess, who'd grown up to be a beautiful maiden of twenty, and said, "Come see if there's some jewel here you like."

Seeing the young merchant, the princess fell in love with him.

"What's the matter, my daughter?"

"Nothing, Papa."

"Do you want any of these things? Speak up."

"No, Papa, I desire neither jewels nor precious stones. I want this handsome youth for a husband."

The king looked at the merchant. "Tell me who you are."

"I am the son of Giumento, an Ismailian merchant. I'm traveling around the world to gain experience in trade, so I'll be fit to succeed my old father."

Considering the vast wealth of the merchant, the king decided to give the youth his daughter in marriage. The boy returned home to invite his father and mother to the wedding. He told them about his meeting with the king and about the marriage engagement. At that, his mother turned pale and began reviling him. "You ungrateful man! So you intend to leave me, do you? You've fallen in love with this princess and can't wait to leave home. Off with you, then, and don't ever let me see you in this house again!"

"But, Mother, what have I done to you?"

"Don't you 'Mother' me! I'm not your real mother, anyway!"

"What! Then who is my mother, if you're not?"

"Goodness knows. You were found in the middle of a forest!" And she told the whole story to the poor boy, who almost fainted.

In the face of his wife's anger, Giumento the merchant was helpless. Deeply grieved, he supplied the youth with money and merchandise and let him go his way.

In despair, the boy came to a forest at night. He threw himself on the ground under a tree and, pounding the earth with his fists, sighed. "Mother, Mother, what is there left for me to do, all alone and miserable. Lovely spirit of my mother, help me!"

As he wept, there appeared beside him an old man in rags, with a long white beard. "What's the matter, my son?" he asked. The youth opened up his heart to him, telling him how he couldn't go back to his betrothed, having found out he was not the real son of the Ismailian merchant.

"What are you afraid of?" said the old man. "Let's go to Spain. I am your father, and I will help you."

The youth looked at the old man in rags and exclaimed, "You—my father? You're imagining things!"

"I assure you, my son, that I am your father. If you come with me, I will bring you prosperity. Otherwise you are doomed."

The young man looked him in the eye and said to himself, "Doomed in any case, what have I to lose by going along with him?" He took the old man onto his horse, and they eventually came to Spain.

He went to the king, who asked, "Where is your father?"

"Right here," said the youth, pointing to the old man in rags.

"That man? And you have the impudence to come and ask for my daughter's hand?"

"Majesty," the old man cut in, "I am the old man who spoke with the planets and announced to you the birth of your daughter and of my son who would marry her. This boy here is none other than that son of mine."

The king was furious. "Get out, uncouth old man! Guards, seize him!"

The guards came forward, at which the old man pulled open the rags covering his chest, and there gleamed the emperor's Golden Fleece.

"The emperor!" cried king and guards in unison.

"May the Holy Crown forgive me!" said the king, kneeling at the emperor's feet. "I knew not to whom I spoke. This is my daughter; do with her as you will."

The emperor was a man who had tired of court life, so he spent his days traveling about the world by himself in disguise, conversing with the stars and planets.

Everyone hugged and kissed, and a date for the wedding was set. The Ismailian merchant and his wife were called to Spain, where the boy welcomed them with open arms, saying, "Mother and Father—for you were a real mother and father to me!—turning me out of the house was the making of me! Although I'm marrying the princess, you will remain with me always."

The two old people were moved to tears of love. The emperor's son married the king's daughter, and there was great rejoicing throughout the city.

> They lived happily ever after,
> While here we are picking our teeth.

(*Palermo*)

153

The Thieving Dove

There was once a king and queen's daughter with such beautiful long hair that she would let no hairdresser touch it, but always combed and arranged it herself. One day while she was dressing her hair, she laid her comb down on the window ledge. A dove lit on the ledge, took the comb in his beak, and flew off with it.

"Oh, my goodness! The dove has taken my comb off!" cried the princess, but by then the dove was already a good distance away.

The next morning the princess was again at the window fixing her hair when the dove returned, seized her hair clasp, and flew off. The third day, she had no sooner done her hair and still had the cloth around her shoulders than down dipped the dove, grabbed hold of the cloth, and made off with it. This time the maiden, truly vexed, climbed down a silken ladder and ran after the dove. But instead of fleeing like all other doves, this one waited for her to approach, then took off only to light a little farther away. The maiden became more and more angry. By a series of short flights, the dove had advanced into the forest, with the princess right behind it. In the heart of the forest stood a solitary hut, and the dove flew inside. The door happened to be open, and the princess caught sight of a handsome youth, whom she asked, "Did you see a dove fly in carrying a cloth?"

"Yes," replied the young man, "I am that very dove."

"You?"

"Yes."

"How can that be?"

"The fairies have cast a spell over me, and I can't go out in human guise until you have sat at the window of this hut for a year, a month, and a day, in sunlight and in starlight, with your eyes fixed on the mountain across the way, where I shall fly as a dove."

Without the least hesitation, the princess took a seat at the window. The dove flew off and came to rest on the mountain. One day went by, then another, then a third, and the princess kept her seat, her eyes trained on the mountain. Weeks passed, and the princess sat on through sunlight, moonlight, and starlight, as though she were made of wood. And little by little she turned dark, ever darker, until she was as black as pitch. Thus passed a year, a month, and a day, and the dove turned back into a man and came down the mountain. When he saw how black the princess had become, he exclaimed, "Phew! What a sight you are! Aren't you ashamed to show yourself after becoming so ugly for the sake of a man? Off with you!" And he spit on her.

The poor girl was mortified. She trudged off and, passing through a field and weeping, she meet three fairies.

"What's the matter?" asked the fairies.

Weeping, she told them her story.

"Don't worry," they said. "You won't stay like that for long." The first fairy stroked the girl on the face, and she was beautiful once again, but far more so than originally; now she was as radiant as the sun. The second fairy clothed her in an empress's gown, while the youngest fairy presented her with a basket of jewels. "Now," announced the fairies, "we will be with you at all times disguised as your maidservants."

Thus they set out and reached the city whose king happened to be that youth. In the twinkling of an eye, the fairies had a palace put up opposite the king's, but a hundred times more beautiful than his. The king looked out, saw the wonder, and thought he was dreaming. At one of the windows appeared a girl who seemed to be an empress, and the king was charmed. "If he starts paying you court," said the fairies, "encourage him." The first day the king stared, the second day he winked, finally he asked if he might call on her. The first couple of times the princess said no, then at last, "Well, Majesty, if you want to visit me, you must prepare a landing stage from my balcony to yours—a carpet of rose petals two inches thick."

The king didn't even let her finish before he'd given orders for the landing stage. Hundreds of women began picking roses, and picked and picked, pulling off petals by the bushels, a thing never before seen.

When the landing stage of rose petals was ready, the fairies said to the princess, "Dress as a grand empress; we shall follow you as your ladies-in-waiting. When you are halfway across the landing stage, make believe you've been stuck by a thorn, and leave the rest to us."

The princess set forth on the rose petals, dressed in the pink gown of an empress. The king eagerly awaited her at the other end, but he was forbidden by the princess to set foot on the landing stage. Halfway across, she screamed, "I'm dying! A thorn has stuck me!" And she pretended to faint. The fairies picked her up and carried her back to her palace. The king wanted to run to her assistance, but was checked by the princess's original order to stay off the landing stage.

From his palace he could see doctors and chemists coming and going, and in the end even a priest came with the viaticum. The king alone was not allowed to go to her. It was rumored that the thorn had caused her legs to swell and that she was fast sinking. Forty days later it was learned that the malady had subsided and the empress was improving. When word went out that she was well, the king renewed his pleas for a meeting with her. So the fairies said to the girl, "Tell him you will visit him, but that you want a landing stage made of three inches of jasmine petals. And when you are halfway across, pretend to be stuck by another thorn."

At once the king had all the jasmine blossoms in the kingdom picked and made into a thick carpet. When all was ready, she started out dressed as an empress. At the other end the king watched with his heart in his mouth, lest she be pricked anew. Halfway across, she screamed, "Ouch! I'm dying! A thorn has gone through my foot!" She swooned, and her ladies picked her up in their arms and carried her back to the palace. The king tore out his hair.

He sent his servants to her repeatedly, but there was no way to see her, much less cross the landing stage, and he knocked his head against the wall in desperation. He ended up sick in bed, but continued to send over messengers to find out how the empress was getting along. Finally he requested permission to come to her, sick as he was, since he wanted to ask for her hand in marriage.

"Tell him," replied the princess, "that I would approach him only if I saw him laid out in a coffin."

Receiving that answer, the king, who by now had lost his mind, had a coffin prepared with candles around it, and pretending to be dead had himself carried past the empress's windows. "Behold, Majesty," they said to her, "our dead king."

The maiden went to the balcony and said, "Phew! Down with you! You did all this for the sake of a woman?" And she spit on him.

Hearing that, the king recalled what he had done to the good maiden as black as pitch, who he then realized wasn't too different from the beautiful empress he had fallen in love with. All of a sudden it dawned on him that the black maiden and the empress were one and the same. You can imagine how upset he was! He almost changed from a false corpse into a real one.

But the three ladies-in-waiting arrived and informed him their mistress was expecting him. The king went in and asked her forgiveness. The royal chapel was immediately opened, and they got married. The king was anxious to keep the three fairies with them, but they bid the couple farewell and departed.

(Palermo)

154

Dealer in Peas and Beans

Once upon a time in Palermo there was a certain Don Giovanni Misiranti, who at noon would dream of dinner and in the evening of supper, and at night he would dream of them both. One day when hunger was gnawing at his stomach, he went outside. "Oh, my luck!" he said to himself. "So you have left me!" Walking along, he spied a bean on the ground. He bent over and picked it up. Sitting down on a roadside post, he studied the bean and thought, What a fine bean! I'll plant it in a pot at once, and a bean plant will come up, with lots of nice pods. I'll dry the pods, then plant the beans in a basin and have many more pods . . . Between now and the next three years, I'll lease a garden, plant the beans, and no telling how many will come up then! The fourth year I'll rent a storehouse and become an important dealer . . .

Meanwhile he had set off on foot again, and gone past St. Anthony's Gate. There was a whole row of stores, with a woman sitting before one of the entrances.

"My good woman, are these stores for rent?"

"Yes, sir," she replied. "Who is interested?"

"My master," he replied. "Whom does one discuss the matter with?"

"With the lady who lives upstairs."

Don Giovanni Misiranti began thinking, and went off to see a friend of his.

"For St. John's sake," he said to his friend, "you mustn't refuse me. Lend me one of your outfits for twenty-four hours."

"Of course, my friend." So Don Giovanni Misiranti got all dressed up, down to gloves and watch. Then he went to a barber to be shaved and, now spruce, passed back through St. Anthony's Gate. He had the bean in the pocket of his waistcoat and glanced at it every now and then on the sly. The woman was still sitting there, and he said, "My good woman, are you the one my servant asked about stores for rent?"

"Yes, sir; have you come to look at them? Follow me, and I'll take you to my master's wife."

With his chest thrust out, Don Giovanni Misiranti followed the woman and introduced himself to the wife of the stores' owner. Seeing a gentleman before her with all the accessories—hat, gloves, and gold watch chain—the lady made a big to-do over him, and then they began discussing the matter at hand. Right in the middle of their conversation, a lovely young lady entered the room. Wide-eyed, Don Giovanni Misiranti asked, "Is she related to you?"

"She's my daughter."

"Single?"

"Yes, she's still single."

"I'm happy to hear that. I too am single."

Shortly afterward, he said, "Now that we've reached an agreement on the stores, I think we ought to come to one regarding the daughter. What does the lady think?"

"We shall see . . ."

Her husband came in. Don Giovanni rose and bowed. "I am a landowner," he said, "and I would like to rent your thirteen warehouses to fill with beans, peas, and all the rest of the harvest. Also, if I may, I'd like to ask for your daughter's hand in marriage."

"Ah! What is your name?"

"I am Don Giovanni Misiranti, dealer in peas and beans alike."

"Well, Don Giovanni, give me twenty-four hours to think it over, and you will have your answer."

That night, mother took daughter aside and told her Don Giovanni Misiranti, dealer in peas and beans alike, wanted her for his wife. The daughter eagerly accepted.

The next day Don Giovanni went back to his friend and borrowed another outfit. The first thing he did was slip the bean into the pocket of the new waistcoat. He went to the residence of the warehouse owner and, receiving the answer, was in seventh heaven.

"I would like to marry as soon as possible, then," he said, "since my many occupations don't give me any time to waste."

"By all means, Don Giovanni," replied the girl's parents. "Would it suit you to draw up the contract in a week's time?"

Throughout that period, Don Giovanni went on borrowing clothes, wearing something different every day, so that his parents-in-law-to-be took him for a very rich man indeed. They signed the contract, and the dowry consisted of two thousand gold crowns cash, sheets, and linen. Seeing so much money at his disposal, Don Giovanni felt himself a new man. He went on a shopping spree, buying presents for his bride and clothes for himself as well as all the trimmings to cut a fine figure.

A week after signing the contract, he got married in fine wedding clothes, with the bean in the pocket of the vest. The newlyweds gave parties and banquets, and Don Giovanni spent money right and left, as though he were a baron. His mother-in-law began to grow uneasy over this unending extravagance. "Don Giovanni," she said, "when do you plan to take my daughter to see your fields? It's harvest time, you know."

That upset Don Giovanni at first, and he could find no excuse. Racking his brains, he took out his good-luck piece, saying, "My luck, you must again help me out."

He had a fine sedan chair readied for his wife and his mother-in-law and announced: "It is time to leave. We shall go toward Messina. I shall ride ahead on horseback, and you will come along behind."

Don Giovanni left on horseback. When he saw a place he thought would serve his purpose, he called to a farmer in the field. "Here are twelve crowns for you. When you see a sedan chair come up with two ladies inside, if they ask you whose fields these are, you are to say, 'They are owned by Don Giovanni Misiranti, dealer in peas and beans alike.'"

The sedan chair passed. "My good man, whose fine lands are these?"

"They are owned by Don Giovanni Misiranti, dealer in peas and beans alike."

Mother and daughter smiled smugly and moved on.

At another estate the same thing happened. Don Giovanni rode on ahead, throwing out twelve-crown pieces; tucked in his pocket was the bean which made up his entire fortune.

When they got to where there was nothing more to see, Don Giovanni said to himself, "Now I'll find an inn and wait for them." He looked around and saw an enormous palace, with a young lady in green standing at the window.

"Pss, pss!" said the young lady, motioning him inside.

Don Giovanni started up the grand staircase which gleamed so he was almost reluctant to walk on the steps, lest he muddy them. The young lady came forward to greet him and, with a sweeping gesture indicating all the lamps, carpets, and gold-sequined walls, asked, "Do you like the palace?"

"Can you imagine my not liking it?" answered Don Giovanni. "I would be happy here even as a corpse."

"Look around, go up to the next floor." And she showed him through all the rooms. Everywhere were jewels, precious stones, fine silks, things Don Giovanni had never even dreamed of.

"Do you see all this? It is yours. Take care of it. Here is the deed. It is a present from me. I am the bean you picked up and kept in your pocket. I shall take my leave now."

Don Giovanni was about to fall at her feet and tell her how grateful he was, but the damsel in green had vanished under his very eyes. The handsome palace, though, remained, and it belonged to him, Don Giovanni Misiranti.

When his mother-in-law saw the palace, she exclaimed, "Ah, my daughter, what luck has come your way! Don Giovanni, dear son, to think you had such a lovely palace and never breathed a word about it to us!"

"That's right! I wanted to surprise you . . ." So he showed them around the palace, seeing it for the first time himself. He pointed out the jewels, the deed to the domain, and then a cellar full of gold and silver, with shovel planted in the midst; then they saw the stables with all the carriages, and finally the lackeys and all the household servants.

They wrote his father-in-law to sell everything and come and live with them at the palace, and Don Giovanni also sent a reward to the good woman he had found seated before the warehouses.

(Palermo)

◈ 155 ◈

The Sultan with the Itch

A fisherman had a little boy who, seeing his father get into his boat, would say, "Take me with you, Father."

"No," replied the fisherman, "a storm might come up."

And if the sea and weather were calm, the man would say, "No, there's danger of sharks."

Or if it wasn't the season for sharks: "No, the boat might sink."

He thus held off for nine years, after which he could object no more: he had to carry his son along to fish in the open sea.

On the open sea the fisherman lowered his nets, while the boy dropped a fishing line. The fisherman pulled up the net, which contained nothing but a minnow; the boy pulled up his line, and hooked to it was an enormous fish. "This, Father, I will take to the king in person." They went back in, the boy donned his Sunday best, slipped the fish into a basket lined with green seaweed, and went off to the king.

At the sight of the fish, the king clacked his tongue. "Come here!" he called to a servant. "Give this little fisherman fifty crowns!" And he asked the boy, "What is your name?"

"I am Pidduzzu, Majesty," replied the little fisherman.

"Well, Pidduzzu, would you like to remain here at the royal palace?"

"Would I!" answered the boy.

So, with his father's approval, Pidduzzu was reared at the palace. He was dressed in fine silk, and had many teachers and professors. He received his education, grew up, and was no longer called Pidduzzu, but "the knight Don Pidduzzu."

Also growing up at the palace at the same time was the king's daughter, Pippina, who loved Pidduzzu better than life itself. When she was seventeen, a king's son showed up to ask for her hand in marriage. Her father, who favored the match, tried to persuade her to marry him. But Pippina was in love with Pidduzzu and informed her father she would either wed Pidduzzu or never marry at all. The king flew off the handle and called in Pidduzzu. "My daughter has lost her head over you, and that cannot be tolerated: you will have to leave the palace."

"Majesty," replied Don Pidduzzu, "are you turning me out like that?"

"It displeases me to do so," said the king, "for you were like a son to me. But have no fear, you will continue to enjoy my protection." So Don Pidduzzu went out into the world, while the princess was shut up in a convent—St. Catherine's, of all places!

Don Pidduzzu took lodgings at an inn. His window overlooked an alley, as did a small window of Pippina's convent. She appeared at the window and, the minute they saw each other, they began comforting one another with gestures and words. Pippina had found a book of magic hidden in her cell by a nun-turned-sorceress, and she passed it down to Don Pidduzzu from her window.

The next day the king went to see his daughter and asked the mother superior for permission to speak to her. As he was king, it was granted him. "Listen, Father," said the princess, "let's settle this matter once and for all. The prince has a brigantine of his own. Give Don Pidduzzu a brigantine. Let both of them sail off, one in one direction, the other in the opposite. Whoever returns with the finest presents will be my husband."

"I like that idea," replied the king. "It shall be done." He called the two suitors to the palace and laid before them his daughter's plan. Both young men were delighted—the prince because he knew Don Pidduzzu hadn't a penny to his name, Don Pidduzzu because, with the book of magic, he was certain of success.

Thus they weighed anchor and departed. Out on the deep, Don Pidduzzu opened the book and read: "Tomorrow, dock at the first land you come to; go ashore with the whole crew and a crowbar." The next morning an island was sighted, so Don Pidduzzu and the crew disembarked, carrying along a crowbar. On land he opened the book and read: "In the very middle of the place you will see a trapdoor, then another, and another; pry them up with the crowbar and descend." That he did. He found the trapdoor in the middle of the island and raised it, using the crowbar as a lever. Underneath was another trapdoor, and under it still another. When he had opened up the last one, he saw a staircase. Don Pidduzzu descended it and found himself in a gold-sequined gallery—walls, doors, floor, ceiling, all gold, and a table laid for twenty-four persons, with gold spoons, salt cellars, and candelabras. Don Pidduzzu looked in the book and read: "Take them." He called the crew and ordered everything carried on board. It took them twelve days to load the treasure on the ship. There were twenty-four gold statues so heavy that a couple of days were needed to load them alone. In the book it was written: "Leave the trapdoors the way you found them." That he did, and the brigantine weighed anchor.

"Hoist your sails and continue your voyage," directed the book. So they sailed an entire month, and the sailors began to grow weary.

"Captain, where are you taking us?"

"Let's push on, boys. We'll be back in Palermo in no time."

Every day he opened the book, but nothing was written in it. At last he saw: "Tomorrow you will sight an island: disembark." On land, the book again said: "In the middle is a trapdoor; raise it. Then two more, and a staircase; descend, and everything you find is yours." This time Don Pidduzzu found a cave hung with hams and cheeses, and countless jars lining the walls. Don Pidduzzu read in his book: "Eat nothing, but take the third jar on the left containing a balm that cures every sickness." So Don Pidduzzu carried the jar on board, where he opened the book: "Go home," it said. "At last!" shouted everyone.

But on the homeward voyage, while they sailed and saw only sky and sea and sea and sky, lo and behold on the horizon loomed ships of Turkish pirates. A battle ensued, and all the men were captured and taken to Turkey. Don Pidduzzu and his pilot were carried before the sultan, who asked his interpreter, "Where are these men from?"

"From Sicily, Majesty," replied the interpreter.

"Sicily! Heaven help us!" exclaimed the sultan. "Chain them up! Put them on bread and water and, for their labor, let them transport boulders!"

So Don Pidduzzu and the pilot began that hard life, and all Don Pidduzzu could think of was his princess waiting for him to come back with gifts.

Note that the sultan suffered from an itch covering him from head to foot and which no doctor was able to cure. Learning about this from the other prisoners, Don Pidduzzu told the guards that, in exchange for his freedom, he would cure the sultan.

The sultan got wind of the statement and sent for the Sicilian. "You'll receive whatever you ask for, if only you cure my itch." A promise wasn't enough for Don Pidduzzu, who insisted on a written agreement and permission to return aboard his ship. The ship had been pulled up on shore, and nothing had been touched or stolen, since these were pirates of honor. Don Pidduzzu filled a bottle with balm from the jar and returned to the sultan. Instructing him to lie down, he took a brush and applied balm to his head, face, and neck. Before nightfall the sultan was shedding his skin like a snake, and underneath that itchy skin appeared new skin, smooth and pink. The next day Don Pidduzzu anointed the sultan's chest, belly, and back, and in the evening his skin changed. The third day arms and legs were anointed, and the sultan was completely well. So Don Pidduzzu sailed off with his crew.

He disembarked at Palermo and jumped into a carriage to go to Pippina, who couldn't contain herself for joy. The king asked him how things had gone. "God only knows, Majesty," replied Don Pidduzzu. "Now I would like a gallery readied for displaying my presents. True, they're trifles, but since I have them here . . ."

And he ordered all the gold objects unloaded. For an entire month they did nothing but unload. When everything was finally in place, he said to the king, "Majesty, I will be ready tomorrow. If you like, go first and view what the prince has brought, then come and see my things."

The next day the king went to see the prince's presents—knickknacks, toiletries, pretty objects, but nothing to rave about. The king heaped praise on him. Then the two of them went to view Don Pidduzzu's display. At the sight of such splendor, the prince gasped, wheeled around, flew down the steps, boarded his ship, and was never seen again.

"Long live Don Pidduzzu!" cried the crowd, while the king embraced him. Together they went to St. Catherine's to fetch Pippina, and three days later the betrothed were joined in matrimony.

Don Pidduzzu sent for his mother and father, of whom he had lost track since leaving home. Poor dears, they were still going barefoot! He had them dressed in a manner befitting a prince's mother and father and from then on they lived at the palace with him.

> They were always happy and content,
> While we are here without a cent.

(Palermo)

156

The Wife Who Lived on Wind

There was, in Messina, a prince as miserly as he was rich, who ate only two meals a day consisting of one slice of bread, one slice of salami as thin as a communion wafer, and one glass of water. He kept only one servant and gave him two pence a day, an egg, and just enough bread to sop up the egg. Thus it happened that no servant could endure more than a week in his service; they all left after only a few days' work. Once he ended up with a servant who was a notorious rogue and, no matter how sly the master, this man could steal his very shoes and socks in a foot race.

When the servant, whose name was Master Joseph, saw how things were, he went to a coal dealer who had her shop next door to the palace, a rich woman and mother of a beautiful maiden, and said, "Neighbor, would you like to marry off your daughter?"

"Please God that some fine young man will turn up, Master Joseph," replied the woman.

"What would you think of the prince as a suitor?"

"The prince? Don't you know how stingy he is? He'd have an eye gouged out sooner than spend a penny!"

"Madam, follow my advice, and I'll make sure there's a wedding. All you need do is say your daughter lives on wind."

Master Joseph went to the prince. "Sir, why doesn't Your Majesty get married? You're growing older, and the passing years will never return . . ."

"Ah! You want to see me dead!" exclaimed the prince. "Didn't you know that with a wife to support, your money runs through your fingers like water? Hats, silk gowns, plumes, shawls, carriages, plays. . . . No, indeed, Joseph, nothing doing!"

"But hasn't Your Majesty heard about the coal vendor's daughter, that lovely maiden who lives on wind? She already has money of her own and cares nothing for luxury, parties, or plays."

"You don't mean it! How can anybody live on wind?"

"Three times a day she takes up her fan, fans herself, and thus fans away her appetite. To look at her plump face, you'd think all she ate was beefsteak."

"Well, arrange for me to have a look at her."

Master Joseph took care of everything, and in a week's time the wedding was celebrated, and the coal vendor became a princess.

Every day she went to the table and fluttered her fan, while her husband looked on as pleased as Punch. Then her mother would smuggle in roast chicken and cutlets, and the princess and the servant would gorge themselves. A month passed, and the coal vendor began to complain to the servant about the heavy expense to which they were putting her. "How much longer must I foot the bill?" she asked. "Let that silly prince of yours contribute something himself!"

Master Joseph said to the princess, "Know what you have to do, my girl?" (In public he addressed her as "Princess," but in private he called her "my girl.") "Tell the prince you would like to see his wealth, just to satisfy your curiosity. If he says he's afraid of a few gold pieces sticking to your shoes, tell him you're willing to go into the treasury barefoot."

The princess asked the prince, but he made a sour face and would not be persuaded. She kept on, saying she was even willing to go into the treasury barefoot, and at last obtained his consent. Then Master Joseph said, "Quick, smear glue all around the hem of your long skirt." The princess did just that.

The prince lifted a plank in the floor, opened a trapdoor, and directed her down the steps. The young woman was speechless with amazement when she saw the heap of gold doubloons. No king in the world was half so rich as her husband! As she gaped and ooed and ahed, she innocently swished her long skirt around, gathering gold pieces by the dozen. When she got back to her boudoir and pulled them off, there was a nice little pile of money, which Master Joseph carried to her mother. So they continued to stuff themselves, while the prince watched the princess work her fan and rejoiced over having a wife who lived on wind.

Once when the prince was out walking with the princess, he met one of his nephews whom he rarely saw. "Pippinu," he said to the youth, "do you know this lady? She's the princess."

"Oh, my uncle, I didn't know you had married!"

"You didn't? Now you do. And you're invited to dine with us one week from today."

After extending the invitation, the prince got to thinking about it and was sorry he'd invited the young man. "There's no telling how much we'll have to spend now! How stupid I was to ask him to dinner!" But there was no getting out of it; he was going to have to plan for dinner.

Then the prince got an idea. "Do you know what, princess? Meat is expensive, and we'll go broke if we have to buy any. Instead of buying meat, I'll go hunting and bring some back. I'll take my gun and after five or six days I'll bring you lots of game without spending a cent."

"Yes, of course, prince," she answered, "but be quick about it."

As soon as the prince was gone, the princess sent Joseph for a locksmith. "Sir," she said to the locksmith, "make me a key right away to this trapdoor. I lost the one I had and now I can't open the door."

In no time she had a key that fitted perfectly. She went underground and returned with a few sacks of doubloons. With that pile of money, she had all the rooms hung with tapestries. She bought furniture, chandeliers, portals, mirrors, carpets, and everything else they have in princes' palaces; she even employed a doorman with livery from head to toe and a stick topped with a gold knob.

The prince came back. "What on earth is this? Where is my house?" He rubbed his eyes, turned around and back, asking, "Where did it go?" And he kept on going round and round.

"Excellency," said the doorman, "what is Your Excellency seeking? Why not go inside?"

"Could this be my house?"

"If not, whose is it? Walk in, Excellency."

"Oh, my goodness!" exclaimed the prince, slapping his forehead. "Every cent of my money has gone into all these things! Wife!"

He flew into the house, saw the white marble stairs and the tapestries on the walls. "Oh, my goodness, every cent! Wife!"

He saw the mirrors and rich frames, sofas, divans, armchairs. "Oh, my goodness! Every cent! Wife!"

He reached his bedroom and threw himself down on his bed.

"What is the matter, prince?" asked his wife.

"Oh, my goodness!" he exclaimed in a whisper, "every cent, wife . . ."

His wife quickly sent for a notary and four witnesses. The notary arrived and asked, "Prince, what is the matter? Tell me, do you want to make your will?"

"Every cent . . . my wife . . ."

"What? How's that again?"

"Every cent . . . my wife . . ."

"You want to leave everything to your wife? Yes, I understand. Is this all right like this?"

"Every cent . . . my wife . . ."

As the notary wrote, the prince gasped once or twice more, then gave up the ghost.

The princess was his sole heir and married Master Joseph when she came out of mourning, and who should get the miser's money in the end but the master swindler.

(Palermo)

157

Wormwood

Over and over it has been told that once upon a time there was a king and queen. Every time this queen had a baby, it was a girl. The king, who wanted a son, finally lost patience and said, "If you have one more girl, I shall kill it."

Just as she had feared, the poor wife ended up bearing another girl, but the prettiest child you ever saw. Lest her husband kill it, she said to its godmother, "Take this infant and do what you think best."

The godmother took it, saying to herself, "What am I to do with a baby girl?" She went into the country and laid it upon a wormwood bush.

There in the country lived a hermit. In his cave he had a doe suckling some newborn fawns. Every day the doe would go outside for something to eat. One evening when she returned to the cave, the fawns attempted to suckle, but the doe's udder was empty and the fawns went hungry. The same thing happened the next day, and the next, and the fawns were starving to death. Feeling sorry for them, the hermit followed the doe and discovered that she was going out every day to nurse a baby girl nestled in a wormwood bush. The hermit picked up the baby and carried her into the cave. He told the doe, "Nurse her here and divide your milk between her and your fawns."

The baby was gradually weaned and grew by leaps and bounds. The older she got, the lovelier she became. She did the hermit's housework, and the hermit came to cherish her as if she had been his very own daughter.

One day another king was out hunting, when right in the middle of everything a fierce storm came up; the wind blew, there was thunder and lightning, and it poured rain. The only available shelter was the hermit's cave. Seeing the king come in soaking wet, the hermit called, "Wormwood! Wormwood! Bring a chair, light the fire, and make His Majesty comfortable!"

"Wormwood?" said the king. "What kind of name is that, good hermit?"

So the hermit told of finding the child in a wormwood bush and naming her after it.

The minute he laid eyes on the girl, the king said, "Hermit, would you like to give her to me to take back to the palace? You are old; how can this child stay by herself like that in the country? I will provide her with teachers who will instruct her . . ."

"Majesty," replied the hermit, "I am devoted to the child, so for her own good I'm happy for her to go to the palace. The education Your Majesty is able to offer her is a far cry from what a poor hermit could give her."

The king bid the hermit goodbye, took Wormwood onto his horse, and rode away with her. At the palace he entrusted her to two noble ladies. Once he was thoroughly acquainted with the girl's merits, he said, "The best I can do by her is marry her and make her queen." He married her, and Wormwood became the realm's queen.

The king was madly in love with her. One day he said, "Wormwood, I am obliged to go away for a while. No matter how short a time I'll be away, it pains me more than I could ever say to leave you."

The king departed. One evening outside the kingdom, he found himself in the company of princes and knights, and each man began singing the praises of his own wife. "Go on, boast all you want," said the king, "but none of you could have a wife as wonderful as mine."

At that, one of the knights turned to him. "Majesty, I bet that if I went to Palermo in your absence, I could make time with your wife."

"Impossible!" replied the king. "Totally impossible!"

"Shall we bet on it?" urged the knight.

"Let's," answered the king.

They agreed on the stakes: a fief. They agreed on the length of time: one month. Then the knight departed. In Palermo he strolled day and night under the windows of the royal palace. The days went by without his getting so much as a glimpse of the queen: the windows were always shut.

Then one day as he was walking there, quite downcast, an old woman approached, begging for alms. "Get away from here," he said, "don't bother me!"

"What's troubling you, sir, and making you so gloomy?"

"Get away, let me alone!"

"Tell me what's bothering you, sir; maybe I can help you."

So the knight told her about the bet and his desire to enter the palace, or at least to find out what the queen looked like.

"Put your mind at rest, sir; I'll see to everything."

The old woman packed a basket with eggs and fruit, went to the palace, and asked to speak to the queen. When she was alone with the queen, she embraced her and whispered in her ear, "My daughter, you don't know me, but I'm a relative of yours, and it gives me joy to bring you these few things."

The queen was unacquainted with her relatives; for all she knew, the old woman could have been one. She therefore trusted her, invited her to live at the palace, and ordered everyone to respect her. At any hour of the day or night the old woman was allowed to go in and out of the queen's room and do whatever she pleased.

One day while the queen was sleeping, the old woman entered her room. She approached the bed, peeped under the cover, and saw that the queen's bare back was graced by a very beautiful mole. Then with a pair of scissors the old woman cut the tiny hairs sprouting from the mole and put them away, after which she left the palace, quite pleased with herself. When the knight had these hairs in his possession and heard the old woman's description of the queen, he could no longer contain himself for joy. He rewarded the old woman with a goodly sum of money and departed. On the appointed day he went before the king and the other knights, who were quite anxious to know who would win the bet. The knight spoke: "Majesty, I apologize for what I'm about to tell you. Is it true, or isn't it, that your wife is such and such—" and he gave a minute description of her face.

"That is correct," replied the king, "but that proves nothing. You could have heard those things without ever actually seeing her in the flesh."

"In that case, Majesty, listen carefully: is it true, or isn't it, that your wife has a mole on her left shoulder?"

The king turned pale. "Well, yes."

The knight handed the king a locket. "Majesty, I hate to tell you, but here is proof that I have won the bet"—and with trembling hands the king opened the locket and saw the hairs from the queen's mole. He hung his head in silence.

Without delay the king returned to his palace. Happy to see him back after such a long absence, the queen came out to meet him, laughing. The king neither greeted nor embraced her. He ordered horses harnessed to a carriage and said to his wife, "Climb up," while he too climbed up and sat beside her, taking the reins himself.

Bewildered, the queen looked at him with apprehension, but the king didn't open his mouth. When they reached the foot of Mount Pellegrino, the king reined in the horses and said "Get down." The queen alighted and the king, without dismounting, dealt her a resounding blow with his whip that knocked her down. Then he whipped the horses to a gallop and disappeared from sight.

That day a doctor and his wife were on their way up to the Sanctuary of St. Rosalie, in fulfillment of a vow made before the birth of their son. Bringing up the rear was their Moorish slave, Alí. When they got to the foot of Mount Pellegrino they heard the sound of moaning. "Who can it be?" said the doctor. Going in the direction of the moans, they found a young woman lying on the ground, wounded and half dead. The doctor bandaged her up the best he could and said to his wife, "Let's put off our trip until another day and try to help this young woman. We'll carry her home and see if we can cure her."

That they did. Lodged and nursed by the doctor and his wife, the young woman got well. But no matter how many questions they put to her, she refused to talk about her past or say how her misfortune had come about. In spite of that, the doctor's wife, pleased over finding a young woman so good and virtuous, grew quite fond of her and engaged her as a maid.

One day the doctor said to his wife, "Dear, it is time we fulfilled our promise to St. Rosalie. We'll leave our little girl with the maid and depart with Alí."

The next morning they left early, while the maid and the little girl were still sleeping. After going a short distance, Alí slapped his forehead. "Master, Alí forget! No basket lunch!"

"Go back at once and get it!" said his master. "We'll wait for you here."

Now this slave, seeing how his owners had taken such a liking to the maid, had developed a mortal hatred for the poor young woman. Forgetting the lunch basket was a mere pretext. Running back home, he found the young woman and the child still sleeping. He approached with a butcher knife and slit the little girl's throat. Then he rejoined his owners.

When the young woman awakened she felt herself drenched with blood, then saw the child with its throat slit next to her. "Oh, my heavens!" she screamed. "Those poor, poor parents! Woe is me! What will I ever tell them?" In a panic, she opened a small window through which she fled into the countryside, running as fast as her legs would carry her. She came to a desolate plain, in the middle of which stood an old palace in partial ruin. The young woman entered it, but there was not a living soul in sight. She spied an old dilapidated sofa on which she sank down and promptly fell asleep, exhausted from fright and running.

Let us leave the young woman sleeping and turn back to that king who didn't want any daughters. In time his wife revealed that the daughter she had borne was not dead but had been entrusted to her godmother and heard from no more. The king couldn't rest after hearing that, and one day he said, "My wife, I'm leaving home and will return only when I have news of my daughter." After traveling far, he was overtaken by night on a desolate plain. He saw an old palace in partial ruin and went inside.

Let us leave this father in search of his daughter and go back to that king who had abandoned his wife at the foot of Mount Pellegrino. The more he thought about it, the more he was assailed with misgivings and remorse. "What if that knight was lying? What if my wife was actually innocent? Could she still be alive? Could she be dead by now? Here in this palace without her there's no peace for me. I shall go to the four corners of the globe and return only when I've had some news of her."

After traveling far, he was overtaken by night on a desolate plain. He saw an old palace in partial ruin and went inside. Another king was already there resting in an armchair. He took a seat nearby.

Let us leave that king and take up the doctor. Back from his pilgrimage, he entered his house expecting to see his little girl; instead, he found the house deserted and the child slain. His first impulse was to go and say to the slave, "Alí, we'll go after that wicked woman, to the ends of the earth if necessary, and we'll slay her the way she murdered our little girl."

So he set out. On a deserted plain night overtook him in the vicinity of an old palace in partial ruin. He entered and found two kings sitting in armchairs side by side. The doctor and Alí sat down in the two armchairs opposite them. So they sat, all four of them silent, each one lost in his own thoughts.

In the middle of the room was a lantern, which said, "I want oil."

Then into the room walked a little oil cruet, which said to the lantern, "Come on down lower."

The lantern let himself down, and the cruet poured oil into him. Then the cruet said to the lantern, "Have you anything of interest to tell me?"

"What would you like to hear? Yes, there is something that might interest you."

"Tell me."

"Listen," began the lantern, "there was a king who, wishing no more daughters, told his wife that if one more girl were born to her, he would kill the baby. To save the child, the wife had her whisked away. Listen to this: the child grew up and married a king. This king, misled by a knight, took her to Mount Pellegrino, struck her, and left her lying unconscious on the ground. A doctor came that way and heard a groan . . ."

Bit by bit as the lantern advanced in the story, the men seated in the armchairs looked up one by one, opened their eyes wide, and nearly jumped out of their seats at all they heard, while Alí shook like a leaf.

"Just listen to this," pursued the lantern. "What should the doctor and his wife see when they approached but a lovely young woman lying wounded on the ground. He took her home and later entrusted his little girl to her care. There was a slave who loathed the young woman, so what did he do but kill the little girl and make the blame fall on the young woman . . ."

"Poor young woman!" sighed the cruet. "And where is she now? Is she living or dead?"

"Sh . . ." said the lantern. "She's upstairs sleeping on a sofa. Here are her father the king and her husband the king who, regretful of the evil turns they have done her, are both out looking for her. And there's the doctor seeking to kill her, thinking she murdered his baby."

The father king and the husband king and the doctor had risen. The doctor immediately seized Alí, just barely in time to prevent his escape. All three of them fell on him and tore him apart.

Then they ran upstairs and knelt before the couch on which Wormwood was sleeping.

"She's mine!" said the father king. "She's my daughter!"

"She's mine!" said the husband king. "She's my wife!"

"She's mine!" said the doctor. "I saved her life!"

In the end she went to the king who was her husband. He invited her father and the doctor to the palace for a gala celebration of her return, and from then on they were all one happy family.

(Palermo)

◈ 158 ◈

The King of Spain
and the English Milord

A king said to his son on the boy's eighteenth birthday, "Time is going by, old age is drawing nigh. Why don't you take a wife? If we die, who will inherit the kingdom?"

The son was not overly impressed by these words, and he said to his father, "We still have much time to think about that, Father."

But the king continued dropping hints to his son about getting married, until the son shut him up by replying, "Father, please understand that I will marry only when I've found a girl as white as ricotta and rosy as a rose."

At that, the king summoned his councilors. "Gentlemen, I have the honor to inform you that the prince will take a wife when he finds a maiden as white as ricotta and rosy as a rose. What is your advice?"

The wise councilors replied, "Majesty, select a few of your grandees, supply each with a portrait painter, servants, coachmen, lackeys, and all the rest, and send them around the world in search of this maiden. At the end of a year's time it will be incumbent upon the prince to marry the best one among the maidens they find."

So the grandees set out from the court, each with a master painter and a battery of servants, coachmen, and lackeys. One went to one kingdom, another to another, and they visited all the kingdoms of the earth.

One of these grandees went to Spain, where the first thing he did was stop and talk with a chemist. In the course of the conversation they became friends.

"Wherefore does Your Worship come to us?" asked the chemist.

"We come," replied the grandee, "in search of a maiden as white as ricotta and rosy as a rose. Would there be one in these parts?"

"Oh, if that is all you are after, we have a rare beauty here—a young lady who is truly as white as a mold of ricotta and rosy as a rosebud. But to get a look at her is no easy matter: she never appears in public. I've not seen her myself, and what I tell you is pure hearsay. She's the daughter of people who've gone down in the world, so you never see her out in public."

"How can we manage to see her?"

"I'll think of a way." The chemist went to the girl's mother. "Madam," he said, "in my shop there's a painter who's traveling about painting portraits of the most beautiful faces in the world. Wanting is the portrait of your daughter. If you allow him to paint it, you will be paid forty gold crowns for the favor."

The mother, who was in dire straits, spoke to her daughter and persuaded her to accept. The painter entered, followed by the grandee and the chemist. Seeing the maiden, they all three exclaimed, "How lovely she is!"

The painter did the portrait, put the finishing touches on it at the chemist's shop, and the grandee had it framed in gold. Thus with the painting hanging by a cord around his neck, he presented himself to the king.

It was the time appointed for the return of all the grandees who had gone around the world, and they all gathered in the audience hall, each with a portrait hanging from his neck. The prince viewed them all and, stopping before the maiden from Spain, said, "If the face itself is like the portrait, it is truly a perfect face."

"Majesty," replied the grandee, "if this face does not please you, none ever will."

The grandee was sent from the court back to Spain to fetch the maiden. First, however, she spent four months at a palace learning to be a queen and, once she had learned, being quite intelligent, she was married by proxy, then departed in a carriage for the prince's realm. Her mother was highly commended for the pious upbringing she had given her daughter, and the chemist was handsomely rewarded for his role in the affair. The prince rode on horseback to meet his bride. When they met, he dismounted and entered her carriage. Just imagine how happy they were.

Her queen mother-in-law also liked her. She whispered in her son's ear, "You've found just the wife for you. I like the purity her eyes bespeak."

No doubt about it, the princess led the life of a saint, keeping to her quarters all the time and never sticking her head outside. She and her mother-in-law got along with one another like a pair of pigeons—a rarity, since mothers-in-law and daughters-in-law have quarreled from the beginning of time. But Beelzebub, as you know, is always on the lookout for a way in, and one day mother-in-law said to daughter-in-law, "My daughter, why do you stay shut up like this all the time? Go out on the balcony and get a little fresh air."

Obediently, the young lady stepped onto the balcony.

At that instant, an English milord happened by, looked up, and thought no more of looking down. The young lady saw him and went back inside, closing the window behind her. But there was no stopping the lord now, and he began walking all around the palace in an effort to see the lady again.

One day a bent old woman asked him for alms, and he replied, "Let me alone, you old hag!"

"What's the matter with Your Lordship?"

"Be off with you, it's none of your business!"

"Do tell me your trouble all the same. Who knows but what I can help."

"The trouble? I long to see the princess, but can't see her."

"That's all? Just give me a ring set with a single diamond and leave everything to me."

The milord had faith in her and bought her the ring. She hastened to the palace.

"Where do you think you're going?" asked the guard.

"To see the princess. I have a ring for sale that she alone can afford."

The message was relayed, and the princess invited her in. At the sight of the ring, she asked, "How much do you want for it?"

"Three hundred crowns, Majesty."

"Give this little old woman three hundred crowns immediately, and ten crowns more for her pains."

Rubbing her hands, the old woman returned to the nobleman, who asked, "What did the princess say to you?"

"She promised me an answer within the next ten days." And without a word to anyone, she tucked the three hundred crowns away in her own coffers.

Ten days later she returned to the milord. "I am to go back to the princess, but do you want me to go empty-handed? Know what you should do? Send her a costly necklace."

Lords, as you know, are kings without crowns. So the old woman received a priceless necklace and went off to sell it to the princess.

"It is beautiful," admitted the princess. "How much is it?"

"For you, Majesty, one thousand crowns."

"Let one thousand crowns be paid her immediately, plus forty crowns for her pains!"

The old woman grabbed the money and ran off to the milord.

"What did the princess tell you?"

"Would you believe it, her mother-in-law was there, and she couldn't talk to me. But she took the gift, and next week for sure she'll have an answer."

"And what will you be wanting to take her next week as a gift?"

"Listen: we've given her a ring, and also a necklace. Next time let's give her a fine gown."

For the coming week, the milord ordered the most beautiful gown you ever saw and handed it to the old woman.

"Majesty," said the old woman to the princess, "this gown is for sale. Do you want to buy it?"

"Exquisite! Magnificent! How much do you want for it?"

"Five hundred crowns."

"Give her five hundred crowns, plus twenty more for her pains."

When the old woman returned, the lord asked, "What did she tell you?"

"She told me that you are to give a grand ball at your palace and invite the prince and princess, and everything will be settled there."

Overjoyed, milord made the grandest of preparations, then sent out an invitation to the prince. The princess exclaimed, "How splendid! A grand ball! I will wear the gown I bought from that old woman!" She also put on the ring and the necklace, and they were off to the ball.

For the first waltz, milord went up and invited the princess to dance with him and, convinced that everything was all set now, he winked at her. At that, the princess wheeled about and returned to her seat beside the prince. Assuming that she merely wanted to put on a few airs, the nobleman went up and invited her to dance with him again, and once more he winked at her. What did she do then but sit back down beside the prince. Milord invited her a third time, again winking at her, and the princess turned her back on him. When the ball was over, prince and princess said farewell and departed, leaving the nobleman very much out of sorts.

"Although she wore gown, ring, and necklace, she refused to dance with me. What is the meaning of that?"

In those days, monarchs followed the practice of disguising themselves as peasants and going around to cafés to hear what the people were saying. In one such café, the prince came face to face with milord. Talking about first one thing and then another, the lord, who failed to recognize the prince in his disguise, said, "Just look at that slut of a princess. I sent her a ring, which she accepted. I sent her a necklace, which she accepted. I sent her a gown, which she also accepted. If you only knew what all that cost me! She promised me heaven and earth, directing me to give a grand ball expressly for that, and then she didn't say one word to me the whole evening long!"

When he heard this account, the king turned crimson. He rushed back to the palace, drew his sword, and lunged for his wife. His mother was standing by and jumped between the couple to shield her daughter-in-law. Angry, the prince summoned a ship captain. "Captain, carry aboard your ship this slut"—from that time on, the princess had no other name—"sail out to sea and then kill her, cut out her tongue, pickle it, and bring it home to me. Throw everything else overboard!"

The captain took the unfortunate girl and left. Her mother-in-law was too broken-hearted for words, and they separated in silence. All around them people did nothing but weep.

The captain had a dog on board with him, so he killed the dog and pickled its tongue. When the boat came in sight of land after days and days of sailing, the captain had the poor princess put ashore with an abundant supply of food and clothing. The ship sailed away again, leaving the princess there all by herself.

She took refuge in a cave, where her provisions slowly dwindled. She was almost out of food, when a frigate appeared on the horizon. The princess signaled to it, and the captain saw her. "Land!" he ordered, and they landed.

"What are you doing here, my lady?" asked the captain, bowing to her.

"I was on a vessel. It was shipwrecked, and I alone survived."

"Where would you like me to take you?"

"To Brazil," promptly replied the young lady, recalling that the prince had an older brother who was emperor of Brazil and that the queen mother always spoke of him with deep affection. "I have relatives in Brazil."

The captain took her on board and set sail for Brazil. Before arriving, she said to him, "Captain, I should like to ask another favor of you: so as not to be recognized by my relatives, I should like to be dressed as a man."

The captain had her disguised and her hair shortened. With her beauty, she now looked like a handsome page.

After disembarking she strolled through the streets looking all about her. Seeing a lawyer's office, she entered and asked, "Sir, could you use me as a clerk?"

"Yes, indeed, I could"—and he engaged her as a clerk. He gave her a piece of work, which she completed in the twinkling of an eye. The lawyer couldn't believe it! He assigned her a more intricate task, which she took care of in less than no time, thus capturing the lawyer's admiration once and for all. For starting wages, he paid the clever clerk twelve crowns a day.

Now the lawyer had a daughter, and he thought to himself, I shall marry her to the young clerk, whom he informed of his plan.

"Please, sir," replied the clerk, "let's put that off for the time being. Let me first get ahead in my profession, and then *I'll* propose."

The fame of the lawyer's young clerk spread, and he was summoned by the royal secretariat. The youth presented himself, and the secretary gave him a document to copy. In no time the task was completed. Word of this youth who exquisitely copied every kind of document reached the emperor—the clerk's own brother-in-law—and the emperor said, "Bring this young man hither!"

The emperor liked him at first sight, retained him at the palace, and engaged him as his squire.

Let us leave them and return to the prince. His anger subsided, and he began to repent of his rashness. "Maybe she was innocent. Oh dear wife, how foolish I was! Now what is left of you? Oh wife, I murdered you!" Constantly plagued by such thoughts, he lost his mind.

At that, the queen sat down and wrote her son, the emperor of Brazil, that his brother had gone crazy and the people were on the verge of rebellion. "Come to us for a while," concluded the letter. The emperor read it and burst into tears. "Squire," he said, "will you go to my brother? I appoint you viceroy and give you carte blanche."

The squire accepted. He assembled a large retinue and two beautiful ships and sailed off. The distance was great, but in due time he arrived.

"Here comes the viceroy!" exclaimed all the people. "Here comes the viceroy!" The cannons fired a salvo, and the viceroy disembarked. The queen came forward and received him with as many honors as if he had been her son in person. "Welcome, Viceroy!"

"Your Majesty's humble servant! Allow me, Majesty, before all else, to see to the affairs of your people." He proceeded to settle the mountain of unfinished public business, and the people found it almost too good to be true that such a viceroy had come to govern them.

One day at last, the viceroy said to the queen, "Now, Majesty, tell me a little about this matter of your lost daughter-in-law."

The queen told him the whole story from beginning to end—about the lord, the conversation in the café, her daughter-in-law's departure, everything—and relating it, her eyes filled with tears.

"Very well. Now let's see," answered the viceroy. "Send for this lord who was the cause of all your misfortune."

The lord arrived, and an audience was granted him. "Milord, this is a matter of life and death. What is your version of the story concerning the princess?" asked the viceroy.

The lord gave as truthful an account as he could, without omissions or additions.

"But, Milord, did you ever talk to her yourself?"

"I never did."

"Did you give her those gifts directly?"

"No, the old woman did." (Meanwhile the queen mother eavesdropped, and with her the prince, who was still half out of his mind.)

"Is the old woman who performed those services for you living or dead?"

"She may still be living."

"Well, lock this lord up in a room," ordered the viceroy, "and send for the old woman."

By order of the viceroy, the old woman was produced.

"Tell me, good old woman, about those sales of yours to the princess." And the old biddy unbosomed herself.

"But tell me, dear, did you ever carry the princess any message?"

"Never, Viceroy."

At that, the king regained his senses. "Oh, my wife, you were innocent!" he began crying. "Oh, dear wife, I killed you without cause!"

"Please calm down, Majesty," said the viceroy. "We may yet find a remedy."

"How can matters be remedied, now that she is dead? Oh, my wife, my wife, I've lost you forever."

The viceroy went behind a screen, dressed up like the princess she actually was, put back her hair that had been cut off, and reappeared before mother-in-law, prince, and court.

"Who are you?" cried the queen.

"Your daughter-in-law! Don't you recognize me?" But the prince already had his wife in his arms, hugging and kissing her.

The sentence had been delivered while she was still disguised as the viceroy: the old woman was to be burned at the stake, and the English lord guillotined. No time was lost in executing the order.

The great queen wrote her son, the emperor of Brazil, about the episode, and he still marvels to this day. "Children, children, just fancy: my secretary was my sister-in-law, and I never suspected it!"

The two captains—the one who killed a dog instead of the princess, and the one who rescued her and took her to Brazil—were elevated to the rank of court grandees. And all the sailors were awarded red pompons for their berets.

(Palermo)

❖ 159 ❖

The Bejeweled Boot

The son of a merchant became an orphan at an early age along with his sister, who was the apple of her brother's eye. He got his education, then put his services at the disposal of the king of Portugal. His penmanship so delighted the eye that the king engaged him as his secretary. Now it happened that certain letters written by him went to the king of Spain, who exclaimed, "What exquisite handwriting! This scribe would make me an excellent secretary." So he wrote to the king of Portugal:

I have read your letter, and I am full of admiration for the beautiful hand your secretary writes. In the name of the friendship that binds us, I beseech you to let me have him to be my secretary, since there is no one in Spain who writes so handsomely.

These kings always made a point of showing one another the utmost courtesy. Therefore the king of Portugal, however much he hated to lose his secretary, told the young man to go to his colleague.

"Majesty," inquired the youth, "what am I to do about my sister? I can't just walk off and leave her."

"Don Giuseppe," answered the king, "I have no idea. All I know is that you must go. Your sister is a good maiden, and stays to herself. Tell your maidservant to keep an eye on her, and you will have nothing to worry about."

The youth had no choice but inform his sister of the situation. "Dear little sister, here's how matters stand: I am obliged to go away, the king of Spain wants me as his secretary. You will remain behind with our maidservant. When I am all settled, I'll send for you to come to Spain too." The sister burst into tears. "So we won't feel so far apart," he continued, "let's have our portraits painted. I'll take yours with me, and you'll keep mine here with you." That they did.

The king of Spain heartily welcomed Don Giuseppe and immediately put him to writing, while he stood by admiring the beautiful script. He became so fond of this new secretary that he would say, no matter what problem arose in the kingdom, "Don Giuseppe, you take care of that . . . Use your own judgment, in which I have complete confidence. Whatever you do is well done!"

As a result, intense jealousy spread among all the highest placed men at the court—the squire, the original secretary, the knight—and they sought some way to tarnish Don Giuseppe's reputation.

The squire went to the king and said, "Good for you, Majesty! You certainly found the right man! I'm referring to Don Giuseppe whose praises Your Majesty is always singing! Goodness knows what he is secretly about while all your trust is in him!"

"What are you saying? What is the matter?"

"What's the matter? Every day in his room he takes out a portrait, contemplates it, kisses it, and weeps. And then he hides it!"

So the king went and surprised Don Giuseppe kissing the portrait. "May one ask whom you are kissing, Don Giuseppe?"

"My sister, Majesty."

The king looked at the portrait and saw such a beautiful maiden that he could not help but be impressed. Her brother then proceeded to relate all her charms.

But also present was the squire, who could never resist putting Don Giuseppe in the wrong. He glanced over the king's shoulder at the portrait and snorted, "Who, this woman? But I know her and have had dealings with the same."

"With my sister?" exclaimed the youth. "But she's never been out of doors! How could you have seen her when no one else has ever laid eyes on her?"

"Yes, I have had to do with your sister."

"Liar!'"

After much arguing back and forth, the king interrupted them. "Let the matter be settled once and for all: if it is true, Squire, that you have had to do with Don Giuseppe's sister, then you have one month to bring in proof of it. If you produce it, Don Giuseppe will be beheaded. Prove nothing and *you* will be beheaded."

It was a royal order, and final. The squire departed. When he reached Palermo, he began sounding out everybody on this maiden, and they all said she was a rare beauty, but that no one had ever seen her, since she never left the house. Days and days went by, and every day the squire could feel the ax a little closer to his neck. He was thus walking around one evening wringing his hands and saying, "What can I possibly do?" when an old woman approached him. "Please give me something, kind sir, I'm starving!"

"Off with you, cursed hag!"

"Give me something, and I will help you."

"I'd like to see the person who could help me right now!"

"Tell me the trouble, and I'll help you."

So the squire told her everything.

"What! Is that all? Leave everything to me and consider youself already in possession of the needed evidence."

It poured down rain that night, and there was much lightning and thunder. The old woman leaned against the front door, shivering with cold and pitifully weeping. At the sound of her wails, the mistress of the house, who was none other than Don Giuseppe's sister, said, "Poor old thing! Bring her inside!"

The front door was opened, and the old woman stepped inside all huddled up. "Brrrrrrrr! I'm freezing to death!"

The lady immediately seated her at the fireside and had food served to her. Sly as could be, the old woman took in everything, noting in which room the mistress slept. When the mistress at last went off to bed and, exhausted by the long evening of stormy weather, fell asleep, the old woman tiptoed into her bedchamber, lifted the covers, and gazed at the maiden from head to foot. On her right shoulder grew three little hairs that were like three golden threads. With a tiny pair of scissors the old woman snipped them and tied them up in the corner of her handkerchief. Then she drew the covers back over the maiden and quietly returned to her own bed.

She huddled up again and started wailing anew. "I can't get my breath! I can't stay here any longer, let me out!"

The mistress of the house woke up and said to her maid, "Let that old woman out, or none of us will get a wink of sleep."

The squire was walking up and down in front of the palace. The old woman gave him the three hairs and walked away with a handsome reward. The next day the squire sailed back to Spain.

"Majesty," he said, going before the king, "here is the sign of Don Giuseppe's sister! Three gold hairs from her right shoulder!"

"Woe is me!" said Don Giuseppe, covering his face with his hands.

"Now I will give you a month's time: defend yourself, or the sentence will be carried out. Guards!"

The guards came forth, surrounded the secretary, and marched him off to prison, where he was given one slice of bread and one glass of water daily. But the jailer, seeing what a good person the prisoner was, began smuggling to him the same food as the other prisoners received. What pained Don Giuseppe most of all, however, was that he couldn't write his sister a single line. He finally appealed to the jailer. "Would you grant me a favor? Would you allow me to write my sister a note, and then post it yourself?"

The jailer had a big heart and said, "Go ahead." So Don Giuseppe wrote his sister, telling her everything that had happened and how he was about to be beheaded on her account. The jailer took the letter and posted it.

The sister, who, having received no word from her brother up to now, was worried, and anxiously read the letter. "My dear little brother!" she cried. "How could such misfortune have befallen us?" She began thinking how she could help him.

She sold all their possessions and property, and with the proceeds bought as many fine jewels as she could. Then she went to a skillful goldsmith and said, "Make me a beautiful boot set with all my jewels." Next, she ordered a mourning dress of solid black, and set sail for Spain.

Upon her arrival in Spain she heard the sound of trumpets, and what should she see but soldiers leading a man blindfolded to the scaffold. Wearing her long black dress, with only a stocking on one foot and the marvelous boot on the other, she began running through the crowd crying, "Have mercy, Your Majesty! Have mercy!"

For this beautiful lady dressed in black, with one foot so magnificently shod and the other one bare, everyone made way. The king heard her. "Don't lay a hand on her," he said to the soldiers. "What is the matter?" he asked her.

"Have mercy, Majesty, and justice be done! Have mercy, Majesty, and justice be done!"

"It is granted. Speak!"

"Majesty, your squire, after enjoying my person, stole my boot that formed a pair with this one"—and she showed the boot set with diamonds and other precious stones.

The king was dumbfounded. He turned to the squire. "And you were able to do such a deed! After taking your pleasure with this young woman, you stole her boot! And you now have the nerve to stand before me!"

The squire fell into the trap. He replied, "But, Majesty, I never saw this lady before!"

"What do you mean you never saw me! Be careful what you say!"

"I swear I never saw you before!"

"If that is so, then why did you claim to have had dealings with me?"

"When did I ever say that?"

"When you swore you had known the sister of Don Giuseppe, so as to send him to his death!"—whereupon she made herself known to the king.

The squire was forced to confess his fraud. Seeing the sister's innocence, the king ordered Don Giuseppe freed and brought to his side, while the squire was blindfolded and led to the scaffold. Brother and sister embraced, weeping for joy. "Off with his head!" ordered the king, and the squire was beheaded then and there. The king returned to the palace with the brother and sister and, seeing how beautiful and virtuous she was, he asked her to marry him.

> They were as happy as happy could be,
> While here we sit, picking our teeth.

(Palermo)

160

The Left-Hand Squire

Once, it is told, there was a king of Spain who had a left-hand squire and a right-hand squire. The left-hand squire was married to a "Madonna," so beautiful, gracious, and modest was she. In all the time he had been at court, the right-hand squire had never laid eyes on that beautiful countenance, and was half angry over this.

He took to telling the king, "Majesty, you can't imagine what a handsome wife the left-hand squire has! A magnificent lady indeed, Majesty!"

On another day, he said, "Majesty, this morning I caught a glimpse of your squire's wife, and the sight left me speechless. There simply aren't words to tell you how lovely she is!"

And still another time. "Would you believe, Majesty, that the left-hand squire's lady grows lovelier all the time?"

Overnight the king was filled with desire to see this beauty for himself. He mounted his horse and rode with his knights up to the left-hand squire's palace. At that very moment the lady happened to be at the window. The king felt his heart skip several beats. He looked at her as they rode by, but that was all he could do, since it was unfitting for a king to stop and stare up at a window, lest the people gossip. He came back by the palace on his way home, but the lady, modest soul that she was, had withdrawn from the window. Unable to let matters rest, the king went home to his palace and ordered no one to leave it until his return: he had got the bright idea of calling on the lady while her husband was under orders to stay inside the royal palace.

He dressed up as a soldier and went to the left-hand squire's palace. He rang the bell, and the door was answered by the maid, who asked, "What do you wish?"

"I must speak to the lady of the house."

"What do you wish of my lady?"

"I have to talk to her."

"My lady is resting and cannot receive you."

"I shall come inside anyway."

"No, you cannot." She gave him a shove and was about to shut the door in his face, when the king unbuckled his soldier-jacket and showed her the Royal Fleece.

The maid fell to her knees. "Pardon me, Majesty! I did not recognize you!"

"That is all right," replied the king. "You prove that you are a faithful maidservant. Now I wish you merely to let me look on the princess's face, and I will leave."

"Of course, Majesty"—and on tiptoe she led him to where her lady was resting. She was in a deep sleep, when one's face becomes rosier, and the king grew weak in the knees at the beautiful sight. He removed one of his gloves, laid it on the canopy, and reached out to caress her; but he checked himself in time.

He stood there contemplating her to his heart's content, then all of a sudden turned away and departed.

When the king got home, the knights and all the court were free to leave. The left-hand squire returned to his house and went to his wife. What should meet his eye as he entered the bedchamber but the glove the king had placed on the canopy and forgotten. The squire might just as well have beheld the Devil. From that day forward, he no longer looked at his wife.

The poor lady, innocent as a lamb, knew not what to make of this change of heart in her husband and, keeping to herself and never complaining, she grew thin and wrinkled.

Her maidservant would say, "My lady, wherefore are you always sad and alone, while other ladies go to balls and the theater?"

One day that evil-hearted right-hand squire chanced to walk by the left-hand squire's residence, and whom should he see on the balcony but the poor princess, now thin as a rail. Even this evil-hearted man was moved to pity and told the king about it. "Would you believe, Majesty, the once exquisitely beautiful wife of the left-hand squire has fallen off and faded beyond recognition."

The king grew thoughtful and, after much pondering, slapped his forehead. "Oh, dear, what have I done!"

Two days later, he gave orders for a court banquet. Every knight was to bring his wife or, if unmarried, his sister or some other lady of his household. The left-hand squire had no choice but to take his wife, since he had neither sister nor anyone else he could bring. He summoned the maidservant and instructed her to tell his wife to get herself the most beautiful outfit conceivable, sparing no expense, since she was invited to the banquet at the court.

At the banquet, the lady was seated beside her husband, who sat on the king's left. The king proceeded to ask his guests about their life, questioning everyone except his left-hand squire and the squire's wife. At last he turned to her. "And how have you spent your life, my lady?"

Softly, the poor lady replied in verse:

> "A vine was I, a vine am I;
> He pruned me once, though now no more.
> I know not why
> My master tends his vine no more."

Then the squire answered her:

> "A vine were you, a vine are you yet;
> I pruned you once, though now no more.
> The reason is the lion's threat,
> And thus your master tends his vine no more."

The king realized that the vine was the lady, who had been deserted by her husband upon finding the glove on the canopy. Now aware of all the harm his curiosity had wrought, he said:

> "About this vine of which you speak:
> I raised its leaves and saw the stalk,
> But touched it not,
> To keep my crown from blot;
> I swear by it the truth to speak."

Now one knows that when kings swear by their crown, they are taking the gravest of oaths, so when the squire heard that his wife was innocent, he was utterly speechless.

After the banquet, the king took the couple aside and told them how the glove had found its way to the lady's bed, and he thus concluded his account. "I admired the maidservant's fidelity to her lady and, even more, the integrity of this lady who never looked at any man but her husband. Forgive me for all the grief I have caused you."

(Palermo)

161

Rosemary

There was once a king and queen who had no children. Strolling in the garden one day, the queen noticed a rosemary bush with many seedlings growing around it, and said, "Just look at that! A mere rosemary bush has all those children, while I am a queen and childless!"

Not long afterward, the queen herself became a mother. But she was delivered not of a baby, but a rosemary bush! She planted it in an exquisite pot and watered it with milk.

They received a visit from a nephew of theirs, who was the king of Spain. "Royal aunt," he asked, "what plant is this?"

"Royal nephew," replied his aunt, "that is my daughter, and I water her four times daily with milk."

The nephew was so charmed with the plant that he planned to steal it. He took it, pot and all, and carried it aboard his yacht, purchased a nanny goat for milk, and ordered the anchors raised. During the voyage he milked the goat and fed the rosemary plant four times a day. When he disembarked at his city, he had the bush planted in his garden.

This youthful king of Spain loved to play the flute, and every day he circled through the garden playing and dancing. As he played and danced, a comely maiden with long hair emerged from the rosemary foliage and began dancing beside him.

"Where do you come from?" he asked her.

"From the rosemary bush," she answered.

When the dance was done, she disappeared into the rosemary foliage and was seen no more. From that day forward, the king would rush through all his official business to go into the garden with his flute. He would play, and the lovely maiden would come out of the rosemary bush; they would dance and converse, holding hands.

At the height of the romance, war was declared against the king, and he had to go off to battle. "Rosemary, my dear," he said, "do not come out of your plant until I return. When I get back I will play three notes on the flute, and then you can come out."

He summoned the gardener and instructed him to water the rosemary bush four times a day with milk. He added that if he found the plant withered upon his return, the gardener would be beheaded. With that, he was off.

Now the king had three sisters, girls with much curiosity, who had been wondering for some time why their brother spent hours on end in the garden with his flute. While he was away at war, they proceeded to inspect his bedchamber and found the flute. They picked it up and carried it to the garden. The oldest girl tried to play it and drew forth one note. The second girl took the instrument from her hands, blew, and produced another note. Then the youngest, in her turn, also sounded one note. Hearing the three notes and believing the king to be back, Rosemary jumped out of the bush. "Ah!" exclaimed the sisters. "Now we understand why our brother spent all his time in the garden!" Malicious girls that they were, they caught hold of the maiden and beat her unmercifully. All but dead, the poor thing fled back to the rosemary bush and out of sight.

When the gardener came by, he found the shrub partially withered, with its leaves fading and drooping. "Woe is me! Now what will I do when the king returns?" He ran inside his house and said to his wife, "Farewell, I must flee for my life. Water the rosemary with milk"—and he was gone.

Mile after mile the gardener walked through the countryside, finding himself in a forest when night fell. Fearful of wild animals, he climbed a tree. At midnight, beneath the tree, a dragon-woman and a dragon-man had agreed to meet. Cold chills went over the gardener crouched in the treetop as he listened to their fierce snorting.

"What's new?" asked the dragon-woman.

"What do you expect?" answered the dragon-man.

"Don't you ever have anything of interest to tell me?"

"As a matter of fact, I do: the king's rosemary bush has withered."

"How did that happen?"

"Well, the king went off to war, his sisters started playing his flute, and out of the rosemary came the enchanted girl. The sisters all but killed her with their blows. So the bush is withering away."

"And there's no way to save it?"

"Yes, there is a way . . ."

"Tell me."

"It's not something to be repeated; the trees around us have eyes and ears."

"Go on, tell me. Who could be out here listening in the middle of the forest?"

"Well, I'll tell you this secret: one would have to take the blood from my windpipe and the fat from the nape of your neck and boil them together in a pot, then grease the whole rosemary bush with the solution. The shrub will dry up completely, but the girl will emerge well and healthy."

The gardener listened, his heart in his mouth. As soon as the dragon-man and dragon-woman fell asleep and began snoring, he ripped a knotty branch from the tree, jumped to the ground and, dealing two hearty blows, sent them both to kingdom come. Then he drew blood from the dragon-man's windpipe, scraped fat from the dragon-woman's scruff, and rushed home as fast as his legs would carry him. He awakened his wife and said, "Quick, boil this stuff!" Then he took it and greased the rosemary shrub, twig by twig. The maiden emerged, and the bush dried up. The gardener took her by the hand and led her into his house, put her to bed, and served her a bowl of tasty hot broth.

The king came back from the war, and the first thing he did was take his flute out to the garden. He played three notes, then another three— yes, he was in the mood to make music! He went up to the rosemary bush and found it all dried up, with every leaf gone.

In bestial fury he rushed up to the gardener's house.

"Your head will roll this very day, wretch!"

"Majesty, calm down and step inside for a minute. I have something wonderful to show you!"

"Something wonderful, my foot! Your head will roll, for sure!"

"Just come inside, and then do whatever you like!"

The king went in and found Rosemary in bed, as she was still convalescing. She looked up and said, her eyes full of tears, "Your sisters beat me nearly to death, but the poor gardener saved my life!"

The king was overjoyed to find Rosemary again; he had only contempt for his sisters and deep gratitude for the gardener. When the maiden was completely well, he decided to marry her, and he wrote his uncle king that the rosemary plant he stole had become a lovely young lady, and he invited him and the queen to the wedding. The king and queen, who had given up all hope of ever hearing of the plant again, went wild with joy when the messenger presented them with the letter stating that the plant was really a beautiful maiden, their daughter. They set sail at once and "Boom! Boom!" went the cannons in salute as they pulled into port where Rosemary stood awaiting them. The wedding took place, and all of Spain rejoiced and feasted.

(Palermo)

162

Lame Devil

Lame Devil lived in Hell. Men were dying and coming straight to Hell and face to face with Lame Devil, who asked, "Well, friends, what brings you here? Why is everyone coming down below?"

"All because of women," the dead would reply.

From hearing this answer over and over, Lame Devil was filled with curiosity and the ever-growing desire to satisfy it: just what was this business involving women all about?

He dressed up in the guise of a knight and betook himself to Palermo. There on a balcony stood a girl, and he found her very much to his liking; so he proceeded to saunter up and down the street. The more he strolled, the more he liked her, and he sent word asking for her hand in marriage. No dowry was required, he would take her with just the clothes on her back, but on one condition: that she would ask for everything she wanted while still engaged, because once they were married he wanted to hear no more requests.

The girl agreed, and the knight showered her with gifts and clothes enough for a lifetime. They got married, and one evening when there was a gala performance at the theater, they went out together for the first time. Now one knows how women do when they get to the theater: she started out eyeing the marquise's outfit and the countess's jewelry, when she spied the baroness wearing a hat totally unlike any of her own three hundred hats, and she was consumed with desire for one like it. But she had agreed to ask her husband for nothing more. The new wife pulled a long face indeed. Noticing it, her husband asked, "Rosina, what's wrong? Something is the matter, I know."

"No, no. Nothing . . ."

"But you don't look well."

"I assure you, nothing is the matter with me."

"If something is wrong, you'd better tell me."

"Well, if you must know, it's quite unfair for the baroness to have a particular kind of hat I don't have and for me not to be able to ask for one like it! That's what the matter is!"

Lame Devil exploded like a firecracker. "Zounds! So it's true that men all go to Hell through the fault of you women! I understand now."

He walked off and left her right there in the theater.

He returned to Hell and proceeded to tell a fellow devil of his everything that had resulted from taking a wife. The fellow devil said he too would like to see what marriage was like, only he wanted a king's daughter, to discover if it was the same old story even in royal families.

"Well, go ahead and try it, brother!" replied Lame Devil. "Do you know how we can work it? I will steal into the body of the king of Spain's daughter. With a devil inside her, she will be taken sick, and the king will decree: 'Whoever cures my royal daughter will be rewarded with her royal hand in marriage.' You will come on the scene dressed as a doctor, I will leave her body when I hear your voice, she will be well, and you will marry her and become king. Does that suit you?"

They went through with the scheme, and everything happened as planned, up to the moment the fellow devil was brought to the sick princess's bedside. As soon as he was alone, he began saying in a low voice, "Brother Lame Devil! Eh, brother, it's me. You can come out now and let the princess get her breath! Do you hear me, Lame Devil?"

But one should always be wary of devils' promises. Lame Devil, as a matter of fact, did hear the voice. "What? What is it? Ah, yes, yes indeed; I'm quite comfortable, so why should I move?"

"Brother, what did you promise me? Are you joking? The king is beheading those who try but fail to cure his daughter! Brother Devil! Eh, brother!"

"Yes, I'm very comfortable right where I am. And you expect me to move out?"

"What are you saying? My very life is at stake!"

"Oh, don't talk to. Go on! I wouldn't even leave here under fire!"

The poor fellow devil begged and pleaded, but all in vain. The time allotted to him was drawing to a close, so the sham doctor went to the king and said, "Majesty, to cure your daughter I need only one thing: for you to order the cannons on your frigates fired."

The king went to the window: "Frigates, fire!"

"Boom! Boom! Boom!" went the cannons on the frigates.

Lame Devil, who saw nothing from his present vantage point inside the princess, asked, "Brother, what's the meaning of all those cannonades?"

"A ship is coming into port, and they are firing a welcome."

"Who's arriving?"

The fellow devil went to the window. "Oh! It's your wife arriving!"

"My wife!" exclaimed Lame Devil. "My wife! I'm getting out of here! I'm clearing out this very instant! I can't even stand the smell of her!"

Out of the princess's mouth shot a streak of fire as Lame Devil fled, and in the same instant the princess was cured.

"Majesty! She's well, Majesty!" called the fellow devil.

"Hurrah!" replied the king. "Daughter and crown are now yours."

So began the fellow devil's woes.

> May you who tell this story, or hear it,
> Stay always clear, clear of the Pit.

(Palermo)

❖ 163 ❖

Three Tales by Three Sons
of Three Merchants

There were once three sons of three merchants, and they all three decided to go hunting together. They went to bed early, and at midnight one of them awakened, saw the moon, and mistook it for the sun. He dressed in his hunting outfit, took his dogs, and went to rouse his friends. They all three set out while it was still night. The sky clouded up and rain came pouring down, but the hunters could find no tree with foliage thick enough to shelter them. They saw a light and discovered a palace.

"Is this any time to be knocking?" asked the maid. "It happens to be the middle of the night."

"Won't you give us shelter?" asked the hunters.

"I'll go ask the mistress of the house," replied the maid. "Madam, at the door are three men, soaking wet; shall I let them in?"

"Yes."

So they came inside and sat down before the lady of the house, a beautiful widow, who said, "Put on these clothes that belonged to my poor husband and give your own a chance to dry out. And have something to eat. Then each of you must tell me a story—something that has happened to you personally. I'll marry the one who tells the most hair-raising, bloodcurdling tale."

The oldest boy began.

"Well, madam, I am the son of a merchant. Once my father sent me out on business. Along the way I was joined by a man all muffled up, whom I had never seen before, but who seemed to know his way around those parts. At nightfall he said, 'Come with me; I know a good place to sleep.' We entered a solitary house, and the door closed behind me. I found myself in a large room, in the middle of which stood an iron cage full of men locked up inside. 'Who are you?' I inquired of them, and they gave me to understand by signs that I too would be caged with them. But they could not speak because a giant stood guard over them, the very one who had men caught and then locked them up. I too was grabbed by the giant and thrown into the cage. 'What happens next?' I asked my fellow prisoners. 'Keep quiet!' they said. 'Every morning the giant eats one of us.' So we lived in silent fear, huddling up together whenever the giant reached into the cage.

"From time to time the giant would get bored and pick up a guitar and play. Once while he was playing, the strings snapped. 'If someone inside the cage can fix a guitar, I'll set him free,' he said. Right away I said in a loud voice, 'Sir, I am a guitar maker and my father is a guitar maker, the same as my grandfather and all my relatives.' The giant said, 'We shall see,' and pulled me from the cage. I picked up the guitar, tightened here, loosened there, and finally got it fixed. Then the giant patted me on the head and handed me a ring. 'Slip the ring on your finger and you will be free,' he said. I did, and found myself at once outside. I started running through the countryside until I ended up once more before the giant's door. 'What! I'm back where I started from?' I began running in the opposite direction, and ran and ran until I again stood before that door. 'I can't get away from here!' I cried. In that instant I heard someone go, 'Pss! Pss!' I looked up and there at one of the upstairs windows stood a little girl who said softly, 'Throw away that ring if you want to escape!' I tried to remove it, but ended up exclaiming, 'I can't get it off!' 'Cut off your finger! Hurry!' 'I have no knife!' I replied. 'Here's one!' said the child, throwing it down to me. Beside the door was the base of a column; I placed my hand on it and, one stroke, cut off the finger with the ring on it. Then I was able to flee and return to my father's house."

Throughout the account the lady had exclaimed, "Oh, you poor dear!" Now she drew a sigh of relief and turned to the second hunter, who began:

"It happened, madam, that once my father, a merchant, gave me a sum of money for a business transaction. I embarked and was sailing on the open sea when a mighty storm arose, forcing us to throw all our cargo overboard. The tempest was followed by total calm, and we stayed put in the middle of the sea. Our provisions ran out quickly, and we had nothing more to eat. 'Gentlemen,' said the captain, 'famine is upon us. We will all write our names now on slips of paper, and every morning there will be a drawing. The man whose name is drawn will be killed and served up to those remaining.' Just imagine, madam, the fear that gripped us on hearing that announcement! But what else could we do, if we didn't want to starve to death in a body? Every morning, then, lots were drawn, and the man whose turn it was to die was cut up and served to all the rest of us.

"Finally only two of us remained—the captain and myself. The next morning we drew lots. I had resolved that if it fell to the captain, I would slay him; but that if I turned out to be the unlucky one, I would fight for my life. It proved to be the captain's turn who, poor man, held forth his hands, saying, 'Here I am, my brother.' It grieved me to do so, but I summoned up the courage and slew him. I quartered him, stringing up one of the quarters on the ropes. An eagle swooped down and made off with that quarter of human flesh. I hung up another quarter; back came the eagle and took it. I was quite upset. The third quarter was also devoured by the eagle. Only the last quarter remained and, as the eagle landed to take possession of it, I seized the bird by the feet. The eagle soared upward and across the sky, with me hanging on for dear life. Coming close to a mountain, I let go. Plunging this way and that, I finally reached flat earth and returned home."

"Oh, you poor, poor thing!" exclaimed the lady. "That too was a frightful experience. Now it is your turn," she said to the third hunter.

"Madam, my story will make your hair stand on end. I too was entrusted by my father, mind you, with a commercial venture. At nightfall I took lodgings at an inn. After dinner I retired to my room and knelt by my bed, as I always do at night, to say my prayers. At a certain place in my prayers, I bent over to kiss the floor, and what should I see under the bed but a man! I looked closer: he was dead. 'This man was killed last night,' I thought to myself, 'and everyone who sleeps in this bed no doubt meets the same fate.' So what did I do? I picked up the corpse and put it in the bed, while I stretched out under the bed and held my breath. One or two hours went by, and I heard the door open. It was the innkeeper with a knife in his hand, and the scullery boy with a hammer; the innkeeper's wife brought up the rear with a lamp.

" 'He's sound asleep,' they said. 'Let him have it!' The innkeeper positioned his knife on the dead man's head, the scullery boy struck it with his hammer, and the woman said, 'Now take him and put him under the bed; we'll throw the one from last night out the window.' Under the window was a deep ravine, and I could just feel every bone in my body broken. But the innkeeper said, 'Let's leave everything as it is for the rest of the night. Tomorrow we'll be able to see what we're doing.' They went off, and I breathed freely once more. I settled down to wait for daylight.

When the sun came up I went to the window and signaled to the towns on the other side of the ravine. Officers of the law were dispatched to the inn to set me free and arrest the innkeeper and his household."

The lady began deliberating as to which of the three stories was the most frightful and, say what you will, she has yet to make up her mind.

(Palermo)

◈ 164 ◈

The Dove Girl

There was once a lad who led a dog's life. One day when he was particularly miserable, having nothing to eat, he went down and sat by the sea, hoping to think of a way out of his plight. After a while he looked up and saw a Greek heading for the same spot. The man asked, "What's troubling you, my lad? You look so worried."

"I'm starving to death, that's what. I've nothing to eat, nor hope of getting anything."

"Oh, lad, cheer up! Come with me, and I will give you food, money, and whatever else you desire."

"What do I do in return for all that?" inquired the young man.

"Nothing. With me, you will work only one time out of the whole year."

The poor boy couldn't believe his ears! They put their signatures to the agreement and, for quite some time, the youth had absolutely nothing to do. Then one day the Greek called him and said, "Saddle two horses; we are leaving." He got everything ready, and they departed. After a long ride they came to the foot of a steep mountain. "Now," said the Greek, "you must scale the mountain, to the top."

"How can I do that?" asked the lad.

"That's my secret."

"But suppose I don't wish to."

"We made an agreement you would work once a year. Like it or not, the time has come. You must go to the top and throw down to me all the stones you find up there."

With that, he took a horse, killed it, flayed it, and ordered the lad inside the hide. An eagle flying overhead at that moment spied the horse, swooped down, seized it in his claws, and soared off with it, with the youth inside. The eagle came to rest on the summit of the mountain, and the boy leaped out of the hide. "Throw me down the stones!" cried the Greek from below. The young man looked about him: a far cry from stones! There lay brilliants, diamonds, and gold ingots as thick as tree trunks! He peered down the slope and saw the Greek who, from that distance, looked no bigger than an ant and continued to order, "Come on, throw those stones down to me!"

The boy thought to himself, Now if I throw him the stones, he will leave me up here on the mountaintop, and I won't have any way to get back down. I'd better hold on to the stones and try to get out of this predicament by myself.

Surveying the mountain summit, his eyes fell on what looked like the opening to a well. He lifted the lid, lowered himself through the opening and lo and behold, there he was inside a magnificent palace! It was the residence of Wizard Savino.

"What are you doing on my mountain?" queried the Wizard when he saw the boy. "I'm going to roast you and have a feast. You came to steal my stones for that thief of a Greek. Every year he tries to pull the same thing on me, and every year I make a meal off his henchman."

Quaking in his boots, the boy fell on his knees before the Wizard and swore he had no stone on his person.

"If you are telling the truth," replied Wizard Savino, "your life will be spared." He went up, counted the stones, and saw that they were all still there. "Very well," said the Wizard, "you were telling the truth. I'm taking you into my service. I have twelve horses. Every morning you will give each horse ninety-nine blows with a cudgel. But make sure I hear those blows from where we are right now. Is that clear?"

The next morning the young man entered the stable with a thick cudgel in his hand. He felt sorry for the horses, though, and couldn't bring himself to beat them. One of the horses then turned around and addressed him. "Please don't beat us. We were once men like yourself, and Wizard Savino turned us into horses. Cudgel the ground instead, and we will neigh as though you are beating us."

The boy followed the suggestion, and the Wizard heard the blows of the cudgel and the neighing, and was satisfied. "Listen," said one of the horses to the youth one day, "would you like to discover your fortune? Go into the garden, where you will see a beautiful pond. Every morning twelve doves come there to drink. They slip into the water, then emerge as twelve beautiful maidens as dazzling as the sun. After hanging their dove clothing on a tree limb, they begin playing. What you must do is hide among the trees and, when they are right in the middle of their game, seize the dress of the most beautiful maiden and hide it under your shirt. She will say to you, 'Give me my dress! Give me my dress!' But don't you dare return it, or she will become a dove once more and fly away with the others."

The young man did as the horse had told him; he crouched in a spot where they couldn't see him and waited for morning. At dawn he heard a flutter of wings that grew louder and louder. Peeping out, he saw a flight of doves. Making himself as small as possible he said, "There they are!" When they reached the fountain, the doves drank, then dived into the water. They returned to the surface as twelve beautiful maidens, who resembled angels from heaven, and began running and frolicking.

When the time seemed right, the youth crept forward, reached out, seized a dress, and stuffed it under his shirt. At that, all the maidens turned back into doves and flew off. Only one, unable to find her dove dress, remained in the youth's presence, and all she could say was, "Give me my dress, give me my dress!" The youth started running, with the girl right behind him. At last, after running some distance along a road the horse had shown him, he arrived home and introduced the maiden to his mother. "Mother, this is my bride. Don't let her out of the house under any circumstances."

Before descending the mountain, he had filled his pockets with precious stones. As soon as he got home, he decided to go out and sell them, leaving the maiden with her mother-in-law. "Give me my dress! Give me my dress!" screamed the girl all day long, making a nervous wreck of the old woman, who said, "Merciful heavens, this is driving me crazy! Let's see if I can find that dress!"

It occurred to her that her son might have put it away in the chest of drawers. She looked and, sure enough, there was a beautiful dove dress. "Could this be the dress, my daughter?" She'd not taken it fully out of the drawer before the girl seized it, threw it on, turned back into a dove, and flew off.

The old woman was terrified. "Now what will I do when my son returns? How will I explain the disappearance of his bride?" The words were no sooner out of her mouth than the bell rang and in walked her son; finding his wife gone, he was fit to be tied. "Mamma," he screamed, "how could you fail me like that!" Then when he had calmed down, he said, "Mamma, give me your blessing; I am going after her." He tossed a morsel of bread into a knapsack and was off.

Crossing a forest, he came upon three brigands engaged in a dispute. They hailed him and said, "Come and be our judge, since you are an outsider. We stole three objects and are now arguing over who should get what. You decide for us."

"What are the objects?"

"A purse that each time you open it is full of money, a pair of boots that carry you faster than wind, and a cloak that makes its wearer invisible."

"Let me verify," answered the lad, "if all that is true." He slipped into the boots, picked up the purse, and wrapped up in the cloak. "Can you see me?" he asked.

"No!" answered the brigands.

"Nor will you see any more of me." He fled in the boots that went like the wind, and arrived on top of Wizard Savino's mountain.

Once more he hid near the pond and saw the doves come to drink, his wife in their midst. He leaped out and made off with her dress, which she had hung on the tree.

"Give me my dress! Give me my dress!" she screamed anew. But this time the youth lost no time in setting it afire and burning it up.

"That's right," said the maiden. "Now I will remain with you and be your bride, but first you must go and behead Wizard Savino, and then turn the twelve horses in the stable back into men. All you need do is pull three hairs out of the mane of each one."

So, wearing the cloak that made him invisible, the young man cut off the Wizard's head, then freed the twelve knights previously transformed into horses, gathered up all the precious stones, and rode home with the maiden, who was none other than the daughter of the king of Spain.

(Palermo)

◈ 165 ◈

Jesus and St. Peter in Sicily

I. Stones to Bread

Once when the Lord was going about the world with the thirteen Apostles, they found themselves out in the country with no bread and they were starving. "Each of you pick up a stone," directed the Lord. The Apostles each picked up a stone, Peter choosing the tiniest one he could find. Then they continued on their way, each one bent under his burden,

594

except Peter, who moved along with ease. They came to a town and attempted to buy bread, but none was to be had. "Well," said the Lord, "I will bless you, and the stones will become bread."

So he did, and all the Apostles had hearty loaves to eat; but Peter who had picked up that pebble, found in his hand a wee, small roll. Crushed, he asked the Lord, "And what about my dinner, Lord?"

"Well, my brother, why did you pick up such a small stone? The others, who loaded themselves down, got bread aplenty."

They set out again, and once more the Lord told them each to pick up a stone. This time Peter, crafty as he was, took up a rock he could scarcely lift and thus walked with great difficulty, while the others all advanced with light stones. The Lord said to the Apostles, "Boys, we'll now have a laugh at Peter's expense."

They came to a town full of bakeries, where bread was just then coming out of the ovens. The Apostles all threw away their stones. St. Peter brought up the rear, bent over double under the weight of his rock. When he saw all that bread he flew into a rage and refused to touch it.

II. Put the Old Woman in the Furnace

As they walked along, they met a man. Peter went up to him and said, "As you can see, here comes the Lord. Ask him a favor."

The man went to the Lord and said, "Lord, my father is old and ailing. Make him strong again, Lord!"

"The burden of old age," replied the Lord, "is something no doctor can do anything about! But listen carefully: if you slip your father into the furnace, he'll come back out as a child!"

No sooner said than done, the man ran his old father into the furnace, and when he drew him back out, he had become a boy.

Peter was tickled pink with this procedure. "Now," he said to himself, "I shall see if *I* can turn some old soul into a child." Just then he met a man on his way to ask the Lord to cure his dying mother. "Whom are you seeking?" asked Peter.

"I'm seeking the Lord, as my mother is advanced in age, sick and infirm, and the Lord alone can restore her to health."

"Very well! The Lord isn't here yet, but Peter is, and he can help you. Know what you have to do? Fire up the furnace, slip your mother into it, and she will be cured."

The poor man, knowing that St. Peter was dear to the Lord, believed him. He flew home and slipped his mother into fiery furnace. What else did he expect? The old woman was burned to a crisp.

"Woe is me!" cried the son. "What a saint for this world and the next! He's had me burn up Mother!"

He returned in search of Peter and found the Lord. Hearing what had happened, the Lord split his sides laughing. "Peter, Peter, what have you done?" Peter tried to apologize, but couldn't get in a word edgewise for the screams of the poor son. "I want my mother! Give me back my mother!"

The Lord then went to the dead woman's home and, pronouncing a blessing, revived and rejuvenated the woman. And he spared Peter the punishment he deserved.

III. A Tale the Robbers Tell

Time and again it's been told that in the days when the Lord roamed the world with the Apostles, he was once overtaken by night on a country road.

"Peter, how will we manage tonight?" asked the Lord.

"Down below are shepherds tending their flock. Come with me," said Peter.

So they made their way in single file down the hill to the flock.

"Greetings! Can you give us shelter for the night? We are poor pilgrims who are exhausted and starving to death!"

"Greetings!" replied the overseer and his shepherds, but none of them budged an inch. They were in the process of rolling out dough on the board, and thought if they offered thirteen persons dinner, they would all go hungry. "Over there is the haystack," they said. "You can sleep there."

The poor Lord and his Apostles drew in their belts and went off to bed without a word.

They'd scarcely gone to sleep, when they were awakened by an uproar —robbers arriving in a band and shouting, "Hands up! Hands up!" There were curses and the sound of blows and the fleeing footsteps of the shepherds scattering into the countryside.

When the robbers were in possession of the field, they made a clean sweep of the flock. Then they took a look in the haystack. "Hands up, every one of you! Just who's in here?"

"Thirteen poor pilgrims, weary and hungry," replied Peter.

"If that's so, come out. Supper's on the board, completely untouched. Eat your fill at the expense of the shepherds, for we must flee for our life!"

Hungry as they were, those poor souls needed no begging. They ran to the board, and Peter exclaimed, "Praised be the robbers! They are more thoughtful of starving poor people than are the rich."

"Praised be the robbers!" said the Apostles, and had a hearty meal.

IV. Death Corked in the Bottle

There was a rich and generous innkeeper who put up a sign that read: WHOEVER STOPS AT MY INN EATS FREE OF CHARGE. The people poured in all day long and he served every one of them for nothing.

Once the Lord and his twelve Apostles came to that town. They read the sign, and St. Thomas said, "Lord, until I see it with my own eyes and feel it with my own hand, I won't believe it. Let's go inside this inn."

So Jesus and the Apostles went inside. They ate and drank, and the innkeeper treated them like royalty. Before leaving, St. Thomas said, "My good man, why don't you ask a favor of the Lord?"

So the innkeeper said to Jesus, "Lord, I have a fig tree in my garden, but I never get to eat a single fig. As fast as they ripen, the boys scamper up and eat every one of them. Now I'd like you to ordain that whoever climbs this tree can no longer come down without my permission."

"So be it!" said the Lord, and blessed the tree.

The next morning, the first one to steal figs stayed hanging to the tree by one hand, the second by a foot, while the third was unable to pull his head through a fork of branches. When the innkeeper found them, he gave them a good dressing down, then released them. Once they heard about the magic property of that tree, the little boys of the town steered clear of it, and the innkeeper was at last able to eat his figs in peace.

Years and years went by. The tree grew old and bore no more fruit. The innkeeper called in a woodcutter to fell the tree, after which he asked, "Could you make me a bottle out of the wood from this tree?" The woodcutter made him the bottle, which retained the magic property of the tree—that is, whoever entered it couldn't come out without the innkeeper's permission.

The innkeeper too grew old, and one day Death came for him. The man said, "By all means, let us be on our way. But first, Death, I'd like to ask you a favor. I have a bottle full of wine, but there's a fly in it, and I am loath to drink it. Would you please jump in and get the fly out, so I can have one last drink before going off with you."

"Oh, if that's all you're asking!" said Death, and jumped into the bottle. Then the innkeeper corked the bottle up and said, "I have you now, and you're not coming out."

With Death trapped and corked up, nobody in the world died any more. Everywhere people were seen with white beards down to their feet.

Taking note of this, the Apostles began dropping hints to the Lord, who finally decided to go and speak with the innkeeper.

"My dear man," said the Lord, "do you think it's appropriate to keep Death shut up all these years? What about those poor old decrepit people who must drag on and on without ever being able to die outright?"

"Lord," answered the innkeeper, "do you want me to let Death go? Promise to send me to Paradise, and I'll unstop the bottle."

The Lord thought it over. What am I to do? If I deny him that favor, there's no telling what a mess I might be in! Therefore the Lord said, "So be it!"

At that, the bottle was uncorked, and Death was free. The innkeeper was allowed to live a few more years, so as to merit Paradise, and then Death returned for him.

V. St. Peter's Mamma

It's been said that St. Peter's mamma was a miser through and through. Never did she give to charity or spend a penny on her fellow man. One day while she was peeling leeks, a poor woman came by begging. "Will you give me a little something, good woman?"

"That's right, everybody comes to me begging. . . . Well, take this, and don't ask for any more!" And she gave her one leaf of a leek.

When the Lord called her into the next life, he sent her to Hell. The head of Heaven was St. Peter, and as he sat on the doorstep, he heard a voice. "Peter! Just look at how I'm roasting! Son, go to the Lord, talk to him, get me out of this misery!"

St. Peter went to the Lord. "Lord," he said, "my mother is in Hell and begging to be let out."

"What! Your mother never did a good turn in her whole life! All she has to her credit is one little leek leaf. Try this. Give her the leek leaf to catch hold to, and pull her up to Paradise by it."

An angel swooped down with the leek leaf. "Grab hold!" ordered the angel, and St. Peter's mamma caught hold of the leaf. She was about to be pulled up out of Hell, when all the poor souls there with her and seeing her rise, latched on to her skirts. So the angel drew up not only her but all the others as well. Then that selfish woman screamed, "No! Not you all! Get off! Just me! Just me! You ought to have had a saint for a son, as I did!" She kicked and shook them from her, jerking about so much to get free, that the leek leaf broke in two and St. Peter's mamma went plummeting to the bottom of Hell.

(Palermo)

598

The Barber's Timepiece

Time and again the story has been told about a barber who owned a clock that had run for centuries without being wound up; never did it stop or lose a minute, but always kept perfect time. The barber had wound it up just once, and from then on, ticktock, ticktock, ticktock . . .

The barber was an old man, so old that he had lost count of the centuries he had lived and the generations of people he had seen. People from all over were accustomed to run to his shop to ask the clock things they needed to know.

There came the burly farmer, weary and out of sorts, in need of rain for planting time. Seeing the sky forever cloudless, he said, "Tell me, clock, when is it going to rain?"

The clock ticked:

> Tick Tock, Tick Tock, Tick Tock, Tick Tock,
> I shine, I shine, I shine, I shine,
> No rain, no rain, the sky is mine.
> Come thunder, thunder, thunder, thunder,
> And next year, water, water, water!

There came an old man leaning on a cane and wheezing from asthma, who asked, "Clock, clock, is there much oil left in my lamp?"

Right away the clock ticked:

> Tick Tock, Tick Tock, Tick Tock, Tick . . .
> Three score, three score, three score,
> Burns low, burns low, little more;
> Three score past, three score past, three score past:
> Poor wick, poor wick, poor wick! Tick . . .

There came a youth in love and up in the clouds, who said, "Tell me, clock, could anyone fare better in love than I?"

> Tick Tock, Tick Tock, Tick Tock

replied the clock:

> Be a fool and you will fall:
> Today you cut a figure at a ball,
> Tomorrow lie you 'neath a pall!

Then came the foremost outlaw in the land, the head of the terrible Camorra, all rigged up in his tasseled beret and long hair and buttons and rings, and muttered in his beard, "Say, clock, how many potentates is there 'at can 'scape my clutches? Speak up, or I'll bust your guts!"

And the clock outdid him, muttering in his beard:

> Ticktockticktockticktock,
> Bewarebewarebeware:
> Tidesturn, tidesturn, tidesturn;
> Burnandonedayyouwillburn!

Next came a poor man, suffering, hungry, half naked, sick all over. "Oh, clock, oh, clock, when will my tribulations end? Tell me, for pity's sake, when to expect Death?"

As usual, the clock provided an answer:

> Tick Tock, Tick Tock, Tick Tock,
> For him who sings no song,
> Life may be very long.

Thus, all sorts of people came to see this wonderful clock, spoke to it, and received an answer. It could say when fruit trees would bear, when winter and summer would arrive, when the sun would rise and set, and how many years people had lived. In short, it was a timepiece without equal, an ingenious creation, and there was nothing under the sun it did not know. Everybody would have liked to have it in his house, but no one could, since it was enchanted; people therefore longed in vain to own it. But everyone, whether he wanted to or not, secretly or openly had to praise the old master barber who had been clever enough to make this unique timepiece and make it run forever and ever, without anyone being able to break it or take possession of it, except the artist who fashioned it.

(Inland vicinity of Palermo)

◈ 167 ◈

The Count's Sister

The tale has been told over and over that once upon a time there was a count just rolling in wealth. He also had an eighteen-year-old sister as lovely as the sun and moon. He jealously guarded the girl, keeping her locked up all the time in a wing of his palace, so that no one had ever seen or talked to her. Now the beautiful little countess, having had quite enough of being shut up, began slowly making a hole in the wall of her room. She worked on it at night, and kept the spot concealed by day with a painting. Flanking the count's palace was the prince's, and the opening made by the countess came through into the prince's chambers, right behind another painting, in such a way that it was not visible.

One night, the little countess pushed the picture slightly aside and looked into the prince's room. She saw a precious lamp burning, and addressed it:

> "Golden light, silver light,
> Does your prince sleep, or watch in the night?"

And the lamp relied:

> "Come in, my lady, have no fear;
> My prince is sleeping soundly here."

She went in and lay down beside the prince. He awakened, took her in his arms, kissed her, and asked:

> "Whence do you come, where do you live, dear lady?
> From what country could you be?"

Laughing, she answered:

> "Prince, dear prince, you ask awry;
> Love me, love me, do not pry!"

When the prince woke up again and found that radiant goddess gone, he dressed in haste and summoned his council. "Council! Council!" The council convened, and the prince related what had taken place, asking in conclusion, "What must I do to keep her with me?"

"Sacred Crown," replied the council, "when you take her in your arms, tie her hair around one of your arms. That way when she's ready to leave she will have to wake you up."

Night fell, and the young countess asked:

> "Golden light, silver light,
> Does your prince sleep, or watch in the night?"

The lamp answered:

> "Come in, my lady, have no fear:
> My prince is sleeping soundly here."

In she went and slipped under the covers.

> "Whence do you come, where do you live, dear lady?
> From what country could you be?"

601

"Prince, dear prince, you ask awry;
Love me, love me, do not pry!"

So they fell asleep, but not before the prince had fastened the countess's lovely hair to his arm. The countess pulled out a pair of scissors, cut off her hair, and left. The prince woke up. "Council! Council! The goddess left me her hair and vanished!"

"Sacred Crown," replied the council, "put your head through the fine gold chain she wears around her neck."

The next night the countess returned:

"Golden light, silver light,
Does your prince sleep, or watch in the night?"

And the lamp answered:

"Come in, my lady, have no fear:
My prince is sleeping soundly here."

When the prince had her in his arms, he again asked:

"Whence do you come, where do you live, dear lady?
From what country could you be?"

As usual, she answered:

"Prince, dear prince, you ask awry;
Love me, love me, do not pry!"

The prince put his head through her gold necklace, but as soon as he was fast asleep she cut the chain and vanished. "Council! Council!" he called when it was day, and related what happened. The council said, "Sacred Crown, take a basin of saffron water and put it under the bed. When she removes her nightgown, throw it into the saffron water. . . . That way, when she puts it back on to go off, she will leave a trail behind her."

At nightfall, the prince prepared the basin of saffron water and went to bed. At midnight she said to the lamp:

"Golden light, silver light,
Does your prince sleep, or watch in the night?"

And the lamp answered:

"Come in, my lady, have no fear;
My prince is sleeping soundly here."

Awakening, the prince put the usual question to her:

602

"Whence do you come, where do you live, dear lady?
From what country could you be?"

And she gave the usual answer:

"Prince, dear prince, you ask awry;
Love me, love me, do not pry!"

When the prince was sound asleep, she eased out of bed without making a sound and made ready to leave, but her nightgown was soaking wet in the saffron water. Silently she wrung the gown out and slipped away without leaving a trace.

From that night on, the prince awaited his goddess in vain, and he was very sad. But nine months later, upon awakening one morning, he found in bed beside him a beautiful baby boy that looked like a cherub. He dressed in haste, crying, "Council! Council!" He showed the council the baby, saying, "This is my son. What can I do now to get his mother back?"

"Sacred Crown," answered the council, "pretend he is dead. Put him in the middle of the church and give orders for all the women in the city to come and mourn him. The one who wails most of all will be his mother."

The prince did that very thing. All sorts of women came, saying, "Son, son!" then departed as freely as they had come. At last appeared the young countess; with tears streaming down her cheeks, she began pulling out her hair and crying:

"Son, dear son!
I was much too beautiful,
And so I cut my locks;
I was much too beautiful,
And so I cut my chain;
I was much too vain,
And so a saffron gown is now my gain."

The king and the council and everybody began crying, "She's the mother! She's the mother!"

At that moment a man pushed forward with his sword unsheathed. It was the count, who raised the blade over his sister. But the prince jumped between them and said:

"Halt, O Count, here is the key:
Count's sister is she, and wedded to me!"

And they got married in that very church.

(Inland vicinity of Palermo)

Master Francesco Sit-Down-and-Eat

Once upon a time, it has been said over and over, there was a cobbler named Master Francesco, and since he was the laziest man alive, everyone called him Master Francesco Sit-Down-and-Eat. He had five daughters, each lovelier than the other, and all as good as gold. But with that father of theirs who worked little and earned even less, they were at a loss to make ends meet. He got up late, dressed, and off to the tavern he went, where he would spend every penny the daughters had earned.

At last they told him that, for better or worse, he had to go to work. So he picked up cobbler's bench, lasts, and hammer, threw them over his shoulder, and went through town crying, "Shoe repairs! Shoe repairs!" But knowing him for the chief lazybones and drunkard in town, people would have nothing to do with him. Realizing he would starve to death in his own town, he went to another town three miles away, where he cried, "Shoe repairs! Shoe repairs! Come one, come all and get your shoes repaired!" He cried himself hoarse, but still no one brought him any work, while his pangs of hunger became ever sharper.

Night fell, and lo and behold a lady called to him from a large mansion. He went inside and found her in bed. "Fix this worn shoe for me."

Master Francesco fixed it for her the best he could, and the lady paid him a groat, saying, "I know you have five daughters. I am sick and need someone to wait on me. Would you let me have one of your daughters for a maid?"

"I certainly will, my lady," replied Master Francesco. "I'll send her to you tomorrow."

Back home, he told his daughters everything and said to the oldest, "You will be the one to go tomorrow."

In the morning, the daughter went to the lady who exclaimed, "Ah, so you did come, my child! Sit down here and give me a kiss. I want you to be happy here with every comfort and joy anyone could ask for. As you can see, I am bedridden, so you will be in charge of the house. Go now, my child, sweep the house, make things tidy, then tidy up yourself and put on your best dress, so that my husband will find everything in order when he returns."

The girl began sweeping and, in order to sweep under the bed, raised the bedspread which came all the way to the floor. What should she then see but a long, long hairy tail that came out from under the sheet and reached all the way under the bed.

Woe is me! she thought to herself. Just look at the mess I'm in now! She's an ogress, not a lady! At that, she backed slowly away from the bed.

"You listen to me!" said the lady, whose voice had already changed, "Sweep everywhere but under the bed. Is that clear?"

The girl pretended she was going to sweep another room, but sneaked out of the house and returned home. "What, you're back already?" exclaimed her father.

"Father, that is an ogress, not a lady; beneath the bed she has a black hairy tail this long. Say what you will, I'm not going back."

"Stay at home, then," said Master Francesco, "and we'll send the second girl there." The second girl got the same attention and words from the lady as the first, but she too spied the tail and went running home.

Now Master Francesco was greedy for the lady's generous pay; with it he could eat and dress without having to do a lick of work himself. So he sent the next daughter to the lady, and then the next, and finally the youngest, and every one of them came flying back home frightened to death by the awful black hairy tail.

"We're better off here," they said, "better off at home working our fingers to the bone day and night and wearing our old rags than scarcely turning a hand for good food and clothing from the ogress, who would eat us in the end! Father, if that appeals to you so much, go to the ogress yourself."

The father knew no peace until he had entered the lady's service himself. The work was so very easy, and he could eat and dress like a prince.

As a matter of fact, the lady treated him like a prince, offering him fine clothes, tasty dishes, gold rings, joys, and comforts. All he had to do was go to market, then come home and tidy the bedchamber, after which he was free to sit down, stretch out his legs, and lounge about for the rest of the day. In no time Sit-Down-and-Eat grew fatter and fatter. When he could get no plumper, the lady called him to her. "Yes, my lady?" he said approaching the bed.

The ogress sneered, grabbed him by the arm, digging her nails into him, and said:

> "Sit-Down-and-Eat, Eat-and-Sit-Down,
> On which part should I first to town,
> Your head, your feet, or under gown?"

Shaking like a leaf, Master Francesco answered in a whisper:

> "Believe your daughters, it is meet,
> Else be eaten, starting with your feet."

So the ogress seized him by the feet and sucked him completely up in one long gulp, without leaving a single bone.

> "The girls were at peace and didn't pine
> For Master Francesco who died like a swine;
> Let whoever tells or hears this story
> Never die a death so gory."

(Inland vicinity of Palermo)

◈ 169 ◈

The Marriage of a Queen and a Bandit

Once, they say, there was a king and queen who had a daughter they wanted to marry off. The king had a proclamation posted for all monarchs and holders of noble titles to assemble at the royal palace to be reviewed. They all assembled, while the king and his daughter watched them parade by. The first one that captured his daughter's fancy was going to be her husband. Filing past in first place were all the kings, next the princes, then the barons, knights, and professors. The king's daughter saw no king she liked, nor any prince. The barons came up, but neither did they appeal to her. It was the same with the knights.

The professors passed, and she pointed at one of them. "Father, my husband will be that one." He was a foreign professor whom no one knew. Since the king had made a promise, he had no choice but give his daughter in marriage to the professor. After the wedding, the bridegroom wished to be off at once. The bride bid her mother and father goodbye, and the couple departed, followed by the army. After half a day's march, the soldiers said to the bridegroom, "Your Highness, let us now have lunch."

"This is no time for lunch," replied the bridegroom.

A bit further on, they repeated the proposal, only to be told a second time, "This is no time for lunch."

Exasperated, the soldiers answered, "In that case, you and your royal bride go on to the country where you're headed."

"And you and the entire military staff may go your way too," replied the man. So the soldiers turned back, and the newlyweds continued on by themselves.

They came to a desolate, rocky terrain covered with wild vegetation. "We are home," announced the bridegroom.

"What! There's no house here!" protested the king's daughter, growing worried.

The bridegroom tapped his stick three times, and an underground cavern sprang open. "Walk in," he ordered.

"I'm afraid."

"Get in there, or I'll kill you!"

The bride went in. The cavern was full of dead people, young and old, piled up on top of each other.

"See these bodies?" asked the bridegroom. "Your job will be this: take each one of them and stand them up in a row against the wall. Every night I'll bring in a fresh cartload of them."

So began the married life of the king's daughter. She picked up the dead people from off the pile and stood them up against the wall, so that they took up less space and left room for more bodies. And every evening her husband came in with a cartload of new dead people. It was hard work, because dead people are particularly heavy. Nor could she ever get out of the cavern, for the opening had even disappeared.

The king's daughter had brought with her a little furniture, including an old chest of drawers, a gift from an aunt who was something of a fairy. One day when the bride opened a drawer, the chest spoke: "At your command, little mistress!"

Without delay she said, "I wish to get out of here and go home."

At that, a white dove flew out of the chest and said, "Write your father a letter and put it in my bill."

The bride wrote the letter, which the dove carried to the king and waited for an answer. The king wrote: "My daughter, find out immediately how to leave your cavern, and trust in my help."

When the dove came back to the girl with her father's answer, she decided to get on her husband's good side that evening, in order to draw the secret out of him. "Do you know what I dreamed?" she said. "That I left the cavern."

"It takes more than a dream to get out!" replied the husband.

"Why? What does it take?" she asked in an innocent manner.

"Well, to begin with, you have to have someone born prematurely like myself, after seven months, to strike the stick three times on the rock. Then the cavern will open."

As soon as the dove relayed to the king the secret about the person born after seven months' time, the king sent soldiers through city and country to find someone born after only seven months. A washerwoman hanging out things to dry saw that bustle of troops and thought to herself, They'll steal my sheets, so she began hurriedly pulling them off the line.

"Don't be afraid, we're not here to rob you," said a corporal. "We're looking for someone born after only seven months. No matter who it is, the king wants him."

"Oh," replied the washerwoman, "as a matter of fact, I have a son born prematurely after only seven months." She went into the house and got him for the soldiers. The young man, thin as a rail, joined the king at the head of the soldiers to go and free the princess. He struck the rock three times with the stick, and the cavern opened. The princess was there waiting for them and rode off with her father, the young man, and the soldiers.

Along the way they saw an old woman in a garden. "Ma'am," they said to her, "should a man come by asking about us, we've not been by—all right?"

"Huh?" answered the old woman. "You want some beans for your soup tonight?"

"Perfect!" they exclaimed, "You're just the person we need."

Soon the brigand himself came by, having found the cavern wide open and his wife gone. "Have you seen a woman go by with the army?" he asked the old woman.

"Huh? You're making marmalade to go with tea?"

"Marmalade my eye! A seven-months' man and the king and his daughter."

"Ah! A pound of parsley and basil!"

"No, no, NO! The king's daughter with soldiers!"

"No, not salty cucumbers!"

The bandit shrugged his shoulders and stormed off. "But, sir," she called after him, "why did you take offense? Whoever heard of salty cucumbers?"

Back in the safety of her father's house, the princess got married again shortly afterward, this time to the king of Siberia. Her first husband, the bandit, however, continued to pursue her and formed a plot. He dressed up as a saint and had himself put into a picture. It was a big picture with a heavy frame fastened by three bolts, and the bandit stood inside, like a saint, behind a thick glass. The picture was taken to the king of Siberia to purchase. When he saw it, he found it so beautiful and lifelike that he bought it to hang over his bed. When no one was in the room, the bandit came out and placed a bewitched paper under the king's pillow. When the queen saw that saint's picture over her husband's bed, she gave a start, for it looked like her first husband, the bandit. But the king took her to task for being afraid of the picture of a saint.

They went to bed. Once they were asleep, the bandit turned the first door bolt to come out. The queen awakened at the sound and pinched her husband to make him listen too, but the king continued to sleep, since the magic property of that paper kept whoever had it under his pillow fast asleep. The bandit turned the second bolt, and the king slept on, while the queen was paralyzed with fear. He turned the third bolt, stepped out, and said to the queen, "I will now cut off your head. Put your neck firmly on the pillow."

To prop her neck up high, the queen took her husband's pillow too, and in the process, the enchanted paper fell on the floor. At once the king awakened, sounded the trumpet he wore around his neck night and day, as is customary with kings, and the soldiers came running from all directions. They saw the bandit, slew him, and that was that.

(Madonie)

◈ 170 ◈

The Seven Lamb Heads

An old woman had a granddaughter. The granddaughter always stayed home and did the housework while the old woman went out shopping. One day she brought home seven little lamb heads. She gave them to her granddaughter and said, "Atanasia, I'm going out. Cook these seven little heads for me, and we'll eat them when I return."

The girl put the heads on to cook. A cat sat nearby and, smelling the aroma that issued from the casserole, said:

> "Mew mew mew mew mew MEW,
> Half for me and half for YOU!"

So the girl took one of the seven lamb heads, divided it, gave half to the cat, and ate the other half herself. The cat ate, then said once more:

"Mew mew mew mew mew MEW,
Half for me and half for YOU!"

The girl cut another head in two, one part for the cat and one for herself. The cat, however, was still not satisfied, and meowed anew:

"Mew mew mew mew mew MEW,
Half for me and half for YOU!"

In short, one by one, the seven heads after being halved ended up in the cat's belly and Atanasia's. When they were all gone the girl started worrying, scratching her head and saying, "Now what will I do when Granny comes back?" Seeing no other solution, she opened the door and fled.

When Granny returned and found the house door wide open, and half the bones from the heads on the floor and the other half in the dish, and no sign of her granddaughter, she began saying, "Every last one, she ate every last one. . . . Every last one, she ate every last one. . . ." And she proceeded to turn the house upside down, saying, "She ate every last one . . ."

She flew out of the house without looking where she was going. She took a few steps, started thinking, then shook her head. "Every last one, she ate every last one . . ." and for the life of her she couldn't calm down.

Meanwhile, Atanasia walked and walked until she came to a forest and saw thousands of roses. "How beautiful!" she exclaimed and, using some cotton thread she had with her, she made herself a crown of roses, a necklace, and two bracelets. Putting them on, she then lay down under a tree and went to sleep. In the morning, the king was hunting in the forest and saw the girl as she slept there. He gazed at her for a long time and liked her so much that he fell in love with her. He awakened her and said, "I am the king. Will you marry me?"

"As you can see," replied Atanasia, "I am a poor girl. How can I ever hope to marry you?"

"If that's all that is bothering you," answered the king, "don't give it a thought. I want you, so you have to become my wife."

The girl blushed, and nodded in agreement.

"Come to the palace with me, then."

"Of course, but I left Granny at home and must go get her."

The king sent a carriage for Granny, and at the wedding banquet seated her by her granddaughter. It was a sumptuous feast, and the old woman leaned over and whispered in her granddaughter's ear, "Every last one, you ate every last one . . ."

"Hush up, will you!" said the granddaughter.

"What," broke in the king, "are the wishes of madam your aunt?"—
which is how he referred to her.

"She wants a dress like mine," replied the bride.

"Let one be made for her at once," ordered the king.

After dinner, conversation began, and the old woman went on whispering to her granddaughter, "Every last one, you ate every last one . . ."

"What does madam your aunt want?" asked the king.

"She wants," explained Atanasia, "a ring like mine."

"Let one be made for her at once," ordered the king.

But the old woman was already starting up again. "Every last one of
them, you ate every last one . . ."

"What does madam your aunt want?"

But Atanasia had had enough by then and replied: "She's a hungry old
skinflint and even in the midst of all this royal splendor she can't take her
mind off those messy lamb heads!"

Outraged by such greed, the king called the guards and ordered her
head chopped off in the middle of the town square.

Where her head fell, a tree sprang up. It was a weeping willow, and at
every wind that swayed it, you could hear: "Every last one, she ate every
last one . . . Every last one, she ate every last one . . ."

(Ficarazzi)

◈ 171 ◈

The Two Sea Merchants

Once there were two friends. One of them had a son, whereas the other
one had no children at all. But they both loved this boy with all their
heart. They were sea merchants, very important ones, who sailed the
seven seas. One day the childless merchant had to sail off for his wares.
As he got ready to leave, his friend's son begged to go along to get some
experience in sailing and trading, and he pleaded with his father to let
him accompany his godfather. Neither father nor godfather wanted him
to go, but the boy kept after them until they finally agreed for him to
leave on a ship that would sail with his godfather's.

While they were on the high seas, a storm came up, so furious that the two ships lost sight of one another. The godfather's ship came through all right, but the young man's vessel went to the bottom of the sea and all its crew drowned. The youth fortunately found a plank to straddle, and floated until he reached land. There he wandered about with little hope, and entered a forest inhabited by wild animals. Fearful of the animals, he spent the night in the top of an oak tree. At daybreak, after making sure there were no wild beasts around, he came down the tree and continued on through the forest until he arrived at a high wall that seemed to have no beginning or end. Climbing a nearby tree, the youth reached the top of the wall; on the other side was a city, and the wall had been built to protect it from wild animals.

The youth somehow made it down the other side and entered the city. "Now I'll go buy something to eat," he decided, and turned into a street flanked by shops. He entered a bakery and asked for bread, but the baker did not reply. He went into a pork butcher's shop and asked for some salami, but the pork butcher did not reply. He went around to all the shops, but no one paid any attention to him.

"I shall go to the king at once and protest!" said the youth to himself, and marched straight to the royal palace. "May I have a word with the king?" he asked the guard. But the guard remained silent. Upset and disheartened because no one would talk to him, the youth entered the palace and proceeded through the rooms. He came to the most beautiful chamber of them all, containing a royal bed, a royal bedside table and a royal washstand, and thought to himself, Since nobody is saying anything to me, I'm going to lie down and go to sleep.

At once two lovely young ladies came tripping out and, in total silence, prepared a table and served him supper. He dined, and then went to bed.

So began a life of ease in that silent city. One night as he slept in the royal bed, there approached, veiled from head to toe and accompanied by two maids of honor, a maiden of marvelous appearance, who asked, "Are you steady and courageous?"

"Yes."

"If so, I will tell you my secret. I'm the daughter of Emperor Scorzone who, before dying, cast a spell over this city and all the citizens, servants, army, and myself. This spell is enforced by a sorcerer. But if you stay with me every night for a whole year without looking at me or revealing my secret to a soul, the spell will be broken, and I'll be the empress and you the emperor, hailed by all the people."

"I am steady and courageous," replied the young man.

But a few days later, he told her that, in order to stay calmly at her side for a whole year, he first had to go and say goodbye to his father, mother, and godfather, and he promised to return to her without delay. The empress wasn't at all sure she should let him go, but he pleaded so much that she had a ship readied for him and a few of her treasures put aboard. She gave him a wand, explaining, "Take this wand and command, and you'll find yourself instantly where you wish to be. But remember to reveal my secret to no one."

The youth boarded the ship, tapped the wand, and found himself in the port of his father's city. He ordered his treasures carried to the best inn, where he took lodgings. "Do you know any sea merchant here?" he asked the people.

"In this city there are two," he was told, "two friends and important merchants, but they've recently been reduced to poverty."

"How's that?"

"The son of one of them, mind you, was lost at sea, but the boy's father refused to believe it was a simple accident. He blamed his friend and brought a lawsuit against him. In the suit both men lost every penny to their name."

Hearing that, the young man sent for his father, who did not recognize him. "I would like," said the boy, "to enter into a commercial transaction with you and your colleague, considering that you are experienced sea merchants."

"Impossible!" replied the father. "My colleague and I have gone bankrupt because of a lawsuit over my son, who met his death through the fault of my colleague."

"That doesn't matter," said the youth. "I'll put up all the capital myself."

And he ordered a fine dinner prepared and also invited his father's friend and the men's wives. When they found themselves face to face at the inn, the two friends and their wives scowled at one another, now being enemies. They tried to eat, but the two friends, angry as they were with each other, couldn't swallow a thing. Then the youth picked up a forkful from his own plate and extended it to his father, saying, "Father, accept this morsel given you by your son, who is right here safe and sound."

They all jumped to their feet. Mad with joy, everybody grabbed and kissed one another, weeping for happiness. The youth divided his treasures between father and godfather so they could continue their trading, then said, "Now I bid all of you farewell, for I must be off again."

"Where are you going?" asked his mother.

"I can't tell you that."

But his mother kept on and on asking him until he finally told her about Emperor Scorzone's daughter, whom he was not allowed to look upon and see how beautiful she was.

"Listen," said his mother, "I'll give you a Tenebrae candle, so that when she goes to sleep, you can light it and see what she looks like."

The young man boarded his ship, tapped the wand, and found himself back in the port of Emperor Scorzone's city. He went to the royal palace, where the emperor's daughter was waiting for him. At night they retired, and he could hardly wait to look upon her beauty. While she was sleeping, he picked up the candle, lit it, and began to uncover the girl. But a drop of molten wax fell on her, scalding her bare flesh. She woke up. "Traitor! You have revealed my secret! Now you won't be able to free me!"

"Woe is me! But I will still try to free you! Is there no other way?"

"You must go into the forest, fight with the sorcerer who enforces the spell, and kill him!"

"Yes. And after I've killed him?"

"Slit open his belly; there you will find a rabbit. Cut open the rabbit and you will find a dove. Cut open the dove and you will find three eggs. Guard those eggs with your life and bring them back here without breaking them. Then the city and all of us in it will be free. Otherwise we'll be under the spell forevermore, and you along with us. Take this wand and go out and fight!"

The young man left, armed with the wand. He came upon a drove of cows and, in their midst, cowherds and the owner of the cattle. "Sir," he said to the owner, "would you give me a piece of bread? I have gone astray in these parts."

The owner of the livestock fed and kept him on as a cowherd. The cowherds were told one day, "Take the cows to pasture, but whatever you do, don't let them stray into the forest, for in there lives a sorcerer who kills not only human beings but cows as well."

The youth went out with the herd and, when the cows were close to the forest, with shouts and whacks he drove them right on in. The owner threw up his hands. "And now," he asked, "who's going into the forest after them?"

None of the herdsmen were willing, so the owner sent the new cowherd, together with another boy. They entered the forest, and the boy was scared to death.

The sight of the cows in the forest infuriated the sorcerer, who came rushing out with an iron club surmounted with six bronze spikes. Gripped with fear, the boy crouched in the underbrush. But the young man stood his ground and waited for the sorcerer to approach.

"Traitor! How dare you come and tear up my forest!"

"I come to destroy not just your forest but you along with it!" the youth replied, and the fight began.

They fought and fought, throughout the day. They were tired at last, but neither of them had yet received a scratch. The sorcerer said:

> "Had I soup of bread and wine,
> I'd quarter you like a swine!"

The young man answered:

> "Had I soup of milk and bread,
> I'd chop off your head!"

So they bid one another farewell, agreeing to continue their fight the next day. The youth rounded up the cows and, together with the boy, drove them back to the barn.

Seeing them back alive, all the men were speechless. The boy told of the fierce combat between the new cowherd and the sorcerer, repeating their exchange of words and the sorcerer's wish for soup of bread and wine and the youth's for soup of milk and bread. So the owner ordered a pail of milk and bread readied for tomorrow and instructed the boy to carry it to the forest and have it all ready for the youth.

They again drove the cattle into the forest, where the sorcerer reappeared, and the fight started all over. Right in the thick of it, the sorcerer said:

> "Had I soup of bread and wine,
> I'd quarter you like a swine!"

But there was no soup of bread and wine. Then the youth said:

> "Had I soup of milk and bread,
> I'd chop off your head!"

And right away the boy passed him the pail of milk and bread. The young man took a ladleful, poured it into his mouth, and immediately dealt the sorcerer a blow on the head that sent him crashing to the ground stone dead.

He slit open the sorcerer's belly and found the rabbit, cut open the rabbit and found the dove, cut open the dove and found the three eggs. He took the eggs and put them away with great care, then drove the cows back to the barn, where he was received in great triumph. The master wanted him to remain on the farm, but the young man declined and, after making the man a present of the sorcerer's forest, took his leave.

When he got back to the silent city, he went at once to the royal palace. The beautiful maiden ran down to meet him, then took him by the hand and led him into the secret chamber of Emperor Scorzone, her father. There she picked up the emperor's crown and placed it on the youth's head, saying, "You are now emperor, and I am empress." She then led him to the balcony. There she took the three eggs and said, "Throw one to the right, one to the left, and one straight before you."

The minute the eggs were thrown, all the people began talking and shouting, and silence gave way to a great clamor; carriages again rolled, the army launched into maneuvers, the guard changed and, all together, people and troops cried: "Long live our Emperor! Long live our Empress!" And they remained emperor and empress their whole life long, while we are still as poor as ever.

(Province of Palermo)

◈ 172 ◈

Out in the World

There was a widow with two daughters and a son named Peppi, who was at a loss to earn his bread. While mother and daughters were spinning one day, Peppi said, "Listen, Mother; with your leave, I am going out into the world."

Along the way, he came to a farm and asked, "Can you use the services of a young man here?" By way of reply, they sicked the dogs on him.

Peppi moved on and reached another farm as it grew dark. "Viva Maria!" he said.

"Viva Maria! What can we do for you?"

"Should you be in need of the services of a young man . . ."

"Oh," he was told, "sit down, sit down. I believe our cowherd is leaving. Wait here while I go ask the master."

The man went upstairs and asked the master, who said, "Yes, give him something to eat, and when I come down we'll discuss the matter."

So they set bread and cheese before Peppi, who began eating. As the master was on his way downstairs, he met the cowherd coming up the steps and asked him, "Is it true you are leaving?" "Yes, sir," answered the cowherd. Then the master went to Peppi and said, "Tomorrow morning, take the cattle out to pasture, my boy, but understand this: if you wish to remain here, all you get is your simple board and nothing more."

"I'll stay," said Peppi. "God's will be done."

He slept through the night and then, in the morning, took bread and a little food and drove the cattle out to pasture. He would keep them out all day and bring them in at nightfall. The carnival season was approaching, and Peppi came home one day with a long face.

"Peppi!" said the steward.

"Oh, me!"

"What's the matter?"

"Nothing!"

The next morning he was on his way out with the herd, as downcast as ever, when he met the master, who said, "Peppi."

"Oh, me!"

"What's wrong?"

"Nothing!"

"Nothing, Peppi? Why don't you tell me what it is?"

"Why should I have to tell you? Carnival is coming, and you couldn't give me a little money just this one time so I could go and celebrate with my mother and sisters?"

"What's that you're saying? Talk to me about anything under the sun, but don't ever mention money. If you want bread, you can have all you want. But money, NO!"

"But what should I do if I had to buy a little meat?"

"I couldn't say. The conditions I laid down in the beginning are still the same."

It was daylight, and Peppi trudged to the pasture with those confounded cattle. He sat down, gloomy as ever. Suddenly he heard his name called. "Peppi?" After looking all around, he concluded, "With my troubled heart, I imagine I hear things, but it's all in my head."

But about that time he heard his name called again. "Peppi! Peppi!"

"Who on earth is calling?"

An ox turned around and said, "I am."

"What! You are talking?"

"Indeed I am. Why are you so long-faced?"

"Why wouldn't I be? Carnival time is approaching, and the master won't give me a cent."

"Listen to me, Peppi; when you go in tonight, you must say to him, 'You couldn't even give me the old ox?' The master can't stand the sight of me, for I've never worked willingly, and he'll make you a present of me. Is that clear?"

When Peppi went home at nightfall with a face longer than ever, the master said, "Peppi, what's troubling you, to be still so glum?"

"I must speak to you: you couldn't give me the old ox that is older than time itself? I could at least slaughter him when I got home and soak his tough meat a little."

"Take him," answered the master, "and I'll even give you a piece of rope to lead him away."

The next morning as soon as it was daylight Peppi took the ox, a knapsack, and eight rolls, and set out for his village, his beret on his head. As he was crossing a plain, two horsemen came galloping up shouting, "Look out, there is a bull coming this way! Look out, the bull will kill you!"

Softly, the ox spoke. "Say to them, Peppi, 'If I capture it, will you give it to me?'"

Peppi repeated those words, and the men replied, "What! You'd never capture it, he would kill you and the ox."

"Peppi," said the ox, "get behind me and don't be afraid." The bull dashed up, snorting furiously, and locked horns with the old ox. They began thrusting at one another, but the old ox was so tough that the bull was stunned after a bit.

"Peppi," said the ox, "Take him and yoke him to me." Peppi obeyed, bid the horsemen farewell, and moved on.

Passing through a certain village, he heard a proclamation: WHOEVER FEELS UP TO PLOWING FIFTY ACRES OF LAND IN ONE DAY'S TIME WILL RECEIVE THE KING'S DAUGHTER IN MARRIAGE; OR, IN THE EVENT HE'S ALREADY MARRIED, TWO PILES OF GOLD. BUT WHOEVER TRIES AND FAILS, DIES.

Peppi put his oxen in a shed and went to present himself to the king. The guards refused to let him in, since he was so ragged, but the king himself appeared and ordered him admitted to the palace.

He entered and said, "Your Majesty's humble servant."

"What can we do for you?"

"I heard the proclamation, I have a pair of oxen, and I would like to see what I can do with those fifty acres."

"Did you hear the whole announcement?"

"I heard it: if I fail, my head falls. Majesty, you will have to furnish me with the plow and a little hay, as I'm just passing through here and have nothing at all."

"Lead the oxen into my barn and feed them," directed the king. Peppi took them there, and the old ox said, "Give me a half sheaf of hay and the bull a whole sheaf." In the morning Peppi took the plow and four sheaves of hay and was off to the field. He had someone show him the land, yoked the oxen, and took his place behind the plow.

The councilors watched from the balcony facing the field, and said to the king, "Majesty, what are you thinking of? Don't you see that fellow finishing the plowing? Do you intend to offer your daughter that ugly peasant for a husband?"

"Just what do you gentlemen advise me to do?" asked the king.

"At noon send him a roasted hen, some tender celery, and a bottle of drugged wine . . ."

They sent a servant to Peppi with this meal. "Come eat while it's hot!" they advised. All he had left to plow was a triangle of land no bigger than a curate's hat. He went to eat his lunch, and fed the old ox a half sheaf of hay and the bull a whole sheaf. Then he began nibbling on the pullet and sipping the wine. He drank it all, finished the chicken, and lay back to sleep. The old ox ate his hay, waited for the bull to finish his, and allowed Peppi to sleep on for a while. When the bull also had finished eating, the ox began to prod Peppi with his hoof.

"Ah . . . ah . . ." mumbled Peppi in his sleep.

"Get up," said the ox, "get up, or your head will roll!"

He arose, washed his face, yoked the oxen and, still half asleep, finished plowing the corner of land, then proceeded to go back over it.

"There wasn't enough opium in the wine, confound it!" commented the councilors from the balcony.

Peppi gave himself wholeheartedly to the task and by ten o'clock in the evening the plowing was all done. He returned to the palace, fed his oxen, and went before the king, saying, "Bless me, Papa."

"Oh, have you finished? What do you want, two piles of gold coins?"

"I'm a bachelor, Majesty. What would I do with gold pieces? I've come for my wife."

They took him and washed him from head to foot and dressed him in princely garb, down to a gold watch. And he got married.

The old ox said to him, "Now that you're married, you must kill me and put all my bones in a basket and go out and plant them one by one in the soil which you plowed. Leave out only one hoof, which you are to put in your mattress. As for my flesh, tell the cook he can cook it any way he likes—as rabbit meat, hare's meat, chicken, turkey, capon, and even fish."

So Peppi slaughtered the old ox. The king did not want him to do so, as even he had grown fond of him, but Peppi said, "No, Father, we will kill him and then we won't have to buy any meat for the wedding feast." And he ordered the cook to prepare the ox meat like that of every kind of animal. A hearty meal indeed resulted; dishes began coming in that delighted everyone present. "This is hare. . . . This is rabbit. . . . This dish here is made from the meat of a young animal. . . . Excellent meat!"

That evening, when the bride fell asleep, Peppi slipped the ox's hoof under the mattress, set the basket of bones on his shoulder, and went out to sow them according to the ox's instructions. Then he came back to bed, without his wife's having heard a thing. Shortly afterward she awakened and said, "Oh, the dream I just had! There seemed to be many cherries and many apples hanging over my mouth. And many roses and jasmines . . . I still feel as though I saw them all. . . ." She reached out and picked an apple.

"I'm not dreaming, this is a real apple!"

"No, you're not dreaming," replied her husband. "These are cherries in my mouth!" And he reached out and picked some cherries.

The king came in to wish them a good day, and found the bedchamber full of flowers and fruit out of season, which he too proceeded to sample.

The councilors stepped out on the balcony, and their gaze fell on the field which Peppi had plowed: it was dense with trees of every variety. They called the king. "Look, Your Majesty, are they not trees out there in the field plowed by Peppi?" Squinting, the king said, "Why, they certainly are, beyond any shadow of a doubt! Let's drive out for a closer look"—and they seated themselves in the carriage.

When they got there, they saw orange trees, lemon trees, plum trees, cherry trees, grapevines, pear trees—all laden with fruit. The king picked some of each and rode home quite content.

Now the king had two other daughters, married to princes' sons, and they began asking their sister, "Does your husband really do all these things?"

"How should I know?"

"Silly girl, ask him how he does them."

"I'll ask him tonight."

"Fine, and then come tell us immediately."

That night, in bed, the bride began putting questions to him; and so that she would be quiet and let him sleep, he told her everything. The next morning she told her sisters, who then told their husbands. When they were all together with the king, the brothers-in-law said, "Shall we make a bet, Peppi?"

"Such as?"

"That we can tell you how you got all these trees to grow."

"Let's make a bet."

"Very well. You stake everything you have acquired here, and we'll stake everything we possess."

They went to a notary, and the agreement was sealed.

Then the brothers-in-law told all. Peppi, who trusted his wife, wondered, "And who told them—the sun?"

He handed over all his possessions and was again as poor as ever. He set out dressed as a peasant and carrying his knapsack. Coming to a hut, he knocked on the door.

"Who is it?"

"It's me, reverend hermit."

"What are you seeking?"

"Could you tell me where the sun rises?"

"Sleep here tonight, my boy, and tomorrow morning I will send you to another hermit older than I am."

The next morning at dawn the hermit gave him a round loaf of bread, and Peppi bid him farewell. After walking a long way he came to another hut occupied by a hermit with a white beard down to his knees.

"God bless you, reverend father."

"What can we do for you?"

"Could you tell me where the sun rises?"

"Ah, my son, keep going until you come to another father older than I am!"

Peppi bid him farewell and continued on to another hut. Kissing the hermit's hand, he said, "Noble father, God bless you . . ."

"What are you seeking?"

"Could you tell me where the sun rises?"

"Ah, my son. . . . Who knows, you might make it there. Here, take this pin and walk on. You will hear a lion roar and you will call out, 'Brother lion, your brother the hermit sends greetings along with this pin to get the thorn out of your paw. In return, you are asked to arrange for me to talk to the sun.' "

Peppi did as he was told and removed the thorn from the lion's paw. "You have given me a new lease on life!" exclaimed the lion.

"Now you are asked to arrange for me to talk to the sun."

The lion guided him to a vast sea of black water. "Here the sun shows himself, but there first appears a serpent, to whom you must say, 'Brother serpent, your brother the lion sends greetings; in return, you are asked to arrange for me to speak to the sun.' "

The lion left him, and Peppi saw the water stir. The serpent appeared, and Peppi repeated word for word what the lion had instructed him to say. "Make haste," replied the serpent, "jump into the water and slip under my wings, or the sun's rays will burn you."

Peppi got beneath one wing. The sun rose, and the serpent said, "Go on, Peppi, tell the sun what you have to tell him before he gets away."

"O treacherous sun, only you could have betrayed me. You shouldn't have done so, traitor!"

"*I* betrayed you?" replied the sun. "No, it was not I. Do you know who did? Your wife, to whom you revealed your secret."

"Well, excuse me, sun," replied Peppi. "But there is a favor only you can do for me: if you would set at half-past twelve tonight, I could recover my possessions."

"Of course, I'll gladly do you this favor."

Peppi thanked him for everything and departed. Back home his wife had his broth waiting for him. He ate and then sat outside for a while. His princely brothers-in-law passed by and he called to them, "Brothers, let us make another bet."

"Just what are you staking? Your property is all gone."

"Well, I'll stake my life, and you two can put up my property."

"Very well, then, you stake your life and we'll stake your property and also our own. But what, by the way, is this bet?"

Then Peppi said, "When does the sun set?"

"Of all things, he's gone mad and no longer knows when the sun sets!" the brothers-in-law told each other; to him they replied, "What? At half-past nine o'clock, that's when it sets!"

"Wrong! I say it sets at a half-past midnight!"

They had the agreement drawn up in writing, then proceeded to watch the sun. At half-past nine the sun was about to go down, when Peppi said to him, "O sun, is this how you keep the promise you made me?"

Then the sun remembered and, instead of setting, lingered on and on, up to half-past midnight.

"What did I tell you?" said Peppi.

"You are right," replied the brothers-in-law, and gave him back his possessions at once along with their own.

"Now," said Peppi, "I intend to show you the heart of a peasant"—as they still called him. He gathered up all their belongings and returned them, saying, "Take all of this back; I have no desire for the property of others, but only for my own."

He went back to his old way of life with his wife. The king insisted on embracing him and, removing his crown, placed it on Peppi's head. The brothers-in-law, as one might expect, were furious, but what could they do? The following day there was a magnificent feast, with all the relatives present. Everyone was happy as course after course followed, concluding at long last with coffee, ices, and cassata. So goes the story of Peppi, who started out as a starving cowherd and ended up the wealthiest and happiest of kings.

(Salaparuta)

A Boat Loaded with . . .

The parents of a certain little boy were very loyal to St. Michael the Archangel. They never let a year pass without celebrating his feast. The father died, and the mother continued to keep St. Michael's day every year with the little money she had left. Then came a year when she found herself penniless and with nothing more to sell in order to observe the feast. She therefore took the child off to sell to the king.

"Majesty," she said to the king, "will you buy this little boy of mine? I'm asking only twelve crowns, or whatever you want to give me for him, just so I can keep St. Michael's day."

The king gave her one hundred gold pieces and kept the little boy. Then he got to thinking. Just imagine, this poor woman has sold her own son so as to be able to honor St. Michael the Archangel, while here I am king and pay him no honor at all. He therefore had a chapel built, bought a statue of St. Michael the Archangel, and celebrated his feast. But once the feast was over, he threw a veil over the statue and thought no more about the saint.

The little boy, whose name was Peppi, grew up in the palace and played with the king's daughter, who was his age exactly. They were always together, and when they got older, they fell in love. The councilors noticed it and said to the king, "Majesty, what's going to happen? You certainly won't marry your daughter to that poor fellow, will you?"

The king replied, "What can I do? Can I send him away?"

"Follow our advice," said the councilors. "Send him on a trading expedition in your oldest and ricketiest boat. Give the order for him to be abandoned on the open sea. He will drown, and our problem is solved."

The king liked the idea, so he said to Peppi, "Listen, you are to go on a trading expedition. You have three days to load your boat."

The boy spent the night wondering what cargo he'd best put aboard the vessel. No ideas came to him the first night nor the second. He thought and thought the third night, and finally called on St. Michael the Archangel. St. Michael appeared and said to him, "Don't be discouraged. Tell the king to have your boat loaded with salt."

The next morning Peppi got up in the best of spirits. The king asked him, "Well, Peppi, did you decide on something?"

"I would like Your Majesty to give me a boatload of salt."

The councilors smiled to themselves. "Perfect! With that cargo, the boat will sink right away!"

The boat sailed off, loaded with salt and towing a smaller vessel.

"What's that one?" Peppi asked the captain.

"Oh, that concerns me," answered the captain.

In point of fact, as soon as they reached the open sea, the captain got into the smaller boat, said "Good night," and left Peppi all by himself.

Peppi's boat started leaking and threatened to sink at any moment in the rough sea. "Dear Mother!" called Peppi. "Dear Lord! St. Michael the Archangel! Help me!" In a flash appeared a solid gold ship, with St. Michael the Archangel at the helm. Peppi grabbed the rope they threw him and tied his boat to St. Michael's, which sped over the sea like a bolt of lightning and glided into an unknown port.

"Do you come in the cause of peace, or in the cause of war?" voices on shore asked.

"In the cause of peace!" said Peppi, who was then allowed to land.

The king of that country extended Peppi and his companion an invitation to dinner (without knowing that the companion was St. Michael).

"Mind you," said St. Michael to Peppi, "they don't know what salt is in this land." Peppi therefore carried along a bag of it to the palace.

They sat down to the king's table and started eating, but nothing had any taste to it. Peppi said, "Why on earth, Majesty, is your food like this?"

"That's what we are used to," replied the king.

Peppi then sprinkled a little salt over everybody's food. "See how it tastes now, Majesty."

The king took a few bites and exclaimed, "Delicious! Delicious! Have you much of this stuff?"

"A whole boatload."

"What are you charging for it?"

"An equal weight of gold."

"In that case I'll buy the entire load."

"Agreed."

After dinner they had the salt unloaded and weighed. In one pan of the scales they put salt, in the other they put gold. So Peppi loaded his boat with gold and, after plugging up the leaks, sailed away.

The king's daughter spent her days on the balcony with her telescope trained on the horizon as she waited for Peppi to come back to her. Catching sight of the boat, she went running to her father. "Papa, Peppi's back! Peppi's back!"

As soon as he landed, Peppi greeted the king and then began unloading all that gold. The councilors were livid with rage and said to the king, "Majesty, this is more than we bargained for."

"What can I do about it?" said the king.

"Send him on another expedition."

So, a few days later, the king told him to be thinking of another cargo, as he was to go out again. After some thought, Peppi called on St. Michael, who said, "Have a ship loaded with cats."

In order to supply Peppi with cats, the king issued a proclamation: LET ALL THOSE PERSONS OWNING CATS BRING THEM TO THE ROYAL PALACE, AND THE KING WILL BUY THEM.

The boat was soon loaded with cats and went meowing across the sea.

When they got farther out to sea than the first time, the captain said "Good night" and took his leave. The boat began sinking, and Peppi called on St. Michael the Archangel. In a flash the gold boat was there and, with lightning speed, towed him to an unknown port, where a delegation met them to see if they had come in the cause of peace or of war. "In the cause of peace," the two said, and were immediately invited to dinner by the king.

By each person's plate lay a whisk broom. "What are they for?"

"You'll see in just a minute," answered the king.

The food was brought in, and in rushed a horde of rats that jumped up on the table and made for the dishes. Each of the guests was supposed to drive them away with the whisk brooms, but that did no good because the rats only came right back and in such numbers that everybody was helpless.

Then St. Michael said to Peppi, "Open the sack you brought along." Peppi untied the sack and let out four cats, which pounced on the rats and tore them to bits.

Overjoyed, the king exclaimed, "What wonderful little creatures! Do you have many?"

"A whole boatload."

"Are you asking a fortune for them?"

"Just their weight in gold."

"Perfect!" The king bought the entire lot of them, and into one plate of the scales went the cats while into the other went the gold. So again, once the boat had been repaired, Peppi sailed home with a boatload of gold.

Waiting for him at the port was the king's daughter dancing for joy. The porters unloaded gold and gold and more gold, the king was in a dilemma, and the councilors were livid with rage. They said to the king, "Twice we've failed. The third time we'll get him. Let him rest a week, then send him off again."

This time when Peppi called him, St. Michael said, "Have them load a boat with beans."

When the boat loaded with beans was about to sink, the usual gold vessel showed up, and before you knew it Peppi and St. Michael were putting in somewhere else.

The ruler of this city was a queen, who invited them both to dinner. After dinner the queen pulled out a deck of cards and said, "Shall we play a round?" So they began a game of lansquenet. The queen was a champion player and had all the men who lost to her imprisoned in a dungeon.

But there was no way St. Michael could lose, and the queen realized that if she continued to play she would forfeit everything she owned.

She therefore said, "I am declaring war on you." They agreed on the time of the war, and the queen drew up all her troops. St. Michael and Peppi were an army of only two, and they rushed into battle with their swords pitted against all the others. But St. Michael the Archangel produced a mighty gust of wind that stirred up a thick cloud of dust. Nobody could see a thing, and St. Michael slipped up to the queen and cut off her head with his sword.

When the dust finally settled and everyone saw the queen's head severed from her body, a shout of joy went up, for she was a queen nobody could stand, and the men said to St. Michael, "We choose Your Honor for our king!"

St. Michael said, "I am king in other parts. Choose your king from among yourselves."

They made an iron cage for the queen's head and hung it up at a street corner, while St. Michael and Peppi descended into the dungeon to free the prisoners. It was crowded with bad-smelling, starving people, and on the ground lay dead bodies alongside the living. Peppi threw out handful after handful of beans, and the men scrambled for them and gulped them down like animals. That way St. Michael and Peppi revived the men, made bean soup for them, and sent them all home.

Beans were unknown in that city, so Peppi sold his for their weight in gold. Then with a boatload of gold and an escort of soldiers at his command, he sailed home, announcing his arrival with a volley of cannon fire.

This time the gold boat as well came into port, and the king welcomed St. Michael the Archangel. At dinner St. Michael said to the king, "Majesty, you have a statue which you honored on a single feast day and then left to gather cobwebs. Why was that? Did you perhaps lack money?"

The king said, "Oh, yes, that's St. Michael the Archangel. I'd completely forgotten."

Then St. Michael said, "Let's go see the statue."

They got to the chapel and found the statue covered with mold. The stranger said, "I am St. Michael the Archangel and I ask you, Majesty, why you have wronged me like this."

The king fell to his knees and said, "Pardon me. Tell me how I can serve you! From now on, your feast will be the most lavishly celebrated of all!"

The saint replied, "You will celebrate the marriage of your daughter and Peppi, since these two young people are meant to become man and wife."

So Peppi married the king's daughter and became king in his turn.

(Salaparuta)

◈ 174 ◈

The King's Son in the Henhouse

It is told that once there was a cobbler with three daughters—Peppa, Nina, and Nunzia. They were as poor as church mice, and although the cobbler went about the countryside to mend shoes, he couldn't make a cent. Seeing him come home empty-handed, his wife cried, "Wretch! What will I put in the pot today?" He was tired and said to his daughter Nunzia, who was the youngest, "Listen, will you come with me to get something to make soup with?"

They went off through the fields to pick herbs for soup. They got to the end of a field and, looking for herbs, Nunzia discovered a fenneltop so big that for all she tugged at it she was unable to uproot it and had to call her father. "Father! Father! Just look at what I've found. But I can't pull it up!"

Her father also tried and tried; the fennel came up and, underneath, a trapdoor stood open. At the door appeared a handsome youth, who said, "Lovely maiden, what are you looking for?"

"Just what do you expect us to be looking for? We are starving to death and therefore picking a few things to make soup with."

"If you are poor, I will make you rich," said the young man to the cobbler. "Leave your daughter with me, and I will give you a sack of money."

"What!" exclaimed the poor father. "Leave you my daughter?" But the youth finally talked him into it, and he picked up the money and left, while Nunzia followed the young man underground.

Down below was a house so sumptuous that the girl thought she had reached Paradise. She began a life that could have been termed blissful, except that she missed her father and sisters.

Meanwhile the cobbler had chicken and beef aplenty every day, and was quite well off. One day Peppa and Nina said to him, "Father, will you take us to see our sister?"

They went to the palace where they had found the fennel, knocked on the trapdoor, and the youth invited them in. Nunzia was delighted to see her sisters again, and showed them around the house. Only one room did she refuse to open.

"Why not? What's in there?" asked the sisters, consumed with curiosity.

"I don't know. Not even I have ever been in there. My husband has forbidden me to enter that room."

Then she went off to arrange her hair, and her sisters insisted on helping her. They undid her tress, inside which they found a key. "This," whispered Peppa to Nina, "must be the key to the room she wouldn't show us!" And pretending to dress her hair, they unfastened the key, then stole off to open up the room.

Inside the room were numerous women: some embroidered, others sewed, and the rest cut out clothes. And they sang:

> "Bundles and clothes we create
> For the king's son they await!"

"Ah, our sister is expecting a baby and didn't tell us!" exclaimed the sisters. But in that instant, the women in the room, realizing they were being observed, went from beautiful to ugly and changed into lizards and green reptiles. Peppa and Nina fled.

Seeing them so upset, Nunzia asked, "What's the matter, sisters?"

"Nothing, we just wanted to tell you goodbye, for we are leaving now."

"So soon?"

"Yes, we must go home."

"But what happened to you?"

"Well, we took the key you had in your hair, and opened that door . . ."

"Oh, my sisters! That will be my undoing!"

As a matter of fact, those women in the room, who were none other than fairies, went to the young man, whom they were holding prisoner there underground, and said, "You must send your wife away. Immediately."

"Why?" he asked, with tears in his eyes.

"You must send her away at once. Orders are orders, is that clear?"

So the poor husband, whose heart was breaking, had to go to her and say, "You must leave the house at once, it's the fairies' order; otherwise I'm done for!"

"My sisters have brought about my downfall!" she said, bursting into tears. "Now where will I go?"

"Take this ball of yarn," he said. "Tie one end to the doorknob and walk off unwinding it. Where the yarn runs out, stop."

Sorrowful, Nunzia obeyed; the ball turned and turned as she walked on and on, and it seemed to have no end. She passed under the balcony of a magnificent palace, and there the ball came to an end. It was the palace of King Crystal.

Nunzia called, and the maids appeared. "Please put me up for tonight," she said. "I don't know where to go, and I'm expecting a baby!" —because in the meantime she had discovered that she was with child.

The maids went to tell King Crystal and the queen, but they replied they would open their door to no one under the sun. Note that many years prior to this, their son had been carried off by the fairies, and they had not seen hide nor hair of him since. They were therefore highly distrustful of strange women.

"Could I just stay in the henhouse, for one night?" asked the poor thing.

Moved to pity, the maids persuaded the king and queen to let her stay in the henhouse, and they took her a little bread, as she was starving to death. They wanted to hear her story, but she only shook her head and repeated, "Ah, if you only knew! If you only knew!"

That very night she gave birth to a fine baby boy, and a maid went at once to tell the queen. "Majesty, you should see the beautiful baby this foreign woman has just had! He looks exactly like your son!"

Meanwhile the fairies said to the youth, who was still underground, "Did you know your wife gave birth to a fine baby boy? Would you like to come see it tonight?"

"If only I could! Will you take me to him?"

That night a knock was heard on the henhouse door. "Who is it?"

"Open up, it's me, your baby's father." And in walked Nunzia's husband, who was the king's son kidnapped by the fairies; they were now bringing him back to see his own son. Behind him came all the fairies, and the henhouse was immediately tapestried and carpeted in gold; the couch was decked with a gold-embroidered counterpane, the baby's cradle turned gold, and everything glittered, making it look like day, while music played and the fairies sang and danced, and the prince rocked the baby and said:

> "If my father knew
> That you are his son's son,
> In clothes of gold would you be wrapped,
> In cradles of gold would you be rocked;
> I would be with you day and night in one,
> Sleep, sleep, O royal son!"

And as they danced, the fairies went to the window and sang:

> "Let the cocks crow not yet,
> The clock strike not yet,
> The time is not yet, not yet, not yet."

Let's leave them and go to the queen. A maid came to her and said, "My queen, my queen, let me tell you! The strangest things you ever saw are going on in the foreign woman's roost! It's no longer a henhouse, but all bright like Paradise. You can hear somebody singing who sounds exactly like your son. Just listen!"

The queen went to the door of the henhouse and listened. But at that moment a rooster crowed, and nothing more was heard, nor did light shine any longer under the door.

That morning the queen herself decided to take the foreign woman her coffee. "Will you please tell me who was here last night?"

"Oh, I'm not at liberty to say, but even if I were, what could I tell you? I wish I knew myself who it is!"

"But who can it be?" said the queen. "What if it were my son?"—and she went on so, until the foreigner finally told her whole story from the beginning, how she'd gone out for herbs, and all the rest.

"Then you are my son's wife?" asked the queen, embracing and kissing her. "Ask him tonight what is needed to free him."

That night, at the same time, the fairies gathered with the king's son. The fairies began dancing, while he rocked his son, singing all the while:

> "If my father knew
> That you are his son's son,
> In clothes of gold would you be wrapped,
> In cradles of gold would you be rocked;
> I would be with you day and night in one,
> Sleep, sleep, O royal son!"

While the fairies danced, the wife said to her husband, "Tell me what is needed to free you!"

"You'll have to see that the cocks don't crow, the clock doesn't strike, nor the bells ring. Cover the window with a dark cloth with the moon and stars embroidered on it, so that you can't see when it's daytime. Once the sun is high in the sky, pull away the cloth, and the fairies will turn into lizards and green reptiles and flee."

The next morning, the king had his crier announce this order: SILENCE ALL BELLS AND CLOCKS, AND BUTCHER ALL YOUR COCKS!

Everything was readied, and that night, at the usual time, the fairies began dancing and making music, while the king's son sang:

> "If my father knew
> That you are his son's son,
> In clothes of gold would you be wrapped,
> In cradles of gold would you be rocked;
> I would be with you day and night in one,
> Sleep, sleep, O royal son!"

And the fairies went to the window singing:

> "Let the cocks crow not yet,
> The clocks strike not yet,
> The time is not yet, not yet, not yet."

They danced and sang the whole night long, and they continued to go to the window and, seeing that it was still night, they repeated:

> "Let the cocks crow not yet,
> The clocks strike not yet,
> The time is not yet, not yet, not yet."

When the sun was directly overhead, the curtain was drawn back. Some fairies became snakes, others green lizards, and they all fled.

The king's son and his wife embraced the king and the queen.

> They were always happy as could be,
> While here we are without a penny.

(Salaparuta)

The Mincing Princess

Time and again the tale has been told of a king who had a marriageable daughter as lovely as lovely could be. One day he called her and said, "My daughter, you've reached the age when it is fitting to marry. I'm going to notify all my fellow monarchs and friends that a grand celebration will be held on a certain day. They'll all attend, and you will take your pick."

The day arrived, and all the monarchs showed up, each one accompanied by his entire family. Of all those present, it was the son of King Garnet that the princess fell in love with. She informed her father of her choice. You are well aware of how news spreads among friends: King Garnet's son found out he was the lucky man, and rejoiced. Noon rolled around and everybody sat down to the king's banquet of fifty-seven courses. For dessert, pomegranates were served. Now pomegranates are not to be found in every country, and they were altogether unknown at King Garnet's court. The prince started in on one, but dropped a seed on the floor. Thinking it might be something very valuable, he bent over to pick it up. When the princess, who hadn't been able to take her eyes off of him, saw that, she rose from the table and, flushed with anger, ran and locked herself in her room. Her father, the king, followed her to see what was the matter. He found her weeping. "Papa, I really liked that boy, but I now see he's a small-minded person and I want nothing more to do with him."

The king returned to the table, thanked all the monarchs for coming, and bid them goodbye. But that was too much for King Garnet's son. Instead of leaving, he disguised himself as a peasant and began skirting the palace. Now at the royal palace they were looking for a gardener. Since he knew something about gardening, he applied for the post. An agreement was reached over the salary, instructions were given him, and he became the royal gardener. He had a cottage in the garden and carried to it a trunk containing gifts intended for his betrothed, pretending the trunk was full of his clothing.

Across the cottage window, he stretched a shawl embroidered in gold. The window of the princess faced upon the garden, and as she looked out, the gleam of the shawl caught her eye. She called the gardener. "Tell me, whose shawl is that?"

"Mine."

"Will you sell it to me?"

"Never."

She then ordered her maids to try to persuade him to sell her the shawl. The maids offered him any amount of money, and even to exchange the shawl for something else of value, but all to no avail. At last the gardener said, "I would give her the shawl, if only she would let me sleep in the first room of her apartment."

The maids burst out laughing and ran off to tell the princess. Then, discussing the matter, they said, "But if he's so foolish as to want to sleep in the first room of your apartment, why not let him? No one will be the wiser for it, it will cost us nothing, no harm can come of it, and you will get the shawl." So the princess consented. At night when the whole house was sleeping, they called him in and left him there to sleep. They woke him up early the next morning and escorted him out. And he handed over the shawl.

A week later, the gardener hung up a second shawl more beautiful than the first. The princess wanted it, but for this one the gardener was asking to sleep in the second room of her apartment. "You let him sleep in the first room, so you might as well put him to bed in the second!" reasoned the maids, and it was granted him.

Another week went by, and the gardener put on display a gown embroidered in gold and adorned with pearls and diamonds. The princess fell in love with it, but to get it there was nothing else to do but let the gardener sleep in the third chamber of her apartment—that is, in the antechamber of the princess's bedroom. What was there to be afraid of, since that poor gardener was certainly half crazy?

The gardener stretched out on the floor, as on the other nights, and pretended to sleep. He waited for the moment when everyone would surely be asleep and then, as if seized with a chill, he began to shake all over, while his teeth chattered noisily. He was propped against the princess's door and, with all his trembling, made it sound like a drum being rolled. The princess woke up and, because of that racket, found it impossible to go back to sleep. She told him to be quiet. "I'm cold!" he moaned, trembling all the more. Unable to calm him down and fearing they would hear him in the palace and learn of her strange pact with the gardener, she finally got up and opened the door. "He's so simple-minded," she thought to herself, "no harm can possibly come of it."

Simple-minded or no, the fact is that, from that night on, the princess began expecting a baby. Her anger and shame were boundless. Worried that everybody would soon know, she told the gardener. "There's nothing else for you to do," he said, "but flee with me."

"With you? I'd rather die!"

"All right, remain at court until everyone finds out."

So she had to resign herself to running away with him. She tied her things up in a small bundle, took a little money, and one night they ran away on foot.

Along the way they met cowherds and shepherds, passed through fields and pastures. And she asked, "Whose flocks are these?"

"They belong to King Garnet."

"Oh poor me!"

"Why? What's the matter?" asked the gardener.

"Poor me, for refusing his son for a husband!"

"Too bad for you!" said the gardener.

"And whose land is all this?"

"King Garnet's."

"Oh, poor me!"

Dead tired, they came to the house of the young man, who had told her he was the son of King Garnet's steward. It was a smoke-blackened hut containing an old bed, a stove, and a fireplace, and next to it were the barn and the henhouse. "I'm hungry," he said. "Wring a chicken's neck and cook me some chicken." The princess obeyed. They spent the night in the hut, and in the morning the youth went out, saying he wouldn't be back before evening.

The princess stayed in that humble house by herself, and all of a sudden a knock was heard. She opened the door, and there stood King Garnet's son dressed from head to toe in royal garb. "Who are you?" he asked. "What are you doing here?"

"I am the wife of your steward's son."

"That may well be. But you don't look like an honest woman to me. What if you were a thief? Somebody's always sneaking up here stealing my chickens."

And the prince called the hens and counted them. "One's missing!" he said. "How's that? They were all here yesterday at this hour." Then he began rummaging through the house. In the stove he found the feathers of the hen which the princess had cooked the night before. "So you're the thief! I've caught you red-handed! Be thankful I was the one to catch you. I won't turn you over to the law!"

At the shouts of the prince, his mother the queen appeared. She saw the young woman in tears and said to her, "Don't worry, my son is a strange boy. You will work for me. I'm expecting a little grandson and have to get his baby clothes ready. You will help me sew." And the queen led her off to make swaddling bands, baby gowns, jackets, and pants.

When the gardener came home in the evening, the young woman wept and told him everything, saying he was to blame and had to take her away from there at once. But he calmed her down and persuaded her to stay there. "But what will we do?" she asked. "Our baby will come and we won't have a stitch to put on him!"

"Tomorrow," he told her, "when the queen gives you more sewing, take a baby gown and hide it in your bosom."

So the next day as she was leaving, the young woman waited until the queen turned her head for a minute, and slipped a baby gown down her bosom. A minute later the prince came in and said to his mother, "Mamma, just who's working here with you? That thief? You know she's capable of stealing everything!" At that, he reached out and pulled the baby gown out of her bosom. The young woman could have gone through the floor. But this time as well the queen sided with her. "These are matters that concern women," she told her son. "What are you doing meddling in them?" She comforted the young woman, who wept as though her heart would break, and told her to come back tomorrow to string a few pearls.

The young woman returned to the hovel that night and told her husband about her latest misfortunes. "Don't give them a thought," he said. "That king is an old skinflint. Just be sure to slip a string of pearls into your pocket tomorrow."

The next day, when the queen wasn't looking, the young woman thrust a string of pearls into her pocket. But when the prince came in, he said, "You're giving this thief pearls? I'll just bet at least one string has already found its way into her pocket!" Rummaging in her pocket, he came up with the pearls, and the young woman fainted. Then the queen held smelling salts to her nose, reviving and consoling her.

The next day while she was working at the queen's, her labor pains began and she had to go to bed. The queen put her in the prince's bed, where she gave birth to a fine baby boy.

In walked the prince. "What, Mamma, this thief in my bed?"

"Enough of this comedy, my son," said the queen. "Dear daughter, this son of mine is your husband, whom you refused all because of a little pomegranate seed, and who became a gardener in order to win you."

Everything was now in the open. The princess's parents were summoned, together with all the neighboring monarchs, and there were three full days of feasting and merrymaking.

(Province of Trapani)

176

The Great Narbone

It is told, ladies and gentlemen, that there was once a king who had one son. This son, eager to marry, sent painters to all the kingdoms to paint portraits of the most beautiful girls of every class. The first painter to return brought back the portrait of a washerwoman's daughter, a maiden of rare beauty indeed. When he saw it, the king's son said, "She's the one I want!" and, escorted by servants and soldiers, he left for the city where the girl lived.

The girl was on her way out to wash and carried a bundle of clothes on her head. With a slap, the prince sent the bundle sailing into the river and said, "I am marrying you, and you will be queen." Taking her by the hand, he said, "Let's go to your father." The girl burst into tears.

Her father was furious. "Go and joke with your own kind, and leave us poor people to our own worries!"

"On my honor," replied the prince, "I want your daughter for my wife, and you will receive a pope's income."

He left them a great sum of money, had the daughter outfitted like a queen, and departed. Back at the palace following the wedding, there was a whole week of gala balls, then the couple settled down in their own quarters and loved each other dearly.

Meanwhile war was declared on the prince's father by the king of Africa. The prince went to the kingdom's defense, leaving his wife in his father's safekeeping. He went to war, and in the first battle, he was the victor.

Let's leave him at war and turn to his wife. One of the king's ministers had begun making eyes at this princess, but the first time he tried to approach her, she slapped him.

Stung to the quick, the minister went flying to the king and said, "Majesty, as Your Highness can see, your daughter-in-law is in league with the cook and certain others . . ."

The king wrote his son, who replied, "Whatever you see fit to do with my wife, do it."

The king showed the minister the letter and asked, "What shall her sentence be?"

"Majesty, let us select two ruffians to take her to the woods and kill her."

Thus was it done. The princess suspected nothing; she only knew she was to go to the country, and had put on her jewels. "But where are we going?" she asked after a certain distance.

"Keep walking and be quiet!" said one of the men, who pulled out his knife and pricked her to keep her moving. Arriving in the darkest part of the woods, they decided it was time to kill her. "Why must you kill me?" wept the poor girl. "Take my jewels in exchange for my life!"

The ruffians took the jewels and spared her life. The princess remained there, alone and bitter. A goatherd came by and, in exchange for a present of money, gave her a man's suit of clothing. She hid her own royal clothing under a mulberry tree and marked the trunk with a cross in order to be able to find it again.

She set out on the road dressed as a man, and ran into four thieves.

"Who goes there?" asked the thieves.

"A fugitive from justice," replied the princess.

"But who are you?"

"The Great Narbone."

"Oh, we have heard of you, we know of your prowesses . . ." and they took her with them into a cave. Other thieves joined them, some twenty in all, and learning that this was the Great Narbone, so illustrious in valor, they made a big to-do over him and named him their chief.

"Since you confer such an honor on me," said Narbone, "what I say from now on, goes. Let's put that in writing, each of us signing in our own blood."

"Yessirree," said the thieves, and they all drew blood from their arms and signed the pledge of obedience to the chief.

While they were doing that, the watch came in and reported that twelve silversmiths were going by with their cargoes. "Who will look after this robbery?" wondered the thieves.

"I'll see to it myself with two of you," said Narbone.

Taken by surprise, the silversmiths fired their guns, but the thieves fired even more guns, so the silversmiths fled, leaving behind twelve loads of gold objects. (They got away, thanks to the Great Narbone.)

The thieves took the treasures and cried, "Long live the Great Narbone!"

The prince came back from the war, closed himself up in his room, and wept. Noblemen went and tried to console him. "Prince, why all this weeping? You'll cry your eyes out. Come to the country with us and forget everything."

They went hunting in the country and were caught by the thieves and carried into the cave. "You are the prince, are you not?" asked the Great Narbone. "Is your father's minister still alive?"

637

"Very much so," said the prince.

"Write him a letter at once," ordered the Great Narbone, "and have him come to the cave of the Great Mountain."

The prince wrote, and the minister had no choice but to come. The thieves were on the lookout; as soon as they saw the minister they seized him and took him to the cave. Chief Great Narbone had a fine feast prepared and invited everyone to dinner—the twenty-four thieves, plus the prince, the minister, and himself, who made twenty-seven. While dining, he said, "Now, honorable minister, explain that business about the prince's wife."

The minister began trembling. "I know nothing about it . . ."

"Don't tremble, tell the prince what happened. Just what designs did you have on his wife?"

When the minister refused to talk, Great Narbone pointed a pistol at him, saying, "Either you tell the whole story, or I'll blow your brains out!"

Stuttering, the minister started his tale.

"Majesty," said the thieves' ringleader to the prince, "did you hear what really happened?" Drawing a large knife, he cut off the minister's head and stood it in the center of the table.

"That takes care of that wicked soul, Majesty! Now we can go on with our dinner! Throw the body out of the cave!" And with bloody hands, he went back to his food.

When dinner was over, he excused himself, went to the mulberry tree for the princess's clothing, and put it on. When the prince saw her come in and realized that it was his wife, he wept tenderly and asked her forgiveness.

The princess had the thieves pardoned and they all rode on horseback, with the carriage of the prince and princess in their midst, to the palace. Just imagine the celebrations! The thieves returned to their towns as rich men and committed no more robberies.

> They lived happily ever after,
> And we are here 'midst friends and laughter.

(Province of Agrigento)

177

Animal Talk and the Nosy Wife

Once there was a young married man who was unable to make ends meet where he lived, so he moved to another town and entered the service of a priest. One day while working in the field, he found a large mushroom, which he carried to his employer. The priest said to him, "Go back to the very same spot tomorrow, dig where the mushroom was growing, and bring me what you find."

The farmer dug down and found two vipers. He killed them and carried them to his master. That same day some eels had been brought to the priest, who said to his servant, "They will be for the young man's dinner. Pick out the two thinnest eels and fry them for him." The servant made a mistake and fried the vipers, which she then served the farmer. He ate them with relish.

When dinner was over, the youth looked down and sat the cat and dog, and he heard them talking. The dog said, "I'm supposed to get more meat than you are." The cat replied, "No, I'm supposed to get the most."

"Since I go out with the master," said the dog, "and you stay at home, I have to have more to eat than you do."

"It's your duty to accompany the master when he goes out," answered the cat, "just as it's mine to remain in the house."

The farmer realized that, by eating the two vipers, he had acquired the ability to understand the speech of animals.

He went downstairs to the stable to give the mules their barley and heard them talking to one another. "He ought to give me more barley than he gives you," the lead mule was saying, "since I carry him on my back."

"He owes me every bit as much," protested the other mule, "since I carry the burdens."

After overhearing that discussion, the farmer divided the barley into equal parts. "What did I tell you?" said the second mule. "He's being very fair."

The farmer went back upstairs and was met by the cat, who said, "Listen, I know you understand us when we talk. The master called for the vipers and was told by the serving woman that she served them to you by mistake. Now the master wants to know if you acquired the magic ability to understand animal talk. He read of such things in a book of magic, and will pressure you to admit that you can now understand our speech. But you must deny that ability, or you will die and it will go to the master."

After that warning, the farmer boy, regardless of all the priest's questioning, was careful to admit nothing. The priest finally gave up and dismissed him. On the road the youth met a flock. The shepherds were worried, because every night a few sheep vanished from the flock. "How much will you give me if I see to it that no more disappear?" asked the farmer. The steward replied, "When we see that no more are missing, we'll give you a mare and a young mule." The farmer remained with the flock and, that night, went to bed outside on the hay. At midnight he heard voices: they were the wolves calling the dogs. "Oh, brother Vitus!"

"Yea, brother Nick!" answered the dogs.

"Can we come after the sheep?"

"No, you can't," replied the dogs, "there's a shepherd sleeping outside."

So for a week the farmer slept outside and heard the dogs warning the wolves not to approach. Therefore no sheep were missing in the morning. On the ninth day he had the faithless dogs killed and new ones put on guard. That night the wolves again cried, "Oh, brother Vitus, can we approach?" And the new dogs answered, "Yes, come right ahead, your friends have been put to death, and we'll make mincemeat of you."

The next morning the shepherds gave the farmer a mare and a young mule, and he departed. When he got home his wife wanted to know whose animals he was leading. "Our own," he replied.

"How did you come by them?"

But the man gave no explanation and remained silent.

In a neighboring town the fair was in full swing, so the farmer decided to take his wife to it. They both climbed on the mare's back, and the mule followed along behind. "Mamma, wait for me!" said the mule. And the horse replied, "Come on, step lively, you are light whereas I have two persons on my back!"

Hearing that exchange, the farmer burst out laughing.

Her curiosity aroused, his wife asked, "What are you laughing at?"

"Nothing at all. I was just laughing."

"Tell me this minute why you're laughing, or I'll dismount and go back home."

"Well," said her husband, "I'll tell you when we get to Santo."

They reached Santo, and the woman started up again. "Now tell me why you were laughing. Just what was so funny?"

"I'll tell you when we get home."

The wife then refused to attend the fair and insisted on going home immediately. The minute they entered the house, she said, "Now tell me."

"Go for the priest," said the husband, "and then I'll tell you."

In all haste the wife threw on her veil and went for the priest.

As the husband waited for him, he thought, Now I'll have to tell her, and I will die. A sad fate! But first I'll confess my sins and receive communion, so I'll die in peace.

As he brooded, he threw a little grain to the hens. The hens hurried up to eat, but the rooster bounded forward flapping his wings and drove them off. The farmer asked the rooster, "Why don't you let the hens eat?"

"The hens must do as I say," replied the rooster, "even if there are great numbers of them. I'm not like you who have only one wife. You let her rule you, and you will now die from telling her you understand our talk."

The farmer thought about it, then said to the rooster, "You have more brains than I do."

He took off his belt, moistened it, made sure it was as flexible as flexible could be, and proceeded to wait. His wife returned and said, "The priest is on his way. Now tell me why you were laughing."

The husband took his belt and lashed the daylights out of her. In walked the priest. "Who wanted to go to confession?"

"My wife."

The priest took the hint and left. In a little while the wife came to, and her husband said, "Did you hear what I was supposed to tell you, wife?"

"I'm not interested any more."

And from that day on, she wasn't the least bit nosy.

(Province of Agrigento)

◈ 178 ◈

The Calf with the Golden Horns

It is told that there was a husband and a wife with two children—a boy and a girl. The wife died, and the husband remarried. His new wife had a daughter who was blind in one eye.

The husband was a farmer and went out to work the fields of a certain estate. His wife couldn't stand the sight of her stepchildren. She baked bread and sent them to her husband with it. But she directed them to another estate in the opposite direction, so as to lose them for good. The children came to a mountain and called their father: "Papa! Papa!" But there was no answer; all they heard was an echo of their calls.

They were lost, and wandered at random through the countryside. Soon the little boy became thirsty. They found a fountain, from which he wanted to drink. But the little girl, who had a sixth sense and knew of the hidden properties of fountains, asked:

> "Fountain, fountain dear,
> What must he fear
> Who would quench his thirst right here?"

And the fountain replied:

> "Whoever drinks of me, be it lad or lass,
> Will for sure become a little ass."

So the little brother remained thirsty, and they moved on. They came to another fountain, and he was all ready to bend over and drink, but his little sister asked:

> "Fountain, fountain dear,
> What must he fear
> Who would quench his thirst right here?"

The fountain replied:

> "Whoever drinks of me, be it lad or lass,
> Will thereafter be a wolf, alas!"

The little brother again refrained from drinking, and they moved on. They came to still another fountain, and the little sister asked:

> "Fountain, fountain dear,
> What must he fear
> Who would quench his thirst right here?"

The fountain replied:

> "Whoever drinks of me, be it lad or lass,
> Will become a little calf, alas!"

The sister forbade her brother to drink, but his thirst by now was so intense that he said, "If I must choose between dying of thirst and becoming a little calf, I'll become a little calf," and he bent over and drank. Right away he turned into a little calf with golden horns.

642

So the little sister continued on with her brother who was now a calf with golden horns. They kept going until they came to the sea. On the shore stood a beautiful cottage where the king spent his holidays. The king's son was at the window and saw the beautiful maiden walking along the strand with a little calf.

"Come up here to me," he said.

"I will," she replied, "if you allow my little calf to come with me."

"Why are you so fond of him?" asked the king's son.

"I love him because I raised him myself, and I don't want him out of my sight a minute."

The prince fell in love with the maiden and married her; the little calf with the golden horns lived with them, and the three were always together.

Meanwhile, the father, who had come home and found his children missing, was deeply grieved. One day, to forget his sorrow, he went out to gather fennels. He came to the seashore and saw the prince's cottage. At the window stood his daughter; she recognized him, but he didn't know who she was.

"Come in, kind sir," she said, and her father went in. "Don't you know me?" she asked.

"To tell the truth, you do look familiar."

"I am your daughter!"

They fell into each other's arms. She told him her brother had become a little calf, but that she had married the king's son. The father was quite pleased to learn that the daughter he had given up for lost had made such a fine match and also that his son was alive, even if he was now a little calf.

"Now, Father, empty your sack of fennels, and I'll fill it with money."

"Goodness knows how happy your stepmother will be!" said the father.

"Why not tell her to come here to live, along with her daughter who's blind in one eye?" said the daughter.

The father agreed, and returned home.

"Who gave you this money?" asked the wife, filled with amazement when he opened the sack.

"Wife! Would you believe I have found my daughter and that she's married to a prince and wants us all to come and live with her—me, you, and your daughter that's blind in one eye."

Hearing that her stepdaughter was still alive, the woman was consumed with rage, but she replied, "Oh, what wonderful news! I can't wait to see her!"

So, while the husband stayed behind to settle his affairs, the wife and the daughter who was blind in one eye went to the prince's cottage. The prince was out, and when the stepmother found herself alone with her stepdaughter, she seized her and threw her out the window, which looked straight down into the sea. Then she dressed the daughter who was blind in one eye in her stepsister's clothes and said, "When the prince returns, you will start crying and tell him: 'The calf with the golden horns put out my eye, and now I'm blind in that eye!'" After giving those instructions, she went back home, leaving the girl there by herself.

The prince returned and found her in bed weeping. "Why are you crying?" he asked, thinking it was his wife.

"The calf, with a thrust of his horns, blinded me in one eye. Ooo, Ooo!"

At once the king cried, "Call the butcher and have him slaughter the calf!"

Hearing that, the little calf ran to the window overlooking the sea and said:

> "Sister, O sister,
> They whet the knife
> To take my life,
> They ready the bowl
> As my blood's last goal!"

And from the sea came this reply:

> "In vain do you wail;
> I lie within a whale!"

Hearing that, the butcher was afraid to slaughter the calf, and said to the prince, "Majesty, come hear what the calf is saying."

The prince approached and heard:

> "Sister, O sister,
> They whet the knife
> To take my life,
> They ready the bowl
> As my blood's last goal!"

And from the sea the voice answered:

> "In vain do you wail;
> I lie within a whale!"

The prince immediately sent for two sailors, and they began fishing for the whale. They caught him, opened his mouth, and out came the prince's wife, safe and sound.

The stepmother and stepsister who was blind in one eye were thrown into prison. As for the calf, they sent for a fairy who changed him into a handsome young man because, in the meantime, he had grown up.

(Province of Agrigento)

◈ 179 ◈

The Captain and the General

Once upon a time in Sicily there was a king who had one son. This son married Princess Teresina. When the wedding festivities were over, the prince sat down in his room very sad and worried. "What's the matter?" asked his bride.

"I was thinking, Teresina dear, that we must take an oath: the first one of us to die must be waked by the other for three days and three nights closed up in the tomb."

"Oh, if that's all that bothering you!" said the bride. She picked up his sword, and they kissed the cross on the hilt to seal their pledge.

One year later, Princess Teresina was taken sick and died. The prince gave her a grand funeral and, that night, picked up sword, two pistols, and a purse on gold and silver coins, went into church, and had the sacristan let him down into the tomb. He said to the sacristan, "Three days from now, come and listen at the tomb. If I knock, open up; if I haven't knocked by nightfall, that will mean I'm coming back no more." Then he gave the sacristan one hundred crowns for his pains.

Closed up in the tomb, the prince lit the torch, opened the coffin, and wept as he gazed upon his deceased wife. Thus the first night went by. On the second, a hissing was heard at the back of the tomb, and out crawled a huge and fierce serpent, followed by a brood of young snakes. Mouth open, the serpent made a lunge for the dead woman, but the prince pointed his pistol at him and planted a bullet in his head. At the shot, the little serpents turned and wriggled away. The prince remained there in the tomb, with the dead serpent at the foot of the coffin. In a short while here came the young serpents back, each carrying in his mouth a wad of grass. They crawled around the dead serpent, putting grass on his wound, in his mouth, over his eyes, and rubbing his body with it. The serpent opened his eyes again, writhed, and was quite sound once more. He turned about and fled, followed by his young.

The prince lost no time in taking up the grass left there by the serpents, putting it in his wife's mouth and strewing it over her body. She began breathing again, color flowed back into her face, and she got up, saying, "Ah, how I've slept!"

They embraced and immediately sought the hole through which the serpents had entered; it was large enough for the couple to crawl through also. They came out into a meadow dense with that serpent grass, and the prince gathered a large sheaf, which they carried off with them. They went to Paris in France and rented a palace beside the river.

A little later the prince decided to become a merchant. He left his wife with a woman of good morals to help with the housework and, once he had bought a ship, he departed. He promised to return in a month and to signal his arrival, when the ship came in sight of the palace, with three blank cannon shots.

No sooner was he gone than a captain of the Neapolitan army came down the street and saw Teresina at the window. He began flirting with her, but Teresina withdrew. Then the captain called to an old woman and said, "Ma'am, if you can arrange for me to meet the lovely young lady who lives in this palace, there will be two hundred crowns waiting for you!"

The old woman went and begged Teresina to please help her, since they wanted to seize her possessions. "I have a chest full of things," she explained, "and they will confiscate it. Would you be so good, madam, as to keep it at your house for me?"

Teresina agreed, and the old woman had the chest brought in. At night, out of the chest jumped the captain. He seized the lady and spirited her away to his ship. They went to Naples where the captain and Teresina, forgetting her husband, lived as man and wife.

One month later, her husband's ship came up the river and fired three cannon shots, but no wife appeared on the balcony. Finding the house empty, with no sign of her, the man sold all his goods and traveled through the world until he reached Naples, where he enlisted as a soldier. One day the king ordered a gala military parade, in which all the soldiers marched. As the captains filed by, arm-in-arm with their wives, the soldier-prince recognized Teresina with her arm in the captain's. Teresina also recognized the prince among the soldiers and said, "Look, Captain, there's my husband there among the soldiers. What shall I do?"

The captain had her point him out. Oh, yes, the soldier was in his company and had only recently been named quartermaster. The captain invited to his house all his noncommissioned officers—corporals and quartermasters. They gave a banquet, at which Teresina did not appear. While they were eating, the captain had a silver knife and fork slipped into the young quartermaster's pocket. The knife and fork were missed after a time and a search was made for them. In whose pocket should they be found but that poor innocent young man's. He was court-martialed and condemned to die before the firing squad. Now the quarter-master had a friend among the soldiers of the squad. He gave the friend a bit of serpent grass and said, "When you fire on me, try to make a great cloud of smoke. While the soldiers are going through the 'Shoulder arms!' put a little of this grass in my mouth and on my wounds and leave me lying there."

He was shot. Screened by the smoke, his friend stole up and filled his mouth with grass. The prince revived, got up, and went off on all fours.

For some time the daughter of the king of Naples had been sick and on the verge of death. The doctors could do nothing for her. The king decreed throughout the kingdom:

WHOEVER COMES AND CURES MY DAUGHTER WILL RECEIVE HER IN MARRIAGE, IF HE IS SINGLE. IF HE IS MARRIED, I WILL NAME HIM PRINCE.

Dressed as a doctor, the prince showed up at the royal palace. He crossed a hall packed with worried doctors, reaching the sick girl just as she drew her last breath and died. "Majesty," said the prince, "your daughter is already dead, but I have a way to cure her yet. Please leave me alone with her."

They left him and he drew out of his pocket a little of the grass and put it in the dead girl's mouth and nose. The king's daughter started breathing again and was completely well. "Fine, doctor," said the king. "you will now be my son-in-law."

"I'm sorry, Majesty," said the prince, "but I'm already married."

"Well, what favor do you ask?"

"Majesty, I wish to be general commander of all the armed forces."

"By all means." And the king ordered two grand celebrations, the first in honor of his daughter's recovery, the second in honor of the new general.

To his celebration, the general invited all the captains, including the one who had carried his wife off. Nor did the general fail to have a gold knife and fork slipped into the captain's pocket. The knife and fork were found, and the captain was thrown into prison.

The general went to question him. "Captain, are you married or single?"

"Honorable General," said the captain, "to tell the truth, I am not married."

"And that lady who was with you?"

In that instant she appeared, handcuffed between two soldiers, and cried, "No, no, the captain kidnapped me from our house. I never forgot you . . ."

But her words were in vain. The general ordered them both coated with pitch and burned to death. So, after all his many trials and tribulations, he ended up alone and general commander of all the regiments.

(Province of Agrigento)

❖ 180 ❖

The Peacock Feather

A king went blind. The doctors had no idea how to cure him. Finally one said that the only way to restore sight to those blind eyes was with a peacock feather.

Now the king had three sons. He called them in and asked, "My sons, do you love me?"

"You're as dear to us, Father, as life itself."

"Well, you must get me a peacock feather, so that my sight will be restored. Whichever one of you brings it to me will have my kingdom."

The sons departed, the two older boys and their little brother. They didn't want the little brother to come along, but he wouldn't hear of being left behind. They entered a forest, and night fell. They all three climbed a tree and went to sleep in the branches. The youngest boy was the first to wake up. It was dawn, and he heard the song of the peacock in the heart of the forest. So he went down the tree and then in the direction of the song. He came to a fountain of clear water and bent over to drink. When he stood up, he saw a feather float down from the sky. He looked up, and there was the bird flying through the air.

When the brothers found out that the youngest had the peacock feather, they seethed with envy, for he would now inherit the kingdom. Then without a moment's hesitation, one of them seized him, the other killed him, and together they buried him and took possession of the feather.

When they got back to their father, they gave him the peacock feather. The king passed it over his eyes, and his sight returned. The minute he could see again, he said, "Where is your little brother?"

"Oh, Papa, if you only knew! We were sleeping in the forest, and an animal came by. It must have carried him off, for that was the last we saw of him."

Meanwhile, in the spot where the boy was buried, a handsome reed came up. A shepherd passed by, saw the reed, and said to himself, "What a fine reed! I shall cut it and make myself a shepherd's pipe." That he did, and when he put his lips to it, the reed sang:

> "O Shepherd holding me,
> Play gently, don't afflict me.
> They slew me for the peacock feather;
> My brother, to be sure, was the traitor."

Hearing this song, the shepherd said to himself, "Now that I have this pipe, I can give up herding sheep! I shall go all over the world and earn my living piping!" So he left his flock and went to the city of Naples. As he played his shepherd's pipe, the king looked out and listened. "Oh, what beautiful music!" he said. "Invite that shepherd inside!"

The shepherd went in and played in the halls of the king. The king said, "Do let me play a little bit myself."

The shepherd handed him the pipe, the king began playing, and the pipe said:

> "O Father holding me,
> Play gently, don't afflict me.
> They slew me for the peacock feather;
> My brother, to be sure, was the traitor."

"Oh," said the king to the queen, "just listen to this pipe. Here, you play a little bit." The queen began playing, and the shepherd's pipe said:

> "O Mother holding me . . ."

and so on. The queen also was dumbfounded and begged her middle son to play too. He shrugged his shoulders and claimed it was all nonsense, but in the end he had to obey; he'd no sooner put his lips to it than the pipe sang out:

"O brother who seized me . . .

but it stopped right there, since the boy was shaking like a leaf and passed the pipe to his big brother, saying, "You play! You play!"

But the elder brother refused, saying, "You've all gone crazy with that shepherd's pipe!"

"I order you to play!" cried the king.

So the oldest boy, pale as a ghost, proceeded to play:

"O brother who slew me,
Play gently, you afflict me.
You killed me for the peacock feather,
You were, to be sure, my betrayer."

At those words the king fell to the floor, stunned with grief. "You wicked boys!" he cried. "To get the peacock feather yourselves, you killed my child!"

The two brothers were burned to death in the town square. The shepherd was named captain of the guards. And the king spent the rest of his days secluded in his palace, sorrowfully playing the shepherd's pipe.

(Province of Caltanissetta)

◈ 181 ◈

The Garden Witch

There was once a cabbage patch. It was a time of famine, and two women were out looking for something to eat. "Friend," said one of them, "let's go into this garden and pick cabbages."

"But someone is surely guarding it," answered the other woman.

The first one went to see. "There's not a soul around. Let's go in."

They went into the garden and each picked an armful of cabbages; they carried them home, prepared a good supper, and the next day returned for two more armfuls.

Now that garden belonged to an old woman, who came home and

discovered that her cabbages had been stolen. "I'll take care of that," she said to herself. "I'll get a dog and tie him to the gate."

The friends saw the dog, and one of them said, "No indeed, I'm not going in this time for cabbages."

"Don't be silly," replied the other one. "We'll get two cents' worth of hard bread and throw it to the dog. That way we can do what we wish."

They bought the bread and, before the dog could go "Bow-wow!" threw it to him. He dived into the bread and remained perfectly quiet. The friends got their cabbages and left.

The old woman appeared and saw what had happened. "So you let them pick cabbages right under your nose! You're not fit to be a watchdog! Get up!" In his place she put a cat. "When it meows, I'll dash out and nab the thieves!"

The two friends returned for cabbages and spied the cat. They bought two cents' worth of lung, and before the cat could go "Mew" they threw him the lung and he kept quiet. The old woman appeared and, finding neither cabbages nor thieves, had it out with the cat.

"Now who will I put here? The rooster! This time the thieves won't get away from me."

"No, indeed, I'm not going in this time," said one of the two friends. "There's the rooster!"

"Throw him some grain," said the other, "and he won't crow."

While the rooster pecked on the grain, they cleaned out the cabbage patch. The rooster finished the grain, then crowed, "Cockadoodledo!" The old woman appeared, found the cabbages missing, grabbed the rooster, and wrung his neck. Then she said to a peasant, "Dig a grave the size of me!" She stretched out in it and had herself covered over with dirt, leaving only an ear above ground.

The next morning the women returned, looked all around, but saw no one in the garden. The old woman had had the grave dug in the path the two women would take. Going in, they didn't notice anything unusual; but coming out with their arms full of cabbages, the first friend saw the ear sticking out of the ground and exclaimed: "Oh, friend, look at this wonderful mushroom!" She bent over and began tugging on the mushroom. She pulled and pulled with all her might. She gave one final jerk, and out jumped the old woman.

"Ah-HA!" cried the witch. "So you're the ones who picked my cabbages! Just let me get my hands on you now." She seized the woman who had yanked her by the ear. The other one took to her heels and escaped.

"Now I'm going to eat you whole," said the old woman, clutching the thief.

"Wait," said the woman. "I'm expecting a baby. If you let me go, I promise that, boy or girl, when it's sixteen, I'll give it to you. Do you agree?"

"I agree!" replied the witch. "Pick all the cabbages you like and be gone. But remember your promise."

Shaking like a leaf, the woman made her way back home. "Ah, friend, you fled to safety, but I got caught, and I promised the old woman that I'll give her the son or daughter born to me when the child turns sixteen!"

Two months later, she gave birth to a baby girl. "Ah, poor daughter!" sighed her mother. "I'll nurse and raise you, and you'll be eaten alive!" And she wept.

When the girl was almost sixteen, she was out buying oil for her mother one day and met the witch. "And whose daughter are you, lass?"

"Signora Sabedda's."

"You've really grown up. . . . I'm sure you're delicious. . . ." Caressing her, she continued, "Here, take this fig home to your mother and, when you hand it to her, say, 'What about your promise?' "

The girl went to her mother and told her everything. ". . . And she told me to say, 'And what about your promise?' "

"My promise?" repeated the mother, and burst into tears.

"Why are you weeping, Mother?"

But the woman made no reply. After weeping a while, she said, "If you run into the old woman again, tell her, 'I'm still quite small.' "

But the girl was already sixteen and ashamed to say she was quite small. So the next time the witch crossed her path and asked, "What did your mother tell you?" she replied:

"I'm a big girl already . . ."

"Well, come along with your grandmother who has so many beautiful presents for you," said the witch and seized the girl.

She took her home and locked her up in the chicken coop and stuffed her full of food to fatten her up. After a short space of time, she decided to see if the girl was fat, and said, "Let me have a look at your little finger."

The girl picked up a mouse that had his nest in the chicken coop and showed the witch the mouse's tail instead of her finger.

"My goodness, you're thin, still too thin, my little one. Keep on eating."

But a little later, the temptation to gobble her up was just too much, so the witch led the girl out of the chicken coop. "My, you're the picture of health! Let's heat up the stove now for me to bake bread."

They made up the bread. The girl heated the oven, swept it out, and got it all ready.

"Now put the bread in," said the witch.

"Granny, I don't know how to put the bread in. I can do everything but that."

"I'll show you how. Slide the bread over here to me."

The girl passed the bread, and the witch put it in the oven.

"Now pick up the large slab that closes the oven."

"How do I lift the slab, Granny?"

"I'll do it myself!" said the witch.

When she bent over, the girl grabbed her by the legs and shoved her inside the oven. Then she picked up the large slab and closed the oven, with the witch inside. She ran home immediately to tell her mother, and the cabbage patch was now all theirs.

(Province of Caltanissetta)

◈ 182 ◈

The Mouse with the Long Tail

It is told that once there was a king who had a daughter beautiful beyond words. Marriage proposals came to her from kings and emperors everywhere, but her father refused to give her to anyone, because every night he was awakened by a voice saying, "Don't marry off your daughter! Don't marry off your daughter!"

The poor girl would look at herself in the mirror and ask, "Why can't I marry, beautiful as I am?" Nor could she stop worrying about it. One day while they were all at dinner, she asked her father, "Father, why can't I marry, beautiful as I am? Listen to me: I'm giving you two days, and if in that time you don't find someone to betroth me to, I shall kill myself."

"If you put it that way," replied the king, "here's what you have to do: dress up in your Sunday best, go to the window, and the first passer-by will be your husband. And that's that!"

The daughter obediently went to the window in her Sunday best, and what should come down the street but a small mouse with a tail a mile long that smelled to high heaven! The mouse stopped and studied the king's daughter at the window. And the instant she felt those eyes on her she drew away screaming. "Father, what have you done to me? The first passer-by to look at me turns out to be a mouse. Surely you don't expect me to marry a mouse?"

Her father stood in the center of the room with his arms crossed. "I do indeed, daughter. What I said goes. You must marry the first interested passer-by." Without delay he wrote and invited all princes and court grandees to his daughter's gala wedding banquet.

With great pomp the guests appeared and took their places at the table. They had all sat down, but the bridegroom was nowhere in sight. Then a scratching was heard on the door, and who should it be but the small mouse with the smelly tail. A lackey in livery opened the door and asked, "What do you want?"

"Announce me," said the mouse. "I'm the mouse who's come to wed the princess."

"The mouse who's come to wed the princess!" announced the butler.

"Bring him in," said the king.

The mouse scampered in, darted across the floor, climbed up the armchair next to the princess's, and sat down.

At the sight of the mouse there beside her, the poor maiden turned her head in disgust and shame. But the mouse pretended not to notice, and the more she turned away, the closer he moved to her.

The king told the story to all the guests who, in approval of the king's whims, smiled and said, "Yes, indeed, the mouse ought to be the princess's husband."

Their smiles gave way to laughter, and they proceeded to laugh right in the mouse's face. Mortified, the mouse took the king aside and said, "Listen, Majesty, either you warn these people not to make light of me, or suffer the consequences."

He scowled so intensely that the king agreed and, upon their return to the table, ordered everyone to respect the betrothed and cease laughing.

The food was brought in, but the mouse, being short and seated in the armchair, didn't reach up to the table. A cushion was placed under him, but that wasn't enough, so he went and sat in the center of the table.

"Any objections?" he asked, glancing about defiantly.

"No, no one objects," the king assured him.

But among the guests was a very fastidious lady who could hardly keep quiet at the sight of the mouse poking his mouth into her food and dragging that long, smelly tail over her neighbors' plates. Once the mouse had finished eating her food and turned to that of the other guests, she blurted out, "How filthy! Who ever saw anything so disgusting! I can't believe my eyes when I see such things at the king's table!"

Whiskers bristling, the mouse leveled his muzzle at her, then leaped furiously up and down the table lashing his tail, flying in the guests' faces, and snapping their beards and wigs: everything his tail hit disappeared immediately—soup tureens, fruit bowls, plates, cutlery and then, one by one, all the guests; the table also disappeared, along with the palace, and all that remained was one vast deserted plain.

Finding herself along and abandoned in the middle of this wasteland, the princess started crying and saying:

"Alas, my mouse!
My loathing has changed to longing!"

Repeating those words, she set out on foot, with no idea where she was going.

She met a hermit, who asked, "What are you doing out in these wilds, my good maiden? Heaven help you if you meet a lion or an ogress!"

"Don't speak to me of such things," said the princess. "All I want is to find my mouse. My loathing has changed to longing."

"I don't know what to tell you, my girl," said the hermit. "Keep on going until you meet a hermit older than I am who might be able to advise you."

She continued on her way, constantly repeating, "Alas, my mouse . . ." until she met the other hermit, who said, "What you must do is dig a hole in the ground, squeeze into it, then see what happens."

The poor girl removed the hairpin from her hair, having nothing else to dig with, and dug and dug until she made a hole in the ground the size of herself. Then she squeezed into it and came out in a dark and spacious cave. "Whatever this leads to," she said, and started walking. The cave was full of cobwebs that stuck to her face, and the more of them she brushed off, the more she then found on her. After a day's walk she heard rushing water and found herself on the edge of a large fishpond. She put one foot in the water, but the fishpond was deep. She could not go forward, nor could she turn back, for the hole had closed behind her. "Alas, my mouse! Alas, my mouse!" she repeated. At that, water began rising all around her. There was no escape, so she plunged into the fishpond.

When she was underwater, she saw that she was not underwater, but in a large palace. The first room was all in crystal, the second all in velvet, and the third all in sequins. So she wandered from room to room over precious carpets and lighted by glittering lamps, constantly repeating:

"Alas, my mouse!
My loathing has changed to longing!"

She came to a sumptuously laid table and sat down and ate. Then she went into a bedroom, where she got in bed and went to sleep. Then she heard the rustle made by a mouse scampering about. She opened her eyes, but all was dark. She heard the mouse running through the room, climbing up on the bed, slipping under the covers, and all of a sudden he was stroking her face, emitting little squeaks as he did so. She dared not say anything, and remained huddled up in a corner of the bed trembling.

The next morning she rose and wandered through the palace again, but still saw no one. That night the table was laid as before, so she ate and went to bed. Once more she heard the mouse running through the room and coming almost up to her face, but she didn't dare say a word.

The third night when she heard the rustle, she took heart and said:

> "Alas, my mouse!
> My loathing has changed to longing!"

"Light the lamp," said a voice.

The princess lit a candle, but instead of the mouse she saw a handsome youth.

"I am the mouse with the smelly tail," said the young man. "To free me from the spell that transformed me, I had to meet a beautiful maiden who would fall in love with me and suffer all that you have suffered."

Imagine the joy of the princess. The couple left the cave immediately and got married.

> They lived happily ever after,
> While here we sit picking our teeth.

(Caltanissetta)

❖ 183 ❖

The Two Cousins

Once, it is told, there were two sisters—one a marquise, the other fallen into straitened circumstances. The marquise had an ugly daughter, the other had three daughters who worked with their hands for a living. One day, having no money to pay the rent, they were all put out in the street. A footman of the marquise happened by and told her about it, pleading until she agreed to lodge the homeless family in a loft over the front door. In the evening the girls sat and worked by lantern light, so as to save the oil in the lamp.

But even that struck their tyrannical marquise aunt as wasteful, so she had the lantern extinguished. The girls then spun by moonlight. One evening the youngest sister decided to stay up and spin until the moon set. As the moon descended, she followed it. Thus moving along and spinning, she was caught in a storm and took shelter in an old monastery.

In the monastery she found twelve monks. "What are you doing here, young lady?" they asked, and she told them.

The oldest monk said, "May you become lovelier than ever!"

The second one added, "When you comb your hair, may pearls and diamonds come pouring out!"

"While you're washing your hands," said the third, "may fish and eels emerge from them."

The fourth monk spoke. "When you speak, may roses and jasmine issue from your mouth!"

"May your cheeks," commanded the fifth monk, "become two lady apples!"

The sixth said, "When you work, may you be done the minute you begin!"

They showed her the way back, instructing her to look behind her when she was halfway there. She looked behind her and became radiant like a star. She got home, where the first thing she did was fill a wash-basin and plunge her hands into it. Out came a pair of eels that wriggled as though they had just been caught. Her mother and sisters, full of astonishment, made her tell everything. They combed her hair, collected the pearls that fell, and carried them to the marquise aunt.

The marquise immediately asked for all the details and decided to send out her own daughter, who stood sorely in need of beauty. She made her wait on the balcony all evening long and, when the moon started down, told her to follow it.

The girl found the monastery of the twelve monks, who recognized her right away as the marquise's daughter. The oldest monk said, "May you grow still uglier!"

The second monk took up from there. "When you comb your hair, may countless serpents crawl from it!"

"When you wash," said the third, "may countless green lizards issue from you!"

"When you speak," added the fourth, "may a world of filth squirt from you!"

With that, they dismissed her.

The marquise was anxiously awaiting her, but when the girl returned uglier than ever, her mother nearly died from the shock. She asked her what had happened, and almost died from the stench that poured from her mouth when she spoke.

Meanwhile, the pretty little cousin was sitting before the door when a king came by. He saw her, fell in love, and asked for her hand in marriage. The marquise aunt consented. The girl left for the king's country, accompanied by the marquise aunt as her most important relative. After going a certain distance, the king rode on ahead to prepare for her arrival at the palace. No sooner was he out of sight than the marquise seized the bride, tore out her eyes, thrust her into a cave, and put her own daughter in the carriage.

When the king saw the ugly cousin step from the carriage dressed as a bride, he took fright. "What is the meaning of this?" he asked scarcely above a whisper. The girl opened her mouth to answer, and her breath nearly knocked him over. The marquise told a tale about sorcery being worked on them along the way, but the king didn't believe a word of it and sent them both to prison.

The poor blind girl there in the cave began calling for help, and a little old man passing by heard her. Seeing how badly off she was, he carried her to his house, which then filled up with pearls, diamonds, roses, eels, and jasmine. He filled two baskets with all those things and went and stood under the king's windows.

"Tell him," the girl had advised, "that you're selling them in exchange for eyes."

The marquise called him immediately, gave him one of her niece's eyes, and took all the beautiful things, with the intention of telling the king her daughter produced them. The old man carried the eye back to the girl, who put it in place once more.

The next day he returned to the palace with two more baskets. The marquise, who was anxious to convince the king that her daughter continuously produced eels and jasmine, immediately paid with the other eye. But the king was not to be taken in, for every time he went up to the girl, her breath was still the same.

Now that she had her sight back, the pretty cousin could embroider. She embroidered a large square of material with her own portrait, displaying it for sale on the boulevard on which the king's palace stood. The king came by, saw the portrait, shuddered, and asked the old man who had embroidered it. The old man revealed everything. The king had the maiden brought to the palace, boiled the marquise and her daughter in oil, and lived happily from then on with his little queen.

(Province of Ragusa)

184

The Two Muleteers

Once, so the tale goes, there were two muleteers who were friends. One of the men put all his trust in God, the other put all his in the Devil. One day as they traveled along together, one of them said to the other, "Friend, it is the Devil who helps us."

"No," replied the other, "whoever trusts in God is helped by God."

They argued back and forth, and finally the man with faith in the Devil said, "Friend, let's bet a mule."

At that moment a knight dressed in black rode by (it was the Devil in disguise), and they asked him which one of them was right. "*You* are right," replied the knight, "it is the Devil who helps you."

"You see?" said the man, and took the mule. But his friend wasn't convinced, so they bet again, submitting this time to the judgment of a knight dressed in white (still the Devil, in a new disguise). And so, staking one mule after the other and constantly running into the Devil in disguise, the man with faith in God lost all his mules. "In spite of everything, I still believe *I* am right," he said. "I would even bet my eyes on it."

"Very well, let's make still another bet," said his friend. "If you win, you'll get back all your mules. If I win, you'll give me your eyes."

They met a knight dressed in green and asked him who was right. "That's easy," answered the knight. "The one who helps you is the Devil." And he spurred his horse onward.

So the man who trusted in the Devil took out the eyes of the one who trusted in God and left him blind and disheartened in the middle of the country.

The poor man began groping his way about and finally found a cave, into which he slipped for the night. It was full of bushes, and the muleteer had just hidden in them when he heard a crowd of people enter. All the Devils in the world were meeting in that cave, and the head Devil questioned them one by one on what they had accomplished. One of the Devils told of disguising himself as a knight and causing a certain poor soul to gamble away everything he owned, down to his very eyes.

"Fine," said the head Devil. "His sight will never return unless he places in the sockets two blades of this grass growing here at the mouth of the grotto."

"Ha, ha, ha!" laughed the Devils. "Can you picture him discovering the secret of that grass?"

The poor muleteer hiding there and trembling was overjoyed, but he held his breath and waited anxiously for the Devils to depart so he could go pick that grass and recover his sight.

But the Devils went on telling stories. "I," said another, "caused a fishbone to stick in the throat of the king of Russia's daughter, and no doctor can get it out, despite the king's promise to heap riches on the man who succeeds. Nobody can do a thing, unaware that all that's needed are three drops of juice from those sour grapes on the vine growing on the girl's balcony."

"Speak quietly," advised the head Devil; "the stones have eyes, and the bushes have ears."

Just before dawn the Devils departed, and the muleteer was able to come out of the bushes and grope his way to the grass that restored lost eyesight. He found it and saw again as well as ever. Without delay, he set out for Russia.

In Russia all the doctors were assembled in conference in the princess's bedchamber. Seeing the muleteer arrive shabby and dusty from all the ground he had covered, they burst out laughing. But the king, who was present, said, "So many have tried, let's give him a chance too." And he had the room cleared, in order to leave the muleteer alone with the princess. The man then went to the balcony, picked three sour grapes, and squeezed them one by one into the princess's mouth. The princess began to stir again and was well in a flash.

Just imagine her father's joy. No reward was too great for the muleteer. The king loaded him down with gold, which the king's men helped him home with. Having given him up for dead, the muleteer's wife took him for a ghost when he walked in.

Her husband told her everything and showed her his treasure. They constructed a large palace. His muleteer friend came by and, seeing him with eyes as good as ever and rolling in wealth, asked, "Dear friend, how did you do it?"

"Didn't I tell you that the man who puts his trust in God is helped by God?" he said, and told him his story.

"Tonight," said the friend to himself, "I shall go to that cave and see if I can't get rich too."

The Devils assembled, and the same Devil as before told of the muleteer listening to their secrets, regaining his eyesight, and curing the king of Russia's daughter.

"Didn't I tell you," said the head Devil, "that the stones have eyes and the bushes have ears? Quick, let's set fire to all this brushwood here."

They burned up the underbrush, and the man hiding in it was reduced to ashes. He thus learned how it is that the Devil helps you.

(Province of Ragusa)

185

Giovannuzza the Fox

There was once a poor man who had an only son, and the boy was as simple-minded and ignorant as they come. When his father was about to die, he said to the youth, whose name was Joseph, "Son, I am dying, and I have nothing to leave you but this cottage and the pear tree beside it."

The father died, and Joseph lived on in the cottage alone, selling the pears from the tree to provide for himself. But once the season for pears was over, it looked as though he would starve to death, since he was incapable of earning his bread any other way. Strangely enough, the season for pears ended, but not the pears. When they'd all been picked, others came out in their place, even in the middle of winter; it was a charmed pear tree that bore fruit all year long, and so the youth was able to go on providing for himself.

One morning Joseph went out as usual to pick the ripe pears and discovered they'd already been picked by somebody else. "How will I manage now?" he wondered. "If people steal my pears, I'm done for. Tonight I shall stay up and keep watch." When it grew dark he stationed himself under the pear tree with his shotgun, but soon fell asleep; he woke up to find that all the ripe pears had been picked. The next night he resumed his watch, but fell asleep right in the middle of it, and the pears were again stolen. The third night, in addition to the shotgun, he carried along a shepherd's pipe and proceeded to play it under the pear tree. Then he stopped playing, and Giovannuzza the fox, who was stealing the pears, thinking Joseph had fallen asleep, came running out and climbed the tree.

Joseph aimed his gun at her, and the fox spoke. "Don't shoot, Joseph. If you give me a basket of pears, I will see to it that you prosper."

"But, Giovannuzza, if I let you have a basketful, what will I then eat myself?"

"Don't worry, just do as I say, and you will prosper for sure."

So the youth gave the fox a basket of his finest pears, which she then carried to the king.

"Sacred Crown," she said, "my master sends you this basket of pears and begs your gracious acceptance of them."

"Pears at this time of year?" exclaimed the king. "It will be the first time I've ever eaten any in this season! Who is your master?"

"Count Peartree," replied Giovannuzza.

"But how does he manage to have pears in this season?" asked the king.

"Oh, he has everything," replied the fox. "He's the richest man in existence."

"Richer than I am?" asked the king.

"Yes, even richer than you, Sacred Crown."

The king was thoughtful. "What could I give him in return?" he asked.

"Don't bother, Sacred Crown," said Giovannuzza. "Don't give it a thought; he's so rich that whatever present you made him would look paltry."

"Well, in that case," said the king, very embarrassed, "tell Count Peartree I thank him for his wonderful pears."

When he saw the fox back, Joseph exclaimed, "But Giovannuzza, you've brought me nothing in return for the pears, and here I am starving to death!"

"Put your mind at rest," replied the fox. "Leave everything to me. Again I tell you that you will prosper!"

A few days later, Giovannuzza said, "You must let me have another basket of pears."

"But, sister, what will *I* eat if you carry off all my pears?"

"Put your mind at rest and leave everything to me."

She took the basket to the king and said, "Sacred Crown, since you graciously accepted the first basket of pears, my master, Count Peartree, takes the liberty of offering you a second basket."

"I can't believe it!" exclaimed the king. "Pears freshly picked at this time of year!"

"That's nothing," replied the fox. "My master takes no account of the pears, he has so much else far more precious."

"But how can I repay his kindness?"

"Concerning that," said Giovannuzza, "he instructed me to convey his request to you for one thing in particular."

"Which is? If Count Peartree is so rich, I can't imagine what I could do that would be fitting."

"Your daughter's hand in marriage," said the fox.

The king opened his eyes wide. "But even that is too great an honor for me, since he is so much richer than I am."

"Sacred Crown, if it doesn't bother him, why should it worry you? Count Peartree truly wants your daughter, and it makes no difference to him whether the dowry is large or not so large, since no matter how big it is, beside all his wealth it will only be a drop in the bucket."

"Very well, in that case, please ask him to come and dine here."

So Giovannuzza the fox went back to Joseph and said, "I told the king that you are Count Peartree and that you wish to marry his daughter."

"Sister, look at what you've done! When the king sees me, he will have me beheaded!"

"Leave everything to me, and don't worry," replied the fox. She went to a tailor and said, "My master, Count Peartree, wants the finest outfit you have in stock. I will pay you, in cash, another time."

The tailor gave her clothing fit for a great lord, and the fox then visited a horse dealer. "Will you sell me, for Count Peartree, the finest horse in the lot? We won't look at prices, payment will be made on the morrow."

Dressed as a great lord and seated in the saddle of a magnificent horse, Joseph rode to the palace, with the fox running ahead of him. "Giovannuzza," he cried, "when the king speaks to me, what shall I reply? I'm too scared to say a word in front of important people."

"Let me do the talking and don't worry about a thing. All you need say is, 'Good day' and 'Sacred Crown,' and I'll fill in the rest."

They arrived at the palace, where the king hastened up to Count Peartree, greeting him with full honors. "Sacred Crown," said Joseph.

The king escorted him to the table, where his beautiful daughter was already seated. "Good day," said Count Peartree.

They sat down and began talking, but Count Peartree didn't open his mouth. "Sister Giovannuzza," whispered the king to the fox, "has the cat got your master's tongue?"

"Oh, you know, Sacred Crown, when a man has so much land and wealth to think about, he worries about it all the time."

So, throughout the visit, the king was careful not to disturb Count Peartree's thoughts.

The next morning, Giovannuzza said to Joseph, "Give me one more basket of pears to take to the king."

"Do as you wish, sister," replied the youth, "but it will be my downfall, you will see."

"Put your mind at rest!" exclaimed the fox. "I assure you that you will prosper."

He therefore picked the pears, which the fox carried to the king, saying, "My master, Count Peartree, sends you this basket of pears, and would like an answer to his request."

"Tell the count that the wedding can take place whenever he likes," replied the king. Overjoyed, the fox returned to Joseph with the answer.

"But, sister Giovannuzza, where will I take this bride to live? I can hardly bring her here to this hovel!"

"Leave that up to me. What are you worried about? Haven't I done all right so far?"

Thus a grand wedding was performed, and Count Peartree took the king's beautiful daughter to be his wife.

A few days later Giovannuzza the fox announced: "My master intends to carry the bride to his palace."

"Fine," said the king. "I will go along with them, so I can finally see all of Count Peartree's possessions."

Everyone mounted horses, and the king was accompanied by a large body of knights. As they rode toward the plain, Giovannuzza said, "I shall run ahead and order preparations made for your arrival." As she raced onward, she met a flock of thousands upon thousands of sheep, and asked the shepherds, "Whose sheep are these?"

"Papa Ogre's," they told her.

"Keep your voice down," whispered the fox. "Do you see that long cavalcade approaching? That's the king who's declared war on Papa Ogre. Tell him the sheep are Papa Ogre's, and the knights will slay you."

"What are we to say, then?"

"I don't know! Try, 'They belong to Count Peartree!' "

When the king came up to the flock, he asked, "Who owns this superb flock of sheep?"

"Count Peartree!" cried the shepherds.

"My heavens, the man really must be rich!" exclaimed the king, overjoyed.

A bit farther on, the fox met a herd of thousands upon thousands of pigs. "Whose pigs are these?" she asked the swineherds.

"Papa Ogre's."

"Shhhhhhhh, see all those soldiers coming down the road on horseback? Tell them they are Papa Ogre's, and they'll kill you. You must say they are Count Peartree's."

When the king approached and asked the swineherds whose pigs those were, they told him, "Count Peartree's," and the king was quite glad to have a son-in-law so rich.

Next the king's party met a vast herd of horses. "Whose horses are these?" asked the king. "Count Peartree's." Then they saw a drove of cattle. "Whose cattle?" "Count Peartree's." And the king felt ever happier over the fine match his daughter had made.

Finally Giovannuzza reached the palace where Papa Ogre lived all alone with his wife, Mamma Ogress. Rushing inside, she exclaimed, "Oh, you poor things, if you only knew what a horrible destiny is in store for you!"

"What has happened?" asked Papa Ogre, scared to death.

"See that cloud of dust approaching? It"s a regiment of cavalry dispatched by the king to kill you!"

"Sister fox, sister fox, help us!" whimpered the couple.

"Know what I advise?" said Giovannuzza. "Go hide in the stove. I'll give the signal when they've all gone."

Papa Ogre and Mama Ogress obeyed. They crawled into the stove and, once inside, pleaded with Giovannuzza. "Giovannuzza dear, close up the mouth of the stove with tree branches, so they won't see us." That was just what the fox had in mind, and she completely stopped up the opening with branches.

Then she went and stood on the doorstep, and when the king arrived, she curtseyed and said, "Sacred Crown, please deign to dismount; this is the palace of Count Peartree."

The king and the newlyweds dismounted, climbed the grand staircase, and beheld such wealth and magnificence as to leave the king speechless and pensive. "Not even my palace," he said to himself, "is half so beautiful." And Joseph, poor man, stood gaping beside him.

"Why," asked the king, "are there no servants around?"

In a flash, the fox answered, "They were all dismissed, since my master wanted to make no arrangements whatever before first knowing the wishes of his beautiful new wife. Now she can command what best suits her."

When they had scrutinized everything, the king returned to his own palace, while Count Peartree remained behind with the king's daughter in Papa Ogre's palace.

Meanwhile Papa Ogre and Mamma Ogress were still closed up in the stove. At night the fox went up to the stove and whispered, "Papa Ogre, Mamma Ogress, are you still there?"

"Yes," they answered in a weak voice.

"And there you will remain," replied the fox. She lit the branches, made a big fire, and Papa Ogre and Mamma Ogress burned up in the stove.

"Now you are rich and happy," said Giovannuzza to Count Peartree and his wife, "and must promise me one thing: when I die, you must lay me out in a beautiful coffin and bury me with full honors."

"Oh, sister Giovannuzza," said the king's daughter, who had grown quite fond of the fox, "why do you talk about death?"

A little later, Giovannuzza decided to put the couple to the test. She played dead. When the king's daughter saw her stretched out stiff, she exclaimed, "Oh, Giovannuzza is dead! Our poor dear friend! We must have a very beautiful coffin built at once for her."

"A coffin for an animal?" said Count Peartree. "We'll just pitch her out the window!" And he grabbed her by the tail.

At that, the fox jumped up and cried, "Penniless man! Faithless, ungrateful wretch! Have you forgotten everything? Forgotten that your prosperity is due to me? You'd still be living on charity, if it hadn't been for me! You stingy thing! Ungrateful, faithless wretch!"

"Fox," begged Count Peartree all flustered, "forgive me, dear friend, please forgive me. I meant no harm, the words just slipped out, I spoke without thinking . . ."

"This is the last you'll see of me"—and she made for the door.

"Forgive me, Giovannuzza, please, remain with us . . ." But the fox ran off down the road, disappeared around the bend, and was never seen again.

(Catania)

◈ 186 ◈

The Child that Fed the Crucifix

One day a God-fearing farmer found a little baby boy abandoned in his field. "Poor innocent thing," he said. "What inhuman soul could have left you here to your fate? Don't be afraid. I'll carry you home and raise you myself."

From that day on, things began going marvelously well for him. The trees were laden with fruit, the wheat shot up and yielded grain galore, the vines provided the richest harvest ever. In a word, the farmer had never before known such prosperity.

The child grew, and the bigger he got, the wiser he became. But living way off in the country like that, he had never seen a church or a holy picture, nor did he know of our Lord and the saints. One day the farmer had to go to Catania. "Will you go with me?" he asked the child.

"As you wish, sir," replied the child, and set out with the farmer for the city.

When they came to the cathedral, the farmer said, "I must now go about my business. Go into church and wait for me there until I finish."

The child entered the cathedral and, seeing the gold-embroidered vestments, the rich altar cloths, the flowers, the candles, he was full of amazement, never having beheld their likes before.

He made his way slowly to the high altar and saw the crucifix. He knelt on the steps and addressed the crucifix: "Dear Friend, why have they nailed you to this cross? Have you committed some crime?"

And the head on the crucifix nodded Yes.

"Oh, poor friend, you mustn't do that any more, seeing how you now have to suffer!"

And the Lord again nodded Yes.

He went on like that for a good while talking to the crucifix, until all the services were over. The sacristan was ready to lock up, but he saw that little country boy kneeling at the high altar. "Hey, there! Get up, it's closing time!"

"No," replied the child, "I'm staying. Otherwise that poor soul will be all alone. First you nailed him to the cross, and now you're going off and leaving him to his fate. Isn't it true, friend, that it will make you happy for me to remain here with you?"

And the Lord nodded Yes.

Hearing the child talk to Jesus, and seeing Jesus respond, the sacristan was terrified and ran and told the pastor everything. The pastor replied, "He is surely a holy person. Leave him in church and take him a dish of macaroni and a little wine."

When the sacristan took him the macaroni and the wine, the child said, "Put everything down right there, and I'll come and eat at once."

Then he turned to the crucifix and said, "Friend, you must be hungry. Goodness knows how long ago you last ate anything. Have a little macaroni." He took the dish, climbed up on the altar, and began tendering the Lord forkfuls of macaroni. And the Lord opened his mouth and ate. Then the child said, "Friend, aren't you thirsty? Have a little of my wine." He held a glass of wine to the Lord's lips, and the Lord opened his mouth and drank.

But once he had divided his food and wine with the Lord, he fell down dead, while his soul flew off to heaven and gave praise to God. But the pastor had hidden behind the altar and witnessed everything. Thus he saw that after sharing his dinner with the Lord, the child crossed his arms, while his soul separated from his body and flew heavenward in song. The pastor rushed over to the child's body lying before the altar: he was dead. Immediately the pastor had it announced throughout the city that a saint lay in the cathedral, and he had him placed in a gold coffin. Everyone rushed in and knelt around the coffin. Even the farmer came, gazed at the little body in the gold coffin, and recognized his son. "Lord," he said, "you gave him to me, and you took him from me and made him a saint!" Then he returned home, and everything he set about was a success; thus, he became rich.

But of the money he earned, he gave amply to the poor. He lived a devout life and, when he died, he had a place in Paradise. May that be the lot of us all!

(Catania)

❖ 187 ❖

Steward Truth

Once there was a king who had a nanny goat, a lamb, a ram, and a steer. Being deeply attached to these animals, he wanted them in the care of only a reliable person. Now the most reliable soul he knew was a farmer everyone called Steward Truth, since he had never in his whole life told a lie. The king sent for him and entrusted the animals to him. "Every Saturday," said the king, "you will come to the palace and report to me on each one of them." So every Saturday Steward Truth came down the mountain, went before the king, removed his cap, and had this conversation:

"Good day to you, Royal Majesty!"

"Good day to you, Steward Verity! How is Nanny?"

"White and rascally!"

"How is Lamb?"

"White and lovely!"

"How is Ram?"

"Fat and lazy!"

"How is Steer?"

"Quite fat, never fear!"

The king took him at his word, and after this conversation, Steward Truth went back up the mountain.

But among the king's ministers was one who envied the esteem in which the king held the steward, and one day he said to the king, "Is it possible that old steward really can't tell a lie? Let's bet that next Saturday he tells one."

"I'll stake my head on it that he won't!" exclaimed the king.

So they made the bet, each one staking his life. Time passed, Saturday was now only three days off, and the more the minister thought about it, the less confident he was of finding a way to make the steward tell a lie.

He pondered morning, evening, and night, and his wife seeing him so worried, asked, "What's the matter? Why this ill-humor?"

"Let me alone," he replied. "To tell you about it would really fix things!"

But she begged so sweetly that she finally wormed it out of him. "Oh, is that it? *I'll* do something about it," she said.

The next morning the minister's wife dressed in her finest outfit, her richest gems, and a star of diamonds on her brow. Then she climbed into her carriage and drove to the mountain where Steward Truth pastured nanny goat, lamb, ram, and steer. When she arrived, she got out of the carriage and began to look around her. At the sight of a beautiful lady the likes of whom he had never seen before, the poor farmer was thrown into a sea of confusion. He bustled about and did his best to give her a fitting welcome.

"Dear steward," she began, "would you do me a favor?"

"Noble lady," he replied, "only ask me. Whatever you wish, I will do."

"As you can see, I'm expecting a baby, and I long for roasted beef liver. I'll simply die if you don't give it to me."

"Noble lady," said the steward, "ask me for anything you like except that one thing, which I cannot give you. The steer belongs to the king and is his dearest animal."

"Poor me!" groaned the woman. "I really will die if you don't satisfy this desire of mine. Steward, dear steward, please! The king will know nothing about it, and you can tell him the steer fell down the mountain!"

"No, I can't say that," said the steward, "nor can I give you the liver."

Then the woman started moaning and groaning, and threw herself down and truly looked as though she were about to die. She was so beautiful that the farmer's heart melted; he killed the steer, roasted the liver, and served it to her. Overjoyed, the woman ate it in two bites, bid the steward a hasty farewell, and drove off in her carriage.

The poor steward remained there by himself, weighed down with shame. "Now what will I tell the king, come Saturday? When he asks me, 'And how is Steer?' I'll no longer be able to say, 'Quite fat, never fear!'" He picked up his staff, planted it in the ground, and hung his cape on it. He moved back, then took a few steps forward, bowed and, facing the staff, began speaking:

"Good day to you, Royal Majesty!"

And then, speaking now like the king, now in his own voice, he continued:

"Good day to you, Steward Variety! How is Nanny?"

"White and rascally!"

"How is Lamb?"

"White and lovely!"

"How is Ram?"

"White and lazy!"

"How is Steer?"

There he remained speechless. He began stuttering to the staff: "Royal Majesty . . . I took him to pasture . . . and he went plunging off a mountain peak . . . and broke every bone in his body . . . and then he died . . ." And he became all flustered.

"No," he reflected, "I won't tell the king that, that is a lie!" He planted his staff in another place, draped his cape back on it, repeated the act, the bow, the conversation, but at the question, "How is Steer?" he again faltered. "Majesty, he was stolen . . . by thieves . . ."

He went to bed, but couldn't sleep a wink. In the morning—it was Saturday—he set out, head bowed, still wondering what to tell the king. Every time he came to a tree, he bowed and said, "Good day to you, Royal Majesty!" He recommenced the dialogue, but could never get through it. At last, after passing one tree after another, he thought of an answer. "That's just the answer!" he exclaimed to himself. Again in high spirits, he saluted every tree he came to and repeated the whole conversation, down to the last line; the more he said the answer, the better he liked it.

At the palace the king was waiting with his whole court to see who would win the bet. Steward Truth removed his cap, and the conversation began:

"Good day to you, Royal Majesty!"

"Good day to you, Steward Verity! How is Nanny?"

"White and rascally!"

"How is Lamb?"

"White and lovely!"

"How is Ram?"

"White and lazy!"

"How is Steer?"

"Royal Majesty,

Here the verity:

There came a lady from high society

So plump and lovely

I fell in love with her beauty
And smote Steer dead out of love for the lady!"

With that said, Steward Truth bowed his head and added, "Now if you wish to send me to my death, master, send me. But I have spoken the truth."

The king, although saddened by the death of the steer, rejoiced over winning the bet and presented Steward Truth with a sack of gold coins. The whole court burst into applause, excepting the minister whose envy cost him his life.

(Catania)

◈ 188 ◈

The Foppish King

There was once a king who thought he was handsome. He had a mirror, to which he was constantly saying:

> "Mirror, mirror so gay and fine,
> I beseech thee, give me a sign
> If anyone has looks surpassing mine."

His wife paid no attention at first; then, unable to stand the thought of her husband's vanity any longer, she said to him the next time he repeated those lines:

> "Hush, King, and hear my view:
> Perchance someone's more handsome than you."

The king sprang to his feet and said, "I'm giving you three days: either you tell me who's handsomer than I, or your head will roll."

The queen immediately regretted her remark, but it was too late, and she could already feel the executioner's ax on her neck. Worried to death, she retired to her chambers and wept for two days straight. On the third day she opened the window to enjoy the sun once more while there was still time. In the street was an old woman who looked as though she were waiting for her. "Majesty, give me alms!" she said.

"Let me alone, good old soul," said the queen. "I have enough troubles of my own."

The old woman, lowering her voice, said, "I know all about it, and I can help you."

The queen looked at her. "Come in," she said.

The old woman entered the palace. "What do you know?" asked the queen.

"I know what the king told you."

"Is there a way out for me?"

"Indeed there is."

"Speak, I'll give you whatever you ask."

"I'm not asking for anything. Listen to me. At noon go to the table with the king. Then ask him to do you a favor. 'Spare your life?' he will ask. 'No,' you will say. 'In that case,' he will answer, 'so be it.' Then you will say, 'Handsomer than you is the son of the emperor of France, hidden under seven veils.'"

The queen followed the old woman's advice to the letter, and the dialogue unfolded exactly as predicted.

The king didn't bat an eyelash. "If the son of the emperor of France is truly handsomer than I," he said to his wife, "you will then deal with me however you choose." Three days later the king set out for France, accompanied by a few soldiers. He went before the emperor and requested to see his son.

"My son is sleeping just now," said the emperor in a hushed voice, "but come with me."

He took him into his son's room and lifted the first veil. They saw a glow filtering through. He lifted the second veil, and the glow intensified. He lifted the third veil and the fourth, and the light grew ever brighter, flooding the room. Now the last veils were removed, and through ever waxing radiance the prince was seen on his throne, scepter in hand and sword at his side, and so dazzling was he that the king fell down in a swoon. Vinegar and smelling salts were held to his nose, and the empress had him carried into her rooms. The king revived, and remained there three days to recuperate.

The prince said to his father, the emperor, "Papa, before that king leaves, I want to talk to him."

The king was brought in, and this time he was stronger and didn't faint. They got to talking, in the course of which the prince asked: "Would you like to see me at your house?"

"If only that were possible!" replied the king.

"If you wish to see me again," said the prince, "take these three gold balls and drop them into a golden basin filled with milk clean and pure. I will then appear to you just as you see me here."

Upon his return the king said to his wife, "I'm back. Now you may deal with me as you choose."

"God bless you!" replied his wife.

The king told her everything and showed her the three gold balls. But the grief over his lost illusion and the impression made on him by the prince's radiance were too painful to endure, and he died a few days later.

After the king was buried, the queen called her most faithful chambermaid and said, "Bring me three gallons of pure milk and leave me by myself." She filled the basin with the milk, threw in the three gold balls; at once surfaced the sword, then the scepter, then the prince himself. They talked together, after which the prince dived back into the milk and disappeared.

The next morning the queen sent for more fresh milk and returned to see the prince, repeating the procedure day after day, until the chambermaid finally got tired and said to herself, "There's witchcraft going on here or some ugly mischief."

So the next day when the queen sent her after milk, she broke a crystal glass and ground it up into fine particles in the mortar, then threw this glass dust into the milk. When the queen dropped in the three gold balls, the scepter rose, but covered with blood. Then the prince appeared, dripping blood from head to foot, for coming through the milk, he had to pass right through those tiny splinters, which cut all his veins. "Ah!" he said, "you have betrayed me!"

"No!" replied the queen. "It is not my fault, forgive me!" But he had already disappeared into the golden basin.

At the royal palace of France, the emperor's son was found covered with wounds from head to foot, and the court doctors were unable to cure him. His father issued a proclamation that any doctor or surgeon who healed his son could name his reward. In the meantime the city dressed in mourning, and the bells tolled constantly.

From the moment she had seen the prince wounded, the queen knew no more peace, so she left for France dressed as a man in shepherd's clothing. The first night she was overtaken by darkness in a forest. She crouched under a tree to say her prayers. Nearby was a circular clearing where, at midnight, all the Devils of Hell converged and sat in council, with their chief in the center; all the Devils in turn told him of their mischief, and finally came Lame Devil's turn to speak.

"What about you, clumsy thing?" everybody said to him. "You always bungle everything."

"Not this time I didn't! After all the years I've worked, I finally did something stupendous!" And he told all about the king and queen and the prince, and what he had caused the chambermaid to do. "But the prince has only three more days to live, and then we'll bring him here with us."

The chief Devil spoke. "But is there no cure for this prince?"

"There is," replied Lame Devil, "but I'm not revealing it."

"You can tell us."

"No. What if someone overheard?"

"Silly! Could anyone come snooping around here at this hour without dying of fright?"

"Well, listen. You'd have to go into the forest of the monastery where the glass herb grows, fill two knapsacks with it, grind it in a mortar, drain off the juice into a glass, and douse him with it from head to foot. He would then be as sound again as ever."

Hearing that, the queen couldn't wait for dawn to break in order to go look for the monastery and the glass herb. After walking a great distance she reached the monastery and called the monks who proceeded to exorcise her without opening the door.

"Don't exorcise, I am baptized."

Hearing that, they opened the door, and she asked them to please give her two knapsacks of the glass herb, and the monks went and picked it for her. The next day she reached the prince's city, where all the houses were draped in mourning. Dressed as a shepherd, she approached the guard, who refused to let her in. The emperor looked out about that time and asked the shepherd what he wanted.

"Dismiss all the surgeons, Majesty, leave me alone with the prince, and tomorrow he will be well."

The emperor, who was by now at his wit's end, agreed and left the shepherd alone with his son, instructing the servants to procure all he asked for. The shepherd called for a mortar and crushed the herb. He requested a glass, into which he drained the juice. He poured the juice over the prince's wounds which, one by one, closed and disappeared.

He sent for the emperor and showed him his son, again well and more handsome than ever. The emperor wanted to load him down with treasure, but the shepherd refused everything and insisted on leaving. "Take this ring, at least, as a souvenir," said the prince, handing him a ring.

The queen went home as fast as she could and, the minute she arrived, went and fetched a little milk clean and pure, obtaining it herself rather than sending the chambermaid for it. She poured it into the basin and dropped in the three gold balls. The prince appeared, but brandished his scepter at her.

"No, I did not betray you," cried the queen, throwing herself at his feet. "On the contrary, I saved your life, and here is the ring you gave me!"

The king lingered in doubt, so she told him the whole tale. A deep love sprang up between them and they married with the consent of the emperor of France, while the chambermaid was condemned to death.

> Their life was happy and long;
> But we, poor we, sing another song.

(Acireale)

◈ 189 ◈

The Princess with the Horns

It is told that once there was a father of three sons, and the man had nothing to his name but a house. The house was sold with the understanding that three bricks in the center of one of the walls would still belong to the father of three sons. When he was about to die, he decided to make a will. "But what will you bequeath?" asked his neighbors. "You have no possessions." His sons didn't even want to send for the notary, but he came all the same, and the dying man dictated this will to him: "To my oldest son I leave the first brick, to the middle boy the second one, to the youngest the third."

With their father dead and gone, the three boys, scapegraces that they were, knew hunger and want. The oldest son said, "I can no longer live in this town. I shall remove the brick my father left me, and go out into the world."

When he went for the brick, the lady who now owned the house said she would pay him if he would leave it where it was and not mar her wall. "No, madam," he replied, "my father left me that brick, and I'm taking it." He dislodged the brick and found a tiny purse which, together with the brick, he carried off with him.

Along the way he grew hungry and pulled out the purse. "O purse, give me two pennies to buy some bread!" He opened the purse, and there were the two pennies.

"O purse, give me one hundred crowns!" he tried saying, and then, in the purse, he found one hundred crowns.

Whatever the sum he named, the purse provided it. In no time he was rich enough to build a palace opposite the king's. He looked out of his palace, and who should look back at him from the king's palace but the king's daughter. They began courting, and he made friends with the king, at whose palace he was always welcome. Seeing that he was so much richer than her father, the princess said, "I'll marry you only when you reveal the source of all your money."

Big fool that he was, he trusted her and showed her the purse. She feigned indifference, but drugged his wine and then replaced his purse with another one that looked just like it. When the poor boy realized this, he was obliged to sell everything he owned to make ends meet, and once more he was as poor as a church mouse, with nothing at all to his name.

Meanwhile, he got word that his middle brother was rich. He looked him up, embraced and kissed him, then asked how he had grown rich. The brother told that, running out of money, he had dislodged the brick left him and found a cloak, which he put on and became invisible to everybody around him. Starving to death, he entered a shop and made off with a loaf of bread, without anyone seeing him. Then he robbed silversmith, haberdasher, and king's messenger, until he was rolling in money.

"Since that's how it is, dear brother," said the oldest boy, "will you do me a favor and lend me your cloak for something special? Then I'll return it." Out of love for his brother, the middle boy lent him the cloak.

He put it on, left the house, and was now invisible to everyone he met. Without a minute's delay and outdoing his brother, he started stealing right and left, taking everything he came across. When his possessions were nicely replenished, he returned to the king's palace. "Where on earth," asked the princess, seeing him now richer than ever, "did all that come from? Tell me, and we'll get married at once."

Still as gullible as ever he again divulged everything, and showed her the cloak. Again she served him drugged wine, and replaced the cloak with another just like it. When he woke up, he wrapped himself in the cloak and, thinking himself invisible, roamed through the palace in search of his purse. But the guards mistook him for a thief, gave him a good thrashing, and threw him out.

Wondering what to do next, the poor boy decided to return to his birthplace and eke out a living the best he could. Upon his arrival he learned that his little brother was a millionaire living in a large palace, with hordes of servants. "I'll go to my little brother," he said to himself. "He certainly won't send me away."

The little brother, who had given him up for dead, welcomed him with open arms and told how he had grown rich. "Just listen to this: our father, as you know, left me the last brick. One day when I was desperate for money I decided to pull the brick out and sell it. Behind the brick I found a horn. As soon as I saw it I had the urge to play it and, when I blew it, out came countless soldiers, saying, 'At your orders, General!' I removed the horn from my lips, and the soldiers retreated. Realizing now what I could do, I visited towns and cities with my soldiers, fighting battles and wars and collecting all the money I could. When I had enough to last me a lifetime, I came back here and built this palace."

When the oldest brother heard all that, he begged to borrow the horn, promising to return it as soon as he had finished with it. He took the horn off to a city renowned for its wealth, played it, and watched the soldiers come pouring out. Once the plain was covered with them, he gave the order to plunder the city. The soldiers needed no coaxing, and returned loaded down with gold, silver, and treasures of all kind. So he reappeared before the princess richer than ever.

But he had fallen into the trap twice, and he fell into it once more. He told his secret, and the princess drugged his wine and exchanged horns with him. When he woke up, the king and queen turned him out quite rudely for getting drunk. Deeply mortified, he took his wealth and set out for another town.

As he made his way through a forest, out rushed twelve robbers. He began playing the horn, expecting the soldiers to come flying to his defense; but blow as he would, the robbers took everything and beat him unmercifully for being so presumptuous. They left him lying on the ground half dead but still holding the horn in his mouth and blowing it. Then he realized that it wasn't the magic horn. Believing that he had ruined himself and his brothers, he decided to jump off a cliff.

He looked for an appropriate cliff, went up to its mossy edge, and jumped. But halfway down, a fig tree was jutting out, and he remained hanging from its branches. The tree was laden with black figs. I'll at least die on a full stomach, he thought to himself, and proceeded to eat fig after fig.

He ate ten, twenty, thirty, only to discover he had sprouted a horn for every fig eaten. They grew on his head, on his face, on his nose, and he now had more branches than the tree that held him. As though his case hadn't been hopeless enough before, he now found himself so monstrous that he was more determined than ever to end his life.

He threw himself out of the fig tree and dropped through empty space, but with all those horns, he caught on another fig tree a hundred feet below. It was laden with still more figs than the other tree, but with white figs. "I couldn't possibly sprout another horn—there's no more room; doomed however you look at it, I might as well eat my fill." At that, he started on the white figs. He'd eaten scarcely three, when he realized he had three horns less. He continued eating and saw that for every white fig he ate, a horn vanished. He ate enough to make them all disappear, and ended up with smoother skin than ever.

When all of his horns were gone, he came down the white fig tree and climbed back up the cliff to the other fig tree. He picked a goodly supply of black figs, tied them up in his scarf, and went off to the city. Disguised as a farmer, he took his figs to the royal palace in a basket. Now this was fruit out of season. The guard immediately called to him and invited him in. The king bought the whole basket, and he took leave, kissing the king's knee.

At noon, the king and his family began eating figs. The fruit especially delighted the princess, who gorged herself with it. They were all too excited to look up from their plates, but when they finally did, they saw each other covered with horns. The princess was a real forest. Terrified, they sent for every surgeon in town, but none of them knew what was wrong with the royal family. The king then issued a proclamation that whoever rid them of these horns would have his every wish granted.

When the fig vendor heard the proclamation, he returned to the white fig tree and picked a heaping basket of fruit, then disguised himself as a surgeon and went before the king.

"Royal Majesty, I can save all of you and remove your horns."

Hearing that, the princess blurted out, "Majesty, you must have them taken off of me first," and the king consented.

The surgeon had himself closed up in a room with the princess and removed his disguise. "Do you recognize me, yes or no? Hear what I have to tell you: if you return the purse that produces money, the cloak that makes its wearer invisible, and the horn that turns out soldiers, I will remove all the horns. But if you refuse, I will cause as many more to grow out on you."

The princess, who couldn't bear those horns a minute longer and who knew this young man still had magic tricks at his command, put her trust in him. "If I give you everything back, you must remove the horns and then marry me," she said, and handed over purse, cloak, and horn.

He gave her as many white figs as she had horns, and thus turned her back to her former self. Then he performed the same cure on the king, queen, and everybody else with horns in the royal palace. The king gave him the princess in marriage, and they became man and wife. The cloak and the horn were returned to his brothers, while he kept the money-producing purse and remained the kings's son-in-law for life. The king died a year later, and he and his wife became king and queen.

(Acireale)

◈ 190 ◈

Giufà

I. Giufà and the Plaster Statue

There was a mother who had a son that was lazy, foolish, and full of mischief. His name was Giufà. The mother, who was poor, had a piece of cloth, and she said to Giufà, "Take this cloth out and sell it. However, don't let it go to any chatterbox, but only to somebody of few words."

Giufà took the cloth and went about the town shouting, "Cloth for sale! Cloth for sale!"

A woman stopped him and said, "Let me see it." She inspected the cloth, then asked, "How much do you want for it?"

"You talk too much," answered Giufà. "My mother doesn't want to sell it to any chatterbox"—and off he went.

He met a farmer, who asked, "How much do you want for it?"

"Ten crowns."

"No, that's too much!"

"Talk, talk, talk! You shan't have it!"

Thus, everyone who called or approached him struck him as too talkative, and he refused to sell the cloth to anyone. Going here, there, and yonder, he slipped into a courtyard, in the middle of which stood a plaster statue. Guifà said to it, "Would you like to buy the cloth?" He waited a moment, then repeated, "Do you want to buy this cloth?" Getting no answer, he exclaimed, "At last I've found someone of few words! You bet I'll sell the cloth now." And he draped it over the statue. "It costs ten crowns. Do you agree? I'll be back tomorrow, then, for the money." At that, he left the courtyard.

As soon as his mother saw him, she asked about the cloth.

"I sold it."

"Where's the money?"

"I'm going back for it tomorrow."

"But is the person trustworthy?"

"She's just the kind of woman you had in mind for the sale. Would you believe she didn't say one word to me?"

The next morning he went to collect the money. He found the statue all right, but the cloth had vanished. "Pay me for it!" said Giufà. The longer he waited for a reply, the angrier he grew. "You took the cloth, didn't you? And now you refuse to pay me? Well, I'll show you a thing or two!" He picked up a hoe and smashed the statue to smithereens. Inside the statue he found a pot of gold. He poured it into his sack and went home to his mother. "Mamma, she didn't want to pay me, so I hit her with the hoe and she gave me all of this."

Mamma, who was a shrewd woman, replied, "Hand it here, and don't breathe a word about it to anyone."

II. Giufà, the Moon, the Robbers, and the Cops

One morning Giufà went out to pick herbs, and before he made it back to town, night had fallen. As he walked along, the moon played hide and seek among the clouds. Giufà sat down on a stone and, watching it appear and disappear, he began saying, "Come out, come out!" followed by "Hide, hide!" Over and over he repeated, "Now come out! "Now hide!"

There by the wayside two thieves happened to be quartering a stolen calf and, upon hearing "Come out!" and "Hide!" they took fright and ran off, thinking the law was after them. And they left the meat right there.

Hearing the robbers flee, Giufà went to see what it was all about and found the quartered calf. He took his knife and went to work on it himself. After filling his sack with meat, he continued on his way.

He got home and called, "Mamma, will you open the door?"

"Is this any hour to be returning?" asked his mother.

"Night overtook me bringing back the meat, and tomorrow you must sell it all, so I'll have some money."

"Go back to the country tomorrow, and I'll sell the meat for you."

Next evening when Giufà came home, he asked his mother, "Did you sell the meat?"

"Yes, to the flies, on credit."

"And when will they pay?"

"When they have something to pay with."

For a week Giufà waited for the flies to bring him the money. When they didn't, he went to the judge and said, "Your honor, I demand justice. I sold the flies meat on credit, and they have not paid me."

"Here is the sentence," said the judge; "when you see a fly, you are authorized to kill it."

At that very moment a fly lit on the judge's nose, and Giufà drew back his fist and squashed it.

III. Giufà and the Red Beret

Work didn't suit Giufà. He would eat and then go out to roam the streets. His mother was constantly telling him, "Giufà, that's no way to get ahead in life! Can't you at least make an effort to do something worthwhile? All you do is eat, drink, and drift! I've had enough; either you go to work and buy your own things, or get out!"

Giufà went off to Càssaro Street to get his own things. He picked up one thing at one shop, another at another, and was soon outfitted from head to foot. He promised all the merchants, "Let me have it on credit, and I'll drop in one day before long and pay you."

For his last purchase, he selected a fine red beret.

When he saw himself all dolled up, he said, "There we are, my mother can't call me a tramp now!" But remembering the bills he had to pay, he decided to play dead.

He threw himself on his bed. "I'm dying! I'm dying! I'm dead!" he cried, and crossed his hands and stretched out his legs. His mother began tearing her hair. "O my son, my son! What a calamity! My son!" At the sound of her cries, people poured in to sympathize with the poor mother. The news spread, and even the merchants came to view the body. "Poor Giufà," they said, "he owed me"—let us say—"six groats for a pair of pants. . . . I'll wipe that off the books, and may he rest in peace!" And all the merchants came and canceled his debts.

But the merchant who had sold him the red beret let the debt stand. "I refuse to mark off the beret." He went to view the dead youth and saw him wearing the brand-new beret. He had a bright idea. When the grave-diggers carried Giufà off to church to bury him, he followed them, hid in church, and waited for nightfall.

It grew dark and some robbers came into church to divide up a sack of stolen money. Giufà remained motionless in his coffin, while the beret merchant hid behind the door. The robbers shook the money out of the sack, all in gold and silver coins, and made as many piles of money as there were robbers in church. One twelve-groat piece was left over, and they had no idea which one of them should get it.

"For the sake of peace," proposed one of the robbers, "let's do this: there's a dead man there who will be our target. Whoever throws the coin right into his mouth wins it."

"Perfect! Perfect!" they all agreed.

And they all got into place to take aim. Hearing this, Giufà stood up right in the middle of his coffin, and thundered, "Dead souls! Rise up, all of ye!"

The robbers left their money and fled.

Finding himself alone, Giufà ran to the heaps of gold, but at the same moment out jumped the beret merchant and reached for the money. They divided it evenly, until only a five-farthing piece remained.

"I'm taking this," said Giufà.

"No you're not, I am!"

"It's mine!" insisted Giufà.

"Hands off, it's mine!"

Giufà picked up a candle-snuffer and waved it at the beret man, screaming, "Put the five farthings right here. I'm taking the five farthings!"

Outside, the robbers patrolled the church on tiptoe, to see what the dead would do; they regretted going off and leaving so much money behind. With their ears to the door, they heard the heated squabble over five farthings.

"Woe to us!" they said. "There's no telling how many dead souls have risen from their tombs! They each get scarcely five farthings, and even at that there's not enough to go around!" And they fled at breakneck speed.

Giufà and the beret man each carried home a sack bulging with money, and Giufà ended up with five farthings more than the man.

IV. Giufà and the Wineskin

Realizing that nothing could be made of her boy, Giufà's mother hired him out as a tavern-keeper's helper. The tavern-keeper said to him, "Giufà, go down to the sea and wash this wineskin for me, but wash it well, if you don't want a beating." Giufà went to the sea with the wineskin, which he washed and washed, all morning long. Then he said to himself, "How am I to know if it has been washed enough? Who can I ask?" There wasn't a soul on shore, but out at sea was a boat that had just left port. Giufà pulled out his handkerchief and began frantically waving it and shouting, "You out there, come back here! Come here!"

"They're signaling to us from shore," said the captain. "Let's go back in; there's no telling what message they have for us. Maybe we left something behind . . ." They rowed to shore in a rowboat, and there was Giufà. "What on earth is it?" asked the captain.

"Tell me, Your Honor, is this wineskin washed enough?"

The captain was fit to be tied; he grabbed up a stick and gave Giufà the walloping of a lifetime.

"But what was I to say?" wailed Giufà.

"Say, 'Lord, speed them up!' so we will make up the time you have caused us to lose."

Giufà slung the wineskin over a back still smarting from the blows and went off through the fields, repeating in a voice loud and clear, "Lord, speed them up, Lord, speed them up, Lord, speed them up."

He met a hunter taking aim at two rabbits, and Giufà went on saying, "Lord, speed them up, Lord, speed them up . . ." Off hopped the rabbits.

"You little dickens! All I needed was for you to come along!" exclaimed the hunter, and hit Giufà on the head with the butt of his gun.

"But what was I to say?" wailed Giufà.

"Say, 'Lord, let them be killed!' "

With the wineskin over his shoulder, Giufà went off, repeating, "Lord, let them be killed . . ." and whom should he meet but two men in a heated quarrel and all ready to fight. "Lord," said Giufà, "let them be killed!" At that, the two men separated and pounced on Giufà, crying, "You cad! You want to add fuel to the fire, do you?" Reconciled at once, they lit into Giufà and beat him black and blue.

"But what must I say?" sobbed Giufà, when he was able to speak.

"What must you say? You must say, 'Lord, separate them!' "

"All right. Lord, separate them, Lord, separate them . . ." began Giufà, continuing on his way.

A bride and groom happened to be coming out of church at the conclusion of their wedding. When they heard "Lord, separate them" the bridegroom blew up; he removed his belt and gave Giufà a thrashing, crying, "You ill-omened buzzard! How dare you think of separating my wife and me!"

Unable to take any more blows, Giufà dropped unconscious to the ground. When they went to pick him up and he opened his eyes, they asked, "What were you thinking of to say such a thing to newlyweds?"

"But what was I supposed to say?"

"You should have said, 'Lord, make them laugh! Lord, make them laugh!' "

Giufà picked up the wineskin again and went on his way, repeating that line. But he passed a house where a dead man lay in his coffin, surrounded by candles and weeping relatives. When they heard Giufà go by saying, "Lord, make them laugh," one man came out with a cudgel and gave Giufà the rest of the blows he had coming to him.

Giufà then realized he'd better keep his mouth shut and run straight to the tavern. But the tavern-keeper who had sent Giufà out first thing in the morning to wash the wineskin had his share of blows to give him for not returning before night. Then he fired Giufà.

V. Eat Your Fill, My Fine Clothes!

Giufà, fool that he was, never got invited anywhere or asked to honor anyone with his company. Once he went to a farm to see if they would give him something, but noticing how slovenly he was, they sicked the dogs on him. His mother then bought him a fine topcoat, a pair of pants, and a velvet vest. Now dressed as a country gentleman, Giufà returned to the same farm. They made a big to-do over him, invited him to sit down to the table with them, and quite turned his head with all their compliments. When they served him, Giufà carried food to his mouth with one hand; with the other he stuffed food into all his pockets as well as his hat, saying, "Eat your fill, my fine clothes, for they invited you, not me!"

VI. Giufà, Pull the Door After You!

Giufà had to go out in the fields with his mother. She left the house first and said, "Giufà, pull the door after you!"

Giufà began pulling, and pulled until he'd ripped the door off the hinges. Then he loaded it onto his back and trailed along behind his mother. After going a little way he said, "Mamma, it's heavy! It's weighing me down, Mamma!"

His mother wheeled around. "What's weighing you down?" Then she saw him with the door on his back.

With such a burden they moved slowly, and night came on while they were still far from home. Fearing bandits, mother and son climbed a tree, with Giufà still carrying the door on his back.

At the stroke of midnight, here came bandits to divide up money beneath the tree. Giufà and his mother held their breath.

In a few minutes, Giufà whispered, "Mamma, I have to make water."

"What!"

"I have to."

"Wait."

"I can't wait another minute."

"Yes, you can."

"No, I can't, Mamma."

"Go on and do it, then."

And Giufà did it. When the bandits heard water coming down, they said, "How about that, it's started to rain!"

A few minutes later, Giufà whispered, "Mamma, I have to do something else now."

"Wait."

"I can't wait another minute."

"Yes you can."

"Mamma, I can't."

"Go on and do it, then!"

And Giufà did it. When the bandits felt it falling on them, they said, "What on earth is this, manna from heaven? Or is it the birds?"

The next thing Giufà whispered (he was still holding that door on his back) was, "Mamma, it's heavy."

"Hold on to it anyway."

"But it's heavy, Mamma!"

"You better not let go of it."

"Mamma, I can't hold it any longer"—and he let it fall right on the bandits.

Without waiting to see what had hit them, the bandits fled like the wind.

Mother and son came down the tree and found a sack full of coins the bandits had been dividing up. They took the sack home, and the mother said, "Don't tell a soul about this, unless you want the law to send us both to prison."

Then she went out and bought raisins and dried figs, climbed to the roof and, when Giufà went outdoors, she pelted him with raisins and figs. Shielding himself, Giufà called into the house, "Mamma!"

"What do you want?" she replied from the roof.

"Raisins and figs are falling!"

"It's obviously raining raisins and figs today, what else can I say?"

When Giufà had gone off, his mother took the gold from the sack and left in its place rusty nails. A week later, Giufà looked in the sack and found the nails. He began shouting at his mother. "Give me my money, or I'll tell the judge!"

"What money?" she replied, and paid no more attention to him.

Giufà went to the judge. "Your Honor, I had a sack of gold, and my mother replaced it with rusty nails."

"Gold? Whoever heard of you having any gold?"

"Yessirree, I did—that day when it rained raisins and dried figs."

And the judge had him sent to the crazy house.

(Sicily)

◈ 191 ◈

Fra Ignazio

Each day Fra Ignazio, the lay brother, had to go out begging for the monastery. He preferred to go where there were poor people, since whatever they gave, they gave cheerfully. One person he would never approach, however, was a certain notary named Franchino, a stingy soul who bled the poor.

One day Notary Franchino, offended with Fra Ignazio for passing him up, went to the monastery to complain to the prior of Fra Ignazio's rudeness. "Do I seem to you, Father, a person of so little consequence?"

The prior urged him not to worry, and promised to speak to Fra Ignazio. Satisfied, the notary took his leave.

When Fra Ignazio returned to the monastery, the prior said, "What do you mean by slighting Notary Franchino? Tomorrow you are to go to him and accept everything he gives you."

Fra Ignazio bowed in silence. The next morning he went to the notary, and Franchino filled his knapsacks with every good thing imaginable. Fra Ignazio heaved the sacks onto his back and set out for the monastery. At the first step he took, a drop of blood fell from the sacks, followed by another and still another. People along the way noticing the sacks dripping blood, said, "Well, well! Today is a meat day for Fra Ignazio! The Fathers will have a fine feast for a change!" Without a word Fra Ignazio continued on his way, leaving a trail of blood behind him.

When he reached the monastery, the brothers, seeing him come in with all that blood, exclaimed, "Fra Ignazio is bringing us meat today! And freshly slaughtered!" They opened the knapsacks, but they contained no meat. "But where did all that blood come from?"

"Don't be afraid," replied Fra Ignazio. "That blood is actually flowing from the knapsacks, since the alms Franchino gave me are not his own earnings, but the blood of the poor people he has robbed."

From then on, Fra Ignazio went no more to the notary for alms.

(Campidano)

❖ 192 ❖

Solomon's Advice

Once there was a shopkeeper with a general store. As he went to open up early one morning, what should he find lying on the doorstep but a dead man. Afraid of being arrested, he took off, leaving his wife and three sons. When he got to a certain town, he looked around for work, but found none. At last he heard about a gentleman who was looking for a servant. For want of anything better, he said, "Let's inquire into that." The gentleman, whose name was Solomon, was a prophet, and all the citizens went to him for advice. The shopkeeper became Solomon's devoted servant and enjoyed his master's highest esteem. He remained with Solomon for twenty years, after which he had the urge to return to his own family, from whom he had heard nothing since his departure. "Master," he said to Solomon, "I have decided to go back and see my people. Let us settle our accounts, and I will leave." In the twenty years he had been a servant, he had never taken a cent of the pay that was due him. The master figured out his wages; he owed him three hundred crowns, which he gave him.

The servant took his leave and was already down the steps, when the master called to him. "Everybody comes to me for advice," said Solomon, "and there you are going off and asking me nothing."

"How much do you charge for a word of advice?" asked the servant.

"One hundred crowns."

The servant thought it over, went back up the steps, and gave him the hundred crowns. "Give me a word of advice."

"Do not abandon the old road for the new," said Solomon.

"What! Is that all? And to think I paid one hundred crowns to hear that!" exclaimed the servant.

"That's so you will remember it," answered Solomon.

The servant was on his way back downstairs, when he had second thoughts and returned to say: "As long as I'm still here, give me another word of advice."

"That will be another hundred crowns," replied Solomon. The servant handed him another hundred crowns, and the master pronounced his advice: *"Don't meddle in other people's business."*

The servant thought to himself, To be going home now with only a hundred crowns, I might just as well go empty-handed, but with a final word of advice from Solomon. So he paid out his last hundred crowns for this advice: *"Postpone anger until tomorrow."*

He turned to go, and again the master called him, gave him a cake, and said, "Don't cut this until you're at the table with all your family."

The servant was walking along the road, when he met a band of travelers, who said, "Would you like to come with us? We are going that way and you can travel with us."

I gave my master one hundred crowns, thought the servant, for the advice not to abandon the old way for the new, so he did not join those men, but went his own way.

As he continued along the same road, he soon heard gunfire, shouts, groans: the travelers were being attacked by bandits, who killed every one of them. Praised be those hundred crowns that went to my master! thought the servant. His word of advice has saved my life.

Darkness descended upon him in a desolate stretch of countryside, and he was unable to find shelter. Finally he saw a solitary house, knocked, asked for a night's lodging, and was invited in. The master of the house prepared supper, readied the table, and they sat down to eat. After they had finished, he opened a door leading underground, and out came a blind woman. The master poured out soup for her in a death's head and gave her a bit of reed to use as a spoon. The blind woman ate, then the man led her back underground and closed the door after her.

Next he turned to the traveler. "What do you have to say about what you just saw?"

Remembering the second word of advice, the servant replied, "I think you must have your own explanations."

Then the master of the house said, "That is my wife. When I used to go away, she would receive another man. Once I came back and found them together. The bowl she eats out of is the man's head; the spoon is the reed I used to gouge out her eyes. Now what do you think? Did I do right, or wrong?"

"If it struck you as right, that means it was right."

"Excellent," answered the master of the house. "Everyone who says I did wrong is put to death."

And the traveler thought to himself, Praised be the second hundred crowns which have saved my life a second time!

The next evening he reached his town. He looked for his street, his house. The windows were lighted, and there stood his wife with a handsome youth, whom she patted familiarly on the face. The sight made the man so furious that he drew his gun at once, but then he thought, I gave my master one hundred crowns for the advice, Postpone anger until tomorrow.

So, instead of shooting, he went to a woman living across the street and asked, "Just who lives in that house over there?"

"That's the house of a woman who is happiness itself, since her boy came home today from the seminary and said his first Mass, and she can't make enough fuss over him."

And the man thought to himself, Praised be the last hundred crowns, which have saved my life for the third time! He ran back to his house, his wife opened the door, his sons did not recognize him, they all embraced. When the neighbors left, they sat down to the table, and the man cut the cake: in it were the three hundred crowns which Solomon had taken so that his advice would be remembered.

(Campidano)

❖ 193 ❖

The Man Who Robbed the Robbers

Six terrible bandits, thriving on murder and robbery, lived in a house on the hill. One of their rooms was packed with money, and whenever they left home, they would hide the house key under a rock.

One day as a farmer and his son were going after wood, they saw the bandits come out, so the two men hid. Thus they learned where the robbers left the key. When the bandits were well out of sight, the farmer and his son took the key from under the rock, entered the house, and filled their pockets with money. Then they locked the door again, replaced the key, and returned to town quite pleased with themselves.

The next day father and son again robbed the bandits, as they did the day after that. On the third day when the son opened the door he fell into a slimy pit which the bandits had dug right next to the threshold. His father tried his best, but couldn't pull the boy out. Fearing the bandits would return and, finding the son, also recognize the father, the man cut off the boy's head and carried it home.

The bandits came back and found a body in the pit, but had no idea whose it was since it was headless. They decided to hang it from a dead tree on the hilltop, with one robber keeping watch to see who came to mourn the dead soul. Wishing to retrieve the boy's body, his father consulted a sorceress, who advised him what to do.

He climbed the hill in the night and hid close to the tree. Another of his sons hid on the other side of the hill making the sound of two rams locking horns by clashing two wooden blocks. The bandit guarding the body hadn't eaten all day long and, at the sound of those clashes, went to capture the rams to roast. As soon as he was gone, the dead boy's father untied the body and fled.

Learning of that, the bandits were determined to take revenge on the dead man's companion, but they could not find him. One day, much later, they were in the village on business and heard that a local man had recently come into a sizable fortune; he was none other than the father of the dead boy. Right away the bandits ordered a cooper to make them six large casks, with lids on top. Fully armed, each bandit climbed into a cask. They sent the cooper to the rich man, who lived nearby, to ask if he would please keep the casks until their owner called for them, as the cooper had no room to store them himself. The rich man agreed and had the casks placed in the wine cellar. That night, before retiring, a serving woman went to get some wine and heard voices in the casks. Someone was asking, "Well, is it time, yes or no, to come out and kill the master of the house?" At that, the servant flew back up the steps, shaking like a leaf. She awakened the master and told him everything. He sent for officers of the law and they all descended into the cellar and slew the bandits. That was the end of them, while the man who had robbed robbers remained rich and lived at home in peace.

(Campidano)

194

The Lions' Grass

There was a carpenter who had a daughter lovely beyond words, but they were quite poor. The girl's name was Mariaorsola and, since she was so beautiful, her father never let her go out of the house or even look out the window. Opposite the carpenter lived a merchant, who was very rich and had one son. The boy heard that the carpenter had a daughter and went over to his house, asking, "Mr. Anthony, will you make me a table?"

"Bring me the lumber, which I have no money to buy myself, and I will make the table for you."

On the sly, the boy carried him lumber belonging to his parents, who didn't want him going in such poor people's houses, and was constantly on the lookout for Mariaorsola. One day when she thought he'd already gone, she came downstairs. Peppino saw her and fell in love.

"Mr. Anthony," he said to the carpenter, "I'm asking you for Maria-orsola's hand in marriage."

"My boy," replied the man, "don't make fun of us. Mariaorsola is just too poor; your mother and father wouldn't have her for a daughter-in-law."

"I'm not joking," said Peppino. "Don't you worry about my mother and father. Mariaorsola suits me, and I shall marry her."

So the marriage agreement was concluded behind the back of Peppino's mother and father.

But Peppino's mother heard from the townspeople that her son had just taken a wife and immediately told her husband.

"What's to be done?" asked the merchant. "We must send him off!"

When Peppino came in that night, his father said to him, "You can see that I'm old, so you must sail to the Continent with the wares."

"Well," said the son, "just let me know when you want me to go."

The day Peppino told Mariaorsola, "I'm going to have to go away," the young wife burst into tears. He left her a handful of money, saying, "Be happy, now; don't worry, and don't forget me at any time."

The next day when he left his house to start on his trip, Mariaorsola peeped out of the window and heard him telling people on the street, "Farewell! I'm going away and will return in one year."

At the sound of Peppino's voice, Mariaorsola went into a swoon. She was put to bed and, from that day on, hovered between life and death.

After a year, Peppino returned to Port Torres and immediately sent a message to his house announcing his arrival and requesting a cart for the wares he was bringing. Mother, father, and friends came to meet him. After greeting them, he asked all of a sudden, "How is everybody on our street?"

"They're all fine," he was told, "except Mr. Anthony's daughter, Mariaorsola. "If she's not already dead, she will be in no time. She's been in bed ever since you went away."

Peppino fainted. They put him in a carriage, carried him home, and sent for the doctor. He was heartbroken over Mariaorsola, but the doctor didn't know what was wrong with him, and his mother went to pieces.

Note that, before going away, Peppino had told two close friends about his secret marriage. These friends went to the doctor and said, "It so happens that the young man was recently married behind the back of his parents, and his wife has been gravely ill from the time he went away. That's his whole trouble, and until he has that young woman back, he won't recover."

The doctor went and told the boy's parents. "What shall we do?" the father asked his wife, who grew even more upset upon hearing of her son wedding a poor girl.

"Rather than have him die, it is better to see him married to the carpenter's daughter," answered the mother, and sent over to find out how Mariaorsola was.

"Mariaorsola is dying," replied the bride's mother. "In all the time she's been sick, you've never asked me about her, and now that she's about to die you think of her!"

"I will take her to my house," said Peppino's mother.

"Leave her be, she's dying, I tell you."

But Peppino's mother insisted and, picking up Mariaorsola, carried her to the merchant's house, where the girl was laid on a sofa in front of Peppino's bed.

"Peppino," called his mother, "look at your Mariaorsola."

At those words, Peppino gradually revived and stepped out of bed. "Mariaorsola!"

Seeing Peppino at her bedside, Mariaorsola also gradually revived.

So they got well. And when they were strong once more they celebrated their marriage and loved each other immeasurably.

After a short period of happiness, Mariaorsola fell ill. "Listen, Peppino," she said, "if I die you must recite the office of the dead before my body." And, lo and behold, she died!

They carried her away, and Peppino had forgotten to recite the office.

That night he thought of it. "Oh dear, I forgot!"—and he ran to church at once and knocked.

"Who is it?" called the sacristan.

"Please come down." And when the sacristan appeared, Peppino said, "Open up the tomb of that dead woman, and I'll give you ten crowns."

"How can I do that? Suppose people hear about it?"

"Nobody will know. It's pitch-dark."

So the sacristan opened the tomb and left him. Peppino knelt and began reciting the office. As he continued, he heard roars, and into church rushed two lions. The lions started fighting. One knocked the other off his feet and bit him to death. The living lion then ran off and snatched some of the grass growing there in the cloister of the church, forced open the dead lion's mouth, and rubbed the grass over his teeth. The dead lion came back to life, and together the two lions ran off.

Peppino, having meanwhile finished reciting the office, said to himself, "Let's see if I can bring Mariaorsola back to life!" He took a little of the grass, rubbed the dead girl's teeth with it, and she got up. "What have you done, Peppino?" she said. "I was very happy with things the way they were!"

Peppino gave her his cloak and took her by the arm.

"What's going on?" asked the sacristan. "What are you doing, taking the dead woman away?"

"Let me go, my wife is alive!"

He took her home, put her to bed, and warmed her up with heavy warm clothing. Then he lay down beside her and slept.

It must have been about seven o'clock in the morning when Peppino's mother knocked at his door. "Who is it?" called Mariaorsola.

Hearing the dead woman's voice, the mother-in-law fell down the steps, struck her head, and died.

A little later, the servant girl went up and knocked. "Who is it?" asked Mariaorsola. "Are you still there knocking?"

The servant girl too was frightened out of her wits, fell down the steps, struck her head, and lay there dead.

When Peppino woke up, Mariaorsola said, "It's impossible to get any sleep in this house. They're always knocking on the door."

"And you answered?"

"Of course I did."

"What have you done? They thought you were dead!"

Peppino opened the door and saw his mother and the servant girl lying dead at the foot of the stairs. "Oh, me, what misfortune has befallen us!" he said to himself. "But I must keep quiet about it and not frighten my wife!" And with the lions' grass he brought the two dead women back to life.

When Mariaorsola was sick, she had made a vow to go to the church of St. Gavino if she got well. "Tomorrow," she said to her husband, "let's go to St. Gavino's."

They set out, but after a while she said, "Peppino, I forgot and left my ring on the windowsill."

"Oh, come, let's not worry about it."

"No, I'm going back for it; a breeze could rise and blow it away."

"I'll fetch it for you, but in the meantime don't you go near the sea, where the king of Muscovy's boat is." At that, he turned back.

Mariaorsola, however, approached the sea, and there was the king of Muscovy, who seized her and carried her off.

When Peppino returned with the ring, he looked everywhere for Mariaorsola, but she was nowhere in sight. Then he dived into the sea and swam off. He spotted a boat and waved a white handkerchief.

"Quick, there's a man in the sea!" said the boat's owner. They took him aboard, and Peppino asked, "Have you seen the king of Muscovy's boat?"

"No, we haven't."

"Please be so good as to take me to Muscovy."

In Muscovy, there was Mariaorsola dressed in queenly attire. When Peppino saw her, he smiled at her, but she looked the other way. There was no possibility of approaching her, so he offered himself to the king as a footman, and was engaged to wait on the table. Finding Mariaorsola alone at the table, he said, "Well, if it isn't my own Mariaorsola! You no longer recognize me?"

She made a wry face and turned her back on him; she already had a plan for wrecking him. To one of the king's pages, she said, "Take all the silver spoons and stuff them into the pocket of that footman."

Once the spoons were missed, she ordered, "Search that footman!"

The spoons were thus found in Peppino's pocket. "So this is the thief! Throw him into prison and then have him hanged before my windows!"

Peppino still had some of the lions' grass and, when they led him to the gallows, he said to his confessor, "I am innocent, so when they hang me, please keep them from breaking my neck and carry me to your house and rub my teeth with this grass. I will then come back to life."

At the appointed hour, the confessor said to the hangman, "Be careful not to break his neck." Then he asked the king's permission to carry the body home with him. The hangman hanged him, taking care not to break his neck, and the confessor carried the body to his monastery. As soon as the grass touched his teeth, Peppino came back to life and, thanking the confessor, set out on the road.

He went to the country of the king of the seven crowns. The king's wife had just died, and the palace was draped in mourning.

"I wish to enter the palace," said Peppino to the guard.

"Do you think they need you right now?" snapped the guard.

"Tell them I wish to enter."

He was so insistent that they finally let him in. "Majesty, I wish to remain alone with the dead queen." The king therefore ordered everybody out of the room.

Peppino closed the doors, removed the queen from her coffin, laid her in bed, placed grass between her lips, and she came back to life. Peppino threw open the room. "Majesty, here is your wife." The palace shed its mourning at once, and the festivities began.

From that day on, the king always kept Peppino at his side. "Peppino," he said one day, "I am old, you are now our son, and I am going to give you my seven crowns."

"What kings," asked Peppino, "will attend the coronation of the king of the seven crowns?"

"The kings of Spain, Italy, France, Portugal, England, Austria, and Muscovy. They are the seven who crown the king of the seven crowns."

"I will accept the seven crowns, then," said Peppino.

The invitations were sent out, and the king of Muscovy got ready for the journey. His wife, Mariaorsola, made herself a special dress for the occasion, and thus they came to the palace of the king of the seven crowns.

In the hall, Peppino spotted Mariaorsola at once among all those kings and queens, but she did not recognize him. The coronation was followed by the banquet. After the banquet, wearing the seven crowns, Peppino said, "Now we must each tell a story."

So, one by one, they told a story. "Now I'll tell mine," said Peppino when it was his turn. "Don't anyone get up from the table until I have finished." He told his whole story, beginning with his marriage to Mariaorsola, who sat there on pins and needles. She pleaded a headache, insisting on leaving, but Peppino waved her down. "Nobody get up!"

At the end of his account, he asked the king of Muscovy, "What does such a woman deserve?"

"Hang her first," replied the king of Muscovy, "then burn her, then throw her ashes to the wind."

"Let it be done," ordered Peppino. "Seize the king of Muscovy's wife" —and she was throttled on the spot.

And he remained king of the seven crowns.

(Nurra)

195

The Convent of Nuns and the
Monastery of Monks

There was a tailor who had a daughter named Jeannie. She was a beautiful girl and attended school. So beautiful was she that a youth named Johnny was always following her around, and she was at a loss to escape him. Finally one day she said to her girlfriends, "Shall we found a convent?"

"Let's do so," replied her companions.

These girls included daughters of kings, knights, and noblemen. Twelve of them got together and told their fathers, "We are going to establish a convent."

"What! You're founding a convent just for yourselves?"

But the girls were determined to have such a convent, for which they chose a site far from town. They took along ample provisions and all twelve of them settled down in their convent. They elected Jeannie to be their abbess.

Now Johnny, who was in love with Jeannie, said to his friends, "I haven't seen Jeannie for ages. Where could she be?"

"You're asking us?"

"Since my beloved has disappeared, I shall become a monk. Why don't we establish a monastery?"

So they founded a monastery of monks.

One night the nuns in the convent ran out of provisions. Abbess Jeannie was in charge of procuring food. Looking out of the window, she saw a light in the distance and set out toward it to get more provisions. She came to a house and entered it, but found no one around. There was a table set with twelve glasses, twelve spoons, twelve napkins, and twelve large bowls of well-seasoned macaroni. Jeannie put the twelve bowls of macaroni in a basket and returned to the convent. She rang the dinner bell, the nuns came in, Jeannie gave each one a bowl, and they ate.

The house where she had procured the macaroni was the monks' monastery. When the monks returned and found their table stripped, the father superior, who was Johnny, said, "What thieving magpie has made off with our dinner? Tomorrow night someone will have to stand guard!"

The next night one of them was appointed to keep watch, and was told, "The instant you whistle, we'll all rush in. Beware of falling asleep." But in no time the monk was snoring up a storm. The abbess returned, saw the twelve plates of macaroni on the table, looked around her, spied the sleeping monk, put the macaroni in her basket, then picked up the pan and rubbed its grease on the sleeping monk's face.

She reached the convent, rang the bell, and they sat down and ate.

When the father superior saw the monk with the black face he said, "A fine guard you made!" The next night he appointed another monk to watch. But this one too fell asleep and woke up with a black face. The same thing happened eleven nights in a row, every monk taking his turn, until it finally fell to the father superior to stand guard.

Johnny only pretended to fall asleep. When Jeannie had filled her basket with macaroni and came over to black his face with the pan, he stood up, saying, "Hold it, you're not getting away this time!"

"Oh!" she said. "Please don't harm me!"

"I'll not lay a hand on you, but you must bring me those eleven nuns of yours."

"All right, but on condition you won't harm a hair on their heads."

"I promise."

So the abbess went off with her basket of macaroni. After she'd given the nuns their supper, she said, "Listen, my sisters, we are obliged to go to the monks' monastery."

"And what will they do to us?"

"They'll not harm us. They promised."

So they went.

"We want a room to ourselves where we can stay."

The superior led them to a room with twelve beds, and the nuns retired for the night.

The other monks returned and found the table stripped. "What! Even with Reverend Father standing guard, supper still vanished?"

"Quiet!" ordered the superior. "We've caught the thieving magpie."

"You don't mean it!"

"Yes, and eleven others, and now they will cook macaroni for us." He went and knocked on the nuns' door, saying, "Quick, wake up! You must cook macaroni for us."

"My nuns," replied the abbess, "cannot cook without music."

"We will play," said the monks.

Although half starved, they picked up trumpets and violins and played. But instead of preparing supper, the abbess and her nuns took the mattresses and threw them out the window, tied the sheets to the window ledge, climbed down them and, one by one, dropped onto the mattresses. Back to their convent they ran and bolted the door behind them.

Meanwhile the monks continued to play, starving to death. "How about that macaroni?" they said. "Isn't it ready yet?" They knocked on the nuns' door, but there was no answer. They broke the door down and found the beds empty, with neither sheets nor mattresses. "Well! They pulled a fast one on us! We'll have to give them a taste of their own medicine!"

They built a cask and closed up the superior in it. They went to the nuns' convent and hid until nightfall, when one of the monks rolled the cask up to the door, knocked, and asked the nun who opened the door, "Would you please keep this cask for us overnight?" And he left it inside the door.

But the abbess knew what was up. "We are done for!" she said to herself. "My sisters," she announced at the supper table, "there's no telling what will happen here, but don't be afraid." As a matter of fact, while they were still eating, the superior climbed out of the cask and knocked at the refectory door.

"Who is it?"

"Open up."

The nuns opened the door, and in walked the father superior.

"Good evening," said the nuns. "Make yourself at home." And the superior sat down and ate with them, talking of this and that. At the end of supper he drew a phial from his pocket, saying, "Have a sip, sisters."

The nuns drank, but the abbess emptied her little glass into her habit. All the nuns fell asleep but the abbess, who only pretended to sleep. When he saw them all sleeping, the superior tied a rope around the waist of each, for the purpose of lowering them from the window, to which he walked and called the other monks. But Jeannie slipped up behind him, grabbed him by the ankles, and threw him out the window headfirst.

Then she roused her companions. "Quick, we have to get away from here. We will write our fathers to fetch us, as we're tired of being nuns!"

So each one went home. The monks also abandoned their monastery and returned home.

Johnny was again in love with Jeannie and, with his head all bandaged, went and asked her to marry him. She finally agreed, but before the wedding she made a sugar doll the size of herself.

On her wedding night, she said to her husband, "Put out the candle when you enter the bedchamber, for at the convent I became accustomed to staying in the dark."

When it grew dark, she went into the bedchamber, put the sugar doll in her bed and hid under the bed, from where she could make the doll move its limbs by means of a string. Her husband came into the room, holding a sword. "Well, Jeannie," he began, "what have you done to me? Do you remember stealing my supper?"

The doll nodded, "Yes."

"Do you remember throwing me out the window and cracking my skull?"

The doll nodded, "Yes, I do."

"And you have the nerve to admit it?"

He lifted his sword and thrust it into the sugar doll's heart.

"There, Jeannie, I've killed you! Now I will drink your blood!" And he passed his tongue over the sword. "Jeannie! You were sweet in life and you are sweet in death!" He was pointing the sword at his own heart, when out leaped Jeannie. "Stop!" she cried. "Don't kill yourself, I am alive!"

They embraced, and from that time on, they were one happy couple.

(Nurra)

◈ 196 ◈

The Male Fern

The proudest young man in Gallura was a bandit, whom not even the law had ever managed to nab. One night after a party, while the whole countryside slept, the bandit, with his gun slung over his shoulder, was crossing a field where a solitary church stood, when suddenly out of the brush sprang a wild boar and started running around the church. The man took aim, fired, and killed it.

The path led right up to the church. When the bandit was a short distance from the door, he heard singing and laughter inside the church. Pausing to listen, he thought to himself, What with this wild boar and so many merry souls, a fine little party could be got up, and I could continue on my way early tomorrow morning. So he entered the church, dragging the dead boar behind him. "A wild boar for this charming gathering!" he cried, and all the people there, men and women, burst out laughing, joined hands, and began dancing in a ring. The bandit was about to give them his hands, when he noticed the empty eye sockets of every one of them, and realized that this was not a ball of the living but of the dead.

Dancing with unflagging gaiety, the dead tried to put him in the middle of the circle, and a female ghost brushing by him said, "Come with me, and I will tell you where the three flowers of the male fern grow!" The youth was eager to join her, having heard that the discovery of the three flowers of the male fern would make everyone invulnerable to gunshot. But at that moment, one of the dead men left the group and approached him. The bandit recognized him: it was his godfather.

"Beware, godson," said the dead godfather; "whoever penetrates the circle of the dead will nevermore leave it. If you don't do your best to get away now, tomorrow you too will be one of the dead. But just as I answered for you in life, so will I save you from death. Come in nonetheless and dance with us, but right in the middle of everything, sing these lines:

> 'Dance and sing, dance and sing,
> Go now, enjoy your fling!
> Come time for our fine fling,
> Then we will dance and sing, dance and sing!' "

The bandit went at once to the woman who had promised to reveal where the male fern grew, and she said to him, "Whoever would have the three flowers must go on the first day of August all the way to the bend in the river, where the cock's crow is never heard. The three flowers will blossom at midnight. No matter what happens, guard against all fear, and pick them."

"I will pick them," declared the bandit, "and no one will ever die again from gunshot."

The dead woman laughed. "That's what you think! You are now in the circle of the dead and will be with us forevermore." Dancing, she held his hand.

At that, the bandit realized the time had come for the lines learned from his godfather, and he sang out:

> "Dance and sing, dance and sing,
> Go now, enjoy your fling!
> Come time for our fine fling,
> Then we will dance and sing, dance and sing!"

Hearing that song, all the dead souls groveled in a heap, crying. In a flash the bandit reached the door, jumped on his horse, and fled. The dead struck out after him, but were never able to catch him.

On the first of August, the bandit set out for the river. The night was fair, but all of a sudden at midnight a storm broke, unleashing hail, lightning, thunder, and tongues of fire. The bandit held his ground, waiting for the flowers to blossom. Lo and behold, during a flash of lightning, he saw one flower of the male fern blossom, and picked it.

Next was heard the sound of thousands of hoofbeats, and out charged droves of wild boars, stags, bulls, cows, and every other kind of animal, maddened by the storm. It looked as though the man would be trampled to death any moment, but he fearlessly held his ground and waited for the fern to blossom. Then, bringing up the rear, crawled a serpent which clutched his ankle, slowly made its way up his leg and trunk, and wrapped around his neck; the man felt himself being choked to death, but still didn't budge. Finally the serpent looked him in the eye, let out a shrill hiss, and vanished. And the bandit noticed that the second flower had blossomed, and picked it.

At this point the bandit was satisfied he had already made man invulnerable to gunfire and, now sure of himself, he waited for the third flower to blossom. All of a sudden the silence was broken by approaching hoofbeats and the din of shots. The bandit was calm at first; but a squad of armed policemen appeared on the ridge, pointed their guns at him, and he said: "The police would have to find me before the third flower has blossomed and while it is still possible to die from a bullet wound!" He thus took fright, trained his gun on the squad, and fired a shot.

Immediately police and horses vanished, and along with them the flowers of the male fern. The third flower never blossomed—so much the worse for the soul of the man who failed to hold out—while bullets continue to go their own sweet and fatal way.

(Gallura)

◈ 197 ◈

St. Anthony's Gift

Once upon a time the world had no fire. Men were cold and went to St. Anthony, who was off in the desert, to ask for help, since they could no longer endure all that cold weather. St. Anthony took pity on them and, as fire was confined to Hell, he decided to go down and get some.

Before becoming a saint, St. Anthony had been a swineherd, and a certain piglet from his herd had always refused to leave him, and followed him wherever he went. So St. Anthony, with his piglet and his fennel staff, showed up on Hell's doorstep and knocked. "Open up, I'm cold and want to get warm!"

Right away the Devils at the door saw that this was no sinner, but a saint, and, they said, "No indeed! We know you! You won't get in here!"

"Let me in, I'm so cold!" begged St. Anthony, while the pig rooted against the door.

"The pig can come in, but not you!" said the Devils, and they cracked open the door just enough for the animal to squeeze through. As soon as St. Anthony's pig was inside Hell, he began running around rooting everywhere, throwing the whole place into an uproar. The Devils scampered behind him picking up firebrands, raking up pieces of cork, standing up the tridents he knocked over, putting pitchforks and torture tools back in their places. It was maddening, but they could neither catch the pig nor drive him out.

At last they turned to the saint, who had remained on the doorstep. "That confounded pig of yours is playing havoc with Hell! Come get him out!"

St. Anthony stepped inside Hell, tapped the pig with his staff, and the animal quieted down on the instant.

"As long as I'm here," said St. Anthony, "I may as well sit down a minute and warm up," and he seated himself on a sack of cork, holding out his hands toward the fire.

Every now and then a devil ran by him, on his way to inform Lucifer of some soul or other on earth whom he had lured into sin. And each time St. Anthony would whack him on the back with his fennel staff.

"We dislike these jokes," said the Devils. "Keep that stick quiet."

St. Anthony held the staff up, letting it lean on him, with the point jutting out on the floor. The first Devil that came by crying, "Lucifer! One more soul for sure!" tripped, and fell on his face.

"That will do! You've vexed us quite enough with this stick," declared the Devils, "so we are now going to burn it up!" They grabbed it and thrust the tip into the flames.

In that instant the pig began to tear up the place once more, sending cords of wood flying into the air along with hooks and torches. "If you want me to quiet him," said St. Anthony, "you'll have to give me back my staff." They handed it to him and the pig calmed down immediately.

Now the staff was made out of fennel, and fennel wood has a spongy pith, so that if a spark of glowing cinder gets into it, it goes on burning in secret, with no outward sign of fire. The Devils were therefore unaware that St. Anthony's staff contained fire. After preaching to the Devils, the saint took his staff and his piglet and left, and the Devils breathed a sigh of relief.

Back on earth, St. Anthony raised the staff with the fiery point and, waving it around, he scattered sparks as though blessing the people. And he sang:

"Fire, fire
For every shire!
Fire to the universe I deliver,
Nevermore may you shiver!"

From that time on, to man's great satisfaction, there was fire on earth. And St. Anthony returned to his desert to meditate.

(Logudoro)

◈ 198 ◈

March and the Shepherd

There was a shepherd who had more sheep and rams than there were grains of sand by the sea. With such a large flock, he was always worrying lest one of the animals die. Winter was long, and the shepherd constantly pleaded with the months: "December, do be kind to me! January, please don't kill off my animals with your freezes! February, be good to me, and I will honor you eternally!"

The months paused to lend an ear to the shepherd's prayers and, highly appreciative of the least show of homage, they granted his requests. They sent down neither rain nor hail nor animal diseases of any kind, and the sheep and rams continued to graze throughout the winter without catching so much as a cold.

Then came March, the most cantankerous month, and things continued to go smoothly. The last day of the month arrived, and the shepherd saw his worries at an end. They were now on the threshold of April, of springtime, and the flock was safe. The shepherd stopped pleading and started hooting and swaggering. "Little old March, March my boy, nightmare of flocks, who's afraid of you now? Lambs? *I* certainly am not. It's spring at last, and you can't harm me now, so you might as well march right off, March, to you know where!"

Hearing that ungrateful shepherd dare address him so disrespectfully,

March felt his blood boil. Buttoned up in his raincoat, he ran to the house of his brother April and said:

> "April, I fain would ask a favor:
> Three days lend thy brother
> To punish yon shepherd
> For being so absurd."

April, who loved his brother March, lent him the three days. The first thing March did was to whip around the world enlisting all the winds, tempests, and pestilences that were abroad, which he then unleashed on the shepherd's flock. The first day all the rams and sheep took sick and died. The second day it was the lambs' turn. By the third day not a single living animal was left in the flock; all the shepherd now had were his eyes for shedding tears.

(Corsica)

❖ 199 ❖

John Balento

Once in a small town there was a very poor cobbler who worked his fingers to the bone mending old shoes. His name was John Balento, and although short in stature, he had brains. One day as he was sewing up a shoe, he accidentally drove the awl through his finger. "Ouch, OUCH!" he cried. "Poor me!" The neighbors heard, but paid no attention; they didn't care what happened to John Balento. But John's cries aroused the curiosity of all the flies in town, who came running to see what was the matter in the cobbler's house. One lit on the wounded finger and sucked the blood oozing from it; the others spied a bowl of noodles ready to serve and swarmed over it.

"What are all these flies doing here?" cried John. "Shoo! Get out!" He tried to fan them away with a piece of leather, but they were stubborn and continued swarming over the bowl of noodles. So John Balento swung his fist through the swarm with such force that a real slaughter resulted. On the ground he counted one thousand dead and five hundred wounded. "A master stroke if there ever was one!" he said to himself.

"People think I'm good for nothing, but if I try, I too can make my mark!"

He took a dry twig, dipped it in ink, and wrote in great big letters on a band of cloth: MEET JOHN BALENTO WHO HAS JUST KILLED ONE THOUSAND AND WOUNDED FIVE HUNDRED. He then attached this banner to the big hat he wore.

Reading it, all the country people burst out laughing and asked, "How many, John?"

"I slew one thousand and wounded five hundred!"

So spread John Balento's fame, from mouth to mouth and town to town. In a year's time John Balento was known far and wide as one of the boldest paladins in the kingdom.

Meanwhile the cobbler had left thread, awl, cobbler's wax, knife, and bench, and gone out into the world to seek his fortune. He rode a donkey that was all skin and bones, and carried neither belongings nor money. After a three days' ride through the forest, he came to an inn. Riding up to the door, he cried, "Here comes John Balento who slew one thousand and wounded five hundred!"

Now the inn was full of robbers. Hearing the name of such a famous hero, the robbers were seized with fear and made a mad dash through doors and windows, fleeing in every direction, leaving their dinner untouched and abandoning shiny arms and sturdy horses. John took his time dismounting and then went to the table. "Eat your fill, illustrious paladin," said the innkeeper. "I'll always be grateful to you. Your mere presence has taken a band of robbers off my hands."

"That's nothing compared to other exploits of mine," replied John Balento, with his eyes on his plate and his mouth full.

When he had eaten his fill, he chose the finest steed—the ringleader's own horse—climbed into the saddle, and said to the innkeeper, "Send me word if you ever need help. No one will wrong you and get away with it as long as John Balento is alive!" He spurred the horse and galloped off, amid the bows of innkeeper and servants.

Now John had never been on horseback before. He hung on with his knees and felt as if he were being bounced into the air at every step. "O my dear awls!" he said to himself; "O my thread, what was I thinking of to put you aside!" But he continued to travel and soon learned to stay in the saddle; everywhere he was received with highest honors.

At length he ended up in the land of the giants. At the sight of him, the giants, stout as chestnut trees and tall as poplars, threw open their oven-sized mouths, smacked their lips, and indicated they planned to eat him whole. John shook like a leaf.

"So you are John Balento, who slew one thousand and wounded five hundred!" cried the chief of the giants. "Do you want to fight me? Come on across the river."

"Listen," said John, "you had better let me go my way. You know what I'm like. . . . Take pepper, for instance; it's little but powerful! If I touch my sword, then heaven help you!"

The giants talked the matter over, then in a milder tone they said, "All right, we will let you go. But first prove your strength to us. See that rocky mass over there? We want you to roll it over here, so we can make a millstone for our mill. If you succeed, we will be your subjects and you will be our king."

John Balento cupped his hands around his mouth and started shouting. "Get out of the way, everybody living in the valley, run for your life! The illustrious John Balento will now set into motion the Giantstone and commit a massacre! Take to your heels, everybody!"

All the poor families began stampeding out of the valley. In the end even the giants took fright; first one and then another fled, and in no time they were all flying as fast as their legs would carry them, shouting as they ran, "Look out for John Balento, slayer of one thousand and wounder of five hundred!"

When the last soul had disappeared from sight, John spurred his horse, forded the river, and made his way through the land of the giants as calmly as you please. His fame preceded him, growing greater by the day.

After riding some distance, he came upon two armies that were ready to clash. The king was there surrounded by his generals, and they were all in low spirits; if the king lost this battle he would forfeit throne and crown, and die. At the sight of John Balento, he felt hopeful once more.

"Famous John Balento," said the king, "Heaven sends you to lead us to victory. Take command of my army."

John felt the time had come to speak the truth. "Majesty," he said, "I'm not the person you think I am; I'm nothing but a poor cobbler. The only things I can handle are awl and thread. . . ."

"Yes, yes," interrupted the king. "But let's save all that talk for later. Time is running short! You are our general. Here is my horse all saddled and ready to go, my armor, and my sword!"

Protest as he would, they forced him into the coat of mail, put him in the saddle, and the king's spirited horse bounded off, neighing. At the sight of the general bearing down on the enemy, all the other horsemen followed, plunged furiously into battle, and wiped out the enemy in a flash.

With the battle won, the soldiers began celebrating their victory, but the general was nowhere in sight. Where could he be? They found him four leagues away: he had galloped clear through the enemy ranks and kept going. The horsemen brought him back in triumph to the king.

"If you had kept up with me," said John to the king, who bowed to him in gratitude, "we would have conquered three kingdoms and three crowns by now. Anyhow, we won the battle, so let's be content with that! Farewell!"

"What! You wish to be on your way already? And here I wanted to give you my daughter in marriage!" said the king.

But John was not to be persuaded. He would accept nothing at all, and resumed his travels through the world.

After going a good way, he came to the kingdom of the Amazons. As everyone knows, the Amazons, famous women warriors, had their own kingdom, headed by a queen, and no man was allowed inside their boundaries. Whoever fell into their hands was cut into pieces and fed to the dogs, while his hide was used to make drums. The queen of the Amazons was a cruel woman who had never smiled or laughed in her whole life.

John Balento landed in their midst. The Amazons captured him, threw him into chains, and dragged him before the queen. The court of the Amazons with all those horses was full of flies. The horses swished their tails, the Amazons fanned themselves with fans; but John, who was in chains and unable to move, had flies all over him.

"You are as good as dead!" announced the queen. "Such is our law. Why did you enter my kingdom?"

Hanging his head, John said to himself, "O my awls, my thread, my bench! If I'd stuck by you, I wouldn't be in all this mess now!"

"Listen," continued the queen, "I don't like to kill a poor youth as though he were a dog. Speak the truth, and your life will be spared: did you really and truly slay a thousand and wound five hundred?"

"In one blow, Majesty."

"How did you do it?"

"Take off my chains, and I'll show you."

The queen ordered the chains removed at once. All around him, on horseback, the Amazons stared at him. There wasn't a sound, save that of the horsetails and the fans and the buzz of the flies.

"How did I do it? Like this!"—and John Balento swung his fist through the flies swarming about him and killed them all. "Count them."

"So they were flies! Oh, me, Oh, my!" And all the Amazons burst out laughing, holding their sides and rocking on the backs of their horses. The one who laughed loudest was the queen. "Ha, ha, ha, ha, ha, ha, ha! Mercy me! Ha, ha, ha, ha, ha, ha, ha! Ouch! I've never laughed so much . . . John Balento, you are the first person in my whole life to make me laugh! And with your skill in killing flies, you **are a** real godsend for my kingdom! Stay with us and you will be my husband."

The nuptials were celebrated amid gala festivities and balls, and the cobbler became king of the Amazons.

> So goes my little tale.
> Now it is your turn
> All of us to regale.

(Corsica)

❖ 200 ❖

Jump into My Sack

Many, many years ago, in the barren mountains of Niolo, lived a father with twelve sons. A famine was raging, and the father said, "My sons, I have no more bread to give you. Go out into the world, where you will certainly fare better than here at home."

The eleven older boys were getting ready to leave, when the twelfth and youngest, who was lame, started weeping. "And what will a cripple like me do to earn his bread?"

"My child," said his father, "don't cry. Go with your brothers, and what they earn will be yours as well."

So the twelve promised to stay together always and departed. They walked a whole day, then a second, and the little lame boy fell constantly behind. On the third day, the oldest brother said, "Our little brother Francis, who's always lagging, is nothing but a nuisance! Let's walk off and leave him on the road. That will be best for him too, for some kind-hearted soul will come along and take pity on him."

So they stopped no more to wait for him to catch up, but walked on, asking alms of everyone they met, all the way to Bonifacio.

In Bonifacio they saw a boat moored at the dock. "What if we climbed in and sailed to Sardinia?" said the oldest boy. "Maybe there's less hunger there than in our land."

The brothers got into the boat and set sail. When they were halfway across the straits, a fierce storm arose and the boat was dashed to pieces on the reefs, and all eleven brothers drowned.

Meanwhile the little cripple Francis, exhausted and frantic when he missed his brothers, screamed and cried and then fell asleep by the roadside. The fairy guardian of that particular spot had seen and heard everything from a treetop. As soon as Francis was asleep, she came down the tree, picked certain special herbs, and prepared a plaster, which she smoothed on the lame leg; immediately the leg became sound. Then she disguised herself as a poor little old woman and sat down on a bundle of firewood to wait for Francis to wake up.

Francis awakened, got up, prepared to limp off, and then realized he was no longer lame but could walk like everyone else. He saw the little old woman sitting there, and asked, "Madam, have you by chance seen a doctor around here?"

"A doctor? What do you want with a doctor?"

"I want to thank him. A great doctor must certainly have come by while I was sleeping and cured my lame leg."

"I am the one who cured your lame leg," replied the little old woman, "since I know all about herbs, including the one that heals lame legs."

As pleased as Punch, Francis threw his arms around the little old woman and kissed her on both cheeks. "How can I thank you, ma'am? Here, let me carry your bundle of wood for you."

He bent over to pick up the bundle, but when he stood up, he faced not the old woman, but the most beautiful maiden imaginable, all radiant with diamonds and blond hair down to her waist; she wore a deep blue dress embroidered with gold, and two stars of precious stones sparkled on her ankle-boots. Dumbfounded, Francis fell at the fairy's feet.

"Get up," she said. "I am well aware that you are grateful, and I shall help you. Make two wishes, and I will grant them at once. I am the queen of the fairies of Lake Creno, mind you."

The boy thought a bit, then replied, "I desire a sack that will suck in whatever I name."

"And just such a sack shall you have. Now make one more wish."

"I desire a stick that will do whatever I command."

"And just such a stick shall you have," replied the fairy, and vanished. At Francis's feet lay a sack and a stick.

Overjoyed, the boy decided to try them out. Being hungry, he cried, "A roasted partridge into my sack!" Zoom! A partridge fully roasted flew into the sack. "Along with bread!" Zoom! A loaf of bread came sailing into the sack. "Also a bottle of wine!" Zoom! There was the bottle of wine. Francis ate a first-rate meal.

Then he set out again, limping no longer, and the next day he found himself in Mariana, where the most famous gamblers of Corsica and the Continent were meeting. Francis didn't have a cent to his name, so he ordered, "One hundred thousand crowns into my sack!" and the sack filled with crowns. The news spread like wildfire through Mariana that the fabulously wealthy prince of Santo Francesco had arrived.

At that particular time, mind you, the Devil was especially partial to the city of Mariana. Disguised as a handsome young man, he beat everybody at cards, and when the players ran out of money, he would purchase their souls. Hearing of this rich foreigner who went by the name of prince of Santo Francesco, the Devil in disguise approached him without delay. "Noble prince, pardon my boldness in coming to you, but your fame as a gambler is so great that I couldn't resist calling on you."

"You put me to shame," replied Francis. "To tell the truth, I don't know how to play any game at all, nor have I ever had a deck of cards in my hand. However, I would be happy to play a hand with you, just for the sake of learning the game, and I'm sure that with you as a teacher I'll be an expert in no time."

The Devil was so gratified by the visit that, upon taking leave and bowing goodbye, he negligently stretched out a leg and showed his cloven hoof. "Oh, me!" said Francis to himself. "So this is old Satan himself who has honored me with a visit. Very well, he will meet his match." Once more alone, he commanded of the sack a fine dinner.

The next day Francis went to the casino. There was a great turmoil, with all the people crowded around one particular spot. Francis pushed through and saw, on the ground, the body of a young man with a bloodstained chest. "He was a gambler," someone explained, "who lost his entire fortune and thrust a dagger into his heart, not a minute ago."

All the gamblers were sad-faced. But one, noted Francis, stood in their midst laughing up his sleeve; it was the Devil who had paid Francis a visit.

"Quick!" said the Devil, "let's take this unfortunate man out, and get on with the game!" And they all picked up their cards once more.

Francis, who didn't even know how to hold the cards in his hand, lost everything he had with him that day. By the second day he knew a little bit about the game, but lost still more than the day before. By the third day he was an expert, and lost so much that everyone was sure he was ruined. But the loss did not trouble him in the least, since there was his sack he could command and then find inside all the money he needed.

He lost so much that the Devil thought to himself, He might have been the richest man in existence to start with, but he's surely about to end up now with nothing to his name. "Noble prince," he said, taking him aside, "I can't tell you how sorry I am over the misfortune that has befallen you. But I have good news for you: heed my words and you will recover half of what you lost!"

"How?"

The Devil looked around, then whispered, "Sell me your soul!"

"Ah!" cried Francis. "So that's your advice to me, Satan? Go on, jump into my sack!"

The Devil smirked and aimed to flee, but there was no escape: he flew head-first into the yawning sack, which Francis closed, then addressed the stick, "Now pound him for all you're worth!"

Blows rained fast and furious. Inside, the Devil writhed, cried, cursed. "Let me out! Let me out! Stop, or you'll kill me!"

"Really? You'll give up the ghost? Would that be a loss, do you think?" And the stick went right on beating him.

After three hours of that shower, Francis spoke. "That will do, at least for today."

"What will you take in return for setting me free?" asked the Devil in a weak voice.

"Listen carefully: if you want your freedom back, you must bring back to life at once every one of those poor souls who killed themselves in the casino because of you!"

"It's a bargain!" replied the Devil.

"Come on out, then. But remember, I can catch you again any time I feel like it."

The Devil dared not go back on his word. He disappeared underground and, in almost no time, up came a throng of young men pale of face and with feverish eyes. "My friends," said Francis, "you ruined yourselves gambling, and the only way out was to kill yourselves. I was able to have you brought back this time, but I might not be able to do so another time. Will you promise me to gamble no more?"

"Yes, yes, we promise!"

"Fine! Here are a thousand crowns for each of you. Go in peace, and earn your bread honestly."

Overjoyed, the revived youths departed, some returning to families in mourning, others striking out on their own, their past misdeeds having been the death of their parents.

Francis, too, thought of his old father. He set out for his village but, along the way, met a boy wringing his hands in despair.

"How now, young man? Do you make wry faces for sale?" asked Francis, in high spirits. "How much are they by the dozen?"

"I don't feel like laughing, sir," replied the boy.

"What's the matter?"

"My father's a woodcutter and the sole support of our family. This morning he fell out of a chestnut tree and broke his arm. I ran into town for the doctor, but he knows we are poor and refused to come."

"Is that all that's worrying you? Set your mind at rest. I'll take care of things."

"You're a doctor?"

"No, but I'll make that one come. What is his name?"

"Doctor Pancrazio."

"Fine! Dr. Pancrazio, jump into my sack!"

Into the sack, headfirst, went a doctor with all his instruments.

"Stick, pound him for all you're worth!" And the stick began its dance. "Help! Mercy!"

"Do you promise to cure the woodcutter free of charge?"

"I promise whatever you ask."

"Get out of the sack, then." And the doctor ran to the woodcutter's bedside.

Francis continued on his way and, in a few days, came to his village, where even greater hunger now raged than before. By constantly repeating, "Into my sack a roasted chicken, a bottle of wine," Francis managed to provision an inn where all could go and eat their fill without paying a penny.

He did this for as long as the famine lasted. But he stopped, once times of plenty returned, so as not to encourage laziness.

Do you think he was happy, though? Of course not! He was sad without any news of his eleven brothers. He had long since forgotten them for running off and leaving him, a helpless cripple. He tried saying, "Brother John, jump into my sack!"

Something stirred inside the sack. Francis opened it and found a heap of bones.

"Brother Paul, jump into my sack!"

Another heap of bones.

"Brother Peter, jump into my sack!" Calling them all, up to the eleventh, he found each time, alas, only a little pile of bones half gnawed in two. There was no doubt about it: his brothers had all died together.

Francis was sad. His father also died, leaving him all alone. Then it was his turn to grow old.

His last remaining desire before dying was to see again the fairy of Lake Creno who had made him so prosperous. He therefore set out and reached the place where he had first met her. He waited and waited, but the fairy did not come. "Where are you, good queen? Please appear one more time! I can't die until I've seen you again!"

Night had fallen and there was still no sign of the fairy. Instead, here came Death down the road. In one hand she held a black banner and, in the other, her scythe. She approached Francis, saying, "Well, old man, are you not yet weary of life? Haven't you been over enough hills and dales? Isn't it time you did as everyone else and came along with me?"

"O Death," replied old Francis. "Bless you! Yes, I have seen enough of the world and everything in it; I have had my fill of everything. But before coming with you, I must first bid someone farewell. Allow me one more day."

"Say your prayers, if you don't want to die like a heathen, and hurry after me."

"Please, wait until the cock crows in the morning."

"No."

"Just one hour more, then?"

"Not even one minute more."

"Since you are so cruel, then, jump into my sack!"

Death shuddered, all her bones rattled, but she had no choice but jump into the sack. In the same instant appeared the queen of the fairies, as radiant and youthful as the first time. "Fairy," said Francis, "I thank you!" Then he addressed Death: "Jump out of the sack and attend to me."

"You have never abused the power I gave you, Francis," said the fairy. "Your sack and your stick have always been put to good use. I shall reward you, if you tell me what you would like."

"I have no more desires."

"Would you like to be a chieftain?"

"No."

"Would you like to be king?"

"I wish nothing more."

"Now that you're an old man, would you like health and youth again?"

"I have seen you, and I'm content to die."

"Farewell, Francis. But first burn the sack and the stick." And the fairy vanished.

The good Francis built a big fire, warmed his frozen limbs briefly, then threw the sack and the stick into the flames, so that no one could put them to evil use.

Death was hiding behind a bush. "Cockadoodledo! Cockadoodledo!" crowed the first cock.

Francis did not hear. Age had made him deaf.

"There's the cock crowing!" announced Death, and struck the old man with her scythe. Then she vanished, bearing his mortal remains.

(Corsica)

713

Notes

English translation edited and slightly revised by Italo Calvino.

For each tale, except the first, Calvino lists, in the following order:
(1) the particular version followed—indicated by the original compiler's name (see Bibliography for complete data on the compilations)
(2) the place where the tale was collected
(3) the narrator's name, when known.

1. "Dauntless Little John" (*Giovannin senza paura*). I begin with a folktale for which I do not indicate, in contrast to my procedure in all the other tales, the particular version I followed. As the versions of it from the various regions of Italy are all quite similar, I let myself be freely guided by common tradition. Not only for that reason have I put this tale first, but also because it is one of the simplest and, in my view, one of the most beautiful folktales.

Italian tradition sharply diverges from the Grimms' "Tale of a Boy Who Set Out to Learn Fear" (Grimm no. 4) which is no doubt closer to my no. 80. The type of tale is of European origin and not found in Asia.

The disappearance of the man limb by limb is not traditional, but a personal touch of my own, to balance his arrival piece by piece. I took the finishing stroke of the shadow from a Sienese version (De Gubernatis, 22), and it is merely a simplification of the more common ending, where Little John is given a salve for fastening heads back on. He cuts his head off and puts it on again—backward; the sight of his rear end so horrifies him that he drops dead.

2. "The Man Wreathed in Seaweed" (*L'uomo verde d'alghe*) from Andrews, 7, Menton, told by the widow Lavigna.

This sea tale transfers to an unusual setting a plot well known throughout Europe: that of the younger brother who goes down into the well to free the princess and is subsequently abandoned there himself (cf. my no. 78). Andrews's collection of tales presents no more than brief summaries in French; for this tale, then, as well as the following, taken from the same compilation, I gave free rein to my imagination in supplying details, while adhering to the basic plot. I chose the name Baciccin Tribordo (Giovanni Battista Starboard) to replace the original name whose meaning is not very clear. In the original text, the princess is abducted by a dragon instead of by an octupus, and the dragon changes into a barnacle, which seemed to me too easy to catch.

3. "The Ship with Three Decks" (*Il bastimento a tre piani*) from Andrews, 2 and 27, Menton, told by Giuanina Piombo *dite* La Mova, and by Angelina Moretti.

Prosperous sea-trading, with unusual cargoes coming into ports where the merchandise is highly prized, is a metaphor of luck in the popular mind. It recurs in diverse folktales and is woven into various plots (cf. my no. 173, from Sicily). In this tale from the Italian Riviera border, the curious motifs of the ship with three decks and of the isles inhabited by animals are incorporated into the widespread type featuring the enchanted filly (in one of Andrews's versions, advice is given by the horse) and grateful animals (cf. my nos. 24 and 79). I have freely rendered the two versions summarized in French by Andrews.

4. "The Man Who Came Out Only at Night" (*L'uomo che usciva solo di notte*) from Andrews, 14 and 21, Menton, told by Iren Gena and Irene Panduro.

A tale full of oddities, the most striking of which is that of women constables, given as a historical fact regarding a particular police system. In Andrews's first variant, the bridegroom turns into a toad.

5. "And Seven!" (*E sette!*) from Andrews, 4, 23, 47. (The first two were collected in Menton, the third near Ventimiglia.)

Marriage anecdotes and fairy-tale initiation motifs (the secret name to remember) are blended in this old story widespread in Europe (of English, Swedish, or German origin, according to scholars), subjected to literary treatment in the seventeenth century in Naples (Basile, IV, 4) and well known throughout Italy.

6. "Body-without-Soul" (*Corpo-senza-l'anima*) from Andrews, 46, Riviera ligure.

This Ligurian Jack differs from fellow heroes and liberators of princesses by his systematic cautiousness bordering on distrust (he is one of the few who, the minute he receives a magic gift, must test it before he is able to believe in it). In that respect he takes after his mother, who will not let him go out into the world until he has given proof of perseverance by felling the tree with his kicks. I have been faithful to the original version while aiming to endow it with a particular rhythm.

7. "Money Can Do Everything" (*Il danaro fa tutto*) from Andrews, 64, Genoa, told by Caterina Grande.

This story, of oriental origin (found in the *Panchatantra*), stresses in its Genoese version a utilitarian and commercial moral all its own. (The final remark of the king was even too harsh along that line, so I decided to give credit also, as is meet, to cleverness. . . .)

8. "The Little Shepherd" (*Il pastore che non cresceva mai*) from Guarnerio (*Due fole nel dialetto del contado genovese* collected by P. E. Guarnerio,

Genoa, 1892), Torriglia, near Genoa, told by the countrywoman Maria Banchero.

A feature of this Genoese variant of the widespread tale of the "three oranges" includes encounters with creatures like those in the paintings of Hieronymus Bosch—tiny fairies rocking in nutshells or eggshells. We meet the same beings in another Genoese version (Andrews, 51).

9. "Silver Nose" (*Il naso d'argento*) from Carraroli, 3, from Langhe, Piedmont.

Bluebeard in Piedmont is Silver Nose. His victims are not wives but servant girls, and the story is not taken from chronicles about cruel feudal masters as in Perrault, but from medieval theological legends: Bluebeard is the Devil, and the room containing the murdered women is Hell. I found the silver nose only in this version translated from dialect and summarized by Carraroli; but the Devil-Bluebeard, the flowers in the hair, and the ruses to get back home were encountered all over Northern Italy. I integrated the rather meager Piedmont version with one from Bologna (Coronedi S. 27) and a Venetian one (Bernoni, 3).

10. "The Count's Beard" (*La barba del Conte*). Published here for the first time, collected by Giovanni Arpino in July 1956, in certain villages of southern Piedmont: Bra (told by Caterina Asteggiano, inmate of a home for old people, and Luigi Berzia), in Guarene (told by Doro Palladino, farmer), in Narzole (told by Annetta Taricco, servant woman), and in Pocapaglia.

This long narrative, which writer Giovanni Arpino has transcribed and unified from different versions with variants and additions from Bra and surroundings, cannot in my view be classified as a folktale. It is a local legend of recent origin in part (I am thinking, for instance, of the geographical particulars given), that is, not prior to the nineteenth century, and containing disparate elements: explanation of a local superstition (the hairpins of Witch Micillina), antifeudal country legend such as one finds in many northern countries, curious detective-story structure à la Sherlock Holmes, many digressions nonessential to the story (such as the trip from Africa back to town—which Arpino tells me also exists as a separate story—and all the allusions to Masino's past and future adventures which lead to the conclusion that we are dealing with a "Masino cycle," Masino being the wily hero and globetrotter from a country whose inhabitants are reputed to be contrastingly slow and backward), verse (of which Arpino and I have presented only as much as we could effectively translate), and grotesque images which seem rooted in tradition, such as the sacks under the hens' tails, the oxen so thin that they were curried with the rake, the count whose beard was combed by four soldiers, etc. . . .

11. "The Little Girl Sold with the Pears" (*La bambina venduta con le pere*) from Comparetti, 10, Monferrato, Piedmont.

I changed the name Margheritina to Perina (Pearlet), and I invented the motif of the peartree and the little old woman (in the original, the magic props come from the king's son, who is under a spell), to reinforce the pear/girl link.

12. "The Snake" (*La Biscia*) from Comparetti, 25, Monferrato, Piedmont.

The luxuriant story from *The Facetious Nights* (III, 3) about Biancabella and the serpent, one of Straparola's finest, is here told, on the contrary, in bare rustic simplicity, in the midst of meadows ready for mowing, fruits, and seasons. The episode of the pomegranate tree with its fruit that cannot be plucked was added by me to fill out a somewhat sketchy passage in the Piedmontese version. I took it from a Tuscan variant (Gradi), based on motifs from this tale and others, where supernatural help comes from a red and gold fish.

13. "The Three Castles" (*I tre castelli*) from Comparetti, 62 and 22, Monferrato, Piedmont.

These two Piedmontese tales are variants of a single type. I took the beginning from one and concluded with the other. Nothing was added; I merely underlined a few elements already in the text (such as the tax collector) and the rhythm.

14. "The Prince Who Married a Frog" (*Il principe che sposò una rana*) from Comparetti, 4, Monferrato, Piedmont.

The tale of the frog bride is common to all of Europe; scholars have counted 300 versions. Comparing it, for instance, with Grimm, no. 63, or with Afanas'ev's "The Frog Prince," this variant which we can classify as distinctly Italian (since it shows up uniformly throughout the Peninsula, even if slinging to locate the bride is rather rare) stands out in its near-geometrical logic and linearity.

15. "The Parrot" (*Il pappagallo*) from Comparetti, 2, Monferrato, Piedmont. See my remarks on this folktale in the Introduction, p. xxx–xxxi. I have taken the liberty of doctoring the two versions published by Comparetti—the Piedmontese one and a Tuscan one, from Pisa (1)—and I heightened the suspense by placing the interruptions at the crucial moments.

16. "The Twelve Oxen" (*I dodici buoi*) from Comparetti, 47, Monferrato, Piedmont.

The folktales about the sister who rescues her brother or brothers changed into animals can be divided into two groups: the one where the seven sons are under a curse (as in Basile, IV, 8, or in Grimm, 9 and 25), and the other where the sole brother is transformed into a lamb (as in Grimm, 11, or in my no. 178). The brothers are most commonly transformed into birds (swans, ravens, doves), and the first literary manifestation of the motif dates back to the twelfth century; the latest is possibly Andersen's "Wild Swans."

17. "Crack and Crook" (*Cric e Croc*) from Comparetti, 13, Monferrato, Piedmont.

This is one of the oldest and most famous tales, which has occupied the attention of scholars for generations. The Piedmontese version I followed is faithful to the oldest tradition and includes the curious character-names and a brisk dose of rustic cunning. Herodotus (*Histories*) tells in detail about Egyptian King Rhampsinitus's treasure, chief source of the vast narrative tradition concerning wily robbers put to the test by a ruler. The beheading of a cadaver so it will not be recognized is also encountered in Pausanias, who presents the myth of Trophonius and Agamedes (*Description of Greece*, IX, 372). Either through the Greeks or through oriental tradition the tale entered medieval literature, in the various translations of the *Book of the Seven Sages* and other Italian, English, and German texts. Literary versions by Italian Renaissance story writers are numerous.

18. "The Canary Prince" (*Il Principe canarino*) from Rua (in *Archivio per lo studio delle tradizioni popolari*, Palermo-Turin, VI [1887], 401), Turin.

This folktale from Turin, with its balladlike pathos, develops a medieval motif, which is also literary. (But Marie de France's *lai, Yonec*, is quite different, being the story of an adultery.) My personal touches here include the prince's yellow suit and leggings, the description of the transformation in a flutter of wings, the gossip of the witches who traveled the world over, and a bit of stylistic cunning.

19. "King Crin" (*Re Crin*) from Pitrè (in *Archivio per lo studio delle tradizioni popolari*, I [1882], 424), Monteu da Po, Piedmont.

Of illustrious origin (since it is certainly related—at least in the motif of the bridegroom who cannot be seen in his true form—to the myth of Amor and Psyche), the folktale about the swine king is one of the most widespread in Italy. This Piedmontese version has a beginning full of brio. The development repeats—with the walnuts to be cracked, spying on the sleeper, etc.—a motif also common to other types and of which my no. 140 presents a richer version.

20. "Those Stubborn Souls, the Biellese" (*I biellesi, gente dura*) from Virginia Majoli Faccio (*L'incantesimo della mezzanotte* [*Il Biellese nelle sue leggende*], Milan, 1941), Valdengo, Piedmont.

This tale is also found in Trieste, starring the Friulians (Pinguenti, 51).

21. "The Pot of Marjoram" (*Il vaso di maggiorana*) from Imbriani, p. 42, Milan.

Gallant banter in verse sustains this thin narrative, the plot of which is known throughout Europe and already in germ in Basile's *Viola* (II, 3) together with other motifs. The verse, which I have handled quite freely, varies from region to region, but always follows the same repertory of jokes.

22. "The Billiards Player" (*Il giocatore di biliardo*) from Imbriani, p. 411, Milan.

The only original part is the beginning, which attests the tale's urban origin—the young man who spends all his time in cafés, the unknown gentleman challenged to a game of billiards—then the action moves onto the supernatural plane. But perhaps under the impetus of that particular opening, the ancient motif of girls in their bath (the dove girls) is charged with a touch of idle urban fantasy. The aborted ending is my own.

23. "Animal Speech" (*Il linguaggio degli animali*) from Visentini, 23, Mantua, Lombardy.

The man who understands animal speech will be pope: it is an old European superstition, related also by the Grimms (no. 33) with a medieval flavor of its own and smacking of half-satanical theology and the wisdom of the "bestiaries." The revelation through the horses' neighing of the father's cruel order and the dog's sacrifice of himself for his master are my own developments of the theme.

24. "The Three Cottages" (*Le tre casette*) from Visentini, 31, Mantua, Lombardy.

This is the famous tale of the three little pigs, which has a new twist in its Mantuan version with three little sisters instead of three animals that build houses. But it retains the humorous and scary elements characteristic of the well-known Walt Disney creation. The three little pigs in Italy are often three geese (see my no. 94).

25. "The Peasant Astrologer" (*Il contadino astrologo*) from Visentini, 41, Mantua, Lombardy.

Here we have the age-old farce about peasant cunning, with the country detective, astrology turned to ridicule (which is rare in folklore), and a slight grudge on the part of people who labor in the fields toward the house servants, a grudge that gives way at once to complicity.

26. "The Wolf and the Three Girls" (*Il lupo e le tre ragazze*) from Balladoro (in *Giambattista Basile*, Naples, IX [1905], no. 6), Pacengo, Lake Garda.

"Little Red Riding Hood" cannot be called popular in Italy. It must have reached Lake Garda from Germany (its close is more like the Grimms' version than Perrault's), but with the variant of three girls. The nonsense rhymes in the dialogue with the wolf are a whim of my own in keeping with the reply at the outset of the original. What I called cakes are, in the original, *spongàde*—a type of rolls with raisins. I also omitted an episode that would have been too gruesome in this meager text: the wolf kills the mother and makes a doorlatch cord out of her tendons, a meat pie out of her flesh, and wine out of her blood. The little girl, pulling on the doorlatch, says, "What a soft cord you've put here, Mamma!" Then she eats the meat pie and drinks the wine, with comments in the same vein.

According to scholars, "Little Red Riding Hood" has never been a "popular" tale in Italy or elsewhere, because its reputation is not based on an oral tradition but on Perrault's vivid "Petit Chaperon Rouge" with its ballet rhythm, or on the Grimms' cruder version. The Italian counterparts (all northern) closely follow the literary tradition. More interesting is the version from Abruzzo (see my no. 116), which combines different motifs but retains the famous dialogue between the little girl and the wolf in bed.

27. "The Land Where One Never Dies" (*Il paese dove non si muore mai*) from Balladoro (in *Lares*, I [1912] fasc. 2–3, pp. 223–26), Verona.

Among the many tales or legends of temporary victory over death, this folktale is outstanding for its particular gothic flavor provided by those old people and their punishments, bones lying on the ground, and the cart loaded with worn-out shoes. Distinctive also are the landscapes transformed in the course of time and the estrangement of the man who comes back to town several generations later. All I added here was the length of the old men's beards.

28. "The Devotee of St. Joseph" (*Il devoto di San Giuseppe*) from Balladoro, 42, Verona.

This folktale shows an almost Voltairean impudence, but that should not mislead one regarding the legend's origins and significance. At its heart is a theological issue and, as such, it is well known all over Italy and in other Catholic countries. Alexandre Dumas *père* in two of his books of Neapolitan recollections, *Le Corricolo* and *L'Histoire des Bourbons de Naples*, relates it as told in the pulpit by a famous preacher as evidence of the supremacy of the cult of St. Joseph among Neapolitan mendicants; the sinful devotee was the famous bandit Mastrilli. A study on the subject (Guido Tammi, *Il devoto di San Giuseppe nella leggenda popolare*, Rome: Edizioni dell'Ateneo, 1955), continuing Pitrè's research and comparing various popular versions, maintains that the legend, reminiscent in structure of medieval "Arguments" between Divine Justice and the Virgin interceding for sinners, is crystallized in this form at the time of the Counterreformation under the impetus of the church's emphasis on devotion to St. Joseph. A certain Father Giovanni Crisostomo of Termini told the tale in a Palermo pulpit in 1775 and was denounced and seized by the Inquisition.

29. "The Three Crones" (*Le tre vecchie*) from Bernoni, 16, Venice.

In this particular Venetian version, the crone takes a wad of aromatic jam into her mouth and spits it on the hand of the passing youth. Instead, I have her drop her handkerchief.

30. "The Crab Prince" (*Il principe granchio*) from Bernoni, III, 10, Venice.

A rather rare tale, in an original version, entirely aquatic with its intricate underwater labyrinth, its courageous heroine who swims, and the little ballet on cliffs with the eight maids of honor. My personal touches here are limited to the following: I bring out the princess's absorption in the customs of fish; I focus on the underwater routes that are a bit hazy in the original; and I add a choreographic note to the scene with the ladies on the rocks.

31. "Silent for Seven Years" (*Muta per sette anni*) from Bernoni, III, 12, Venice.

A motif widespread in Europe, yet rare in Italy—heroic persistence in pretending to be mute in order to save one's brothers—gives singular narrative power to this folktale including family and childhood fictional elements, as well as aspects of demonic legends and tales of adventure. The tale also contains the well-known medieval theme of the persecuted wife, enhanced here by the victim's silence.

32. "The Dead Man's Palace" (*Il palazzo dell'Omo morto*) from Bernoni, III, 13, Venice.

This tale is identical to the beautiful story that opens the *Pentameron*. Basile's narrator need only say "canal" in place of "street," and the melancholy of the tale would have a well-nigh natural setting in Venice, while the Dead Man's palace becomes one of those decaying palaces on the lagoon, with the Moorish slave evoking booty from the Levant. The tale closes on an unusual note of melancholy.

33. "Pome and Peel" (*Pomo e Scorzo*) from Bernoni, III, 2, Venice.

Moments of grand tragic theater are presented in the story (most likely of Indian origin) of the brother turned to stone in Basile's *Cuorvo* (IV, 9). Gozzi uses its subject for his finest play, *Il Corvo* (The Raven). More moving is the Grimms' "Faithful John" (no. 6). Peel (Scorzo or Bella Scorza), brother-servant, is in the popular versions that open with the queen and her handmaiden simultaneously conceiving as they respectively eat an apple and its peel.

34. "The Cloven Youth" (*Il dimezzato*) from Bernoni, III, 9, Venice.

The story of Pietro Pazzo (Peter the Mad), who can realize every desire through the power transmitted to him by a fish and even impregnate the princess that laughed at him, was told in a lively manner by Straparola (III, 1), and scholars tend to consider it of Italian origin. Similar to it is Basile's story about Peruonto (I, 3), who receives magic powers because of shading three sons of a fairy as they slept. Peculiar to this Venetian version is the halved hero in place of the usual simpleton.

35. "Invisible Grandfather" (*Il nonno che non si vede*) from Bernoni, III, 14, Venice.

The punishment of the girl, in the Venetian version, is the transformation of her head into a goat's, with the ears of a hare. But since that makes the tale too much like another one I have transcribed (no. 67, "Buffalo Head"), I substituted for those metamorphoses the appearance of a beard on her as in other variants (Tuscany, Comparetti, 3, and Abruzzo, Finamore, 1). The Tuscan "Buffalo Head" and the many other versions of that type are among the most mysterious and "ethnological" Italian folktales. But the Venetian variant has a distinct logic all its own. Thus the girl's punishment for leaving something behind is accounted for: her oversight has prevented the breaking of an evil spell.

36. "The King of Denmark's Son" (*Il figlio del Re di Danimarca*) from Sabatini (*El fio del re de la Danimarca*, popular Venetian tale edited by Francesco Sabatini, in *Gli Studi in Italia*, Rome, III [1880], vol. II, fasc. 2), Venice.

Goodness knows how this folktale, woven entirely of traditional motifs, got the title, "King of Denmark's Son." Perhaps it is from some theatrical memory associated with an atmosphere of vague melancholy.

37. "Petie Pete versus Witch Bea-Witch" (*Il bambino nel sacco*) from Gortani, p. 118, Cedarchis, Friuli.

Regarding characteristics of tales for children, see my remarks in the Introduction (p. xxx). I have attempted to give a typical example here, taking all the liberties to which the rudimentary quality of the original texts entitled me. I chose the names "Pierino Pierone" (Petie Pete) and "Strega Bistrega" (Witch Bea-Witch) while "Margherita Margheritone" comes from the Friulian verses. Throughout, I have put key nonsense rhymes into the mouth of all the characters. Other touches of mine include the quail's call and the hunter (in the original, the child yells, and some boys shooting marbles come and free him) and the stepladder of pots (in the original the witch tries to climb up on the mantelpiece, using spoon, knife, and fork for a ladder). The story about the little boy in the sack is known throughout Europe, and in Northern and Central Italy.

38. "Quack Quack! Stick to My Back!" (*Quaquà! Attaccati là!*) from Zorzut, II, p. 140, Cormons, Friuli, told in 1911 by Giovanni Minèn, 66-year-old parish organist.

In many versions, all a bit loosely constructed, this old and mysterious tale (common to all Europe) has come down to us about the princess who does not laugh. One of the first literary treatments of it is in an English poem of the fifteenth century, "The Tale of a Basyn." This Friulian version is not very different from the Grimms' no. 64, but it is richer and more amusing. The protagonist here has scalp disease. Scalp disease in folktales is at times a sign of good luck, at others a sign of wickedness.

39. "The Happy Man's Shirt" (*La camicia dell'uomo contento*) from Zorzut, II, p. 47, Cormons, Friuli, told in 1912 by Orsola Minòn, housewife.

A story with a famous literary source. Starring Alexander the Great, it figures in the *Pseudo-Callisthenes* and from there passes into medieval Latin legends and oriental narratives.

40. "One Night in Paradise" (*Una notte in Paradiso*) from Zorbut, p. 169, Cormons, Friuli, told in 1913 by Giovanni Minèn.

This legend incorporates the grand medieval motifs—Death, the Hereafter, Time. But here, the contemporary narrator brings modern times into the picture, with the town grown into a large city with tramways, motorcars and airplanes.

41. "Jesus and St. Peter in Friuli" (*Gesú e San Pietro in Friuli*).

The cycle of popular legends about Jesus and the Apostles who go about the world is common throughout Italy, and almost always these short narratives pivot around St. Peter, with whom the people are on very familiar terms. Popular tradition makes of Peter a lazy man, glutton, and liar, whose elementary logic is always contrary to the faith preached by the Lord, whose miracles and acts of mercy never fail to put Peter to shame. Peter, in this sort of common man's gospel, is the human opposite of the divine, and his relationship with Jesus is somewhat like Sancho Panza's with the *hidalgo*. I found the largest number of legends of this cycle in Friuli and Sicily. I have given a selection from Friuli and, further on, a selection from Sicily; but almost all of them are common to both regions as well as to Christendom on the whole, together with another large cycle of legends and tales about the Hereafter, where St. Peter officiates as doorkeeper to Heaven. What prompted me to linger over the Friulian tradition was not only the wealth of material collected (as early as the middle of the last century by Caterina Percoto, and later by Gortani and Zorzùt, who has published an ample selection of it), but also the harmonization of the tales' episodes and religious moral with the rugged landscape which is always present or implied in the Friulian oral narrative—rough, completely concrete, devoid of all mysticism, yet not forbidding.

I. "How St. Peter Happened to Join Up with the Lord" (*Come fu che San Pietro è andato col Signore*) from Zorzùt, p. 22, Cormons, told in 1909 by the widow Caterina Braida Minèn, 40-year-old housewife.

I have tried to re-create a popular narrative pace in place of Zorzùt's modern literary dialogue, which is too "poetic" at times.

II. "The Hare Liver" (*La coratella di lepre*) from Zorzùt, p. 105, Cormons, told in 1913 by Giovanni Minèn.

III. "Hospitality" (*L'ospitalità*) from Gortani, p. 12, Camia.

IV. "Buckwheat" (*Il grano saraceno*) from Percoto (in *La Ricamatrice*, Milan, 1 September 1865, p. 223), Friuli.

One of the popular legends included among the writings in Friulian dialect of the "peasant countess" Caterina Percoto (1812–87), lady of letters, patriot, and outstanding woman of her time.

42. "The Magic Ring" (*L'anello magico*) from Schneller, 44, Trentino.

Many traditional motifs (of Asiatic origin) are interwoven here, giving an impression of improvisation, with intervals of moralizing, as if to make up for the uneven pace.

43. "The Dead Man's Arm" (*Il braccio di morto*) from Schneller 35, Trentino.

Macabre mountain story with a wealth of gothic details, to which I naturally made my own contribution.

44. "The Science of Laziness" (*La scienza della fiacca*) from Pinguentini, 30, Trieste.

A Triestine proverbial joke, linked to the ancient theme of striving to excel in laziness and to the traditional satire of the lax morals of the Levant. A whole landscape is conjured up by just a few touches—the garden shaded by the fig trees, the cushions, the immobility of the men.

45. "Fair Brow" (*Bella Fronte*) from Ive, 3, Rovigno d'Istria.

The grateful dead man is a motif of diverse medieval legends, including *L'Histoire de Jean de Calais*, which has come into the narrative tradition of our seacoast regions, drawing inspiration from misfortunes endured at the hands of the Turks, an inevitable element in Italian sea stories.

46. "The Stolen Crown" (*La corona rubata*) from Forster, 10, Zara.

This tale is full of old motifs pieced together, as if told by someone narrating without remembering the story exactly. But at the same time it never flags in imagination, which makes it a pleasure both to read and to transcribe. Though I added a touch here and there, I have preserved the tale's carefree incoherence.

47. "The King's Daughter Who Could Never Get Enough Figs" (*La figlia del Re che non era mai stufa di fichi*) from Bagli, 5, Castelguelfo, Emilia.

I found examples of strange contests to win the princess's hand (such as the pasturing of hares) also in Tuscany and in Lombardy. They appear, moreover, in the folktales of all of Europe (see, for instance, Grimm, no. 165).

48. "The Three Dogs" (*I tre cani*) from Bagli, p. 40, Imola, Emilia, told by Teresa Ronchi.

Specialists have devoted extensive studies to this folktale, as well as to a related one about the two brothers (see my no. 58, "The Dragon with Seven Heads"). Ranke found 368 versions in all Europe, 14 of them in Italy. Its origins may well be anterior to the "two brothers" type, and its primary center of diffusion was France. I reduced to a minimum the episode of the liberation of the princess from the dragon, as it is fully related in "The Dragon with Seven Heads." The ending with the transformation of the dogs into kings is taken from a Tuscan version (Pitrè, T. 2), Siena, told by Umiltà Minucci, dressmaker. There are other versions in Northern, Central, and Southern Italy and in Sardinia.

49. "Uncle Wolf" (*Zio Lupo*) from Toschi, Faenza, Emilia.

Uncle Wolf, Barbe Lof, Barbe Zucon, Nonno Cocon: it is the simplest folktale for children in the popular tradition, widespread in Northern and Central Italy, with its rudimentary elements of gluttony and excrement and with its progression of fear. This extremely simple type—and I followed one of the richest versions—will lead to the perfect grace of "Little Red Riding Hood."

50. "Giricoccola" from Coronedi, S. 2, Bologna.

The moon's journey across the sky gives this Bolognese Cinderella or Snow White a melancholy refinement all her own.

51. "Tabagnino the Hunchback" (*Il gobbo Tabagnino*) from Coronedi, S. 38, Bologna.

Common to all of Italy (as to all of Europe, especially the North), this folktale about the successive ruses in the ogre's house was told by Basile as *Il corvetto* ("The Jackdaw") in *The Pentameron* (III, 7). But Basile's version does not fully reflect the richness and cleverness of popular tradition, which gives free rein to its fantasy for the meeting of the tests.

52. "The King of the Animals" (*Il Re degli animali*) from Coronedi, S. 26, Bologna.

Among the many folktales about the enchanted palace, this one differs from the others with its strange oriental atmosphere (teeming with animals and mysterious as a tapestry). It is also rather incoherent, almost like a vernacular *Alice in Wonderland*, candid and full of amazement. I threw a little light on the maid of honor, who, in the original, remains in the shadows. And I let the ring, moreover, be a gift from the good aunt rather than from the king of the animals as it is in the original version.

53. "The Devil's Breeches" (*Le brache del Diavolo*) from Coronedi, S. 28, Bologna.

This is a tale about a bargain with the Devil, but differing from other tales of the same type. Curious details of the present Bolognese version: the protagonist oppressed by his good looks, dodging employers' wives who are in love with him and thus keep him from working for anyone in the long run; the revolting description of filth, in contrast with the bridal trousseau (the trousseau is an elementary fairytale motif, quite appropriate when—as in this case, I believe—the narrator is a woman). I presented evidence not in the original of the sisters' actually offering their souls to the devil; otherwise damnation for a simple envious impulse would have been too severe a punishment, with all the cruelty that customarily results from envy in folktales.

√ 54. "Dear as Salt" (*Bene come il sale*) from Coronedi, 3, Bologna.

This tale opens, exactly like *King Lear*, with a "test of love" imposed by a king on his three daughters. But similarities to Shakespeare's masterpiece stop right there. The rest of it, in general, is related to Perrault's *Peau d'asne*, in which the daughter flees from home to avoid her father's unnatural passion for her. I picked, among versions all over Italy, this brisk and elegant Bolognese variant because of the particular logic of the girl's repartee—"You were as dear as salt to me." As Pitrè has noted, "in the Bolognese dialect, *amôur* means not only 'love' but also 'savor.' " I took the beginning with three different-colored thrones from a *Cinderella* of Parma.

55. "The Queen of the Three Mountains of Gold" (*La Regina delle Tre Montagne d'Oro*) from Coronedi, S. 31, Bologna.

Among the folktales (very widespread in Italy) of the enchanted castle, the beautiful maiden to be freed, her disappearance and the subsequent search for her, this type is distinguished by the motif of the pond from which the girl is liberated by degrees and by the carrousels of animals.

56. "Lose Your Temper, and You Lose Your Bet" (*La scommessa a chi primo s'arrabia*) from Coronedi, 18, Bologna.

Be patient, never get angry; the old rule preached by the rich to the poor backfires in the face of the man who wishes to take advantage of it, in this ancient motif of the bet which takes shape here in a jovial country tale.

57. "The Feathered Ogre" (*L'orco con le penne*) from Pitrè, T. 24, Garfagnana Estense, Tuscany, told by Rosina Casini.

Just as there are tales of distrust, so are there tales of generosity, almost independently of the event narrated, thanks to the rhythm imparted by the narrator. This curious and lively Tuscan tale is the triumph of the obliging man who knows that helping others is a small matter and need not originate above. The title "Feathered Ogre" is my own invention; the original speaks of a vague "beast" (whose traits are nonetheless the ogre's). Also the ending with the retention of the ogre on the ferry is mine, but it does not strike me as arbitrary, since the same thing happens in the Grimms' tale no. 29.

58. "The Dragon with Seven Heads (*Il Drago dalle sette teste*) from Nerucci, 8, Montale Pistoiese, Tuscany, told by the girl Elena Becherini.

One of the most widespread folktales in Europe and in Italy, and the object perhaps of the most detailed folkloristic studies. Ranke records 800 versions of it, including 27 Italian ones (in addition to the 1,100 versions of the similar "Three Dogs"; see my no. 48). In its most complete form (aside from the derivations of the simple motifs such as the liberation of the princess from the monster, which has flourished ever since the myth of Perseus and Andromeda, down to the medieval legend of St. George and the Dragon), it appears to be of European origin; Northern France, according to Ranke, would be its principal center of radiation at the outset of the Middle Ages. The Italian versions are quite rich and harmonious (better, I would say, than Grimm nos. 60 and 85). With this Tuscan version I consolidated details from others.

59. "Bellinda and the Monster" (*Bellinda e il Mostro*) from Nerucci, 16, Montale Pistoiese, Tuscany, told by the widow Luisa Ginanni.

To produce the richest version possible of this very famous folktale common to all Italian regions, I combined this Tuscan version with one similar in Roman dialect (Zanazzo, 27) and I added a motif from an Abruzzese version—the tree of sorrow and laughter (from De Nino, 29).

60. "The Shepherd at Court" (*Il pecoraio a Corte*) from Nerucci, 7, Montale Pistoiese, Tuscany, told by the girl Elena Becherini.

The tale of magic turns into a story of cunning with a certain raciness and also includes a fearful note, such as the encounter with the stranger whose bed is the rock, and a touch of grotesque humor with the dance the shepherd compels the entire court to dance. Nerucci's text is, as always, the most verbose; I toned it down and added only a few particulars, dramatizing the boy's bad manners at court and the spell of the dance; I also had the bean soup overturned on the prison floor. Variants of the tale are found throughout Italy.

✓ 61. "The Sleeping Queen" (*La Regina Marmotta*) from Nerucci, 46, Montale Pistoiese, Tuscany, told by Pietro di Canestrino, laborer.

Concerning the narrative style of this folktale, see Introduction, p. xxiv.

62. "The Son of the Merchant from Milan" (*Il figlio del mercante di Milano*) from Nerucci, 19, Montale Pistoiese, Tuscany, told by Ferdinando Giovannini, tailor.

In the Introduction (p. xxv), I took this tale as an example of how, in Nerucci's book, one moves from the folktale to the short story of fortune or the bourgeois novel of adventure. Note that this is one of the rare folktales with an unhappy ending, which brings it right into line with many modern narratives. (One must remember, however, that when desires are satisfied in folktales, the tales as a rule close with the loss of riches gained through magic.) The riddle motif is common to tales of all regions.

63. "Monkey Palace" (*Il palazzo delle scimmie*) from Nerucci, 10, Montale Pistoiese, Tuscany, told by the tailor Ferdinando Giovannini.

The episode, quite common in folklore, of the prince who weds an animal is carried to the extreme here, with an entire population changed into monkeys and a display of grotesque effects typical of a grand ballet. I have accented this display by describing the monkeys in the city and their transformation.

64. "Rosina in the Oven" (*La Rosina nel forno*) from Nerucci, 32, Montale Pistoiese, Tuscany, told by the widow Luisa Ginanni.

In the country tales the beauty of girls is described as radiantly white. This Cinderella, thanks to the spell cast over her by toads, shines in the night; a ray of sunshine destroys her; she will come back to life in the fire.

65. "The Salamanna Grapes" (*L'uva salamanna*) from Nerucci, 40, Montale Pistoiese, Tuscany, told by the widow Luisa Ginanni.

One of the richest tales of *The Arabian Nights* of Galland ("Histoire du Prince Ahmed et de la fée Pari-Banou"), full of descriptions of marvels and treasures, in a bare Tuscan version. The sequel, or the story of the fairy Pari-Banou, lengthy and rich in Galland, is reduced in Nerucci's version to an unnecessary appendix; I thus chose to conclude the tale with general disappointment, a type of close traditionally common to some popular stories about contests.

66. "The Enchanted Palace" (*Il palazzo incantato*) from Nerucci, 59, Montale Pistoiese, Tuscany, told by the farmer Giovanni Becheroni.

One of the finest folktales about the enchanted palace, which can be classed as a variant of the Amor and Psyche type, with the invisible wife lost and regained by the husband, rather than the reverse. The charm of this Tuscan version derives from the person of the solitary prince absorbed in his books; I have aimed to accent his character by making him inept at hunting, with his pursuit of the hare leading him to the palace (here I followed a Piedmontese version—Comparetti, 27; in Nerucci, the guardian of the palace is an ill-defined "monster"). The plot is often incoherent, as in the case of the hermit's strange behavior. Nerucci fails to account for the innkeeper's sly drugging of the wine; I justified the act, following the version from Monferrato, by the daughter's claim to Fiordinando's love.

67. "Buffalo Head" (*Testa di Bufala*) from Nerucci, 37, Montale Pistoiese, Tuscany, told by the widow Luisa Ginanni.

One of the most suggestive and mysterious folktales in Italy. Different details appear in its different versions, but the plot is substantially the same, from Venice to Sardinia. The supernatural creature that rears the protagonist can be a lizard, snake, or dragon in the service of fairies, a monster, an ogre or ogress, an old woman, a woman with a bull's head, or someone invisible except for his hands (as in the most rational and civilized version—Bernoni's, source of my no. 35). The offense responsible for the transformation of the girl's face is usually ingratitude, her failure to say thank you when she goes away; sometimes it is leaving an object behind. Usually ingratitude and oversight go hand in hand; rarely is the offense curiosity (the customary forbidden door). The supernatural being always takes revenge by transforming the protagonist's face into the head of some animal (buffalo, goat, cat, or donkey); or else a beard grows out of her face, or a sheep's fleece on her neck; or she may simply become ugly, or even end up with no head at all.

68. "The King of Portugal's Son" (*Il figliolo del Re di Portogallo*) from Nerucci, 25, Montale Pistoiese, Tuscany, told by the farmer Giovanni Becheroni.

Part-folktale and part-romance, and in the end like a piece of news told at fairs, it most likely comes from an old popular poem, *Istoria di Ottinello e Giulia.*

69. "Fanta-Ghirò the Beautiful" (*Fanta-Ghirò, persona bella*) from Nerucci, 28, Montale Pistoiese, Tuscany, told by the widow Luisa Ginanni.

In Basile's *Pentameron* (III, 6), a father feels disgraced to have only daughters and no sons, which inspires Belluccia to pretend to be a man. This theme is better developed in the popular tradition, particularly in Tuscany and in Southern Italy. The humor of the present brisk and precise Tuscan version stems altogether from an affirmation of feminine pluck and resolution—the attitude that always determines the ups and downs of women in men's garb so common in the stories and comedies of the sixteenth and seventeenth centuries.

70. "The Old Woman's Hide" (*Pelle di vecchia*) from Nerucci, 13, Montale Pistoiese, Tuscany, told by the widow Luisa Ginanni.

The image of the beautiful girl who steps out of the old woman's skin (well portrayed in Grimm no. 179) lends charm to this tale, which is one of many variants of the "dear as salt" type, like my no. 54. This Tuscan version in the original begins in a similar way, so for the sake of variety, I took the beginning of an Abruzzese version (Finamore, 26).

71. "Olive" (*Uliva*) from Nerucci, 39, Montale Pistoiese, Tuscany, told by the widow Luisa Ginanni.

The name Olive is encountered in a mystery play (*Rappresentazione di Santa Uliva*) and in a popular poem (*Istoria de la Regina Oliva*). Both tell of a woman with mutilated hands and cursed with the cruelest of vexations. Those motifs are also common to the present Tuscan tale, where cruelty and religious intolerance (manifested in the inhuman Jew) and a fundamentally cheerful nature (dramatized by the girl eating the pears) are calmly and strikingly blended as in Paolo Uccello's predella at Urbino. The story of the persecuted girl with her hands lopped off is common to quite a few versions all over Europe (cf. Grimm no. 31) and Asia (*The Arabian Nights*) and is also found in every region of Italy. According to Pitrè, it "figures in popular narratives blending sacred and profane elements and giving rise to Genoveffa, Orlanda, Florencia, Saint Guglielma, to the daughter of the king of Dacia, the queen of Poland, Crescenzia, and Saint Olive."

72. "Catherine, Sly Country Lass" (*La contadina furba*) from Nerucci, 3 and 15, Montale Pistoiese, Tuscany, told by the widow Luisa Ginanni and the tailor Ferdinando Giovannini.

The tale of a country lass's feats of cunning is common to all of Europe (the oldest written version seems to be a Norwegian saga of the thirteenth or fourteenth century), interspersed with oriental motifs. A few of the witticisms (such as "neither naked nor clothed") also appear in the popular Italian book of the seventeenth century, *Bertoldo, Bertoldino e Cacasenno.*

73. "The Traveler from Turin" (*Il viaggiatore torinese*) from Nerucci, 48 Montale Pistoiese, Tuscany, told by Benvenuto Ginanni, decorator.

A country Robinson Crusoe. Not only is the tale in its general plan like Defoe's novel, it also presents all the marginal similarities such as the father's opposition to the son's seafaring vocation, the shipwrecked man's industry, reflections on the vanity of wealth for the single man. But the source of the story is to be found in that great store of seafaring tales, "The Voyages of Sinbad" (*The Arabian Nights*), particularly the fourth voyage, which the Tuscan tale faithfully repeats, tacking on the sentimental story of the beloved who also ends up in the cavern. (In the Arabian narrative, Sinbad kills all his companions, so as to survive on their victuals.) The oriental motif of the husband buried with the dead wife also comes up, with various justifications, in other popular Italian narratives (see my no. 179, Sicily, and no. 197, Sardinia). About the mysterious bull, *The Arabian Nights* is very vague, speaking only of an animal that feeds on cadavers. The gratuitous naming of places includes the landlocked city of Turin as the birthplace of the seafarer.

74. "The Daughter of the Sun" (*La figlia del Sole*) from Comparetti, 45, Pisa, told by an old woman of the people.

The myth of Danaë is very much alive in Italian folklore and usually serves to introduce the later vicissitudes of the daughter engendered by the Sun. The tale, I believe, can be considered truly Italian, or pretty much so; it is actually found only in Italy, Spain, and Greece, with its crude stories of magic, mutilations, and self-prepared banquets. The bean field is not in the Pisan text but another Danaë tale, from Rufina (Florence), *Faina* (Pitrè, T. p. 9). All the magic feats are in the original, except the girl's passing through the wall and walking on the cobweb—my own inventions.

75. "The Dragon and the Enchanted Filly" (*Il Drago e la cavallina fatata*) from Comparetti I, 17, Pisa, told by the narrator of the preceding tale.

Into a rather unusual plot are woven motifs of the serpent king (cf. my no. 144), Fanta-Ghirò (cf. my no. 69), and the false report of the monstrous childbirth (cf. my nos. 31, 71, 87, 141).

76. "The Florentine" (*Il Fiorentino*) from Comparetti, 44, Pisa, same narrator as in the preceding.

This is the story of Ulysses and Polyphemus translated into a Tuscan peasant story with farmer and priest, local satire, the misery of the Florentine who has nothing to boast of, and the petty cautionary moral of always staying close to home. I have accentuated the character satire in the way the story seems to demand; hence my inclusion of the tale's last sentence.

77. "Ill-Fated Royalty" (*I Reali sfortunati*) from Comparetti, 42, Pisa, same narrator as in the preceding.

This folktale presents several unusual aspects: double plot, intertwining of the different loves and matrimonial exchanges, precision of political plot with conspiracies, coup d'état, and well-defined international relations. In addition, there is the serious tone of a tear-jerker. All those aspects suggest that a romance could be the source of the tale. The geography, on the other hand, is imaginary, with that splendid mountain passage that leads directly into Scotland.

78. "The Golden Ball" (*Il gobbino che picchia*) from Comparetti, 40, Pisa, same narrator as before.

North European (see Grimm no. 91) and also Italian tradition is full of stories about the world underground, with the youngest of three brothers or comrades going below and freeing the princess and subsequently finding himself abandoned down there by his traitorous brothers. I chose this strange and carefree Pisan version, dressing it up a bit toward the end.

79. "Fioravante and Beautiful Isolina" (*Fioravante e la bella Isolina*) from Nuti (tale in Pisan vernacular, recorded and annotated by Oreste Nuti, Milan 1878), Pisa, told by Tonchio di Pitolo, "farmer and schoolmaster."

In the grand line of stories about the prince disguised as a servant, who frees a princess from a spell with the help of a talking filly and three grateful animals (cf. my no. 6), this Pisan folktale is in a vernacular charged with color (which I have toned down in keeping with the book's overall tone) and interspersed with lines from Tasso and moral explanations. It presents rare fanciful details (the two dolphins, for instance, wrangling over the tress) and an even rarer subplot about the weaver in love, the poor girl who sacrifices herself to save her beloved and consequently brings about his marriage to a princess. Are these variants "literary," or a part of genuine popular tradition? An episode with a sad dénouement is unusual indeed in folktales, even if overshadowed by the usual happy ending of the princely wedding. Also, there is a bitter resignation to class barriers, which are cynically shrugged off by the decision to provide a dowry for the unfortunate girl in love, so that she can marry someone else.

80. "Fearless Simpleton" (*Lo sciocco senza paura*) from Pitrè, T. 39, Florence, told by Paolina Sarta, "who had heard it in Leghorn and told it in a very scatterbrained way."

Unlike Dauntless Little John of my no. 1, this tale's hero, the titular protagonist of the Florentine original, takes no account of dangers or strange occurrences but is an irresponsible prattler. The notable feature of this version is the protagonist's speech, which makes him a full character—and that is rare in oral narrative (cf. Grimm no. 4). Pitrè's text closes with the hero's head cut off and put on backward (see note on my no. 1); but since that brings an element of fantasy into an otherwise realistic narrative, I thought it best to exclude it.

81. "The Milkmaid Queen" (*La lattaia regina*) from Pitrè, T. II, 25, Leghorn.
The girl prisoner brought by the wind or a bird to an ogress's house is already in Basile (IV, 6).

82. "The Story of Campriano" (*La storia di Campriano*) from Giannini, 1, Tereglio, near Lucca, Tuscany.

This tale has been identically preserved all over Italy and with the same title probably from the fourteenth century. A. Zenatti, publishing a verse version of 1572 (*Storia del Campriano contadino*, Bologna 1884) lists fourteen printed editions covering the period from 1518 down to the popular editions of his generation. One of the first editions specifies "composed for a Florentine." Straparola relates the same jokes in the story of

Pre' Scarpacifico (I, 3). Every version whether literary or popular, with the exception of the present version from Lucca followed by me, contains the final episode of Campriano closed up in a bag and the trick to bag someone else. I did not reinsert it, because, with the three traders swindled, the story is already complete, and the ending—although highly traditional—strikes me as arbitrarily tacked on. The proof is that it exists also separately in literary and oral versions.

83. "The North Wind's Gift" (*Il regalo del vento tramontano*) from Comparetti, 7, Mugello, Tuscany.

Known in all Europe and Asia, the folktale about magic gifts, dispensers of food and wealth successively taken from their rightful owner and later regained by means of another magic gift that delivers blows (see my no. 127), came into this Tuscan variant on a ripple of rustic rebellion—a very slight breeze, not a north wind like the one Geppone recognizes as the sole cause of his hardships and also as his only possible rescuer. Note that the prior-landlord is never explicitly condemned as the thief he is in reality; the farmers blames his misfortunes exclusively on the wind, or on his talkative wife. But submission gives way in the end to fury, in the blows that rain on the prior.

84. "The Sorceress's Head" (*La testa della Maga*) from Pitrè, T. 1, "told by Beppa Pierazzoli, of Pratovecchio in the upper Val d'Arno, longtime resident of Florence."

This is obviously derived from the myth of Perseus, of which several motifs are repeated: Danaë's isolation, which still does not prevent her from conceiving; the Medusa enterprise imposed on Perseus by King Polydectes; the power to fly (in the myth, with winged sandals); the three Gray Women, daughters of Phorcys, with only one eye and one tooth among the three of them; the silver shield that serves as a mirror for viewing Medusa; blood that changes into snakes; the freeing of Andromeda from the monster, the petrification of King Polydectes. I found no other versions of this in collections of popular tales, and that would lead to the conclusion that we are dealing here with a recent popular remolding of the classical myth.

85. "Apple Girl" (*La ragazza mela*) from Pitrè, T. 6, Florence, told by Raffaella Dreini.

Raffaella Dreini is the best storyteller recorded in 1876 by Giovanni Siciliano, one of Pitrè's collaborators. As I said in the Introduction (p. xxix), the secret of the narrative lies in the metaphor—the evocation of the girl's freshness by that of the apple. (A near-surrealistic effect is produced by blood flowing from the punctured pome.) I also touched on the matter of kings living next door to one another (Introduction, p. xxxi), with regard to how Tuscan storytellers see kings.

86. "Prezzemolina" from Imbriani, 16, Florence.

A folktale of well-known motifs resolved in a manner somewhat too simple, but told with vitality, entirely in dialogue, and with that cheerful figure of Memè, cousin of the fairies. It is one of the best-known folktales, found throughout Italy.

87. "The Fine Greenbird" (*L'uccel bel-verde*) from Imbriani, 6, Florence, and from other versions.

For this tale, known all over Italy, I supplemented Imbriani's version with details from many others, so as to come up with the richest text possible of the story. The tale is known all over Europe (see the Grimms' sprightly no. 96) and also in western Asia (to which scholars say it spread from Europe; see the last narrative, which is quite similar to our tradition, of Galland's *Arabian Nights*, "Histoire de deux sœurs jalouses de leur cadette"). The first literary version is the story of Ancillotto, king of Provino, in Straparola (IV, 3), very much like the tale as it has come down to us by word of mouth. Modeled on the story in *The Facetious Nights* in Madame d'Aulnoy's "La Princesse Belle-Etoile et le Prince Chéri." Gozzi used it for one of his most polemical theatrical fables (*L'augellino Belverde*), subsequently adapting it into *L'amore delle tre melarance* ("The Love of the Three Oranges").

88. "The King in the Basket" (*Il Re nel paniere*) from Imbriani, 3, Florence.

From Basile's *Sapia Licarda* (III, 4), story of the seduction of three sisters and of the wiles of the third to escape and get even with a tyrannical and rakish king, this Florentine tale retains only a childish taste for raids and acts of vandalism, with no real maliciousness. For the sake of variety, I made the wardrobe the goal of the girl's third foray. Thus I came up with a pair of silver pumps as "convicting evidence," rather than the pot of *verdèa* (name of a plant unfamiliar even to the narrator). In the Florentine original, the name of the third sister is Leonarda, which I refined to Leonetta.

89. "The One-Handed Murderer" (*L'assassino senza mano*) from Imbriani, 17, Florence.

This is one of the most romantic folktales in Italy, and this particular version unfolds in an atmosphere of obsessive fear, yet with no recourse to the supernatural. The suggestive detail of the hand lopped off is not in Imbriani; I took it from a Pisan folktale (Comparetti, 1). The pistol concealed in the towel is my own invention; in Imbriani the one to fire is the husband, at last awakened, and he does so unjustifiably. The original ends with the miserly father's repentance, which I omitted.

90. "The Two Hunchbacks" (*I due gobbi*) from Pitrè, T. 22, Florence, told by Raffaella Dreini.

This is one of the legends pertaining to the famous "walnut tree of Benevento" (like the tale about the two friends, one of whom blinds the other—see my no. 184), but it is of ancient European diffusion, and widespread in Italy.

91. "Pete and the Ox" (*Cecino e il bue*) from Pitrè, T. 42, Florence.

This tale has a style slightly different from usual (and not only this version but other Italian versions of the tale as well, and even the foreign ones—cf. Grimm, 37 and 45): it resembles a child's drawing, with tiny little men and great big moo-cows and figures inside one another with no sense of perspective. Common to all the versions is extreme coarseness, which I endeavored to preserve in my draft. I deviated from the Florentine version only in the beginning, preferring the versions which present the transformation of peas into children as a curse rather than a blessing. I left out Pete's drowning in a pond at the end, to close in a better way. And I retained the slightly scatological overtone characteristic of children's stories.

92. "The King of the Peacocks" (*Il Re dei Pavoni*) from Marzocchi, 34, Siena.

Of literary origin, from a French tale at the end of the seventeenth century, *La Princesse Rosette* by Madame d'Aulnoy. This folk version is faithful to its literary source as far as plot is concerned, but its style is quite simple in comparison with the mannerism of the French text, albeit refined and ornate for a popular narration. I avoided transcribing sentences which struck me as too exalted. The strangest image, that of voices coming from the trees as feathers swirl in the air, is not in the French text. The theme of the bride replaced along the way by the hideous daughter of the nurse is among the most widespread in Italy (see my no. 101), but the peacock décor is probably Madame d'Aulnoy's own invention.

93. "The Palace of the Doomed Queen" (*Il palazzo della Regina dannata*) from Marzocchi, 25, Siena, told by "a certain Smida, an old woman who makes hose."

A rather sinister folktale, with that cold determination of a heroine too lacking in pity to attract the admiration which courageous girls in tales always inspire. Medieval touches abound, such as the chained queen's vision of Hell and the romantic fatality of love's revenge. It seems that this type of tale exists only in Italy, from Venice to Sicily. The beginning, with the dog running off with the girl's food and entering the unoccupied palace, is common to all the versions.

94. "The Little Geese" (*Le ochine*) from Signora Olga Cocchi, who kindly provided me with the manuscript, Siena.

The three little pigs have here become three geese—or, more exactly, one single goose: in this Sienese version, there is no mention of the other two, who built their houses out of flimsier material and whom the wolf (or, in this case, the fox) ate. We see only the third, the wisest, with her iron cottage and her successive foils to the greedy beast. Here we have keener zoological and geographical perceptions than usual with the Maremma landscape and the seasonal migration of palmipeds. Other goose stories in Northern Italy are closer to the "three pigs" type.

95. "Water in the Basket" (*L'acqua nel cestello*) from Comparetti, 31, Jesi, Marche.

This is one of the versions which best correspond to the most widespread Italian tradition of the folktale about two sisters or stepsisters—one who is kind, the other ill-willed—with supernatural beings burdened with human sufferings and squalor.

96. "Fourteen" (*Quattordici*) from Gianandrea, 6, San Paolo di Jesi, Marche.

The little digger with the strength of an ox is the hero of this coarse rustic epic, accomplishing something akin to the feats of Hercules. He has no supernatural help, however, only the strength of his own arms. A king is not the one who imposes the tasks, but a farmer who wants to get rid of Fourteen because of the pay demanded by the boy. The tasks culminate with a descent into Hell, whereby the hero regains his liberty. I have freely mixed the folktale from Marche with a similar one from Abruzzo (Finamore, 27). Stories about strong peasants in Northern Italy present elements in common with "Fourteen."

97. "Jack Strong, Slayer of Five Hundred" (*Giuanni Benforte che a cinquecento diede la morte*) from Gianandrea, 7, San Paolo di Jesi, Marche.

Boasts of strength and the ruses of the little man against the big man are eternal narrative motifs from David to Chaplin. The version I followed has the vitality and irony required by the theme. The death of the giant was somewhat insipid in the Marche text (an iron chest for a trap); I replaced it with the episode of the sheep intestines from a Bolognese version (Coronedi, S. 32), putting in wolves where the original speaks of lions, for the sake of regional accuracy. (For the close, I turned back to the Marche version.) I also transcribed another version (my no. 199, "John Balento"), from Corsica, with different and curious episodes such as that of the Amazons.

98. "Crystal Rooster" (*Gallo cristallo*) from Gianandrea, 5, Jesi, Marche.

Fables featuring animals in a group are always entertaining. This is more rigmarole than fable, very comical, thanks to the letter repeatedly consulted with the utmost seriousness.

99. "A Boat for Land and Water" (*La barca che va per mare e per terra*) from Zanazzo, 21, Rome.

This is in the grand line of literary swashbuckling, made up as it is of those "quantitative" skills at the root of *Gargantua*. Many variants all over Italy (see my no. 126).

100. "The Neapolitan Soldier" (*Il soldato napoletano*) from Zanazzo, 11, Rome.

The semblance of bravery and mutual mockery of the soldiers lend a modern touch to this Roman folktale. The original shows the first two soldiers deriding the third because he is from Naples. I made those two Roman and Florentine respectively, to accentuate the spirit of the barracks.

This type, according to Stith Thompson, is quite rare. It is found here and there in Europe, but I saw no other Italian version of it.

101. "Belmiele and Belsole" (*Belmiele e Belsole*) from Zanazzo, 29, Rome.

This is one of the most widespread types in Italy. To produce a richer version, I combined the Roman with one from Florence (Imbriani, 25). The escape of the nurse and her daughter at the end is my own invention (in place of the customary tarring of a transgressor).

102. "The Haughty Prince" (*Il Re superbo*) from Menghini (*"Due favole romanesche"* in *Volgo di Roma*, vol. 1, fasc. 2, Rome 1890) and from Zanazzo, 1, Rome.

Love for the haughty prince with the seven veils (see my no. 36, Venice) serves here only as a starting point for a series of tales of sorcery and cures, in which the Italian tradition is quite rich. Nor is the famous "walnut tree of Benevento," meeting place of witches, wanting. The "magic counter-formula" is an addition of my own.

103. "Wooden Maria" (*Maria di Legno*) from Zanazzo, 24, Rome, and from other versions.

One of the most widespread folktales in all of Italy. I have drawn on several versions, in order to come up with the most complete text possible. As early as the sixteenth century the tale about the father who wants to marry his daughter and about her flight from him in disguise was told by Straparola (I, 4: Doralice, daughter of King Tebaldo of Salerno, flees in a cupboard, weds a king, etc.). The seventeenth century finds the tale in Basile's *Orza* (II, 6: the daughter of the king of Rocc'Aspra runs away, magically changing into a she-bear, etc.), and in Perrault's *Peau d'asne*, which is closer to the popular versions (cf. also Grimm no. 65).

104. "Louse Hide" (*La pelle di pidocchio*) from Zanazzo, 6, Rome.

A folktale motif—the louse hide—and a story motif—hunchbacks thrown into the river—are deftly combined in this Roman version, as also in Doni's sixteenth-century short story, where the object to be guessed is a lizard lung.

105. "Cicco Petrillo" from Zanazzo, 10, Rome.

This little story about the boundlessness of human stupidity is the same the world over; only the dramatizations of it vary. I followed the carefree Roman version, drawing illustrations from others as well.

106. "Nero and Bertha" (*Nerone e Berta*) from Zanazzo, 417, Rome.

This little tale serves as an explanation for two popular sayings at once: that of the old woman who wept for Nero, and "times aren't what they used to be."

107. "The Love of the Three Pomegranates" (*L'amore delle tre melagrane* [*Bianca-come-il-latte-rossa-come-il-sangue*]) from Finamore, 54, Montenero-domo, Abruzzo, told by the illiterate Domenica Rossi.

As far as this being apparently one of the few folktales that can be called distinctly Italian, I refer the reader to my remarks on it in the Introduction (p. xxviii). Italy in any case saw the first literary version of it, "The Three Lemons," by Basile (V, 9), with its plot of metamorphoses like a baroque flourish. Carlo Gozzi transposed the tale for Commedia dell'Arte masques in his *Love of the Three Oranges*. The many popular versions are in large part faithful to the tradition that inspired Basile. In the Abruzzese version I followed, among forty other Italian versions, the fruits containing the girls consist of a walnut, a hazelnut, and a chestnut; elsewhere they are watermelons, lemons, oranges, apples, pomegranates, or *melangole* (which means in some places "oranges," in others "bitter oranges"). I took the pomegranates, as in a Pisan version (Comparetti, 11), because they were already at the end of the present Abruzzese version as metamorphoses of the dove (this last part, which I find in various southern versions, is not in Basile), and I wanted to make a cycle of transformations that ended as it began. For the verses interspersed in the text, I drew on different versions, a Campanian one in particular (Amalfi, 9, Avellino); but the first (*Giovanottino dalle labbra d'oro*, etc.) are Umbrian, from Spoleto (Prato, p. 28).

108. "Joseph Ciufolo, Tiller-Flutist" (*Giuseppe Ciufolo che se non zappava suonava lo zufolo*) from De Nino, 62, Sulmona, Abruzzo.

The medieval legend about the grateful dead man, which appears in a French poem of the thirteenth century, *Richars li biaus*, and then in *Novellino* and in Straparola, is here told with a tiller in place of the noble protagonist, peasant rather than knightly prowesses, and a beggar instead of the mysterious savior-knight.

109. "Bella Venezia" (*La Bella Venezia*) from De Nino, 50, Lama dei Peligni and other villages of Abruzzo.

The folktale about Snow White in Southern Italy puts bandits in place of the dwarfs that figure in the Grimms' classic version. And quite often in the rest of Italy also, the envious mother or stepmother is not a queen but an innkeeper, as in the Abruzzese version of "Bella Venezia," which differs from the best-known versions in having no magic looking glass for the stepmother to ask if anyone exists lovelier than herself; but the question comes up in the course of the chitchat of the travelers who stop at the inn. The dwarfs appear only in a version from Piedmont and another from Calabria, probably as late literary acquisitions, since dwarfs are practically nonexistent in Italian oral tradition. Northern and Tuscan versions are often devoid of both bandits and dwarfs, thus lacking the tale's most suggestive element.

110. "The Mangy One" (*Il tignoso*) from Finamore, 17, San Eusanio del Sangro, Abruzzo.

It is probably medieval and Nordic in tradition, and I have chosen to give it a melancholy tone perhaps not so evident in the original from Abruzzo, but which nevertheless results from the circumstances, as if the diabolical apparition at the outset cast a shadow over the entire story, including its loves and victories.

111. "The Wildwood King" (*Il Re selvatico*) from Finamore, 19, San Eusanio del Sangro, Abruzzo, told by an illiterate man of the people.

The touching figure of this misanthrope, half-ogre and half-king-in-exile, who keeps to the woods and shelters the abandoned girl, lends a gentle touch to this obscure southern tale. In other versions, the wildwood king (or hairy man) asks to be killed and cut up in pieces and buried in the various rooms of the house. He is a figure indeed worthy of ethnological speculations; we merely emphasize his manner of a dethroned cannibal, of a defeatist Esau.

112. "Mandorlinfiore" from Finamore, 71, Atri, Abruzzo.

This Abruzzese version of the widespread medieval story about the pre-destined man stands out with its opening dramatization of a superstition centering around the birth of a child.

113. "The Three Blind Queens" (*Le tre Regine cieche*) from De Nino, 51, Canzano Peligno and elsewhere in Abruzzo.

Nonchalant and cruel, but never cynical.

114. "Hunchback Wryneck Hobbler" (*Gobba, zoppa e collotorto*) from De Nino, 70, Acciano and elsewhere in Abruzzo.

This folktale, so precise, rational, and moralistic as to resemble a literary creation, was apparently collected only in Abruzzo. A rare instance in folk-lore: the witch or evil sorceress, who is condemned in the end to be tarred and burned, escapes, and the narrator seems to be on her side. I just barely accentuated the literary character, at the outset.

115. "One-Eye" (*Occhio-in-fronte*) from De Nino, 61, Pratola Peligna and elsewhere in Abruzzo.

The myth of Ulysses and Polyphemus has remained in Italian oral tradition as a separate tale of fear. Ulysses's companions are friars in this version which De Nino collected here and there in Abruzzo (and in Apulia and Sicily); in other Abruzzese versions they are shepherds, students, or beggars. The end, with the ring, is common to all the Abruzzese versions and also to the strange Pisan transposition that is the source of my no. 76. We can say that the story tends to preserve its mythical elements in localities of shepherds; thus it is also found in the mountainous areas of Lombardy.

116. "The False Grandmother" (*La finta nonna*) from De Nino, 12, Bugnara, San Sebastiano, and elsewhere in Abruzzo.

Like the version from Garda, this is one of the rare popular versions collected in Italy of "Little Red Riding Hood" (see note on my no. 26). It presents all the characteristics of children's tales in the popular tradition: cruelty, mention of bodily needs, and a rigmarole of questions and answers. A realistic detail: the house is a real peasant house, with only one bed, and the barn on the ground floor.

117. "Frankie-Boy's Trade" (*L'arte di Franceschiello*) from Finamore, 24, San Eusanio del Sangro, Abruzzo, told by an illiterate man of the people.

Another tale of bets with the thief, differing from the one about the "treasure of King Rhampsinitus" (cf. my no. 17) but nevertheless widespread in Europe (compare with Grimm no. 192 and Afanas'ev's "The Thief"). Here we are in the climate of artful bandits, of cattle stealing, of mortmain.

118. "Shining Fish" (*Pesce lucente*) from De Nino, 10, Sulmona and elsewhere in Abruzzo.

Found also in *The Arabian Nights*, but so varied as to lead one to believe we are here dealing with an independent tradition.

119. "Miss North Wind and Mr. Zephyr" (*La Borea e il Favonio*) from Francesco Montuori (in *Rivista delle Tradizioni popolari*, Rome, I [1894], p. 761), Pesche, Molise.

A meteorological fable slightly suspect of literary influence.

120. "The Palace Mouse and the Garden Mouse" (*Il sorcio di palazzo e il sorcio d'orto*) from Eugenio Cirese (*Tempo d'allora, figure, storie e proverbi* in Molise dialect, Campobasso, 1939), Campobasso, Molise.

This classical theme (see Horace, *Satires*, II, 6, vs. 79–117) already handled by eighteenth-century Italian authors, appears here in a vivid variant, the work of a writer in Molise dialect, but of unquestionable popular extraction.

121. "The Moor's Bones" (*Le ossa del moro*) from Corazzini, 5, Benevento, Campania.

A cruel folktale of oriental origin, rather widespread in Italy as in all of Europe. I combined the Benevento version with others, especially one from Abruzzo (Finamore, 7, Ortona a Mare).

122. "The Chicken Laundress" (*La gallina lavandaia*) from D'Amato, 6, Avellino, Campania.

A curious tale that very well constitutes a type in itself, notwithstanding affinities here and there with "The Prince Who Married a Frog," "Rosemary," the "Cinderella" type, and the "Dove Girl" type.

123. "Crack, Crook, and Hook" (*Cricche, Crocche e Manico d'Uncino*) from Amalfi, 13, Avellino, Campania.

We have already come across "Cric" and "Croc" [Crack and Crook] as names of famous thieves in Piedmont (see my no. 17). Here is another story about the brazenness of robbers, which I take from the above version, since it has its own Neapolitan vitality, although the source could be a school book by a Tuscan author (Gradi, p. 105).

124. "First Sword and Last Broom" (*La prima spada e l'ultima scopa*) from Vicenzo della Sala (in the journal *Giambattista Basile*, Naples, I [1883] no. 1, pp. 2–3), Naples.

Among the variants of "Fanta-Ghirò the Beautiful" (my no. 69), this tale is connected to the Neapolitan tradition of Basile (III, 6) both in its bourgeois ambience and in the disgrace experienced over having only daughters and no sons. The enchanted filly plays an important role in the original (a theme developed in my no. 75). For the sake of variety, I made the filly a silent counselor here, and I expanded the episodes on the road, which were very sketchy in the original.

125. "Mrs. Fox and Mr. Wolf" (*Comare Volpe e Compare Lupo*) from Benedetto Croce (in *Giambattista Basile*, Naples, I, no. 6 [15 July 1883]), Naples.

I give this little animal fable in homage to Benedetto Croce's early enthusiasm as a collector of Neapolitan popular traditions. In the first issues of the journal *Giambattista Basile*, we come across quite a few little tales, ballads, and traditions compiled by him.

126. "The Five Scapegraces" (*I cinque scapestrati*) from Pellizzari, p. 89, Maglie, Apulia.

The story of the extraordinary companions is told by Basile in a very lively tale (III, 8), which is quite similar to this folk version, even down to their names, with foot race and all. It probably spreads from Basile over Europe (it is essentially the same as Grimm no. 71).

127. "Ari-Ari, Donkey, Donkey, Money, Money!" (*Ari-ari, ciuco mio, butta danari!*) from Pellizzari, p. 19, Maglie, Apulia.

One of the liveliest Italian versions of this very widespread narrative with a taste for hunger, tavern deceptions, and family quarrels. Versions all over Italy.

128. "The School of Salamanca" (*La scuola della Salamanca*) from Pellizzari, p. 111, Spongano, Apulia.

The folktale about the sorcerer's pupil is of Indian origin, and in all truth seems to come to us from a world so conversant with marvels as to be able to represent the most arbitrary metamorphoses with the swiftness and rhythm of a ballet. The widespread Italian tradition is gay and rich and has its oldest testimony in a highly entertaining story by Straparola (VIII, 5). The Apulian version I followed has a medieval tint, with its initial recall of the famous university of Salamanca. Concerning the transformations, I took account of other versions, especially a Lucanian one (Comparetti, 63), mainly for the sake of the rhythm.

129. "The Tale of the Cats" (*La fiaba dei gatti*) from Pellizzari, p. 37, Maglie, Apulia.

The story of the two sisters, tale of kindness, sometimes presents in place of the verminous witch (see my no. 95) a community of cats, a type of perfect society, industrious and just. Other versions are found all over Italy.

130. "Chick" (*Pulcino*) from Pellizzari, p. 53, Maglie, Apulia.

Perrault's "petit-Poucet" and the Grimms' "Hansel and Gretel" have here become the far too numerous progeny of an Apulian peasant in times of famine; at the head of them is a hunchback, traditionally shrewd and lucky. I made the crowns worn by the ogre's children into crowns of flowers, as in the striking Bolognese version (Coronedi, 17).

131. "The Slave Mother" (*La madre schiava*) from Pellizzari, p. 127, Maglie, Apulia.

This is not a folktale, but a "sad romance" in the oral tradition with a happy fairy-tale ending, and it belongs to the particular category of stories predominant in maritime localities, especially in the South, about Turkish pirates and abductions. This tale has its own particular tone of sorrow: the person kidnapped is not a youth or maiden, but a mother. In popular narratives, compensation for suffering usually comes before the end of youth; here a woman suffers and grows old before a happy solution is reached. The story also brings to mind another popular variety—the discovery of treasure. The narrative stands out in its realism and wealth of local color. There is, moreover, a suggestion of a *Bourgeois Gentilhomme* situation when the country family takes up residence in the city of Naples.

132. "The Siren Wife" (*La sposa sirena*) from Gigli, 5, Taranto.

Giuseppe Gigli translated the folktales he compiled into a lyrical Italian which overshadowed the spirit in which they were originally narrated. Consequently his text is the least suited to a project like mine. I was nevertheless charmed by the unusual characteristics of this tale (the return to the classical tradition of Sirens, the motif of the adulteress's reinstatement), and I attempted to retell the story in a simpler language. Considering these successive touches determined by varying tastes, who can say how far we may be from the original popular spirit, if indeed the tale is of popular origin? Let us therefore accept this tale with more reserve than the others, as far as its "popularity" goes. I composed all the lines of the Sirens' songs. The fairy and the rescued wife fly off at the end of the original on a broom; as I found no other instance of flying brooms in Italian folklore, I replaced the broom with an eagle.

133. "The Princesses Wed to the First Passers-By" (*Le Principesse maritate al primo che passa*) from Comparetti, 20, Potenza, Lucania; from Lombardi, 23, Santo Stefano d'Aspromonte, Calabria, told by the late Crea Domenico, charcoal burner; and from Pitrè, 16, Casteltermini, Sicily, told by Vicenzo Midulla, sulfur miner.

The three kings of animal realms who marry three princesses and sisters and help their brother-in-law free a beautiful girl from a spell constitute a widespread Italian folk motif. It is found in a tale by Basile (IV, 3) containing animals like a hunting tapestry (and retold in German in the late eighteenth century by Musäus). In the three similar Southern versions I followed, the brothers-in-law are not animals, but men who have dominion over animals, like the swineherd and the fowler, or over the world of the dead, like the gravedigger.

134. "Liombruno" from Comparetti, 41, Potenza, Lucania.

The source is a Tuscan ballad of chivalry. Transplanted in Lucania, it has absorbed something of the somber religiosity of that region. The *Bellissima istoria di Liombruno*, in verse and dating from the end of the fourteenth century, is a complete story of human destiny, in the tradition of medieval romance: the birth ensuing from a vow to the Devil, rescue by a fairy, indoctrination in love and knighthood, homecoming and benefit for the parents, the joust of the unknown knight, the "boast," loss of the beloved, and then a series of typical folktale motifs such as the seven pairs of iron shoes, the three magic objects for which thieves compete, the house of the winds.

135. "Cannelora" from Comparetti, 46, Potenza, Lucania.

This is essentially like the story of Fonzo and Canneloro in Basile (I, 9), with the addition of the episode of the quarreling gardeners and that of the fairy changed into a snake. There is no mention in the Lucanian version of furniture giving birth to other furniture (after swelling up—a detail I omitted); I took it from the similar story in the *Pentameron* (I, 9), making an exception to my rule to use only popular motifs, and no doubt it is Basile's own invention. Both Basile's opening and the Lucanian beginning of the tale are subtly or outright lascivious.

136. "Filo d'Oro and Filomena" (*Filo d'Oro e Filomena*) from Comparetti, 33, Potenza, Lucania.

This tale is of the same family as Amor and Psyche (cf. my no. 174). I named the girl Filomena myself (in the original text she has no name) and also specified Filo d'Oro's transformations into a man with a beard, a man with whiskers, and a man with sideburns (the original simply says he assumed the form of another man). The mother who impedes the birth of the child until she puts her hands on her head repeats the myth of the birth of Hercules (a widespread motif in Sicily).

137. "The Thirteen Bandits" (*I tredici briganti*) from La Rocca, 6, Pisticci, Lucania.

The motif of "Open sesame!" may be of modern literary origin, deriving from one of the most successful narratives of Galland's *Arabian Nights*, "The Story of Ali Baba and the Forty Thieves." It now belongs to European folklore (it is also in Grimm no. 142) and often, as in this Lucanian version, takes on a strong regional character. In the widespread Italian tradition, no light is thrown on the slave Nirguaba who, in *The Arabian Nights*, plays a key role in the extermination of the thieves.

138. "The Three Orphans" (*I tre orfani*) from Lombardi, 41, Tiriolo, Calabria.

A religious allegory of rare beauty, with the mysterious simplicity of a rebus. Calabrian tales are often intertwined with Christian motifs, but nearly always as a distortion of an old pagan plot of magic. Here, on the contrary, we find only the rhythm of the magic tale, with everything converging into the composition of liturgical symbols. But the tale opens on a realistic note— the day laborer offering his services by means of the rather somber lines, "Whoever would have me as his helper,/Him do I want for a master!"

139. "Sleeping Beauty and Her Children" (*La bella addormentata ed i suoi figli*) from R. de Leonardis (in *La Calabria*, VIII [1896], no. 12, p. 93), Rossano, Calabria.

The Italian Sleeping Beauty is quite distinct from Perrault's, since—like Basile's Neapolitan tale about Sun, Moon, and Talia (V, 5)—it concentrates, above all, on what happens *after* the prince has found Beauty, and this succession of events is cruel beyond anything the French version hints of; it is indeed one of the cruelest of all Italian folktales. Scholars attribute to this type rather late literary origins (as to the Grimms' no. 50, "Hawthorn Blossom," derived from Perrault), and as a matter of fact, nearly all the Italian popular versions, from Tuscany to Sicily, are like Basile, even down to the names of the characters. Thus, in the Calabrian version I followed, Sleeping Beauty bore the name Talia. I called her Carol, for the sake of assonance.

140. "The Handmade King" (*Il Reuccio fatto a mano*) from Di Francia, 5, Palmi, Calabria, told by Concetta Basile; and from Lombardi, 13, Feroleto Antico, Calabria, told by Maria Muraca.

This tale was also found in Naples (Basile, V, 3), Abruzzo, and Sicily.

141. "The Turkey Hen" (*La tacchina*) from Di Francia, 10, Palmi, Calabria, told by Annunziata Palermo.

The second part is the very widespread story about the woman with amputated hands (cf. my no. 71, "Olive"), but of the whole beginning— down to the marriage to the beggar girl—I found no other versions, not even partial ones. It is probably of recent tradition, with realistic episodes like the riot, the introduction of nineteenth-century figures such as the English lord (regarding the fate of this character in southern folklore, see note on my no. 158), and the supernatural exemplified only as the miracle worked by a saint. In the Calabrian text, the account of the miracle—the regaining of the amputated hands—was a bit too meager (just an encounter with St. Joseph, who causes a pond to appear and tells the woman to immerse her arms); I chose to follow the commonest version in Italy, with the babies slipping from the mother's arms into the water and the course pursued from there on. St. Joseph's verses are my own, but based on similar lines in a Tuscan version (Pitrè, T. II, 13).

142. "The Three Chicory Gatherers" (*Le tre raccoglitrici di cicoria*) from Di Francia, 27, Palmi, Calabria, told by Annunziata Palermo.

This is a variant of the Bluebeard type (see my no. 35), and I found such a cannibalistic folktale also in Tuscany, Abruzzo, and Sicily.

143. "Beauty-with-the-Seven-Dresses" (*La Bella dei Sett'abiti*) from Di Francia, 23, Palmi, Calabria, told by Pasquale Di Francia.

Although this tale belongs to the well-known group in Italy about the enchanted palace and the supernatural wife lost and magically recovered, it is quite rich in rare and original motifs, and also capriciously and elaborately incoherent (excepting the motif of the grass that revives the lizard, which is quite common).

144. "Serpent King" (*Il Re serpente*) from Di Francia, 1, Palmi, Calabria, told by Di Francia's sister, Teresa.

Composed altogether of well-known motifs, this folktale is distinguished by its animalistic details in a vein either gothic or oriental, from the opening procession of lizards and snakes through the fields up to the queen, to the end of the tale where the enchanted palace is meticulously described with the gold animals that inhabit it. I omitted the episode in the Calabrian version of the flea-skin test, which is already familiar to the reader (see my no. 104); correctly guessing what it is, the snake marries the empress. In place of it I put the snake's transformation into a man after sloughing off seven skins, as in various other versions—Tuscan, Campanian, Sicilian, and Piedmontese.

145. "The Widow and the Brigand" (*La vedova e il brigante*) from Luigi Bruzzano (in *La Calabria*, VII [1894], nos. 2–5), Roccaforte, Calabria, in Greek dialect.

The story of the mother of questionable morals who consorts with bandits or giants while her son is out hunting, and of her schemes to bring about her son's death (and then of the son's revenge) is one of the most pungent and obscure folktales circulating in Italy, apparently originating in Eastern Europe. It is also one of the most psychologically suggestive tales with its amoral mother. I followed a Greek version from Calabria, in which the theme is introduced directly, in the realistic setting of a poverty-stricken countryside where we see mother and son roaming in search of work, the son bringing down birds with his slingshot, and passing bandits tempting the widow. I ended the tale with the son's revenge, omitting all the rest of his adventures which resemble those of "The Dragon with Seven Heads" and "The Three Dogs," with three ferocious animals instead of the dogs. In other versions the mother is almost always a queen rather than a poor peasant; sometimes she gives birth in prison to a son who frees them both, whence the adventures begin.

146. "The Crab with the Golden Eggs" (*Il granchio dalle uova d'oro*) from Luigi Bruzzano (in *La Calabria*, X [1897] nos. 1–3), Roccaforte, Calabria, in Greek dialect.

Only in this Greek version from Calabria are the golden eggs laid by a crab instead of a bird, as in the oriental motif known throughout Italy and

Europe. But is it really a crab? The original is contradictory: it first speaks of a crab (*caridaci*), then of a cockerel (*puddhaci*). The bricklayer fires on the crab, and it falls, still alive (I omitted this detail). Where I spoke of the shell and the claws, the original speaks of "the front half" and the "back half."

147. "Nick Fish" (*Cola Pesce*) from Pitrè (*Studi di leggende popolari in Sicilia e Nuova raccolta di leggende siciliane* [vol. XXII of "Biblioteca delle tradizioni popolari siciliane"], Turin 1904), Palermo, told by a sailor.

This is the finest of the seventeen Sicilian popular versions of the famous legend of Nick Fish, published by Pitrè in an appendix to his detailed study. Among the scholars to write on the legend was Benedetto Croce. His article, "La leggenda di Niccolò Pesce," based on a Neapolitan tradition, appeared in *Giambattista Basile*, III, (1885), no. 7, and was reprinted separately in Naples, 1885. A controversy ensued, and Pitrè and Arturo Graf expanded the study. The first literary mention of the legend is by a Provençal poet of the twelfth century, Raimon Jordon. A rich repertory of literary versions, including Schiller's ballad, *Der Taucher* ("The Diver"), is to be found in the above-mentioned study by Pitrè. Regarding Nick Fish and Benedetto Croce, see Carlo Levi's fine page in *L'Orologio* (pp. 343 ff.).

148. "Gràttula-Beddàttula" from Pitrè, 42, Palermo, told by Agatuzza Messia, seventy-year-old seamstress of winter quilts.

Of all the Italian variants of the famous "Cinderella," the most colorful and Mediterranean is this tale about the date-palm trees, told by the great illiterate narrator of Palermo, Agatuzza Messia (see Introduction, pp. xxi–xxiv). There is no moralizing here as in Perrault and Grimm; all is one grand play of fantastic marvels. The Cinderella motif of the lost slipper is not retained in "Gràttula-Beddàttula," but is found in all the other Italian versions.

149. "Misfortune" (*Sfortuna*) from Pitrè, 86, Palermo, told by Agatuzza Messia.

One of the most touching southern folktales is this one about the girl pursued by her evil luck, which brings misfortune to herself and others. Contrary to the custom of ostracizing the bearer of ill-luck, one takes pity on her here, in the framework of an individual cult to Fate, to whom tribute is paid in the form of vows and petitions. Men are at the mercy of the erratic psychology of the Fates. Messia superbly sketches the character of the protagonist's wicked and mad Fate. But the finest characters of Messia emerge from types like the charitable washerwoman, who is mistress of the Fates' cult and viewed with affection. (If she refrains from telling the prince about Misfortune, it is to protect her from snares and in no wise indicates a dislike for the girl.) Note how the customary generality of folktales gives way to linguistic and technical precision when Messia speaks about the washerwoman's work.

150. "Pippina the Serpent" (*La serpe Pippina*) from Pitrè, 61, Palermo, told by Agatuzza Messia.

I saw versions compiled in Emilia (with the girl changed into an eel the first time she lays eyes on water), Tuscany (see my no. 64), Abruzzo, Calabria, and Sicily. But other versions are centered on my no. 101, which is quite similar.

151. "Catherine the Wise" (*Caterina la Sapiente*) from Pitrè, 6, Palermo, told by Agatuzza Messia.

The intelligent woman, both cultured and honored, is frequent in Italian folklore. The present Sicilian popular version is far richer than the Neapolitan literary one by Basile (V, 6) and contains curious reminiscences of such medieval institutions as the "free school" and allusion to a pedagogy that we will call democratic, with equality between the sexes.

152. "The Ismailian Merchant" (*Il mercante ismaelita*) from Pitrè, 100, Palermo, told by Agatuzza Messia.

Pitrè quotes as a source a Venetian edition (1555) of a popular romance (which is also a source of my no. 112). The story assumes biblical echoes in Agatuzza Messia's narration: I refer to the threatened massacre of innocents and the strange mention of an "Ismailian" merchant. The emperor's disclosure of the Golden Fleece beneath his rags is a grand bit of theater. His role, as he sadly roams the world questioning the planets, carries us into a vaguely Shakespearean atmosphere.

153. "The Thieving Dove" (*La colomba ladra*) from Pitrè, 101, Palermo, told by Agatuzza Messia.

Widespread in all the South. Other versions in Campania, Calabria, Sicily, and Sardinia.

154. "Dealer in Peas and Beans" (*Padron di ceci e fave*) from Pitrè, 87, Palermo, told by Agatuzza Messia.

This is the story of Puss-in-Boots, but without cat, fox, or any other animal suggestive of wiles to obtain credit in the eyes of others. In the present story, the poor protagonist comes up with the maneuvers himself as he gives free rein to his imagination regarding the bean found on the ground. Only at the end does the bean change into a fairy (but such supernatural intervention is not really indispensable), and a poor soul's dream of easy wealth becomes miraculous reality. Whereas in the cat story virtuous poverty and venturesome shrewdness collaborate as two distinct persons, the present tale combines them in a single character not nearly as appealing: he represents triumphant bluff, the dream of a poverty-stricken world devoid of prospects.

155. "The Sultan with the Itch" (*Il Balalicchi con la rogna*) from Pitrè, 69, Palermo, told by Agatuzza Messia.

Not so much a folktale as a tale of adventure, with certain geographical notions and, above all, a clear-cut idea of differences between one civilization and another, of relationships with the Moslem world, all of which is particularly characteristic of oral narrative in the South. Here Messia, who never set foot on a ship, gives vent to her marine fantasies.

156. "The Wife Who Lived on Wind" (*La sposa che viveva di vento*) from Pitrè, 92, Palermo, told by Agatuzza Messia.

A tale that closes with a proverb—"And who should get the miser's money in the end but the master swindler"—and which is marked by Messia's rich taste for description.

157. "Wormwood" (*Erbabianca*) from Pitrè, 73, Palermo.

The boast about the wife's fidelity, the bet with a swindler who brings in false proof of seduction, the adventurous ways the wife takes to prove her innocence—they are all elements of legends of chivalry (see the ballad, *Madonna Elena*, dating from the end of the fourteenth century) that subsequently pass into stories about merchants (as in Boccaccio, II, 9) and from there, through diverse versions, go all the way to Shakespeare, who drew on them for *Cymbeline*. This version stands out among the many popular ones because of its romantic details and strange dénouement. The popular versions are very numerous in Northern, Central, and Southern Italy, and only Pitrè's Sicilian compilation gives five of them (73–77), all excellent and included in my compilation (nos. 158, 159, 160, 176).

158. "The King of Spain and the English Milord" (*Il Re di Spagna e il Milord inglese*) from Pitrè, 74, Palermo, told by Agatuzza Messia.

Out of the sententious speech of Messia and all her proverbs and expressions comes a romantic story starring a woman who exemplifies various virtues such as the Spanish or Moslem ideal of chastity typical of the sheltered woman, and intellectual and political bravura. Her mother-in-law, diametrically opposed to the cruelty represented by her counterparts in other tales, is the affective center of the narrative. There is also the English Milord, the equal in Southern folklore of all legendary kings—rather, he is superior to the kings because of his wealth—with a touch of romantic perversity. Another notable element is the pressure exercised by an ill-governed people toward solving difficulties. The surrounding geography is realistic: there is Spain, sister country of Sicily, and nineteenth-century Brazil, the empire to which men unjustly persecuted flee and make their fortune. Messia lets herself go in this tale, with all her flair for dramatic narration. One minute she is talking like a sailor; the next, her tone is very genteel. I translated quite faithfully, adding no touches of my own, except the red tassel at the conclusion.

159. "The Bejeweled Boot" (*Lo stivale ingioiellato*) from Pitrè, 75, Palermo, told by Rosa Brusca, a forty-five-year-old blind woman.

This is a subcategory of the slandered wife (or sister) type, common to Europe (cf. Afanas'ev's "The Merchant's Slandered Daughter").

160. "The Left-Hand Squire" (*Il Bracciere di mano manca*) from Pitrè, 76, Palermo, "from a woman to whom Messia told it."

Regarding the Sicilian popular storyteller's conception of the court of kings and court ethics, see my remarks in the Introduction, p. xxxi. These ethics gave rise to the present version of the famous tale supposedly featuring Pier delle Vigne, as transmitted by Jacopo d'Acqui in Latin, with verse in Piedmontese dialect. D'Ancona observes, "The image of the vine must have come from attributing this adventure to Frederick the Second's prime minister." The story goes back farther (D'Ancona ascertained) and appears in Greek, Hebrew, and Arabic versions. It was also recounted by Brantôme (*Vie des dames galantes*, II), in reference to the Marquis of Pescara, with verses in Italian, which I followed in part, in preference to the frequently altered lines in the popular versions.

161. "Rosemary" (*Rosmarina*) from Pitrè, 37, Palermo, told by a woman.

Another folktale about the plant-woman. This one repeats "The Mulberry," one of Basile's finest Neapolitan tales (I, 2), with a few additional details such as the watering with milk and the prince's flute-playing. The girl's dancing to the flute music is the only thing I added, but a dance rhythm is already in the Palermo original.

162. "Lame Devil" (*Diavolozoppo*) from Pitrè, 54, Palermo, told by a blind man by the name of Giovanni Patuano.

Machiavelli's tale *Belfagor* comes from a popular tradition, as shown by the fact that Straparola also uses it (II, 4). I decided to give this Sicilian "Lame Devil" a more or less stylized translation which accents the rudimentary vitality of the narration.

163. "Three Tales by Three Sons of Three Merchants" (*I tre racconti dei tre figli dei tre mercanti*) from Pitrè, 103, Palermo, told by Rosa Vàrrica.

The frame, with the unresolved ending, is most often found in literary stories.

164. "The Dove Girl" (*La ragazza colomba*) from Pitrè, 50, Palermo, told by a woman.

The swan girl or dove girl, whose bird costume the hero takes away, thereby compelling her to remain a woman, is a universally known motif and often combines with the motif of the sorcerer's servant who must climb a mountain of precious stones. I began with the Palermo version, showing the "lad who led a dog's life" in search of work, and the Greek from the Levant. I departed from the text by having the boy go up the mountain, not on a winged horse, but in a horse's hide carried upward by an eagle, as in other Southern versions (taking into special account one from Lucania—La Rocca, 9). I also borrowed the final episode of the invisible cloak, a very widespread motif, to make the plot complete.

165. "Jesus and St. Peter in Sicily" (*Gesú e San Pietro in Sicilia*). Same type of legends as in the series, "Jesus and St. Peter in Friuli." (See my no. 41.) St. Peter presents the same characteristics here as in the Friulian compilation—laziness and gluttony.

I. "Stones to Bread" (*Le pietre in pane*) and II. "Put the old woman in the furnace" (*La vecchia nel forno*), from Pitrè, 123, Bagheria, Palermo, told by a certain Gargano.

III. "A Tale the Robbers Tell" (*Una leggenda che raccontano i ladri*) from Pitrè, 121, Borgetto, Palermo.

This tale is traditional among thieves, who claim to have received Jesus Christ's blessing. Sending me a version from Santa Ninfa, the Honorable Antonino Destefani-Perez mentioned that his uncle was retained for a time by robbers who tried to convince him they were not as black as the world made them out to be, referring to God's blessing of them in the gospel and also telling this little tale, by way of additional proof.

IV. "Death Corked in the Bottle" (*La morte nel fiasco*) from Pitrè, 124. Palermo, told by Gioacchino Ferrara, butler in a Sicilian home.

One of the many tales about death checkmated, in the frame of popular tradition regarding encounters with Jesus and the Apostles. In the original, the innkeeper bears the strange name of Accaciúni, meaning "cause," and the story ends on the proverb, "No death without cause."

V. "St. Peter's Mamma" (*La mamma di San Pietro*) from Pitrè, 126, Palermo, told by Agatuzza Messia.

An illustrious popular legend, known in most of Europe (the oldest known literary version being a German poem of the fifteenth century).

166. "The Barber's Timepiece" (*L'orologio del Barbiere*) from Pitrè, 49, Borgetto, Palermo, told by Rosa Amari.

"Who could possibly fail to see that this wonderful timepiece is the Sun?" writes the worthy compiler, Salomone-Marino. "And the Master who made it, the old man who wins everyone's praise for his divine work, is none other than God. His creations reveal His existence. What wisdom is contained in this tale beneath its modest simplicity!" Modest simplicity? Although I do not always champion oral and popular poetry over literary poetry, here is truly a case where a miracle must be proclaimed: we are on the level of the great moments of allegorical poetry. And more striking than the symbolism—which is unquestionably interesting, along with the cultural and oracular importance of the sun—is the poetic interpenetration of metaphysical space and the human comedy in so precise and harmonious a construction, with a language so rich in invention, nobility, and characterization. Rosa Amari has turned out a little masterpiece, which I wished to include, although fully realizing that much is inevitably lost in translating from dialect a text that relies mainly on the spoken word (and the rustic assonance of those lines, "nearly all proverbial," as Pitrè notes).

167. "The Count's Sister" (*La sorella del Conte*) from Pitrè, 7, Borgetto, Palermo, told by Francesca Leto.

The most beautiful Italian folktale of love, in its finest popular version, so touchingly told that I should have liked to retain it just as it was, in dialect.

168. "Master Francesco Sit-Down-and-Eat" (*Mastro Francesco Siedi-e-mangia*) from Pitrè, 127, Borgetto, Palmero, told by Francesca Leto.

Salomone-Marino compiled this tale for Pitrè, giving it a moralistic-allegorical interpretation (with a quasi-Freudian overtone). But it stands out principally as a comedy of manners (resulting from the experience of girls serving in wealthy households), with contempt for the old sick lady and her fussy ways, and with the character of the town loafer so strikingly depicted.

169. "The Marriage of a Queen and a Bandit" (*Le nozze d'una Regina e d'un brigante*) from Pitrè, 21, Polizzi-Generosa, Palermo.

A few capital burlesque details are tacked onto the theme of wedding a bandit (see note on my no. 89): the professor-husband, the seven-months' offspring, and the old deaf woman. The power of the seven-months' man is extraordinary; for instance, it is said that persons plagued with intermittent and stubborn fevers need only go to any man born after being carried only seven months and say to him instantly, *"Settimu di Maria, fammi passari lu friddu a mia!"* and they will be cured.

170. "The Seven Lamb Heads" (*Le sette teste d'agnello*) from Pitrè, 94, Ficarazzi, Palermo, told by Giuseppe Foria.

This lies between the folktale and the character story: the miserly, whiny soul that delights more in complaining than in rejoicing, and whose little losses are never forgotten in the face of later gains. I chose the Ficarazzi variant because of the dialogue with the cat, in preference to a Calabrian one which nevertheless has a more pronounced moral, since it opposes the old woman's stinginess to her niece's generosity; the niece gives bread and fish to an old beggar (who is St. Joseph). But I did turn to the Calabrian version from Di Francia (8) for the meeting with the king in the woods and the ending with the beheading and the willow; and I used a Sicilian variant from Pitrè (89) for the old woman's nagging during the banquet.

171. "The Two Sea Merchants" (*I due negozianti di mare*) from Pitrè, 82, Palazzo-Adriano, Palermo.

A tale combining elements of adventure and magic, as in certain ancient ballads.

172. "Out in the World" (*Sperso per il mondo*) from Pitrè, 27, Salaparuta, Palermo, told by Antonio Loria.

A masterpiece of Italian popular narrative. The traditional magic repertory boils down to the peasant's actual experience: the search for work from farm to farm, his entering into bondage, the solidarity of the old animal, and the necessity for sacrificing him without a word of regret or pity. And the contest to win the princess is no longer equestrian, but a show of peasant strength in plowing a piece of land. Miracles can be none other than plants that spring up in haste, fruit out of season, or daylight prolonged through the intercession of the Sun, omnipotent lord and friend. I found no precise counterpart to the tale in its entirety. The type is found here and there in Europe and also in India.

173. "A Boat Loaded with . . ." (*Un bastimento carico di . . .*) from Pitrè, 116, Salaparuta, Palermo, told by Calogero Fasulo.
Several versions, literary and folk.

174. "The King's Son in the Henhouse" (*Il figlio del Re nel pollaio*) from Pitrè, 32, Salaparuta, Palermo, told by Rosa Cascio La Giucca.
Related to the most illustrious of all tales, "Amor and Psyche," that is, the first purely fairy-tale narrative, a written version of which has come down to us in Apuleius' *Metamorphoses* (second century A.D.). Scholars count seventy-one Italian oral variants of it.

175. "The Mincing Princess" (*La Reginotta smorfiosa*) from Pitrè, 105, Erice, Sicily, told by eight-year-old Maria Curatolo.
Told in the sixteenth century by Luigi Alamanni (almost exactly as it is here, with pomegranate seed and all) in the story about Blanche of Toulouse and the Count of Barcelona, in a solemn and precise style typical of a historian. But it is one of the oldest "romantic" stories in existence, and scholars seem to believe that it originated in the Italian Middle Ages. Basile came out with a very similar tale in the seventeenth century (IV, 10), except for the pomegranate seed. Other European popular versions (cf. Grimm, 52) attribute the princess's objection to some physical feature of the suitor—often the twisted hair in his beard—as do almost all the other Italian versions I examined.

176. "The Great Narbone" (*Il Gran Narbone*) from Pitrè, 77, Cianciana, Sicily, told by Master Vincenzo Restivo, shoemaker.
See note on my no. 157.

177. "Animal Talk and the Nosy Wife" (*Il linguaggio degli animali e la moglie curiosa*) from Pitrè, 282, Cianciana, Sicily, told by Rosario di Liberto, miner.
An old oriental fable ("Story of the Ox and the Donkey with the Farmer" from *The Arabian Nights*) which assumes the tone of a peasant anecdote here with the nosy wife. Remarkable in this version are the calls between wolves and dogs in the night, with a sort of lawless complicity: "Oh, Brother Vitus!" "Yea, Brother Nick!"

178. "The Calf with the Golden Horns" (*Il vitellino con le corna d'oro*) from Pitrè, 283, Casteltermini, Sicily, told by Dame Vicenza Giuliano, weaver.
I found only childish and rudimentary versions of this folktale known

throughout Europe (and related to my nos. 16 and 101). Here and there I introduced elements from other Italian versions (such as the verse) into the Sicilian original, and I toned down the cruelty of the ending.

179. "The Captain and the General" (*Il Capitano e il Generale*) from Pitrè, 202, Casteltermini, Sicily, told by Agostino Vaccaro.

An old Buddhist legend—Indian and Chinese—(with the husband buried alive, according to custom, with his dead wife) and directly absorbed into European folklore (cf. Grimm, 16) with various adaptations via the medieval *exempla*. In Sicily it became a barracks story in which the secret of success is linked to advancement in a military career.

180. "The Peacock Feather" (*La penna di hu*) from Pitrè, 79, Vallelunga, Sicily, told by Elisabetta Sanfratello, 55 years of age, maidservant.

One of the most moving tales there is on the theme of sacrifice of the youngest. It exists all over Europe (Grimm 28, 57, 97) and in all of Italy as tale and ballad, and contains the melancholy of the laments that come from the reed pipe in which the soul of the slain boy resides. Such melancholy is already in the somber, ugly cry of the peacock, the bird created to be viewed, whose tail feathers contain the eyes of Argus. I followed this Sicilian version with an unhappy ending (no resurrection of the boy) which seems in keeping with the spirit of the tale. But I replaced the pipe made from a bone of the dead boy with a reed sprung from the grave, as in many other versions.

181. "The Garden Witch" (*La vecchia dell'orto*) from Pitrè, 20, Vallelunga, Sicily, told by Elisabetta Sanfratello.

Of all the variants of "Prezzemolina" (see note on my no. 86), this Sicilian one has the most unusual opening with the mushroom-ear, and that is why I include it in spite of the common childish plot that follows. (Of its narrator, Pitrè writes: "The *sancta simplicitas* of the poor in spirit is her particular gift and at the root of the narrative's naïveté.") The girl who is ashamed to say "I am still little" is my invention.

182. "The Mouse with the Long Tail" (*Il sorcetto con la coda che puzza*) from Pitrè, 40, Caltanissetta, Sicily, told by little Maria Giuliano.

Concerning the opening, see note on my no. 133. The plot is close to types where the supernatural spouse is lost and then found again.

183. "The Two Cousins" (*Le due cugine*) from Pitrè, 62, Noto, Sicily.

Of a popular tradition that includes my nos. 2, 95, and 129.

184. "The Two Muleteers" (*I due compari mulattieri*) from Pitrè, 65, Noto, Sicily.

Of remote oriental origin (there are versions of it that go back 1,500 years), it also appears in compilations of the Grimms' (no. 107) and Afanas'ev's ("Honesty and Deceit"), but in versions less synthetic and forceful than those of the Italian tradition. There was no putting out of eyes in the Sicilian text, but I included the episode on the basis of almost all the other versions, as the story requires it, in my opinion. Many versions mention witches gathering round a tree (the famous "walnut tree of Benevento"), but the motif is common to other folktales (cf. my no. 90, very similar to this one, and nos. 18 and 161).

185. "Giovannuzza the Fox" (*Lo volpe Giovannuzza*) from Gonzenbach, 65, Catania, Sicily.

Puss-in-Boots in Sicily is a fox (in Pitrè's Palermo version, "la vurpi Giuvannuzza," nickname of the fox in popular tradition); but the plot of the tale is closer to its Italian literary versions—Straparola, XI, 1 (the cat of Costantino Fortunato) and Basile, II, 4 (the cat of Gagliuso)—with the assisted man's ingratitude toward the providential animal, a pessimistic ending that occurs in almost all the Italian versions, in contrast to Perrault's. (The Catanian version I followed showed the fox pardoning the protagonist and thus a happy close, but this seemed unjustified, so I omitted the episode altogether.) For my transcription I also kept before my eyes Pitrè, 88, told by Angela Smiraglia, eighteen-year-old country girl.

186. "The Child that Fed the Crucifix" (*Il bambino che diede da mangiare al Crocifisso*) from Gonzenbach, 86, Catania, Sicily.

This is the widespread legend that came back into the limelight with the Spanish film, *Marcelino, pan y vino*. But this Sicilian version stands out from all the others with their excessive mysticism, thanks to its tone of popular and nonconformist religiosity as expressed in the child's solidarity with Christ betrayed by man. The original attributes the child's ignorance of Jesus and the church to his simplicity. For the sake of realism, I let it ensue from the characters' isolation in the country, in a remote and desolate part of Sicily. The original also speaks of a second miracle worked by the child: he makes a rosary without ever having seen or heard of one.

187. "Steward Truth" (*Massaro Verità*) from Gonzenbach, 8, Catania, Sicily.

I followed Gonzenbach's version, dressing it up with a few livelier passages from Pitrè's Palermo version, 78. The story about the man who tries to lie, but with no success, is quite old. It appears in the *Gesta Romanorum* and Arabic compilations, but is especially striking in Straparola's Bergamasque story (III, 5) about Travaglino the cowherd and the bull with the golden horns. The oral tradition is more forceful and supple.

188. "The Foppish King" (*Il Re vannesio*) from Pitrè, 38, Acireale, Sicily.

A strangely morbid popular tale with the king's narcissism, his love and envy of the handsome prince, and the queen's ritual to summon the prince by means of milk, basin, and golden balls.

189. "The Princess with the Horns" (*La Reginotta con le corna*) from Pitrè, 28, Acireale, Sicily.

This came down to us in a fifteenth-century ballad, but only in the present

Sicilian version do I find the beginning with the three bricks and the boy's eventual attempts to take his own life.

190. "Giufà" from Pitrè, 190.

The large cycle about the fool, even if we are not dealing with the folktale properly speaking, is too important in popular narrative, Italian included, to be omitted. It comes from the Arabic world and is appropriately set, subsequently, in Sicily, which must have heard it directly from the Arabs. The Arabic origin is seen in the very name of the protagonist—Giufà, the fool for whom everything turns out well.

I. "Giufà and the Plaster Statue" (*Giufà e la statua di gesso*) Casteltermini, Sicily, told by Giuseppe Lo Duca.

One of the finest and most widespread stories about fools, with grand theatrical gags (for instance, the dialogue with the statue, and the exchange of few words).

II. "Giufà, the Moon, the Robbers, and the Cops" (*Giufà, la luna, i ladri e le guardie*), ibid.

Remarkable for the return at night through the fields and the hide-and-seek with a moon in harmony with the sleepy rhythm of the passage.

III. "Giufà and the Red Beret" (*Giufà e la berretta rossa*) Palermo, told by Rosa Brusca.

One of the most Sicilian of tales, with the red beret and the mother's lament.

IV. "Giufà and the Wineskin" (*Giufà e l'otre*) Palermo, told by a worker from the Oretea Foundry.

V. "Eat Your Fill, My Fine Clothes!" (*Mangiate, vestitucei miei!*) Palermo, told by Francesca Amato.

VI. "Giufà, Pull the Door After You!" (*Giufà, tirati la porta!*) Palermo, told by Rosa Brusca; and Trapani, told by Nicasio Catanazaro, nicknamed Baddazza.

191. "Fra Ignazio" from Bottiglioni, 108, Cagliari, Sardinia, told by Bonatia Carlucciu.

Bottiglioni notes: "Fra Ignazio was born in Làconi, and his name is still very popular in Cagliari and Campidano. He was the alms-seeker of the Capuchin monastery, where one can still view his bed, rosary, and crucifix. The people of Calgiari call him venerable and have the same devotion for him as for a saint."

192. "Solomon's Advice" (*I consigli di Salomone*) from Mango, 11, Campidano, Sardinia.

In various parts of Italy, and not just in Sardinia, one encounters the story of the three pieces of advice with the same fatalism, the same truculence, and the same hint of lasciviousness (the caress given the young priest). But only here do I find the name "Solomon," which seems to link the tale with its Eastern source. (The oldest Indian, Arabic, and Persian collections include it, and from there it passes into the books of Christian *exempla* and the stories of the Middle Ages and the Renaissance.) Peculiar to the Sardinian version, I believe, is the fear at the opening of being an innocent victim of the law.

193. "The Man Who Robbed the Robbers" (*L'uomo che rubò ai banditi*) from Pietro Lutzu (in *Due novelline popolari sarde {dialetto campidanese} quale contributo alle leggende del tesoro di Rampsinite Re di Egitto*, Sassari 1900) Oristano, Sardinia, told in 1874 by a certain Beppa Rosa Massa di Santa Giusta.

This is not the "Rhampsinitus" type, as Lutzu believed, but the "Ali Baba" type, however much we would like to see the age-old story of a pharaoh reset in a rocky Sardinian landscape along with the replacement of the king's treasury by a bandits' cottage and the wiles to get in by means of a hidden key. But whatever the tale's origin, every detail here is newly invented and darkly realistic—the beheaded body hung from a dead tree, the sorceress's advice, the clashing of the rams on the mountain. Even the ending, which is closer to "Ali Baba," takes on local color—the village scene with the cooper and the house of the farmer grown mysteriously rich, then the appearance of casks and the prompt arrival of officers of the law.

194. "The Lions' Grass" (*L'erba dei leoni*) from Loriga, 7, Porto Torres, Sardinia.

The plot of this tale is widespread (see note on my no. 179), but the beginning is Sardinian with the thwarted love of the young people and the precariousness of life which brings sickness or death at every turn.

195. "The Convent of Nuns and the Monastery of Monks" (*Il convento di monache e il convento di frati*) from Loriga, 8, Porto Torres, Sardinia.

Well-known motifs wind in and out of this very rudimentary, but witty and graceful little tale. I altered the following details of the original: the unmotivated decision to become monks and nuns; the father superior's tying up the nuns to set fire to them. And I added the name "Johnny" (Gianni) myself. Cf. my no. 151, which is quite similar to this tale.

196. "The Male Fern" (*La potenza della felce maschio*) from Bottiglioni, 13 and 15, Tempio, Sardinia, told by Anna Rosa Ugoni and Nicoletta Atzena.

The Sardinian legends published by Bottiglioni are quite short and meager in narrative development, but infused with local color. Here I combined two tales, giving as an introduction one of the many elementary stories about encounters with a multitude of dead people and then proceeding with the beautiful legend of the male fern. Dying from gunfire is almost equivalent to a particular sickness, and the bandit as a generous hero wants to liberate man from it. But only courage can free man from such a death, and that is the whole point of the legend (in short, it takes more courage *not* to fire than to fire). Man proves weak, incapable of dominating his fear, and the story ends sorrowfully.

197. "St. Anthony's Gift" (*Sant'Antonio dà il fuoco agli uomini*) from Bottiglioni, 29, Nughedu S. Nicolò, Sardinia, told by Adelasia Floris; and from Filippo Valla (in *Rivista delle tradizioni popolari*, I [1894], 499) Osieri, Sardinia.

St. Anthony in Sardinia takes the role of Prometheus. Fire is a diabolical element, but stealing it and delivering it to man is a holy use, and we have a high-spirited comedy indeed. "Close by Nughedu," notes Bottiglioni, "is the chapel of St. Anthony of the Fire, where his feast is celebrated annually." And Filippo Valla describes the big bonfires that are lit on January 17; included in the wood that goes into these fires is cork, which, according to Sardinian tradition, fed the fire of Hell. The accounts of both Bottiglioni and Valla are extremely brief and rudimentary. I aimed to heighten the narration and bring out the saint's shrewdness, taking a hint from Valla regarding the pig which throws Hell into such chaos (a Hell strangely well-ordered). The concluding verse is composed of words incomprehensible in part even to Sardinians.

198. "March and the Shepherd" (*Marzo e il pastore*) from Ortoli, 1, Olmiccia, Corsica, told in 1882 by A. Joseph Ortoli.

"March and the Shepherd," a famous Tuscan apologue, is based on the wiles of the shepherd who goes to the hills when he says he is going to the plain and vice versa; the shepherd and the month vie with one another in cunning and peasant mockery. Here, though, the relationship smacks of a religious cult: the shepherd prays to the months, occasionally losing faith and blaspheming one, which is then unleashed against him with the fury of an irate deity. I retold the tale from Ortoli's French translation.

199. "John Balento" (*Giovan Balento*) from Carlotti, p. 187, Corsica.

A Corsican variant of the famous story about the braggart, of which I gave a version from Marche in my no. 97. Strange is the closing episode with the Amazons; I slightly modified it by presenting a country invaded by flies.

200. "Jump into My Sack" (*Salta nel mio sacco!*) from Ortoli, 22, Porto-Vecchio, Corsica, told in 1881 by Madame Marini.

One of the many variants of a very old theme (cf. my no. 165, IV), which here becomes almost a local legend, thanks to the place names. I omitted one of Francis's statements in the original, as it struck me as oratorical and out of keeping with the rest of the story: in one of his last wishes, Francis tells the fairy: "I want to see Corsica happy, and invaded and plundered no more by Saracens." The original becomes obscure toward the end. In keeping with the vague allegorical atmosphere, I did a bit of rearranging; in the beginning, for the fairy's first apparition, I placed her in a tree.

And with this wise and stoical folktale I bring my book to a close.

Bibliography

The Bibliography does not list the following classics, to which frequent reference is made solely by the author's name and the number, if not the title, of a particular tale:

Basile. *Il Pentamerone* (Eng. tr., Sir Richard Burton, 1893).
Boccaccio. *The Decameron*, tr. G. H. McWilliam, 1972.
Grimm. *The Grimms' German Folk Tales*, tr. Magoun & Krappe, 1960.
Perrault. *Contes* (*Complete Fairy Tales*, tr. A. E. Johnson et al., 1961).
Straparola. *Le Piacevoli Notti* (*The Facetious Nights*, tr. W. G. Waters, 1894).

AMAL.   Amalfi, Gaetano. *XVI conti in dialetto di Avellino.* Naples, 1893.

ANDER.   Anderson, Walter. *Novelline popolari sammarinesi* (3 fasc.). Tartu, 1927, 1929, 1933.

ANDR.   Andrews, James Bruyn. *Contes ligures* (recueillis entre Menton et Gênes). Paris, 1892.

"ARCH."   "Archivio per lo studio delle tradizioni popolari" (Tri-annual journal directed by G. Pitrè and S. Salomone-Marino). Palermo-Torino, 1882–1906.

BAB.   Babudri, F. *Fonti vive dei Veneto-Giuliani.* Milan: Trevisini, n.d. (In "Canti, novelle e tradizioni delle regioni d'Italia," collection edited by Luigi Sorrento. My numbers refer to the section "Fiabe.")

BAGLI   Bagli, Giuseppe Gaspare. *Saggio di novelle e fiabe in dialetto romagnolo.* Bologna, 1887.

BALD.   Baldini, Antonio. *La strada delle meraviglie.* Milan: Mondadori, 1923.

BALL.   Balladoro, A. *Folk-lore veronese: Novelline.* Verona-Padova, 1900.

BERN. I   Bernoni, Dom. Giuseppe, comp. *Fiabe e novelle popolari veneziane.* Venice, 1873.

BERN. II   ———. *Tradizioni popolari veneziane.* Venice, 1875. (My numbers concerning this collection refer to the pages.)

BERN. III   ———. *Fiabe popolari veneziane.* Venice, 1893.

BOLOGN.   Bolognini, Nepomuceno. "Fiabe e leggende della Valle di Rendena nel Trentino," essay in *Annuario della Società degli alpinisti tridentini.* Rovereto, 1881.

759

BOTT. Bottiglioni, Gino. *Leggende e tradizioni di Sardegna* (dialectal texts written phonetically). Geneva, 1922 (vol. V, series II of the "Biblioteca dell' 'Archivium Romanicum'" under the direction of Giulio Bertoni).

CARL. Carlotti, Dom Domenico, *Racconti e Leggende di Cirnu bella* (in Corsican dialect). Leghorn, 1930.

CARR. Carraroli, D. *Leggende, novelle e fiabe piemontesi*, in "Arch.," XXIII. Turin, 1906.

CAST. Castelli, Raffaele. *Leggende bibliche e religiose di Sicilia*, in "Arch.," XXIII. Turin, 1906.

COMP. Comparetti, Domenico. *Novelline popolari italiane*. Turin, 1875.

CONTI Conti, Oreste. *Letteratura popolare capracottese*, with preface by Francesco D'Ovidio, 2nd ed. Naples, 1911.

CORAZ. Corazzini, Francesco. *I componimenti minori della letteratura popolare italiana nei principali dialetti*, or *Saggio di letteratura dialettale comparata*. Benevento, 1877. (Contains Tuscan, Venetian, Bolognese, Bergamasque, Vicenzan, and Beneventan tales.)

CORON. Coronedi-Berti, Carolina, comp. *Novelle popolari bolognesi*. Bologna, 1874.

CORON. S. ———. *Al sgugiol di ragazú* (Bolognese popular tales). Bologna, 1883.

D'AM. D'Amato, A. "Cunti irpini." (MS in Museo Pitrè, Palermo, of seven short tales in Irpino dialect with translation.)

DEGUB. De Gubernatis, Alessandro. *Le tradizioni popolari di S. Stefano di Calcinaia*. Rome, 1894.

DEN. De Nino, Antonio. *Fiabe* (vol. III of *Usi e costumi abruzzesi*). Florence, 1883.

DIFR. Di Francia, Letterio. *Fiabe e novelle calabresi*. Turin: "Pallante," fasc. 3–4, December 1929, and fasc. 7–8, October 1931.

FARIN. Farinetti, Clotilde. *Vita e pensiero del Piemonte*. Milan: Trevisini, n.d.

FERR. Ferraro, Giuseppe. "Racconti popolari monferrini," Rome: MSS 131–140 in Museo arti e trad. pop., 1869.

FINAM. Finamore, Gennaro. *Tradizioni popolari abruzzesi*. Lanciano, 1882 (vol. 1, 1st part); Lanciano, 1885 (2nd part).

FORS. Forster, Riccardo. *Fiabe popolari dalmate*, in "Arch.," X. Palermo, 1891.

GARG.   Gargiolli, Carlo. *Novelline e canti popolari delle Marche.* Fano, 1878.

"GIAMB. BAS."   "Giambattista Basile," Archivio di letteratura popolare, edited by Luigi Molinaro del Chiaro. Naples (started in 1883).

GIANAN.   Gianandrea, Antonio, comp. *Novelline e fiabe popolari marchigiane.* Jesi, 1878.

GIANN.   Giannini, G. *Novelline lucchesi.* 1888.

GIGLI   Gigli, Giuseppe. *Superstizioni, pregiudizi e tradizioni in Terra d'Otranto* (supplemented by popular poems and tales). Florence, 1893.

GONZ.   Gonzenbach, Laura, comp. *Sicilianische Märchen* (2 vols.). Leipzig, 1870.

GORT.   Gortani, Luigi, comp. *Tradizioni popolari friulane,* vol. I. Udine, 1904.

GRADI   Gradi, Temistocle. *Saggio di letture varie.* Turin, 1865.

GRIS.   Grisanti, Cristoforo, comp. *Usi, credenze, proverbi e racconti popolari di Isnello.* Palermo, 1899.

GRIS. II   ———. Idem, vol. II, 1909.

GUAR.   Guarnerio, P. E. *Primo saggio di novelle popolari sarde* (in "Arch.," II, pp. 18–35, 185–206, 481–502; III, 233–40).

IMBR.   Imbriani, Vittorio. *La Novellaja Fiorentina.* Leghorn, 1877.

IMBR. P.   ———. *XII Conti pomiglianesi (con varianti avellinesi, montellesi, bagnolesi, milanesi, toscane, leccesi, ecc.).* Naples, 1877.

IVE   Ive, Antonio. "Opuscolo per nozze Ive-Lorenzetto," containing four Istrian tales. Vienna, 1877.

IVE II   ———. *Fiabe popolari rovignesi.* Vienna, 1878.

IVE D   ———. *I dialetti ladino-veneti dell'Istria.* Strasbourg, 1900.

LAR.   La Rocca, L. *Pisticci e i suoi canti.* Putignano, 1952.

LOMB.   Lombardi Satriani, Raffaele, ed. *Racconti popolari calabresi,* vol. I. Naples, 1953.

LOR.   Loriga, Francesco, comp. "Novelle sarde" (MS 59, Museo arti e trad. pop., Rome).

MANGO   Mango, Francesco, comp. *Novelline popolari sarde* (vol. IX of Pitrè's "Curiosità popolari tradizionali"). Palermo, 1890.

MARZ.   Marzocchi, Ciro, comp. "130 novelline senesi" (MS 57, Museo arti e trad. pop., Rome).

MOR.   Morandi. "5 fiabe umbre" (MS 179, Museo arti e trad. pop., Rome).

NER. Nerucci, Gherardo, comp. *Sessanta novelle popolari montalesi.* Florence, 1880.

NIERI Nieri, Idelfonso. *Cento racconti popolari lucchesi (e altri racconti).* Florence: Le Monnier, 1950.

ORT. Ortoli, J. B. Frédéric. *Les Contes populaires de l'île de Corse* (vol. XVI, "Les littératures populaires de toutes les nations"). Paris, 1883.

PELL. Pellizzari, Pietro, comp. *Fiabe e canzoni popolari del contado di Maglie in Terra d'Otranto.* Maglie, 1881.

PING. Pinguentini, Gianni, comp. *Fiabe, Leggende, Novelle, Storie paesane, Storielle, Barzellette in dialetto triestino.* Trieste, 1955.

PITRÈ Pitrè, Giuseppe, comp. *Fiabe, novelle e racconti popolari siciliani* (vols. IV–VII of "Biblioteca delle tradizioni popolari siciliane"). Palermo, 1875. (The 300 narratives are consecutively numbered, from one volume to the next. I have used the designation "Pitrè alb." for the seven tales in Alban dialect from Piana de' Greci, given in the appendix of vol. IV.)

PITRÈ T. ———. *Novelle popolari toscane,* part one (vol. XXX of "Opere complete di Giuseppe Pitrè," edizione nazionale). Rome, 1941.

PITRÈ T. II ———. Idem, part two. Rome, n.d.

PRATI Prati, Angelico. *Folklore trentino.* Milan: Trevisini, n.d.

PRATO Prato, Stanislao, comp. *Quattro novelline popolari livornesi.* Spoleto, 1880. (I use the number on the Leghorn tales and the page number for the Umbrian tales.)

"RIV. TRAD. POP." "Rivista delle tradizioni popolari," directed by Angelo De Gubernatis, Rome (begun in 1893–94).

SCHN. Schneller, Christian. *Märchen und Sagen aus Wälschtyrol.* Innsbruck, 1867.

TARG. Targioni-Tozzetti, Giovanni, ed. *Saggio di novelline, canti ed usanze popolari della Ciociaria* (vol. X of "Curiosità popolari tradizionali"). Palermo, 1891.

TIRAB. Tiraboschi, Antonio, comp. "Sei quadernetti manoscritti di fiabe in dialetto bergamasco." Bergamo: Biblioteca Civica.

TOSCHI Toschi, Paolo. *Romagna solatia.* Milan: Trevisini, n.d.

VECCHI Vecchi, Alberto. *Testa di Capra.* Modena, 1955.

VISEN. Visentini, Isaia, comp. *Fiabe mantovane* (vol. VII of "Canti e racconti del popolo italiano"). Turin, 1879.

VITAL.   Vitaletti, Guido. *Dolce terra di Marca*. Milan: Trevisini, n.d.

VOC.   Vocino, Michele and Nicola Zingarelli. *Apulia Fidelis*. Milan: Trevisini, n.d.

ZAG.   Zagaria, Riccardo. *Folklore andriese*. Martina Franca, 1913.

ZAN.   Zanazzo, Giggi, comp. *Novelle, favole e leggende romanesche* (vol. I of "Tradizioni popolari romane"). Turin-Rome, 1907.

ZORZ.   Zorzùt, Dolfo. *Sot la nape . . . (I racconti del popolo friulano)*, 3 vols. Udine, 1924, 1925, 1927.

Books by Italo Calvino available from
Harcourt Brace & Company
in Harvest paperback editions

The Baron in the Trees

The Castle of Crossed Destinies

Cosmicomics

Difficult Loves

If on a winter's night a traveler

Invisible Cities

Italian Folktales

Marcovaldo, or The seasons in the city

Mr. Palomar

The Nonexistent Knight and *The Cloven Viscount*

t zero

Under the Jaguar Sun

The Uses of Literature

The Watcher and Other Stories